AF823894

# Modern Development Paths of Agricultural Production

Volodymyr Nadykto
Editor

# Modern Development Paths of Agricultural Production

## Trends and Innovations

Springer

*Editor*
Volodymyr Nadykto
Tavria State Agrotechnological University
Melitopol, Ukraine

ISBN 978-3-030-14917-8 ISBN 978-3-030-14918-5 (eBook)
https://doi.org/10.1007/978-3-030-14918-5

Library of Congress Control Number: 2019935551

© Springer Nature Switzerland AG 2019, corrected publication 2024
This work is subject to copyright. All rights are reserved by the Publisher, whether the whole or part of the material is concerned, specifically the rights of translation, reprinting, reuse of illustrations, recitation, broadcasting, reproduction on microfilms or in any other physical way, and transmission or information storage and retrieval, electronic adaptation, computer software, or by similar or dissimilar methodology now known or hereafter developed.
The use of general descriptive names, registered names, trademarks, service marks, etc. in this publication does not imply, even in the absence of a specific statement, that such names are exempt from the relevant protective laws and regulations and therefore free for general use.
The publisher, the authors and the editors are safe to assume that the advice and information in this book are believed to be true and accurate at the date of publication. Neither the publisher nor the authors or the editors give a warranty, expressed or implied, with respect to the material contained herein or for any errors or omissions that may have been made. The publisher remains neutral with regard to jurisdictional claims in published maps and institutional affiliations.

This Springer imprint is published by the registered company Springer Nature Switzerland AG
The registered company address is: Gewerbestrasse 11, 6330 Cham, Switzerland

# Contents

# Part I

# Improving the Efficiency of Fruit Tree Sprayers

Nikolai Struchaiev, Larysa Bondarenko, Olexandr Vershkov and Andrei Chaplinskiy

## 1 Introduction

Current trends in the formation of strategies for providing the population with high-quality fruit products include several components. According to the Food and Agriculture Organization, world losses from plant pests and diseases each year account for about 20–25% of the potential world crop yield [1]. The development of spraying methods is based on the mechanics and physics of the formation and deposition of liquid droplets, whose sizes range from units and tens of micrometers to several millimeters [2]. However, currently, there are no designs of sprayers with absolutely monodisperse spraying, for example, with a diameter of 200 mm [3].

As a result, during spraying, unproductive losses of pesticides are always observed due to the drift of very small droplets (20–80 mm) and the flow of very large droplets (360–1000 mm) from the treated object to the soil. Therefore, the search for effective methods for influencing droplets in flight and during deposition on plants remains an urgent task.

### *1.1 Analysis of Recent Studies and Publications*

The mechanism of putting drops on the leaves of plants is diverse and complex. Large drops with a weak wind are deposited on the leaves under the action of gravity, from top to bottom, mainly on the upper side of the leaves. Smaller droplets are deposited on the leaves under the action of inertial forces when captured by the wind, mainly on the windward side of the leaves [4]. The possibility of protection against wind exposure using tunnel sprayers [5] and the implementation of controlled aerosol

N. Struchaiev · L. Bondarenko (✉) · O. Vershkov · A. Chaplinskiy
Tavria State Agrotechnological University, B. Khmelnitsky ave. 18,
Melitopol 72310, Ukraine
e-mail: larbond@ukr.net; larysa.bondarenko@tsatu.edu.ua

© Springer Nature Switzerland AG 2019

V. Nadykto (ed.), *Modern Development Paths of Agricultural Production*,
https://doi.org/10.1007/978-3-030-14918-5_1

deposition using an electro-aerosol installation [6] is noted. However, the issues of increasing the efficiency of sprayers of fruit plantations by improving the deposition of droplets through the use of thermodynamic methods used in the operation of sprayers are virtually unknown [7].

### *1.2 Statement of the Objective and Tasks of the Study*

The purpose of the study is to establish the possibility of increasing the efficiency of fruit plant sprayers by reducing the entrainment of small droplets, by improving the deposition of droplets by using thermodynamic and chemical methods used in the operation of sprayers.

To achieve this goal, the following tasks were solved:

- Determine the dependence of condensation growth of water droplets on thermodynamic parameters;
- Determine the effect of hygroscopic additives on the process of growth and deposition during spraying;
- Propose a tunnel-type sprayer scheme with a protective spraying effect.

## 2 The Basic Part of the Study

The objectives of our work are an experimental study of the dependence of the condensation increment of free-falling water droplets on thermodynamic parameters and the effect of the addition of hygroscopic substances. Thermal deposition of droplets, thermoprecipitation, in the pale of the temperature gradient, occurs as a result of such a phenomenon as thermophoresis, that is, the movement of fluid particles along the lines of the temperature field in the direction of lowering the temperature. In this case, the growth of droplets begins when the dew point reaches 100% relative humidity. This can be achieved in two ways: either by increasing the temperature of the main part of the solution used or by reducing the temperature of the additional part of the solution (pure or with added hygroscopic substances), the drops of which will become condensation centers.

### *2.1 Materials and Methods of Research*

The experiment was conducted on a pilot plant on the basis of an improved monodisperse aerosol generator [2, 8, 9] (see Fig. 1), which consists of a test chamber (glass tube) 150 cm long, a droplet generator with a thermocouple, two dedicated volumes in which measurements are made, two pulsed illuminators of these volumes, upper

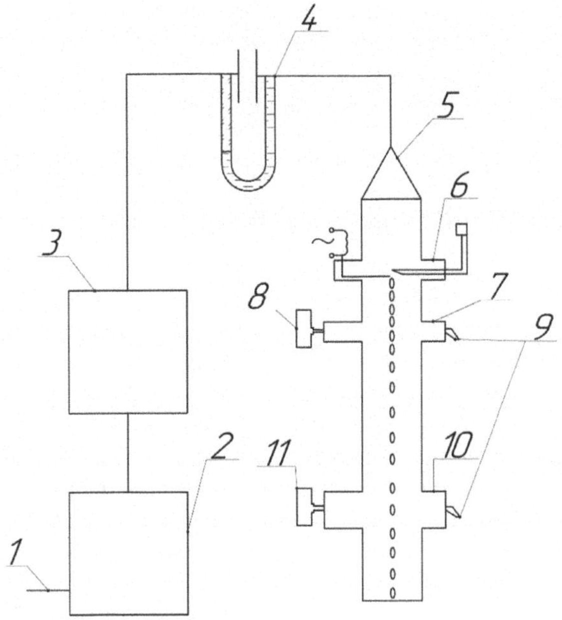


**Fig. 1** Experimental setup: 1—air inlet, 2—evaporator, humid air generator, placed in a thermostat, 3—damper, 4—flow meters, 5—test chamber, 6—droplet generator with thermocouple, 7—upper dedicated volume, 8—upper camera, 9—pulsed illuminators, 10—lower allocated volume, 11—lower camera

camera and lower camera, an evaporator, a moist air generator, placed in a thermostat, and a damper.

The dynamics of the process of condensation growth of free-falling water droplets from thermodynamic parameters and the influence of additives of hygroscopic substances was estimated using the dependence [10]:

$$r^5 = r_0^5 + A \cdot \tau, \tag{1}$$

where

- $\boldsymbol{r}$ final radius of the drop, m;
- $\boldsymbol{r_0}$ the initial radius of the drop, m;
- $\boldsymbol{\tau}$ time of a drop in a humid environment, s;
- $\boldsymbol{A}$ parameter ($m^5/s$), which is determined by the formula:

$$A = \frac{5 \cdot D \cdot \mu}{\rho \cdot R_c \cdot T} \cdot \Delta p_0 \cdot r_0^3 \tag{2}$$

where

- $\boldsymbol{D}$ the diffusion coefficient of vapor to the surface of the drop, $m^2/s$;
- $\boldsymbol{\mu}$ molecular weight of steam, kg/kmol;
- $\boldsymbol{R_c}$ universal gas constant, J/(kmol K);

$\Delta\, p_0$ the initial decrease in the elasticity of the saturated vapor above the drop introduced into the flow, Pa;
$\rho$ fluid density, kg/m$^3$;
$T$ absolute temperature, K.

## 3 Results and Discussion

Let us calculate the expected size increase for a drop of a saturated $NH_4NO_3$ solution with a radius of 30 microns for the duration of the passage of the chamber section 90 cm at a speed of 30 cm/s, which consists of a sedimentation rate of ~10 cm/s and an associated flow rate of ~20 cm/s, $\tau = 3$ s. We obtain $r_{\text{end}} = 35$ mm, and the change in the radius of the drop $\Delta r = 5$ mm. The difference in velocity according to Stokes will be $\Delta V = V_{н} - V_{к} = 14.4 - 10.8 = 3.6$ cm/s. With a frequency of 100 s$^{-1}$, the distance between the particles is 0.036 cm. If we measure the distance between the first and tenth particles, it will be 0.36 cm; i.e., it can be measured, in the worst case, to 0.1 mm, with an error in the measurement of the radius of about 3%.

Figure 2 shows the results of experimental studies of condensation growth of droplets in single-component and multicomponent aerosols of hygroscopic substances. The experiments were carried out at temperatures of a droplet introduced into an aerosol equal to the dew point temperature. For example, at a temperature of 25 °C and a relative humidity of 65%, to begin the condensation of water, the aerosol was cooled to 17.5 °C. The temperatures of the aerosol and the cooled drop were set and controlled by an evaporator, a humid air generator placed in a thermostat, and a droplet generator with a thermocouple.

As can be seen from Fig. 2, the rate of condensation growth increases with increasing concentration of the second component of hygroscopic substances. At the initial stage, the rate of condensation growth of droplets for all experiments is almost linear, and later, it slows down according to the power dependences presented in the table, which also lists the single-component and multicomponent additives of hygroscopic substances in aerosols.

Table 1 shows one-component and multicomponent additives hygroscopic substances in aerosols.

Empirical dependences of condensation growth of droplets are obtained.

For clean water:

$$d = -0.01\,\tau^2 + 0.9\,\tau + 23.00 \tag{3}$$

where

$d$ diameter of a drop, micron

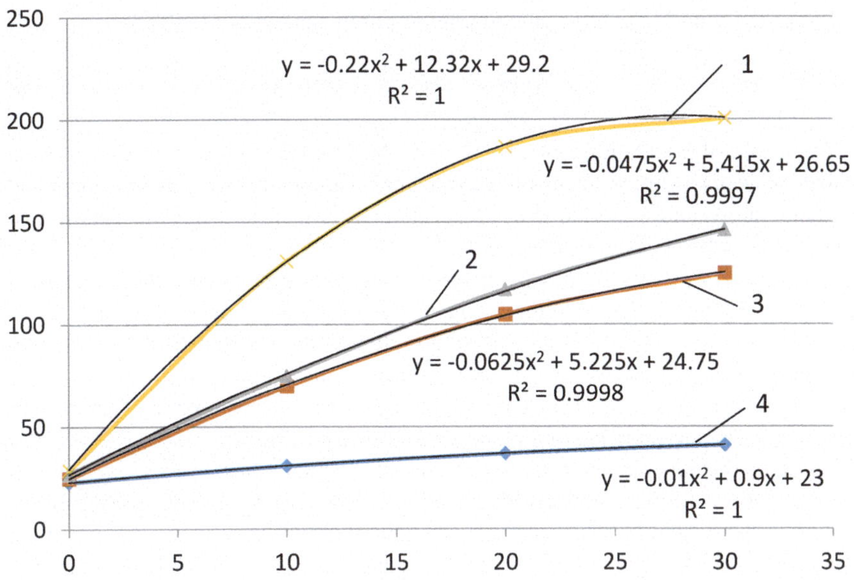


**Fig. 2** Rate of condensation growth of droplets in single-component and multicomponent aerosols of hygroscopic substances: up to 30 min, along the *X*-axis time in minutes, along the *Y*-axis the diameter of the droplet, mm; 1—water, 2—urea additive, 3—ammonium nitrate additive, 4—ammonium nitrate + urea additive (50/50%)

**Table 1** Single and multicomponent additives hygroscopic substances in aerosols

| No. | Substance | Chemical formula | Solution concentration % |
|---|---|---|---|
| 1 | Water | $H_2O$ | 100 |
| 2 | Urea additive | $H_2O + CO(NH_2)_2$ | 55 |
| 3 | Ammonium nitrate additive | $H_2O + NH_4NO_3$ | 55 |
| 4 | Ammonium nitrate + urea additive 50/50%) | $H_2O + NH_4NO_3 + CO(NH_2)_2$ | 55 |

For area supplemented water:

$$d = -0.0625\,\tau^2 + 5.225\,\tau + 24.75 \tag{4}$$

For water with ammonium nitrate:

$$d = -0.0475\,\tau^2 + 5.415\,\tau + 26.65 \tag{5}$$

For water with the addition of ammonium nitrate and urea (50/50%):

$$d = -0.22\,\tau^2 + 12.32\,\tau + 29.20 \tag{6}$$

The results of calculations and experiments show the effectiveness of condensation growth of water droplets from thermodynamic parameters, such as humidity and air temperature and a cooled drop, respectively, as well as from the addition of hygroscopic substances, during spraying;

Injection of cooled drops into the stream of sprayed pesticides and other substances used for spraying garden plantings reduces the entrainment of small drops due to their condensation on cooled drops and their return to the sprayed area. The addition of hygroscopic substances significantly improves the condensation effect; at the same time, it is possible to pick up supplements of hygroscopic substances that simultaneously serve for feeding plants or for other purposes.

Based on the data obtained, we proposed a conceptual scheme of a sprayer with a cooling thermal curtain to improve the design of the existing garden sprayer. Introduction to the design of new structural elements that will lead to an increase in the efficiency of chemical treatment of plants, reduction of chemical losses, improvement of the environment and sanitary and hygienic working conditions will make the sprayer competitive in the market of chemical plant protection devices.

Reducing the entrainment of aerosols and chemical vapors is possible due to the formation of a cooling air curtain and separation from the environment of collectors on which nozzles are installed for spraying the working fluid, tunnel-forming shields, and also due to condensation and return of chemical vapors to the spray area when they are cooled with an air curtain which passed through the installed cooler.

The conceptual diagram of a sprayer with a cooling air curtain contains a casing with a fan placed in it and an outlet nipple located above the collectors on which the nozzles are installed for spraying the working fluid. The nozzles are separated from the environment by tunnel-forming shields; a cooler is installed in the suction inlet of the fan (see Fig. 3). The working fluid from the tank, pump, under pressure is fed to the collectors and enters the nozzles, with which it is sprayed on the crown of the plant.

When the fan rotates, the air is drawn into the suction nozzle; its temperature and the temperature of the spray solution are reduced when in contact with the cooler. Further, the cooled air and the solution enter the upper part of the sprayer, between the tunnel-forming shields, and form a cold air curtain to separate the collectors on which the nozzles for spraying the working fluid are installed from the environment.

The air curtain and tunnel-forming shields significantly reduce the loss of working fluid from blowing and dispersion. Evaporation losses, which are unavoidable with traditional spraying methods, are reduced due to condensation and the return of chemical vapors to the spray area when cooled by air that has passed through the installed cooler.

**Fig. 3** Conceptual diagram of a sprayer with a cooling air curtain: 1—a casing with a fan placed in it, 2—an outlet, 3—a collector, 4—sprays, 5—tunnel-forming shields, 6—a cooler, 7—pipes for supplying the working fluid

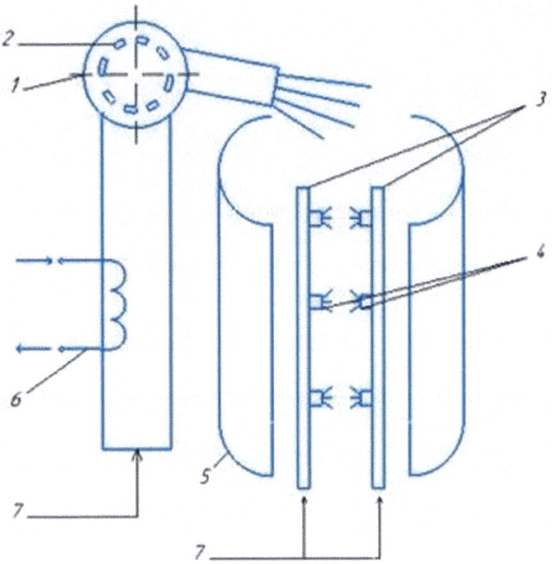


## 4 Conclusions

One of the reserves for improving the efficiency of spraying fruit plantations is to improve this process after dropping the droplets from the sprayer and precipitating the droplets using the thermodynamic and chemical methods used in the operation of the sprayers. Condensation of small droplets due to condensation growth of water droplets from thermodynamic parameters by cooling them and introducing centers of droplet formation with the addition of hygroscopic substances into this zone increases the intensity of their growth.

We determined:

- dependence of condensation growth of water droplets on thermodynamic parameters;
- the effect of the addition of hygroscopic substances on the process of growth and deposition during spraying;
- proposed conceptual scheme of a tunnel-type sprayer with a vortex spraying effect.

The results can be used in systems of various types of sprayers as a way to increase the efficiency of spraying fruit plantations, by improving the process after dropping the droplets from the sprayer and precipitating the droplets using thermodynamic and chemical methods.

## References

1. Solovyova, N.F.: Modern technical means for spraying crops. M.: Rosinformagrotiex, 59p (2002)
2. GOST ISO 5682-1-2004 Spray equipment. Spray Nozzle Test Methods. M.: Standardinform, 16p (2006)
3. Shamaev, G.P.: Handbook of machines for the control of pests and diseases of agricultural crops, M.: Kolos, 143p (1980)
4. Oripov, M.O.: Investigation of the deposition of sprayed liquid with a low-volume ground spraying of cotton, depending on the elements of the microclimate of the surface air layer. Tashkent, 19p (1974)
5. Dogoda, P.A.: Mounted single-row chamber sprayer. Doslidnitske **12**(26), 368–372 (2008)
6. Dogoda, P.A.: Controlled deposition of aerosols using charge during chemical plant protection, Melitopol 44–51 (2008)
7. Seraja, E.M.: Intensification of the deposition of pesticides when spraying crops. Melitopol **9**(8), 119–125 (2009)
8. Wolf W.R.: Rev. Sci. Instr. **32**, 1124 (1963)
9. Binek, B., Dhnalova, B.: Staub **27**, 11 (1967)
10. Todes, O.M., Fedoseev, V.A.: Colloid J. **24**(4) 1962

# Refractory Materials Manufacturing Based on Sewage Sludge

**Ludmyla Chernyshova, Serhii Kiurchev, Oleg Peniov and Vitaliy Cherkun**

## 1 Introduction

The protection of the environment is a pressing issue due to the growth of the world's population. The volume of sewage from machine-building as well as other enterprises with galvanizing lines is growing. It is reasoned by the increase in the degree of wastewater treatment, the growth of industrial production as well as urban population. It takes vast land areas to store raw sludge that creates a serious threat of the secondary pollution [1]. Therefore, the main task of the sludge treatment technology is to obtain a product that is harmless in sanitary terms. Its quality composition and properties should ensure the possibility of further use in the national economy. More significant problem is the processing and utilization of wastewater from galvanizing and pickling divisions of the machine-building industry. Since such wastewater contains heavy metal ions, its storage in urban landfills is prohibited. Therefore, the development of wastewater sludge disposal process with the subsequent application of the reaction product is topical.

L. Chernyshova (✉) · S. Kiurchev · O. Peniov · V. Cherkun
Tavria State Agrotechnological University, B. Khmelnitsky ave. 18, Melitopol 72310, Ukraine
e-mail: liudmyla.chernyshova@tsatu.edu.ua; ludmyla.chernyshva@tsatu.edu.ua

S. Kiurchev
e-mail: serhii.kiurchev@tsatu.edu.ua

O. Peniov
e-mail: oleh.penov@tsatu.edu.ua

V. Cherkun
e-mail: vitalii.cherkun@tsatu.edu.ua

© Springer Nature Switzerland AG 2019

V. Nadykto (ed.), *Modern Development Paths of Agricultural Production*,
https://doi.org/10.1007/978-3-030-14918-5_2

## 2 Review of Literary Sources

Forming sludge consists mainly of metal hydroxides and mechanical impurities. There are several methods of disposal and burial of sludge of galvanic shops. One of them is the method of dehydration of sludge to 75–85% humidity with the subsequent burial [2, 3]. The authors [4] offer a method of decontamination of sludge, which is formed after treatment of galvanic production sewage. The method includes adding of phosphoric acid, with the subsequent use as a pigment. As a result, solventless orthophosphate of heavy metals turns into compounds of bright coloration. Another paper presents [5] studies of the quantitative and phase composition for the wastes of galvanic production. Iron, chromium and nickel leaching in solution of 10% sulfuric acid at temperature of 50–60 °C 0.5 h. The recovery of iron, chromium and nickel proved to be 99.7, 99.46 and 98.0%, respectively. Three methods of neutralization of the galvanic sludge on the basis of zinc and iron were compared. They were quite effective in conversion of dangerous waste into inert material. The methods included calcium oxide neutralization, conversion to inert material by adsorption of organic and inorganic pollutants into activated carbon, mobile waste component transformation into insoluble phosphates. Thus, the concentration of zinc in filtrate in sludge decreased by 99.7% compared to raw silt. After the treatment of sludge with activated carbon, zinc retention was 99.9%. After the treatment by phosphoric acid, it is made up 98.7%. The advantages of the galvanic method of sludge treatment by activated carbon comparing with other two methods are high sorptioning ability, low pH, volumetric changes of sludge and easy application [6]. The analysis of scientific periodical publications showed different ways of utilization of the galvanic sludge. We offer our own method of neutralization that allows to obtain a waterproof material on the basis of the galvanic production waste.

## 3 Research Methodology

The work is aimed at production of materials based on an inorganic adhesive made from heavy metal hydroxides (HMH) sludge and alundum—waste of an abrasive production.

## 4 Results

As a test material, we studied the waste galvanic production. The basis of green color sludge was heavy metal hydroxides of chromium, iron, zinc, calcium (up to 80%) associated by crystallization water. There were also carbonates, sulfites of those metals, mineral impurities and surface-active substances with other organic substances. Chemical composition of the investigated waste is shown in Table 1.

The sludge, which had been dehydrated on the filter press to the humidity of 75–80%, was transferred into the chamber with a heater, where concentrated sulfuric and phosphoric acids were added. Acids react with crystallization water, as well as with carbonates and organic substances which the sludge contained. The crystal structure of sludge and organic substance was destructed when polymer–metal compounds were formed. After that, the obtained solution was processed at 100 °C for 60–90 min. Dark green liquid of high viscosity with a density of 1.67 g/cm$^3$ formed.

Hundred grams of alundum was thoroughly mixed with a different quantity of adhesive (40 g; 45 g; 50 g; 60 g; 70 m; 80 g) and was put into the forms. At an ambient temperature, it was soaked for 10 h and dried at a temperature of 180 °C in a drying cabinet 2 h. Then, the samples were heat-treated within 1 h at a temperature of 300 °C. Produced samples were tested for durability (Table 2).

Unit weight of obtained samples increased from 0.57 to 0.9 g/cm$^3$ with increasing of the adhesive content when the strength decreased. Higher durability has the compositions with a low content of the adhesive. The more the binder was, the more the reaction of phosphate formation was as well as the evolving of gases. As a result,

**Table 1** Chemical composition of the investigated sludge

| Components of sludge | $Cr(OH)_3$ | $Fe(OH)_3$ | $Ca(OH)_2$ | $Zn(OH)_2$ | Mineral impurities | Organic substances |
|---|---|---|---|---|---|---|
| Content, % | 53.6 | 3.2 | 10.0 | 14.1 | 10 | 9.1 |

**Table 2** Testing the strength of wet and dry fused alumina samples

| Sample composition | Cross-sectional area, m$^2$ | Excessive load, kg | Voltage, mPa |
|---|---|---|---|
| 40 g adhesive + 100 g fused alumina<br>Dry fused alumina<br>Wet fused alumina | 3.8 × 3.7<br>0.0014<br>3.9 × 3.8<br>0.0015 | 1400<br>950 | 10<br>6.3 |
| 45 g adhesive + 100 g fused alumina<br>Dry fused alumina<br>Wet fused alumina | 0.0014<br>0.0015 | 1520<br>1070 | 10.9<br>7.1 |
| 60 g adhesive + 100 g fused alumina<br>Dry fused alumina<br>Wet fused alumina | 0.0016<br>0.0016 | 850<br>400 | 5.8<br>2.4 |
| 70 g adhesive + 100 g fused alumina<br>Dry fused alumina<br>Wet fused alumina | 0.0016<br>0.0015 | 450<br>220 | 2.8<br>1.4 |
| 80 g adhesive + 100 g fused alumina<br>Dry fused alumina<br>Wet fused alumina | 0.0014<br>– | 360<br>– | 2.6<br>– |

we obtained a more porous structure with low strength. More durable samples were obtained if the filler was dry fused alumina. In case the wet water molecules prevent the formation of polymeric chains of chromium polyphosphate, aluminum, the result was sampled with low strength. The next aim of the research was to test the obtained samples for water resistance and acid resistance and to process the samples with a solution of cement. For the experiment, samples with dry corundum were taken. The samples were being soaked for 24 h in neutral, acidic and alkaline environments. Then, the concentration of $Cr^{3+}$ in solution was determined. After 6 days, the concentration of chromium was also measured. The results of the experiment are shown in Table 3. The solution volume was 200 ml, and the sample weighted up to 20 g. The pH of the samples was measured after 20 min of continuous stirring at an ambient temperature. Later, the pH of the environment stabilized (Table 3).

Thus, the optimal composition for the manufacture of durable and water-resistant samples is the composition of 40 g of adhesive (from heavy metal hydroxides) with 100 g of fused alumina.

Figure 1 shows the change in the concentration of chromium in the samples for acidic, neutral, alkaline solution 1 day.

As a result of the experiments, a significant change in pH was observed in a neutral environment. In alkaline environment, the maximum pH change was 0.5.

**Table 3** Change in the environment the soak of the sample in different environments

| Sample composition | $H_2O$ dist. ΔpH | $[Cr^{3+}]$, mg/l after 24 h | Cement solution Δ pH | $[Cr^{3+}]$, mg/l after 24 h | Cement solution, double processing | $[Cr^{3+}]$, mg/l after 24 h |
|---|---|---|---|---|---|---|
| 1 | 2 | 3 | 4 | 5 | 6 | 7 |
| 40 g of adhesive + 100 g of dry fused alumina | 3.08 | 0.1 | 0 | 0.05 | 0.05 | 0.3 |
| 45 g of adhesive + 100 g of dry fused alumina | 0.5 | 1 | 0.4 | 0.1 | 0.4 | 0.8 |
| 50 g of adhesive + 100 g of dry fused alumina | 3.7 | 0.5 | 0.4 | 0.1 | 0.5 | 0.83 |
| 60 g of adhesive + 100 g of dry fused alumina | 3.85 | – | 0.35 | | 0.4 | – |

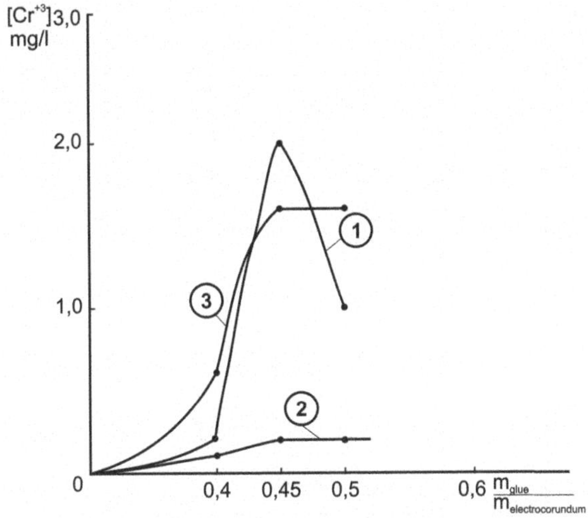


**Fig. 1** Change of chromium concentration while the samples are processed by: 1—acidic, 2—neutral, 3—alkaline solution 1 day

That was 7 times less than in a neutral environment. In the acidic environment, the pH practically did not change, and the maximum $\Delta pH = 0.15$. A release of iron on the sample surface was observed as well. Chromium content after 1 day remained small. With increasing of the weight ratio of the adhesive to the fused alumina mass, the concentration of chromium increased. In addition, the high porosity of the material promoted water penetration into the sample. It saturated the solution with ions $Cr^{3+}$, since the adhesive m/m fused alumina equated to 0.45. There is dependence between the chrome (III) concentration and the ratio of m adhesive to fused alumina weight processing of the sample 6 days. The dependence is presented in the diagram (Fig. 2).

## 5 Conclusions

Our research is useful because it solves the problem of neutralization of the galvanic production sludge. Its crystalline structure is destroyed, and the mass in the form of adhesive is formed. By adding various fillers (fire clay, aluminum oxide and others), it is possible to obtain a durable and waterproof material. That solves two problems: the problem of utilization of harmful waste of the galvanic production and manufacturing of material on the basis of an inorganic adhesive. Thus, we defined the optimal composition for the manufacture of durable and water-resistant samples. It is the composition of 40 g of adhesive (from heavy metal hydroxides) with 100 g of fused alumina.

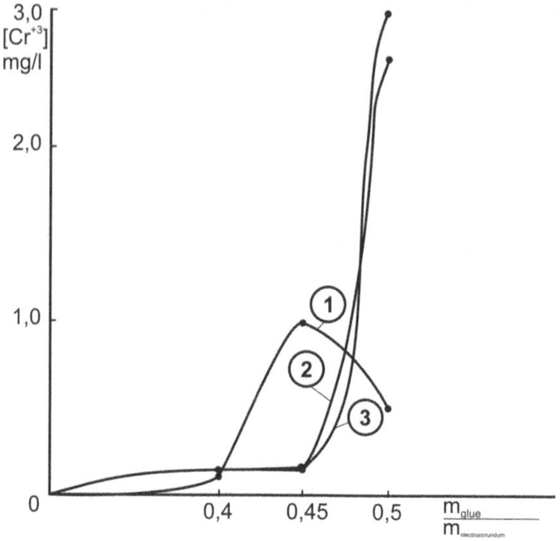


**Fig. 2** Dependence between the chrome (III) concentration and the ratio of m adhesive to fused alumina weight processing of the sample 6 days in: 1—acidic, 2—neutral, 3—alkaline environments

## References

1. Pinaev, A.V.: Migracija ionov tjazhelyh metallov v pochvu pri zahoronenii osadkov stochnyh vod gal'vanicheskogo proizvodstva (avtoref. dis. … kand. biol. nauk: 03.00.16, Tol'jatti, 2006)
2. Ljubarskij, V.M.: Osadki prirodnyh vod i metod ih obrabotki. Strojizdat, Moscow (1980)
3. Sander, K., Jepsen, D., Schilling, S., Tebert, C.: Definition of waste recovery and disposal operations. Institute for Environmental Strategies, Hamburg, Germany (2004)
4. Dobrovol'skij, I.P., Denisov, S.E., Abyzov, V.A., Selihov, A.B.: Vestnik JUUrGU. **17**, 48–50 (2012)
5. Garole, D.J., Garole, V.J., Dalal, D.S.: Research. J. Chem. Sci. **2**(3), 61–63 (2012)
6. Višnja, O., Mikelić, I.L., Kollar, R., Mikulić, N., Medunić, G.: Research gate: science network. Article (PDF Available). Arch. Ind. Hyg. Toxicol. **63**(3), 337–344 (2012)

# Increasing the Efficiency of the Technological Process of Processing Castor-Oil Seeds into Castor Oil

**Volodymyr Didur, Volodymyr Kyurchev, Andriy Chebanov and Anatoliy Aseev**

## 1 Introduction

Oilseed crops have a major national economic importance, since they are a source of valuable food and technical products. Among oilseeds, an important place in the southern part of Ukraine is taken by castor-oil plant and its main product is castor oil, used for the chemical, electrical, medical, aviation, and other industries [1–3]. It should be noted that castor oil can be used for the production of biofuels [4–7].

In production of vegetable oils from oil-bearing raw materials (sunflower, rapeseed, soybeans, etc.), including from castor seeds, oil extraction pressing methods use presses or extraction with help of solvents [8–10]. The solvents used to extract oil by the extraction method [10] should satisfy the requirements imposed on them by the technique and technology of the extraction process (dissolve the oil well and quickly, be completely removed from oil and meal, do not impart an extraneous smell and taste, be safe for health of service staff, be cheap and not rare, etc.). However, such a solvent does not yet exist, and all solvents used in industry satisfy only some of the requirements. In this connection, during extraction, quality parameters of oil and meal reduce, and the cost of the process and the harmful effect on the human body increase.

The pressing method of obtaining vegetable oils does not have such drawbacks. But in case of using this process in its pure form (without additional steps) occurs not completed extraction of oil, thereby the loss of oil in production increases. It is connected with the fact that oil in the meal of castor oil is distributed in the form of thin membranes on the surface of the particles of crushed seeds (rushanka) and is retained by the forces of intermolecular interaction. Their magnitude is much higher than the pressure developed by modern presses used for its spinning. Therefore, extracting oil by pressing from an unprepared meal gives a small amount of the final

V. Didur (✉) · V. Kyurchev · A. Chebanov · A. Aseev
Tavria State Agrotechnological University, B. Khmelnitsky ave. 18, Melitopol 72310, Ukraine
e-mail: didurva@gmail.com

© Springer Nature Switzerland AG 2019

V. Nadykto (ed.), *Modern Development Paths of Agricultural Production*,
https://doi.org/10.1007/978-3-030-14918-5_3

product. The task of meal's preparing (moisture–heat treatment or frying) before pressing is to weaken the forces that hold the oil in the seed. This is achieved by moistening the seed. But the moistened seed becomes very plastic, and therefore, after pressing, oil separates poorly from it. To give a seed a certain elastic properties, moisture removes by drying [11]. However, the indicators of moisturizing and drying of castor meal by different sources differ a lot [12]. This makes it difficult to select valid technological regimes of squeezing oil out on the experimental line. The search for such technological regimes is an urgent task.

The purpose of this work is to increase the efficiency of processing castor-oil seeds by justifying the technological regimes of moisture–heat treatment of the meal in the roaster.

## 2 Materials and Methods

The main task of this method is to establish integrated indicators of the efficiency of technological processes for processing castor oil.

Indicators of efficiency in the processing of castor beans will be the quantity (specific quantity) of the obtained oil and its quality. The quality of castor oil must comply with the requirements of GOST 6757-96 [13]

Thus, a set of criteria for determining the optimal levels for oil dehydration can be achieved using the following set of indicators:

$$\mu_0 \rightarrow \max;\ \eta_\Phi \rightarrow \max, \quad (1)$$

where $\mu_0$—is the relative amount of oil; $\eta_\Phi$—the relative number of oilseed sludge.

The relative amount of oil is determined by the formula

$$\mu_0 = \frac{m_0}{m_n}, \quad (2)$$

where $m_0$—the mass of the oil pressed from the prescribed weight of crushed seeds, kg; $m_n$—predetermined weight of crushed seeds of castor-oil plant, kg.

The relative amount of oil sludge is determined by the formula

$$\eta_\Phi = \frac{m_\Phi}{m_n}, \quad (3)$$

where $m_\Phi$—mass of an oil sludge obtained from a given mass of crushed castor seeds, kg.

For experimental researches were used castor seeds of castor-oil plant Khortitskaya variety 7. Crushing of castor-oil seeds in a rusk was carried out on a special single-roll machine. The space between the grinding rolls was set depends on physical and mechanical characteristics of the seeds (length, width, and thickness).

Before the experiments, the initial moisture of the crushed seeds was found according to GOST 5947-68 [13].

The mass of water in the material was determined by the formula

$$g = g - g_{c.вещ},\tag{4}$$

where $g$—total mass of wet material, kg; —mass of dry material, kg.

The moisture content in the material to the total mass (%) was determined by the Formula (5)

$$w = \frac{g_{вл}}{g_{c.вещ.}} \cdot 100\tag{5}$$

Determination of complex indicators of the efficiency of technological processes for the processing of castor-oil plant was carried out in an experimental setup; the technological scheme is shown in Fig. 1.

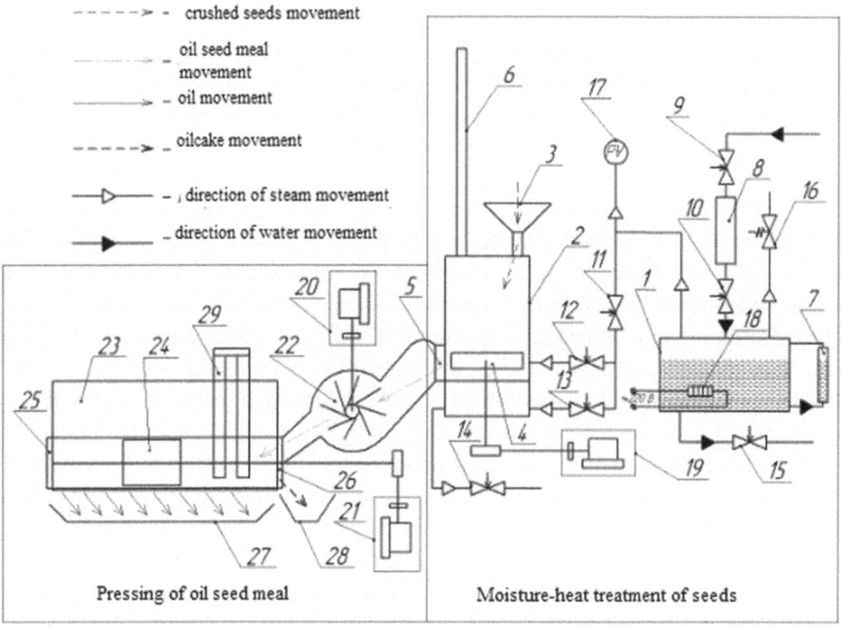


**Fig. 1** Scheme of a technological experimental complex for oil extraction from oilseeds: 1—steam generator; 2—roaster; 3—loading hopper; 4—mixer; 5—unloading window; 6—a pipe for evacuation of vapors; 7—dimensional glass of the steam generator; 8—additional capacity for water; 9 to 15—through-flow control valves 16—non-return valve; 17—manometer; 18—heating element; 19 to 21—electric motor with reducers; 22—beater; 23—press; 24—the piston of the press; 25 to 26—insertion of the channel; 27—capacity for collecting oil; 28—container for harvesting cake; 29—heating element

Frying (moistening and drying) of the meal of castor seeds occurs in the roaster 2. According to the technological process, the water in the steam generator 1 is fed through an open straightway control valve 9, an additional water tank 8, and a through-flow control valve 10. Water level monitoring is performed using a measuring glass 7. After reaching the required level of water in the steam generator 1, the valves 10 and 9 overlap. The water is heated by means of three heating elements 18 located in the tank of the steam generator. The pressure in the steam generator 1 is controlled by means of a pressure gauge 17. The emergency pressure is controlled by a safety valve 16.

The crushed castor-oil seeds are fed to the roaster 2 through the hopper 3. The necessary level of the crushed seeds is checked by means of a measuring ruler which is inserted into the hopper. Mixing during the frying process is carried out by means of a stirrer 4, driven by a motor with a reducer 19. The steam is roasted off by means of a steam extraction pipe 6. The humidification of the oilseed meal is carried out by the hot steam, which is fed into the roaster tank 2 through the open valves 11 and 12. The valve 13 is closed at the same time. Moreover, the valve 12 is fully open, and the valve 11 is open to such a position that the vapor pressure remains unchanged, which is controlled by a pressure gauge 17. The drying of the meal is carried out by a dead steam that is fed to the bottom of the roaster through the valves 11, 13, and 14. To do this, the valve 12 supplying the hot steam is closed, and the valves 13 and 14 are opened completely. The position of the valve 11 does not change for the preset pressure value. Frying of the meal ends with the unloading of material through the unloading window 5. The valve 15 is necessary for draining water from the steam generator.

The oil is squeezed out in the press (Fig. 1). To do this, the meal is fed to the beater 22 through the unloading window; the beater is rotated by the electric motor with the gear 20. In turn, the beater 22 pushes the meal into the press channel 23 in the required amount. At this time, the piston 24 is moved to the extreme left position. In order to maintain a constant temperature of the meal after roasting, the heating channel 29 preliminarily heats the press channel to a temperature of 85 … 90 °C. After the press channel 23 is filled with a meal, the channel insert 26 is closed, the electric motor with the reduction gear 21 is turned on, which causes the piston 24 to move. Piston 24, moving to the right, squeezes the meal. The oil is squeezed out by squeezing through the curb plate (located at the bottom of the press) and enters the oil collection tank 27. Next, the squeezed pulp should be removed from the press channel 23. To do this, the insert 26 should be removed; an electric motor with a gear 21 is moved on to move the piston 24. The spent pulp is squeezed out by the piston and exits into the bottom of the beater 22, which has a discharge slot.

While preparing cleaned seeds for frying, the roaster's determining parameters are: steam pressure, $p_{гр}$, height of the oilseed meal layer $H$, initial moisture of the oilseed meal $W_{н}$, final moisture of the oilseed meal separately for the wetting periods $W_{к.з}$ and drying $W_{к.с}$. The roaster of the experimental setup has one tank, the speed of the stirrer 4 does not change, and it is 32.5 rpm.

In various sources of the literature, the final moisture at humidification $W_{к.з}$ and drying $W_{к.с.}$ of oilseed meal of castor oil has a contradictory character [12]. That is, the final moisture content of the meal for the wetting period was controlled to the level $W_{к.з} = 13\%$; the final moisture of the meal for the drying period to the level $W_{к.с} = 5\%$. Steam pressure at researches is accepted $p_{гр} = 3, 4$ и 5 kPa. The height of the layer of crushed seeds in the roaster can vary from H = 0 mm to $H = 240$ mm. Carrying out preliminary experiments allowed to determine the parameters of the loading of the crushed seeds into the roaster. So, to provide the necessary amount of oilseed meal, which is fed to the press for squeezing castor oil, the sufficient height of the layer of the crushed seeds in the roaster is 80 mm.

In order to obtain a constant amount of steam, it is necessary to calibrate the control valve 11 (Pic. 1) for different pressures. To do this, the steam generator, filled with water, includes all the heating elements 18 and creates a steam pressure $p_{гр} =$ 5 kPa. The valves 13 and 14 are fully open. And the valve 11 is open to such an angle of rotation that the vapor pressure controlled by the pressure gauge 17 does not change. When this position of the valve is fixed, a mark should be made. The same calibration marks are made for pressure values $p_{гр} = 4$ kPa and $p_{гр} = 3$ kPa.

For the period of moistening of the oilseed meal, it is also necessary to know the capacity of the steam generator (kg/s). For this, a container with water was taken and weighed. The hose after valve 14 was installed in this container with water and sealed. In the steam generator, a steam pressure $p_{гр} = 5$ kPa was generated. For the period of moistening of the oilseed meal, it is also necessary to know the capacity of the steam generator (kg/s). For this, a container with water was taken and weighed. The hose after valve 14 was installed in this container with water and sealed. In the steam generator, a steam pressure pgr = 5 kPa was generated. The valves 13 and 14 were fully opened, and the valve 11 was opened to the required mark. The time was controlled, and it was 30 s. After 30 s, valves 11, 13, and 14 were closed, and the container with water was again weighed.

The productivity of the steam generator was determined as:

$$Q_{пг} = \frac{m_e - m_{e.п}}{t}, \tag{6}$$

where $m_e$—mass of the container with water before carrying out the researches, kg; $m_{e.п}$—mass of the container with water and steam after the research, kg; $t$—the time during which the studies were conducted, sec.

The same experiments were carried out for pressure values $p_{гр} = 4$ kPa and $p_{гр} = 3$ kPa.

To ensure the height of the layer of crushed castor-oil seeds, served in the roaster, it is necessary to calibrate the tank of the roaster. To do this, a measuring ruler through the hopper 3 measured the depth of the roaster's tank and measured the values. Then they poured crushed seeds into the tank of the roaster. Periodically, the electric motor with a reducer 19 was switched on and the crushed seeds were mixed with a stirrer 4. The crushed seeds were poured so that the measured value on the ruler was reduced by 80 mm, and the mark was placed on the ruler according to the level of the upper rim of the hopper.

In order to determine the amount of steam for moistening the crushed seeds to the final moisture content (13%), it is necessary to determine the weight of the crushed seeds for the corresponding layer height. For this purpose, a weighed crushed seeds were poured into the roaster. At the same time, the height of the filling layer of the crushed seeds was controlled. Mass of a layer with height of 80 mm was fixed.

The amount of steam necessary to moisten the crushed seeds to the final moisture content was determined by the formula

$$\Delta g_{вол} = g_1 \cdot \frac{w_2 - w_1}{100 - w_1}. \tag{7}$$

The time required to moisten the crushed seeds to the final moisture $w_2$ at a given steam pressure $p_{гр}$ was determined as:

$$T = \frac{\Delta g_{вол}}{Q_{пг}}. \tag{8}$$

The moisture difference obtained during wetting and drying the oilseed meal of castor-oil seeds was determined by the formula

$$\Delta W = W_{к.з} - W_{к.с.}. \tag{9}$$

To carry out research on the experimental setup (Fig. 1), it is necessary to have the following equipment and materials: a thermo cabinet, a scale with a scale value up to the third decimal place, and aluminum weighing bottles for samples with the material being examined.

The algorithm for carrying out the experiment is as follows:

Weigh each empty weighing bottle on the scale with a scale division of 0.001 g.

Create a steam pressure in the steam generator 1 equal to $p_{гр} = 5$ kPa, which is controlled by a pressure gauge 17. If a steam pressure is higher than necessary, it can be vented. For that, valves opens 11, 13, and 14 (Fig. 1).

Fill the weighed crushed seeds into the roaster 2 to a height of 160 mm. Determine by the Formulas (7) and (8) the amount of steam and the time required to moisten the rush to the final moisture content $W_2$. Turn on the electric motor with the reducer 19, what rotates the stirrer 4 of the roaster 2.

For the humidification period, valve 12 was opened completely and on the corresponding label—valve 11 (valve 13—closed). After the time required for moistening the crushed seeds to the final moisture content $W_2$ [the time was found through the Formula (8)], the shutter of the unloading window 5 was opened and the crushed seeds were sampled.

For the drying period, the valve 13 was fully opened, and the valve 12 was closed. The position of the valve 11 was not changed. The drying time to the required humidity was determined experimentally so that the final moisture content after drying was at a level from 5 to 10%.

After the drying period with the corresponding time, the electric motor with reducer 20 of the beater drive gear 22 was turned on, the unloading window 5 was fully opened, and the oilseed meal was fed to the pre-heated to 80–85 °C press channel 23. With full filling of the press channel 23, the motor 20 with the beater drive gear 22 turned off. The channel was closed by a shutter 26, and the electric motor with a reduction gear 21 of the press piston 24 was turned on. The load was monitored with an ammeter that was installed in the motor circuit 21. The oil in the tank 27 was discharged to the weighing bottle. Each weighing bottle was weighed. Humidity of the samples was determined by drying them to absolutely dry mass according to the procedure given in GOST 5947-68 [13]. In this case, the mass of water in the experimental material and the moisture content of the material with respect to the total mass were determined from Formulas (4) and (5).

Experiments 1–9 were repeated at a steam pressure in the steam generator 1 $p_{гp} = 3$ kPa, $p_{гp} = 4$ kPa, $p_{гp} = 5$ kPa.

The efficiency of the oil extraction process was determined on the basis of the evaluation of efficiency criteria.

## 3 Results and Discussion

As a result of the experimental studies, the relative amounts of oil and oil sludge are obtained as a function of the final moisture content of the pulp after the drying process, when the oilseed meal was moistened to $W_{к.з.} = 11\%$ and the heating steam pressure was $p_{гр} = 3$ kPa, $p_{гр} = 4$ kPa и $p_{гр} = 5$ kPa (Fig. 2). When the pulp was moistened to $W_{к.з.} = 11\%$ and dried to $W_{к.с} = 6.25\%$, the steam generator flow rate $Q = 0.194$ kg/min, the greatest relative amount of oil $\mu_o = 0.17$, was obtained, with the relative amount of the oil sludge has the smallest value and is $\eta_ф = 0.06$.

The regression equations of the dependences shown at Fig. 2 have the form:

– for the relative amount of oil at a heating steam pressure $p_{гр} = 3$ кPa

$$\mu_0 = -0.0095W_{к.с.}^2 + 0.1202W_{к.с.} - 0.2098, \quad (10)$$

– for the relative amount of oil at a heating steam pressure $p_{гр} = 4$ кPa

$$\mu_0 = -0.0122W_{к.с.}^2 + 0.1835W_{к.с.} - 0.5695, \quad (11)$$

– for the relative amount of oil at a heating steam pressure $p_{гр} = 5$ кPa

$$\mu_0 = -0.0087W_{к.с.}^2 + 0.0749W_{к.с.} - 0.0081, \quad (12)$$

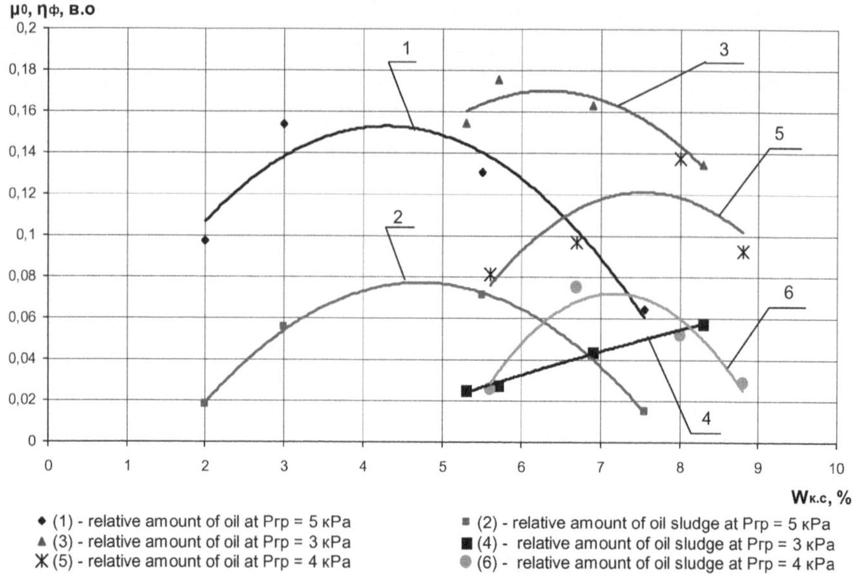


**Fig. 2** Dependence of the relative amount of oil and oil sludge on the final moisture content of the pulp after the drying process at $W_{к.з} = 11\%$; $p_{гр} = 3$ кПа, $p_{гр} = 4$ кПа, and $p_{гр} = 5$ кПа

– for the relative amount of the oil sludge at the heating steam pressure $p_{гр} = 3$ кPa

$$\eta_{ф} = -0.0005W_{к.с.}^2 + 0.018W_{к.с.} - 0.0573, \tag{13}$$

– for the relative amount of the oil sludge at the heating steam pressure $p_{гр} = 4$ кPa

$$\eta_{ф} = -0.0177W_{к.с.}^2 + 0.2544W_{к.с.} - 0.8402, \tag{14}$$

– for the relative amount of the oil sludge at the heating steam pressure $p_{гр} = 5$ кПа

$$\eta_{ф} = -0.0078W_{к.с.}^2 + 0.0738W_{к.с.} - 0.0974. \tag{15}$$

To determine the influence of moistening of the pulp on the final product ($\mu_o$ та $\eta_ф$) taking into account the preliminary experiments (Fig. 3), investigations were carried out at a heating steam pressure of $p_{гр} = 3$ kPa and for different initial humidity $W_{к.з} = 9$, 11, and 13% (Fig. 3).

It has been established by investigations that pulp's moistening to $W_{к.з.} = 9\%$ and $W_{к.з} = 13\%$; the maximum relative amount of oil is $\mu_o = 0.162$ and $\mu_o = 0.114$, respectively. With this amount of oil, the relative amount of oil sludge is

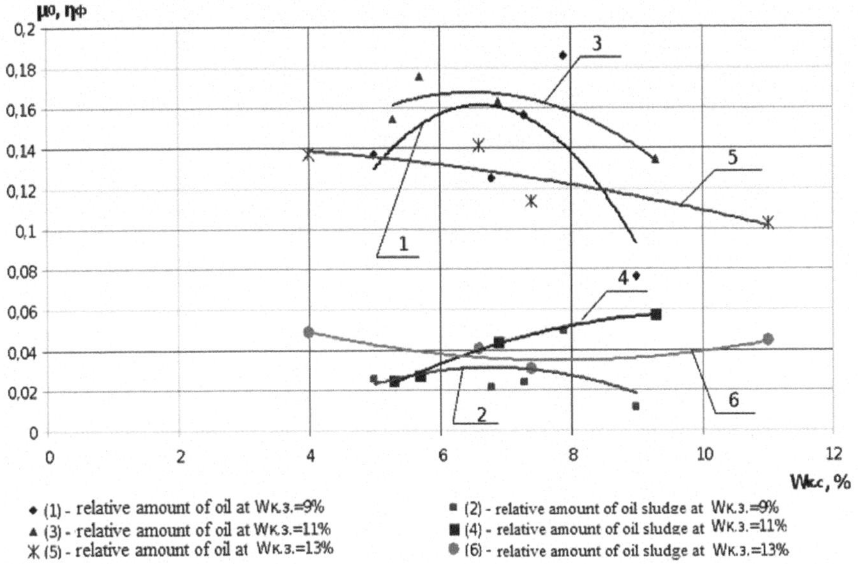


**Fig. 3** Dependence of the relative amount of oil and oil sludge on the final moisture content of the pulp after the drying process at $W_{к.з} = 9\%$; $W_{к.з} = 11\%$; $W_{к.з} = 13\%$, and $p_{гр} = 3$ kPa

$\eta_ф = 0.03$ и $\eta_ф = 0.05$, respectively. The resulting relative amount of oil when the pulp is moistened to $W_{к.з} = 9\%$ and $W_{к.з} = 13\%$ has a lower value (4.7 and 8.8%, respectively) than the amount of oil ($\mu_o = 0.17$), obtained by moistening the oilseed meal to $W_{к.з} = 11\%$. This is the predominant factor in selecting the necessary regimes for moisture–heat treatment of the oilseed meal of castor-oil plant during the production of castor oil.

The regression equations of the dependences shown at Fig. 2 have the form:

– for the relative amount of oil at a heating steam pressure $p_{гр} = 3$ kPa

$$\mu_0 = -0.0095W_{к.с.}^2 + 0.1202W_{к.с.} - 0.2098, \tag{16}$$

– for the relative amount of the oil sludge at the heating steam pressure, $p_{гр} = 3$ kPa

$$\eta_ф = -0.0005W_{к.с.}^2 + 0.018W_{к.с.} - 0.0573, \tag{17}$$

– for the relative amount of oil at a heating steam pressure $p_{гр} = 3$ kPa and humidification of the pulp to $W_{к.з} = 9\%$:

$$\mu_o = -0.0121W_{к.с.}^2 + 0.1601W_{к.с.} - 0.3683, \tag{18}$$

– for the relative amount of oil at a heating steam pressure $p_{гр} = 3$ kPa and humidification of the pulp to $W_{к.з.} = 13\%$:

$$\mu_o = -0.0004W_{к.с.}^2 + 0.0004W_{к.с.} + 0.1431, \quad (19)$$

– for the relative amount of the oil sludge at the heating steam pressure $p_{гр} = 3$ kPa and the humidification of the pulp to $W_{к.з.} = 9\%$:

$$\eta_ф = -0.0027W_{к.с.}^2 + 0.036W_{к.с.} - 0.0904, \quad (20)$$

– for a relative amount of a oil sludge at a heating steam pressure $p_{гр} = 3$ kPa and moistening the pulp to $W_{к.з.} = 13\%$:

$$\eta_ф = 0.001W_{к.с.}^2 - 0.0152W_{к.с.} + 0.0944. \quad (21)$$

## 4 Conclusions

Experimental studies have established the main modes of moisture–heat treatment of the oilseed meal of castor-oil plant before pressing: The heating steam pressure is $p_{гр} = 3$ kPa; the final moisture content of the pulp for the humidification period is $W_{к.з.} = 11\%$; the final moisture content of the pulp for the drying period is $W_{к.с} = 6.25\%$. This made it possible to obtain the highest efficiency indicators: The relative amount of oil was $\mu_o = 0.17$, and the relative amount of the oil sludge was $\eta_ф = 0.06$. At the same time, the oil yield increased by 4.7 and 8.8% compared to the initial moistening of the pulp up to $W_{к.з.} = 9\%$ and $W_{к.з.} = 13\%$, respectively. Statistical processing of the experiments made it possible to obtain a regression equation designed to predict performance indicators over a wider range.

## References

1. Barnes, D.J., Baldwin, B.S., Braasch, D.A.: Degradation of ricin in castor seed meal by temperature and chemical treatment. Ind. Crops Prod. **29**(2–3), 509–515 (2009). https://doi.org/10.1016/j.indcrop.2008.09.006
2. Mutlu, H., Meier, M.A.: Castor oil as a renewable resource for the chemical industry. Eur. J. Lipid Sci. Technol. **112**(1), 10–30 (2010). https://doi.org/10.1002/ejlt.200900138
3. Malajowicz, J., Kusmirek, S.: Characteristics and possibilities of industrial use of castor oil. Przem. Chem. **95**(9), 1756–1760 (2016). https://doi.org/10.15199/62.2016.9.20

4. Deb, A., Ferdous, J., Ferdous, K., Uddin, M.R., Khan, M.R., Rahman, M.W.: Prospect of castor oil biodiesel in Bangladesh: process development and optimization study. Int. J. Green Energy **14**(12), 1063–1072 (2017). https://doi.org/10.1080/15435075.2017.1357558
5. Zhu, Q.L., Gu, H., Ke, Z.: Congeneration biodiesel, ricinine and nontoxic meal from castor seed. Renew. Energy **120**, 51–59 (2018). https://doi.org/10.1016/j.renene.2017.12.075
6. Hadiyanto, H., Yuliandaru, I., Hapsari, R.: Production of biodiesel from mixed waste cooking and castor oil. In: MATEC Web of Conferences, vol. 156, p. 03056. EDP Sciences (2018)
7. Conejero, M.A., César, A.D.S., Batista, A.P.: The organizational arrangement of castor bean family farmers promoted by the Brazilian Biodiesel Program. Energy Policy **110**, 461–470 (2017). https://doi.org/10.1016/j.enpol.2017.08.036
8. Bernesson, S.: Influence of some physical press parameters on capacity and oil extraction efficiency of small oil expeller. Swed. J. Agric. Res. **28**(3), 147–155 (1998)
9. De Lima, M.S., Rocha, L.A., Molina, E.F., Caetano, B.L., Marçal, L., Mello, C., Nassar, E.J.: Thermoanalysis of soybean oil extracted by two methods. Quím Nova **31**(3), 527–529 (2008). https://doi.org/10.1590/S0100-40422008000300012
10. Dasari, S.R., Goud, V.V.: Effect of pre-treatment on solvents extraction and physico-chemical properties of castor seed oil. J. Renew. Sustain. Energy **6**(6), 063108 (2014). https://doi.org/10.1063/1.4901542
11. Didur, V., Chebanov, A., Didur, V., Aseev, A.: Fundation of operating practices of seed meal moisture and heat treatment on oil extraction from castor beans. J. Agric. Environ. **1**(1), 9–15 (2017)
12. Didur, V.A.: Rozrobka tekhnolohii, eksperymentalnoho ustatkuvannia tekhnolohichnoi linii po hlybokii pererobtsi nasinnia rytsyny v kastorovu oliiu dlia vyrobnytstva mastyl dlia silskohospodarskoi tekhniky: zvit pro naukovo-doslidnu robotu (Melitopol, 2005) [Дідур В. А. Розробка технології, експериментального устаткування технологічної лінії по глибокій переробці насіння рицини в касторову олію для виробництва мастил для сільськогосподарської техніки: звіт про НДР]
13. GOST 5947-68. Semena hlopchatnika. Tehnicheskie usloviya. Standartinform, Moskow (1968) [ГОСТ 5947-68. Семена хлопчатника. Технические условия]

# Mechanism for the Maintenance of Investment in Agriculture

**Halyna Hrytsaienko, Igor Hrytsaienko, Andriy Bondar and Dmitry Zhuravel**

## 1 Introduction

Increasing production and improving the quality of agricultural products, strengthening food security countries are based on the most advanced technologies of machinery and equipment in agribusiness. For their successful functioning It is necessary to solve a number of complex tasks, one of which is to develop an effective mechanism for investing technical services in agriculture.

### *1.1 Analysis of Recent Research and Publications*

Various aspects of the technical service study were conducted in their scientific papers Berggren [1], Voitjuk [2], Lips [3], Mazzetto [4], Northern [5], Rublevo [6], Sundin [1], Sidorchuk [7], Fontanini [4], Jiao [5], Tsalsant [4], and others. They considered the possibility use of funds for the cost of repairs and maintenance, and also suggested ways to improve the efficiency business service. At the same time, the question of the investment mechanism of the technical service in agriculture remains underdeveloped, what determines the relevance of the chosen research topic.

H. Hrytsaienko (✉)
Department of Entrepreneurship, Trade and Exchange Activities, Tavrichesky State Agrotechnological University, B. Khmelnitsky Ave., 18, Melitopol, Ukraine
e-mail: halyna.hrytsaienko@tsatu.edu.ua

I. Hrytsaienko · A. Bondar · D. Zhuravel
Department of Maintenance and Systems in AIC, Tavrichesky State Agrotechnological University, B. Ave. Khmelnitsky, 18, Melitopol, Ukraine

© Springer Nature Switzerland AG 2019

V. Nadykto (ed.), *Modern Development Paths of Agricultural Production*,
https://doi.org/10.1007/978-3-030-14918-5_4

### *1.2 Setting Goals and Objectives of the Study*

The main purpose of the article is to develop a mechanism for investment in the development of technical services in agriculture.

### *1.3 Materials and Research methods*

The publications of Ukrainian and foreign scientists on the investment mechanism and maintenance were theoretical and methodological basis of the study.

The interdisciplinary and multidimensional nature of the scientific problem led to the use of such a complex in the research process methods: general scientific—abstraction—to identify the essence of the technical service; economic and statistical—association qualitative and quantitative aspects; systematic approach—justification of the investment mechanism for the development of technical services in agrarian production.

## 2 The Main Part of the Study

Previous studies using the hierarchical analysis method have led to the conclusion that the mechanization of agriculture farm is a major factor in the intensive development of agricultural production, which allows to increase the amount of quality production per unit area of land (livestock of cattle) [8]. Formation and operation of technical trading requires the proper development of the agricultural machinery market.

In Fig. 1 shows its components, among which the central place belongs to the technical service.

One of the main functions of the technical service is to provide consumers of equipment (agrarians) with the technical means. According to the State Statistics Service of Ukraine, the number of tractors of all brands (without tractors on which cars are mounted) in general for Ukraine in 2011–2016 increased by 8.4% and amounted to 339.8 thousand units, and the number of combine harvesters decreased by 2.5% to 52.7 thousand units (Table 1).

During the study period, the number of the tractors increased annually at an average by 1.6%, and the number of the combine harvester decreased by 0.5%. At the same time, in 2016, the loading of arable land per a tractor was 107 hectares (for comparison: in the USA—28 hectares, in France—14 hectares per tractor [9]), grain and pod sowing per 1 grain harvester—269 hectares (for comparison: In the USA, France, and Germany, the load on the combine is about 55 hectares [9]).

Insufficient provision of the agrarian production of the agricultural machinery prevents for the implementation of the necessary technological operations in due time and leads to significant losses of the crop.

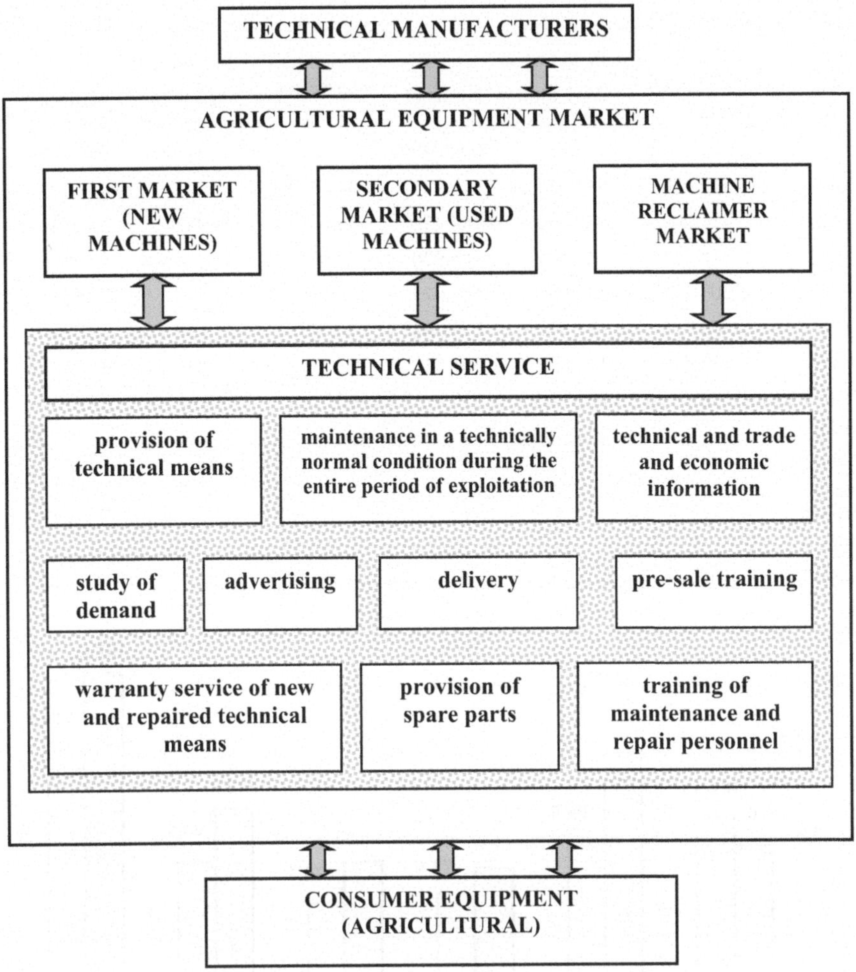


**Fig. 1** Components of the agricultural machinery market. *Source* Compiled using [12]

The analysis of agricultural machinery movement for 2011–2016 shows a lack of stability in its updating (see Fig. 2).

For 2011–2016, the number of tractors received during the year exceeded the number of those who dropped out only in 2012, 2015, and 2016, combine harvesters—accordingly in 2011 and 2016. According to calculations [10], the tractor fleet today is 45% from the needs of agriculture, combine harvesters—48%, and other types of machinery—from 35 to 60%. The average depreciation rate of agricultural machinery is 70%, including tractors—78% and combine harvesters—71%.

For the exploration period, the total cost of equipment bought by agricultural enterprises from UAH 4725.5 million in 2011 grew 3.5 times and increased to 16505.6 million UAH in 2016 (the average annual growth rate was 28.4%). Unfortunately,

**Table 1** Dynamics of the tractor and combine harvester availability in agriculture of Ukraine, units

| Years | Tractors of all brands, total | Including in | | Combine harvester, total | Including in | |
|---|---|---|---|---|---|---|
| | | Agricultural enterprises | Farms of the population | | Agricultural enterprises | Farms of the population |
| 2011 | 313,480 | 134,554 | 178,926 | 54,074 | 32,062 | 22,012 |
| 2012 | 322,209 | 137,958 | 184,251 | 54,651 | 31,997 | 22,654 |
| 2013 | 315,261 | 129,341 | 185,920 | 52,065 | 29,364 | 22,701 |
| 2014 | 309,111 | 120,638 | 188,473 | 50,019 | 27,196 | 22,823 |
| 2015 | 309,716 | 127,852 | 181,864 | 56,328 | 37,537 | 18,791 |
| 2016 | 339,829 | 132,686 | 207,143 | 52,714 | 37,946 | 14,768 |
| 2016% by 2011 | 108.4 | 98.6 | 115.8 | 97.5 | 118.4 | 67.1 |
| Average annual growth rate, % | 1.6 | −0.3 | 3.0 | −0.5 | 3.4 | −7.7 |

*Source* Calculated according to the State Statistics Service of Ukraine

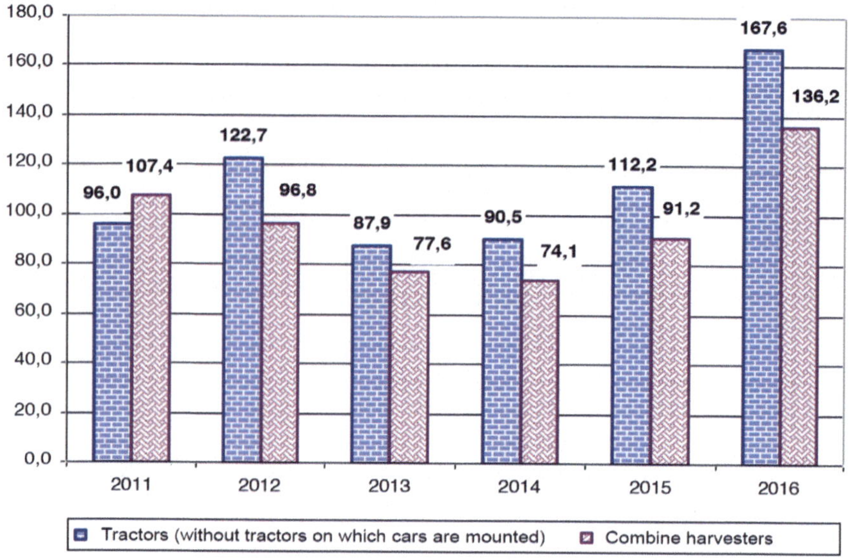


**Fig. 2** Ratio of the number of tractors and combine harvesters in agricultural enterprises received during the year to those who dropped out, %. *Source* Compiled according to the State Statistics Service of Ukraine

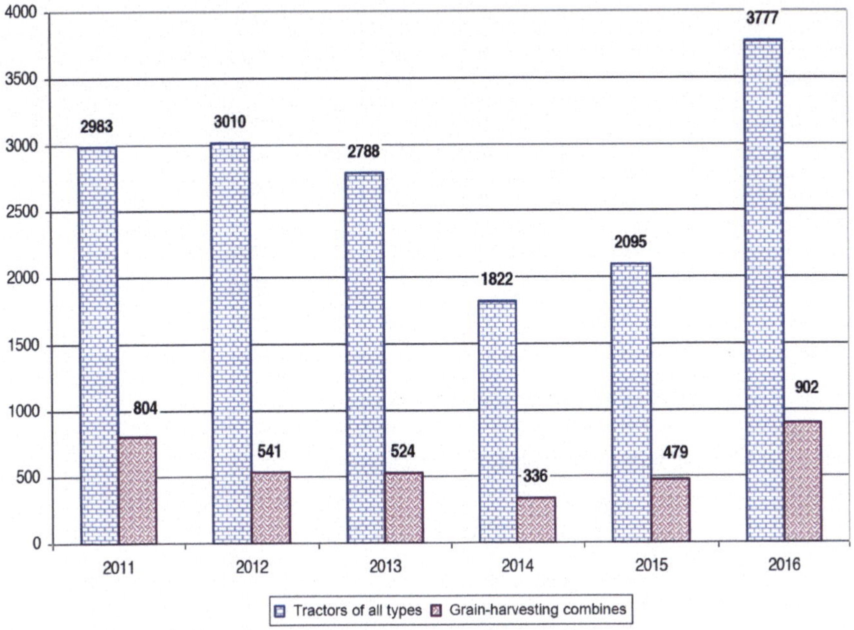


**Fig. 3** Dynamics of purchase of the new tractors and combine harvesters by agricultural enterprises, units. *Source* Compiled according to the State Statistics Service of Ukraine

these changes occurred mainly due to a significant increase in average prices. In Fig. 3, the dynamics of the tractor purchase of all types and combine harvesters is showed.

For 2011–2016, the number of new plows, harrows, and harvester rolls increased at an annual rate (at an average annually by 8.9, 10.7, and 9.3% accordingly). But at the same time, the fastest rates of reduction in the number of purchases were demonstrated by beet harvesters, milking machines, and apparatuses, as well as manure pickup trucks (on average annually by 25.6, 15.1, and 16.4% accordingly).

The domestic production of machinery and equipment for agriculture affects the provision of agricultural machinery (see Table 2).

For 2011–2016, the actual production of agricultural machinery in Ukraine has significantly decreased. Thus, the combine harvester production was reduced almost 6 times, plows and boring—almost three times, ramming and cultivators—twice, tractors—almost half time. At the same time, the production of seeders and mowers on the contrary increased more than twice.

Insufficient quantity and quality of the domestic production agricultural machinery prompts agrarians to give preference to import analogues. According to the calculations, import accounts for almost 80% of the funds spent on technical equipment of agriculture. The largest demand in Ukraine is for tractors, harvesters from “Case,”

**Table 2** Domestic production of machinery and equipment for agriculture, pcs

| Machinery and equipment | 2011 | 2012 | 2013 | 2014 | 2015 | 2016 | Average annual growth rate, % |
|---|---|---|---|---|---|---|---|
| Tractors | 6847 | 5280 | 4273 | 4121 | 4206 | 4894 | −6.5 |
| Plows | 6115 | 15,403 | 6203 | 4446 | 3672 | 2068 | −19.5 |
| Row cultivators and cultivators | 7396 | 4652 | 4358 | 3739 | 3440 | 3607 | −13.4 |
| Disk harrows | 7421 | 1645 | 1975 | 2044 | 2252 | 2799 | −17.7 |
| Seeders | 2133 | 823 | 1208 | 787 | 705 | 4520 | 16.2 |
| Mowers | 1710 | 2182 | 2307 | 2609 | 2463 | 3729 | 16.9 |
| Combine harvesters | 399 | 59 | 68 | … | 100 | 65 | −30.4 |

*Source* Compiled according to the State Statistics Service of Ukraine

**Table 3** Dynamics of tractor and combine harvester import

| Years | Tractors | | | Combine harvesters | | |
|---|---|---|---|---|---|---|
| | Quantity, thousand pcs. | Average price, USD US per pc. | Million dollars USA | Quantity, pcs. | Average price, USD US per pc. | Million dollars USA |
| 2011 | 146.4 | 3819.0 | 559.1 | 2619 | 10,3932.8 | 272.2 |
| 2012 | 104.9 | 5384.2 | 564.8 | 1647 | 11,3721.9 | 187.3 |
| 2013 | 156 | 3606.4 | 562.6 | 1570 | 11,6751.6 | 183.3 |
| 2014 | 108.8 | 2515.6 | 273.7 | 935 | 10,2139.0 | 95.5 |
| 2015 | 39.7 | 5775.8 | 229.3 | 1285 | 82,957.2 | 106.6 |
| 2016 | 58.9 | 8789.5 | 517.7 | 3139 | 86,492.5 | 271.5 |
| 2016% by 2011 | 40.2 | 230.2 | 92.6 | 119.9 | 83.2 | 99.7 |
| Average annual growth rate, % | −16.6 | 18.1 | −1.5 | 3.7 | −3.6 | −0.1 |

*Source* Compiled according to the State Statistics Service of Ukraine

"Claas," "John Deere," "New Holland," "Massey Ferguson," and others. In Table 3, the dynamics of tractor and combine harvester import is shown.

In 2016, the tractors were imported for a total amount of 517.7 million dollars. The USA is 2.3 times more than in 2015 but by 7.4% less than in 2011. The situation with combine harvesters is similar. In 2016, they imported $ 271.5 million. The USA is 2.5 times more than in 2015 but by 0.3% less than in 2011. In quantitative terms in 2016, tractors were imported 2.5 times less than in 2011, combine harvesters—on the contrary, accordingly 1.2 times more.

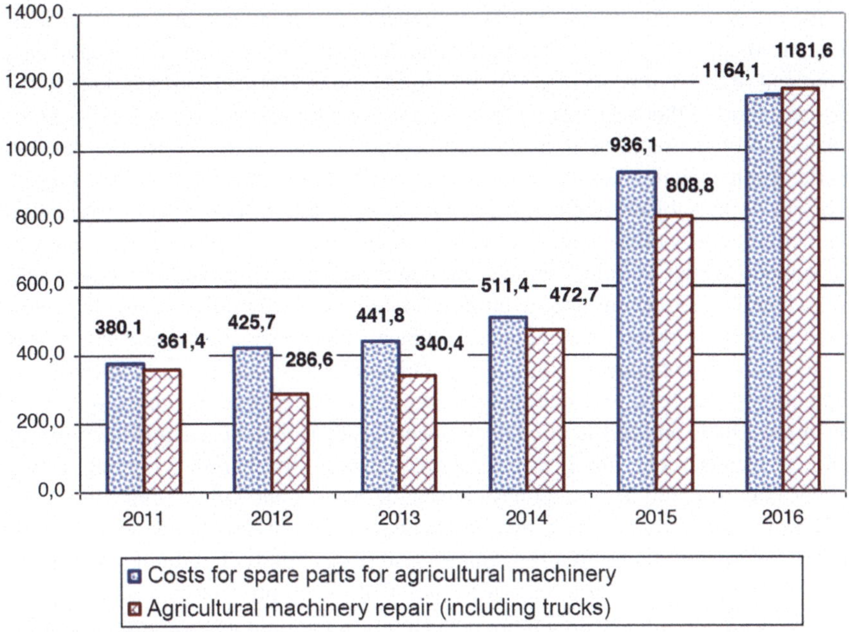


**Fig. 4** Dynamics of expenses for spare parts and repair of agricultural machinery, million UAH. *Source* Calculated according to the State Statistics Service of Ukraine

In 2011–2016, the average price of a tractor in the dollar equivalent increased 2.3 times and the combine harvester, on the contrary, decreased by 16.8%, but in the UAH, the tractor price on average increased almost 7 times, the combine harvester 2.5 times. Of course, the high cost of imported machinery is a significant obstacle to modernizing and updating of the agrarian production technical support.

The analysis of the market for agricultural machinery shows that agricultural producers are mostly turning to the secondary market (used cars) and the market for the restoration of machinery. The technical equipment maintenance in a technically good condition during the whole period of exploitation is another important function of the technical service. Unfortunately, the prices for spare parts and repairs are increasing substantially in the country, the special need for which is observed in used and restored machinery. In Fig. 4, the dynamics of expenses for spare parts and repair of agricultural machinery shows.

For 2011–2016, the cost of spare parts for agricultural machinery has tripled, and an average annual increase was 25.1%. At the same time, the number of engines purchased for tractors, combines and self-propelled machines, as well as trucks, on the contrary, decreased annually by 5.1, 10.7, and 8.0%, accordingly. That is, the overall increase in the cost of spare parts for agricultural machinery was due to the rapid increase in prices for them.

According to calculations, in order to renew the machine-tractor park of agricultural enterprises at the level of technological demand at prices that were established by the end of 2016, it is necessary to invest more than 200 billion UAH. Due to the crisis phenomena that have taken place in agriculture for many years, as well as accumulated problems with the industry technical provision, there is a need to develop an effective mechanism for the technical service investing in agriculture (see Fig. 5).

In Fig. 5, the subjects of technical service in agriculture, sources of investment in their activities, as well as means of the investment source replenishment, are shown. The subjects of the technical service include technical and technological research, research and development organizations and car testing facilities, tractor factories, agricultural and food engineering, agrotechnical enterprises, specialized enterprises in trade, leasing companies, engineering and technical subdivisions of agricultural commodity producers, information advisory services, training institutes, qualification improvement and certification of engineering and technical staff and operators of technical facilities, etc. The investments are necessary for their development. Their sources may be equity funds, moneys raised, borrowed, and public funds.

The analysis of sources of the investment and innovation activity financing of agricultural enterprises shows that they mainly use equity funds. To complement them, we propose paying special attention to the need of the production structure profitability increasing for self-financing opportunities, the fight against price disparity, and also the depreciation policy improvement. We believe it is necessary to increase the production profitability by improving the production organization, payment and labor motivation, logistics, marketing, the innovative technologies use, etc.

A very thoughtful analysis of the price ratio for agricultural products and the industries providing its material and technical resources revealed the existence of disparity, which, according to calculations, significantly influences the profitability of agricultural production. In order to eliminate the disproportions that arise in this case, it is necessary to apply a system of state regulation administrative instruments.

Improvement of depreciation policy is the next important step in the formation of its own investment resources. Unfortunately, there are problems that prevent the enterprises timely and effective update of fixed assets—the misuse of funds from depreciation funds, outdated depreciation rates. We believe that the introduction of a controlling system at the enterprises and the use of accelerated depreciation of agricultural machinery will facilitate the purchase of more productive modern machines and equipment.

Despite the fact that enterprises have significant own funds, their size is not enough to invest the development of the technical service subjects of agriculture. That is why it is necessary to look for opportunities to replenish external sources of investment financing (borrowed, moneys raised, and state funds). Among them, special attention is required for issuance of securities for the wide attraction of production structures funds and population for realization of concrete investment projects, creation of conditions for direct foreign investments activization and use of international financial organizations loans, mortgage lending expansion.

One of the most effective ways of investment external financing is a financial leasing and a bank lending. In Table 4, the data for a comparative analysis of finan-

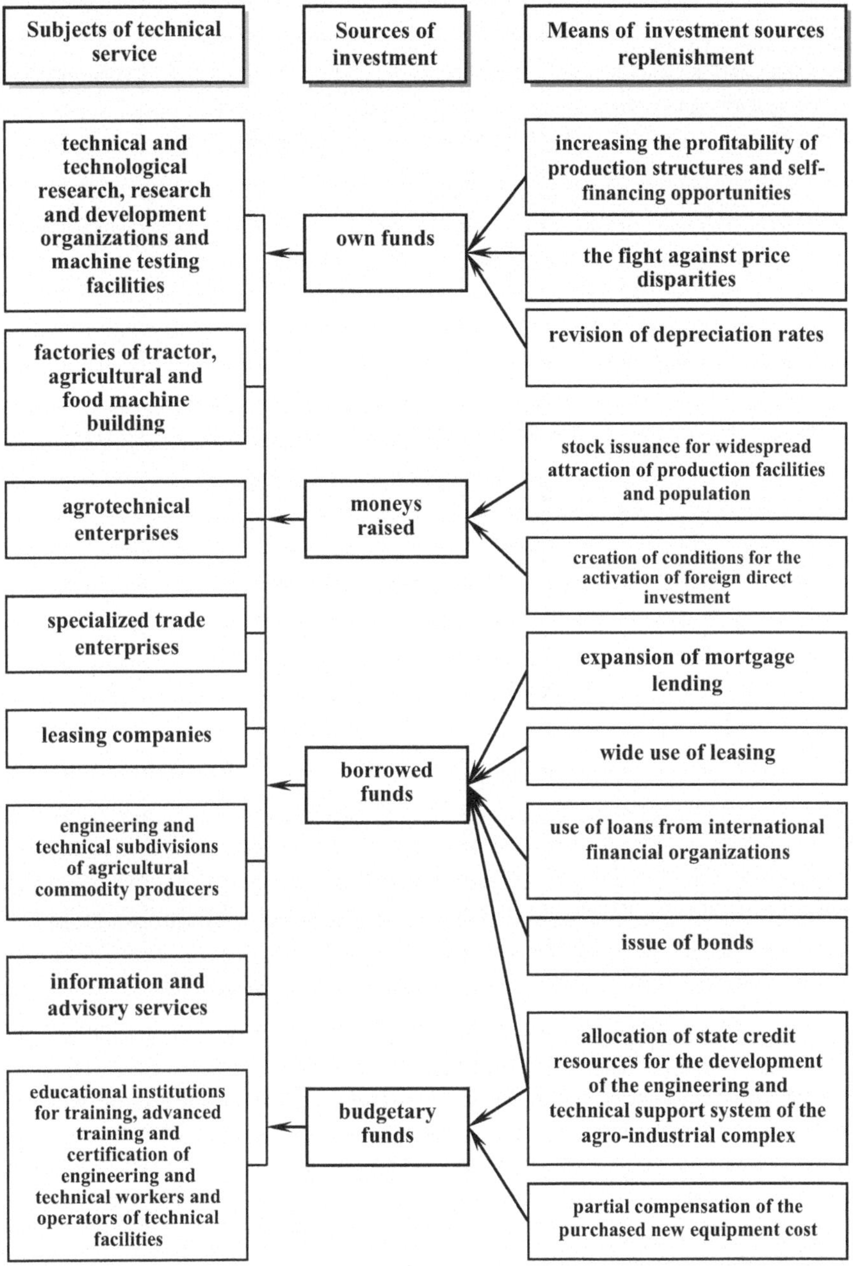


**Fig. 5** Mechanism of technical service investment of agriculture. *Source* Compiled using [12]

**Table 4** Comparative analysis of financial leasing and bank lending terms when purchasing a new agricultural machinery

| Indexes | Tractor KhTZ-242K | |
|---|---|---|
| | NJSC "Ukragroleasing" | Ukreksimbank |
| Cost of a tractor, UAH | 1,734,000.00 | |
| Lending period, years | 3 | |
| Debt repayment | Monthly | |
| Annual rate, % | 11 | 17.5 |
| Advance payment, % | 15 | |
| UAH | 260,100.00 | |
| Monthly payment, UAH | 56,768.44 | 40941.67 |
| Repayment of the principal amount of debt, total, UAH | 1,734,000.00 | |
| Maintenance of debt (payment of interest), UAH | 309,679.00 | 397,645.94 |
| Other payments, UAH | 86,700.00 | 68,799.00 |
| Total cost of acquiring equipment, UAH | 210,379.00 | 2,200,444.94 |
| Increase in the credit service, UAH | 396,379.00 | 466,444.94 |
| % | 22.9 | 26.9 |

*Source* Calculated according to NJSC "Ukragroleasing" and JSC Ukreksimbank

cial leasing and bank lending terms for the purchase of a new tractor KhTZ-242K (production of PJSC "Kharkiv Tractor Plant") are provided.

NJSC "Ukragroleasing" provides the opportunity of financial leasing with an interest rate 11% per annum, the amount of the previous lease payment in respect of the cost of equipment reimbursement from 15%, for a term up to 3 years, with the monthly payment order. In JSC "Ukreksimbank" for the purchase of new agricultural machinery lending terms up to 5 years, monthly or quarterly repayment of debt, the size of the down payment from 15%.

The analysis shows that financial leasing terms are more attractive than bank lending. The rise in the credit service at NJSC "Ukragroleasing" is 22.9%. At the same time, the rise in credit service of UEB is 26.9%, which is more by 4%.

LLC "Advance-Leasing," VAB Leasing, Alfa-Leasing, OTP Leasing, Raiffeisen Leasing Aval, Tekom-Leasing, ESKA Capital provide a full range of financing services for the motor vehicle, agricultural machinery, and special equipment purchase. We have analyzed and recognized the attractive terms for providing their leasing services.

According to the decision of the Cabinet of Ministers of Ukraine [11], agricultural enterprises have the opportunity to submit documents for partial compensation of the cost of purchased agricultural machinery in the amount of 25%. In addition, farm enterprises can receive on a nonrefundable basis, through the state bank, a compensation by 40% of the cost of purchased equipment and domestic equipment: 25% at the expense of the budget program of the CPCMK 2801580 "Financial support of agricultural producers" in the direction "Partial compensation of the cost agricultural machinery and equipment of domestic production," as well as 15%—due to the budget program of the CPCMC 2801230 "Financial support for the development of farm enterprises."

## 3 Conclusion

Sustainable development of domestic agricultural production requires an increase in the volume of investment in the technical service, which will have a leading role in the technical provision of agricultural producers. Unfortunately, the current state of the industry technical potential for a long time has a steady tendency to deteriorate due to many reasons, including the lack of an effective mechanism for investing in the development of the technical service subjects.

The proposed sources of investment source replenishment reveal the conceptual directions of solving existing problems, one of which is the lack of sufficient own funds for investing. As an effective external source of financing for investments in agricultural technical support, we propose a more extensive use of financial leasing.

The prospects for further research on this issue are the identification of the institutional reform directions that will improve the investment attractiveness and technical support of agricultural production.

## References

1. Sandin, J., Berggren, C.: A simplified service engineering approach used by an industrial service solutions provider. Procedia CIRP **30**, 30–35 (2015)
2. Voitiuk, V.: Strukturno-funktsionalne obgruntuvannia systemy tekhnichnoho servisu [Structural and functional maintenance of the system to technical service]. Silskohospodarski mashyny [Agric. Mach.] **27**, 15–24 (2014)
3. Lips, M.: Repair and maintenance costs for nine agricultural machine types. Trans. ASABE **56**(4), 1299–1307 (2013)
4. Calcante, A., Fontanini, L., Mazzetto, F.: Repair and maintenance costs of 4WD tractors in northern Italy. Trans. ASABE **56**(2), 355–362 (2013)

5. Xiao, L., North, D.J.: The graduation performance of technology business incubators in China's three tier cities: the role of incubator funding, technical support, and entrepreneurial mentoring. J. Technol. Transf. **42**(3), 615–634 (2017)
6. Rublov, V.: Innovatsiini napriamy zabezpechennia zapasnymy chastynamy tekhnichnoho servisu silskohospodarskoi tekhniky [Innovative directions of maintenance of spare parts of technical service of agricultural machinery]. Tekhnichnyi servis ahropromyslovoho, lisovoho ta transportnoho kompleksiv [Tech. Serv. Agroindustrial Forest. Transp. Complexes] **2**, 34–40 (2014)
7. Sydorchuk, O.: Kontseptualni zasady rozvytku rynku tekhnichnoho servisu silskohospodarskoho vyrobnytstva [Conceptual basis of development of the market of technical service of agricultural production]. Visnyk ahrarnoi nauky [Bull. Agrarian Sci.] **8**, 48–52 (2014)
8. Didur, V., Hrytsaienko, I., Hrytsaienko, H.: Metod analizu iierarkhii u vyznachenni priorytetnykh napriamiv intensyfikatsii zemlerobstva [Method of analysis of hierarchies in determining the priority directions of intensification of agriculture]. Naukovyi visnyk TDATU [Sci. Herald TDATU] **6**(1), 22–34 (2016)
9. Skotsyk, V.: Kontseptualni zasady rozvytku rynku silskohospodarskoi tekhniky v Ukraini [Conceptual framework for the development of the agricultural machinery market in Ukraine]. Ekonomika ta upravlinnia APK [Econ. Manage. Agro Ind. Complex] **2**, 114–118 (2014)
10. Skotsyk, V.: Problemy vidtvorennia tekhnichnykh zasobiv, yak osnovy intensyfikatsii vyrobnytstva silskohospodarskoi produktsii [Problems of reproduction of technical means, as the basis for the intensification of agricultural production]. Stalyi rozvytok ekonomiky [Sustain. Dev. Econ.] **4**, 229–233 (2013)
11. Cabinet of Ministers of Ukraine: Resolution of the Cabinet of Ministers of Ukraine "On approval of the procedure for using the funds provided for in the state budget for partial compensation of the cost of agricultural machinery and equipment of domestic production" (2017). Available at http://zakon2.rada.gov.ua/laws/show/130-2017. Accessed 22 Apr 2018
12. The Verkhovna Rada of Ukraine: The Law of Ukraine "About the system of engineering and technical support of the agro-industrial complex of Ukraine" (2006). Available at http://zakon3.rada.gov.ua/laws/show/229–16. Accessed 22 Apr 2018

# Location of Social Capital in the Labor Protection of the Enterprise

Mykola Hrytsaienko, Yuri Rogach and Mykhailo Zorya

## 1 Introduction

The state of labor protection is an important socioeconomic characteristic of both an individual enterprise and society as a whole. The presence and growth of the proportion of hazardous and harmful production factors, the low level of industrial and technological discipline of workers, the absence of adequate control over the safe conduct of the work by the heads in accordance with the regulatory acts on labor protection, inconsistency of reforms, lack of investment, poor conditions and labor safety are one of the main reasons for the unsatisfactory state of occupational safety. Formation of socioeconomic conditions for the creation of safe and harmless working conditions is the most important task for any enterprise, regardless of the form of ownership, the organizational–legal form of management and the sphere of activity. Therefore, the problem of finding new tools for the system of occupational safety modernizing, one of which may be the social capital of the company.

### *1.1 Analysis of Recent Studies and Publications*

Social capital is considered by scientists as a complex of able to bring income of connections and socioeconomic relations that arise in a particular social network based on existing norms and trust. The connection of the labor protection system with the social capital of the company has already attracted the interest of researchers. So,

M. Hrytsaienko (✉)
Department of Entrepreneurship, Trades and Exchange Activities, Tavria State Agrotechnological University, B. Khmelnytsky Ave., 18, Melitopol, Ukraine
e-mail: nick.gric@gmail.com

Y. Rogach · M. Zorya
Department of Civil Security, Tavria State Agrotechnological University, B. Khmelnytsky Ave., 18, Melitopol, Ukraine

© Springer Nature Switzerland AG 2019

V. Nadykto (ed.), *Modern Development Paths of Agricultural Production*,
https://doi.org/10.1007/978-3-030-14918-5_5

N. Breinegaard, J. H. Jensen, and J. P. Bond on the basis of the study concluded that the unsatisfactory psychosocial working environment, which includes such concepts as social capital, organizational justice and quality of management, contributes to the early retirement of civil servant disability [1]. C. Dura, A.-P. Păun, and R. I. Moraru revealed the link between corporate health, life-saving, and financial performance in socio-responsible companies [2]. M. Jarvis, A. Viverere, and P. Tint defined the differences between formal security and real security in small- and medium-sized enterprises in Estonia, on which, among other factors, is influenced by social capital [3]. T. Oksanen, A. Kouvonen, Y. Vachter, M. Virtanen, and M. Kivimaki investigated the influence of social capital on the development of depression in the working place [4]. T. Osho, A. Inoue, and A. Tsutsumi studied the effect of social capital on the connection between unfavorable working conditions and psychological disorder among Japanese workers [5]. S. Rao analyzed the culture of labor safety and the accidents at the enterprise in terms of an available social capital [6]. J. J. Tang, S. Leka, N. Hunt, and S. MacLennan investigated the impact of the social capital on occupational safety and health in the Chinese education sector [7]. M. Tei-Tominaga and M. Nakanishi analyzed the impact of the social capital, favorable and ethical working conditions on accidents at manufacturing site, injuries, and incidents of serious psychological disorders among Japanese nurses in health care [8].

Despite the existence of research in this direction, the issues of the influence of social capital on the system of labor protection at enterprises of Ukraine remain poorly understood, which determines the relevance of the chosen research topic.

## 1.2 Statement of the Objective and Tasks of the Study

The purpose of the article is to highlight the place of social capital in the company's labor protection.

## 1.3 Materials and Methods of Research

Solving the set of research tasks was due to the application of general scientific and special methods: theoretical generalization, system analysis and synthesis, comparative analysis, abstract-logical and economic-statistical analysis, method of analysis of hierarchies on the basis of expert assessments, which were used to develop conceptual provisions with improvement of the development mechanism of the labor protection system at the enterprise on the basis of the formation and use of social capital. Research theoretical basis is the scientific work selection of domestic and foreign scientists, and the toolkit for research has become the system analysis, factor analysis and regression analysis, the method of analysis of hierarchies. The information base of the study is legislative and normative acts of Ukraine that regulate labor protection issues, materials of the State Statistics Service of Ukraine, scientific articles and monographs of scientists, results of surveys and analytical work of authors, etc.

## 2 The Basic Part of the Study

The issue of labor protection is an urgent problem in all countries of the world. The legislation of Ukraine recognizes the labor protection as a system of legal, socioeconomic, organizational and technical, sanitary and hygienic and therapeutic and preventive measures and means aimed at saving the life, health, and efficiency of a person in the process of labor activity [9].

The state of labor protection at the enterprise characterizes many indicators, including indicators of injuries related to production. In Table 1, indicators of occupational injuries at Ukrainian enterprises for 2010–2016 are presented.

Data analysis (Table 1) shows that for 2010–2016, indicators of occupational injuries at enterprises of Ukraine tend to be reduced. Thus, the total number of accidents during this period decreased annually by 15.7% and amounted to 4287 units

**Table 1** Dynamics of injury indicators connected with production at enterprises of Ukraine

| Indexes | 2010 | 2011 | 2012 | 2013 | 2014 | 2015 | 2016 | Average annual growth rate, % |
|---|---|---|---|---|---|---|---|---|
| Number of accidents, units | 11,947 | 10,705 | 9732 | 1855 | 4805 | 4254 | 4287 | −15.7 |
| Including with a fatal outcome | 543 | 561 | 548 | 687 | 356 | 305 | 340 | −7.5 |
| Per 100 thousand employees, persona | 5.6 | 5.9 | 5.8 | 5.1 | 4.3 | 4.0 | 4.5 | −3.6 |
| The number of accident victims[a], total, persona | 12,234 | 10,914 | 10,067 | 1867 | 4973 | 4444 | 4429 | −15.6 |
| Per 1000 employees, persona | 1.2 | 1.1 | 1.0 | 0.9 | 0.6 | 0.6 | 0.6 | −10.9 |
| Number of days of disability, total, thousand people days | 492.1 | 444.4 | 412.4 | 28.9 | 200.7 | 211.8 | 222.4 | −12.4 |
| Per 1000 employees, people days | 47.4 | 43.3 | 39.9 | 38.6 | 22.4 | 26.3 | 28.3 | −8.2 |

[a]Which resulted in disability for 1 working day or more, and from fatal accidents
*Source* Calculated according to the State Statistics Service of Ukraine

**Table 2** Expenses of Ukrainian enterprises due to accidents related and not related to production

| Indexes | 2010 | 2011 | 2012 | 2013 | 2014 | 2015 | 2016 | Average annual growth rate, % |
|---|---|---|---|---|---|---|---|---|
| Expenses of the enterprise, caused by accidents, total, mln | 21.2 | 22.8 | 22.4 | 22.3 | 18.8 | 18.8 | 15.0 | −5.6 |
| Accrued cost of spoiled tool equipment, destroyed buildings, constructions, UAH million | 4.3 | 5.5 | 6.5 | 5.1 | 3.2 | 7.4 | 4.1 | −0.8 |
| The amount of fines paid by the officials of the enterprise for violation of the legislation on labor protection related to an accident, including for its concealment, thousand UAH | 181.3 | 95.8 | 247.6 | 199.7 | 145.3 | 102.9 | 115.7 | −7.2 |

*Source* Calculated according to the State Statistics Service of Ukraine

in 2016. The number of accidents with a fatal outcome per 100 thousand employees decreased annually by 3.6%, and in 2016, it was 4.5 units. At the same time, there are many factors that cast doubt on the contradiction of the given data: the physical and moral depreciation of fixed capital of many enterprises, the unsatisfactory from the point of view of labor protection, the state of technological processes, loose industrial and technological discipline, insufficient level of personnel training, lack of responsible attitude of employers to occupational safety issues, etc.

That is, we can assume that there is a high probability of concealing accidents from registration, which significantly distorts the real state of business. In favor of this view are data on the costs of enterprises caused by accidents that are related and not related to production (see Table 2).

During 2010–2016, the expenses of enterprises, caused by accidents, decreased annually by 5.6%, the cost of the spoiled tool equipment, the destroyed buildings, constructions—accordingly 0.8%, the amount of fines paid by the officials of the enterprise for violation of the legislation on labor protection related to an accident, including for its concealment—by 7.2%, respectively.

We believe that the size of the costs of enterprises caused by accidents does not reflect the real state of affairs in the sphere of occupational safety.

According to the US Department of Labor [10], in 2016, the number of injuries in the workforce with a fatal outcome of 100,000 was 3.6 units (a quarter more than in Ukraine). According to the International Labor Organization [11], in Romania this index was 3.8 units, Bulgaria—3.0 units, Czech Republic—2.9 units, Belarus—2.3 units, Japan and Slovakia—2.0 units, Spain—1.8 units, Sweden—1.0 units. That is, we can conclude that the level of occupational injuries in Ukraine is very high compared to other countries of the world.

The most traumatic in Ukraine is the processing industry, as well as extractive industry and quarrying. In 2016, the number of victims of industrial accidents in the processing industry amounted to 990 people (22.4% of the total in Ukraine), including the number of victims in the production of food products was 188 people, or 19.0% of victims in the metallurgical industry—157 people, respectively, or 15.9%, in the manufacture of machinery and equipment—respectively, 117 people or 11.8%. In the extractive industry and development of quarries, the number of victims of industrial accidents in 2016 amounted to 969 people (21.9% of the total in Ukraine), of which 81.4%, or 789 people—in the extraction of stone and brown coal.

To determine the influence of factors on the effectiveness of the company's occupational safety and health system, we use the method of hierarchy analysis (MAI), which is used to solve complex-structured multicriteria tasks with a hierarchical structure. Its essence consists in decomposing complex tasks into several simple components (quantitative and qualitative), for further processing by methods of matrix algebra of experts successive judgments, giving estimates in pairwise comparisons of criteria and alternatives. At the same time, the method is systemic, economical in terms of time for its application, requires the involvement of a small number of experts, allows obtaining new knowledge about the subject of evaluation, and checks the validity and consistency of the expert judgments received.

The hierarchical model is based on the data of Table 3, which contains a list of factors that determine the effectiveness of the occupational safety system at the enterprise.

At the first level, the purpose of the hierarchy analysis is determined—an assessment of the labor protection system effectiveness at the enterprise (see Fig. 1).

Achievement of this goal is conditioned by the importance of the relevant factors (elements of Level 2): trust and respect for the employees of the company, trust and reciprocity among staff members of the same hierarchical level, personnel training on safety and health regulations, compliance with labor and technological discipline, external security controls and labor protection at the enterprise, allocation of funds for material and technical means of labor protection, material and technical supporting of sanitary and hygienic and therapeutic and preventive at the enterprise. Finally, social capital, organizational measures and material and technical supporting of occupational safety at the enterprise, which generalize the above-mentioned factors, form elements of the third level of the hierarchy.

The levels of trust and respect of employees to the management personnel of the enterprise (both public and private), and trust and reciprocity among employees

**Table 3** List of factors that determine the effectiveness of labor protection system at the enterprise

| Indicator | Level of hierarchy | Conditional mark |
|---|---|---|
| Labor protection at the enterprise | 1 | 1 |
| Social capital of the enterprise | 3 | 2.1 |
| Trust and respect of employees to the management of the enterprise | 2 | 2.1.1 |
| Trust and reciprocity between employees of one hierarchical level | 2 | 2.1.2 |
| Organizational measures on occupational safety at the enterprise | 3 | 2.2 |
| Personnel training on safety and health regulations | 2 | 2.2.1 |
| Compliance with labor and technological discipline | 2 | 2.2.2 |
| External control of safety and labor protection at the enterprise | 2 | 2.2.3 |
| Material and technical support of the labor protection system at the enterprise | 3 | 2.3 |
| Allocation of funds for material and technical means of labor protection | 2 | 2.3.1 |
| Allocation of funds for material and technical support of sanitary-hygienic measures at the enterprise | 2 | 2.3.2 |
| Allocation of funds for medical and preventive measures | 2 | 2.3.3 |

*Source* Compiled based on the research

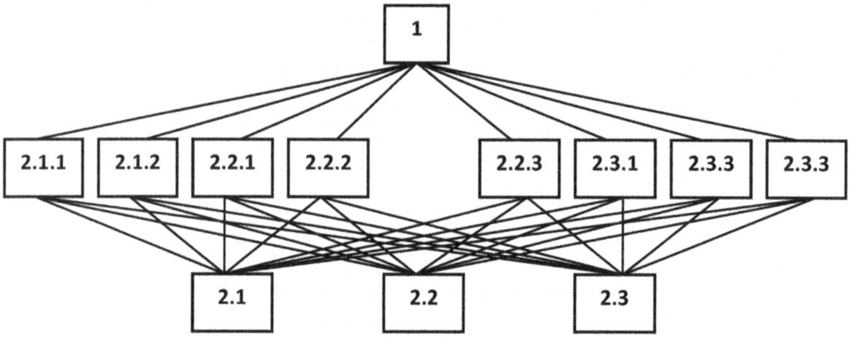


**Fig. 1** Hierarchy to assess the factors affecting the effectiveness of the occupational safety system at the enterprise. *Source* Compiled based on the research

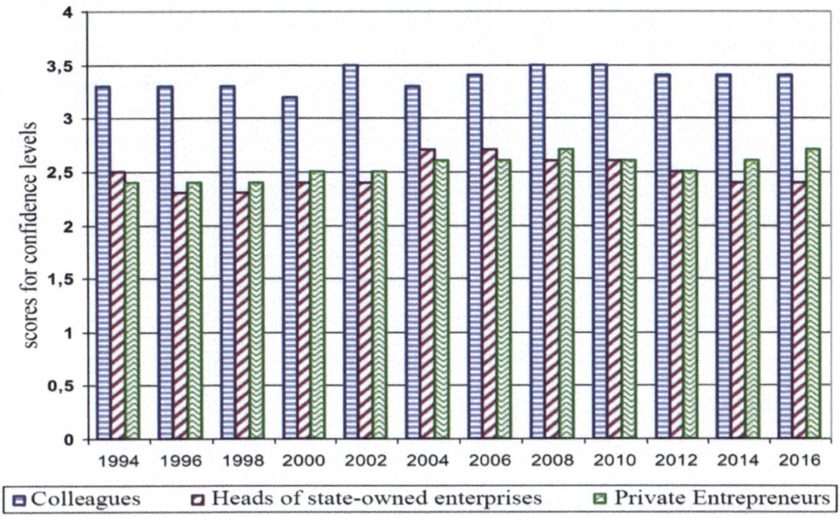


**Fig. 2** Dynamics of confidence levels for colleagues, heads of state enterprises and private entrepreneurs, scores*. *The average score was calculated for the 5-point scale, where 1 is complete distrust, and 5 is full confidence. *Source* Compiled based on the results of national annual monitoring surveys [13, pp. 148–149]

of one hierarchical level (colleagues) are determined on the basis of the results of national annual monitoring surveys (see Fig. 2).

Personnel training of on safety and labor protection rules, compliance with labor, and technological discipline are estimated taking into account the data of the Social Insurance Fund of Ukraine. External control of safety and labor protection, material and technical support financing of labor protection, sanitary and hygienic and preventive measures at enterprises are determined on the basis of data from the State Statistics Service of Ukraine.

As experts during the analysis of the efficiency of the labor protection system, leading teachers of the Department of Civil Security of the Tavria State Agrotechnological University have been involved.

More detailed features of the hierarchy analysis method are given in [12].

In Table 4, it shows the matrix of pairwise comparisons and priority vectors of elements of the second level of the hierarchy regarding the effective use of the occupational safety system at the enterprise (object 1 hierarchical model).

Local vectors of priorities are defined by the formula:

$$W_i = \frac{A \times B}{\lambda_{\max}}, \tag{1}$$

where

$\lambda_{\max}$ the maximum actual value of the matrix;

**Table 4** Matrix of pairwise comparisons and priority vectors elements of the second level of the hierarchy relative to the object 1

| | 2.1.1 | 2.1.2 | 2.2.1 | 2.2.2 | 2.2.3 | 2.3.1 | 2.3.2 | 2.3.3 | Local priority vector $W_i$ |
|---|---|---|---|---|---|---|---|---|---|
| 2.1.1 | 1 | 1/5 | 1/7 | 1/7 | 5 | 1/7 | 1/7 | 1/7 | 0.0240 |
| 2.1.2 | 5 | 1 | 3 | 1/5 | 5 | 1/7 | 1/5 | 1/5 | 0.0596 |
| 2.2.1 | 7 | 1/3 | 1 | 1/3 | 5 | 1/3 | 1/3 | 1/3 | 0.0636 |
| 2.2.2 | 7 | 5 | 3 | 1 | 7 | 3 | 5 | 5 | 0.3171 |
| 2.2.3 | 1/5 | 1/5 | 1/5 | 1/7 | 1 | 1/7 | 1/7 | 1/7 | 0.0167 |
| 2.3.1 | 7 | 7 | 3 | 1/3 | 7 | 1 | 9 | 7 | 0.2823 |
| 2.3.2 | 7 | 5 | 3 | 1/5 | 7 | 1/9 | 1 | 1/3 | 0.1002 |
| 2.3.3 | 7 | 5 | 3 | 1/5 | 7 | 1/7 | 3 | 1 | 0.0211 |
| $\lambda_{max} = 10.0945$; CI = 0.2992; CR = 0.2122 | | | | | | | | | |

$A$ is the matrix of weight judgments;
$B$ is a matrix of normalized priorities.

Maximum actual value of the matrix is determined by the formula:

$$\lambda_{max} = R_1 \sum_{i=1} a_{i1} + R_2 \sum_{i=1} a_{i2} + \cdots + R_n \sum_{i=1} a_{in},$$
$$R = \sqrt[n]{\prod_{j=1}^{n} a_{ij}} \times \left( \sum_{i=1}^{n} \sqrt{\prod_{j=1}^{n} a_{ij}} \right) \tag{2}$$

where

$i$ index of the line of matrix A;
$j$ column index of matrix A;
$a_{ij}$ expert assessments.

To assess the consistency of the data in the matrices of pairwise comparisons, we use the values:

CI consistency index:

$$\text{CI} = \frac{\lambda_{max} - n}{n - 1}, \tag{3}$$

where

$n$ the number of columns and rows of the matrix;
CR consistency ratio:

$$\text{CR} = \frac{\text{CI}}{\text{CIS}}, \tag{4}$$

where CIS is the average value of the consistency index as a random variable, which is obtained experimentally as a result of processing of a large number of pair comparison matrices that were generated randomly.

According to calculations, the most significant object 2.2.2—compliance with employees technological and labor discipline (local priority vector 0.3171).

In Table 5, it shows the matrix of pairwise comparison and priority vector of elements of the third level of the hierarchy with respect to the object 2.1.1. At the same level of the hierarchical model, eight similar matrices are composed.

According to the statistical characteristics, all matrices of pairwise comparisons are consistent.

The following formula is used to perform hierarchical synthesis:

$$W' = \sum_{i=1}^{n} W_{ji} \times W_i \tag{5}$$

where

$W'$ global priority vector of each alternative;
$W_{ij}$ local priority of the $j$th alternative on the $i$th criterion;
$W_i$ is the local priority of the $i$th criterion.

The results of hierarchy analysis are presented in Table 6.

Thus, on the basis of the hierarchy analysis, one can conclude that the global priority of the highest efficiency of the labor protection system at the enterprise determines

**Table 5** Matrix of pairwise comparisons and priority vectors elements of the third level of the hierarchy with respect to the object 2.1.1

| | 2.1 | 2.2 | 2.3 | Local priority vector $W_i$ |
|---|---|---|---|---|
| 2.1 | 1 | 5 | 1/5 | 0.2344 |
| 2.2 | 1/5 | 1 | 1/5 | 0.0801 |
| 2.3 | 5 | 5 | 1 | 0.6854 |
| $\lambda_{max} = 3.2947$; CI = 0.1473; CR = 0.2541 | | | | |

**Table 6** Summary of the analysis of the hierarchy regarding the integrated assessment of the influence of factors on the effectiveness of the company's occupational safety and health system

| Group of factors | Global priority vector of each group of factors $W'$ |
|---|---|
| Social capital of the enterprise (2.1) | 0.2066 |
| Organizational measures on labor protection at the enterprise (2.2) | 0.3648 |
| Material and technical support of the system of labor protection at the enterprise (2.3) | 0.4284 |

its logistical support (global vector of priorities 0.4284), the next factor—organizational measures for the protection of labor (global priority vector 0.3648).

The dependence of the number of accidents (in) on the costs of enterprises on the labor protection system ($x$), which is calculated on the basis of the indexes of the investigated enterprises for 2010–2016, has the following form:

$$y = -1.2436x + 23312; \quad D = 0.8113; \quad R = 0.9007 \tag{6}$$

That is, every one million of hryvnias, which is invested by enterprises in the system of labor protection, reduces the number of accidents by 1.24 units. The correlation coefficient indicates a close relationship between the indicators ($R = 0.9007$), the determination coefficient—the significance of the factor included in the model (the variation of the dependent variable due to the variation of the independent variable $x$ by 81.13%, the other 18.83%—the influence of factors that are not included in the model).

At the same time, the social capital of the enterprise, which was assessed as the level of trust in management personnel, as well as the trust and relationships between employees of one hierarchical level, has a rather high impact on the effectiveness of the system of labor protection—the global priority vector is 0.2066.

In addition to trust, social capital of the enterprise is characterized by social networks and norms. Social networks are created through partnerships, collaborations, and relationships between staff at different hierarchical levels. The standards of social capital of an enterprise related to the system of labor protection combine the formal rules of its regulation, generally accepted informal rules and norms of labor protection at the enterprise, as well as interpersonal and group informal norms of interaction, existing codes of conduct (see Fig. 3).

The social capital of the company contributes to reducing the uncertainty and risk in the system of labor protection through social control over compliance with the formulated rules and procedures, which serves to create a positive image of the enterprise as a whole.

## 3 Conclusion

The question of the formation and the use of the enterprise social capital need further study, but the results of the conducted research indicate that it is an important factor to be taken into account when improving the system of occupational safety at the enterprise.

Positive changes in the system of labor protection should be achieved not only at the expense of improving its organization and financing mechanism. In the complex of urgent measures, it is expedient to include other mechanisms of its development aimed at developing social interaction, social cohesion and partnership, increasing the level of trust, spreading positive social norms and values, as well as the formation

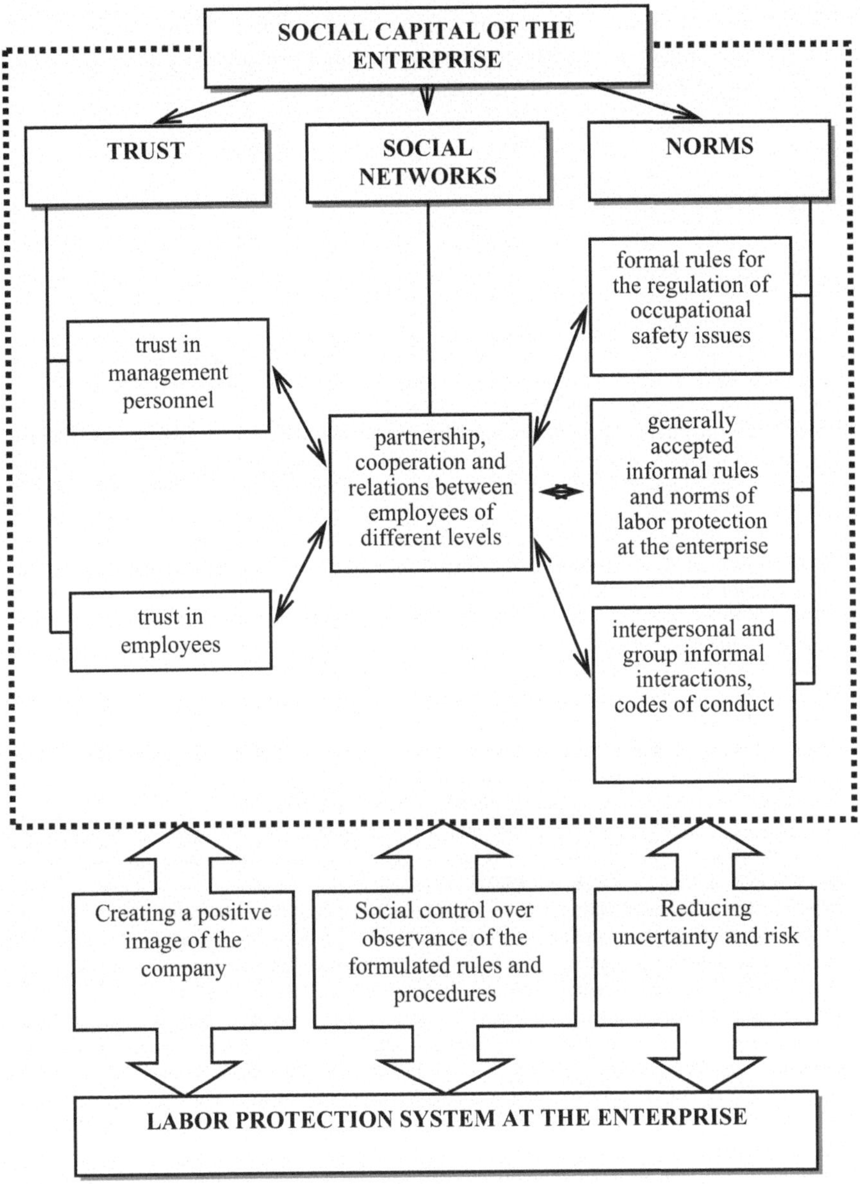


**Fig. 3** Mechanism of influence of social capital on the system of labor protection of the enterprise. *Source* Compiled based on the research

of institutions that promote the accumulation of social capital. These measures will have a positive impact on the effectiveness of the labor protection system, as well as on the economic development of the country as a whole.

## References

1. Breinegaard, N., Jensen, J.H., Bonde, J.P.: Organizational change, psychosocial work environment, and non-disability early retirement: a prospective study among senior public employees. Scand. J. Work Environ. Health **43**(3), 234–240 (2017)
2. Dura, C., Păun, A.-P., Moraru, R.I.: Empirical analysis on the relationship between corporate health and safety performance and the financial outcome within socially responsible companies. Quality Access to Success **19**(162), 155–160 (2018)
3. Järvis, M., Virovere, A., Tint, P.: Formal safety versus real safety: quantitative and qualitative approaches to safety culture—evidence from Estonia. In: Proceedings of the Latvian Academy of Sciences, Section B: Natural, Exact, and Applied Sciences, vol. 70, issue 5, pp. 269–277 (2016)
4. Oksanen, T., Kouvonen, A., Vahtera, J., Virtanen, M., Kivimäki, M.: Prospective study of workplace social capital and depression: are vertical and horizontal components equally important? J. Epidemiol. Commun. Health **64**, 684–689 (2010)
5. Oshio, T., Inoue, A., Tsutsumi, A.: The mediating and moderating effects of workplace social capital on the associations between adverse work characteristics and psychological distress among Japanese workers. Ind. Health **52**(4), 313–323 (2014)
6. Rao, S.: Safety culture and accident analysis—a socio-management approach based on organizational safety social capital. J. Hazard. Mater. **142**(3), 730–740 (2007)
7. Tang, J.J., Leka, S., Hunt, N., MacLennan, S.: An exploration of workplace social capital as an antecedent of occupational safety and health climate and outcomes in the Chinese education sector. Int. Arch. Occup. Environ. Health **87**(5), 515–526 (2014)
8. Tei-Tominaga, M., Nakanishi, M.: The influence of supportive and ethical work environments on work-related accidents, injuries, and serious psychological distress among hospital nurses. Int. J. Environ. Res. Public Health **15**(2), 240 (2018)
9. The Verkhovna Rada of Ukraine: The Law of Ukraine "About the protection of labor" (1992). Available at http://zakon5.rada.gov.ua/laws/show/2694-12. Accessed 8 Aug 2018
10. Census of Fatal Occupational Injuries News Release: United States department of labor. Available at https://www.bls.gov/news.release/cfoi.htm. Accessed 8 Aug 2018
11. Fatal occupational injuries per 100,000 workers by economic activity: International Labour Organization. Available at http://www.ilo.org/ilostat/faces/oracle/webcenter/portalapp/pagehierarchy/Page27.jspx?_adf.ctrl-state=17otw2qayi_45#! Accessed 8 Aug 2018
12. Gry'czayenko, M.I.: Integral evaluation of effectiveness of innovation in the agricultural sector basis of the analytic hierarchy process, Odessa National University Herald. Economy **19**(3/1), 147–150 (2014)
13. Shul'ga, M.O.: Stan suchasnogo ukrayins'kogo suspil'stva: cy'vilizacijny'j vy'mir [The state of modern Ukrainian society: the civilization dimension], Institute of Sociology of the National Academy of Sciences of Ukraine, Kyiv, Ukraine (2017)

# Defining Stability of Technological Process of Growing Fruit Crop Seedlings

**Oleksandr Karaiev, Lyudmyla Tolstolik, Ivan Chyzhykov and Tetiana Karaieva**

## 1 Introduction

Concerning Ukraine entering into WTO and its intention for more close integration with European Union, the current system of technical regulation should be improved in accordance with international experience, being based on risk-oriented approach [1]. It makes the country to adapt its own system of the technical regulation, including standardization and certification, to international and European approaches.

The fruit culture nursery management is being conducted by the Law of Ukraine "On Seed and Seedling Material," according to which a producer of seedling material must be included into the "Register of Nursery Agents" [2]. The annual field evaluation of the long-term nursery plantations is being provided by the law as well. Due to that, technical feasibilities of nursery are being defined in relation to providing of stable production of the certificated planting material. Under such requirements, a nursery should register the results of finished products' quality control as well as the methodology for state control of plants growing and development in the course of technological process.

Thus, the requirements of law pull out before a producer rather intricate tasks in production management. These tasks may be solved by means of the quality control systems application for planting stocks in the process of their growing. The above control systems are currently unavailable.

The solution of this problem is possible due to development and application of the standardized methods of the statistical quality control of planting stocks. Such a practice enables to obtain the certificated planting material and form of the proper terms for nurseries registering in the Register of planting material producers "Registrer of Nursery Agents" [3].

---

O. Karaiev (✉) · L. Tolstolik · I. Chyzhykov · T. Karaieva
Tavria State Agrotechnological University, B. Khmelnitsky av. 18, Melitopol 72310, Ukraine
e-mail: sgm@tsatu.edu.ua

© Springer Nature Switzerland AG 2019

V. Nadykto (ed.), *Modern Development Paths of Agricultural Production*,
https://doi.org/10.1007/978-3-030-14918-5_6

**The Purpose of Research and the Tasks.** The purpose and the tasks of research are to minimize the risk of economic value loss by planting material producers due to providing of the stability of technological process of nursery seedlings growing by the timely realization of operational control of the grafts state and application of corresponding correcting actions.

# 2 Materials and Methods of Research

## 2.1 *Defining of Technological Process Stability*

For realizing of operational control, the method of statistical control of technological process stability of nursery seedlings growing has been worked out and proposed by us [4].

The method has such an algorithm:

(1) The arrays of the values of the controlled parameters $x_i^{(k)}$ are being formed from the sample.
(2) The mean selective arithmetic value and standard deviation are being calculated.
(3) The lower and overhead limits of the confidence interval of *k*-parameter are being calculated as:

$$x_{\min,d}^{(k)} = \overline{x^{(k)}} - \frac{t_M}{\sqrt{M}} s^{(k)} - 3\, s^{(k)} z_M \geq x_{\min}^{(k)};$$
$$x_{\max,d}^{(k)} = \overline{x^{(k)}} + \frac{t_M}{\sqrt{M}} s^{(k)} + 3\, s^{(k)} z_M \leq x_{\max}^{(k)}, \quad (1)$$

where

$x_i^{(k)}$ valid *k*th parameter;
$M$ sample size (calculation is given in Sect. 2.2);
$\overline{x^{(k)}}$ mean arithmetic selective value;
$s^{(k)}$ selective mean square deviation;
$x_{\min,d}^{(k)}$ lower bound of the confidence interval;
$x_{\max,d}^{(k)}$ upper limit of the confidence interval;
$x_{\min}^{(k)}$ lower boundary of the interval of normative values;
$x_{\max}^{(k)}$ upper boundary of the interval of normative values;
$t_M$ distribution coefficient of the arithmetic average in the sample;
$z_M$ distribution coefficient of the mean square deviation in the sample

(4) The process stability is determined by comparing the actual values of parameters of the state of grafts to normative ones:

$$x_{\max,d}^{(k)} \leq x_{\max}^{(k)}, \quad x_{\min,d}^{(k)} \geq x_{\min}^{(k)}. \quad (2)$$

If the confidence intervals of the controlled parameter values belong to the corresponding intervals of normative ones and the condition (3) is adhered, then the process is considered to be stable. Otherwise, the disorder of the process takes place; its level is determined by setting the norms of the controlled parameters values by the formula:

$$y_i^{(k)} = \frac{2x_i^{(k)} - x_i^{\min} - x_i^{\max}}{x_i^{\max} - x_i^{\min}}, \tag{3}$$

where

$x_i$ the value of the $i$th quality parameter of the $k$th product unit;
$x_i^{\max}, x_i^{\min}$ normative values of $i$th parameter,

and by calculating the modules of normalized average values:

$$\overline{y_i} = \frac{1}{M}\sum_{k=1}^{M}\left|y_i^{(k)}\right| \tag{4}$$

## 2.2 *Calculation of the Optimum Sample Size* M

The sample size is determined according to the method, having been worked by us and being set in [5], where for defining the optimal sample size, the expected a priori risk $R_{\text{anp}}$ of economic loss by the producer is calculated by the formula:

$$R_{\text{anp}} = \text{iss} \cdot M + \frac{1}{M+1}\sum_{k=0}^{M}\min\left(\text{pr}\sum_{j=k}^{N-M+k}\frac{j(M+1)!j!(N-j)!C_{N-M}^{j-k}}{(N+1)!k!(M-k)!}, (\text{pr}-\text{cs})\cdot N\right) \tag{5}$$

where

iss controlling cost for one graft;
cs production cost for graft growing;
pr expected price of sales of the unit of final product (seedling);
$N$ the number of units of production in the inspection lot;
$j$ the number of defective units of production in the inspection lot;
$k$ the number of defective units of production in the sample;
$n$ the maximum number of defective units of production that is satisfactory for the receipt of the inspection lot;
$C_{N-M}^{j-k}$ binomial coefficient, which is calculated by the formula:

$$C_{N-M}^{j-k} = \frac{(N-M)!}{(j-k)!(N-M-j+k)!}.$$

A sample size is optimal if its value corresponds to global function minimum (5) and it is determined according to the algorithm:

(1) Accept $M = 3$, calculate the risk of the producer by the Formula (5), and the resulting value is added to an array of risks.
(2) Increase a sample size by $M + 1$, and repeat a calculation.
(3) If the last value from the array of risks is less than the previous one, $M$ is increased by unit and the item (2) is performed. In the opposite case, the $M$ value is reduced by unit and obtained value of sample size is taken as an optimal one.

# 3 Results and Discussions

## *3.1 Methodology for Production Verification of the Method*

**Object of control**. The verification of the proposed method for the control of the stability of technological process of nursery seedlings growing was carried out under productive conditions of the nursery of the research farm "Melitopolske" when growing cherry seedlings (Melitopol, Ukraine). The control over technological processes took place in certain phases of plants growth and development by the values of the controlled parameters, being defined in a sample of an inspection lot. The control has been conducted according to the method that had been worked out by us [6].

The productive verification methodology for the proposed method provided to conduct control over cherry grafts state on generative rootstocks at the certain phases of their growth and development.

Operational control was subject to the samples of cherry sort grafts *Valery Chkalov* in the quantity of $N = 1230$ pcs., *Krupnoplidna*—$N = 1227$ pcs., *Melitopolska chorna* of the first year of cultivation—$N = 1240$ pcs. Grafts were in "phenological phase"—active growth of grafts (Fig. 1a).

The stability of technological process of grafts growing was being controlled by the parameter of the inoculum shoot length with the number of leaves per graft. Thus, according to [6] the grafts of sorts *Valery Chkalov* and *Krupnoplidna* with the number of leaves on grafts from 14 to 16 pcs., the length of their inoculative shoots should be in the range of 45–50 cm (Fig. 2b), and at the grafts of *Melitopolska chorna* sort having the number of leaves from 8 to 10 leaves, the length of the shoots should be from 20 to 25 cm.

**Sampling Inspection Conducting**. For convenience in sample size defining directly in production, the maps for sample volume zones have been worked out according to Formula (1) for the values of the inspection lot $N$, being within the range from 1000 to 15,000 pcs.

The optimum sample amount of $M$ sample grafts at the cost of measuring one graft iss $= 0.157$ UAH, productive prime price of nursery seedling cs $= 4.61$UAH, and the expected price per seedling $r = 50$ UAH made up $Y = 32$ pcs. (Fig. 2).

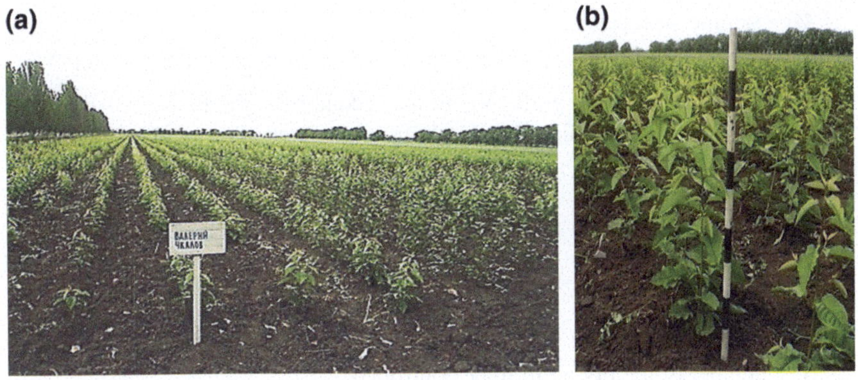


**Fig. 1** General view of Valery Chkalov cherry grafts (**a**) and their control (**b**) in a phenological phase "active graft growth" in the nursery of the research farm "Melitopolske" (Melitopol, Ukraine)

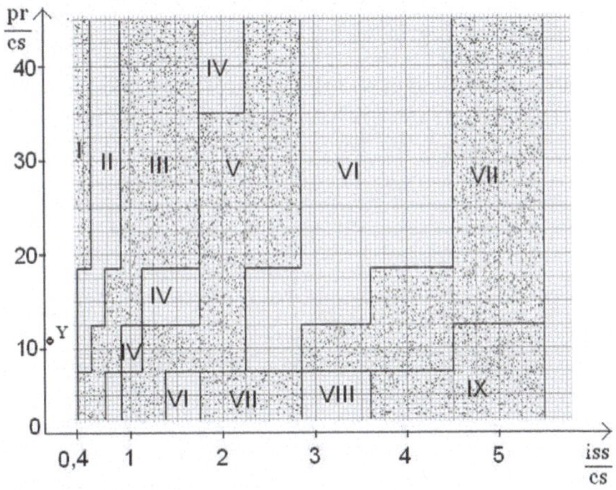


**Fig. 2** Map of volumes of samples *M* for inspection graft lot *N* = 1250 pcs.: I = 32 pcs.; II = 25 pcs.; III = 20 pcs.; IV = 15 pcs.; V = 13 pcs.; VI = 10 pcs.; VII = 8 pcs.; VIII = 6 pcs.; IX = 5 pcs.; *Y*—optimum sample size *M* from a controlled chip lot *N*, pcs.

The choice of grafts for control within the prescribed sample size was carried out according to the following rules [6]:

(1) The plants, located along the perimeter of the field (plot), were not the subject of measurement;
(2) The number of the next graft to be measured is greater than the number of the current controlled plant on a size.

$$\tau = \frac{N_ч}{M}, \tag{6}$$

where

$\tau$ numbers interval (the number is rounded to a smaller integer);

$M$ sample size, pcs.;
$N_ч$ number of grafts, except for those, located along fields perimeter, pcs.

(3) The number of the first graft ($N_1$) in the second row (the first row belongs to the perimeter) is determined randomly from the interval of numbers [1, $\tau$].
(4) The number of the next graft being the subject of control is determined by the formula:

$$N_i = N_1 + \tau(i - 1), \tag{7}$$

where

$N_1$ number of the first graft to be controlled;
$\tau$ numbers interval;
$i$ the sequence number of chips in the sample,

(5) Movement in the course of measurement is carried out according to the scheme shown in Fig. 3.

For a controlled graft lot $N = 1250$ pcs. and the sample size $M = 32$ pcs. according to Formula (6), the step of realizing control is calculated:

$$\tau = \left[\frac{1250}{32}\right] = 39.$$

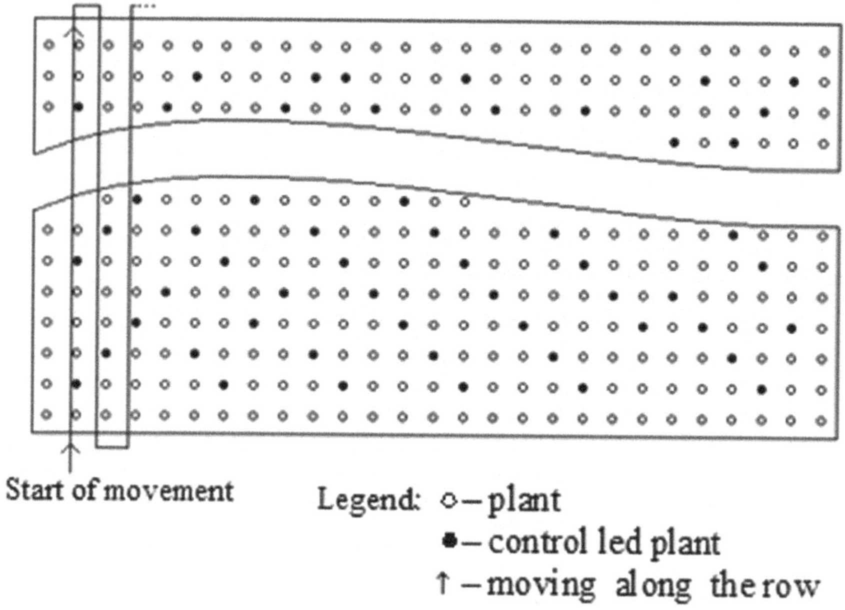


**Fig. 3** Scheme of defining grafts for selective control

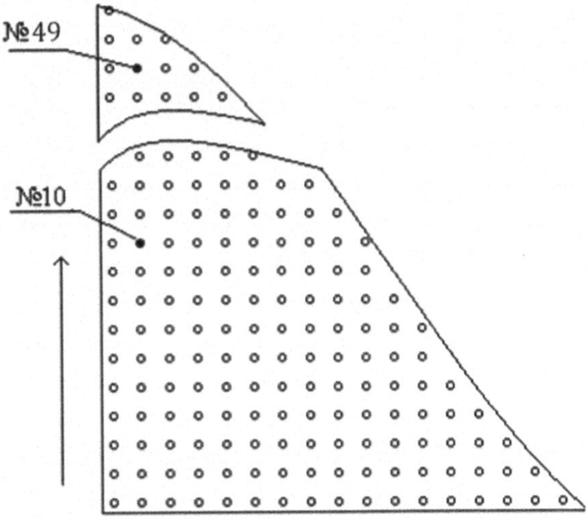


**Fig. 4** Scheme for defining the controlled grafts

Id est every 39th graft was the subject of control. The number of the first graft was elected randomly from an interval of 1–39. The control started from 10th graft, the location of which is given in Fig. 4. The numbers of the following grafts, being under control, were defined by the formula:

$$N_i = 10 + 39(i - 1).$$

## 3.2 *Results of Method Verification*

**Defining Stability of Technological Process.** As the example the measurements results of inoculative shoot length for the sort of Valery Chkalov are given.

The values of inoculant shoot length were as follows, cm: 55; 40; 20; 47; 42; 60; 54; 48; 28; 25; 53; 40; 25; 30; 27; 43; 44; 26; 5; 43; 47; 37; 33; 45; 50; 43; 10; 18; 47; 45; 46; 33. At the same time:

- the average arithmetic value $\overline{x^{(1)}} \approx 37.31$ cm;
- mean square deviation $s^{(1)} \approx 14$ cm;
- coefficient of distribution of the mean arithmetic deviation $t_{32} = 2.356$;
- coefficient of distribution of the mean square deviation $z_{32} = 1.329$.

The values of the coefficients are given in Table 1.

Then, according to (1), we have the value of the lower and upper limits of the confidence interval: $x^{(1)}_{\min,d} = 24.4\,\text{cm}$; $x^{(1)}_{\max,d} = 98.9\,\text{cm}$, under which the conditions (2)

$$x_{\max,d} = 98.9\,\text{cm} < 50\,\text{cm}, \quad x_{\min,d} = 24.4\,\text{cm} < 45\,\text{cm}$$

**Table 1** Values of $t_M$ and $z_M$ coefficients for the $M$ sample size

| *M* sample size, pcs. | Value of coefficients | |
|---|---|---|
| | $t_M$ | $z_M$ |
| 10 | 2.685 | 1.826 |
| 13 | 2.56 | 1.651 |
| 15 | 2.51 | 1.577 |
| 20 | 2.434 | 1.461 |
| 25 | 2.391 | 1.391 |
| 32 | 2.356 | 1.329 |
| 40 | 2.331 | 1.284 |
| 50 | 2.312 | 1.246 |
| 65 | 2.295 | 1.209 |
| 80 | 2.285 | 1.184 |
| 100 | 2.276 | 1.162 |
| 125 | 2.269 | 1.142 |
| 150 | 2.264 | 1.128 |

are not executed that testifies about disorder in technological process.

For *Krupnoplidna* and *Melitopolska chorna* sorts, the disorder of technological process takes place as well.

**Defining the Level of Technological Process Disorder**. Such disorder level has been defined in accordance with Formulas (3) and (4). Thus, the mean values of the modulus of the normalized graft length were as follows:

- *Valery Chkalov*—5.05;
- *Krupnoplidna*—4.1;
- *Melitopolska chorna*—5.4.

These values are given at Shewhart control chart, having been modified by us (Fig. 5) [7].

It is obvious from Fig. 5 that the mean values of the modules of the normalized values of grafts length at all the sorts are greater than 1 (area of inadmissible values); that is, the corrective actions, aimed at stimulating grafts growth (irrigation or fertilization), should be applied.

The results of calculations, carried out according to [3], showed that for *Valery Chkalov* and *Melitopolska chorna* grafts, it is expedient to apply fertilizers instead of irrigation. And for the grafts of *Krupnoplidna,* it is not recommended to conduct irrigation.

The output of the first commercial grade of seedlings, obtained without the use of control and corrective actions, did not exceed 60%, while when applying the method of control having been worked out, it made up 82% (by the parameter of the craft roots length, Fig. 6).

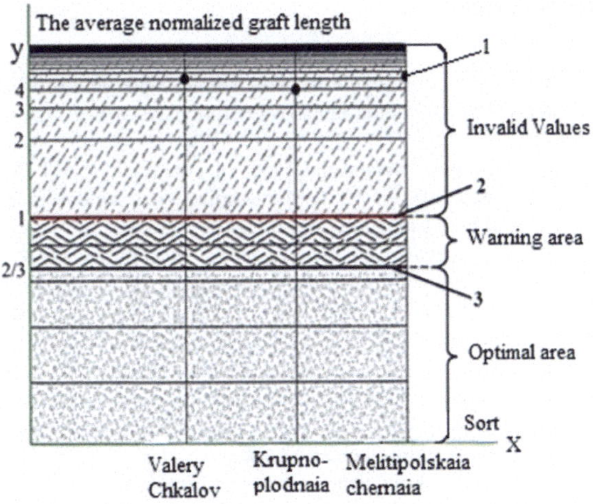


**Fig. 5** Map of average values of the length of sweet cherry variety inoculums sprout: 1—valid parameter values; 2—the upper limit of the optimal area ($y < 2/3$); 3—limit warning area ($2/3 < y < 1$)

**Fig. 6** Root system of *Valery Chkalov* cherry seedlings, being cultivated by using the operational method of control and appropriate corrective actions

## 4 Conclusions

It has been proved that at the stage of planting material growing, the stability of the technological processes in certain phases of the growth and development of plants should be determined by the method of calculating the confidence intervals of the values of the controlled parameters and comparing them with the corresponding intervals of the normative values.

To find out the level of stability or process disorder, a modified Shukhart's map has been proposed (DSTU ISO 8258). On the basis of the above method, the disruption of the technological process of cultivating cherry grafts in the phenological phase "active graft growth" was detected, the inspection lot of which was made up of 3697 pcs., that enabled to introduce the corrective actions timely and to increase the yield of first-grade seedlings by 22%.

## References

1. Zvit: "Tehnichne reguliuvanniav Ukraini: jak zabezpechyty rozvytoki zahyst prav spozhyvachiv"/ Mizhnarodna finansova corporatsiya IFS. Grypa svitovogo banku. – Canada. – 2008, s.80
2. Zakon Ukrainy: "Pronasinniai sadyvnyi material" № 411-IVvid 26.12.2002/VR Ukrainy// Vidomosti VR, 2003, № 13, s. 92; Izzmin., vnes. zgidno iz Zakonom №2505-IV (2505-15) vid 25.03.2005, VVR, 2005, № 17, 18-19, s. 267
3. Yakist productsii rozsadnytstva plodovyh kultur/ O.G. Karaiev, L.M. Tolstolik. – Melitopol: Vydavnytstvo-poligrafichnyi tsentr «Lux», 2014. – 150 s
4. Kultury plodovi. Metod statystychnogo controliu stabilnosti tehnologichnogo protsesy vyroschuvannia: DSTU ISO 7897:2015. – [Chynny vid 2016-01-07]. – K.: Derzhspozhyv standart Ukrainy, 2015. – 9 s.– (Natsionalni Standarty Ukrainy)
5. Kultury plodovi. Kontrol yakosti productsii rozsadnykiv i sadiv metodom statystychnyh vybirok: DSTU ISO 8315:2015. – [Chynny vid 2017-01-07]. – K.:Derzhspozhyv standart Ukrainy, 2015.– 17 s.– (Natsionalni Standarty Ukrainy)
6. Kultury kistochkovi. Schepy. Vymogy ta metody controliuvannia: DSTU 7639:2014. – [Chynny vid 2015-01-07]. – K.: DerzhspozhyvstandartUkrainy, 2014.– 15 s.– (Natsionalni Standarty Ukrainy)
7. DSTU ISO 8258-2001 Statystychnyi control. Controlni karty Shuharta (ISO 8258:1991, IDT)

# Design of Functional Surfaces in CAD System of SolidWorks via Specialized Software

**Yuliia Kholodniak, Yevhen Havrylenko, Iryna Pykhtieieva and Viktor Shcherbyna**

## 1 Introduction

The items which are limited to the composite surfaces are processed on machines with numerical control tool (NCT). Technology of production on machines with NCT allows processing surfaces of any form. Control program for a machine with NCT is created in the automated mode in CAM system. The initial data for CAM system is a three-dimensional computer model of the item, which is formed in CAD system (SolidWorks, AutoCAD, NX CAD et al.) [1].

Processing accuracy is determined by accuracy of the formation of three-dimensional model, accuracy of the determination of trajectory of the cutting tool by CAM system, and accuracy of the machine tool performing the programed operations [2].

The correct formation of three-dimensional model is a necessary condition of the high-quality making of the item. Design technology in CAD system assumes formation of surfaces on the basis of linear frameworks [3, 4]. Geometry of the curved lines (elements of the framework of surface) determines functional properties of the designed surfaces. Required properties of many surfaces are provided by using certain curved lines as elements of the framework. These can be evolvents, trochoids,

Y. Kholodniak (✉) · Y. Havrylenko · I. Pykhtieieva · V. Shcherbyna
Tavria State Agrotechnological University, B. Khmelnitsky Ave 18, Melitopol, Zaporizhia Obl., Ukraine
e-mail: yuliya.kholodnyak@tsatu.edu.ua

Y. Havrylenko
e-mail: yevhen.havrylenko@tsatu.edu.ua

I. Pykhtieieva
e-mail: iryna.pykhtieieva@tsatu.edu.ua

V. Shcherbyna
e-mail: viktor.shcherbyna@tsatu.edu.ua

© Springer Nature Switzerland AG 2019

V. Nadykto (ed.), *Modern Development Paths of Agricultural Production*,
https://doi.org/10.1007/978-3-030-14918-5_7

or curves which do not have analytical presentation and formed on the basis of algorithm.

Existing CAD systems use the limited set of curved lines. For the majority of CAD systems, it is a straight line, second-order curve, B-spline. When the lines are used, which CAD system does not form, they are replaced by the compound curve. The areas of the compound curve are formed by curves from the library of CAD system. Main condition of the formation of compound curves in CAD system is providing required accuracy of presentation of the initial elements of the framework. Modern CAD systems do not have facilities to allow controlling fulfillment of the indicated condition.

Development of computer-aided technologies of design of composite functional surfaces on the basis of the curves which are absent in the libraries of CAD systems is an actual task of geometrical design.

### *1.1 Analysis of Recent Studies and Publications*

Method of the formation of computer models of surfaces is suggested in works [4–10]. Models are created on the basis of frameworks which consist of two families of lines. Lines which circumscribe a surface (generating lines) in its motion are included in the first family. The lines of the second family determine spatial motion of the generating line (directional lines).

Model framework is created on the basis of surface determinant, which comprises analytically presented generating and directional lines. In works [5, 6], the determinant comprises B-spline; in work [4], these are second-order curves. Formation of framework in CAD system for such surfaces does not require additional operations at generation of elements of the framework.

In works [7, 8], the determinant of surface comprises lines which are absent in the library of CAD system (cycloid and trochoid). The linear elements of the model are formed as follows: positions of points belonging to the curve are being determined analytically. The obtained row of points is formed in CAD system and is interpolated by B-spline.

In works [9, 10], the task of reverse engineering is being solved. The positions of points, on which basis the linear elements of the model are formed, are measured on the surface of the real items. Model elements are formed in CAD system by interpolation of the row of points with B-spline.

Works [11, 12] suggest method of presentation of plane and spatial trajectories of moving of the cutting tool with an accompanying polyline or torispherical line of circles. The trajectory is formed on the basis of the row of points, which belongs to the surfaces of three-dimensional computer model, obtained in CAD system.

Exactness of presentation of the initial curve with the contour depends on the number of nodes of interpolation, which are assigned on the initial curve. In CAD system, the curve, which interpolates a row of points, can be formed manually. For example, at formation of B-spline, the sequence of node points of the contour is specified on the monitor screen with the mouse. Applications allowing to form contours of rows of points in the automated mode are absent in modern CAD systems. The manual mode limits the number of initial nodes, and therefore exactness of presentation of lines with the contour. Therefore, estimation of accuracy, with which formed in CAD system lines present theoretical trajectories, is one of the important problems of design of surfaces.

The method of determination of maximal absolute error with which formed contour presents an initial curve is suggested in works [13, 14]. The error of interpolation is determined on the basis of area of possible location of the curve with known geometrical properties. For plane curves, such property is the direction of growth of radiuses of curvature along the curve. For spatial curves, it is the motion of curve, direction of growth of radiuses of curvature and radiuses of contacting spheres. The area of possible location of parts of the initial curve and contour, which interpolates the row of points belonging to the curve, is determined by distance between points, limiting the part.

Solving the issue of using any curves which are set analytically or structurally in CAD system is an important stage of developing the technology of creating control program for a machine with NCT.

### *1.2 Statement of the Objective and Tasks of the Study*

The objective of the study is to develop a method of the formation of computer models of composite surfaces on the basis of the framework, consisting of the curves, which are absent in the libraries of CAD systems.

For achieving the set objective, it is necessary to solve the following tasks:

- to work out the algorithms of the formation of contours which represent the analytically or structurally set curves, which are the linear elements of determinant of the surface with set accuracy;
- to develop software for automated formation of the frameworks in CAD system, consisting of contours, which represent the linear elements of determinant of the surface;
- to test the worked out method at formation of functional surfaces of planetary-rotor compressor.

## 2 Basic Part of the Study

An algorithm, allowing defining the area of possible location of the curve, is based on control of appearing oscillation (change of bulge-concavity) on its parts [15]. For a convex curve, in every point of which there is a single position of tangent, any part is located inside the triangle, which we will name a basal. A basal triangle is limited to the chord, connecting points belonging to the curve, and by tangents to the curve in these points.

Any contour, which interpolates the row of points, assigned on the convex curve, on the parts of which there is no change of bulge-concavity, is located inside the sequence of basal triangles with tops in the node points of the contour. The height of basal triangle of the biggest length can be taken for the maximal absolute error of interpolation.

In case when the initial points of the contour are determined on an unknown curve, the tangents to the contour are determined coming from its supposed properties. This property is the direction of growth of radiuses of curvature along the curve. Direction of growth of radiuses is determined on the basis of adjoining circles, passing through three successive points of row. Direction of increase of radiuses of adjoining circles and direction of increase radiuses of curvature coincides along the curve. A tangent to the contour is determined as a mid-position of tangents to two adjoining circles, passing through the given point. In Fig. 1, the position of tangent to the contour ($t_i$) is determined inside the sector, limited to the lines $'t_i$ and $t_i'$. Line $'t_i$ is a tangent to the adjoining circle, passing through the points of $i - 1$, $i$, $i + 1$ (AC($i - 1$, $i$, $i + 1$)), and a straight line $t_i'$—the nearest to $'t_i$ from tangents to AC($i - 2$, $i - 1$, $i$) or AC($i$, $i + 1$, $i + 1$)).

On the basis of the indicated algorithm, the technology of the automated formation of lines has been worked out in CAD system of SolidWorks. At the first stage, the positions of points, belonging to the convex part of the designed curve and sequence of basal triangles, formed on their basis are determined. Condition of formation of row of points is the maxheight of basal triangle does not exceed the admissible absolute error of formation of the line in CAD system.

For the calculations of positions of nodes of designed curve, the math software Maple is used. The positions of the obtained points in the automatic mode are written down in text files.

On the second stage, formation of the line is performed in SolidWorks. The program for the formation of the lines is written in the environment of Delphi 10.2. With

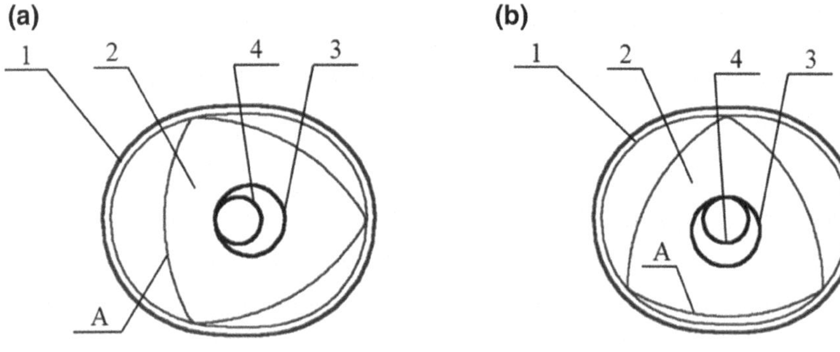


**Fig. 1** Position of operating elements of the planetary-rotor compressor: **a** at the maximal volume of the working chamber; **b** at the minimum volume of the working chamber

the use of tool of application program interface (API), the program is computer-integrated with CAD system—SolidWorks. Raw data for the program work are text files with the positions of points, which are calculated in Maple. The program automatically creates B-spline which interpolates the initial row of points. On the basis of the obtained spline curve, the computer model of the surface is built by means of standard functions of SolidWorks.

The worked out technology is applied for the design of functional surfaces of planetary-rotor compressor.

The flowchart of the compressor is shown in Fig. 1. The compressor consists of the casing (1) and rotor (2). At running by the gearwheel (3) which is connected with the rotor, of the fixed gearwheel (4) which is connected with the casing, the rotor performs a planetary rotation inside of the casing. Tops of the rotor are in permanent contact with the internal surface of the casing. The surfaces of the rotor and casing are limited with three working chambers of the compressor. When the rotor rotates, the volume of the working chambers changes constantly. Figure 1 shows the position of the rotor, at which volume of the working chamber limited with the side A, is maximal (Fig. 1a) and minimum (Fig. 1b). The productivity of compressor operation is determined by the difference of diameters of gearwheels of the toothing of the planetary-rotor mechanism (positions 3 and 4) and sizes of the rotor.

A working surface of the casing is a cylindrical surface, which is determined by an epitrochoid—curve, obtained with motion of the top of the rotor (point of A), rigidly bound by distance $d$ and corner $\varphi$ to the circle of radius $R$. This circle rolls without sliding along the fixed circle of radius $r$ (Fig. 2). Relation between the radius of movable ($R$) and immovable ($r$) circles (the gear-ratio of the toothing) is 3/2.

A rotor cross-section is a curvilinear triangle of Reuleaux, which sides are created with the arcs of circles. Rotor sizes are determined by distance d between the center and top of the triangle.

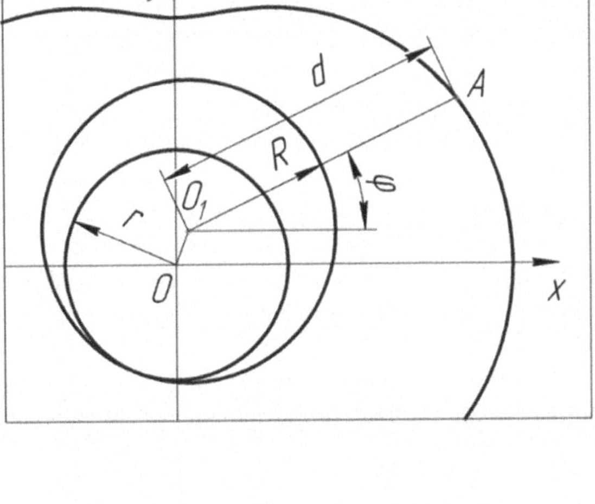


**Fig. 2** A chart of formation of epitrochoid

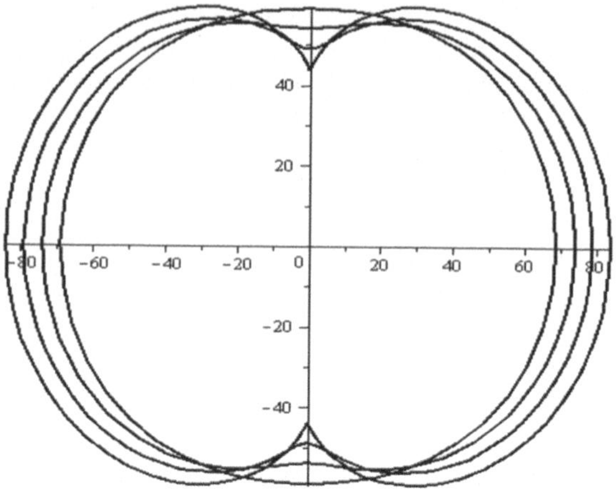


**Fig. 3** Family of epitrochoids for different values of $R$ at $R/r = 3/2$

At unchanging distance $d$, an increase of radiuses of circles results in the form of casing becoming more prolate and the narrowest part of the casing diminishes (Fig. 3). It results in increase of amplitude of vibrations of the rotor and of difference between maximal and minimum volumes of the working chambers. Productivity of the compressor increases. A moment comes at the increase of radiuses of circles, when the profiles of casing and rotor intersect (the rotor is wedged in the casing). For the rotor with a section as a triangle of Reuleaux, which is determined with the size $d = 60$ mm maximal from possible radius of pitch circle of the fixed gearwheel of $R = 23.4$ mm.

For the further increase of volume of working chambers, it was necessary to optimize the profile of the rotor. The side of the optimized profile of the rotor is formed by B-spline, which interpolates a row of points, which node positions are determined structurally. Position of initial points is determined on the compound curve, which parts are formed as follows. Position of the initial profile of the rotor is fixed. The contour of casing is rigidly bound to the circle of radius $r$. This circle revolves around the fixed circle $R$, and the profile of the casing performs planetary motion. At this motion, part of the rotor within the limits of different positions of the casing contour (Fig. 4) is determined. Optimization of formation of the rotor consists in moving away of the indicated part.

To form the initial row of points, the sequence of positions of the rotor contour is determined. The number of these positions is equal to the amount of initial nodes. A compound curve is formed, every part of which belongs to the epitrochoid in one of its positions. Parts are limited to the intersections of successive positions of epitrochoids. In the middle of every part, an initial point is assigned. On the basis of the obtained row of points, B-spline is formed, which determines optimized contour of the rotor.

The optimized profiles of the casing and rotor are formed for the radius of movable circle $R = 30$ mm (reference diameter of the movable gearwheel of the planetary-

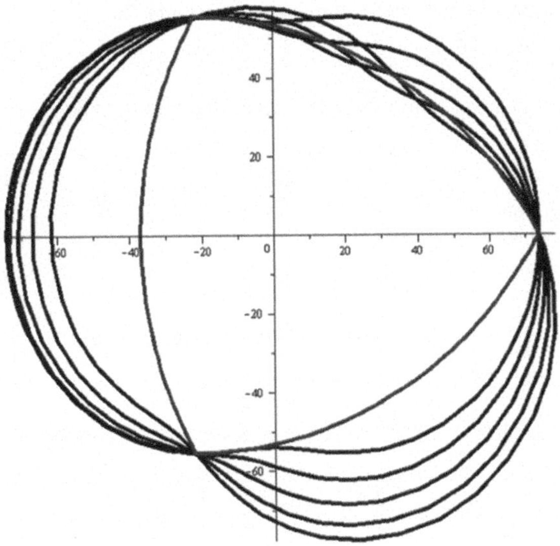


**Fig. 4** Moving of epitrochoid in relation to the profile of the rotor

rotor mechanism) and distances $d = 60$ mm (distance from the center of gravity of the rotor section to its top).

The profile of the rotor side is formed on the basis of 32 points, which positions are expected according to the suggested method. The maximal absolute error of interpolation ($\delta_i$) made up $3.16 \times 10^{-4}$ mm (this value is highlighted of bold in the Table 1).

The chosen parameters of the planetary-rotor machine defined the profile of the casing as an epitrochoid which is determined by equations:

$$\begin{cases} x = 10\cos\varphi + 60 \cdot \cos\frac{\varphi}{3}, \\ y = 10\sin\varphi + 60 \cdot \sin\frac{\varphi}{3}. \end{cases} \tag{1}$$

On the obtained curve, the position of nodes of initial row of points, on which basis B-spline is formed, which is a linear element for formation of the surface model in SolidWorks. Characteristics of the row of points, belonging to one-fourth of epitrochoid, are given in Table 2. The maximal absolute error of presentation of epitrochoid ($\delta_i$) made up $2.94 \times 10^{-4}$ mm (this value is highlighted of bold in the Table 2).

Initial rows of points, on which basis the contours of the rotor and casing are designed, are formed according to the requirements of required accuracy of interpolation. This accuracy must exceed exactness with which machine with NCT provides processing of the programed trajectories. For, especially, exact machine tools (class C), this exactness is $1.6 \times 10^{-3}$ mm [16].

The three-dimensional model of the item, created in CAD system (SolidWorks), is imported into CAM system (PowerMill). The model is imported using the formats

**Table 1** Initial data for forming of the rotor profile

| Number of point, $i$ | 1 | 2 | 3 | 4 | 5 | 6 | 7 | 8 |
|---|---|---|---|---|---|---|---|---|
| Length of chord, mm | 3.8425 | 3.6597 | 3.3215 | 3.1096 | 3.0634 | 2.8952 | 2.9436 | 2.6421 |
| $\delta_i$, $10^{-4}$ mm | 2.19 | 2.31 | 2.23 | 2.65 | 2.18 | 2.96 | 2.48 | 2.46 |
| Number of point, $i$ | 9 | 10 | 11 | 12 | 13 | 14 | 15 | 16 |
| Length of chord, mm | 2.5423 | 2.4871 | 2.6574 | 2.3985 | 2.8415 | 2.9546 | 2.8145 | 2.9821 |
| $\delta_i$, $10^{-4}$ mm | 2.92 | 3.08 | 3.02 | 2.94 | 3.02 | **3.16** | 3.12 | 3.06 |
| Number of point, $i$ | 17 | 18 | 19 | 20 | 21 | 22 | 23 | 24 |
| Length of chord, mm | 2.5648 | 2.8416 | 2.6547 | 2.3698 | 2.8541 | 2.6318 | 2.8423 | 2.7236 |
| $\delta_i$, $10^{-4}$ mm | 2.96 | 2.83 | 3.04 | 2.83 | 2.67 | 2.91 | 2.84 | 2.65 |
| Number of point, $i$ | 25 | 26 | 27 | 28 | 29 | 30 | 31 | 32 |
| Length of chord, mm | 3.2154 | 3.4598 | 3.5618 | 3.6794 | 3.8126 | 3.6421 | 3.8412 | 3.6925 |
| $\delta_i$, $10^{-4}$ mm | 2.89 | 2.92 | 2.58 | 2.71 | 2.86 | 2.67 | 2.59 | 2.51 |

**Table 2** Initial data for formation of the casing profile

| Number of point, $i$ | 1 | 2 | 3 | 4 | 5 | 6 | 7 | 8 | 9 |
|---|---|---|---|---|---|---|---|---|---|
| Length of chord, mm | 3.654 | 3.541 | 3.528 | 3.479 | 3.651 | 3.219 | 3.873 | 3.462 | 3.391 |
| $\delta_i$, $10^{-4}$ mm | 2.46 | 2.63 | 2.41 | 2.16 | 2.36 | 2.82 | 2.46 | 2.65 | 2.81 |
| Number of point, $i$ | 10 | 11 | 12 | 13 | 14 | 15 | 16 | 17 | 18 |
| Length of chord, mm | 3.725 | 3.682 | 3.126 | 2.981 | 3.84 | 3.967 | 3.612 | 3.118 | 3.364 |
| $\delta_i$, $10^{-4}$ mm | 2.63 | 2.31 | 2.64 | 2.87 | **2.94** | 2.64 | 2.51 | 2.69 | 2.74 |
| Number of point, $i$ | 19 | 20 | 21 | 22 | 23 | 24 | 25 | 26 | 27 |
| Length of chord, mm | 3.845 | 3.671 | 3.651 | 3.942 | 3.624 | 3.735 | 3.642 | 3.816 | 3.524 |
| $\delta_i$, $10^{-4}$ mm | 2.46 | 2.32 | 2.45 | 2.56 | 2.74 | 2.62 | 2.34 | 2.71 | 2.63 |

iges, x_b, step, sat et al., which save data about surfaces that limit the item. Control program for processing of the item on a machine with NCT is created by means of standard functions of CAM system.

## 3 Results and Discussion

The suggested in the work method of design in CAD system of composite surfaces is based on the formation of the contours that present lines from the determinant of the surface with set accuracy.

The worked out algorithms allow defining an initial row of points belonging to any curve and provide set accuracy of interpolation at formation of the contour by B-spline or arcs of second-order curves. The software created on the basis of the worked out method is tested on designing functional surfaces of the planetary-rotor compressor.

The computer models of surfaces of the casing and rotor are designed for Melitopol Compressor Plant "Melcom." The models of surfaces are formed according to the gear ratio of the toothing of the planetary-rotor mechanism and sizes of the rotor. In order to increase productivity of the compressor, the working surfaces of the rotor had been optimized. The maximal volume of the working chamber is increased due to the increase of radius of the movable gearwheel of the planetary-rotor mechanism. To avoid wedging of the rotor in the process of compressor operation, the contour of the rotor was changed. The initial contour of the generated circle was substituted by the contour that interpolates the row of points, which nodes are determined according to the specially worked out algorithm. The algorithm is based on the determination of relative position of the contour of casing and rotor in different moments of compressor operation. The design of working surfaces of the compressor required forming of linear elements of the framework on the basis of the row of points that was obtained from analytical presentation of the curve and structurally obtained row of points.

The drawback of the suggested method is that it is based on the formation of solely plane contours. The task of further researches is interpolation with the set accuracy of rows of points, which belong to the space curves.

Solving such task will allow increasing accuracy in designing directional curved lines from the determinant of the surface. Such curves are axial lines of surfaces, which function is transportation of the medium. Such surfaces limit interblade spaces of turbines, channels, and pipelines. Besides, it will make possible in CAD system to form trajectories of moving of processing tools, coming from the set accuracy of location of these trajectories on the surfaces, which limit the model of the item created in CAD system.

## 4 Conclusion

The work suggests the method of formation of computer models of composite surfaces on the basis of the framework consisting of the curves which are absent in the libraries of CAD system.

The following results are received as a result of researches.

1. The algorithms had been worked out to form the contours that represent the analytically or structurally set curves with set accuracy. Generated contours are used in CAD system as linear elements of the model of surfaces.
2. Software had been worked out for the automated formation of contours in CAD system, presenting the curves from the determinant of the surface. The contours are used for the creation of linear frameworks, on which basis the model of the surface is formed by means of standard functions of CAD system.
3. The worked out method is tested on designing functional surfaces of the planetary-rotor compressor. Formation of profiles of the casing and rotor of the compressor was optimized in order to increase its productivity. The models of surfaces are obtained according to the gear ratio of the toothing of 3/2, distance from the center of the cross-section of the rotor to its top $d = 60$ mm. Radius of the pitch circle of the movable gearwheel is increased from $R = 23.4$ mm of the initial model to $R = 30$ mm. Indexes were compared to descriptions of the traditional model of the planetary-rotor compressor, which rotor is made in the formation of a triangle of Reuleaux. The initial design has difference between maximal and minimum volumes of the working chamber that makes up 302,600 $\text{mm}^3$ and after optimization, the difference between volumes increased to 345,400 $\text{mm}^3$.

## References


1. Light, R., Gossard, D.: Modification of geometric models through variational geometry. Comput. Des. **14**, 209–214 (1982). https://doi.org/10.1016/0010-4485(82)90292-5
2. Czerech, L., Kaczynski, R., Werner, A.: Machining error compensation for objects bounded by curvilinear surfaces. Acta Mech. Autom. **6**, 26–30 (2012)
3. Korotkij, V.A., Usmanova, E.A., Hmarova, L.I.: Kompyuternoe modelirovanie kinematicheskih poverhnostey. Geometrija i grafika **3**, 19–26 (2016). [in Russian]
4. Chekalin, A.A., Reshetnikov, M.K., Shpilev, V.V., Borodulina, S.V.: Design of engineering surfaces using quartic parabolas. In: IOP Conference Series: Materials Science and Engineering, vol. 221, UNSP 012015 (2017). https://doi.org/10.1088/1755-1315/221/1/012015
5. Pérez-Arribas, F., Pérez-Fernández, R.: A B-spline design model for propeller blades. Adv. Eng. Softw. **118**, 35–44 (2018). https://doi.org/10.1016/J.ADVENGSOFT.2018.01.005
6. Hoschek, J., Müller, R.: Turbine blade design by lofted B-spline surfaces. J. Comput. Appl. Math. **119**, 235–248 (2000). https://doi.org/10.1016/S0377-0427(00)00381-2
7. Lai, T.-S.: Design and machining of the epicycloid planet gear of cycloid drives. Int. J. Adv. Manuf. Technol. **28**, 665–670 (2006). https://doi.org/10.1007/s00170-004-2423-x
8. Chang, Y.J., Kim, J.H., Jeon, C.H., Kim, C., Jung, S.Y.: Development of an integrated system for the automated design of a gerotor oil pump. J. Mech. Des. **129**, 1099 (2007). https://doi.org/10.1115/1.2757629

9. Kruth, J.P., Kerstens, A.: Reverse engineering modelling of free-form surfaces from point clouds subject to boundary conditions. J. Mater. Process. Technol. **76**, 120–127 (1998). https://doi.org/10.1016/S0924-0136(97)00341-5
10. Saini, D., Kumar, S., Gulati, T.R.: NURBS-based geometric inverse reconstruction of free-form shapes. J. King Saud Univ. Comput. Inf. Sci. **29**, 116–133 (2017). https://doi.org/10.1016/j.jksuci.2014.12.010
11. Zhang, Y., Ye, P., Wu, J., Zhang, H.: An optimal curvature-smooth transition algorithm with axis jerk limitations along linear segments. Int. J. Adv. Manuf. Technol. **95**, 875–888 (2018). https://doi.org/10.1007/s00170-017-1274-1
12. Pessoles, X., Landon, Y., Rubio, W.: Kinematic modelling of a 3-axis NC machine tool in linear and circular interpolation. Int. J. Adv. Manuf. Technol. **47**, 639–655 (2010). https://doi.org/10.1007/s00170-009-2236-z
13. Gavrilenko, E.A., Kholodnyak, Y.V.: Discretely geometrical modelling of one-dimensional contours with a regular change of differential-geometric characteristics. In: 2014 Dynamics of Systems, Mechanisms and Machines (Dynamics), pp. 1–5. IEEE (2014). https://doi.org/10.1109/dynamics.2014.7005654
14. Havrylenko, Y.A., Kholodniak, Y.V.: Formirovanie geometricheskih harakteristik monotonnoy krivoy linii. Vestnik Hersonskogo natsionalnogo tehnicheskogo universiteta **3**(58), 492–496 (2016). [in Russian]
15. Havrylenko, Y., Kholodniak, Y., Vershkov, O., Naidysh, A.: Development of the method for the formation of one-dimensional contours by the assigned interpolation accuracy. East. Eur. J. Enterp. Technol. **1**, 76–82 (2018). https://doi.org/10.15587/1729-4061.2018.123921
16. Rzhevskiy, V.F., Sechkarev, G.A.: Spravochnik po proektirovaniyu avtomaticheskih liniy. Nauka, Moscow (1966). [in Russian]

# Seed Material Size Influence on Its Uniform Sowing Unit Delivery

Volodymyr Kiurchev, Yevhenii Serbii and Svitlana Shevchenko

## 1 Statement of the Problem

Diversity of crop seed material forms and sizes, schemes and standards of their sowing makes creating of universal seeding machine, which would satisfy all the needs of a landowner impossible. Seed material of the most part of row crops is characterized by high looseness, form coefficient up to one, size variation within ±0.25 mm. All these characteristics make it possible to realize precise seed sowing in production quantities.

Traditionally for precise seed sowing, working elements are used for unit choice of a seed from bulk in general, in the form of scoops in mechanical sowing units and cups in compressed units. Arrangements for unnecessary seed material moving off are installed in sowing units additionally. After that, the task of gentle unnecessary seed material moving off is set up. It makes sowing units not reliable, hard in use, their power consumption and value increase, which set rigid requirements for seed size calibration. It is known that the best world technical findings guarantee uniform seed material moving off with variation coefficient on the level of 20–30% [1–3].

Another realization of precise sowing can be made up of continuous regular dense one-seeded flow with constant speed from sowing unit. Such kinds of technical findings are not new and their realization can be represented in the form of vibration sowing units [4, 5] which are used for grain seeds sowing. The sowing units of such design have linear sizes of seed supply track which exceeds seed material size greatly (Fig. 1a).

V. Kiurchev (✉) · Y. Serbii · S. Shevchenko
Tavria State Agrotechnological University, Melitopol, Ukraine
e-mail: office@tsatu.edu.ua

Y. Serbii
e-mail: eugeniyserbii@gmail.com

S. Shevchenko
e-mail: svitlana.shevchenko@tsatu.edu.ua

© Springer Nature Switzerland AG 2019

V. Nadykto (ed.), *Modern Development Paths of Agricultural Production*,
https://doi.org/10.1007/978-3-030-14918-5_8

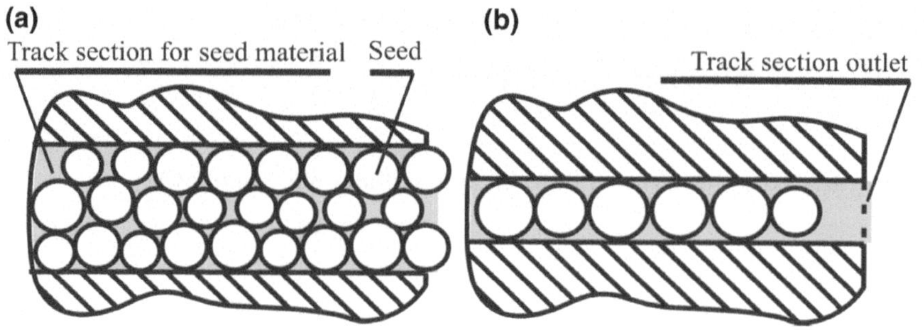


**Fig. 1** Track section sizes for seed material: **a** for total sowing, **b** for precise sowing

It is necessary to have not more than one seed in track section for precise sowing (Fig. 1b). So, investigation of seed material size characteristic influence on its uniform sowing unit delivery, in the form of dense regular one-seeded flow is relevant investigation.

## 2 Purpose of the Article

The objective of the paper is to investigate the seed material size characteristic influence on its uniform sowing unit delivery, in the form of dense regular one-seeded flow.

## 3 Statement of the Main Material

A seed is considered to be delivered to the furrow when the center of its gravity will pass track section outlet. Usually, the seeds of any crop are of different forms and sizes and each of them can be bullet-shaped form. The main statistical characteristics of such stray parameter are mean diameter, standard deviation, variation coefficient, density of distribution. According to previous findings [6, 7], seed size distribution is described by the normal law [8].

As far as the track section size exceeds the maximum diameter of a seed that to avoid self-locking and simplify its filling, some ideal conditions of dense mutual position can be marked out.

1. **Coaxial seed material location in the track** (Fig. 2), which helps to locate the center of its gravity on the axial line of the track section. The distance between centers of gravity of adjoining seeds equals $s = d$.

Difference of seeds feed time under coaxial seed material location in the track section

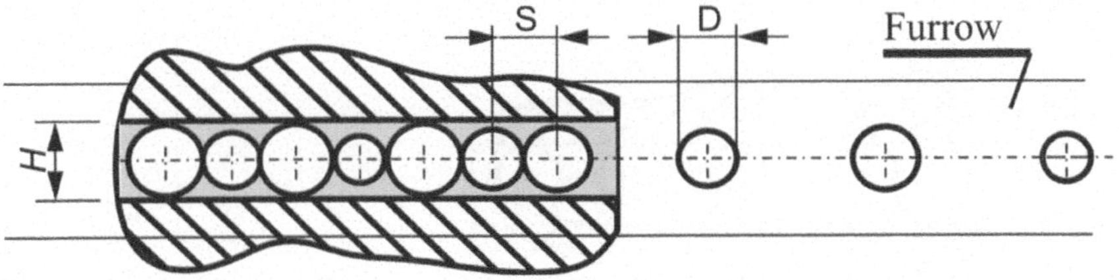


**Fig. 2** Coaxial seed material location in the track section

$$t_1 = \frac{s}{V_D}, \tag{1}$$

will be the function of stray parameter $s = d$, therefore, it will be a stray parameter itself with the main characteristics of mean value $T_1$, standard deviation, and density of distribution $p(t_1)$.

As far as continuous stray parameter $d$ with distribution density, $p(d)$ is connected with functional dependence $t_1 = t_1(s)$ stray parameter (1), which is monotonous and rising then density of distribution $p(t_1)$ can be found with the following form [9]:

$$p(t_i) = p(s(t_i)) \cdot s(t_i)', \tag{2}$$

where $s(t) = t_1 \cdot V_D$ is inverse function up to (1), $s(t)' = V_D$,

then $p(t_1) = \frac{1}{\sqrt{2\pi}\sigma_D} e^{-\frac{(t_1 \cdot V_D - D)^2}{2\sigma_D^2}} \cdot V_D$,

$$\text{or } p(t_1) = \frac{1}{\sqrt{2\pi}\sigma_D / V_D} e^{-\frac{(t_1 - D/V_D)^2}{2(\sigma_D/V_D)^2}} \tag{3}$$

Form (3) is normal law of distribution with mathematical expectation $T_1 = D/V_D$ and standard deviation $\sigma_{t1} = \sigma_d / V_D$.

So, seed material delivery from sowing unit with constant speed $V_D$ will be characterized by uniformity with variation coefficient $\nu_{t1}$:

$$\nu_{t1} = \frac{\sigma_{t1}}{T_1} \cdot 100\% = \frac{\sigma_d}{D} \cdot 100\% = \nu_d, \tag{4}$$

which equals to the variation coefficient of seed sizes $\nu_d$.

2. **Dense location of seed material** (Fig. 3) where seeds touch the opposite sides of track section by turns.

With the width of $H = k \cdot D$ where coefficient $k > 1$ shows the difference between the track width and seed's mean diameter $D$, the distance along the track between centers of gravity of adjoining seeds will be $s = d \cdot \sqrt{2 \cdot k - k^2}$, then feed time between adjoining seeds will be equal to

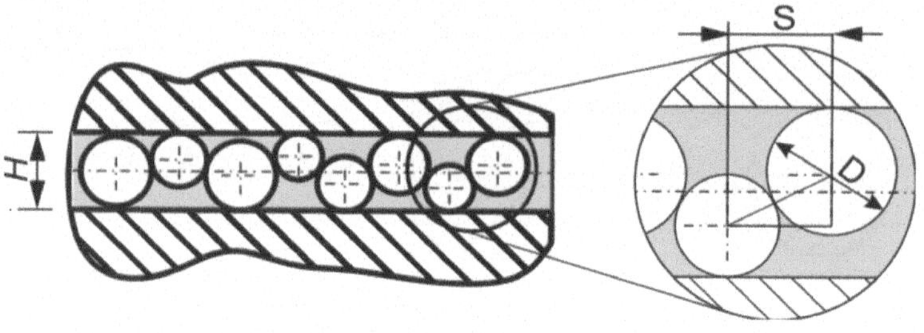


**Fig. 3** Dense location of seed material

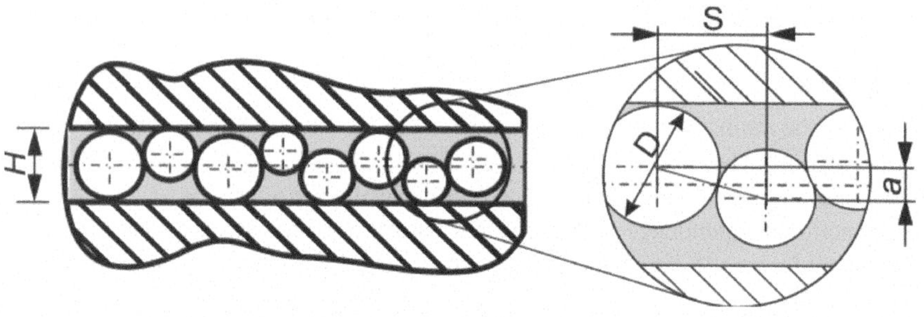


**Fig. 4** Random location of seed material in the track

$$t_2 = d\frac{\sqrt{2 \cdot k - k^2}}{V_D}. \tag{5}$$

Density of its distribution $p(t_2)$ in accordance with (2) under $s(t_2) = t_2 \cdot V_D/\sqrt{2 \cdot k - k^2}$ та $s(t_2)' = V_D/\sqrt{2 \cdot k - k^2}$ and will be equal to

$$p(t_2) = \frac{1}{\sqrt{2\pi}\sigma_D \cdot \sqrt{2 \cdot k - k^2}/V_D} e^{-\frac{\left(t_1 - D \cdot \sqrt{2 \cdot k - k^2}/V_D\right)^2}{2\left(\sigma_D \cdot \sqrt{2 \cdot k - k^2}/V_D\right)^2}}, \tag{6}$$

Form (6) is normal law of distribution with mathematical expectation $T_2 = D \cdot \sqrt{2 \cdot k - k^2}/V_D$ and standard deviation $\sigma_{t2} = \sigma_D \cdot \sqrt{2 \cdot k - k^2}/V_D$, so variation coefficient $\nu_{t2}$ of uniform seed material delivery will be

$$\nu_{t2} = \frac{\sigma_D}{D} \cdot 100\% = \nu_d. \tag{7}$$

Uniform seed material delivery $\nu_{t2}$ with constant speed with dense location within track also equals to variation coefficient of seed sizes $\nu_d$.

3. **Random location of seed material in the track** (Fig. 4), it means disorder of seed location.

In such kind of location, seeds can occupy any position along track and will touch adjoining seeds. Taking into account that seed material location in a seed vessel is random then it means that without additional units for ordering their location along the track also will be occasional and equiprobable. So, the distance along the track between centers of gravity will be even value and will change within the limits of 0 and $H - D = D(k - 1)$:

$$a = D \cdot b, \tag{8}$$

where $b = [0 \ldots k - 1]$ uniformly distributed value with mathematical expectation $B = \frac{k-1}{2}$, mean-squared departure $\sigma_b = \frac{k-1}{2\sqrt{3}}$, and density of distribution $p(b) = \frac{1}{k-1}$ [8].

Then the distance $s$ along the track between centers of seed material gravity will be a function of two independent chance quantities: $s = d\sqrt{1 - b^2}$, and feed time between adjoining seeds will be equal to

$$t_3 = \frac{d\sqrt{1 - b^2}}{V_D},$$

If we will make a replacement $c = \sqrt{1 - b^2}$, then we will have:

$$t_3 = \frac{d \cdot c}{V_D}. \tag{9}$$

Mathematical expectation $C$ of chance quantity $c = c(b)$, we will find with form [9]: $C = \int_{-\infty}^{\infty} c(b) \cdot p(b)db$, then:

$$C = \frac{1}{2}\left(\sqrt{2k - k^2} + \frac{\arcsin(k - 1)}{k - 1}\right). \tag{10}$$

As far as $t_3$ is the product of chance quantities, then its mathematical expectation we will find with form [9]: $T_3 = \frac{D \cdot C}{V_D}$, or:

$$T_3 = \frac{1}{2 \cdot V_D} \cdot D \cdot \left(\sqrt{2k - k^2} + \frac{\arcsin(k - 1)}{k - 1}\right). \tag{11}$$

Dispersion $\sigma_c^2$ of chance quantity $c = c(b)$ we will find with form [9]: $\sigma_C^2 = \int_{-\infty}^{\infty} (c(b) - C)^2 \cdot p(b)db$, then

$$\sigma_C^2 = \frac{-3 \cdot \arcsin(k - 1)^2 - k^4 + 4k^3 + 3k^2 - 14k + 8}{12(k - 1)^2} - \frac{\sqrt{2k - k^2}\arcsin(k - 1)}{2(k - 1)}. \tag{12}$$

Mean-squared departure $\sigma_{t3}$ of time $t_3$ as product of chance quantities $d$ and $b$ [9] we will define with form:

$$\sigma_{t3}^2 = \frac{1}{V_D^2} \cdot \left(\sigma_d^2 + \sigma_C^2 + D^2 \cdot \sigma_C^2 + C^2 \cdot \sigma_d^2\right). \quad (13)$$

Then uniformity of seed delivery $\nu_{t3}$ will be equal to:

$$\nu_{t3} = \frac{\sigma_{t3}}{T_3} \cdot 100\%. \quad (14)$$

Reduction of combined Eqs. (10), (11), (12), (13), and (14) up to one gets dependence $\nu_{t3} = \nu_{t3}\left(k, \sigma_d^2, D\right)$ with no sense, due to its inconvenience and unreasonableness of further algebraic transformation or analyzes. Substitution of numeric data $k, \sigma_d^2, D$ in (10)–(14) allows to get numeric data as $\nu_{t3}$.

## 4 Analysis of the Theoretical Investigation Results

Seed forms and sizes of row and vegetable crops, which need precise sowing, have a wide range of changing and depend of a kind of crop, variety and cultivation conditions: sizes from 2 up to 12 mm, seed form can be bullet shaped (soybean, pea, cabbage, beet, mustard, garden radish, black radish, asparagus) and long shaped (corn, sunflower, carrot, cucumber).

According to Eqs. (4) and (7), it is obvious that uniform delivery from the track depends on seed material diameter variation coefficient during its coaxial and dense location. In disorder location, seed material size in the track will influence the uniform delivery under constant variation coefficient (Fig. 5), but as the graph shows we can neglect this influence without further research damages on account of its insignificance.

It is obvious that the seed form will influence on uniform delivery from the track.

The soybean variety "Almaz" [10] was chosen for analysis of uniform delivery a bullet-shaped seed from the track. This kind of seed has the following geometrical size characteristics: average value—6.27 mm, mean-squared distance—0.47 mm and variation coefficient—7.5%. According to Eqs. (10)–(14), graphic dependence of width track influence was got. It is expressed through coefficient $k = [1.0 \ldots 1.8]$, on the uniform seed delivery $\nu_t$ (Fig. 6).

The graph shows that variation coefficient of seeds delivery is constant and makes up 7.5% during coaxial and dense location. The variation coefficient of seeds delivery $\nu_t$ is raised under disorder seed material location as the result of its chaotic location in the seed vessel, uneven surface, and complex form with width track coefficient increasing. It is possible to highlight two parts of the graph: in the range $k = [1.0 \ldots 1.5]$ variation coefficient $\nu_t$ linear raises from 10.7 up to 12% but the

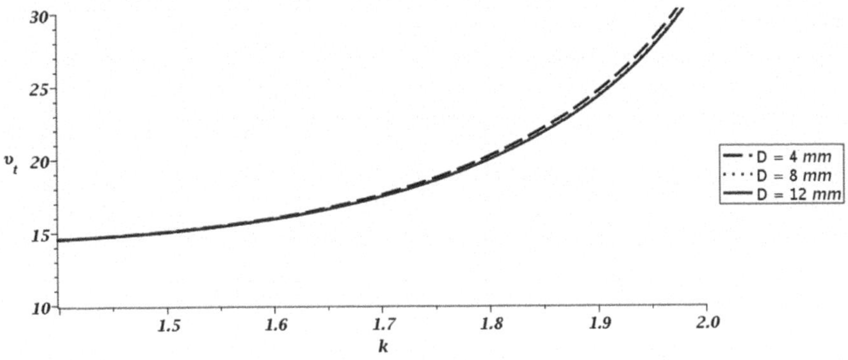


**Fig. 5** Seed sizes influence on the uniform delivery from the track with $\nu_d = 5\%$

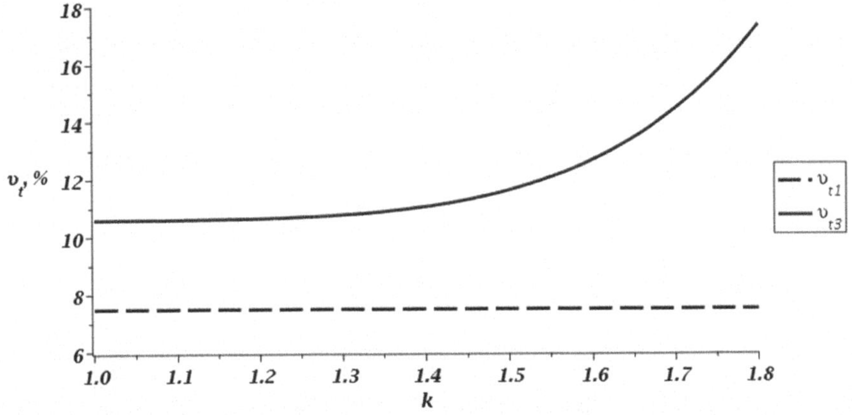


**Fig. 6** Dependence of width track coefficient $k$ influence on the uniform delivery of circle-shaped seeds (soybean variety "Almaz", $D = 6.27$ mm, $\sigma_d = 0.47$ mm, $\nu_d = 7.5\%$)

further raise $k$ is nonlinear, it is a step function. So, for the perfect sphere shaped seed delivery:

- Seeds should have coaxial and dense location in the track;
- Track width coefficient should not exceed 1.5.

Seed material of all crops is different from the bullet-shaped seeds, especially long-form seeds. Maximum (seed length) and minimum (thickness) size ratio of light fraction seed hybrids of corn "Ushytskyi", "P'iatykhatskyi", "Borozenskyi", "Solonianskyi", "Zbruch" [11] is 200% (Table 1).

If seed location ("on the blade", "upright" or "prone position") in the track is probable, then the middle size of seed corn hard fraction hybrids which was calculated on the value of length, width and thickness will be $D = 8.37$ mm, mean-square distance $\sigma_d = 2.0$ mm, and variation coefficient $\nu_d = 23.84\%$. In such conditions,

**Table 1** Middle size characteristics of corn hybrids

| Linear seed sizes, mm | Fractions | | | | | |
|---|---|---|---|---|---|---|
| | Light | | | Hard | | |
| | $D$, mm | $\sigma_d$, mm | $\nu_d$, % | $D$, mm | $\sigma_d$, mm | $\nu_d$, % |
| Length | 10.43 | 0.20 | 1.92 | 9.96 | 0.27 | 2.71 |
| Width | 8.85 | 0.17 | 1.92 | 9.01 | 0.17 | 1.89 |
| Thickness | 5.03 | 0.19 | 3.78 | 6.13 | 0.27 | 4.40 |
| Middle size | 8.1 | 2.78 | 34.26 | 8.37 | 2.0 | 23.84 |

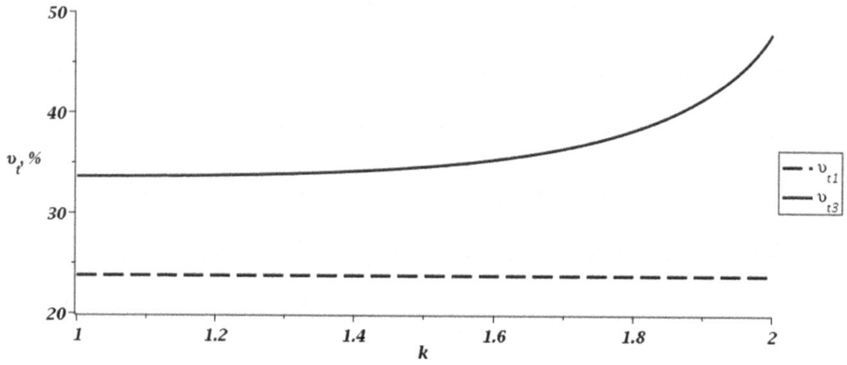


**Fig. 7** Uniform corn seed hybrids of light fraction ($D$ = 8.37 mm, $\sigma_d$ = 2.0 m, $\nu_d$ = 23.84%) delivery dependence of width track coefficient $k$ influence

uniform seed delivery from the track is prohibitive for precise sowing even during coaxial and dense location (Fig. 7).

Even seed material location in the track is advisable for uniform delivery of no bullet-shaped seeds, e.g., corn hybrid seeds, which have too high variation coefficient regarding to middle size. In this case, the distance between seeds $s$ (Fig. 2) will define only by one side of the seed: length, width or height, and size variation coefficient will not exceed 5% (Table 1). As a result, it will be possible to vary the track's width, keeping variation coefficient of seed material delivery from the track up to 20%.

## 5 Conclusions

1. Promising decision of precise sowing is making regular continuous dense one-seeded flow with constant speed from sowing unit.
2. Uniform delivery equal to the variation coefficient of seed sizes during coaxial and dense location in the track.

3. With disorder seed material location in the track, their size will not influence the uniform delivery under constant variation coefficient.
4. For the perfect uniform seed delivery, it is advisable that seeds should have coaxial and dense location in the track, width coefficient should not exceed 1.5, nonround seed sizes will have even track location.

## References


1. Mirovoi rekord Tempo: https://www.vaderstad.com/ru/seyalki-propashnie/seyalki-tempo/mirovoi-rekord-tempo/. Мировой рекорд Tempo. https://www.vaderstad.com/ru/seyalki-propashnie/seyalki-tempo/mirovoi-rekord-tempo/
2. Kozhushko, M.: Analiz rezultativ doslidzhen' prosapnykh sivalok v umovakh zahidnogo region. Pratsi UkrNDIPVt im. L. Pogorilogo, 114–119 s (2016). Кожушко М., Аналіз результатів досліджень просапних сівалок в умовах західного регіону / М. Кожушко, Я. Ценюх // Праці УкрНДІПВт ім. Л. Погорілого, 2016, - с. 114–119
3. Potapov, A.: Parametry I rezhimy raboty mnohophunktsionalnoi I soshnikovoi gruppy propashnoi seialki. Stavropol, p. 23 (2011). Потапов А.А. Параметры и режимы работы многофункциональной сошниковой группы пропашной сеялки. – а.-реф. ... к.т.н., - Ставрополь, 2011 г. – 23 стр
4. Pastukhov, V.: Vykorystannia eksperymentalnoii sivalki z vibratsiino-dyskovymy vysivnymy aparatamy pry sivbi nasinniaovochevykh kultur na kintsevu gustotru. Visnuk KHTUSG, vypusk 135 (Tekhnichni nauky). Kharkiv (2013). Пастухов В.І. Використання експериментальної сівалки з вібраційно-дисковими висівними апаратами при сівбі насінняовочевих культур на кінцеву густоту // В.І. Пастухов, М.В. Бакум, Р.В. Кириченко, В.В. Живолуп / Вісник ХНТУСГ. Випуск 135 (Технічні науки). «Механізація сільськогосподарського виробництва». – Харків – 2013 р
5. Zaiika, P., Bakum, M.: Vysivnyi vibratsiino-dyskovyi aparat. Pat. 37998 Ukraina, MPK A 01 C7/00, Bul. 24 (15.12.2008). Пат. 37998 Україна, МПК A 01 C7/00. Висівний вібраційно-дисковий апарат / П.М. Заїка, М.В. Бакум, Р.В. Кириченко. - № 200802501; заявл. 18.02.2008; опубл. 15.12.2008, Бюл. № 24
6. Yatsun, S.: Mekhaniko-tekhnologichni vlastyvosti sil'skogospodars'kykh materialiv. Kuyiv, Meta, 448 s (2003). Механіко-технологічні властивості сільськогосподарських матеріалів: Підручник [для студ. вищ. навч. закл.] / [Царенко О.М., Войтюк Д.Г., Швайко В.М. та ін.]; За ред. С.С. Яцуна. – К.: Мета, 2003. – 448 с
7. Kyrychenko, R.: Vyznachennia osnovnykh mekhaniko-tekhnologichnykh vlastyvostei nasinnia ripaku, liutserny ta morkvy. Visnyk KHTUSG, vypusk 107 (Tekhnichni nauky). Kharkiv (2011). Кириченко Р.В. Визначення основних механіко-технологічних властивостей насіння ріпаку, люцерни та моркви / Вісник ХНТУСГ. Випуск 107 (Технічні науки). «Механізація сільськогосподарського виробництва». Т.1. - Харків – 2011 р
8. Gmurman, V.: Teoriia veroiatnostei i matematicheskaia statistika. Vysshaia shkola, Moskva, 308 s (1972). Гмурман В. Е. Теория вероятностей и математическая статистика. - М., «Высш. школа», 1972, - 308 с
9. Pechinkin, A.: Teiriia veroiatnostei. Izdatelstvo MGTU im. N.E. Baumana, Moskva, 456 s (2004). Печинкин А.В. Теория вероятностей / А.В. Печинкин, О.И. Тескин, Г.М. Цветкова и др. - М.: Изд-во МГТУ им. Н.Э. Баумана, 2004. - 456 с
10. Kopets', K.: Rozrobka ta obgruntuvannia parametriv prystroiu pidgotovky zeren soii do sushinnia. LNTU, Luts'k, 189 s (2016). Копець К. Є. Розробка та обгрунтування параметрів пристрою підготовки зерен сої до сушіння. - дисс...к.т.н. - ЛНТУ, Луцьк, 2016 р. - 189с

11. Kyrpa, M.: Osoblyvosti aerodynamichnogo separuvannia odnokomponentnykh nasinnievykh sumishei na prykladi kukurudzy. Nasinnytstvo, 45–50 s (2013). Кирпа М. Я. Особливості аеродинамічного сепарування однокомпонентних насіннєвих сумішей на прикладі кукурудзи / М. Я. Кирпа, С. О. Скотар, О. І. Лупітько, О. О. Рослик // Насінництво, 2013. – с. 45–50

# Agrobiological as Well as Mechanical and Technological Framework of Development of the Harvesting Technology with the Method of Grain Crops Combing in Standing Position

**Oleksandr Lezhenkin, Ivan Lezhenkin, Oleksandr Vershkov and Serhii Kolomiiets**

## 1 Introduction

Combine harvesting technique is the main technique for harvesting of grain crops. However, to date, the mechanization of the grain crops harvesting has entered the stage when further increase in the capacity of harvesters has become economically inexpedient, since, basically, it led to an increase in energy costs, an increase in the mass of combines, and also an increase in their cost. In general, the idea of combine harvesting should be revised from numerous points of view, namely due to:

- high transportation costs;
- limited thrashing productivity;
- problem of lodged grain harvesting;
- impossibility to harvest wet mass.

At the same time, there is a way out of this situation. Stationary techniques are sufficient alternative to the combine harvesting of grain crops.

### *1.1 Analysis of Recent Studies and Publications*

Let us consider some of the stationary techniques. The first type of stationary techniques is a three-phase technique. The three-phase method is based on grinding the entire grain mass at the same time as mowing or picking rolls, followed by final refinement of the heap obtained under stationary conditions [1]. The further development

O. Lezhenkin (✉) · I. Lezhenkin · O. Vershkov · S. Kolomiiets
Faculty of Engineering and Computer Technology, Tavria State Agrotechnological University, 18, Bogdana Khmelnitskogo Avenue, Melitopol 72312, Ukraine
e-mail: oleksandr.lezhenkin@tsatu.edu.ua

© Springer Nature Switzerland AG 2019 
V. Nadykto (ed.), *Modern Development Paths of Agricultural Production*,
https://doi.org/10.1007/978-3-030-14918-5_9

of the three-phase technique was the Kuban industrial technique [2]. The essence of this technique is to mow down and grind the bread mass with its subsequent loading into the trailer and transportation to the stationary facilities for finalization. The refinement includes: separation, drying, and finish grinding.

For the zones of high humidity, a stream technique was developed for harvesting grain and herbal crops with threshing at the stationary facilities, which was the continuation of the three-phase and Kuban industrial techniques [3].

These techniques make it possible to achieve the main thing, that is, significantly reduce the loss of grain, withdraw from the field and use the entire biological crop, lengthen the working day. The main disadvantages of these techniques are low productivity, high transportation costs and, as a result, high production costs.

The increase of the stationary techniques' efficiency can be achieved through the use of the method of combing plants. Scientific foundations of this method were provided by Prof. Shabanov [4]. The constructions of combing devices were substantiated by his students: Golubev [5], Danchenko [6], Goncharov [7]. However, the practical implementation of this technique requires additional research.

### *1.2 Statement of the Objective and Tasks of the Study*

The purpose of the study is to substantiate the technology of harvesting grain by combing plants in standing position by studying the agrobiological as well as mechanical and technological properties of grain crops. To achieve this goal, it is necessary to solve the following tasks:

- to study the agrobiological as well as mechanical and technological properties of grain crops, as applied to their combing in standing position;
- to develop a structural scheme of the technological process of grain crops harvesting by the method of their combing in standing position.

## 2 Results and Discussion

An important factor affecting the process of harvesting grain crops is the moisture content of grain and straw. Long-term studies [4] revealed a number of patterns of moisture change in individual crops. However, these studies do not make it possible to provide a comparative analysis of the dynamics of changes in the moisture content of grains and plants currently cultivated in Ukraine.

In order to determine the moisture content of plants in accordance with their height during the harvesting period, field experiments were carried out on different crops. The objects of the research were the spiky crops (wheat, barley, triticale) and panicle crops (millet and oats). The results of the experimental studies are shown in Fig. 1 [8].

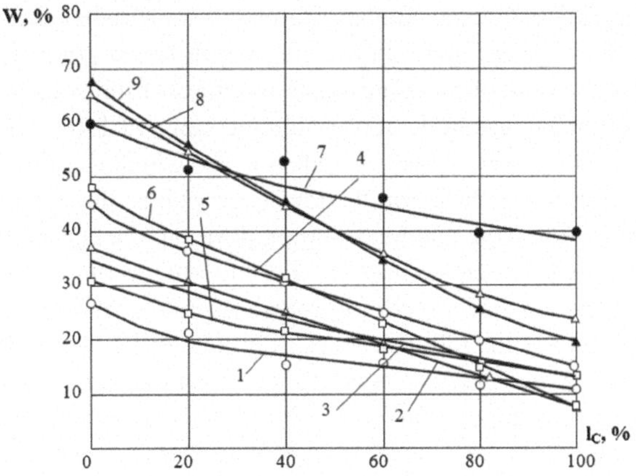


**Fig. 1** Change in the moisture content of plants in height: 1—wheat "Albatross Odessa"; 2—wheat "Odessa-267"; 3—wheat «Victoria Odessa»; 4—barley "Prairie"; 5—barley "Stalker"; 6—triticale "Zenith of Odessa"; 7—millet "Kharkivskoye-31"; 8—millet "Start"; 9—oats "Kubanskiy"

Analyzing the results obtained, it should be noted that the moisture content of the stems increases in the lower part, while for the spiky and panicle crops this pattern manifests itself unequally. So, for example, in the wheat cultivar "Albatross Odessa" the humidity in the lower part of the plant was 27%, and in the upper part—12%, while in oats cultivar "Kubansky" the humidity ratio was, respectively, 67 and 22%. As can be seen from the data given, the moisture content of plants in height is quite significant. Investigations of grain moisture in southern regions of Ukraine showed that the moisture content of the grain does not fluctuate as much as the moisture content of the plants and is 11.4 … 15.6% depending on the crops [8].

At the same time, the difference between the moisture content of the grain and the moisture content of the non-grain part of the plants is quite significant, for the spiky crops, it is 13.5 … 34.2%, and for the panicle crops, it is 41.1 … 53.4%. This property predetermines the effectiveness of the method of combing plants in standing position because the dry grain does not mix with the wet non-grain part.

To substantiate the technique of combing plants in standing position, long-term studies of the size and mass characteristics of grain crops were carried out in Tavria Agrotechnological University [9]. The length of the plant and inflorescence, the diameter of the stem, the diameter of the inflorescence as well as the mass characteristics of the plant parts were determined. It was found that the spread of plant lengths and the lengths of inflorescences relative to average values are insignificant. In general, this is a positive factor when combing plants in standing position. The average values of the diameter of the compressed inflorescence in the spiky crops vary insignificantly and do vary within the range of 7.1 … 7.7 mm, in the panicle crops the mean values of this parameter are somewhat lower being in the range of 4.6 … 6.4 mm. The insignificance of fluctuations in mean square deviations testifies to the stability of the diameter values of the compressed inflorescence for grain crops, which greatly simplifies the design of the combs, as there is no need to develop combs with variable clearance.

The undertaken studies of mass characteristics make it possible to assert that the mass of plants and inflorescences of different grain crops varies within wide limits, but the ratio of plant masses to the mass of inflorescences is more stable and for the cultures under study, it is within the range of 1.76 … 1.99, with the exception of the wheat cultivar "Albatross Odessa," for which this ratio is 2.65.

To justify the combing device, as well as the dynamic calculation of the process of combing the plant in standing position, the task is to determine the strength characteristics of grain crops. To this end, studies of efforts of plucking from the soil and efforts of detachment of inflorescences from the stem, as well as efforts to separate grain from the inflorescences of grain crops cultivated in the South of Ukraine were carried out. The results of the investigations are shown in Fig. 2.

The average values of the efforts of plucking from the soil of different cultures fluctuate within the wide range of 36.4 … 75.3 H. In the effort to detach the inflorescences of the studied cultures, the spread of mean values is much smaller, their average values range from 21.5 to 34.5 H. The ratio of the mean values of the detachment forces of the inflorescence to the mean values of plucking forces is 0.44 … 0.54. The forces of detachment of individual grains from the inflorescences amount to 11.0 … 18.1 H. At the same time, these efforts are 1.7 … 2.2 times less than the efforts of detachment of the inflorescences from the stem.

In this case, the inflorescence is combed, and the stem remains in the soil. If this condition is not met, then the stripping process is disturbed, the plants are pulled out

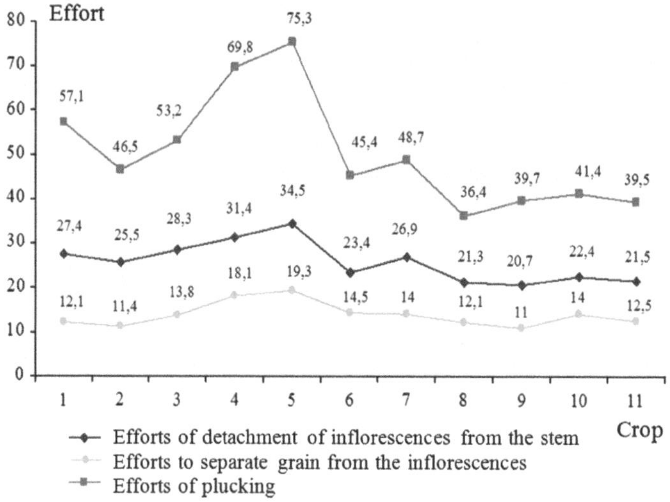


**Fig. 2** Diagram of the average values of the detachment forces of the inflorescences from the stalk and plucking from the soil: 1—wheat "Albatross Odessa"; 2—wheat "Nikonia"; 3—wheat "Odessa-267"; 4—triticale "Hurricane"; 5—winter triticale (grain); 6—spring barley "Stalker"; 7—barley "Prairie"; 8—millet "Kharkov Fodder"; 9—millet "Kharkov-31"; 10—millet "Start"; 11—oats "Kuban"

of the soil when force is applied to them, they are wound onto the working parts and clog them.

Proceeding from the studies of agrobiological as well as physical and mechanical properties of grain crops, a structural scheme of the harvesting grain crops technique was developed (Fig. 3) [10].

The field harvesting machine performs the first operation in the general technological chain of the harvesting process. Its purpose is to comb the plants in standing position, pick up a heap into the trailer and harvest the straw.

The heap obtained as a result of combing is a four-component grain-straw mixture of free grain, tattered spikelets (panicles), straw particles, and spiky crops.

For different cultures, the content of the individual components is different. So, the winter wheat contains 39 … 81% of free grain, 12 … 54% of straw particles, 5 … 14% of chaff, and 1.2 … 7.8% of dangling spikelets [11].

The combed heap collected in the trailer is delivered to a stationary point for its completion. The main purpose of the device for finishing the combed heap is an allocation of free grain and preparation of fodder briquettes from unmalted ears (panicles), coarse straw impurities, and chaff.

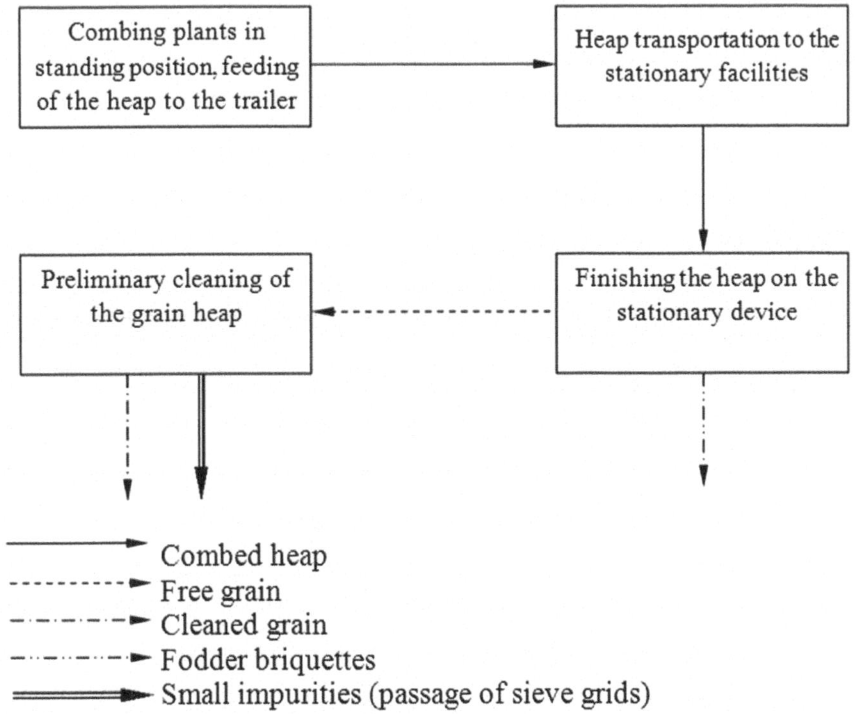


**Fig. 3** Structural scheme of stationary technique by combing plants in standing position

## 3 Conclusion

Analysis of agrobiological properties of cereals made it possible to establish that the moisture content of the stems of plants is not the same, in the upper part of spiky crops, it is 13 … 25%, and in the lower part, it is 27 … 48%, in the upper part of panicle crops, it fluctuates within the limit of 20 … 45%, and at the bottom, it is 60 … 67%, at the same time, the moisture content of the grain is stable and amounts to 11.4 … 15.6%, this pattern of moisture distribution along the length allows combing of the dry grain mixed with damp stem grout.

Studies of the grain crops' biometrics showed that the plants length and the length of the inflorescence fluctuate insignificantly, and the coefficients of variation are 4.0 … 8.4% and 2.4 … 7.2%, which makes it much easier to justify the construction of the combing device. In addition, the stability of the values of compressed inflorescences diameters is 6.2 … 7.2 mm, which will make it possible to get rid of the need to design combs with variable clearance.

It has been revealed that the detachment forces of the inflorescences are half that of the plants peeling from the soil, and the detachment forces of individual grains are 1.7 … 2.2 times less than the detachment forces of the inflorescences from the stem. In this case, the inflorescence is combed, and the stem remains in the soil.

Experimental studies of agrobiological as well as mechanical and technological properties of grain crops allowed to formulate the concept of harvesting, the essence of which consists in combing the plants in standing position, followed by finishing the combed heap at the stationary facilities.

## References

1. Zhuk, Y.M.: Mekhanyzatsyia i elektryfykatsyia sotsyalystycheskoho selskoho khoziaistva. **5**, 4–8 (1959)
2. Kanarev, F.M.: Mekhanyzatsyia i elektryfykatsyia selskoho khoziaistva. **8**, 17–19 (1986)
3. Malta, A.O.: Tekhnyka v selskom khoziaistve. **5**, 6–8 (1988)
4. Shabanov, P.A.: Mekhanyko-tekhnolohycheskye osnovы obmolota zernovыkh kultur na korniu. Doctor's thesis, Melytopol (1988)
5. Holubev, I.K.: Obosnovanye osnovnыkh parametrov y rezhymov rabotы dvukhbarabannoho ustroistva dlia ochesa rysa na korniu. Candidate's thesis, Moskva (1989)
6. Danchenko, N.N.: Obosnovanye parametrov shchetochnoho ustroistva dlia ochesыvanyia metelok rysa na korniu. Extended abstract of candidate's thesis, Cheliabynsk (1983)
7. Honcharov, B.I.: Yssledovanye rabocheho protsessa ochesыvaiushcheho ustroistva dlia obmolota rysa na korniu s tseliu umenshenyia poter zerna. Candidate's thesis, Moskva (1982)
8. Lezhenkin, O.M., Hryhorenko, S.M.: Pratsi TDATA. **36**, 25–28 (2006)
9. Lezhenkin, O.M., Hryhorenko, S.M.: Pratsi TDATA. **26**, 152–158 (2005)
10. Lezhenkyn, A.: Selskyi mekhanyzator. **11**, 27 (2004)
11. Kiurchev, S.V., Lezhenkyn, I.A.: Konstruiuvannia, vyrobnytstvo ta ekspluatatsiia silskohospodarskykh mashyn. **43**(1), 297–302 (2013)

# Simulation of Cereal Crops Harvesting Using the Method of Grain Crops Combing in Standing Position in Conditions of Farming Enterprises

**Oleksandr Lezhenkin, Ivan Lezhenkin, Oleksandr Vershkov and Serhii Kolomiiets**

## 1 Introduction

At present, there are 45,035 farms in Ukraine, while the area of agricultural land is 4.43 million hectares, of which the arable land includes 4.3 million hectares. However, combine harvesters are available only at 15% of farms. To buy new combines for many farmers is an insoluble task, while it should be noted that existing combines are complex, material-intensive, expensive machines. On a pro rata basis to the increase in the capacity of the produced combine harvesters, the possibility of their effective use decreases, which makes the harvesting process more expensive.

In our opinion, the solution to the problem of harvesting grain crops is in the introduction of stationary technology of combing plants in standing position. The introduction of this technology will permit:

- to reduce the harvesting losses;
- to solve the problem of harvesting drooping and laid grain;
- to lessen grain damage;
- to increase the productivity of the harvesting process;
- to exclude the effect of the combine harvester systems on the soil;
- to decrease the energy intensity of harvesting, and, consequently, reduce the cost of harvesting.

O. Lezhenkin (✉) · I. Lezhenkin · O. Vershkov · S. Kolomiiets
Faculty of Engineering and Computer Technology, Tavria State Agrotechnological University, 18, Bogdana Khmelnitskogo Avenue, Melitopol 72312, Ukraine
e-mail: oleksandr.lezhenkin@tsatu.edu.ua

© Springer Nature Switzerland AG 2019

V. Nadykto (ed.), *Modern Development Paths of Agricultural Production*,
https://doi.org/10.1007/978-3-030-14918-5_10

### 1.1 Analysis of Recent Studies and Publications

We shall consider some of the stationary technologies. First, stationary technologies are represented by a three-phase technology. The three-phase method is based on grinding the entire grain mass simultaneously with mowing or selecting rolls, followed by the final refinement of the obtained heap in a stationary facility [1, 2]. The further development of the three-phase technology was the Kuban industrial technology [3, 4]. The essence of this technology is to mow down and grind the grain mass with its subsequent loading into the trailer and transportation to the stationary facility for finalization. The refinement of the mass includes: separation, drying, and finishing.

The efficiency increase of stationary technologies can be achieved through the use of the method of combing plants. Scientific foundations of this method were formed by Prof. Shabanov [5]. The constructions of combing devices were substantiated by his students Golubev [6], Danchenko [7], and Goncharov [8]. However, the practical implementation of this technology requires additional research.

### 1.2 Statement of the Objective and Tasks of the Study

The purpose of the research is to increase the efficiency of harvesting grain crops in farming conditions by justifying the rational interaction of harvesting and transportation complex machines. To achieve this goal, it is necessary to solve the following tasks:

- to develop a graph of states and transitions of the harvesting and transportation complex;
- using the theory of mass service to build a mathematical model for the functioning of the harvesting and transportation complex links in the form of a system of linear equations;
- to solve the resulting system of equations with respect to the probability of finding links in each of the states and carry out a computer experiment with the results obtained;
- to determine the efficiency coefficients of the harvesting-transportation complex links and analyze the result obtained.

## 2 Results and Discussion

Based on the technological process of harvesting grain, portions of the combed heap with the direction of their movement from the field to the stationary facility were accepted for the flow of requirements in the production line. It was taken into account that the flow of portions of combed heap (requirements) is created by field harvesters,

**Fig. 1** General view of a trailed harvester: 1—a wheeled tractor; 2—a sweeper with working elements of combing type; 3—a trailer cart for collection of combed heap

accumulating it in trailer containers (carts) (Fig. 1) [9]. A tractor moves a trailer cart from the field to the stationary finishing facility and thereby creates a flow of portions of the combed heap (requirements) for their completion (maintenance) by a stationary unit.

Possible conditions of the technological chain of the harvesting process, which take into account functioning, location and interaction of machines, both in the field and on the thrashing floor, can be represented as a graph of states (Fig. 2) [10].

$$\left.\begin{aligned}
&-\left(\lambda_{y.M.}+\lambda_{m.x.}+\lambda_{\partial}\right)\cdot p_1+\lambda_{en}\cdot p_{15}=0;\\
&-(\lambda_m+\lambda_{\partial})\cdot p_2+\lambda_{y.M.}\cdot p_1+\lambda_{e.n.}\cdot p_{16}=0;\\
&-(\lambda_m+\lambda_{\partial})\cdot p_3+\lambda_{m.x.}\cdot p_1=0;\\
&-\left(\lambda_{y.M.}+\lambda_{m.x.}\right)\cdot p_4+\lambda_{\partial}\cdot p_1=0;\\
&-(\lambda_{\partial}+\lambda_{e.n.})\cdot p_5+\lambda_{y.M.}\cdot p_3+\lambda_{m.x.}\cdot p_2=0;\\
&-\lambda_{m.x.}\cdot p_6+\lambda_{\partial}\cdot p_2+\lambda_{y.M.}\cdot p_4=0;\\
&-\lambda_{y.M.}\cdot p_7+\lambda_{\partial}\cdot p_3+\lambda_{m.x.}\cdot p_4=0;\\
&-\left(\lambda_{y.M.}\cdot\lambda_{m.x.}+\lambda_{\partial}\right)\cdot p_8+\lambda_{m3}\cdot p_5=0;\\
&-\lambda_{T3}\cdot p_9+\lambda_m\cdot p_5+\lambda_{m.x.}\cdot p_6+\lambda_{y.M.}\cdot p_7=0;\\
&-(\lambda_m+\lambda_{\partial})\cdot p_{10}+\lambda_{y.M.}\cdot p_8=0;\\
&-\left(\lambda_{y.M.}+\lambda_{\partial}\right)\cdot p_{11}+\lambda_m\cdot p_8=0;\\
&-\left(\lambda_{y.M.}+\lambda_m\right)\cdot p_{12}+\lambda_{\partial}\cdot p_8+\lambda_{m3}\cdot p_9=0;\\
&-\lambda_{\partial}\cdot p_{13}+\lambda_m\cdot p_{10}+\lambda_{y.M.}\cdot p_{11}=0;\\
&-\lambda_m\cdot p_{14}+\lambda_{\partial}\cdot p_{10}+\lambda_{y.M.}\cdot p_{12}=0;\\
&-(\lambda_{\partial}+\lambda_{en})\cdot p_{15}+\lambda_{\partial}\cdot p_{11}+\lambda_m\cdot p_{12}=0;\\
&-\lambda_{en}\cdot p_{16}+\lambda_{\partial}\cdot p_{13}+\lambda_m\cdot p_{14}+\lambda_{y.M.}\cdot p_{15}=0;\\
&\sum_{j=1}^{16}P_j=1.
\end{aligned}\right\}\qquad(1)$$

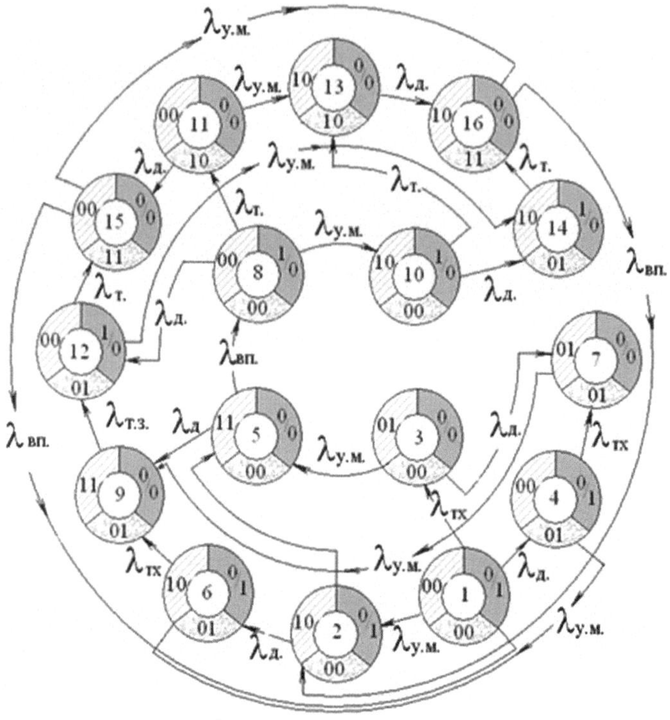


**Fig. 2** Graph of states and transitions of harvesting and transportation complex

Based on the constructed graph of states and transitions, a mathematical model of the harvesting process was obtained in the form of a system of algebraic Eq. (3) describing the joint functioning of HTC links.

The stream intensity of a combed heap ($\lambda$уМ and $\lambda$Т), of a tractor with an empty trailer in the process of its movement from a thrashing floor on the field $\lambda$ТХ, and of the servicing of a portion of the heap by machines ($\lambda$ТЗ, $\lambda$ВП, $\lambda$Д) were determined based on dependencies:

$$\lambda_{y.м.} = \frac{1}{t_{y.м.}};\ \lambda_{m.з.} = \frac{1}{t_{m.з.}};\ \lambda_{m.} = \frac{1}{t_{m.}};\ \lambda_{m.x.} = \frac{1}{t_{m.x.}};\ \lambda_{в.п.} = \frac{1}{t_{в.п.}};\ \lambda_{\partial.} = \frac{1}{t_{\partial.}}, \tag{2}$$

where $\lambda_{y.M.}$ and $t_{y.M.}$ shall mean the intensity and average time of filling the cart with a combed heap the intensity and average time of filling the cart with a combed heap; $\lambda_{m.3.}$ and $t_{m.3.}$ shall mean the intensity and average time of replacing a full cart with an empty one; $\lambda_m$ and $t_m$ shall mean the intensity and average time of transportation of a cart with a heap to the thrashing floor; $\lambda_{m.x.}$ and $t_{m.x.}$ shall mean the intensity and time of an idle tractor movement with an empty trailer; $\lambda_{в.п.}$ and $t_{в.п.}$ shall mean the intensity and time unloading the full trailer into the hopper; $\lambda_{\partial}$ and $t_{\partial}$ shall mean the intensity and time for finishing the heap.

The solution of the system of linear Eq. (1) made it possible to determine the significance of the probabilities of the state of the links in the harvesting and transportation complex in a general form. To find the numerical values of state probabilities, a computer experiment was carried out for the three factors. The following were chosen as variable factors: the intensity of filling the harvesting machine of the trailer trolley, the intensity of finishing the combed heap, and the intensity of tractor movement.

The adequacy of the model was evaluated according to the values of the multiple regression coefficients and the variance of inadequacy. According to the calculations, the multiple regression coefficients are in the range of 0.9977 … 1.000, and the dispersion of inadequacy equals zero. It follows from the data given that the model obtained has sufficient accuracy to describe the change in the probabilities of finding machines of the harvesting complex in various states.

After decoding, the mathematical model took the following form:

$$\begin{aligned}
p_1 &= 0.118248 - 31.2267\cdot\lambda_{\text{Д}} - 30.16\cdot\lambda_{\text{T}} + 8533.33\cdot\lambda_{\text{Д}}\cdot\lambda_{\text{T}} + 175.975\cdot\lambda_{\text{УМ}} - 30285.7\cdot\lambda_{\text{Д}}\cdot\lambda_{\text{УМ}} - 6349.21\cdot\lambda_{\text{T}}\cdot\lambda_{\text{УМ}};\\
p_2 &= 0.0140236 - 1.404444\cdot\lambda_{\text{Д}} + 0.568889\cdot\lambda_{\text{T}} - 711.111\cdot\lambda_{\text{Д}}\cdot\lambda_{\text{T}} + 55.0889\cdot\lambda_{\text{УМ}} - 6857.14\cdot\lambda_{\text{Д}}\cdot\lambda_{\text{УМ}} + 7936.51\cdot\lambda_{\text{T}}\cdot\lambda_{\text{УМ}};\\
p_3 &= 0.0644853 - 18.3289\cdot\lambda_{\text{Д}} - 11.1467\cdot\lambda_{\text{T}} + 3377.78\cdot\lambda_{\text{Д}}\cdot\lambda_{\text{T}} + 104.346\cdot\lambda_{\text{УМ}} - 29142.9\cdot\lambda_{\text{Д}}\cdot\lambda_{\text{УМ}} + 7936.51\cdot\lambda_{\text{T}}\cdot\lambda_{\text{УМ}};\\
p_4 &= 0.0644689 + 5.06667\cdot\lambda_{\text{Д}} - 20.7378\cdot\lambda_{\text{T}} - 2133.33\cdot\lambda_{\text{Д}}\cdot\lambda_{\text{T}} + 41.2698\cdot\lambda_{\text{УМ}} + 14285.7\cdot\lambda_{\text{Д}}\cdot\lambda_{\text{УМ}} - 11111.1\cdot\lambda_{\text{T}}\cdot\lambda_{\text{УМ}};\\
p_5 &= -0.00290133 + 1.40444\cdot\lambda_{\text{Д}} + 2.26667\cdot\lambda_{\text{T}} - 888.889\cdot\lambda_{\text{Д}}\cdot\lambda_{\text{T}} + 16.0603\cdot\lambda_{\text{УМ}} - 6285.71\cdot\lambda_{\text{Д}}\cdot\lambda_{\text{УМ}} + 7936.51\cdot\lambda_{\text{T}}\cdot\lambda_{\text{УМ}};\\
p_6 &= -0.0218262 + 15.1022\cdot\lambda_{\text{Д}} + 13.0044\cdot\lambda_{\text{T}} - 7644.44\cdot\lambda_{\text{Д}}\cdot\lambda_{\text{T}} + 157.397\cdot\lambda_{\text{УМ}} + 31428.6\cdot\lambda_{\text{Д}}\cdot\lambda_{\text{УМ}} - 65079.4\cdot\lambda_{\text{T}}\cdot\lambda_{\text{УМ}};\\
p_7 &= 0.730871 - 35.6444\cdot\lambda_{\text{Д}} - 114.462\cdot\lambda_{\text{T}} + 21688.9\cdot\lambda_{\text{Д}}\cdot\lambda_{\text{T}} - 737.733\cdot\lambda_{\text{УМ}} + 25142.9\cdot\lambda_{\text{Д}}\cdot\lambda_{\text{УМ}} + 138095\cdot\lambda_{\text{T}}\cdot\lambda_{\text{УМ}};\\
p_8 &= 0.00466578 - 1.36\cdot\lambda_{\text{Д}} + 0.444444\cdot\lambda_{\text{T}} + 44.1841\cdot\lambda_{\text{УМ}} - 12571.4\cdot\lambda_{\text{Д}}\cdot\lambda_{\text{УМ}} + 3174.6\cdot\lambda_{\text{T}}\cdot\lambda_{\text{УМ}};\\
p_9 &= 0.0295476 - 4.01778\cdot\lambda_{\text{Д}} - 8.23111\cdot\lambda_{\text{T}} + 1955.56\cdot\lambda_{\text{Д}}\cdot\lambda_{\text{T}} + 19.4032\cdot\lambda_{\text{УМ}} + 15428.6\cdot\lambda_{\text{Д}}\cdot\lambda_{\text{УМ}} + 7936.51\cdot\lambda_{\text{T}}\cdot\lambda_{\text{УМ}};\\
p_{10} &= -0.00122844 - 0.186667\cdot\lambda_{\text{Д}} - 0.871111\cdot\lambda_{\text{T}} + 533.333\cdot\lambda_{\text{Д}}\cdot\lambda_{\text{T}} + 21.3365\cdot\lambda_{\text{УМ}} - 5142.86\cdot\lambda_{\text{Д}}\cdot\lambda_{\text{УМ}} - 1587.3\cdot\lambda_{\text{T}}\cdot\lambda_{\text{УМ}};\\
p_{11} &= -0.0164431 + 6.59556\cdot\lambda_{\text{Д}} + 21.0222\cdot\lambda_{\text{T}} - 7111.11\cdot\lambda_{\text{Д}}\cdot\lambda_{\text{T}} + 24.5016\cdot\lambda_{\text{УМ}} - 12571.4\cdot\lambda_{\text{Д}}\cdot\lambda_{\text{УМ}} + 11111.1\cdot\lambda_{\text{T}}\cdot\lambda_{\text{УМ}};\\
p_{12} &= 0.134399 + 3.08444\cdot\lambda_{\text{Д}} - 42.6222\cdot\lambda_{\text{T}} - 888.889\cdot\lambda_{\text{Д}}\cdot\lambda_{\text{T}} + 87.3778\cdot\lambda_{\text{УМ}} + 30857.1\cdot\lambda_{\text{Д}}\cdot\lambda_{\text{УМ}} - 12698.4\cdot\lambda_{\text{T}}\cdot\lambda_{\text{УМ}};\\
p_{13} &= -0.0309062 + 14.2578\cdot\lambda_{\text{Д}} + 8.124444\cdot\lambda_{\text{T}} - 5155.56\cdot\lambda_{\text{Д}}\cdot\lambda_{\text{T}} + 77.946\cdot\lambda_{\text{УМ}} - 35428.6\cdot\lambda_{\text{Д}}\cdot\lambda_{\text{УМ}} + 22222.2\cdot\lambda_{\text{T}}\cdot\lambda_{\text{УМ}};\\
p_{14} &= -0.027972 + 0.728889\cdot\lambda_{\text{Д}} + 11.2\cdot\lambda_{\text{T}} - 1777.78\cdot\lambda_{\text{Д}}\cdot\lambda_{\text{T}} + 232.857\cdot\lambda_{\text{УМ}} + 14285.7\cdot\lambda_{\text{Д}}\cdot\lambda_{\text{УМ}} - 85714.3\cdot\lambda_{\text{T}}\cdot\lambda_{\text{УМ}};\\
p_{15} &= -0.055648 + 43.2267\cdot\lambda_{\text{Д}} + 168.16\cdot\lambda_{\text{T}} - 8533.33\cdot\lambda_{\text{Д}}\cdot\lambda_{\text{T}} - 324.133\cdot\lambda_{\text{УМ}} + 4571.43\cdot\lambda_{\text{Д}}\cdot\lambda_{\text{УМ}} - 33333.3\cdot\lambda_{\text{T}}\cdot\lambda_{\text{УМ}};\\
p_{16} &= 0.00122 + 1.15556\cdot\lambda_{\text{Д}} + 1.68\cdot\lambda_{\text{T}} - 711.111\cdot\lambda_{\text{Д}}\cdot\lambda_{\text{T}} - 2.58413\cdot\lambda_{\text{УМ}} + 3428.57\cdot\lambda_{\text{Д}}\cdot\lambda_{\text{УМ}} + 11111.1\cdot\lambda_{\text{T}}\cdot\lambda_{\text{УМ}}.
\end{aligned} \tag{3}$$

As an evaluation criterion for the functioning of the harvesting and transportation complex, the coefficient of efficiency of harvesting machines, vehicles and stationary aggregate was accepted.

$$\eta_{y.м.} = \frac{1}{Y}\sum Y_j p_{yj};\ \ \eta_m = \frac{1}{T}\sum T_j p_{mj};\ \ \eta_д = \frac{1}{Д}\sum Д_j p_{дj;} \tag{4}$$

where $Y_j$ shall be the number of working harvesting machines; T shall mean the number of tractors for transportation of combed heap; Д shall mean the number of aggregates for finishing the combed heap; and $p_{Y_j}$, $p_{T_j}$, $p_{Д_j}$ shall mean the correspondent conditions probabilities.

The obtained mathematical models (3), as well as formulas (4), make it possible to find an expression for determining the efficiency coefficient as a function of changes in the intensity of the flows, which in general depend on the productivity of the harvesting complex links.

During the calculation, the number of tractors, necessary to transport the scrubbed heap, was taken as equal to the number of harvesting units, and it was also accepted that there was one stationary aggregate for finishing the combed heap on the farm.

Then taking into account the formulas (4) as well as the graph of states and intensities of the transitions of the harvesting and transport complex (Fig. 2), the efficiency coefficients of the harvesting machine, the tractor for transporting the combed heap, and the stationary aggregate of refinement are determined from the relations

$$\begin{aligned} \eta_{y.M.} &= p_1 + p_3 + p_4 + p_7 + p_8 + p_{11} + p_{12} + p_{15} \\ \eta_g &= p_1 + p_2 + p_3 + p_5 + p_8 + p_{10} + p_{12} + p_{13} \\ \eta_T &= p_1 + p_2 + p_4 + p_6 + p_8 + p_{10} + p_{12} + p_{14} \end{aligned} \tag{5}$$

Factors of efficiency shall be the variable values, since they depend on the numerical values of the state probabilities of the links in the harvesting and transportation complex, which in turn are measured depending on the values of the intensity of the flows.

Equations characterizing the change in the efficiency coefficient have the following form:

$$\begin{aligned} \eta_{УМ} &= 0.62564 - 10.20576 \cdot \lambda_Д + 23.36674 \cdot \lambda_Т + 14934.1 \cdot \lambda_Д \cdot \lambda_Т; \\ \eta_Д &= 0.30179 - 14.360615 \cdot \lambda_Т + 242.564 \cdot \lambda_{УМ} + 28572.3 \cdot \lambda_Т \cdot \lambda_{УМ}; \\ \lambda_Т &= 0.25802 - 17.76225 \cdot \lambda_Д + 498.01 \cdot \lambda_{УМ} + 36000 \cdot \lambda_Д \cdot \lambda_{УМ}. \end{aligned} \tag{6}$$

The calculation of the efficiency coefficient according to the formulas (6) has shown that the efficiency factor of the harvesting machine varies in the range of 0.746 … 0.884, that is, its values are insignificantly dependent on the productivity of the harvesting machine. Based on the result, the harvesting machine is almost completely loaded. Its downtime is due to technological reasons (changing the filled trolley).

In turn, the efficiency of the use of the tractor directly depends on the hauling distance. With increasing distances from field to grain flow, the coefficient of efficiency of the tractor increases and amounts to 0.512 … 0.520, whereas at minimum distances, its value is in the range of 0.18 … 0.361. In the calculations, the distance from the field to the grain path was assumed to be 2.5 … 4.0 km. Increasing the hauling distance of a combed heap up to 8 km can lead to the downtime of the harvesting unit because there is no empty cart. Therefore, it is most expedient for farms to locate a thrashing floor at a distance of 4.0 … 5.0 km from the field.

The numerical values of the efficiency factor of the aggregate for finishing the combed heap vary widely within 0.07 … 0.401 and depend on the intensity. The low values of the efficiency factor of the finishing aggregate play a generally positive role since there is no downtime of the harvesting unit in the field, and the portions of the combed heap delivered from the field are processed in a timely manner.

## 3 Conclusion

1. A graph of states and transitions of the harvesting and transportation complex has been developed which makes it possible to analyze the functioning, location, and interaction of the harvesting complex.
2. A mathematical model of the functioning of the harvesting-transportation complex links has been constructed in the form of a system of linear equations which describe each state.
3. Solution of a system of linear equations, relative to the probability of finding the links of the system in each of the states, allowed to conduct a computer experiment.
4. The calculation of the efficiency factor of the harvesting machine varies in the range of 0.746 … 0.884; thus, the harvesting machine is almost completely loaded. Its downtime is due to technological reasons. In turn, the efficiency of the use of the tractor depends on the hauling distance and varies in the range of 0.18 … 0.52. Therefore, it is most expedient for farms to have a stationary shield at a distance of 4.0 … 5.0 km. The numerical values of the efficiency factor of the heap completion unit range from 0.07 to 0.401 and that means that the whole heap delivered from the field is processed in a timely manner, and there is no downtime of the harvesting unit.

## References

1. Runchev, M.S.: Traktoryi i selhozmashinyi. **9**, 32–36 (1959)
2. Zhuk, Y.M.: Mehanizatsiya i elektrifikatsiya sotsialisticheskogo selskogo hozyaystva. **5**, 4–8 (1959)
3. Kanarev, F.M.: Mehanizatsiya i elektrifikatsiya selskogo hozyaystva. **8**, 17–19 (1986)
4. Kanarev, F.M.: Obosnovanie rabochih organov pochvoobrabatyivayuschih i uborochnyih mashin. 3–9 (1985)
5. Shabanov, P.A.: Mehaniko-tehnologicheskie osnovyi obmolota zernovyih kultur na korny. Doctor's thesis, Melitopol (1988)
6. Golubev, I.K.: Obosnovanie osnovnyih parametrov i rezhimov rabotyi dvuhbarabannogo ustroystva dlya ochesa risa na kornyu. Candidate's thesis, Moscow (1989)
7. Danchenko, N.N.: Obosnovanie parametrov schetochnogo ustroystva dlya ochesyivaniya metelok risa na kornyu. Extended abstract of candidate's thesis, Chelyabinsk (1983)
8. Goncharov, B.I.: Issledovanie rabochego protsessa ochesyivayuschego ustroystva dlya obmolota risa na kornyu s tselyu umensheniya poter zerna. Candidate's thesis, Moscow (1982)
9. Lezhenkin, A.N.: Selskiy mehanizator. **12**, 2 (2004)
10. Lezhenkin, A.N.: Pratsi TDATA. **40**, 195–204 (2006)

# Software Development for the Security of TCP-Connections

**Dmytro Lubko, Sergii Sharov and Oksana Strokan**

## 1 Introduction

Nowadays, cybersecurity can be considered as an important aspect of the policy of any state in the conditions of the existence of a global information space, widespread communication and interaction through the Internet. Despite constant development of new methods of protecting information and computer systems from unauthorized access as well as their updating, cybercriminals continue finding the ways of bypassing the information security systems and carrying out their devastating activities. In turn, the vulnerability of information systems to certain actions (conscious or unconscious) of end users also raises a significant issue in protecting personal data and information flows. The protection of information involves the achievement and saving the security properties in user resources aimed at preventing relevant cyberthreats. As a result, the development of qualitative software products and digital equipment will give a possibility to increase the level of information security for commercial enterprises, government agencies and the state. A number of system and application software designed to provide information security can be found in the software market today. The development and dissemination of the Internet led to a common use of special programs, called sniffers, which are intended for interception and analysis of the network traffic.

---

D. Lubko (✉) · O. Strokan
Tavria State Agrotechnological University, B. Khmelnitsky ave. 18, Melitopol 72310, Ukraine
e-mail: dmytro.lubko@tsatu.edu.ua; di75ma@gmail.com

S. Sharov
Bohdan Khmelnytsky Melitopol State Pedagogical University, Hetmanska St, 20, Melitopol 72312, Ukraine

© Springer Nature Switzerland AG 2019

V. Nadykto (ed.), *Modern Development Paths of Agricultural Production*,
https://doi.org/10.1007/978-3-030-14918-5_11

## 2 Analysis of Recent Research and Publications

Issues of computer networks, their monitoring and analysis, as well as the development of appropriate specialized network traffic analyzers, were studied by such well-known scientists, researchers and programmers: Michael W. Lucas, Oliver M. Heckmann, Ed Wilson, N. Mendkovich, G. Konakhovich, V. Chuprin and other scholars. M. Doinea researched software products for analyzing open-source network traffic [1], M. A. Qadeer studied the packet sniffer principles [2], and Z. Xiao, L. Guo and J. Tracey used sniffers to analyze the traffic of instant messaging systems [3]. Among Ukrainian scientists, we can mention D. Dubov, O. Baranov and others who deal with the problem of cybersecurity. At the same time, the problem of developing software products for analyzing network traffic that perform specific functions is still considered to be relevant.

## 3 Results and Discussion

There is a tendency today toward the active use of information resources in all spheres of economic, social, educational, political and other activities. Along with the rapid growth of hardware capacity, cases of unauthorized data collection, use, distribution of personal data and important information, fraud in the Internet are spreading. It is due to its widespread and massive nature that cybercrime has become transnational, capable of causing a severe harm to the interests of the individual, society and the state [4]. The mentioned circumstances explain the relevance of effective mechanism development and use to ensure the cybersecurity of the state and resolving of the related problems [5].

Currently, there are several issues of information security and protection of personal data on the Internet. Primarily, it concerns the unification of the cybersecurity notion. Today, there are different definitions of this notion, namely: It is a condition of security of the vital interests of the individual, society and the state in the use of computer systems and/or telecommunication networks [6]; it is a status of a person's ability, society and the state to prevent and avoid the directed, first of all, unconscious, negative influence of information [7]. Since national and international documents differ significantly in cybersecurity determination, then the approaches to cybersecurity are different.

As Bellovin and Cheswick [8] point out, the situation with the information security of a computer connected to the network is complicated by several factors at once:

1. The mail program, the network file system and the distributed database are potential sources of danger. In addition, the authentication used by some protocols may be inadequate. However, they must be loaded into RAM to ensure that users are properly served.
2. There are many entry points from which you can start an attack. If computer users are limited to one task, it will be difficult for the outsider to try to penetrate

the security system. On the other hand, a computer connected to the network can be attacked from anywhere in the world, connected to the Internet.
3. Connecting to hazardous computers connected to the Internet causes problems of transitive trust, which also require some actions to prevent them.

Another problem related to providing information security at least at a minimum level is due to the lack of awareness of users about cyberthreats. The constant use of the Internet by ordinary users, who often do not know about the potential risks of losing personal information and cyberattacks by unauthorized third parties, often leads to cases of Internet fraud, theft of passwords and files, loss of money. In this case, the solution to the problem is to inform users of the existing risks regarding the use of the Internet and the importance of security of personal information [9].

Effective cybersecurity is associated with the development of appropriate mechanisms, hardware and software tools to protect against unauthorized access to information (firewall, sniffer, cryptographic protocols, etc.).

The main solution for a bulk of information protection issues is in use of firewall that monitors and analyzes the connection of your computer to the Internet, which is the basis for deciding whether to allow this connection on the basis of the settings of the program. That is, the network filter passes the traffic of only those programs that the user allows. As a result, the effectiveness of the firewall depends on the experience of the user or the system administrator who installs the software setup. There are two types of firewall: personal, installed on a personal computer, and corporate software installed on the gateway between the Internet and the local network. In the first case, the setting is done by the user and the other by the system administrator. In both cases, it is recommended to allow the traffic to only trusted applications and only through the ports on which they work. By all means, in order to prohibit access to certain programs, you need to know how they harm the computer or the corporate network.

One of the tools for analyzing network traffic is sniffers, which means a program or hardware and software device designed for interception, storage and analysis of network traffic. This type of software is often used by system administrators in a legitimate way to address certain problems when transmitting data over a network [2].

Sniffers are designed for a specific type of network, usually for the Ethernet; that is why it is often called network analyzer or the Ethernet Sniffer. This program passively intercepts data addressed to other computers at the level of the network adapter NIC (OSI network interface card). Trapping is carried out using several methods:

- Usual «interception» of the network interface. This method is effective when used in a segment of hubs instead of commutators;
- Connection of a sniffer to a channel gap;
- Branching (software or hardware) of traffic and directing its copy to the sniffer;
- Analysis of side electromagnetic emissions and traffic restoration.

Analysis of network traffic that has passed through a sniffer allows:

- to identify viral and/or looped traffic that increases the download of network equipment and communication channels;
- to identify malicious and unauthorized software, such as network scanners, flooders, Trojans, peer network clients and others;
- to capture any unencrypted (sometimes encrypted) network traffic in order to obtain passwords or other important information;
- to locate a network failure or network agent configuration error.

All the sniffers can be divided into two categories: sniffers that support download and work from the command line and sniffer with graphical interface. Some hybrid variants combine both modes of operation. There are plenty of examples of existing network traffic analyzers for various operating systems: Wireshark, Iris, WinDump, Sniffit, Ultra Network, sniffer, analyzer, Packetyzer, IPDump2, Ferret, LanGrabber Ethereal Network Analyzer Wireshark and others. All these software features are similar in function, but differ in the protocols they support, the depth of the analysis of intercepted packets, filtering capabilities, as well as compatibility with other software products, the user interface and the ability to generate statistical reports. It should be noted that since in the "classic" sniffer traffic analysis is done manually, using only simple automation tools (protocol analysis, TCP stream restoration), it is suitable for analysis of only small volumes of data.

To work with a sniffer, you need to know several recommendations:

- If the sniffer is used on the corporate network, then the rules and procedure for its use must be described [1].
- It is desirable to use filters when intercepting data packets in order to increase the efficiency of analytics.
- The use of sniffer does not make any sense if the data intercepted by it are not used for further analysis.
- Make sure that a specific sniffer is able to intercept network traffic (incompatibility of protocols, operating systems, applications).

According to Androschuk [4], Ukrainian users are highly vulnerable to infections owing to refusal to update software or use of pirated copies of programs. The situation is aggravated by the fact that some users use outdated operating systems, such as Windows XP, where protection from modern cyberattacks is almost absent.

In view of this, we set ourselves the task of developing a software tool to intercept and analyze outbound TCP-connections in 32-bit programs. A software tool allows you to perform tasks that often have to be solved by system administrators, for example, to identify specific causes of client network software failures, to test programs for suspicious activity and intercept the transmitted information. It also provides the ability to control the need for the entire data transfer process in arbitrary software.

The designed sniffer should be fast, reliable, free and stable, and store the data received for further monitoring. Its development was carried out in the following stages:

1. Analysis of the operation of the protocol TCP/IP, specifics of TCP-connections, review of the functional capabilities of the existing network traffic analyzers;
2. Selecting the format of data storage that will be needed to run the software;
3. Selecting the tool environment for the development of the sniffer;
4. Creating software modules;
5. Testing, validation of the software in real conditions, error correction and re-testing.

Let us consider the basic requirements for the developed software.

*Requirements for interface design.*

Graphical user interface software should be easy to use, flexible and intuitively comprehensible, zoomable, looking the same on computers with different localizations. The user interface language is English. The main window should have a minimum number of controls.

Access to basic operations should be implemented through the main menu. Transactions over data packets must be implemented through the context menu, and the ability to view the contents of the package in text as well as in the 16-year form should be implemented.

The download window of the new process should be able to select the executable file by using the dialog or manually entering the path to it. You must also be given the option to load the process parameters and delete them. The connection window to the active process should contain a list of processes and a block of control buttons.

The list of processes needs to be implemented in the form of a table with two columns. The first column contains the digital ID of the program and the icon. If the program does not have icons, you should use the default icon. The second column contains the name of the executable process file.

The manual window of the data packet must contain packet contents and a data management block. The function of sending a package several times in a row with a given interval must be implemented.

The search box should have an intuitive interface with the ability to select a search method (text, using a regular expression or 16-bit sequence).

Requirements for software and hardware.

The main language of software development was the C++ programming language.

To develop the graphical user interface, a cross-platform Qt software development library was used. The programming language Lua [10] was used as a tool for writing and executing scripts.

Software functionality requires the following software and minimum hardware: operating system: 32-bit Windows XP, 7; installed Microsoft Visual C++ 2010 Redistributable Package (×86) software; RAM—2 GB or more; hard drive—500 GB or more. Software Restriction: The sniffer only works with 32-bit software.

Software aspects of sniffer development.

The main software module consists of five main parts. The general scheme of the software is shown in Fig. 1.

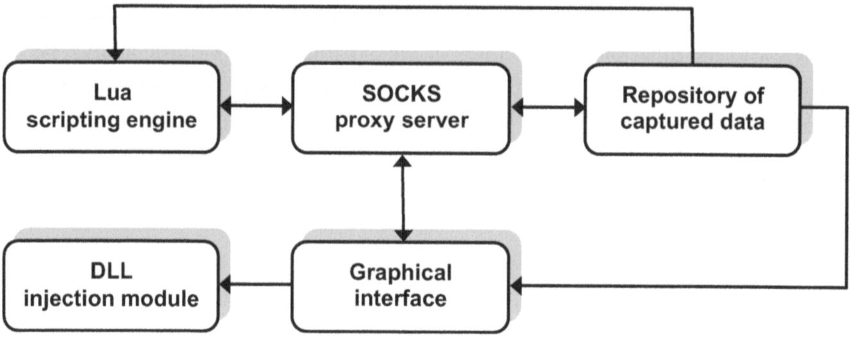


**Fig. 1** General scheme of the developed sniffer

Developing a software tool for analyzing outbound TCP-connections involves the development of a dll-module that redirects all new TCP-connections of the analyzed software to the main module. Before starting to develop this module, the following issues should be solved:

1. How to implement the method of intercepting ws2_32.connect and ws2_32.WSAconnect? There are four basic methods of intercepting the target function call located in the dll. The first one is to replace the function pointer in the library export table. The second involves replacing the pointer to the function in the import table of the main module of the program.

The essence of the third method is to replace the first few instructions (or one if its size is $\geqq 5$ bytes) to the instructions for the unconditional jump to the address (jmp long addr), the parameter of which is the address of the handler's function. The fourth method is most often used in cases where the function code structure does not allow to replace its first instructions with the five-byte instruction jmp long. When developing the software, the third variant of the interception was used, as the most versatile, and has the least number of shortcomings. The scheme of the method is shown in Fig. 2.

2. How to implement the method of loading a working library into the target process? The loading of the working library to the address space was as follows: The target process opens with kernel32.OpenProcess; it allocates the memory area with access rights PAGE_EXECUTE_READ, where using the kernel32.WriteProcessMemory function writes the loading code of the library with the dll extension; a stream that executes this code is loaded.
3. How to implement the protocol of information exchange with the main module of the software? SOCKS version 5 was chosen to exchange information with the main software module, but only part of the protocol that was needed to process the output TCP-connections was implemented.

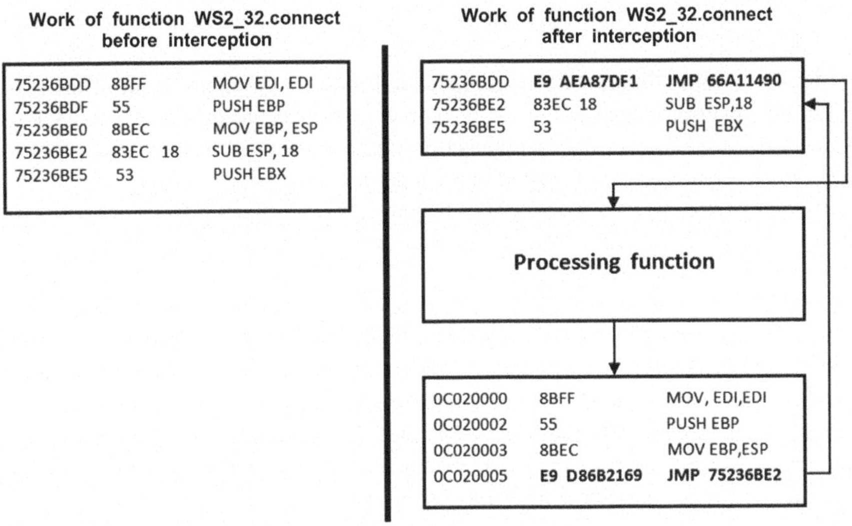


**Fig. 2** Logic of the function working before and after the interception

This allowed intercepting traffic from part of client network applications (those that support work through an intermediate SOCKS server) without the use of a work library.

4. How to integrate the Lua scripting language into a software tool? This task was divided into several stages.

The first stage involved creating an interpreter control class that runs in a separate flow. Due to the fact that the Qt platform provides the ability to implement multi-threaded operation and has a built-in functionality for transmitting data between them using signals, this stage was executed fairly quickly.

To achieve this, two classes were implemented—luaEngineWorker, which encapsulates the entire low-level interpreter control code, and the luaEngine control class, which implements the interface between flows and manages the execution of the interpreter's flow.

The second stage consisted of writing classes that implement the interface between the main code and the interpreter of the Lua programming language. In this step, classes were written that provide full access to the data repository and allow you to manage intercepted connections. For the possibility of full control of the data transmission process, a system of events was implemented, the handlers of which can arbitrarily control the process of data transmission in intercepted connections.

The final stage involved the integration of the Lua interpreter with the software using the SWIG toolkit. This approach frees the programmer from writing a huge amount of code for the interface and significantly reduces the time of development and adjustment of the software.

Let us consider the most significant moments in the development of the main module.

Access to most functions was implemented through the main menu of the program, but functions that directly work with a particular package (e.g., the "Close connection" function for forced connection failure) were resolved to be brought to the context menu that can be called in the data tree.

When developing a software tool, it became clear that the Qt Designer software used to design forms lacked the ability to create and edit a context menu. In view of that fact that the structure and appearance of the menu do not change during the work of the sniff, it was decided to create objects of the context menu when initializing the main window of the software using the procedure initContextMenu. Then we developed, the onPacketListContextMenu (Q Point p) function handler was created for the customContextMenuRequested signal, which is automatically generated when the right mouse button is pressed. This function in the process of running a software tool displays a screen built-in window initialization context menu. The binding of the handler function to the signal was done using the Qt Designer slot editor.

The event log is implemented using the QTextBrowser widget, which has the following benefits: fast work, small overhead for saving text, no flicker when you add a large number of entries in a small amount of time. The function of adding new entries to the event log was to be implemented similarly to the function of the standard printf library. This approach has made it easy to format event log entries without creating any excessive software code.

It was decided to implement the content of the intercepted data packet in text and 16-bit views. The first format of the data presentation was implemented using the standard QPlainText widget, which allows you to view a data packet as text, which is very useful in analyzing text data protocols. The second data viewer mode was implemented using the QHexEdit2 widget, which is freely licensed under the GNU Lesser General Public License.

To have the developed software tool for analysis of network traffic properly used, we will give a brief user's manual.

Work with the program tool (executable file ReqPacketTool.exe) is carried out using the main menu items, the context menu of the data tree and the controls. In addition, for the automation of executable actions, the sniffer developed can accept command line parameters that can be combined arbitrarily. Figure 3 shows the analysis of network traffic on the home page of Wikipedia at https://uk.wikipedia.org.

The main menu contains several sections, which, in turn, consists of sub-items. The section "Main" includes the following items:

- The "Start process" item allows the user to download an arbitrary software tool for further analysis. To do this, manually or by using the dialog (button «...»), select the main executable file of the target software, set all necessary parameters for loading it, if necessary, and click on the "Start" button.
- The "Inject to process" item allows the user to intercept all new connections of the working process. After this, a process selection window appears that contains

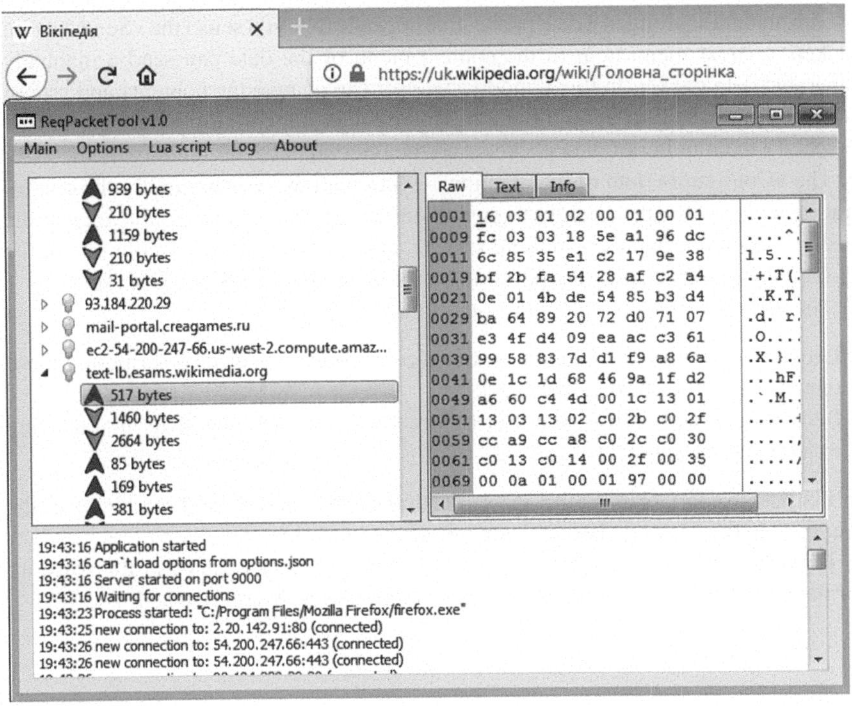


**Fig. 3** Interface of the main window of the software

a list of all processes in the operating system. The user must select the desired process and click on the "Inject" button. It should be noted that with the help of the developed software tool it is impossible to intercept connections of processes that are started by users with a higher level of rights than that of the user who launched this sniffer.

- «Pause sniffing/Start sniffing» allows the user to pause, if necessary, the accumulation of intercepted data and to continue this process at any time.
- The «Clear captured packet list» option allows you to clear the tree of the intercepted data if necessary.
- The "Find" item allows the user to search for the requested information in the intercepted data. The search box has three main modes of operation, namely: searching for text data (which is sensitive to the case), searching through regular expressions and searching for 16-year-old sequences. For the convenience of the user in the search mode of 16-bit sequences, the search according to the template is realized. To do this, all data packets that have an undefined value are marked with a "?". For example, the query "FF??" will find all two-byte sequences that start with the FF code.
- The "Load packets queue"/"Save packets queue" items allow you to download previously exported data/export intercepted data.

The manual sending of a data packet that appears after you select the «Send custom packet» context menu item in the context menu of the data can send a manually generated data packet. In this window, the user can change the contents and size of the sent packet, as well as its direction. To send a package multiple times, you need to specify the number of data packets and the interval between sending them.

The «Configure» item of the «Options» section allows you to modify the configuration of the software, namely to switch on/off logging events, disable/enable the receipt of domain names of remote servers, change the number of columns in the 16-bit data viewer, change the port number of the embedded SOCKS server.

The «Lua script» section contains the following items:

- Item «Start script» allows the user to execute arbitrarily selected script or choose one of the scripts used by the latter.
- Item «Stop script» allows the user to forcefully terminate the script that is being executed at the moment.

Items «Save»/«Clear» in the «Log» section allow you to save the contents of the event log to the file and clear the contents of the event log.

In order to speed up the operation and for the user convenience of the software for the main operations of the main menu, shortcut keys were assigned.

## 4 Conclusion

Consequently, in the conditions of the existence of a global information space cybersecurity takes a significant place among the important issues of many states. For its provision, appropriate technologies of information security are being developed, and state-level legislative documents, hardware and software are being worked out. Consequently, sniffers are designed to intercept traffic for further analysis.

In the process of research based on a technical specification, a software tool was developed for interception and analysis of output TCP-connections and a user's manual was written. The object of automation of the developed sniffer is the system administrators' work, who often have to solve the tasks of providing information security and the correct operation of the internal and external network. Among such tasks, one can distinguish interception, interpretation and storage of data transmitted through the network by various programs. Further analysis of the received data packages helps to analyze suspicious activity of some programs or to identify certain problems in the early stages when it is possible to find the ways of their solution faster and to avoid negative consequences.

The prospects for further research lie in increasing the functionality of the software tool, specifically in modifying it for use in networked software with 64-bit architecture.

## References

1. Doinea, M.: Open source security tools. Open Source Sci. J. **2**(2), 131–144 (2010)
2. Qadeer, M.A., Iqbal, A., Zahid, M., Siddiqui, M.R.: Network traffic analysis and intrusion detection using packet sniffer. In: International Conference on Communication Software and Networks ICCSN'10, pp. 313–317 (2010)
3. Xiao, Z., Guo, L., Tracey, J.: Understanding instant messaging traffic characteristics. In: International Conference on Distributed Computing Systems ICDCS'07, pp. 51–59 (2007)
4. Androshchuk, H.O.: Cybersecurity: trends in the world and Ukraine. In: International Conference on Cybersecurity and Intellectual Property: Problems of Legal Enforcement, pp. 30–36 (2017)
5. Dubov, D.V.: Strategic aspects of cybersecurity of Ukraine. Strat. Priorities **4**, 119–127 (2013)
6. Baranov, O.A.: About the interpretation and definition of «cybersecurity». Leg. Inform. **2**, 54–62 (2014)
7. Furashev, V.: Key aspects of the draft law of Ukraine «On information security». Veche **6**, 29–30 (2012)
8. Bellovin, S.M., Cheswick, W.R.: Network firewalls. IEEE Commun. Mag. **32**(9), 50–57 (1994)
9. Kritzinger, E., von Solms, S.H.: Cyber security for home users: a new way of protection through awareness enforcement. Comput. Secur. **29**(8), 840–847 (2010)
10. Ieruzalimski, R.: Programming in Lua language. DMK Press, Moscow (2014)

# The Formation of Orthogonal Balanced Experiment Designs Based on Special Block Matrix Operations on the Example of the Mathematical Modeling of the Pneumatic Gravity Seed Separator

**Vira Malkina, Serhii Kiurchev, Viacheslav Osadchyi and Oksana Strokan**

## 1 Introduction

The experiment is the most important element of scientific research. The qualitative experiment allows obtaining the information about the object of the research. Therefore, the skill full arrangement of the experiment is crucially important. The task of experiment designs is urgent for the mechanism operation process modeling, particularly for the determination of optimal operation parameters.

However, for accurate determination of parameter values based on mathematical equations, it is necessary to obtain the most complete information. For this purpose, the experimental studies with the largest quantity of factor levels (parameters) affecting the separator operation have to be carried out.

In this way, the problem of formation of such design of experimental studies of the separator, which is going to provide the most accurate and complete model of its operation, is of current importance and will allow determining the optimal work parameters.

In the present paper, solved the task of the mathematical model formation of the of a pneumatic gravity separator by means of the organization of the effective experiment conduction scheme with a sufficiently large number of factor levels and simultaneously small number of experiments operation.

V. Malkina (✉) · S. Kiurchev · O. Strokan
Tavria State Agrotechnological University, B. Khmelnitsky Ave.18, Melitopol 72310, Ukraine
e-mail: vira.malkina@tsatu.edu.ua

V. Osadchyi
Bogdan Khmelnitsky Melitopol State Pedagogical University, Getmanska 20, Melitopol 72312, Ukraine

© Springer Nature Switzerland AG 2019 
V. Nadykto (ed.), *Modern Development Paths of Agricultural Production*,
https://doi.org/10.1007/978-3-030-14918-5_12

The formation of the planning matrix, which possesses the specified features that ensure the sufficiently high level of the experimental research efficiency, is the most important stage of the research.

The range of criteria exists that assesses the quality of the experiment design and the criteria of the design optimality. Depending on requirements imposed to the process of the model formation and analysis chosen one or another criterion.

The choice of a criterion depends on the research task. Thus, while studying the influence of separate factors on the object behavior, used the *E*-criterion, while searching the response function optimum chosen the *D*-optimality criterion.

Since the tasks of our research are the effective design formation of experiment conduction with a large number of factor levels and the following determination of the optimal values of the factors based on the resulting model, therefore the most appropriate in this case is the *D*-optimality criterion.

One of the approaches to the formation of *D*-optimal designs is numerical algorithms. Such algorithms are numerical procedures, so-called exchange algorithms. The number of papers [1–3] deals with the design formation algorithms.

In [4], examined the issue of comparison of the experimental design formation algorithms providing the *D*-optimality criteria.

The methodology of the *D*-optimal design formation based on Kronecker operations of special block matrix operations suggested in [5]. In this paper, the algorithm of the orthogonal balanced design formation of the $m^n$ experiment, where $m$ is the number of factor levels, and $n$ is the number of factors is proposed.

In [6, 7], initiated the mathematical model that describes the operation of the pneumatic gravity seed separator. The experiment carried out according to the Box-Bank design with three two-level factors. Based on the experiment, formed the mathematical model, according to which the rational values of the parameters (factors) of the separator operation are determined. The optimal values of the remaining separator parameters are determined experimentally. In this paper, examined the mathematical model of the separator operation formed based on the conduction of experiments with five factors having five levels. For the model, used formation the scheme of orthogonal balanced design formed on special block matrix operations. The optimal values of separator operation parameters found based on the mathematical model formed according to the results of the experiment.

The objective of the article is the algorithm of the formation of orthogonal balanced experiment designs, which are optimal according to the *D*-efficient criterion, the mathematical model development of the pneumatic gravity seed separator, and determination of the optimal separator parameters based on the developed model.

## 2 Research and Discussion of the Result

It is commonly known that the full factor experiment gives the most complete information about the object of the research with the given number of factors and their levels. In such design, the matrix columns do not correlate; the design is orthogonal and balanced. However, the essential drawback of such design is the necessity of conduction of large number of experiments. Thus, for example, with five factors, each of which varies at five levels, it is necessary to perform $5^5 = 3125$ experiments. This significantly limits the possibilities of application of such designs.

One of the methods of obtaining sufficiently complete information about an object with the reduction of the number of necessary experiments is the fractional design formation. In this case, it is possible to use not the whole matrix of the design of the full factor experiment, but only some of its lines. The matrix of the design, at the same time, formed in such a way that even the part of information sufficiently fully reflects the researched object, in other words, the design must be efficient.

One of the quality indicators of the design completeness is the fact that each factor at each level repeated in the design matrix the same number of times; consequently, the matrix of such design forms an orthogonal array and the design is balanced.

As it is known from [8], the orthogonal design or the orthogonal table $OA_\lambda(t, k, v)$ of size $\lambda v^t \times k$ over the alphabet of $v$ letters is the matrix in which each of the symbol combinations occurs exactly $\lambda$ times for any set of $t$ columns under the limitation of matrix lines for these $t$ columns.

The design formed based on the orthogonal design is balanced, since it contains the same amount of each level of each factor, each pair of factors occurs the same number of times. If the design is balanced and its matrix is orthogonal, then the criterion of the $D$-efficiency of such design is of the greatest importance.

The $D$-efficient criterion minimizes the determinant of the corresponding normal system of linear equations. According to this criterion, the expected forecast error based on the response function is minimal. The value of the $D$-optimality criterion calculated by the formula:

$$D_{\text{eff}} = \frac{\left|M^{\mathrm{T}} \times M\right|^{\frac{1}{m}}}{n \times m}, \tag{1}$$

where

$M$ the matrix of the experiment;
$M^{\mathrm{T}}$ the transposed experiment array;
$n$ the number of factors;
$m$ the number of levels.

Based on the methodology [5], offered the algorithm for the formation of matrix of the orthogonal $m^n$ type design. For the formation of such a matrix, Kronecker operations are used [9].

If $A$ is a matrix of the $m \times n$ size, $B$ is a matrix of the $p \times q$ size, then the Kronecker product is a block matrix of the $mp \times nq$ size

$$A \otimes B = \begin{bmatrix} a_{11}B & \dots & a_{1n}B \\ \dots & \dots & \dots \\ a_{m1}B & \dots & a_{mn}B \end{bmatrix} \tag{2}$$

Let us examine the Kronecker $A$ matrix product $\begin{bmatrix} 0 \\ 1 \\ 2 \\ 3 \end{bmatrix}$ and $A^{\mathrm{T}} = \begin{bmatrix} 0 & 1 & 2 & 3 \end{bmatrix}$:

$$A \otimes A^{\mathrm{T}} = \begin{bmatrix} 0\,0\,0\,0 \\ 0\,1\,2\,3 \\ 0\,2\,4\,6 \\ 0\,3\,6\,9 \end{bmatrix} \tag{3}$$

We perform the "division by remainder" operation with the elements of matrix obtained as the Kronecker product $k$, that is, the calculation of the remainder from the division by $k$.

For example, the Kronecker product $A \otimes A^{\mathrm{T}}$ "by remainder 3" is the matrix:

$$D_3 = \begin{bmatrix} 0 \\ 1 \\ 2 \\ 3 \end{bmatrix} \otimes \begin{bmatrix} 0 & 1 & 2 & 3 \end{bmatrix} = \begin{bmatrix} 0\,0\,0\,0 \\ 0\,1\,2\,3 \\ 0\,2\,1\,0 \\ 0\,0\,0\,0 \end{bmatrix} \tag{4}$$

On the analogy of the Kronecker product, the operation of the block matrix summation considered in [5]. The block sum of matrices $A = \left[a_{ij}\right]$ and $B = \left[b_{ij}\right]$ is the matrix of the type:

$$A \oplus B = \begin{bmatrix} a_{11} + B & \ldots & a_{1n} + B \\ \ldots & \ldots & \ldots \\ a_{m1} + B & \ldots & a_{mn} + B \end{bmatrix}$$

$$A \oplus B = \begin{bmatrix} a_{11}+b_{11} & a_{11}+b_{12} & a_{11}+b_{1n} & \ldots & a_{1n}+b_{11} & a_{1n}+b_{12} & \ldots & a_{1n}+b_{1n} \\ a_{11}+b_{21} & a_{11}+b_{22} & a_{11}+b_{2n} & \ldots & a_{1n}+b_{21} & a_{1n}+b_{22} & \ldots & a_{1n}+b_{2n} \\ \ldots & \ldots & \ldots & \ldots & \ldots & \ldots & \ldots & \ldots \\ a_{11}+b_{m1} & a_{11}+b_{m2} & a_{11}+b_{mn} & \ldots & a_{1n}+b_{m1} & a_{1n}+b_{m2} & \ldots & a_{1n}+b_{mn} \\ \ldots & \ldots & \ldots & \ldots & \ldots & \ldots & \ldots & \ldots \\ a_{m1}+b_{11} & a_{m1}+b_{12} & a_{m1}+b_{1n} & \ldots & a_{mn}+b_{11} & a_{mn}+b_{12} & \ldots & a_{mn}+b_{1n} \\ a_{m1}+b_{21} & a_{m1}+b_{22} & a_{m1}+b_{2n} & \ldots & a_{mn}+b_{q2} & a_{mn}+b_{22} & \ldots & a_{mn}+b_{2n} \\ \ldots & \ldots & \ldots & \ldots & \ldots & \ldots & \ldots & \ldots \\ a_{m1}+b_{m1} & a_{m1}+b_{m2} & a_{m1}+b_{mn} & \ldots & a_{mn}+b_{m1} & a_{mn}+b_{m2} & \ldots & a_{mn}+b_{mn} \end{bmatrix} \tag{5}$$

The obtained matrix is the main block for the formation of the orthogonal balanced design scheme for the $m^n$ kind experiment ($n$ is the number of factors, $m$ is the number of levels).

Let us examine the Kronecker product of matrices $L_n = \begin{bmatrix} 0 \\ 1 \\ \ldots \\ n \end{bmatrix}$ and $L_n^{\mathrm{T}} = \begin{bmatrix} 0 & 1 & \ldots & n \end{bmatrix}$:

$$L_n \oplus L_n^{\mathrm{T}} = \begin{bmatrix} 0 & 1 & 2 & \dots & n \\ 1 & 2 & 3 & \dots & n+1 \\ \dots & \dots & \dots & \dots & \dots \\ n & n+1 & n+2 & \dots & 2n \end{bmatrix} \quad (6)$$

Let us form the $D_m$ matrix as the result of applying the "remainder from division by $m$" of the $L_n \oplus L_n^{\mathrm{T}}$ matrix elements.

Let us form the $F$ matrix as the result of performing the operation of the block summation of the column-matrix $\begin{bmatrix} 0 \\ 1 \\ \dots \\ n-1 \end{bmatrix}$ and the $D_m$ matrix:

$$F = \begin{bmatrix} 0 \\ 1 \\ \dots \\ n-1 \end{bmatrix} \oplus D_m = \begin{bmatrix} 0 + D_m \\ 1 + D_m \\ \dots \\ n + D_m \end{bmatrix}. \quad (7)$$

Then using operation "remainder of the division by $m$" for each element of the result matrix.

For example, the matrix of the orthogonal design of the $5^5$ type experiment ($m = 5$, which is the number of factor levels, $n = 5$ and is the number of factors), is formed according to the offered scheme and has the type:

$$[l_5, l_5 \oplus D_5] = \begin{bmatrix} l_5,\ 0 + D_5 \\ l_5,\ 1 + D_5 \\ l_5,\ 2 + D_5 \\ l_5,\ 3 + D_5 \\ l_5,\ 4 + D_5 \end{bmatrix}, \quad (8)$$

where $l_5$—the matrix $\begin{bmatrix} 0 \\ \dots \\ 4 \end{bmatrix}$.

After the matrix formation row sorting according to the first column, we obtain the experiment design type $5^5$ matrix:

0 0 0 0 0 0
0 1 1 1 1 1
0 2 2 2 2 2
0 3 3 3 3 3
0 4 4 4 4 4

$$\begin{array}{ll}
1 & 0\,1\,2\,3\,4 \\
1 & 1\,2\,3\,4\,0 \\
1 & 2\,3\,4\,0\,1 \\
1 & 3\,4\,0\,1\,2 \\
1 & 4\,0\,1\,2\,3 \\
2 & 0\,2\,4\,1\,3 \\
2 & 1\,3\,0\,2\,4 \\
2 & 2\,4\,1\,3\,0 \\
2 & 3\,0\,2\,4\,1 \\
2 & 4\,1\,3\,0\,2 \\
3 & 0\,3\,1\,4\,2 \\
3 & 1\,4\,2\,0\,3 \\
3 & 2\,0\,3\,1\,4 \\
3 & 3\,1\,4\,2\,0 \\
3 & 4\,2\,0\,3\,1 \\
4 & 0\,4\,3\,2\,1 \\
4 & 1\,0\,4\,3\,2 \\
4 & 2\,1\,0\,4\,3 \\
4 & 3\,2\,1\,0\,4 \\
4 & 4\,3\,2\,1\,0
\end{array} \tag{9}$$

After the formation of the array, the orthogonal coding should be performed in according to [5]. The coding is the process or replacement of our calculated factors by a set of indicators or coded variables. Applied the standardized orthogonal contrast coding:

| | | | | |
|---|---|---|---|---|
| 1 level | 1.58 | −0.91 | −0.65 | −0.50 |
| 2 level | 0 | 1.83 | −0.65 | −0.50 |
| 3 level | 0 | 0 | 1.94 | −0.50 |
| 4 level | 0 | 0 | 0 | 2.00 |
| 5 level | −1.28 | −0.91 | −0.65 | −0.50 |

Thus, we obtain the orthogonal balanced design, which has the best *D*-efficient indicator among designs of the same dimension [5].

So, the following algorithm for the formation of the orthogonal fractional design of the $m^n$ type experiment ($m$ is the number of factor levels, $n$ is the number of factors):

1. To form the matrix—the result of the Kronecker product of two matrices $l_n = \begin{bmatrix} 0 \\ \ldots \\ n \end{bmatrix}$ and $l_n^{\mathrm{T}} = \begin{bmatrix} 0 \ldots n \end{bmatrix}$, ($n$ is the quantity of the factors).

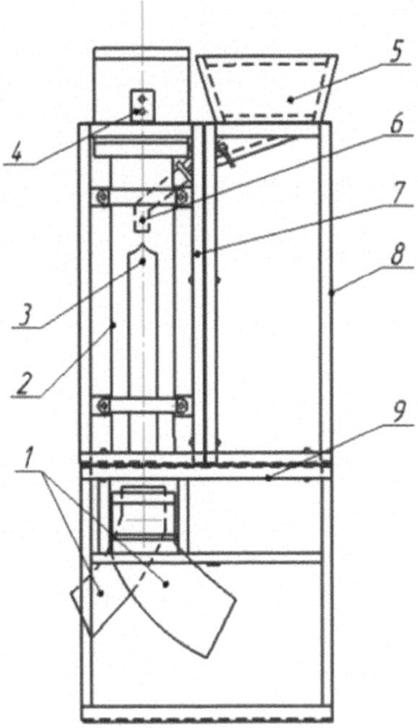


**Fig. 1** Pneumatic gravity seed separator: 1—two distributors; 2—the aspirating channel; 3—the divider; 4—the ventilator; 5—the bunker; 6—the supply branch pipe; 7—the left frame part; 8—the upper frame part; 9—the lower frame part

2. To perform the operation “remainder from dividing” by $m$ ($m$—the number of factor levels) toward the formed matrix and obtain the $D_m$ matrix;
3. Using the block summation operation, to form the matrix $l_m \oplus D_m$;
4. To add the $l_n$ column to the formed matrix and obtain the $[l_n, l_m \oplus D_m]$ matrix;
5. To sort the matrix from step 4 according to the first column;
6. To perform the orthogonal coding operation.

The offered algorithm of the design matrix formation applied while conducting the experimental studies of the optimal parameter values of the pneumatic gravity separator operation [6, 7].

The scheme of the pneumatic gravity seed separator presented in Fig. 1.

Based on the preliminary studies, the degrees of the influence of distinct factors have been determined and established the levels of their variation. The factors that have significant influence on the separation process include:

- the seed drop speed (m/s), $v_0$;
- the angle of seed input (deg), $\alpha$;
- the airspeed (m/s), $v_v$;
- the vertical aspirating channel length (m), $l$;
- the main aspirating channel diameter (mm), $d$.

As a response, the size of the average deviation of the seed motion path in the off-loading point ($m$), $y$, can be considered.

**Table 1** Factor levels

| Factors | The real notations | The coded notations | The variability interval | Factor levels | | | | |
|---|---|---|---|---|---|---|---|---|
| | | | | 5 | 4 | 3 | 2 | 0 |
| The seed drop speed (m/s) | $v_0$ | $x_1$ | 0.2 | 1 | 0.8 | 0.6 | 0.4 | 0.2 |
| The angle of seeds input (deg) | $\alpha$ | $x_2$ | 20 | 100 | 80 | 60 | 40 | 20 |
| The airspeed (m/s) | $v_v$ | $x_3$ | 0.5 | 6 | 5.5 | 5 | 4.5 | 4 |
| The vertical aspirating channel length (m) | $l$ | $x_4$ | 0.25 | 0.8 | 0.65 | 0.5 | 0.35 | 0.2 |
| The main aspirating channel diameter (mm) | $d$ | $x_5$ | 50 | 200 | 175 | 150 | 125 | 100 |

Each factor modified at five levels. The level values of the factor variation is given in Table 1.

For the determination of rational separator operation parameters, the series of experiments conducted according to the design (9). As the result of the experiment conduction, the following model formed, taking into account only the significant factors:

$$
\begin{aligned}
y = & -64.829 + 0.067v_0 + 0.002\alpha + 0.92 \times 10^{-5} v_v \\
& - 0.042v_0^2 - 9.2 \times 10^{-6} v_v^2 - 2.5 \times 10^{-5}\alpha^2 + 1.115l + 0.32d
\end{aligned}
$$

Based on the formed model, the following optimal values of the separator operation parameters were determined in the range of a variety of factors:

- the seed drop speed $v_0 = 0.8$ M/C;
- the angle of seed input (deg) $\alpha = 40°$;
- the air speed = 5 M/C;
- the vertical aspirating channel length $l = 0.7$ m;
- main aspirating channel diameter $d = 200$ mm.

Achieved the maximum separating capacity of the separator is at the value of the average seed motion path deviation in the off-loading point $y = 0.018$ m.

The results of the full-scale experiments, given in the paper [6, 7], confirm the found theoretical values.

## 3 Conclusion

The article describes the algorithm of the formation of the orthogonal balanced experiment designs, which are optimal by the $D$-efficient criterion. The algorithm based on the Kronecker matrix product operations the block matrix summations. On the grounds of the offered algorithm, formed and implemented the experiment for the investigation of the optimal parameter values of the pneumatic gravity seed separator. The optimal values of the parameters found from the mathematical model confirmed experimentally.

## References

1. Mitchell, T.J.: The algorithm for the construction of D-optimal experimental designs. Technometrics **16**(5), 203–210 (1974)
2. Atwood, C.L.: Sequences converging to D-optimal designs of experiments. Ann. Statist. **1**, 342–352 (1973)
3. Yonchev, H.: New computer procedures for constructing exact D-optimal designs. In: Optimal Design and Analysis of Experiments, pp. 71–80. Holland (1988)
4. Cook, R.D., Nachtsheim, C.J.: A comparison of algorithms for constructing exact D-optimal designs. Technometrics **22**(8), 315–324 (1980)
5. Kuhfeld, W.F.: Experimental Design, Efficiency, Coding, and Choice Designs. SAS Technical Documents TS-694C. SAS Institute Inc., Cary, NC (2010)
6. Kyurchev, S.V., Kolodij, O.S.: The methodology of studying of the parameters of the seeds separator of the offered type. In: Conference 2013. Motorization and Energetic in Agriculture, vol. 15(2), pp. 205–213. Lublin-Rzeszow (2013)
7. Kyurchev, S.V., Kolodij, O.S.: The results of research of rational sizes of the vertical aspirating channel of the seeds separator of agricultural crops. In: Conference 2013. Motorization and Energetic in Agriculture, vol. 15(2), pp. 169–175. Lublin-Rzeszow (2013)
8. Dey, A: Construction of nested orthogonal arrays. Discrete Math. **310**(21), 2831–2834 (2010)
9. Graham, A.: Kronecker Product and Matrix Calculus With Applications. Ellis Horwood Limited, Chichester (1981)

# The Parameters Substantiation of Seed Drill Capacity for Stone Crop Seeds

Oleksandr Matkovskyi, Oleksandr Karaiev, Serhii Sankov and Tetiana Karaieva

## 1 Introduction

The sectoral program for gardening development in Ukraine for the period until 2025 provides an increase by 4.5% in the areas occupied by gardens [1]. Under such conditions, the increase in the areas on average should be 4.5 thousand hectares, that, in its turn, requires about 12,500 thousand pieces per year. According to the State Register of Ukraine producers of gardening material, there are 182 farmsteads, their capacity makes up about 8680 thousand seedlings pcs. per annum [2], which provides 3.5 thousand hectares of gardens. That is, we have the shortage of gardening material of domestic production in the amount of 3820 thousand pcs. a year.

The effectiveness of fruit stonecrops seedling production can be increased by growing them without transplanting the rootstocks. Such a technology enables to reduce the cost of seedlings by reducing manual labor to 100 man-h/ha [3] and increase the yield of seedlings of the first commercial grade.

The above technology production application requires the availability of seeding machine with precise stones sowing. Unfortunately, such machines are currently not available in Ukraine, while those, being available, do not meet the agro-technical requirements. That is why the proposed technology doesn't find the widespread use [4].

The purpose of research is to provide sowing of fruit crop calibrated stones directly in the first field of seedlings nursery keeping distance of 15–17 c min a row between them in accordance with parameters substantiation for the seeding machine of precise sowing.

O. Matkovskyi (✉) · O. Karaiev · S. Sankov · T. Karaieva
Tavria State Agrotechnological University, B. Khmelnitsky ave. 18, Melitopol 72310, Ukraine
e-mail: oleksandr.matkovskyi@tsatu.edu.ua; sgm@tsatu.edu.ua

© Springer Nature Switzerland AG 2019

V. Nadykto (ed.), *Modern Development Paths of Agricultural Production*,
https://doi.org/10.1007/978-3-030-14918-5_13

## 2 Results and Discussion

The most suitable to realize sowing of fruit crop calibrated stones are cell disk drives with a horizontal axis of the disk rotation. Structural and technological scheme of such a sowing machine is given in Fig. 1 [5].

According to Fig. 1, the seeding machine consists of two principal parts:

- containers for seeds (stones);
- disk with cells.

Since these parts of the seeding machine are functionally interconnected, the quality of seeding can be achieved by reconciling their constructive and mode parameters.

**Seed Capacity Parameters Substantiation**. Seeding process modeling and parameters substantiation for seeding machine of precise seeds and stonecrops sowing, namely the parameters of the capacity for seed and disk with the cells, are carried out according to the following working hypotheses:

- The first hypothesis: we assume that the mass moving of the stones in a container m (hereinafter—a substance) is equivalent to the motion of liquid with physical parameters that are close to the magnitude to similar parameters of the substance, namely density and dynamic viscosity; the probability of the particle falling into the cell is a growing function of filling the cell with the substance for the period of their contact. The condition of this hypothesis is that the container for a substance is formed by two pairs of planes being in general position, their intersections are cross-lines $a$ and $b$ (Fig. 2);
- The second hypothesis: we assume that the movement of elementary particle in a container substance takes place along the straight lines, crossing the lines $a$ and $b$;
- The third hypothesis: we assume that the speed of the particle of the substance at the moment of leaving the container is proportional to $\sqrt{p}$ (where $p$—the pressure of the substance is greater than atmospheric one). That is, at any point of intersection of substance particle in the domain $d_1$, $d_2$, its velocity depends only on the substance pressure–that is, the condition of the motion continuity;

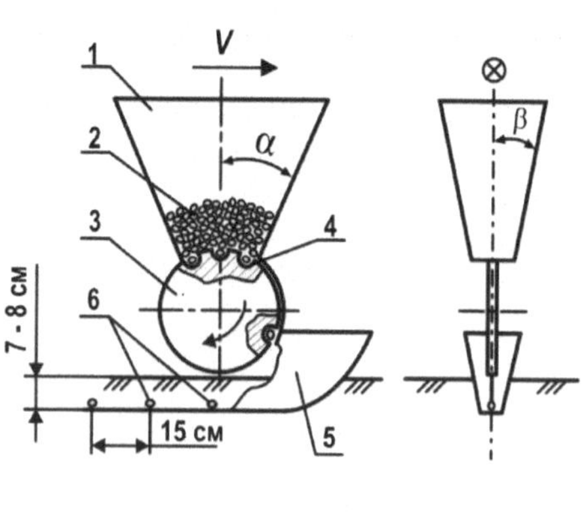


**Fig. 1** Structural and technological scheme of seeding machine for precise sowing stonecrops seeds: 1—stones container; 2—stones; 3—disk; 4—cell; 5—coulter; 6—stones in the furrow

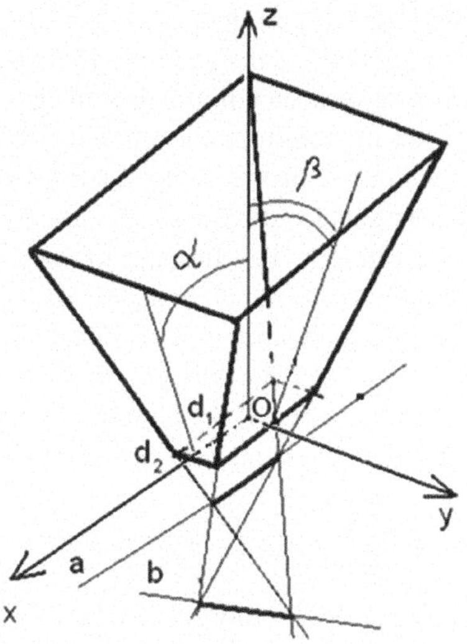


**Fig. 2** Capacity scheme for solving the problem of optimizing the manhole parameters

- The fourth hypothesis: we assume that substance moving in the container is in close to the $d_1 d_2$ manhole, being subordinated to the law of Bernoulli.

Based on accepted hypotheses, the specific consumption of substance $q$ can be calculated by the formula:

$$q = \frac{\Delta Q}{\Delta S} = C\sqrt{p} \tag{1}$$

where $\Delta Q$—substance cost; $\Delta S$—cross section area; $C$—coefficient of proportionality;

$p$—substance pressure greater, than atmospheric one.

We will assume that the tangential pressure of the stones rubbing on the container planes $\tau$ is being conformed to Newton's law of internal friction:

$$\tau = -\mu \frac{\partial u}{\partial n}, \tag{2}$$

where $\mu$ is the coefficient of internal friction (viscosity) of substance; $u$—speed of stones movement; $n$—the normal to the surface.

In accordance with the law of Bernoulli [6] and taking into account formula (1), we have the equality:

$$\rho g h - \frac{\rho v^2}{2} - P_B = \frac{q^2}{C}, \tag{3}$$

where $\rho$—liquid density; $v$—seeding machine wheel speed; $P_B$—pressure loss to overcome the substance internal friction forces; $g$—free fall acceleration; $h$—the level of substance in the container.

Let us introduce the Cartesian coordinate system in such a way, that its center coincides with the center of the lower part of the container, and the axis of the abscissa and ordinates coincide with the axis of the symmetry of the lower part of the container. The moving of substance will be described by the function of the coordinate $q(x, y, z)$, being equal to the mass of the substance, that moves through the unit of area per unit of time.

To calculate the pressure projections of the internal friction forces on the horizontal plane, let us recall the elementary values increase of the coordinates d$x$ and d$y$ and consider the elementary body limited by planes passing through the straight lines:

$$X = x, Y = y, X = x + \mathrm{d}x, Y = y + \mathrm{d}y$$

and the corresponding lines, being creative capacities.

Let us determine the dependence between the differential and the coordinate of abscissa or ordinate from the applique. For this purpose, let's use the scheme given in Fig. 3.

According to the scheme, the length of the segment of the $OC$ will be determined by the formula:

$$OC = OF \cdot ctg\beta = \frac{d_2}{2} \cdot ctg\beta,$$

and the length of the segment $AB$ is determined, basing on the fact, that the triangles of the $CDE$ and $CAB$ are homothetic.

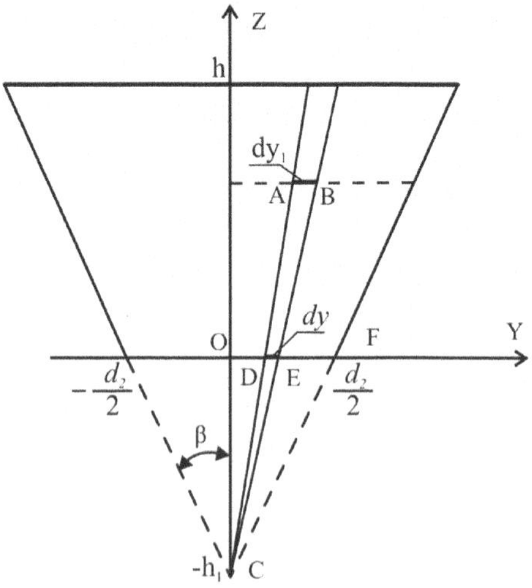


**Fig. 3** Scheme for defining the dependence of the differential from the applique

Then for

$$\mathrm{d}y_1 = AB = \frac{h_1 + z}{h_1} DE = \frac{h_1 + z}{h_1}\mathrm{d}y = \frac{z + \frac{d_2}{2} ctg\beta}{\frac{d_2}{2} ctg\beta}\mathrm{d}y = \frac{2z + d_2 ctg\beta}{d_2 ctg\beta}\mathrm{d}y,$$

but for

$$\mathrm{d}x_1 = \frac{2z + d_1 ctg\alpha}{d_1 ctg\alpha}\mathrm{d}x.$$

Let us find out the projections of the friction forces on the applique, taking into account, that forces of the thrust act on the lateral faces of the elementary body, being under consideration.

For this purpose, we denote the difference between the forces, acting on the planes $Y = y, Y = y + \mathrm{d}y$ as $\Delta F_y$, and on the planes $X = x, X = x + \mathrm{d}x$ *as* $\Delta F_x$. Considering the differentials of d$x$ and d$y$ as functions from applique and taking into account, that substance amount, passing through the lower part of the body under consideration as well as the area, having been formed by the intersection of the body with any horizontal plane, are equal.

Then we have for

$$\Delta F_y = \mathrm{d}x \int\limits_0^h (F(y + \mathrm{d}y) - F(y))\mathrm{d}z = \mathrm{d}x \int\limits_0^h \left(\mu \frac{\partial q(y + \mathrm{d}y, z)}{\partial y} - \mu \frac{\partial q(y, z)}{\partial y}\right)\mathrm{d}z$$

$$= \mu \mathrm{d}x\mathrm{d}y \frac{\partial^2 q}{\partial y^2} \int\limits_0^h \left(\frac{d_2 ctg\beta}{2z + d_2 ctg\beta}\right)^3 \mathrm{d}z,$$

but for

$$\Delta F_x = \mu \mathrm{d}x\mathrm{d}y \frac{\partial^2 q(x, y, 0)}{\partial x^2} \int\limits_0^h \left(\frac{d_1 ctg\alpha}{2z + d_1 ctg\alpha}\right)^3 \mathrm{d}z.$$

Then the resulting projection of friction forces on the lower part

$$P_B = \frac{\Delta F_x + \Delta F_y}{\mathrm{d}x\mathrm{d}y} = \mu\left(I_1 \frac{\partial^2 q}{\partial x^2} + I_2 \frac{\partial^2 q}{\partial y^2}\right), \quad (4)$$

where

$$I_1 = \int\limits_0^p \left(\frac{d_1 ctg\alpha}{2z + d_1 ctg\alpha}\right)^3 \mathrm{d}z; \; I_2 = \int\limits_0^p \left(\frac{d_2 ctg\beta}{2z + d_2 ctg\beta}\right)^3 \mathrm{d}z. \quad (5)$$

Thus, taking into account the formulas (1) and (2), we have that the specific consumption function of a substance satisfies the differential equation in partial derivatives

$$-\mu\left(I_1\frac{\partial^2 q}{\partial x^2}+I_2\frac{\partial^2 q}{\partial y^2}\right)+\frac{q^2}{C}=\rho g h-\frac{\rho v^2}{2} \tag{6}$$

Since it is necessary to maximize the probability, that the stone will fall into the cell during the movement of the cell within the lower part of the container, then the filling of the substance with the cell should be maximal (Fig. 3). Therefore, the optimization problem is to find the maximum of the function

$$F(d_1, d_2, \alpha, \beta, h, \omega, R)=\frac{1}{\omega R}\int\limits_{-d_1/2}^{d_1/2} q(x,0)\mathrm{d}x \tag{7}$$

Assuming that the velocity of the substance particles, being in contact with the space capacitance is zero, then the optimization problem of function (7) must be in its maximum by the function $q$ (1), satisfying the conditions

$$\begin{cases} -\mu\left(I_1\frac{\partial^2 q}{\partial x^2}+I_2\frac{\partial^2 q}{\partial x^2}\right)+\frac{q^2}{C}=\rho g h-\frac{\rho v^2}{2} \\ q\left(x,-\frac{d_2}{2}\right)=q\left(x,\frac{d_2}{2}\right)=q\left(-\frac{d_1}{2},y\right)=q\left(\frac{d_1}{2},y\right)=0 \end{cases} \tag{8}$$

The optimization of function (7) is carried out with the parameters $\alpha$, $\beta$ for given:

| | |
|---|---|
| $h$ | height of substance in the container; |
| $v=\omega R$ | circular speed of the disk; |
| $\omega$ | disk rotation frequency; |
| $N_1$ | cells number per disk; |
| $C$ | coefficient of proportionality; |
| $\mu$ | dynamic viscosity coefficient; |
| $\delta_x$ | cell size; |
| $R$ | disk radius; |
| $d_1, d_2$ | length and width of the container outlet |

**Substantiation of Disk Parameters with Cells for Stones**. Disk parameters have been substantiated for the experimental drill in the unit with the tractor with a minimum speed $v = 1.9$ m/s. The number of cells on disk $N$ will be defined, taking into account the fact, that the average distance between the stones after sowing them into a furrow should be equal to 0.15 m. The scheme of forces, acting on the movement of the stone at the moment of its falling out of the cell (*t. K*), is shown in Fig. 4.

Angle between adjacent cells $\gamma=\frac{2\pi}{N_1}$.

Where $N$ is the number of cells per disk. In order to determine the angle $\alpha_{d,}$ we form the equation of the acting forces

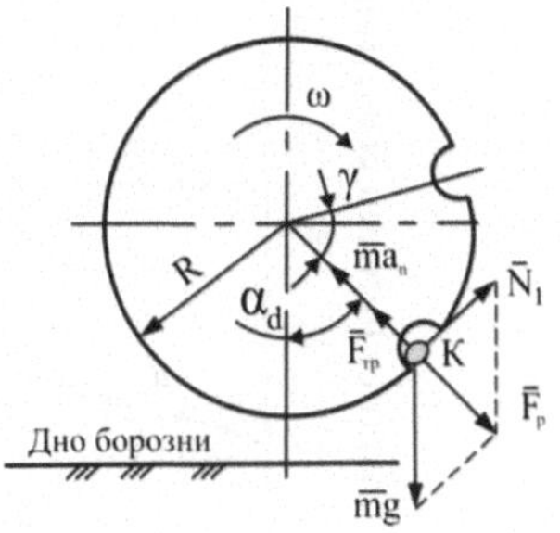


**Fig. 4** Scheme of forces, acting on the stone (*p. K*) at the moment of its falling out of the cell

$$\overline{ma_n} = \overline{F_{mp}} + \overline{F_p}, \tag{10}$$

where $a_n = \omega^2 R$—centrifugal acceleration; $F_{mp} = fN_1$—friction force (where $f$—coefficient of external friction of the stone on the material); $F_p = mg\cos\alpha_d$—an eminent force.

Then Eq. (10) takes the form: $m\omega^2 R = mg\cos\alpha_d - fmg\sin\alpha_d$.

After the simplification we have:

$$\alpha_d = arctg\frac{1}{f} + \arcsin\frac{\omega^2 R}{g\sqrt{1+f^2}}. \tag{11}$$

Then, taking into account the time, for which the next cell takes the position of the previous one, $t = \frac{\gamma}{\omega} = \frac{2\pi}{\omega N}$ we have an equation for calculating the number of cells:

$$N = \left[\frac{2\pi(v - \omega R\cos\alpha_d)}{\omega \cdot \delta s}\right], \tag{12}$$

where $\delta_s$—the distance between the stones in the furrow; $[x]$—is the integer part of the number.

In this case, the distance between the adjacent cells $\delta_l = \frac{2\pi R}{N}$, and its dependence on the speed of the disk is shown in Fig. 5.

For cherry stones a $R = 0.2$ m, $f = 0.5$, $\omega R = 0.38$ m/s, $\delta_s = 0.15$ m, we have $\alpha_d = 77.3°$, $n = 38$ pcs. $\delta_l = 33$ mm. (Fig. 5), and for almond stones: $\delta_l = 54$ mm, $\omega R = 0.605 - 0.625$ m/s, $n = 23$ pcs.

## 3 Materials and Methods

To solve the system of Eq. (8), the laboratory experiments were carried out to determine the values of the coefficients of proportionality $C$ and the dynamic viscosity of $\mu$.

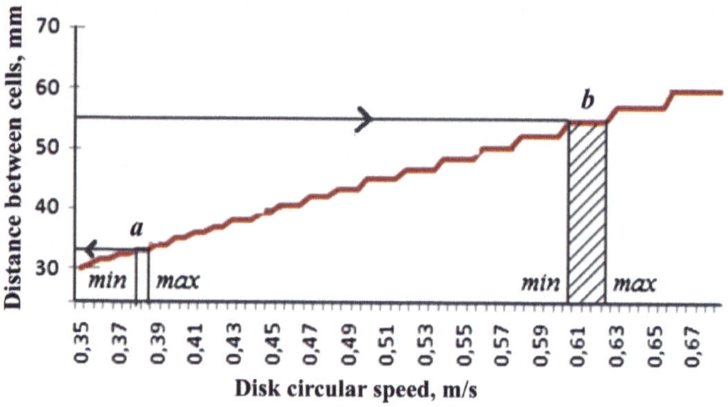


**Fig. 5** Dependence of the distance between the cells on the disk from its speed: (a) the range of speed for the seeds of sweet cherries; (b) the range of disk speed for almonds

**Table 1** Parameters, methods, and means of measurement

| Value | Unit | | Means of measurement | |
|---|---|---|---|---|
| Name of measured value | Name | Marking | Method | Device |
| Change in the volume of the tubes in the container under the influence of its oscillations at intervals of time | The volume of stones | $m^3$ | Mechanic | Bi-harmonic vibrator with a frequency of $10^{-15}$ $s^{-1}$; stopwatch |

**Defining the Proportionality Coefficient**. The proportionality coefficient characterizes the fluidity of the sowing material. In the laboratory experiment, the proportionality factor was determined for cherry and almond. The characteristics of the seed mass moving through the manhole opening were defined, having been mounted on the stand and carried out oscillatory motion. The change in the mass of the stones in the container was determined under the influence of its bi-harmonic oscillations at certain intervals of time.

*Measurement methods.* Methods and instruments for measuring parameters are given in Table 1.

*Performing measurements.* In order to determine the change in the mass of stones in the container, the following parameters should be applied: changes in stones height in the container at intervals: 1, 5, 10 min; changes in stones volume in the container for 1, 5, 10 min.

*Results of measurements.* The following values of the coefficients proportionality have been defined on the basis of data processing results:

- for cherry $C = 0.52 \times 10^{-6}$ $m^3/s$;
- for almond $C = 0.57 \times 10^{-6}$ $m^3/s$.

**Table 2** Parameters, methods, and means of measurements

| Value | Unit | | Means of measurement | |
|---|---|---|---|---|
| Name of measured value (parameter) | Name | Marking | Method | Device |
| Time to advance the metal ball through the stones mass in the container | Time of moving | c | Mechanic | Bi-harmonic vibrator with a frequency of $10^{-15}$ $s^{-1}$; stopwatch |

**Defining the Dynamic Viscosity Coefficient**. In the laboratory experiment, the coefficient of dynamic viscosity was defined for sweet cherries and almonds seeds. The time interval for moving the metal ball through the stones mass under the influence of the container vibrations was determined.

*Measurement methods*. Methods and instruments for parameters measuring are given in Table 2.

*Performing measurements*. In order to determine the time for moving the metal ball through the mass of stones in the container, the following parameters were to be measured:

- height of stones in the container;
- density of stone mass in the container;
- the weight of the ball.

*Results of measurements*. The following values of the proportionality coefficients were defined on the basis of the data processed:

- for cherries $\mu = 0.065$ N s/m$^2$;
- for almonds $\mu = 0.074$ N s/m$^2$.

The accuracy of measuring the parameter values for determining the given coefficients is presented by three significant numbers after a comma. The values of parameters were determined in three replicates, and the obtained results were based on the method of variation statistics.

## 4 Research Results

**Results of theoretical calculations**. The optimization of the function (7) has been carried out at the angles $\alpha$ and $\beta$ at values from 0° to 45° for the seeds of sweet cherry at $n = 38$ pcs., $\delta_x = 0.013$ m, $C = 0.52 \times 10^{-6}$ m$^3$/s, $\mu = 0.065$ N s/m$^2$ for almond seeds at $n = 23$ pcs., $\delta_x = 0.049$ m, $C = 0.5710^{-6}$ m$^3$/s, $\mu = 0.074$ N s/m$^2$. In this case, the size of the cell and the distance between the adjacent cells must satisfy the condition $\delta_l > \delta_x + \Delta s$, where $\Delta s = a/2$ (where $a$—the minimum size of the stone).

**Fig. 6** Cherry stones in a furrow

The system of Eq. (8) was solved by means of the finite differences method for the corresponding parabolic equation of reckoning prior to the establishment of the static regime, and the optimization of the functional was conducted by means of the gradient method.

The calculations were made using the C++ Builder software package Embarcadero RAD Studio 2009 and the worked out computer program; the optimal values of the angles of inclination of the lateral planes of the container $\alpha = 45°$ and $\beta = 45°$ were obtained. It is for these values of the inclination of the lateral capacities of the plane that the highest speed in stones mass moving is achieved and reached the greatest probability for reaching cells by stones.

**Research Tests Results**. The experimental sample of a seeding machine for precise sowing of seed of fruit crops investigation was carried out when seeding cherry stones in the first nursery field. The seeding machine disk had the following parameters: the radius $R = 0.2$ m, the diameter of the cell $\delta_x = 13$ mm, the distance between the cells—33 mm, the number of cells $n = 38$ pcs., the angular speed of the disk $\omega R = 0.38$ m/s (Fig. 6).

According to the results of experimental data processing, it was found out that the average step of sowing exceeded the normative value by 2% and equaled 17.9 cm (coefficient of variation 11.3%), the amount of sown seeds decreased from the normative by 69% and made up 46.6 kg/ha, while the deviation of stones seeding from the axis of the row did not exceed ±1.5 cm.

## 5 Conclusion

For realizing the method of precise seeding, disk-type devices are the most suitable, as the probability of their cells filling with seed material represents a growing function for the period of disk contact with stones within the container manhole.

It was defined that the greatest speed of the stones mass moving through the container's manhole, in the course of which the probability of stones to penetrate into the cells is the greatest, is achieved at deviating plane of the container from the vertical axis by 45°.

For the stones (cherries) being of the least size, the distance between disk cells should make up 33 mm, cells number—38 pcs., and circular speed of seeding drill should make up 0.38 m/s. For the largest stones (almonds), these parameters should be 54 mm, 23 pcs., 0.605–0.625 m/s, respectively. The average seeding step of the experimental model of the seeding machine was 17.9 cm that enabled to increase the output of seedlings of the first commercial grade by 27.4% and decrease the cost of seedlings by 24%.

## References

1. Haluzeva prohrama rozvytku sadivnytstva Ukrayinyna period do 2025 roku. - K.: SPD "Zhytelyev S.I.", 2008. - 76s
2. Derzhavnyy reyestr sub'yektiv posadky ta rozsadnytstva na 2018 rik. http://www.minagro.gov.ua/node/25851
3. Vasilenko, R.K., Rybalov, L.N.: Typical tehnologichni karty dlia vyroschuvannia sadyvnogo materialu plodovykh ta yagidnyh kulturvUkrayinskyi SSR. Zaporozhye: Communar, 1989. 62s
4. Karayev, O.H., Tolstolik, L.M.: Yakist produktsiyi rozsadnytstva plodovykh kultur. Melitopol: Vydavnytstvo-polihrafichnyitsentr «Liuks», 2014. 150s
5. Mukhin, S.P.: Sovremennyietendentsyirazvitiyaposevnojtekhniki. Traktory I selskokhoziaystvennyie mashiny. № 6. S. 16–18 (1993)
6. Landau, L.D., Lifshits Ye.M.: Hidrodinamika. M., 2003. – («Teoreticheskaya fizika», tom VI)

# Methodological Aspects of Determining Parameters of a Scalper-Type Air-Sieved Separator Airflow

**Evgeniy Mikhailov, Marina Postnikova, Natalia Zadosnaia and Oleg Afanasyev**

## 1 Introduction

The preliminary grain cleaning consists in separating at least 60% of the large and air-separable foreign materials from the grain. This increases their safety in temporary storage. Seed and all food grain are subject to pre-cleaning after harvesting.

There is an analysis of the work of preliminary cleaning (PC) means and the methods of their investigation. Studies on the procedure for determining parameters and operating modes of scalper-type air-sieved separators are presented to a lesser extent.

Devices and technological processes of operation of scalper-type air-sieved separators with a closed air system, providing preliminary cleaning of the grain, are known [1–4].

A special novelty in the study of the airflow parameters of air-sieved separators is provided by a two-stage sedimentation chamber and a suction channel of the fan, which influence the validation of the airflow parameters in the fluidization zone of grain materials (GM) and their air separation [5].

The blowing airflow, which diametrically permeates a cylindrical sieve, plays an important role here.

The state of the airflow structure and the analysis of factors ensuring the quality of grain material purification form the main task of the study of the air-sieved separators' working process.

E. Mikhailov (✉) · M. Postnikova · N. Zadosnaia · O. Afanasyev
Tavria State Agrotechnological University, B. Khmelnitsky ave. 18, Melitopol 72310, Ukraine
e-mail: evgenii19508v@gmail.com

© Springer Nature Switzerland AG 2019

V. Nadykto (ed.), *Modern Development Paths of Agricultural Production*,
https://doi.org/10.1007/978-3-030-14918-5_14

### 1.1 Analysis of Recent Investigations and Publications

Grain heap of high humidity and dockage has a negative effect on operability and productivity of pre-cleaning machines [1–3].

At the beginning of the cleaning process, separation of grain heap constituents takes place bearing in mind its aerodynamic properties and especially the critical velocities of winding. At the same time, the phenomena of motion of material particles connected with the grain heap constituents, separation is widely used. However, the quantitative characteristics of the bodies' movement considering the resistance of the air environment still need additional research today [4–9].

An air-sieved separator of a grain heap with a closed air system [5–9] is of a special interest. This separator has a device having a significant importance and allowing to divide the airflow into two components—separating and feeding.

The studies on the methodology of determining the parameters and operating modes of the scalper-type air-sieved separators are presented to lesser extent that determines the topicality of the research.

### 1.2 Goal of Research

Development of the method for determining the airflow parameters of a scalper-type air-sieved separator by studying the structure and air velocities in the air distributor sections.

## 2 Results and Discussion

Based on researches carried out, an air-sieved separator with the technological scheme shown in Fig. 1 [10, 11] is proposed.

The technological process of operation of an air-sieved separator is carried out in the following way. The airflow from the diametrical fan 1 is directed to the air distributor 5 and is distributed to the louvers of the air distributor 7 and the tray-intensifier 10. By changing the position of the extension of the rear movable wall 9, the intensity of the fluidized state of the grain heap on the tray-intensifier is controlled. Segregation is carried out here—the grain, being the heavier fraction, descends in the lower layer, the light impurities (chopped straw, underthreshed spikes)—into the upper layer. Due to the rotation of the cylindrical sieve 13, the grain is spilled through a sieve and through the channel of purified grain material 29 enters the hopper 30. Large impurities enter the hopper for large impurities 27.

Air-separable contaminants travel through channel 20, enter the zone of the sedimentation chamber of the first stage of purification 22, then into the inlet channel of the second stage of purification 23, and at last, settle in the hopper 25.

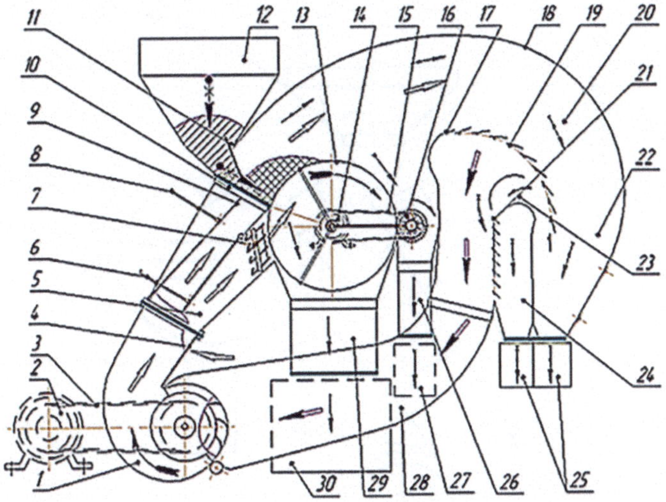


**Fig. 1** Technological scheme of an air-sieved separator, 1—diametric fan; 2—direct current motor; 3—fan drive; 4—louvers for additional airflow inlet in the fan; 5—air distributor; 6—middle movable wall control lever; 7—air distributor louvers; 8—rear movable wall control lever; 9—extension of the rear movable wall; 10—tray-intensifier; 11—charging hopper flap; 12—hopper; 13—cylindrical sieve; 14—reducing gear motor; 15—brush cleaner drive; 16—brush cleaner; 17-airflow cutter; 18—shell of the air channel of a sedimentation chamber; 19—working surface of the first stage cleaning louver; 20—air channel of the first stage cleaning; 21—surface of the second stage cleaning louver; 22—sedimentation chamber of the first stage of purification; 23–input channel of the second stage of purification; 24—sedimentation chamber of the second stage of purification; 25—hopper for impurities of the first and second stages of the sedimentation chamber; 26—the channel and the valve for the large impurities outlet; 27—hopper of large impurities; 28—the suction channel of the fan; 29—channel and valve of the purified grain material; 30—hopper of purified grain material

In the air distributor zone, it is suggested to note and change the following factors (Figs. 1 and 2):

- fan speed;
- the inclination angle of the airflow leaving the air distributor louver to the horizontal;
- the inclination angle of the rear movable wall of the air distributor to the rear immovable wall;
- the inclination angle of the moving middle wall to the immovable middle wall of the air distributor;
- length of the tray-intensifier;
- value of the hopper flap opening.

Let us refer to Fig. 2 showing the example of the cross-sections location for measuring airflow parameters in the air distributor. Here is a scheme of the laboratory and production construction for determining the main parameters and operating modes of supplying and separating constituents of the air distributor.

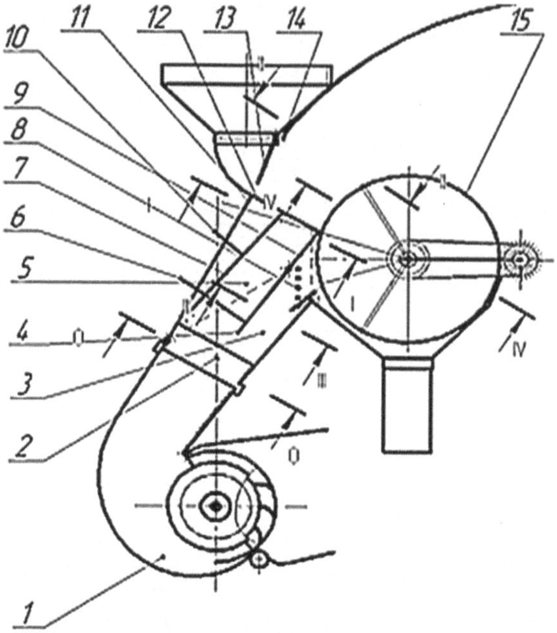


**Fig. 2** Scheme of the laboratory and production installation for determining the main parameters and operating modes of supplying and separating constituents of the air distributor. 1—diametric fan; 2—air distributor; 3—separating component of the air distributor; 4—middle movable wall; 5-feeding component of the air distributor; 6—regulator changing an inclination angle of the middle movable wall; 7—rear movable wall; 8—louvered air distributor; 9—immovable middle wall; 10—regulator changing an inclination angle of the rear movable wall; 11—rear wall; 12—tray-intensifier; 13—feeding device; 14—hopper; 15—cylindrical sieve

The structure and airspeed diagrams in the air distributor are studied in five sections. The values of dynamic pressure and airflow are determined. When the "flow section" of the air distributor is changed, different values of inclination angles of the middle-moving and rear-moving walls and the coefficient of the tray-intensifier flow section is taken into account.

Measurements of the airflow parameters are carried out in accordance with [10, 11] in the following sections:

0-0—the section for measuring the dynamic pressure at the inlet to the distributor;
I-I—the section for measuring the dynamic pressure under the tray-intensifier;
II-II—the section for measuring the dynamic pressure above the tray-intensifier;
III-III—the section for measuring dynamic pressure under the louvered air distributor;
IV-IV—the section for measuring the dynamic pressure in the separating zone of a cylindrical sieve.

Based on the data obtained, the operating airflow rates are determined, velocity diagrams are drawn, and the structure of hydraulic resistances of the pneumatic system is formed. This allows to make adjustments in determining the rational and optimal values of the parameters and operating modes of the pneumatic system. As a result, it is planned to increase the efficiency of the pneumatic separator and reduce its energy capacity.

## 3 Conclusions

The construction and the technological process of the operation of an air-sieved separator with a closed air system are proposed that makes it necessary to develop methods for their investigation.

Methodology for determining airflow parameters of a scalper-type air-sieved separator is developed by studying the structure and airspeed diagrams in the air distributor sections.

Based of the data obtained, the operating speeds of the airflow are determined, and the structure of the hydraulic resistances of the pneumatic system is formed. This allows to make adjustments in determining the rational and optimal values of the parameters and operating modes of an air-sieved separator.

## References


1. Podpryatov, G., Skaletsky, L., Senkov, A., Khilevich, V.: Storage and processing of crop production, 495 p. K: Goal (2002)
2. http://www.nocak.ru/predvaritelnaya-ochistka-zernovogo-voroha.html
3. Nachinov, D.: Tractors and agricultural machines **1,** 17–22 (2005)
4. Mikhailov, E., Zadosnaya, N., Bilokopitov, O.: In: Practice Conference on Questions of Natural Agroproduction in Ukraine, pp. 60–62. Dnipropetrovsk (2015)
5. Mikhailov, E., Zadosnaya, N.: Motrol **17**(9), 43–49 (2015)
6. Mikhailov, E., Zadosnaya, N.: In: Practice Conference Problems of Construction, Production and Operation of Agricultural Machinery, pp. 96–98. Kirovograd (2015)
7. Mikhailov, E., Zadosnaya, N.: Proceedings of the Tavria State Agrotechnological University, **15**(14) 167–173 (Melitopol, 2015)
8. Mikhailov, E., Zadosnaya, N., Afanasyev, O.: Visnyk of Sumy National Agrarian University **10**(2) 96–99 (Sumy, 2016)
9. Mikhailov, E., Zadosnaya, N., Afanasyev, O.: Modern problems of agricultural mechanics: XVII International Sciences Conference, pp. 56–58 (Sumy, 2016)
10. Mikhailov, E., Zadosnaya, N., Afanasyev, O.: Pat. № 116021 U Ukraine, МПК B07B1/28. Pneumatic separator No. 09901; stated. 09/26/2016; has published May 10, 2017, bulletin No. 9
11. Mikhailov, E., Zadosnaya, N., Afanasyev, O.: Pat. № 11892 U Ukraine, МПК B07B1/ 28. Pneumatic separator with closed air system No. 12417; statement 06/06/2016; has published September 11, 2017, bulletin No. 17

# Theoretical Aspects of Plant Material Sealing in a Wedge-Shaped Canal

**Dmytro Milko, Viacheslav Bratishko, Volodymyr Kuzmenko and Oleksandr Kholodiuk**

## 1 Introduction

The problem of nutrients preservation is very relevant at the present stage. In particular, it concerns the storage of haylage components in the diet at the autumn–winter period.

However, this problem can be solved by creating the latest sealing technology and equipment for its implementation. The development of technologies for planting raw materials in post-Soviet countries is carried out from 60 to 70 years.

The project examines the process of sealing plant stem materials in a wedge-shaped canal analyzes the density of the monolith in the height of the layer, which allows us to draw conclusions on further improvement of the technological process in relation to increasing the stored nutrients amount in the continuation of the conservation period.

D. Milko (✉)
Tavria State Agrotechnological University, Melitopol 72310, Ukraine
e-mail: dmitro.milko@tsatu.edu.ua

V. Bratishko
National University of Life and Environmental Sciences of Ukraine, Kiev 03041, Ukraine
e-mail: vbratishko@nubip.edu.ua

V. Kuzmenko
National Scientific Centre "Institute for Agricultural Engineering", Hlevakha 08631, Ukraine
e-mail: vfkuzmenko@ukr.net

O. Kholodiuk
Vinnytsia National Agrarian University, Vinnytsia 21008, Ukraine
e-mail: holodyk@vsau.vin.ua

© Springer Nature Switzerland AG 2019

V. Nadykto (ed.), *Modern Development Paths of Agricultural Production*,
https://doi.org/10.1007/978-3-030-14918-5_15

## 2 Problem Statement

We propose a mathematical model describing the effect of hydrostatic stresses and components of shear viscoelastic deformations on the output density of a plant monolith after passing the rectilinear and inclined sections.

We also want to emphasize that for constructing a mathematical model we used the Kelvin–Voigt model and the generalized Hooke's law. On the other hand, it should be noted that the application of sealing technology in the wedge-shaped canal is an intermediate chain in creating a new energy-efficient sealing technology, which should take into account the physical and mechanical properties of plant stem materials. Exactly, the same research was already done by Ashour [1], who presented the results of compression tests performed on wheat and barley straw bales laid both flat and on-edge and found that wheat bales are stiffer than barley bales. Compression tests were performed by Zhang [2] and Brojan and Clouston [3] on flat and on-edge straw bales and they found a nonlinear stress–strain behavior. Four stages for straw bales compression were identified and an explanation for the mechanical behavior of bales laid flat was proposed by Zhang [2]. Zhang also performed a cyclic test in which a straw bale was subjected to three complete loading and unloading cycles, although neither the maximum stress on the bale nor the maximum bale deformation was kept constant from cycle to cycle.

Attempts to establish accordance between the required density of vegetative raw materials when placed on storage and the quality of nutrients preservation did not lead to positive results due to insufficient consideration of the physical and mechanical properties of plant material.

## 3 Solution

Considering the fact that the study of vegetative raw materials sealing with high humidity is rather specific due, is primarily, to the need to maintain initial humidity in the range of 65–75%, it is necessary to create a mathematical model that would allow to take into account the efforts that arise in particles of plant material. It will allow to manage the process of compaction and release the moisture contained in plant raw materials that ensure the silage process.

In many tasks related to the analysis of the interaction of working bodies and any deformers with materials and environments, it is necessary to use physical equations of stress–strain relations. In this case, such a connection may be manifested in the form of a significant influence of elastic, viscous, and plastic properties [4]. The fundamental laws and physical equations of the relationship of stresses with deformations (deformation rates) are used for the formalization of materials and environments in the form of models with a significant manifestation of certain types of these properties With complex types of properties influence, in particular, when all three properties appear in equal, the models of the connection of stresses with defor-

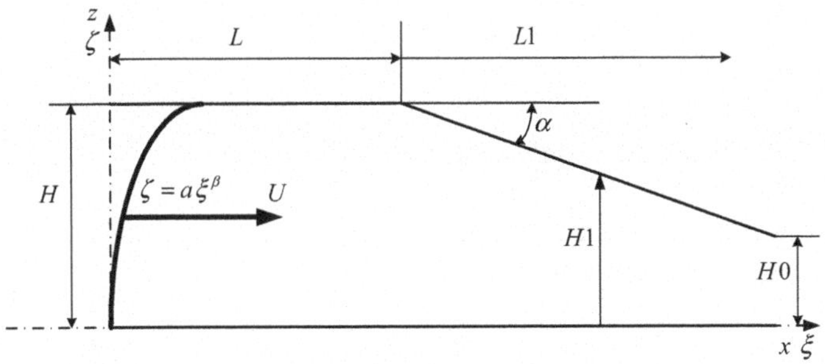


**Fig. 1** Calculation scheme of the process of compaction of plant raw materials in the press with wedge-shaped canal. *H*, *H*1, *H*0—respectively, the height of the layer on a straight line, in the wedge-shaped section and at the exit from the press of the previous seal, m; *L*, *L*1—respectively, the length of the rectilinear and wedge-shaped press, m; $\alpha$—tilt angle of the upper wall of the sealing chamber, rad.; *U*—speed of the piston, mm/s; ζ, ξ—coordinates of the surface of the press

mations (strain rates) are constructed for some cases, that is, for normal or landslide deformations only.

The calculation scheme of the press with a wedge-shaped canal is shown in Fig. 1.

Unfortunately, such a representation of a piston shape in the curvilinear surface form does not allow further integrating the biharmonic potential functions, so we will simplify the shape of the piston to a flat surface.

For calculation of material density, it is necessary to consider two stages of consolidation. The first stage is a seal in a rectilinear section of the press and the second stage is direct sealing in the wedge-shaped section of the press.

The sealing of the material at the end of the rectilinear part of the press can be determined by the dependence

$$\rho = \rho_0 + b\ln\left[\sqrt{\sigma_m^2(1+\tau_{xz})}\right], \tag{1}$$

where

$\sigma_m = \frac{\sigma_x+\sigma_z}{3}$ hydrostatic stress;
$\rho_0$ initial material density;
$b$ an empirical coefficient, determined for a particular material in conditions of uneven compression;
$\sigma_x, \sigma_z, \tau_{xz}$ components of normal stresses;

$$\sigma_x = -\frac{2e^{\frac{Gt}{\mu}}\mu((-5+7\nu)\dot{\varepsilon}_x - 2(-2+\nu)(\dot{\varepsilon}_y+\dot{\varepsilon}_z))}{9(-1+\nu)}; \tag{2}$$

$$\sigma_z = \frac{2e^{\frac{Gt}{\mu}}\mu(2(-2+\nu)\dot{\varepsilon}_x + 2(-2+\nu)\dot{\varepsilon}_y + (5-7\nu)\dot{\varepsilon}_z)}{9(-1+\nu)}; \tag{3}$$

$$\tau_{xz} = 2\mathrm{e}^{\frac{Gt}{\mu}} \mu \dot{\gamma}_{xz}. \quad (4)$$

where

$\dot{\varepsilon}_x$ $\dot{\varepsilon}_z$, $\dot{\gamma}_{zx}$ shear strain rates of the material;
$\mu$ module of material viscosity;
$G$ modulus of vegetation shift;
$\nu$ Poisson's coefficient.

$$\dot{\varepsilon}_x = \frac{\partial u}{\partial x}; \dot{\varepsilon}_z = \frac{\partial w}{\partial z}; \dot{\gamma}_{zx} = \frac{\partial w}{\partial x} + \frac{\partial u}{\partial z}. \quad (5)$$

$u$, $w$ projection of the speed of the press on the axis of the coordinates;

$$u = \Phi_x - \frac{1}{4(1-\nu)} \frac{\partial}{\partial x}(x\Phi_x + z\Phi_z); \quad (6)$$

$$w = \Phi_z - \frac{1}{4(1-\nu)} \frac{\partial}{\partial z}(x\Phi_x + z\Phi_z). \quad (7)$$

where

$\Phi_x$, $\Phi_z$ the Papenkich-Neuer function on the $x$- and $z$-axes;

$$\Phi_x = \int_0^H \frac{ku_{x0}(H+z+\delta-\zeta)}{(H+z+\delta-\zeta)^2 + (L+x+\delta-\xi)^2} \mathrm{d}\zeta;$$

$$\Phi_z = \int_0^{\xi\max} \frac{ku_{z0}(H+z+\delta-\zeta)}{(H+z+\delta-\zeta)^2 + (L+x+\delta-\xi)^2} \mathrm{d}\xi. \quad (8)$$

After integrating the reduced functions, we obtain a solution

$$\Phi_x = \frac{aHkU(-H^2 + H(ax - z) + 2axz)}{(H^2 + x^2 + 2Hz + z^2)(H^2 - 2aHx + a^2(x^2 + z^2))};$$

$$\Phi_z = \frac{aHkU(-a(x-z)(x+z) + H(x+az))}{\sqrt{1+a^2}(H^2 + x^2 + 2Hz + z^2)(H^2 - 2aHx + a^2(x^2 + z^2))}. \quad (9)$$

To calculate the final density of plant material at the exit from the wedge-shaped canal, it is necessary to repeat the same operations as for a straight line plot and substitute them in Eq. (10)

$$\rho L1 = \rho_0 + b' \ln[\sqrt{\sigma_{mL1}^2}(1 + \tau_{xzL1})]; \quad (10)$$

where $\sigma_{mL1} = \frac{\sigma_{xL1} + \sigma_{zL1}}{3}$, hydrostatic stress.

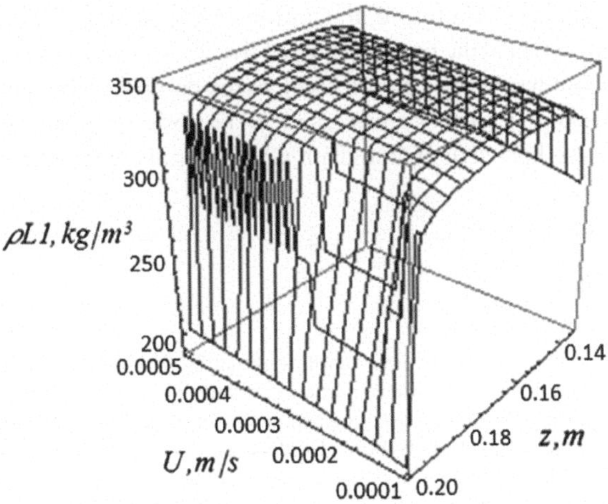


**Fig. 2** Graphic representation of the density distribution by height and speed of the sealing press *U*

In a similar way, the stresses are calculated for the narrowing area $L1$ of the wedge-shaped canal.

Graphical interpretation of the calculation of plant raw material density at the exit from the wedge-shaped canal press is presented in Fig. 2.

The presented graphic dependencies are constructed according to the following initial conditions: $\alpha = \pi/4$, $\rho_0 = 200\ \text{kg/m}^3$, $H = 0.5$ m, $L = 1$ m, $U = 0.0003$ m/s.

## 4 Conclusion

Analyzing the foregoing, we should note the uneven distribution of the compacted material output density at the layer layout height. Since this fact will affect the subsequent distribution of air in the material mass, it is necessary to align the density in the subsequent layout process before the sealing operation.

In the case of storage of compacted plant material in the wedge-shaped canal, there will be problems associated with excessive sealing in the upper layers of plant material and with insufficient sealing in the lower layers. However, the overall density of the package will be within the recommended range. In turn, in the upper layers, there will be processes of destruction of shells of cells with the flow of juice of plants. And in the lower layers, there will be layers with excessive air, which will oxidize crushed plants and juice that stands out from the upper layers. In the pile, all this will reduce the nutritional value of feed materials and reduce the taste appetite of the feed. The consequence of this will be the reluctance of animals to eat food, a decrease in diet, reduced incline, increased cases of animal diseases, deterioration of the sanitary and epidemiological situation in the farm in general.

In general, poor quality sealing of plant material leads not only to the loss of nutrient elements of feed raw materials, but also to many negative economic consequences. Therefore, the uniformity of sealing of plant material for storage is very important at the present stage of livestock development. This also applies to the packing of haymaking in rolls, where the density in the inner part is less than the outside density. Also, these processes can be observed when creating rectangular bales. Very often this occurs when plant raw materials are loading in the trench storage.

## References

1. Ashour, T.: The use of renewable agricultural by-products as building materials (Doctoral thesis). Faculty of Agriculture, Benha University Moshtohor. Toukh, Kaliobia, Egypt (2003)
2. Zhang, J.Q.: Load carrying characteristics of a single straw bale under compression. In: International Straw Bale Building Conference, pp. 1–8. Western Sydney University, Sydney, Australia (2000)
3. Brojan, L., Clouston, P.L.: Effect of plaster type and loading orientation on compression behavior of straw bales for construction. ARPN J. Eng. Appl. Sci. **9**, 1645–1652 (2014)
4. Molari, L., Maraldi, M., Molari, G.: Non-linear rheological model of straw bales behavior under compressive loads. MRC **81**, 32–37 (2017)

# Study of Hydromechanical Parameters Part of the Water Solutions Household in Running Flows

Serhii Movchan, Olena Dereza, Serhii Mazilin and Serhii Dereza

## 1 Introduction

The study of two-component aqueous solutions, sewage and other similar, transparent technical liquids is an important water management problem for the vast majority of industries in the country's water management complex. It is a question of technical means and measures for the treatment of waste water in systems of re-circulating and recycling water supply. It is in these engineering solutions that several important and interrelated technological processes need to be addressed. First of all, it concerns the determination of the quality of treatment of wastewater and, if possible, to carry out the processes of management and management of the above-mentioned technological operations.

Purification of water supply systems is an important component of the unit, which depends not only on safe working conditions on a separate production site, but also on the level of environmental safety in a particular region.

In carrying out laboratory research and practical determination of the quality of treatment of industrial wastewater of widespread use, devices and equipment that measure and determine a significant number of parameters characterizing the state of the entire aqueous solution have been acquired. In the vast majority of cases, the latter circumstance determines not only the technological orientation of the equipment, but also takes into account the current requirements for water treatment equipment for industrial water supply systems.

Such an integrated approach and determines the relevance of the chosen direction of scientific research, which consists in the effective research of transparent technical liquids, sewage, etc.

S. Movchan (✉) · O. Dereza · S. Mazilin · S. Dereza
Tavria State Agrotechnological University, B. Khmelnitsky ave. 18, Melitopol 72310, Ukraine
e-mail: serhii.movchan@tsatu.edu.ua; sgm@tsatu.edu.ua

© Springer Nature Switzerland AG 2019

V. Nadykto (ed.), *Modern Development Paths of Agricultural Production*,
https://doi.org/10.1007/978-3-030-14918-5_16

## 2 Analysis of Literary Sources and Problem Statement

Optical devices and equipment, the principle of which is the use of this or that optical effect of water flow, are called. There are several types of devices using optical effects. The most widespread devices were found using the Doppler effect, based on the measurement of the frequency difference that occurs when the reflection (reflection) of the light flux of particles moving in the flow of fluid is observed. This is due to a number of advantages. First of all, it is simplicity, reliability and relative accuracy of measurements. The use of newly developed and improved optical-mechanical systems is the basis for the automation and control of the processes of determining the individual hydromechanical parameters of particles of impurities of aqueous solutions.

A known method for measuring particle velocity, which is used for static tests in laboratory conditions. The essence of this method is that the moving part is observed using a microscope equipped with a mesh for measurement. In this case, the velocity of the particle is determined as the result of the division of the path traversed by the particle into a time which is measured by a stopwatch. The disadvantage of this method of measurement and further definition is the low accuracy of measurements and the lack of conditions for automation of the measurement process [1].

In a method for measuring the velocity and effective diameter of a particle, the test medium is illuminated by two coherent beams, which are directed at an angle to each other in the probing zone. Then a Doppler signal is fed from the photodetector to a digital memory oscilloscope, determines the frequency of the Doppler signal and calculates the particle velocity. One of the disadvantages of this method is the impossibility of measuring the vertical component of the particle velocity. And this somewhat restricts the functionality of the optical measurement circuit [2].

A well-known design of the optical scheme of the motion of particles in a vertical plane, which uses two optical prisms Dove. Simultaneous use of another prism Dove, installed mirror 180° against the first, not only extends the functionality of the scheme of measurement. However, low accuracy with respect to the vertical component of the particle velocity somewhat restricts the use of this optical circuit [3].

The results of refractive index measurements, concentration of sodium chloride solution and hydrogen peroxide are low phase error. Using the prism Dove creates conditions for preserving the direction of the light beam, which results in the test light being fully internalized. Optical circuit is simple, high-precision with measured refractive index and concentration of liquid solutions. However, functional limitations do not allow measuring sewage and individual parameters in dynamic water flow conditions [4].

The authors propose a new laser doppler vibrometer scanning head, developed on the basis of a pair of rotating optical wedges. The design involves the use of double mirrors and Dove prisms, which allows you to determine low speeds of motion at low rotational speeds in the case of research around the construction of the head for scanning [5].

The work, which uses the modified Mach-Tsander (MZ) interferometer with a built-in Dove prism, is considered. The macroscopic motion of the wedge is driven by a compact motorized rate of translation and is reflected by the rotation of the interference patterns. This method describes the accuracy, simplicity and subsequent digital processing of the image in the region of measuring the speed of the temperature regime of the sensed zone [6].

The problem of stable interference on the optical path of interference bands is solved using a mirror of the preventive feedback. To do this, in an optical scheme, a mirror with a backlink is established, the corresponding beam of light enters the inner plane of the laser and the interference of the feedback mirror occurs [7].

The development of simple, precise, and universal atomic interferometers for interference sounding is used to measure acceleration, rotation and inclination by guiding Romanov beams on separate faces of a pyramidal mirror. The specified application task is solved by using Doppler-sensitive Romanovian transitions. Optical circuit opens the way of deployment of multiaxial interferometers, the use of which is possible in geodesy, geology or inertial navigation [8].

The basis of the development of Laser Photon Rev authors laid the known linear Doppler effect, arising from the linear motion between the source and the observer. The authors in a visual form illustrate the close relationship between the rotational and linear Doppler effects, compare the theoretical preconditions of this connection, and also consider the conditions of motion of other surfaces [9].

Thus, the development of new and improved existing optical-mechanical systems solves not only a purely technical problem, but also scientific and applied, which consists in studying the motion of hydromechanical parameters of particles of impurities of aqueous solutions in countercurrent flows.

## 3 Purpose and Tasks of the Research

The purpose of the work is to create conditions for measuring the velocity of particles of impurities of aqueous solutions with the help of a rotating prism Dove.

The purpose of the work is to develop an opto-mechanical system for simultaneous measurement of the horizontal and vertical particle velocity for different aqueous solutions.

To achieve this goal, the following tasks must be solved:

1. To develop an opto-mechanical system for measuring the hydromechanical parameters of particles of impurities of aqueous solutions due to the vertical component of the speed of motion.
2. Using the theory of electrophoresis to develop a method for calculating the individual components of the motion of particles of impurities of water flow on an example of the value of effective diameter.

## 4 Materials and Methods of Research of the Process of Determining the Measurement of the Velocity of Particles of Impurities of Aqueous Solutions Using Prisms Dove

At the present stage of development of the level of scientific knowledge of the definition of dangerous and harmful factors is an important task. A characteristic feature of modern production is the exploitation and accumulation of very harmful production factors, which play an important role in the system of "man - production environment". Determination of harmful and dangerous substances at an early stage is an integral part of safe working conditions. Therefore, the development of high-performance management tools for assessing the quality of wastewater treatment is a promising area of production activity.

It is common knowledge that electric converters or sensors with electric converters differ in their exceptional versatility. Using the use of electric converters, the particle speed, acceleration, individual hydrodynamic characteristics of the particles (diameter), etc., are measured.

A characteristic feature of their application is that they can be used both in conditions of conducting field experiments, as well as in the static test and in dynamic conditions. The output signal of the electric converters is favorable for further transformation (amplification, integration), at transmission at distances and registration.

In Fig. 1 shows a structural diagram of a universal information and measurement system. The constituent elements and units of the structural scheme, which make it, provide the whole complex of the transformation of the electric signal into the information form, from the perception of the measured value to its automatic processing and registration.

In addition, systems with electric transformation of the output signal provide not only a rather high level of automation of the technological process, but also management of the assessment of the quality of treatment of industrial waste water at a level that provides safe working conditions.

Precisely such effective methods include the determination of hydromechanical parameters, which are considered in methods of measurement and determination of hydro-mechanical parameters of particles of aqueous solutions [6–9].

### *4.1 Method of Measuring the Velocity of Particles in Aqueous Solutions Using Prisms Dove*

The developed optical-mechanical system for determining the velocity of particles of aqueous solutions consists of a source of coherent radiation, two mirrors 2 and 3, a light separator 4, prisms Dove 5, a lens 6, a measurement chamber 7, a photodetector (FEP 84-5), an oscilloscope (C 9-8) It is based on the use of pride Dove, which

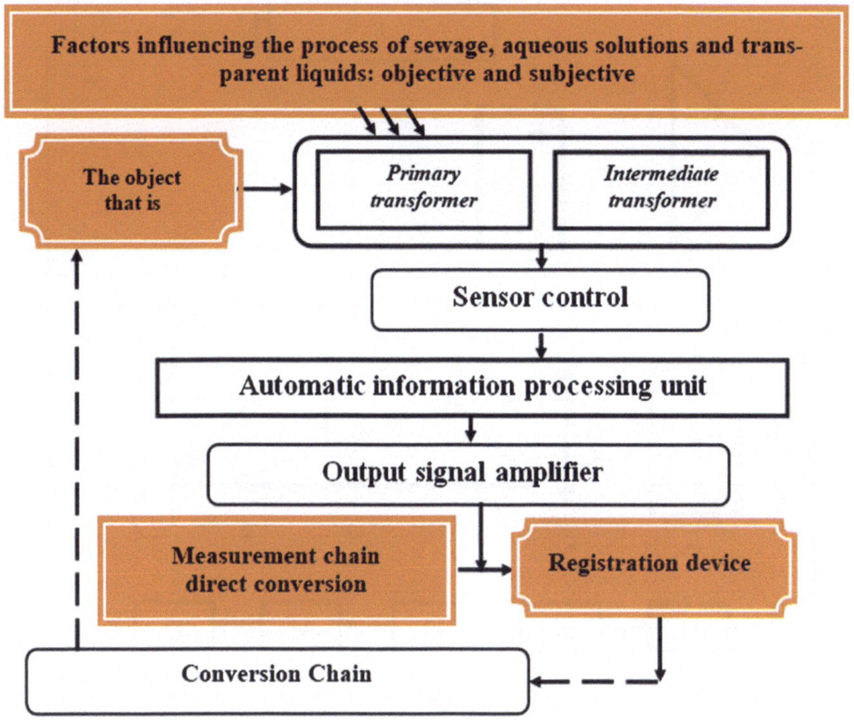


**Fig. 1** Structural diagram of indication of measuring systems

allows you to improve the accuracy of the determination of individual parameters of particles of aqueous solutions [10].

In the developed method for measuring the particle velocity, the coherent laser radiation is sent to the zone of probed aqueous solution. In this case, a vertical system of interference bands is formed, in which the frequency of the Doppler signal is measured and the horizontal component of the particle velocity is determined. With the help of the rotary prism Dove the system of interference bands are rotated to an angle of 90°, measure the frequency of the Doppler signal for this position and determine the vertical component of the particle velocity.

A distinctive feature of the developed technical solution is the simultaneous determination of the horizontal and vertical components of the particle velocity during electrophoresis and sedimentation.

According to the developed technical solution, the measurement is as follows. Two coherent light beams are formed with the help of mirrors 2, 3 and the light divider 4, which are guided by the lens 6 at an angle to each other in the zone of the sensing measuring chamber 7 and form a vertical system of interference bands. The part moves in a horizontal plane at a speed $v_1$. In electrophoresis, the intensity of the

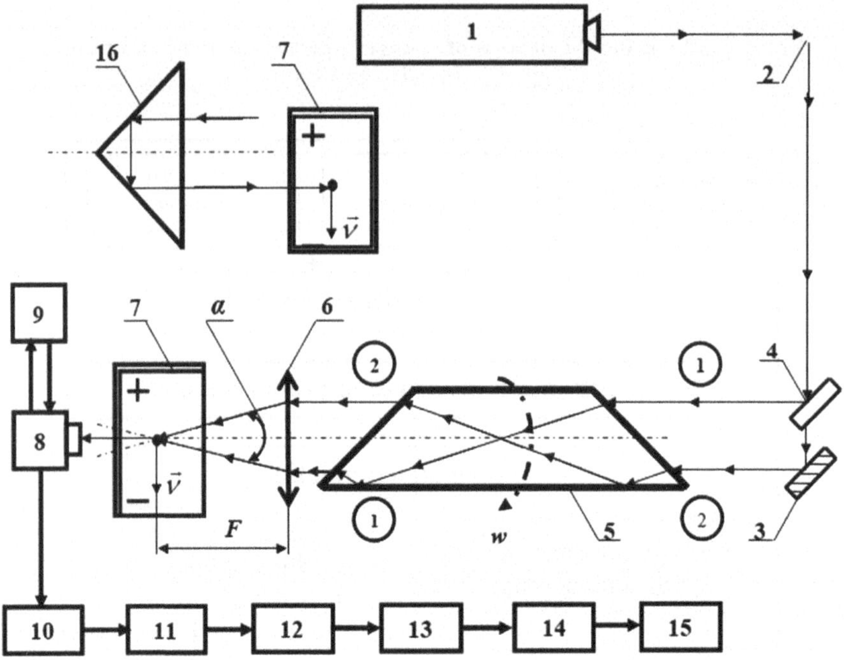


**Fig. 2** Block diagram of a method for measuring the velocity of particles in aqueous solutions using the Dove prism [Patent No. 89040 Ukraine]: 1—the jerel of coherent radiation; 2, 3—mirrors; 4—light separator; 5—prism Dove; 6—lens; 7—measuring cell; 8—photodetector (FEP 84-5); 9—camera with reference solution; 10—signal demodulator; 11—oscilloscope (C 9-8); 12—signal amplifier; 13—digital frequency demodulator signal; 14—analog-digital converter (ADC); 15—personal computer (PC); 16 is a rectangular prism [11]

**Fig. 3** Optical circuit of reflected beam

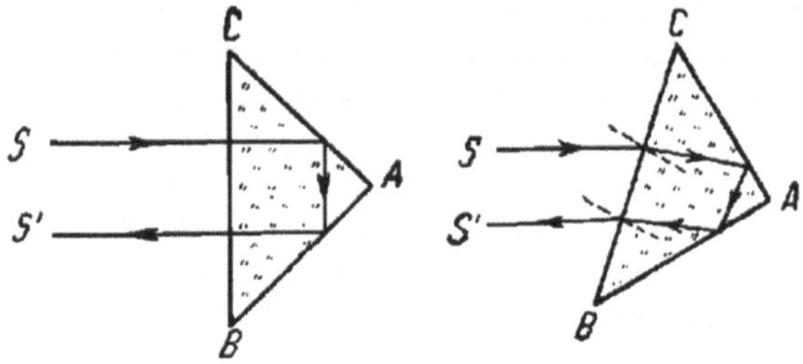


radiation scattered by a particle is recorded using a photodetector 8, whose output is electrically connected to a digital memory oscilloscope 9.

Block diagram of the device for measuring the velocity of particles in aqueous solutions using the prism Dove is presented in Fig. 2.

In the proposed optical scheme, a rectangular prism (16) is used, which is used as a rotary mirror (Fig. 3).

The direction of the reflected beam does not change when the prism is rotated in-on the edge of the right angle—it remains parallel to the beam that gives it.

Depending on the angle of motion of the particle, the rectangular prism also moves. Using a rectangular prism extends functionality.

### 4.2 Combined Scheme for Determining the Speed of Particles of the House-Shock

Determining the frequency $f_1$ of the Doppler signal, calculate the horizontal component $v_1$ of the particle velocity by the formula:

$$v_1 = \frac{\lambda \cdot f_1}{2 \cdot \sin \alpha/2}, \quad (1)$$

where λ is the wavelength of radiation in the solution being investigated, μm.

Then rotate the prism Dove 5 to a 45° angle so that the system of interference tapes returns to an angle of 90° and a horizontal system of interference tapes is formed.

In this case, the frequency $f_2$ of the corresponding Doppler signal is recorded and the vertical component $v_2$ of the particle velocity is determined by the formula:

$$v_2 = \frac{\lambda \cdot f_2}{2 \cdot \sin \alpha/2}. \quad (2)$$

Using the Pivot Rotate, provides a measurement of the vertical component of the particle velocity during sedimentation or pressure flotation. Conditions are created for real-time measurements and automation of the measurement process, which extends the functional capabilities of the developed optical-mechanical system.

Using the Doppler effect and the prism Dove, a combined-on-block diagram was developed (Fig. 4) to determine the electrokinetic zeta potential, effective diameter, electrophoretic velocity of particles of aqueous solutions, and also the number of particles of impurities of aqueous solutions [3].

### 4.3 Measurement Method

To determine the electrophoretic velocity υ of the particles motion of the house-shock of aqueous solutions, the period $T = 1/\nu$ Doppler signal is measured and the particle velocity is determined by the formula:

$$\upsilon = \frac{\lambda}{2 \cdot T \cdot n \cdot \sin \alpha/2}, \quad (3)$$

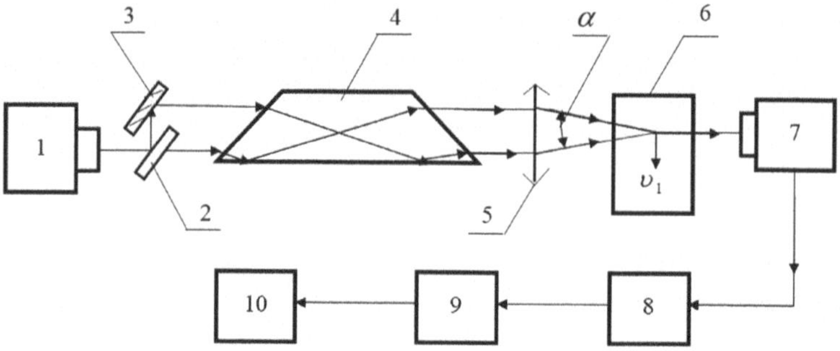


**Fig. 4** Block—scheme of the optical method of measuring the speed of particles of impurities of aqueous solutions: 1—laser LHN-222; 2—light separator; 3—a mirror; 4—prism Dove; 5—lens; 6—measuring chamber; 7— photodetector (FEP 84-5); 8—oscilloscope (C 9-8); 9—analog-to-digital converter (ADC); 10—personal computer (PC)

where $\lambda$ is the wavelength of the coherent radiation of the helium-neon laser, $\lambda = 0.6328\ \mu\text{m}$;

$\alpha$ is the angle between two probing rays, degrees;

$n$ is the refractive index of the liquid, $n = 1.3$.

To determine the diameter of a particle $D$, it is necessary to measure the time $t_1$ of increasing the intensity of the Doppler signal. Then the effective diameter is determined by the formula:

$$D = t_1 \cdot \upsilon \cdot \cos \alpha/2. \tag{4}$$

Thus, an experimental determination of the velocity and diameter of particles of impurities of aqueous solutions is provided in a real-time mode by a remote method.

The optical circuit of the opto-mechanical system allows the formation of two beams for sensing, which, with the help of an optical lens, 5 are rotated at an angle to each other to the measuring chamber 6 and form the system of interference tapes. The intensity of light scattered by a particle moving horizontally at a rate at an electrophoresis is recorded using a photodetector 7. Its output is electrically connected to a digital memory oscilloscope 8. An analog-to-digital converter 9 is used to process the Doppler signal waveform.

The method of measurement and calculation is as follows. First, determine the frequency of the Doppler signal and calculate the horizontal component of the speed $\nu_1$ particle by the formula:

$$\upsilon_1 = \frac{\lambda \cdot f_1}{2 \cdot n \cdot \sin \alpha/2}. \tag{5}$$

In accordance with the formula of Smoluchovsky determine the zeta potential of the particle:

$$\zeta = \frac{3 \cdot \eta \cdot \upsilon}{2 \cdot \varepsilon \cdot \varepsilon_0 \cdot E}, \tag{6}$$

where

| | |
|---|---|
| $\varepsilon$ | is the relative dielectric constant; |
| $\eta$ | is the viscosity of a liquid; |
| $E$ | Is the electric field strength; |
| $\varepsilon_0 = 8.85 \times 10^{-12}$ F/m | Electric became. |

Then, the prism of Dove is turned to an angle of 45° so that the installed interference tape system is turned 90°. The frequency of the Doppler signal is recorded and the vertical component of the particle velocity is calculated by the formula:

$$\upsilon_2 = \frac{\lambda \cdot f_2}{2 \cdot n \cdot \sin \alpha / 2}. \tag{7}$$

Measured by the laser Doppler anemometry rate of sedimentation allows you to determine the value of the diameter $D$ particles of impurities of aqueous solutions using the Stokes law:

$$D = \sqrt{\frac{18 \cdot \eta \cdot \upsilon_2}{g \cdot (\rho - \rho_1)}}. \tag{8}$$

Thus, the application of the rotating prism Dove provides the definition of the vertical component of the velocity of the particle during sedimentation or flotation in real-time and automation of the measurement process, which extends the functional capabilities of the developed optical circuit.

To determine the concentration of particles of impurities of aqueous solutions, use a reference solution with a known concentration of $n_2$ and conduct a parallel measurement for the aqueous solution after treatment with concentration $n_1$ for the reference solution.

In the case of measurement and determination of the optical circuit of the counter flow, the prime Dove is used, which is additionally installed and provided for structural execution. An optical scheme for studying the motion of particles in a vertical plane is given in the work entitled "Electrophoretic Camera Lighting Device" [12].

To calculate the concentration of n particles of impurities in water, the number of extra pulse ks per unit time and particle velocity is measured. Then the concentration of particles n is determined by the formula:

$$n = a \cdot \frac{k}{\upsilon}, \tag{9}$$

where $a$ is the coefficient of proportionality.

Table 1 shows the values of the concentration of impurity particles in drinking water, depending on the value of Doppler pulses per unit time.

**Table 1** The value of the concentration of particles of impurities in drinking water from the value of Doppler pulses per unit time for the selected optimal parameters of the optical laser scheme ($\lambda = 0.6328\ \mu$m, $\alpha = 7°$)

| Quantity | Repeatability of measurements | | | | | | | | | |
|---|---|---|---|---|---|---|---|---|---|---|
| | 1 | 2 | 3 | 4 | 5 | 5 | 7 | 8 | 9 | 10 |
| $C_{Д}$ | 120 | 115 | 107 | 102 | 100 | 102 | 112 | 109 | 110 | 107 |
| $C_f$ | 234 | 225 | 245 | 240 | 225 | 220 | 205 | 200 | 208 | 210 |
| $\upsilon_{П}$ | $4.45 \times 10^{-4}, \ldots, 5.5 \times 10^{-4}$ m/s | | | | | | | | | |

*Note* (1) $C_{Д}$—concentration of impurities in drinking water (number of particles with a maximum diameter value of 300–400 μm), particles/dm$^3$; (2) $C_f$—number of Doppler pulses per unit time; (3) $v_{П}$—amount of water flow

The proportionality factor is determined by using a reference solution with known particle concentration $n_0$:

$$a = \frac{n_0 \cdot \upsilon_0}{k_0}, \tag{10}$$

where $k_0$—the number of Doppler pulses per unit time for reference solution.

Then the calculation formula for the concentration has the following form:

$$n = \frac{n_0 \cdot \upsilon_0}{k_0 \cdot \upsilon} \cdot k. \tag{11}$$

Thus, the technological and optical measurement schemes are available to measure the concentration of impurity particles and to ensure the quality control of drinking water in real time and to automate the measurement process.

## 5 Discussion of the Results of the Study of the Determination of the Value of the Effective Diameter of the Particle

The verification of the results of the measurements of the research was carried out in accordance with the indicated directions of intensification in the work of the systems of recycling water supply (Fig. 5).

The purpose of analyzing the shape of the Doppler signal is to **calculate** the value of the deviation of the diameter from the size of the effective diameter **determined** using optical methods of measurement and processed using computer programs (MathCAD, MathLAV).

One of the tasks of the verification is to minimize the (functional) values of the effective diameter determined by the research, and its value is obtained by the formula:

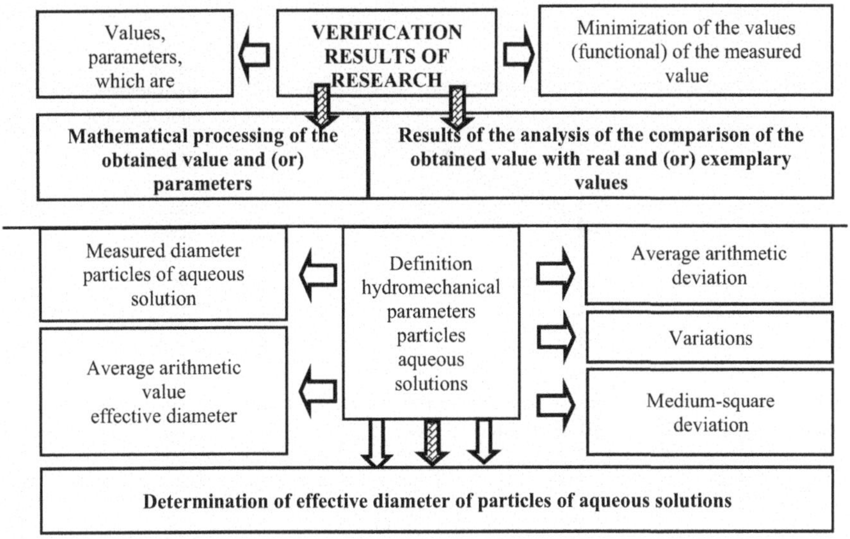


**Fig. 5** Algorithm for checking the results of measurements of research results

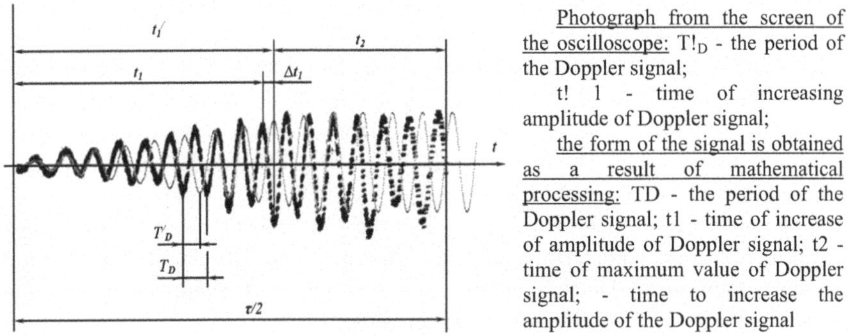


**Fig. 6** Combined graph of Doppler signal shape

$$J = \sum_{i=1}^{n} (D(t) - D_{\text{ex}})^2, \tag{12}$$

where $D(t)$ is the estimated value of the effective diameter, which is determined by the developed mathematical model, *m*;

$D_{\text{ex}}$ is the experimental value of the effective diameter, which is determined in accordance with the well-known method, *m* [6, 7].

Combined graph of the Doppler waveform: from the screen of the oscilloscope and obtained as a result of mathematical processing is presented in Fig. 6.

The finding of the optimal value of the effective diameter is possible under the condition that its value depends on the known parameters, that is, the function:

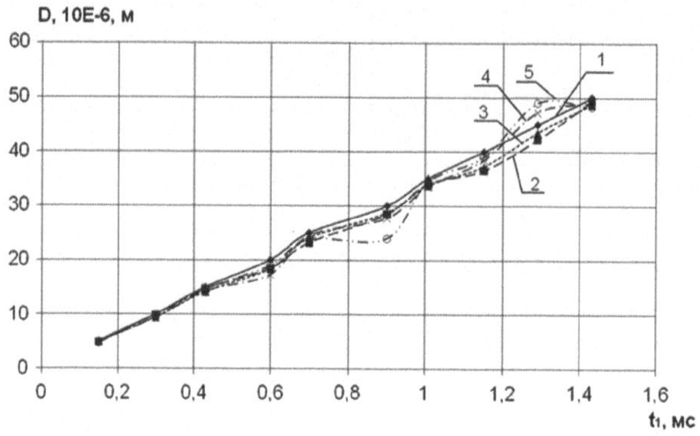


**Fig. 7** Dependence of the time of amplification of the Doppler signal $t_1$, (ms) on the particle diameter, measured ($D$, $10^{-6}$ m) of the aqueous solution

$$D = f(v, t). \tag{13}$$

Thus, the value of the effective diameter of particles $D$ of aqueous solutions depends on the time of increasing the amplitude of the Doppler signal $t$, and the electrokinetic zeta potential—on the particle velocity.

According to the results of the calculations, the time dependencies of the amplitude of the Doppler signal $t_1$ (ms) were obtained depending on the value of the determined diameter of the particles $D$ ($10^{-6}$ m) of the aqueous solution at the fivefold repetition of the measurement. The results of the obtained mathematical calculations are presented in the form of graphic dependencies on Fig. 7.

The graphic dependencies shown (Fig. 7) clearly show that the value of the values of the effective diameter determined by the experimental measurements has the same tendency to increase, and the numerical values are found in the same numerical range with the real values of the effective diameter value.

Values that have a slight deviation from the level of the amplitude of the pre-Pler signal: the first case with values $t_1 = 0.85$–$0.90$ ms is random. At other values of $t_1 = 1.20$–$1.40$ ms indicates that when the maximum value of illumination of two adjacent particles of maximum dimensions, at the intersection of two beams of irradiation, their size significantly affect the determination of particle parameters.

Mathematical calculations were carried out in the next range (Table 2). The angle between two probed beams of the measuring circuit was selected in the range of 5–30° (fixing every 5°) at different values of the electrophoretic velocity of particles: (1) $v = 0.35$–$0.40$ m/s; (2) $v = 0.40$–$0.50$ m/s; (3) $v = 0.45$–$0.50$ m/s.

The results obtained for the mathematical processing of the effective diameter according to known formulas are given in Table 3.

Thus, the verification of the results of measurements of the effective diameter is carried out in several stages. It is assumed that the form of particles (droplets) of concentrated emulsified impurities (particles) of aqueous solutions in the form of a sphere. Therefore, the definition of effective diameter most clearly illustrates

**Table 2** Results of measurement of the effective diameter of particles of aqueous solutions

| Experiment No | 1 | 2 | 3 | 4 | 5 | 6 | 7 | 8 | 9 | 10 |
|---|---|---|---|---|---|---|---|---|---|---|
| Measured diameter particle $D$. $10^{-6}$ m | 4.92 | 9.95 | 14.76 | 18.56 | 23.25 | 28.56 | 34.06 | 36.58 | 42.34 | 49.40 |
| | 4.87 | 9.42 | 14.18 | 18.38 | 24.06 | 28.73 | 33.78 | 37.23 | 43.56 | 48.78 |
| | 4.89 | 9.40 | 14.32 | 17.32 | 24.52 | 27.78 | 33.56 | 38.45 | 47.28 | 49.04 |
| | 4.78 | 9.56 | 14.67 | 19.01 | 23.87 | 24.06 | 34.45 | 39.05 | 49.03 | 48.39 |
| Діаметр, $D$, $10^{-6}$m | 5 | 10 | 15 | 20 | 25 | 30 | 35 | 40 | 45 | 50 |

**Table 3** Results of mathematical processing of the effective diameter value with the use of statistical parameters

| No. series | Rounded diameter particles, mm | Particle diameter $D$, μm | | | | Average arithmetic value effective diameter (μm) | Arithmetic mean deviation (Δ) | Variation (η) | Mean square deviation (σ) |
|---|---|---|---|---|---|---|---|---|---|
| | | 1 | 2 | 3 | 4 | | | | |
| 1. | 5 | 4.95 | 4.90 | 5.00 | 4.95 | 4.95 | 0.05 | 0.1 | 0.071 |
| 2. | 10 | 9.95 | 9.90 | 9.85 | 9.90 | 9.9 | 0.1 | 0.1 | 0.015 |
| 3. | 15 | 14.90 | 14.95 | 15.00 | 15.00 | 14.9625 | 0.0375 | 0.1 | 0.00417 |
| 4. | 20 | 19.95 | 19.95 | 20.00 | 19.90 | 19.95 | 0.05 | 0.1 | 0.00425 |
| 5. | 25 | 24.90 | 24.95 | 24.95 | 25.00 | 24.95 | 0.05 | 0.1 | 0.005 |
| 6. | 30 | 29.90 | 29.90 | 29.95 | 29.90 | 29.9125 | 0.0875 | 0.05 | 0.01083 |
| 7. | 35 | 34.85 | 34.90 | 34.95 | 34.95 | 34.9125 | 0.0875 | 0.1 | 0.0125 |
| 8. | 40 | 39.95 | 39.90 | 39.85 | 40.00 | 39.925 | 0.075 | 0.15 | 0.00867 |
| 9. | 45 | 44.90 | 44.95 | 45.00 | 45.00 | 44.9625 | 0.0875 | 0.1 | 0.004167 |
| 10. | 50 | 49.80 | 49.85 | 50.00 | 49.90 | 49.8875 | 0.0125 | 0.15 | 0.0242 |

the change of particle parameters in the process of electrophoresis. The characteristics that allow the most accurate determination of the effective diameter value are determined:

The presence of mathematical models, improved and re-developed technical solutions creates the conditions for further scientifically-based improvement of technologies, methods and structures of complex intensification of industrial water supply systems.

$$\begin{cases} D = f(V_{\Gamma}, V_B); \\ D = f(t_1, t_2, t_3, T_1, \Delta T_1). \end{cases} \quad (14)$$

Thus, the ecological and economic potential characterizes the degree of environmental balance and the safety of the objects of the environment, namely, such objects include systems that use water, water resources and their constituents. In addition, their potential in the above-mentioned directions is determined by solving several interconnected components. First of all, it concerns: technical, technological, ecological—economic efficiency, use in the development of water treatment technologies and further practical implementation, which is applied in nature.

## 6 Conclusions

1. A method for measuring not only the values of the velocity vector, but also the direction is developed. In this method of laser Doppler anemometry, the rotating Dovde prism is used and the position of the system of interference bands in the probing zone is observed, in which the maximum Doppler frequency is observed. Thus, the direction of the velocity vector is determined.
2. The use of additional automation devices extends the functionality of the water treatment equipment and allows for this process of control and control of the determination of hydromechanical parameters in real time.
   In addition, the developed control and control systems minimize or exclude preparatory operations, the time required to determine the hydromechanical parameters due to optimal physical and chemical processes occurring in water treatment equipment.
3. One of the directions of research is simulation using the Doppler signal shape, when the components of this movement are determined. In addition to increasing the amplitude of the Doppler signal of the particle, its motion is characterized by the time constant of the signal and the time attenuation of this signal. The derivative of the signal is the period and increment of the increase of the period of this signal.
4. During the conducted laboratory investigations and theoretical calculations, a regularity was found between the parameters of the opto-mechanical systems and the values obtained when using the measuring optical circuits.

## References


1. Lobachev, P.V., Krivov, M.N.: Instruments for determining the electrophoretic mobility of particles in dispersed systems. Water Supply Sanit. Eng. **9**(S), 4–6 (1979)
2. Morozov, M.V., Movchan, S.I.: Pat. № 50226A Ukraine, МПК7 G01 N15/ 00. Method for measuring the velocity and effective particle diameter. No. 2001118059; stated. November 26, 2001; has published October 15, 2002, Bul. No. 10
3. Morozov, M.V., Nikiforov, L.E., Movchan, S.I.: Patent for utility model No. 89040 Ukraine, МПК7 (2014.01) G01 N 15/00. Method of measuring the particle velocity in a solution. Application No. u 2013 12593; stated. Oct. 28, 2013, Published 04. 04. 2014, Bull. No. 7
4. Hsu, C.C., Lee, C.Y., Chu, K.H., Wu, T.C.: Dove prism heterodyne refractometer. Opt. Laser Technol. **73**, 6–11 (2015). https://doi.org/10.1016/j.optlastec.2015.04.004
5. Rothberg, S.J., Tirabassi, M.: Development of a scanning head for laser Doppler vibrometry (LDV) using dual optical wedges. Rev. Sci. Instrum. **84**(12), 121704 (2013). https://doi.org/10.1063/1.4845555
6. Jia, Q.K., Qiu, X.D., Wu, Z.W., Zhang, W.H., Chen, L.X.: Transferring linear motion of an optical wedge to rotational frequency shift in an orbital angular momentum interferometer. Appl. Phys. Lett. **111**(9), 091102 (2017). https://doi.org/10.1063/1.4995366
7. Zhu, W., Chen, Q.H., Wang, Y.H., Luo, H.F., Wu, H., Ma, B.W.: Improvement on vibration measurement performance of laser self-mixing interference by using a pre-feedback mirror. Opt. Lasers Eng. **105**, 150–158 (2018). https://doi.org/10.1016/j.optlaseng.2018.01.022
8. Wu, X.J., Zi, F., Dudley, J., Bilotta, R.J., Canoza, P., Muller, H.: Multiaxis atom interferometry with a single-diode laser and a pyramidal magneto-optical trap. Optica **4**(12), 1545–1551 (2017). https://doi.org/10.1364/optica.4.001545
9. Fang, L., Padgett, M.J., Wang, J.: Sharing a Common Origin Between the Rotational and Linear Doppler Effects. Laser Photon Rev. **11**(6), 1700183 (2017). https://doi.org/10.1002/lpor.201700183
10. Morozov, M.V., Nikiforov, L.E., Movchan, S.I.: Patent for Utility Model No. 89040 Ukraine, IPC7 (2014.01) G01 N15/00. Method of measuring the velocity of particles in a solution. Application No. u 2013 12593; stated. October 28, 2013, Published April 10, 2014, Bull. No. 7
11. Movchan, S.I., Dereza, O.O., Gereza, S.V.: Method for measuring the velocity of particles of impurities of aqueous solutions in counter flows. Application No. a 2018 02675. Date of application setting 10.04.2018
12. Movchan, S.I.: Patent for Utility Model № 123331 Ukraine, МПК7 (2018.01) G01 N15/00. Electrophoretic camera illumination device. Application № u 2017 08378; stated. August 14, 2017, published. Feb 26, 2018, Bul. No. 4

# The Efficiency of Tractor Application with Articulated Frame for Cultivating Arable Crops

**Volodymyr Nadykto, Oleksandr Karaiev, Volodymyr Kyurchev and Hristo Beloev**

## 1 Introduction

Until recently there was steady opinion that articulated tractors semi-frames of which are connected by vertical hinge are not suitable for work with mounted tilling machines (seeding machines and cultivators). The reason was and still is in cultivating by such dynamic systems of managing influence—the folding angle of the semi-frames in the horizontal plane—with an amplification factor considerably exceeding 1.0 [1–3]. As a result, mounted machine (seeder or cultivator) at relatively small managing influence can accomplish such angular rejections under which the linearity of its moving does not correspond to agro-technical requirements [4]. That results in substantial rows curvature of tilled crops in the course of inter-row cultivation—to considerable damage of cultural plants.

According to existing in Ukraine agro-technical requirements, a linearity of the cultivated culture is considered to be satisfactory if deviation from rows on the length of 100 m does not exceed 5 cm [5].

Due to practical inconvenience in using the above indicator, we have worked out more profound disperse and frequency one [6]. Its algorithm is the following:

- considering that trajectory oscillations in standing row crops are stationary process, the row not less than 100 m long is chosen on the field;
- the straight baseline parallel to its imaginary axis is being laid, and the oscillations of cultivated plants from it are being measured with the step of 0.5 m.

On the basis of data obtained, the dispersion ($D$) and normalized spectral density are being calculated. The cutoff frequency ($\omega_{cp}$) of the latter is being defined. Rows

V. Nadykto (✉) · O. Karaiev · V. Kyurchev
Tavria State Agrotechnological University, Melitopol, Ukraine
e-mail: volodymyr.nadykto@tsatu.edu.ua

H. Beloev
University of Ruse "Angel Kanchev", Ruse, Bulgaria
e-mail: hbeloev@uni-ruse.bg

© Springer Nature Switzerland AG 2019 
V. Nadykto (ed.), *Modern Development Paths of Agricultural Production*,
https://doi.org/10.1007/978-3-030-14918-5_17

**Table 1** Brief technical description of MTA

| | |
|---|---|
| Tractor engine power (kW) | 132 |
| Mass of tractor (kg) | 8300 |
| Tires: front axle | 23.1R26 (doubled) |
| Rear axle | 23.1R26 (doubled) |
| Tires pressure: front axle (MPa) | 0.9 |
| Rear axle (MPa) | 0.7 |
| Seeder grasp width (m) | 8.4 |

linearity of cultivated crops is considered acceptable under implementation of the following terms:

$$D \leq 12.5\ \text{cm}^2;$$
$$\omega_{cp} \leq 0.25\ \text{m}^{-1}. \quad (1)$$

It should be noted that from these two terms, more important is the first one—dispersion $D$. This is explained by the fact that dispersion $D$ characterizes the energy of the oscillatory process and ($\omega_{cp}$)—only its frequency.

As the analysis of cultivated machine and tractor aggregate (MTA) operations showed, these requirements (1) are quite strict. It often happens that they are impracticable even for MTA on the basis of universally cultivated tractors with guided fore wheels and rigid frame.

To guarantee terms implementation (1) for cultivated machine and tractor aggregates, the GPS system with automatic mechanism of influence at steering wheel of power aggregates is needed. Under the lack of such system, the required result as for cultivated aggregate straightforward movement can be reached by means of mechanization driver-expert high mastery. This is true to MTA as well on the basis of the tractor with articulated frame by proof what this article is.

## 2 Materials and Methods

As a physical object for research, the series of HTZ-170 tractor with articulated frame as well as 12-row cultivated seeder "Kverneland Optima" having 70 cm step between rows was chosen (Fig. 1). Brief technical description of MTA is presented in Table 1.

To provide seeds drilling of the sown culture (sunflower) along the tractor wheelmark, it was equipped with 21, 3R26 doubled tires, and reduced air pressure in them (see brief technical description, Table 1).

Choosing the mechanic driver of several persons was being done on the aggregate including tractor with rigid frame and guided front wheels as well as cultivated seeder. Each of mechanic drivers of the given MTA exercised control pass on the field. After

**Fig. 1** Seeding aggregate including HTZ-170 tractor and "Kverneland Optima" seeder

that, the track trajectory, being left by outermost part of seeder section, was being estimated. The choice of the machine driver was given in favor of the one, who had managed to reach the MTA movement linearity corresponding to the requirements (1).

In the process of field experiments after two contiguous passes of the seeding aggregate (Fig. 1) had been measured:

- deviation the seeder outermost coulter track movement trajectory from the straight baseline—250 measurements at the step of 0.5 m;
- guess row width—250 measurements at the step of 0.5 m.

The data obtained were used to calculate such statistical characteristics as dispersions $D$ and normalized spectral densities.

The cutoff frequency ($\omega_{cp}$) of the normalized spectral densities was determined in accordance with methodology outlined in [7].

Field soil humidity at the layer of 0–10 cm was defined by common thermostate-weight method, while the soil bulk density was defined by the patented in Ukraine method [8]. According to this method, the electronic scale immediately shows the bulk density of the soil.

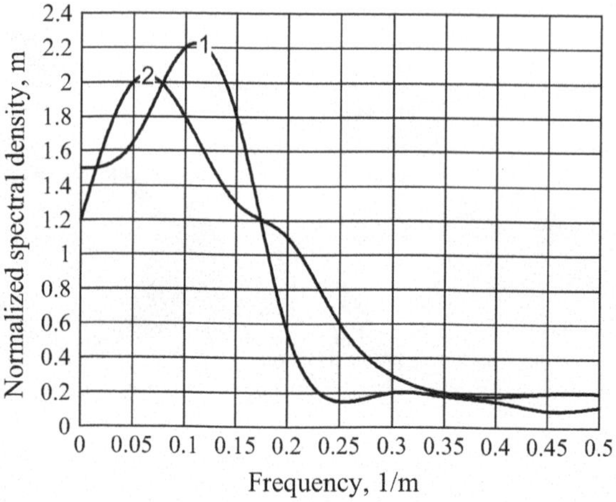


**Fig. 2** Normalized spectral density in off-pass trajectory jogging in the seeder outermost coulter track (1) and guess row (2)

## 3 Results and Discussion

Laboratory and field researches of seeding aggregate were conducted at the soil humidity mean value of which in the layer of 0–10 cm was 17.6%. Its bulk density at the same layer equaled to 1.30 g $cm^{-3}$.

The seeder was tuned at the 6 cm of sunflower sowing depth. The seeding aggregate was moving at the speed of 12.2 km $h^{-1}$.

The analysis of seeding aggregate movement linearity showed that the oscillation dispersions in the seeder outermost coulter track (process 1) and the guess row width (process 2) did not exceed 9 $cm^2$ that fully met the requirement system (1).

Normalized spectral densities in oscillation of both processes differed only by the cutoff frequency ($\omega_{cp}$). In the process 2 (Fig. 2, curve 2), it was approximately 0.35 $m^{-1}$. This is beyond the requirements (1), but they are not produced to the guess row.

The cutoff frequency of normalized spectral density in off-pass trajectory jogging in the seeder coulter track makes up approximately 0.23 $m^{-1}$ (curve 1, Fig. 2). Together with oscillation dispersion (8.56 $cm^2$) it fully corresponds to the requirements (1).

But that was reached due to the high-level mastery of mechanic driver of seeding aggregate made upon the basis of the tractor with articulated frame. We should also emphasize that when the driver having lower qualification operates, the conditions (1) may become unachievable.

The variant of seeding machine–tractor aggregate under consideration can be referred to wide-coverage ones. Its operating performance was equal to 10.2 ha. Hence, it follows that even under using the shift time coefficient 0.65–0.70 by this aggregate for the period of 10-h work 65–70 ha can be sown.

But the more its cutting width is, the more time may be spent for turnings, and thus, the less its output will be. Finally, it is undesired result.

Practice of using wide-cutting MTA shows that it is possible to decrease unproductive time waste for turnings by means of the proper choosing the working run length. By way of evaluative parameter at that the coefficient of using aggregate working passes is used. The inner nature of this index can be disclosed by the following function [9]:

$$\varphi = \frac{\sum L_p}{\sum L_p + \sum L_x}, \tag{2}$$

where $\sum L_p$, $\sum L_x$—the overall length of MTA working passes and its movement on the turning lines correspondingly.

Taking into account that

$$\begin{aligned} \sum L_p &= n_p \cdot L_g; \\ \sum L_x &= (n_p - 1) \cdot L_f; \\ n_p &= \frac{B_f}{B_p}; \\ B_f &= \frac{S_f}{L_g}. \end{aligned}$$

Equation (2) after corresponding transformation acquires the following form:

$$\varphi = \frac{S_f \cdot L_g}{S_f \cdot L_g + (S_f - L_g \cdot B_p) \cdot L_f}.$$

In this expression, $S_f$, $L_f$—area and length of cultivated field; $L_g$ is the length of working run of the field; and $B_p$ is the working width of sowing aggregate.

The basic type of turn of MTA under consideration is pear-shaped one. Way of its motion ($L_t$) on a turning stripe is being defined from the following formula [9]:

$$L_t = (6.6 \div 8.0) \cdot R_a \cdot 2e,$$

Taking for the calculations

$$L_t = 7 \cdot R_a \cdot 2e = 14 \cdot R_a \cdot e,$$

finally receive:

$$\varphi = \frac{S_f \cdot L_g}{S_f \cdot L_g + 14 \cdot (S_f - L_g \cdot B_p) \cdot R_a \cdot e}, \tag{3}$$

where $R_a$, $e$—minimal turning radius and the length of seeding aggregate exit running.

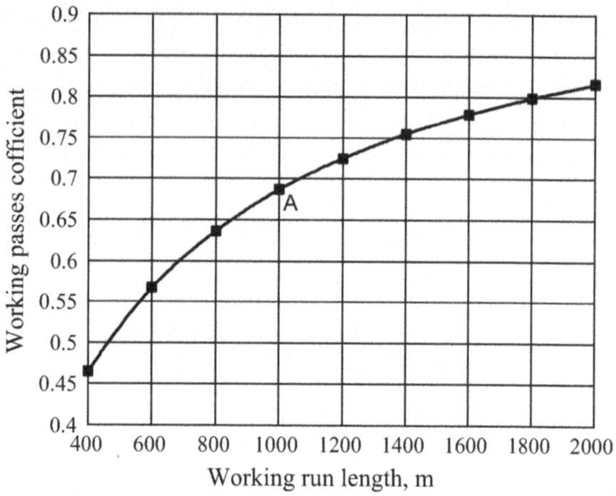


**Fig. 3** Dependence of seeding aggregate working passes coefficient on the field working run length

For the aggregate composed of HTZ-170 tractor and "Kverneland Optima" seeder $R_a = 5.5$ m, exit running length $e = 6$ m, and operating cutting width $B_p = 8.4$ m.

It was already mentioned that our aggregate enables to drill nearly 70 ha for a 10-h working day. In this connection, let us take $S_f = 70$ ha for further calculations.

Subject to assumed values of quantity entering into Eq. (3), the dependence of seeding aggregate working passes coefficient on the field working run length acquires the form as given in Fig. 3.

It is obvious from its analysis that the function $\varphi = f(L_g)$ does not have optimum. At the same time on the graph of Fig. 3, there is a point dividing it conditionally into two parts. The first one is characterized by more intensive increase in estimated figure $\varphi$ from $L_g$ value, and the other one—by less intensive increase.

The methodology of defining the coordinates of this point for the function of such kind is known [10]. Using it for the given case, we determine (see Fig. 3) that this is point A, having the coordinates $L_g = 1000$ m and $\varphi = 0.68$.

Hence, the conclusion follows that for effective operation of wide-coverage seeding aggregate under consideration, the operating field run length should be not less than 1000 m. Otherwise (i.e., under increasing the value of $L_g$), relatively sharp lowering in MTA working passes, coefficient values will take place. It is extremely undesirable to admit it in practical work as it can lead to decreasing the operating performance of machine and tractor aggregate.

## 4 Conclusions

When using a highly qualified mechanic driver for row crops drilling by means of wide-coverage implement including the tractor with articulated frame, the evaluation of nonlinearity of its moving may correspond to the requirements of new frequency and dispersion index. Thus, under conditions of the experiments being conducted, the fluctuation dispersion in deviation amplitude of the track after sowing machine seeding section from the straight line was less than 12.50 $cm^2$ and equaled to 8.56 $cm^2$. The cutoff frequency being of normalized spectral density of these oscillations made up at that 0.23 $m^{-1}$, corresponding to the second condition of the system (1).

For sowing unit, effective operation composed of tractor with minimal turning radius at the level of 5.5 m and 12-row crop seeder having 70 cm inter-row spacing, and working length of the field run should be not less than 1000 m.

## References

1. Pozhidaev, S.: To the problem of indicator choice for estimating the row crops linearity. Visnyk sil's'kogospodars'koyi nauky **11**, 38–42 (1980)
2. Oida, A.: Turning behavior of articulated frame steering tractor. 1. Motion of tractor without traction. J. Terrramech. **20**(3–4), 153–165 (1983). https://doi.org/10.1016/0022-4898(83)90046-0
3. Oida, A.: Turning behavior of articulated frame steering tractor. 1. Motion of tractor with draw-bar pull. J. Terrramech. **24**(1), 57–73 (1987). https://doi.org/10.1016/0022-4898(87)90059-0
4. Nadykto, V., Kyurchev, V.: Applying the tractors of series HTZ-160 and HTZ-170 for tilled crops growing. APC Tech Technol **1**, 8–13 (2013)
5. Ormandghi, K.: Field Works Quality Control. Rosagropromizdat, Moscow (1991)
6. Nadykto, V., Nazarova, O., Chorna, T.: Frequency and dispersion index of tilled crops non-straight linearity. Tract. Agric. Mach. **8**, 18–22 (2009)
7. Lurie, A.: Automation of Agricultural Aggregates. Kolos, Leningrad (1976)
8. Nadykto, V., Kotov, O.: The method for determining the soil density. Patent of Ukraine № 97828 G01 N 1/34 (2015)
9. Iofinov, S., Lyshko, G.: Machine and Tractor Fleet Operation. Kolos, Moscow (1984)
10. Bulgakov, V., Nadykto, V., Velichko, I., Ivanovs, S.: Investigation of draft coefficient of wheeled tractor. Eng Rural Dev. **15**, 1036–1042 (2016)

# Operating Conditions' Influence on the Change of Functional Characteristics for Mechatronic Systems with Orbital Hydraulic Motors

**Anatolii Panchenko, Angela Voloshina, Irina Milaeva and Petro Luzan**

## 1 Introduction

Currently, the creation of self-propelled machines for special purposes (road building, construction, agricultural, etc.) with increased mobility is an important issue. Mobility means the ability of a self-propelled machine to move and maneuver on the ground when performing technological operations with maximum (permanent) power take-off from the drive engine. One of the main factors determining the improvement of the machine's mobility is the characteristics' improvement for the mechatronic systems of the chassis of the self-propelled machine. There is a firm tendency to use continuously variable hydraulic transmissions [1–3]. The application of wheel hubs in transmissions of this type makes it possible to abandon the traditional layout of running gears of self-propelled vehicles and use active hook-on modules with drive axles [4, 5]. Therefore, the study of changes in the output characteristics of mechatronic systems with hydraulic drive of the chassis of self-propelled vehicles, depending on the operating conditions, is a pressing issue.

## 2 Review of Literary Sources

The analysis of the studies showed [6, 7] that the mobility of self-propelled machines can be significantly increased by using a hydraulically driven mechatronic system in the undercarriage. Such a mechatronic system, as a rule, is represented by a pump

A. Panchenko (✉) · A. Voloshina · I. Milaeva
Tavria State Agrotechnological University, B. Khmelnitsky ave. 18, Melitopol 72310, Ukraine
e-mail: tia_tgatu@ukr.net

P. Luzan
Institute of Vocational Education, National Academy of Pedagogical Sciences of Ukraine, Vito-Litovskogo str, 98-a, Kiev 03045, Ukraine

© Springer Nature Switzerland AG 2019

V. Nadykto (ed.), *Modern Development Paths of Agricultural Production*,
https://doi.org/10.1007/978-3-030-14918-5_18

and hydraulic motors [8]. Mathematical models of a pump and a hydraulic motor are proposed. They describe static and dynamic properties of the mechatronic system. Tests of the mechatronic system with two models have been carried out: One was simplified, containing efficiency constants, and another was extended with different values of efficiency. The disadvantage of such a mechatronic system is a small range of load transfer ratio and the mandatory use of onboard gearboxes with a large gear ratio. This complicates the design of the transmission as a whole, making it more laborious, expensive to manufacture, and less reliable in operation.

The use of mechatronic systems with a hydraulically driven undercarriage of self-propelled machines which include orbital hydraulic motors requires the study of changes in the functional characteristics of mechatronic systems [9]. The dynamic characteristics of the hydraulic drive for mechatronic systems with orbital hydraulic machines were considered. The design features of their displacement [10] and distribution [11] systems were substantiated. The influence of the constructive feature of the orbital hydraulic machines on their output characteristics was investigated [12]. Alongside with that the change in the output characteristics of mechatronic systems with a hydraulically driven undercarriage of self-propelled vehicles, according to the operating conditions, has not been studied.

Orbital hydraulic machines are reliable in operation [1–3], have high efficiency in a wide range of operating pressures of the fluid and shaft rotational frequencies, develop large torques, and have a high moving torque, small overall dimensions, and specific gravity per unit of working volume.

## 3 Research Methodology

To improve the functional characteristics of mechatronic systems of self-propelled machines with hydraulic drive chassis, it is necessary:

- to conduct experimental studies of the mechatronic system of a self-propelled machine with a hydraulically driven chassis with an adjustable pump and wheel hubs on the basis of orbital hydraulic motors;
- to justify the rational modes for the operation of an adjustable pump and orbital hydraulic motors that improve the functional characteristics of the mechatronic system of a self-propelled machine with a hydraulically driven chassis.

## 4 Results

To study the influence of operating conditions on changes in the functional characteristics of the mechatronic system with a hydraulic drive of the undercarriage of self-propelled machinery, methodological recommendations were developed for conducting experimental studies. Methodical recommendations included a method for determining the number of repetitions for experiments, a method for measuring the output (input) parameters of hydraulic machines that were the part of the mechatronic system [4, 5]. In this research, we defined operating conditions (modes of operation) of the mechatronic system as variable load in the driving gear of self-propelled machinery and the variable speed of propulsion (wheels). The change in load in the hydraulic system of the undercarriage under the operating conditions is characterized by a pressure differential of the working fluid $\Delta p$.

The mechatronic system of self-propelled machinery with hydraulic drive of the undercarriage consists of an adjustable pump and two mechatronic modules (wheel hubs) with orbital hydraulic motors. The adjustable axial piston pump has a built-in safety valve design and a working volume of 89 $cm^3$.

The executive unit of the mechatronic system under study is the mechatronic module (motor wheel) based on an orbital hydraulic motor represented by a unified range with the working volumes of 160, 320, and 630 $cm^3$ [4, 5].

The mechatronic system is equipped with a regulator that ensures the operation of the engine of a self-propelled machine in constant power mode. The pump power was kept constant by changing the pump regulation parameter $e_1$ when the load was changed by the differential pressure $\Delta p$ of the working fluid in the hydraulic drive of the mechatronic system.

Thus, the power at the pump shaft is determined by the expression

$$N_1 = k \cdot \Delta p \cdot e_1 \cdot n_1 = \text{const}, \quad (1)$$

where $k$ is a coefficient of proportionality; $n_1$ is frequency of the pump shaft rotation.

When the calculated load range of the system was $D = 2$, the pump control parameter varied within the range of $e_1 = 0.5$–$1.0$, and the pressure dropped to $\Delta p = 0.5 \cdot \Delta p_n - \Delta p$. At the same time, the expression $\Delta p \cdot e_1 = 0.5 \cdot \Delta p_n = \text{const}$ was true. In that case, the power at the pump shaft was equal to

$$N_1 = 0.5 \cdot k \cdot \Delta p_n \cdot n_1, \quad (2)$$

where $\Delta p_n$ is the nominal pressure drop in the mechatronic system.

When the load range of the transmission was changed to $D = 3$, the pump control parameter varied too ($e_1 = 0.33$–$1.0$), and the pressure drop $\Delta p = 0.33 \cdot \Delta p_n - \Delta p$. At the same time, the expression $\Delta p \cdot e_1 = 0.33 \cdot \Delta p_n = \text{const}$ was true, and the power at the pump shaft, in that case, was equal to

$$N_1 = 0.33 \cdot k \cdot \Delta p_H \cdot n_1. \quad (3)$$

During the operation of the self-propelled machine, the rotational speed of the internal combustion engine (diesel) was kept constant by means of an all-mode regulator. This allowed us to study the changes in the output characteristics of the mechatronic system hydraulic drive with two fixed values of the rotational speed of the pump driving shaft (1500 and 2500 $min^{-1}$).

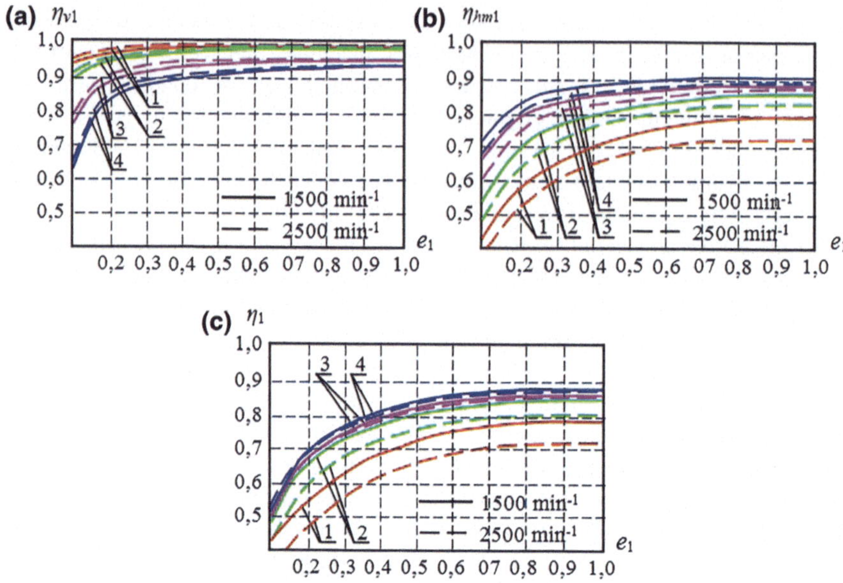


**Fig. 1** Dependence of the change in the pump efficiency on the pump control parameter $e_1$: **a** volumetric $\eta_{v1}$; **b** hydromechanical $\eta_{\mathrm{hm}1}$; **c** total $\eta_1$; 1—$\Delta p = 3.5$; 2—$\Delta p = 7$; 3—$\Delta p = 14$; 4—$\Delta p = 21$

It is known [1, 2] that with increasing pump control parameter $e_1$, volumetric $\eta_{v1}$, hydromechanical $\eta_{\mathrm{hm}1}$, and total $\eta_1$ pump efficiency also increase. The analysis of the experimental studies carried out according to the developed method [4, 5] allowed us to conclude that when the pump control parameter $e_1$ changes from 0.5 to 1.0, the volumetric $\eta_{v1}$, hydromechanical $\eta_{\mathrm{hm}1}$, and total $\eta_1$ pump efficiency vary in a small range from 0.5 to 1.5% (Fig. 1a–c).

In the ongoing studies, the imitation of the load change when driving over rough terrain in the hydraulic drive of the mechatronic system of the undercarriage of self-propelled machinery was carried out by changing the pressure drop $\Delta p$ of the working fluid in a sufficiently large range (7–21 MPa). It should also be noted that in the whole range of variation of the pump regulation parameter $e_1$, an increase in the frequency of rotation of the pump $n_1$ shaft leads to an insignificant (up to 3%) increase in volumetric $\eta_{v1}$ efficiency (Fig. 1a). With the increase in the pump speed $n_1$ from 1500 to 2500 min$^{-1}$, hydromechanical $\eta_{\mathrm{hm}1}$ (Fig. 1b), and total $\eta_1$ (Fig. 1c), the pump efficiency decreases, varying in the range of 2–11%.

The increase in the load on the undercarriage system of self-propelled machinery causes the growth in the pressure differential $\Delta p$ in the mechatronic system which was under study. That leads to a decrease in volumetric $\eta_{v1}$ efficiency by 5–20% (Fig. 2a). The decline in the pump regulation parameter $e_1 = 1.0$–$0.18$, caused by a change in load, also has a significant effect on reducing the volumetric efficiency $\eta_{v1} = 0.99$–$0.8$ and increasing the hydromechanical $\eta_{\mathrm{hm}1}$ (Fig. 2b) pump efficiency.

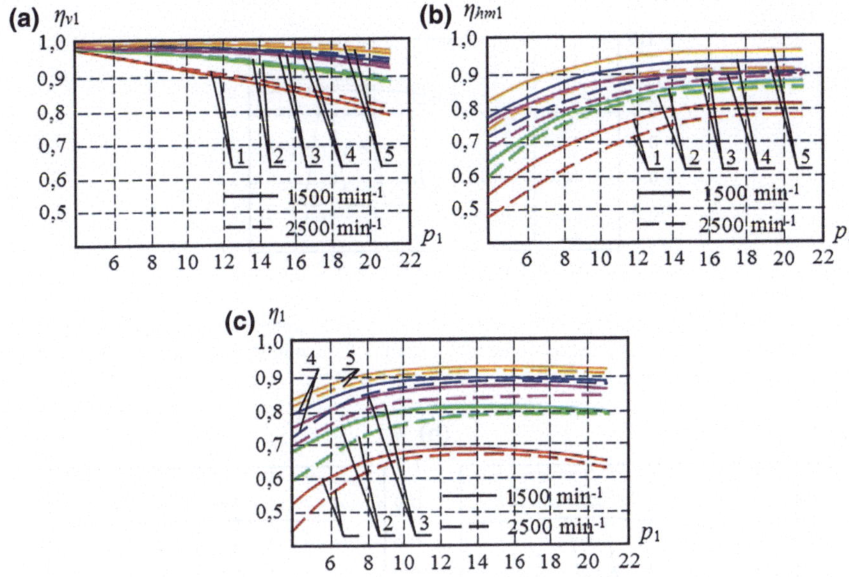


**Fig. 2** Dependence of the change in pump efficiency on the pressure drop $\Delta p$ of the working fluid: **a** volumetric $\eta_{v1}$; **b** hydromechanical $\eta_{\mathrm{hm}1}$; **c** total $\eta_1$; 1—$e_1 = 0.18$; 2—$e_1 = 0.34$; 3—$e_1 = 0.5$; 4—$e_1 = 0.69$; 5—$e_1 = 1.0$

The overall $\eta_1$ efficiency of the pump remains almost unchanged (Fig. 2c) with the increase in pressure drop $\Delta p$ in the mechatronic system from 7 to 21 MPa and with a change in the pump control parameter $e_1$ from 0.5 to 1.0. It should be noted that an increase in the frequency of rotation of the pump shaft $n_1$ leads to a decrease in the hydromechanical $\eta_{\mathrm{hm}1}$ (Fig. 2b) and the total $\eta_1$ efficiency of the pump (Fig. 2c). At the same time, the volumetric $\eta_{v1}$ pump efficiency slightly increases (Fig. 2a). Reducing the pump control parameter ($e_1 < 0.5$) leads to a decrease in the total $\eta_1$ efficiency.

When the pump control parameter is changed in the range $e_1 = 0.5$–$1.0$ at an engine speed of 2500 min$^{-1}$, the total $\eta_1$ pump efficiency remains almost unchanged and is $\eta_1 = 0.86$–$0.88$ (Fig. 3, curve 1). With a decrease in the engine shaft rotation frequency to 1500 min$^{-1}$, the total $\eta_1$ pump efficiency increases and is $\eta_1 = 0.8$–$0.9$ (Fig. 3, curve 2) in the range of variation of the control parameter $e_1 = 0.5$–$1.0$. This could be explained by the fact that the decreasing of the engine speed causes the hydromechanical $\eta_{\mathrm{hm}1}$ pump efficiency increase.

When the load range of the transmission increases in the whole range of variation of the pump control parameter ($e_1 = 0.33$–$1.0$) with a motor shaft rotation frequency of 2500 min$^{-1}$, the overall efficiency of the pump decreases and makes up $\eta_1 =$ 0.8–0.83. Moreover, the highest value of the overall efficiency of the pump ($\eta_1 =$ 0.83) is achieved when the value of the control parameter is equal to $e_1 = 0.65$ (Fig. 3, curve 3). Reducing the rotational speed up to 1500 min$^{-1}$ at the values of

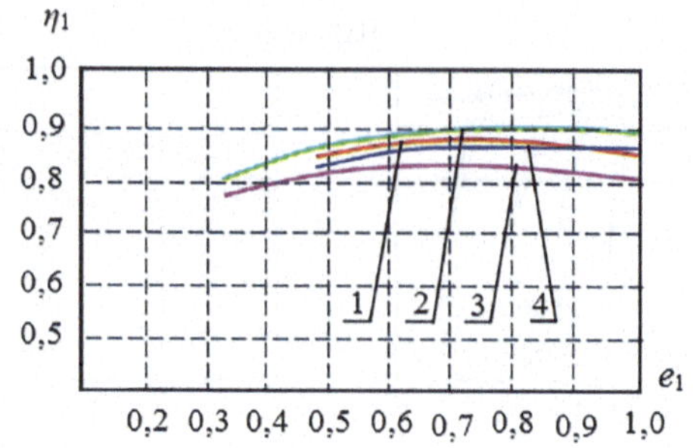


**Fig. 3** Dependence of the change in the overall efficiency of the pump on the regulation parameter $e_1$: 1, 3—with a frequency of rotation of the motor shaft of 2500 $\text{min}^{-1}$; 2, 4—with an engine shaft rotational speed of 1500 $\text{min}^{-1}$

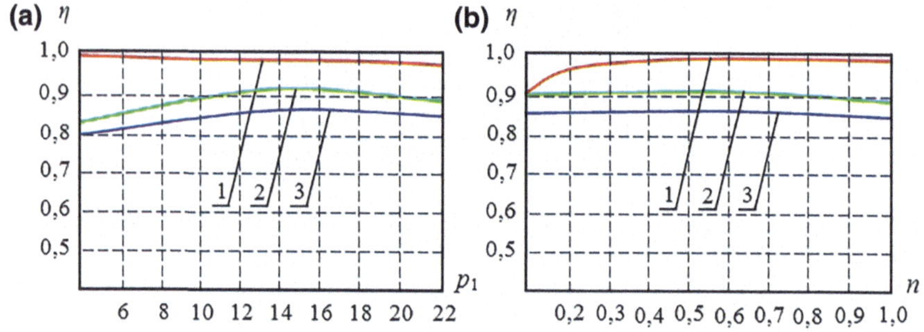


**Fig. 4** Dependence of the change in the efficiency of the hydraulic motor on **a** pressure difference $\Delta p$ and **b** relative frequency of rotation $n$: 1—volumetric $\eta_{v2}$; 2—hydromechanical $\eta_{hm2}$; 3—total $\eta_2$

the regulation parameter $e_1 = 0.33$–$1.0$ leads to the growth in the total $\eta_1$ efficiency of the pump (Fig. 3, curve 4). When the pump control parameter is changed in the range $e_1 = 0.6$–$1.0$, the value of the total $\eta_1$ efficiency increases to $\eta_1 = 0.87$.

Thus, for the considered mechatronic system with a hydraulically driven chassis, the most appropriate operational modes (referring to the power) are changes in the pump control parameter in the range $e_1 = 0.5$–$1.0$, and the differential pressure of the working fluid is $\Delta p = 7$–$21$ MPa. It should be noted that under these operating conditions of the mechatronic system, the life span of the pump also increases.

The volumetric $\eta_{v2}$ efficiency of orbital hydraulic motors under study (Fig. 4a) with a working volume of 320 $\text{cm}^3$ is linearly dependent on the differential pressure $\Delta p$ of the working fluid. In the range of pressure drop $\Delta p$ of the working fluid from 4 to 16 MPa, the volumetric $\eta_{v2}$ efficiency decreases slightly from 0.98 to 0.97 (curve 1). When the pressure differential $\Delta p$ grows, the hydromechanical $\eta_{hm2}$ efficiency of the hydraulic motor increases and, at $\Delta p = 14$ MPa, reaches its maximum value $\eta_{hm2} = 0.92$ (curve 2). The dependence of the change in total $\eta_2$ efficiency (curve 3) is similar to the change in the hydromechanical $\eta_{hm2}$ efficiency of the hydraulic motor.

The increase in the relative frequency of rotation of the output shaft of the hydraulic motor $p$ and, consequently, the speed of a self-propelled machine with hydrostatic transmission (Fig. 4b) leads to the growth of the volumetric $\eta_{v2}$ efficiency of a hydraulic engine (curve 1) and a decrease in its hydromechanical efficiency $\eta_{\mathrm{hm2}}$ (curve 2). The overall efficiency $\eta_2$ of the motor is practically independent referring to the rotational speed $n_2$ of its output shaft (curve 3). The relative frequency $n = n_{2i}/n_{2n}$ of rotation of the output shaft of the hydraulic motor under study is the ratio of the relative frequency each $n_{2i}$ and the nominal values $n_{2n}$ of the rotation speed of the motor shaft, respectively.

Thus, it is possible to determine the range of variation of parameters for the operation of a mechatronic system with a hydraulically driven chassis that ensures high efficiency. High efficiency of the mechatronic system is achieved by changing the relative frequency of rotation of the hydraulic motor shaft $n = 0.25$–$1.0$ and changing the pressure difference of the working fluid $\Delta p = 7$–$21$ MPa. In the specified range of changes in the parameters of the mechatronic system, the total $\eta_2$ efficiency of the hydraulic motor can be taken equal to $\eta_2 = 0.87$. When operating the mechatronic system in the specified range of parameters, its total $\eta$ efficiency can be taken as $\eta = 0.77$.

One of the stages of the experimental studies was testing of a mechatronic system with a hydraulic drive chassis with an increased load range up to a value of $D = 4$ and 8. The studies were conducted on a mechatronic system with two mechatronic modules (wheel hubs), performed on the basis of two and three orbital hydraulic motors with working volumes of 320 and 160 $cm^3$, and 630, 320 and 160 $cm^3$, respectively. During the tests, similar results were obtained; the hydromechanical efficiency $\eta_{\mathrm{hm2}}$ of these hydraulic motors decreased more intensively with increasing rotational speed of the shaft $n_2$ of the hydraulic motor. In the range of low-pressure differences $\Delta p$, the hydromechanical efficiency $\eta_{hm2}$ was also significantly reduced. When operating orbital hydromotors in the range of relative rotational frequencies $n = 0.2$–$1.0$ and differential pressure is $\Delta p = 8$–$22$ MPa, the total efficiency $\eta_2$ of the hydraulic motor can be taken as $\eta_2 = 0.85$.

## 5 Conclusions

Experimental studies were carried out using the mechatronic system of a self-propelled machine with a hydraulically driven chassis which included an adjustable pump and two wheel hubs based on orbital hydraulic motors. The mechatronic system was equipped with a regulator that ensured the operation of the engine of a self-propelled machine in the constant power mode. When the load changed in the hydraulic drive of the mechatronic system, determined by the differential pressure of the working fluid, the pump power was kept constant by changing the pump control parameters.

Experimental studies have substantiated the rational modes of operation of the pump and hydraulic motors that improve the functional characteristics of the mechatronic system of a self-propelled machine with a hydraulic drive of an undercarriage.

It has been established that the most rational modes of pump operation are changes in pump control parameters in the range of 0.5–1.0, and the differential pressure of the working fluid is 7–21 MPa. The condition for ensuring high efficiency of orbital hydraulic motors is their operation in the range of relative rotational frequencies of 0.25–1.0 when the pressure drop of the working fluid varies in the range of 7–21 MPa.

## References

1. Biryukov, B.N.: Rotary Piston Hydraulic Machines. Mechanical engineering (1977)
2. Bashta, T.M.: Mechanical Engineering Hydraulics. Mechanical engineering (1971)
3. Erasov, F.N.: New Planetary Hydraulic Drive Machines. Mechanical engineering, Kiev (1969)
4. Panchenko, A., Kyurchev, V., Voloshina, A., Obernihin, P.: Increase in passability of cars by upgrading the drive chassis. Proc. Tavria State Agrotech. Acad. **8**(8), 28–37 (2008)
5. Panchenko, A.I., Voloshina, A.A.: Planetary Rotary Hydraulic Motors. Calculation and Designing. Publishing and Printing Center "Lux", Melitopol (2016)
6. Liang, Z., Jin, W., Zhengwu, Z.: Research on travel control system of hydrostatic transmission chassis. In: MATEC Web of Conferences, vol. 139, p. 00212 (2017). https://doi.org/10.1051/matecconf/201713900212.
7. Liang, Z., Jin, W., Zhengwu, Z.: Research on parameter design of multi–axis hydrostatic transmission vehicle. In: MATEC Web of Conferences, vol. 139, p. 00215 (2017). https://doi.org/10.1051/matecconf/201713900215
8. Stojek, J., Pluta, J., Jêdrzykiewicz, Z.: Research on the properties of a hydrostatic transmission for different efficiency models of its elements. Acta Montanistica Slovaca **2**(4), 373–380 (1997)
9. Panchenko, A., Voloshina, A., Kiurchev, S., et al.: Development of the universal model of mechatronic system with a hydraulic drive. East.-Eur. J. Enterp. Technol. 4 **7**(94), 51–60 (2018). https://doi.org/10.15587/1729-4061.2018.139577
10. Panchenko, A., Voloshina, A., Milaeva, I., Panchenko, I., Titova, O.: The influence of the form error after rotor manufacturing on the output characteristics of an orbital hydraulic motor. Int. J. Eng. Technol. **7**(4.3), 1–5 (2018). https://doi.org/10.14419/ijet.v7i4.3.19542
11. Voloshina, A., Panchenko, A., Boltyansky, O., Panchenko, I., Titova, O.: Justification of the kinematic diagrams for the distribution system of a planetary hydraulic motor. Int. J. Eng. Technol. **7**(4.3), 6–11 (2018). https://doi.org/10.14419/ijet.v7i4.3.19544
12. Panchenko, A., Voloshina, A., Boltyansky, O., et al.: Designing the flow-through parts of distribution systems for the PRG series planetary hydraulic motors. East.-Eur. J. Enterp. Technol. 3, **1**(93), 67–77 (2018). https://doi.org/10.15587/1729-4061.2018.132504

# Increase in Durability of Motor Crankshaft Pin Surface by Vibrorolling

**Oleksii Novyk, Valeriia Panina, Halyna Dashyvets and Andriy Bondar**

## 1 Introduction

The usual ways of an increase in the crankshaft pin durability by an increase in their superficial hardness by means of superficial training implementation or chemical heat treatment are known. However in these ways, an increase in superficial hardness is reached by a detailed heating with the subsequent shock cooling (usually the microwave oven) or its surface saturation, for example, with the help of nitrogen at a long heating. It is followed by structural changes that quite often cause the emergence of cracks. As a result, there are considerable radial and axial deformations of engine shafts. They reach the good sizes occasionally measured by millimeters.

### *1.1 The Last Research and Publication Analysis*

Emergence of the specified defects leads to detailed rejection or to additional technological way implementation [1].

Such way is additional hot elimination that is negative for high-quality production or detailed restoration, because of the additional thermal deformations. It leads to an increase in detailed cost and, first of all, iron crankshafts.

Increase in durability of crankshaft pins by means of increasing their hardness enables the plastic deformation implementation.

As a result of such processing strips, the microcracks move away from preliminary processing and increase hardness, surface durability and corrosion resistance, its tired hardness. One of the ways of carrying out the plastic deformation is a cold crankshaft pin vibrorolling. The shaft surface vibrorolling reduces the tendency to

O. Novyk · V. Panina (✉) · H. Dashyvets · A. Bondar
Tavria State Agrotechnological University, 18 B.Khmelnytsky Ave, Melitopol 72310, Zaporizhia obl, Ukraine
e-mail: valeriia.panina@tsatu.edu.ua

© Springer Nature Switzerland AG 2019

V. Nadykto (ed.), *Modern Development Paths of Agricultural Production*,
https://doi.org/10.1007/978-3-030-14918-5_19

scuff formation at the expense of a bigger surface oil absorption power. It significantly increases a resource of the engine operation and reduces oil consumption and "oil starvation" at engine start and winter operation.

The research results [1–3] have shown that the working surface consumption after the strengthening processing during the run-in period is 1.1–1.8 times less, and the rate of consumption in the period of the operational consumption is twice less. The blanket rolled at the optimum modes has the microhardness increased for 18–27% [3]. Its greatest increase is observed at the iron, in which graphite inclusions have smaller length and are more isolated. Mat thickness with the increased microhardness fluctuates within 0.05–0.5 mm. The more the diameter of the deforming element is, the thicker a mat with the increased microhardness is.

Besides, at expansion there is a crushing of the graphite inclusions, and the perlite grains after deformation have other orientations in comparison with the initial one [4].

The grain form becomes flattened in the direction of deformation radial forces. The undoubted positive moment should be considered that the vibrorolling is a final operation and can be used both in industrial and in repair productions.

## 2 The Research Main Part

The task of the crankshaft pin working surface vibrorolling is giving of the additional durability and the oil-retaining property. The knurled surface has to have pockets for lubricant deduction and the increased quantity of locations that occurs at the full crossing of flutes (Fig. 1).

It is necessary for receiving such surface that the ratio of a detailed rotation frequency and frequency of the double course spheres is equal to an integer number and a half. Longitudinal machine supply has to be equal to a half of amplitude of the sphere fluctuations.

It is necessary for an increase in durability and the oil-retaining property that the surface processed by a vibrorolling made 50% of the crankshaft pin general working surface. This condition will be satisfied at an appropriate combination of the modes. There are tabular and analytical methods of the processing mode calculation which allow to define the modes taking into account many surfaces of vibrorolling process, but they are quite labor-consuming that limits their implementation in individual and limited production. It is expedient to carry out mode definition in such cases by

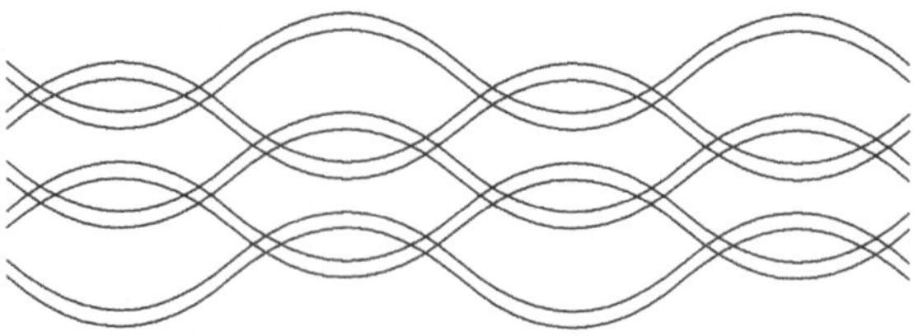

**Fig. 1** Microrelief with the full crossing for giving of the additional durability and the oil-retaining property

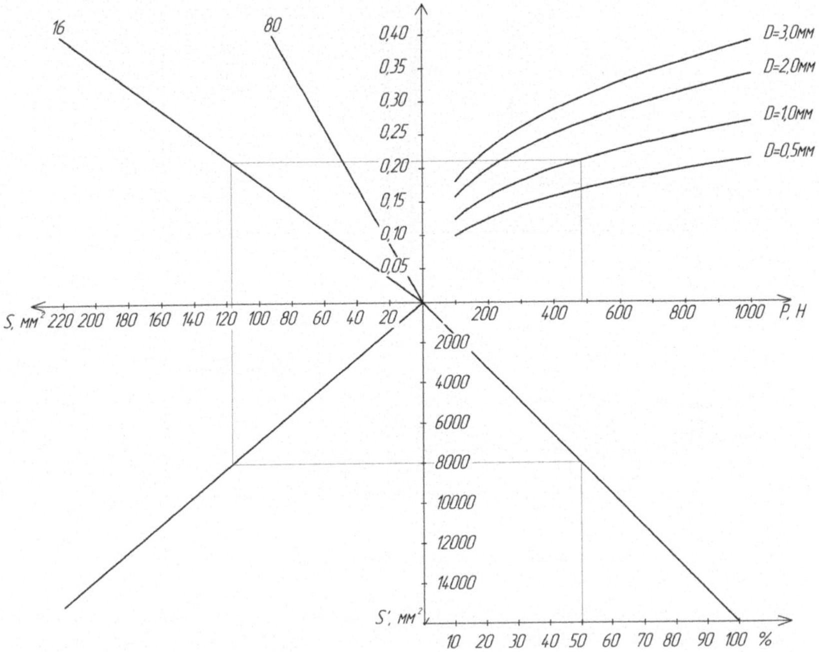


**Fig. 2** Nomograph for the definition of the crankshaft pin vibrorolling modes. The key: % surface covering → the area of the processed surface **S**, the area of the processed surface for one a detailed circulation **S** → flutes width **h** → effort of a sphere clip to a detailed surface **P**

a tabular method—according to the nomograph which graphically reflects process regularities. The nomograph allows to determine one of the dependence parameters on two known parameters.

The nomograph for the definition of the crankshaft pin vibrorolling modes consists of four quadrants (Fig. 2).

In the first quadrant, the dependence of width of a flute on the tool pressing effort to a surface is constructed. The flute width will be equal to a print diameter. Calculations were carried out for spheres from T30K4 alloy by diameters 0.5; 1.0; 2.0; 3.0 mm. The detailed material is the pig iron.

In the second quadrant, the dependence of the flute area for one detailed circulation when processing from its width is graphed. This area will be equal to the length of one wave increased by a flute width and by the wave quantity.

In the third quadrant, the dependence of the area processed depending on the flute area for one detailed circulation is graphed.

In the fourth quadrant, the ratio schedule of the area processed by the tool to the total area of a detailed surface is constructed.

So, for example, if it is necessary to process 50% of a crankshaft pin surface superficial and plastic deformation, then the processing modes have to be such: The area of the processed surface will be 8000 mm$^2$; the area of the processed

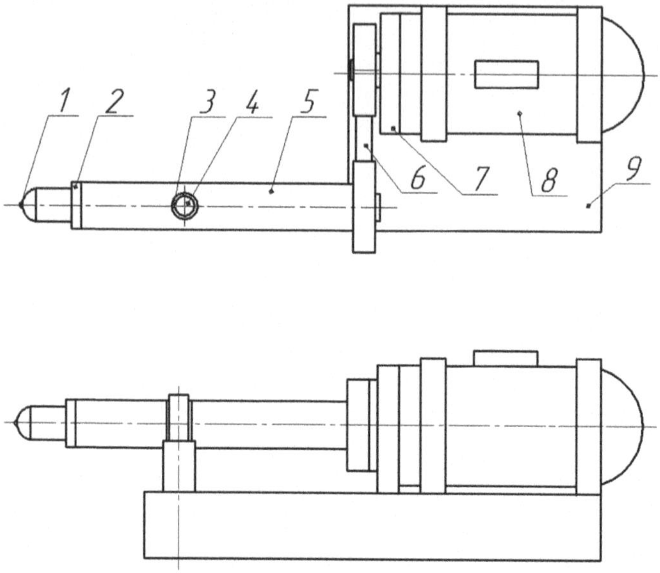


**Fig. 3** A device for the crankshaft pin vibrorolling (1—a sphere; 2—a variable extension; 3—a liner; 4—axis of oscillation; 5—a lever; 6—a crank rod; 7—a column; 8—an electric motor; 9—a plate)

surface for one detailed circulation is 118 mm$^2$, with a frequency of the detailed rotation—16 min$^{-1}$; the knurled flute width is 0.22 mm; with a sphere diameter of the device for the vibrorolling 1.0 mm, the effort of a sphere clip will equal to 480 N.

Similar nomographs have been developed for details of the engine cylinder-piston group and the slide valve of the hydraulic valves [5, 6].

The crankshaft pin surface vibrorolling has the certain difficulties caused by the protruding pins and fillets, which at a shaft rotation make it impossible to use the ordinary equipment for the vibrorolling [7]. Therefore, the special device design for the vibrorolling with a radial supply of the tool (Fig. 3) has been developed.

The plate 9 is a device basis and is intended for its installation on a support of the turning screw cutter. On the plate, the electric motor 8, an axis of oscillation 4 and a column 7 are affixed. For reduction of vibration influence, a bearing inside the column 7 is installed, in which the electric motor 8 shaft rotates. On an axis of oscillation 4 through the liner 3, the lever 5 is installed. An out-shot part of the electric motor shaft is pierced with the eccentricity on which a sphere fluctuation amplitude 1 will depend.

Two bearings which transfer to-and-fro motion from the electric motor eccentric workpiece to the lever 5 are installed in the crank rod 6. In the lever butt end, the variable extension 2 which has sprung a sphere 1 is established. Pressure of a spring is regulated by the screw. Extensions 2 differ on the sphere opening diameter and on length for the lever change.

Thus, the developed device can be used for the surface cylindrical part vibrorolling of both radical and crank pins (without rounded) by means of the turning screw cutter.

## 3 References

1. Effective way of an increase in durability of the crankshaft pin surfaces in the conditions of boundary friction is superficial plastic deformation.
2. By means of the developed device and the corresponding modes, we determine by the offered nomograph, receiving of a microrelief with the full crossing is possible.
3. A certain microrelief allows to increase location quantity twice that gives the additional durability and the oil-retaining ability of the detailed surface.

## References

1. Plehun, D.S., PanIna, V.V.: Metodika viznachennya optimalnogo sposobu vIdnovlennya kolInchastogo valu. ZbIrnik naukovih prats magIstrantIv ta studentIv TDATU, Vip. 16 T.1 MehanIko-tehnologIchniy fakultet. – MelItopol: TDATU, 2016. [Плехун Д.С. Методика визначення оптимального способу відновлення колінчастого валу/ Д.С. Плехун, В.В. Паніна/Збірник наукових праць магістрантів та студентів ТДАТУ, Вип. 16 Т.1 Механіко-технологічний факультет. – Мелітополь: ТДАТУ (2016)]
2. Kudryavtsev, I.V., Grudskaya, R.E.: Novyie sposobyi poverhnostnogo plasticheskogo deformirovaniya. Mashinostroitel. – # 7 – S. 28–29 (1984). [Кудрявцев И. В. Новые способы поверхностного пластического деформирования/И. В. Кудрявцев, Р. Е. Грудская. Машиностроитель. – 1984. – № 7 – С. 28—29.]
3. Kuznetsov, A. M.: Tehnologicheskie i fizicheskie os-novyi deformiruyuschih metodov obrabotki. Avtomobilnaya promyishlennost. – # 12 – S. 36—37 (1974). [Кузнецов А. М. Технологические и физические ос-новы деформирующих методов обработки/ А. М. Кузнецов. Автомобильная промышленность. № 12 – С. 36—37 (1974)]
4. Dehtyar, L.I., Andreychuk, V.K., Muravev, A.I.: Vliyanie PPD na vyinoslivost detaley iz vyisokoprochnogo chuguna s kontsentratorami napryazheniya. Vestnik mashinostroeniya. – # 4 – S. 55—57 (1977). [Дехтярь Л. И. Влияние ППД на выносливость деталей из высокопрочного чугуна с концентраторами напряжения/Л. И. Дехтярь, В. К. Андрейчук, А. И. Муравьев. Вестник машиностроения № 4 – С. 55—57 (1977).]
5. NovIk, O.Yu., PanIna, V.V., Dashivets, G.I.: Obgruntuvannya rezhimIv vIbronakatuvannya poverhon detaley tsilIndro-porshnovoYi grupi dviguna. VIsnik UkraYinskogo vIddIlennya MIzhnarodnoYi akademIYi agrarnoYi osvIti – Vip. 5.–Herson: OLDI-PLYuS, 2017.– S.50–55. [Новік О.Ю. Обгрунтування режимів вібронакатування поверхонь деталей циліндро-поршньової групи двигуна/О.Ю. Новік, В.В. Паніна, Г.І. Дашивець. Вісник Українського відділення Міжнародної академії аграрної освіти – Вип. 5.–Херсон: ОЛДІ-ПЛЮС, С.50-55 (2017)]
6. NovIk, O.Yu., PanIna, V.V.: Pobudova nomogrami dlya viznachennya rezhimIv vIbronakatuvannya zolotnikIv. Naukoviy vIsnik TDATU [Elektronniy resurs]. – MelItopol: TDATU, 2015. – Vip.5, T.1. – Rezhim dostupu. http://nauka.tsatu.edu.ua/e-journals-tdatu/e-index.html. S. 145–149. [Новік О.Ю. Побудова номограми для визначення режимів

вібронакатування золотників/О.Ю. Новік, В.В. Паніна/Науковий вісник ТДАТУ [Електронний ресурс]. – Мелітополь: ТДАТУ Вип.5, Т.1 (2015)]
7. Shestakov, V.V., Kryuchkov, L.P., Kuzmichev, B.P.: Faktoryi, vliyayuschie na stepen poverhnostnogo plasticheskogo uprochneniya pri obrabotke stalnyimi sharikami. Vestnik mashinostroeniya. # 5 – S. 67—68 (1984). [Шестаков В. В. Факторы, влияющие на степень поверхностного пластического упрочнения при обработке стальными шариками/В. В. Шестаков, Л. П. Крючков, Б. П. Кузьмичев. Вестник машиностроения. № 5 – С. 67—68 (1984)]

# The Coefficient Determination of a Damper Washer Hydraulic Resistance for Reducing a Technical Module Oscillation Amplitude

**Volodymyr Bulgakov, Oleksandr Parakhin, Vasil Mitkov and Tetiana Chorna**

## 1 Introduction

Prospects for improving the work productiveness, as well as reducing the fuel-specific consumption by MTA based on drawbar-pull category tractors are practically over [1]. This problem can possibly be solved only by implementing the drawbar-pull concept on the basis of modular power units (MPU) [2].

Developed in Ukraine, the MPU (see brief technical description, Table 1) consists of power (1) and technological (2) modules (Fig. 1). The power module (PM) is a tractor whose nominal traction effort amounts to 14–16 kN. The technological module (TM) is an additionally connected axle with a wheel drive activated from the PM ground-speed power take-off shaft.

Theoretical and practical aspects of using MPU-80 with different agricultural machines have been developed [2–8]. The results of the research have provided a justification of structural-technological schemes of the block-modular machine-tractor aggregates (MTA). It has been determined that the MPU-80 technological module should be located behind the power module. Their direct connection in the vertical longitudinal plane is activated with the help of a four-chamber hinged mechanism located on the TM. The latter is configured as a three-point fastening scheme and constantly remains in the "floating" position.

In order to ensure satisfactory stability and controllability of motion, as well as maneuverability of modular machine-tractor aggregates in a horizontal plane, the MPU-80 technological module must be non-rotatable during the operation motion on a run of cultivated land, and movable in relation to the power module on a headland.

V. Bulgakov
National University of Life and Environmental Sciences of Ukraine, Kyiv, Ukraine

O. Parakhin · V. Mitkov · T. Chorna (✉)
Tavria State Agrotechnological University, Melitopol, Ukraine
e-mail: tetiana.chorna@tsatu.edu.ua

© Springer Nature Switzerland AG 2019

V. Nadykto (ed.), *Modern Development Paths of Agricultural Production*,
https://doi.org/10.1007/978-3-030-14918-5_20

**Table 1** Brief technical characteristic MPU-80

| | |
|---|---|
| *Power module* | |
| Operating weight (kg) | 3700 |
| Engine power (kW) | 58.8 |
| Wheel base (m) | 2.37 |
| Wheel track (m) | 1.4 |
| Distance between a trailing wheel and a hinge pivot of TM (m) | 1.22 |
| Tires: | |
| front wheels | 7.5-20 |
| trailing wheels | 16.9R38 |
| *Technological module* | |
| Operating weight (kg) | 2600 |
| Wheel track (m) | 1.4 |
| Distance between wheels and alpha hinge (m) | 1.22 |
| Tires | 16.9R38 |

**Fig. 1** Modular power unit MPU-80

The turnability of the technological module relative to the power one in the horizontal plane is provided by means of a vertical hinge positioned on the TM. It allows the technological module to rotate ±30° relative to the power module.

Relative rotational fixing of EM and TM during the operation motion on a run of cultivated land must be provided by auto locking of the vertical hinge of the latter. However, such a design concept is very difficult and is characterized by a low reliability. And the latter is especially needed for hinged aggregates, and the speed rate of which is relatively equal to the road transport speed rate.

The practice of using MTA based on MPU-80 has revealed that the connection of its modules in a horizontal plane does not necessarily have to be rigid. But in any case, it must be both simple and very reliable. This article is dedicated to the solution of this practical task.

## 2 Materials and Methods

Laboratory and field tests have analyzed the travel movement of plowing unit based on MPU-80 (Fig. 2). In actual practice, the plowing unit movement takes place during the MTA removing from one field onto the other, or to a home station.

As was mentioned above, the rotation of the MPU-80 technology module relative to the energy one is carried out through the vertical hinge 1 (Fig. 3).

The limiting of the rotation was provided by the hydraulic cylinder 2, the bottom-end and the head-end of which were connected by a hose with a mounted damper in the form of a washer with a hole of diameter *d*.

This design parameter is easy to determine if its influence on the damper washer hydraulic resistance ($K_m$, N m s) is a prior knowledge. To solve this practical problem, a special laboratory facility was developed (Fig. 4).

Let us consider the schematic of the laboratory device (Fig. 5). Here, the hydraulic cylinder hinges in contact with the beam OA.

At the outer point of it (*A*), there is a weight with a value equal to $G_1$. The bottom-end and the head-end of the hydro cylinder are interconnected by a washer with a certain hole diameter *d*, which causes the resistance force ($G_2$) to the transfer of fluid.

In reality, the force that forms the reaction $G_2$ drives the rod of the hydraulic cylinder down. As a result, within a certain time ($t_o$), the beam turns through angle

**Fig. 2** The plowing unit based on MPU-80

**Fig. 3** The placement of the hydraulic cylinder and the damper washer on a frame of MPU-80: 1—vertical hinge of TM; 2—hydraulic cylinders; 3—damper washers

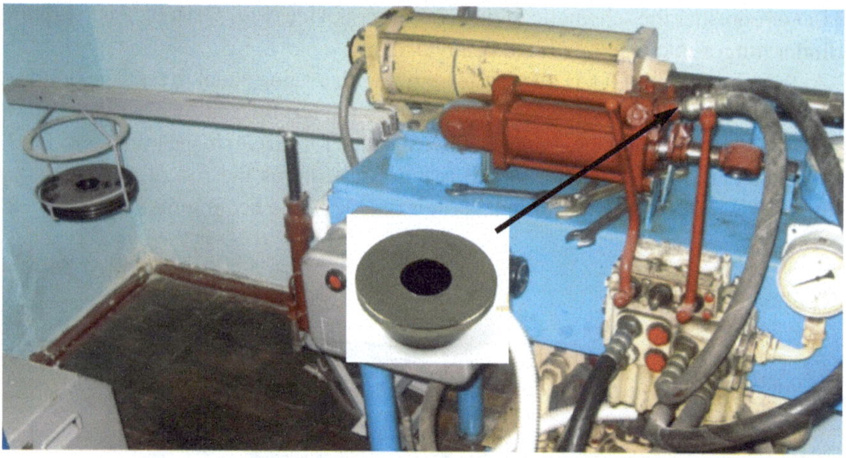

**Fig. 4** The laboratory installation to determine the influence of the damper washer hole diameter on the coefficient of its hydraulic resistance

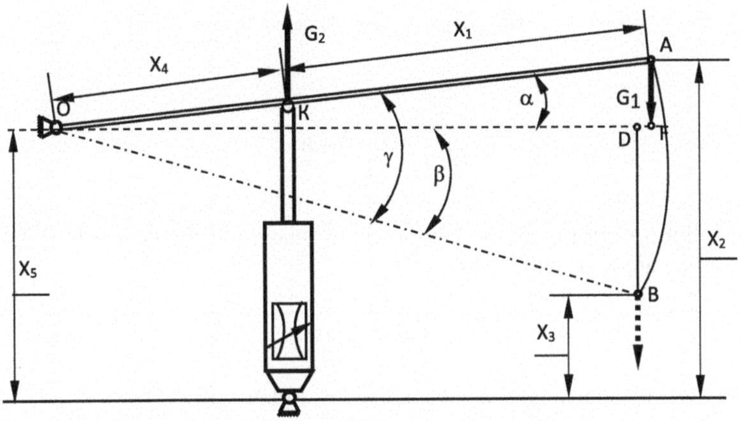


**Fig. 5** Schematic construction of the installation to determine the influence of the damper washer hole diameter ($d$) on the coefficient of its hydraulic resistance ($K_m$)

$\gamma$, and the force point $G_1$ moves from point $A$ to point $B$. It follows that the coefficient $K_m$ can be represented as follows:

$$K_m = \frac{G_2 \cdot h \cdot t_o}{\gamma}, \tag{1}$$

where $h$ is an arm of force $G_2$.

During the movement of the hydraulic cylinder rod down, the point $K$ moves not to a true vertical, but arc-wise and equidistantly to $AB$. When passing, arm of force $G_1$ changes from $h_1$ to $h_2$. At the same time (Fig. 5):

$$h_1 = OF = (X_1 + X_4) \cdot \cos\alpha;$$
$$h_2 = OD = (X_1 + X_4) \cdot \cos\beta.$$

In general, we can assume that force $G_1$ with respect to point $O$ acts on the middle value of arm ($h_{\text{mid1}}$), which is equal to

$$h_{\text{mid1}} = \frac{(h_1 + h_2)}{2} = \frac{(X_1 + X_4) \cdot (\cos\alpha + \cos\beta)}{2}.$$

The same can be said for arm of force $G_2$. The middle value of it ($h_{\text{mid2}}$) will be as follows:

$$h_{\text{mid2}} = X_4 \cdot \frac{(\cos\alpha + \cos\beta)}{2}.$$

The magnitude of the force $G_2$ can be determined from the sum of the moments that act with respect to point $O$. Namely

$$G_1 \cdot h_{\text{mid1}} - G_2 \cdot h_{\text{mid2}} = 0,$$

where

$$G_2 = \frac{G_1 \cdot h_{\text{mid1}}}{h_{\text{mid2}}}.$$

As it follows from Fig. 5, the total angle of rotation of the beam *OA* ($\gamma$) equals the sum of the angles $\alpha$ and $\beta$, that is,

$$\gamma = \alpha + \beta.$$

Again, the indicated angles can be defined as follows (Fig. 5):

$$\alpha = \arcsin\left(\frac{X_2 - X_5}{X_1 + X_4}\right);$$
$$\beta = \arcsin\left(\frac{X_5 - X_3}{X_1 + X_4}\right).$$

Therefore,

$$\gamma = \arcsin\left(\frac{X_2 - X_5}{X_1 + X_4}\right) + \arcsin\left(\frac{X_5 - X_3}{X_1 + X_4}\right).$$

After substituting the desired values into (1), we will deduce

$$K_m = \frac{G_1 \cdot (X_1 + X_4) \cdot \left\{\cos\left[\arcsin\left(\frac{X_2 - X_5}{X_1 + X_4}\right)\right] + \cos\left[\arcsin\left(\frac{X_5 - X_3}{X_1 + X_4}\right)\right]\right\}}{2 \cdot \left[\arcsin\left(\frac{X_2 - X_5}{X_1 + X_4}\right) + \arcsin\left(\frac{X_5 - X_3}{X_1 + X_4}\right)\right]} \cdot t_{o.} \quad (2)$$

The methodology of laboratory research was as follows. Hydraulic cylinder rod was set in the upper position with the help of hydraulic distributor. A weight was attached to the beam, and then, the lever of the hydraulic distributor was set in the "floating" position. In a double retry, the hydraulic cylinder lowering time was recorded, and the lowering was carried from the upper position to the bottom through the angle $\gamma$ that is equal to 70° (Fig. 5) (i.e. 1, 2 rad.). Such measurements were made for each of the washer hole diameters *d*, which was mounted into the hydraulic cylinder.

Diameter *d* was changed as follows: 8, 4, 2 1 mm. A value of the weight was constant, and the gravity force amounted $G_1 = 125$ N.

The geometric parameters of the laboratory installation were as follows:

$X_1 = 0.60$ m; $X_2 = 0.90$ m; $X_3 = 0$; $X_4 = 0.20$ m; $X_5 = 0.50$ m.

The time measurement of the hydraulic cylinder rod lowering motion was taken out by a stopwatch timer with the accuracy of 0.2 s.

## 3 Results and Discussion

When performing the laboratory tests, as already noted above, the inside diameter of the damper washer has four values: 8, 4, 2, and 1 mm. The first of these (8 mm) was conventional. In this variant, the washer was not installed at all, and its inside diameter was determined by the connection hose inside diameter of the hydraulic hose pipe, which connected the bottom-end and the head-end of the cylinder. It was obtained that with the inside diameter increase, the hydraulic resistance of the damper washer calculated by the formula (2) decreases accordingly (Fig. 6).

The obtained dependence is nonlinear and with satisfactory accuracy for practice ($R^2 = 0.94$) is described by the following parabolic dependence:

$$K_m = 0.025 \cdot d^2 - 0.36 \cdot d + 1.67. \qquad (3)$$

Equation (3) has a great practical relevance. Thus, according to the results of mathematical modeling, it has been determined that the examined machine-tractor unit will possess the satisfactory stability and controllability of motion in that case when the value of the damper hydraulic resistance is more than 1,000,000 N m s. To a completely logical question how to actually put this requirement into practice, the dependence (3) is precisely a desired answer.

When $K_m$ = 1,000,000 N m s, a damper washer with a hole diameter not more than 2.2 mm should be inserted into the hose, which connects the bottom-end and head-end of the hydraulic cylinder of the MPU technical module.

Besides, during the movement of the block-modular MTA, the oil transmission through a throttle (from the head-end of the hydraulic cylinder to the bottom-end) will be accompanied by the heating of the hydraulic fluid, and as a result, its density and viscosity decrease. However, in the throttle used, the length of the inside hole is

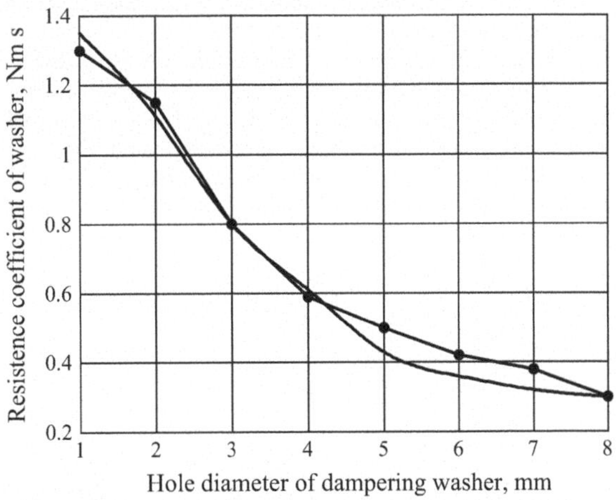


**Fig. 6** Dependence of the damper washer hydraulic resistance coefficient ($K_m$) on its hole diameter ($d$)

practically equal to its diameter. And in this case, as the research shows, the value of the throttle hydraulic resistance practically does not depend on the viscosity of the hydraulic fluid (oil).

## 4 Conclusions

Accepting accuracy which is sufficient for production application (determination coefficient $R^2 = 0.94$), it can be stated that the damper washer hydraulic resistance ($K_m$) dependence on its hole diameter ($d$) is described by the parabolic dependence $K_m = 0.025 \cdot d^2 - 0.36 \cdot d + 1.67$. Its use enables affirmation that for a satisfactory stability and controllability of the modular MTA movement, the hose connection linking the bottom-end and the head-end of the hydraulic cylinder of the MPU technological module should be supplied with a damper washer, and the inside diameter of which does not exceed 2.2 mm.

## References


1. Kutkov, G.: Tractors and Automobiles. Theory and Technological Properties. KoloSS, Moscow (2004)
2. Nadykto, V.: Role tractor's energy saturation rate in formation their draft classification. Tract. Agric. Mach. **3**, 12–18 (2012)
3. Bulgakov, V., Kyurchev, V., Nadykto, V., Olt, J.: Structure development and results of testing a novel modular power unit. Farm Mach. Process. Manag. Sustain. Agric. **7**, 40–44 (2015)
4. Padureanu, V., Lupu, M., Canja, C.M.: Theoretical research to improve traction performance of wheeld tractors by using a supplementary driven axle. In: Computational Mechanics and Virtual Engineering, pp. 410–415. Brasov, Romania (2013)
5. Bulgakov, V., Adamchuk, V., Nadykto, V., Kyurchev, V., Nozdrovicky, L.: Theoretical consideration of the controllability indicator of machine-tractor unit movement. Acta Technol. Agric. **20**(1), 11–18 (2017)
6. Golovach, I.V., Adamchuk, V.V., Bulgakov, V.M., Nadykto, V.T.: Theory of motion controllability of a wheel machine-tractor aggregate. Agric. Sci. Pract. **2**, 3–10 (2016)
7. Adamchuk, V., Bulgakov, V., Nadykto, V., Ihnatiev, Y., Olt, J.: Theoretical research into the power and energy performance of agricultural tractors. Agron. Res. **14**(5), 1511–1518 (2016)
8. Bulgakov, V., Adamchuk, V., Arak, M., Nadykto, V., Kyurchev, V., Olt, J.: Theory of vertical oscillations and dynamic stability of combined tractor-implement unit. Agron. Res. **14**(3), 689–710 (2016)

# Generalization of Factors of Milk Homogenization

Kyrylo Samoichuk, Nadiia Zahorko, Vadym Oleksiienko and Serhii Petrychenko

## 1 Introduction

The main obstacle on the way of developing high-efficiency equipment for milk homogenization is not enough profound researches on the mechanism of dispersion of the milk emulsion fat phase [1]. Despite the considerable quantity of the existent hypotheses on homogenization and possible mechanism of breakup of the fat globules of milk, an up-to-date world practice does not offer a universal homogenization theory and a generalized factor of milk fat dispersion for the most types of equipment for dispersion [2].

## 2 Analysis of Recent Studies and Publications

Analysis of breakup mechanism of the fat globules surrounded by milk plasma allows marking out the prevalent factor of this process—slip velocity of a fat globule relative to surrounding plasma [3]. The criterion of deformation and disruption is Weber number which is closely connected to the Rayleigh–Taylor instability of a fat droplet and thus acceleration. Slip velocity and emulsion flow acceleration are closely connected values [4]. In such machines for emulsion dispersion as valve, rotor, rotor–pulsation, impulse, ultrasound, and eddy, the slip velocity is difficult to calculate. Also, the use of the emulsion flow velocity instead of the slip velocity of a fat globule considerably distorts the results of researches. Unlike slip velocity, the flow acceleration is easy to calculate. Thus, one of the main directions in increasing homogenization degree is to increase the emulsion movement acceleration. Still, modern methods of

K. Samoichuk (✉) · N. Zahorko · V. Oleksiienko · S. Petrychenko
Tavria State Agrotechnological University, B. Khmelnitsky ave. 18, Melitopol 72310, Ukraine
e-mail: samoichuk.k@ukr.net

© Springer Nature Switzerland AG 2019 
V. Nadykto (ed.), *Modern Development Paths of Agricultural Production*,
https://doi.org/10.1007/978-3-030-14918-5_21

**Table 1** Prevalent factor of milk fat dispersion in main types of equipment for homogenization

| Type of homogenizer | Turbulence | Gradient of liquid flow | Slip of fat globule | Electro-hydraulic impact | Cavitation dispersion | Subcavitation dispersion | Boiling in vacuum |
|---|---|---|---|---|---|---|---|
| Valve (spinneret, spiral) | • | • | •■ | | • | | |
| Impulse | | • | •■ | | • | | |
| Rotor-pulsation | • | • | •■ | | • | | |
| Ultrasound | • | | •■ | | • | | |
| Opposite-jet | | • | •■ | | | | |
| Eddy (stream-eddy) | | • | ■ | | | • | |
| Stream with separate homogenization | • | • | •■ | | | | |
| Colloid mill | • | • | ■ | | | | |
| Mixers | • | • | ■ | | | | |
| Electro-hydraulic | | | ■ | • | | | |
| Vacuum | | | | | | | • |
| | Dispersion due to the emulsion flow acceleration | | | | | | |

*Annotation*: • —classical representation of factors of dispersion; ■ —new view on dispersion, ▭ —types of homogenizers which main factor of dispersion can be emulsion flow acceleration

the homogenization equipment calculations are based on determining other factors: cavitation, turbulence, velocity gradient, creating areas of subcavitation dispersion, etc.

Let us analyze the main factors of dispersion in modern homogenizers of dairy industry (Table 1).

Developers and researchers of homogenization consider the main factors of dispersion, apart from fat globule slip velocity, to be turbulence, gradient of liquid flow (in lengthwise and crosswise directions) and cavitation. Electrohydraulic impact combines cavitation and hydraulic impact (high gradient of flow). But all these factors can be unified into a more universal factor—the emulsion flow acceleration. Indeed, turbulence, flow gradient, and streamline of a fat particle result in appearing fat globule slip velocity which is proportional to liquid flow acceleration. Meanwhile, acceleration factor promises to be a more universal factor for many types of homogenizers, and it allows developing designs of high-efficiency machines with low energy consumption.

The most popular and researched are valve homogenizers. For this type of homogenizers, the theoretical and empirical dependences are got to connect its quality and design and process-dependent parameters [5].

Among the most efficient for creating a high acceleration of emulsion are such homogenizers as pulsation, reciprocating, and rotor–pulsation types.

In a rotor–pulsation machine at periodical overlapping of rotor holes and stator, the liquid movement becomes nonsteady and considerable reverse sign pulsations occur. Energy dissipation becomes even with excitation of the additional oscillations occurring due to a vibrating rotor, and as a result of compliance of rotor oscillations with holes overlap, the resonance of pulsations is created which additionally increases homogenization efficiency [6].

In a pulsation reciprocating homogenizer, high values of flow's acceleration and fat globule slip velocity are created due to the reverse sign pulsations at oscillatory movements of the piston. In addition, multiplicity of passing of the emulsion through the piston holes equals 12 and more times [6].

## 3 Statement of the Objective and Tasks of the Study

Objective of the study is to determine the connection between the emulsion flow acceleration and the dispersion of milk emulsion homogenized in valve, pulsation with a vibrating rotor, and reciprocating pulsation machines.

To achieve the set objective, the following tasks were solved:

1. to determine analytically the interrelation between design and process-dependent parameters of the valve, reciprocating pulsation, and pulsation with a vibrating rotor homogenizers and the emulsion flow acceleration;
2. to determine experimentally the presence of correlation between flow acceleration and milk emulsion dispersion;
3. to compare the efficiency of the valve, reciprocating pulsation, and pulsation with a vibrating rotor homogenizers using the factor of the emulsion flow acceleration.

## 4 Results

A pulsation homogenizer with a vibrating rotor consists of the rotor 3 and the stator 6 which have the holes 4 for passing of the emulsion (Fig. 1a). The impeller 5 creates the required head of the fluid which exits the channel 1. The rotor shaft 7 is connected to the crank gear which produces axial oscillations of the rotor up to 3 mm. Thus, when the machine is in use, the rotor makes oscillation movements (vibrates) with the frequency up to 3000 $min^{-1}$ in addition to rotating.

The main part of the pulsation reciprocating homogenizer (Fig. 1b) is a cylindrical chamber 1 with branches for the supply and discharge of the product 4 and 5. Inside of the chamber, the piston 2 makes impulse movements with the frequency n with the help of the rod which is connected to the crank gear. The piston has holes 6, and the emulsion passes through those holes. The product is supplied to the chamber with the help of the outside pump. Pump performance rates the multiplicity of passing of the emulsion through the piston holes.

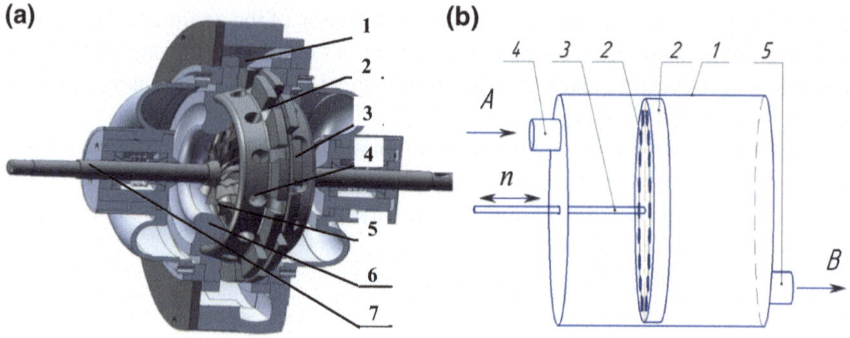


**Fig. 1** Construction: **a** rotor of the pulsation machine with a vibrating rotor: 1—pressure channel of the stator; 2—rotor slot; 3—rotor; 4—cross-holes of the rotor; 5—impeller, 6—stator with holes, 7—rotor shaft. **b** Process chamber of the reciprocating homogenizer: 1—chamber; 2—piston with holes; 3—rod, 4, 5—branches for the supply and discharge of the emulsion, 6—piston holes, A—emulsion supply, B—discharge of the homogenized product

To solve the first task of the study, the theoretical researches had been conducted which result was obtaining the formula of emulsion acceleration. For valve homogenization, the formulas for determining an average size of the fat globules, $d$, $\mu$m, an average emulsion acceleration $a$, m/s$^2$, emulsion velocity $\upsilon$, m/s are of the form [5]

$$d = \frac{3.8 \times 10^3}{\sqrt{P}}, \tag{1}$$

$$a = \frac{\upsilon^2}{L}, \tag{2}$$

$$\upsilon = \varphi\sqrt{\frac{2P}{\rho_{\mathcal{M}}}}, \tag{3}$$

where

P pressure of the valve homogenization, Pa;
$\varphi$ velocity coefficient of the valve;
$\rho_{\mathcal{M}}$ density of milk, kg/m$^3$;
$L$ length of the valve gap, mm.

From these formulas, it is possible to get an expression which relates an average size of a fat globule to emulsion acceleration

$$d = \frac{\varphi}{\sqrt{\rho_{\mathcal{M}} L}} \frac{5.37 \times 10^3}{\sqrt{a}}. \tag{4}$$

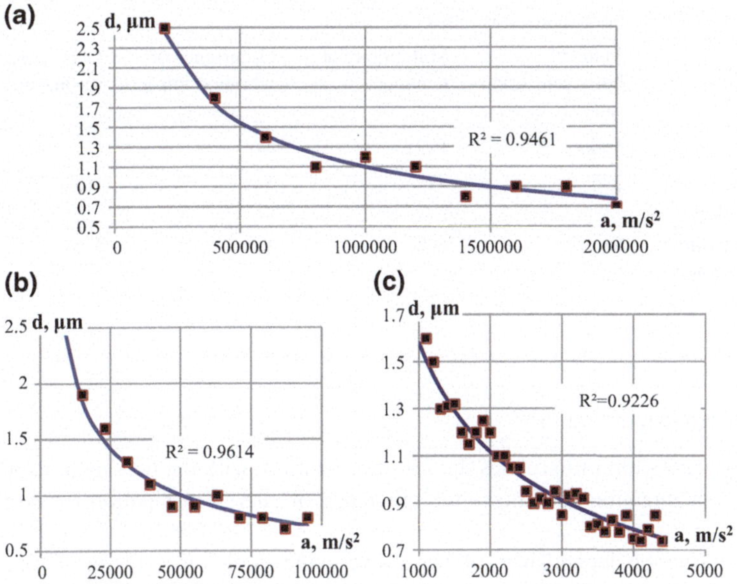


**Fig. 2** Dependence of the average size of fat globules on the average emulsion flow acceleration for **a** valve homogenizer, **b** pulsation reciprocating homogenizer, **c** pulsation homogenizer with a vibrating rotor

The obtained formula (4) shows that the emulsion dispersion after the treatment in a valve homogenizer is evaluated with the acceleration of the emulsion movement in a valve gap. It proves the assumption about the possibility to evaluate dispersion from the emulsion flow acceleration for this type of homogenizers.

The equation of interrelation of the emulsion acceleration in the holes of the pulsation machine with a vibrating rotor with its design and process-dependent parameters is got in the article [7]. For the reciprocating homogenizer—in the work [8].

The subject of the experimental research was cow's milk with fat content of 3.2–4.5% with an average size of fat globules 2.4–2.8 μm. Sizes of milk fat globules after homogenization were found with the help of an optical microscope equipped with a digital camera applying computer programs of analysis and image processing.

According to the formula (4), the curves were built to show the dependence of the average size of fat globules on the emulsion movement acceleration for $\varphi = 0.64$, $L = 3$ mm (Fig. 2a). The results of evaluating milk dispersion after the treatment both in a valve and rotor homogenizers indicate a high degree of correlation between the average size of fat globules and flow acceleration (Fig. 2b, c).

The dependences showed in Fig. 2 for all types of the homogenizers are approximated with the formula

$$d = \frac{\mathcal{K}_h}{\sqrt{a}}, \tag{5}$$

**Table 2** Comparing homogenization ratios

| Type of homogenizer | Value of homogenization ratio ($K_T$) | Multiplicity of treatment | Occurrence of vibration impact | Occurrence of resonance |
|---|---|---|---|---|
| Valve | 1600 | 1 (2) | No | No |
| Pulsation reciprocating | 225 | >12 | Yes | No |
| Pulsation with a vibrating rotor | 68 | 1–8 | Yes | Yes |

where

$\mathcal{K}_h$ homogenization ratio.

Homogenization ratio relates the flow acceleration with the emulsion dispersion after homogenization and shows the efficiency of energy dissipation for homogenization.

Comparing the dependences of dispersion on the emulsion acceleration for reciprocating pulsation and rotor–pulsation homogenization, it is possible to see that treatment in a rotor–pulsation machine is more efficient. It is explained with the influence of the resonance of the axial oscillations of the rotor and overlap of the holes. Meanwhile, the amplitude of the emulsion oscillations grows substantially, whereupon the slip velocity of a fat globule increases. The influence of resonance in the formula (5) is ignored. Thus, homogenization ratio at processing emulsion in a pulsation machine with a vibrating rotor is considerably lower than in other types of homogenizers.

Homogenization ratios and additional influence factors on the process of homogenization are shown in Table 2.

The results of the data of Table 2 show the influence of multiplicity, occurrence of vibration, and resonance on the homogenization ratio. Treatment in valve homogenizers does not provide for the usage of vibration and resonance. Multiplicity equals 1 or 2 for the two-level heads (although the pressure of the second level is much lesser than of the first one). At the vibration impact on the emulsion which occurs in a reciprocating pulsation machine, the energy dissipation is concentrated on the interphase boundary. Besides, the multiplicity of passing of the emulsion through the piston holes equals 12 and more times. It considerably increases the degree of milk fat dispersion as compared to the valve homogenizer. A pulsation machine with a vibrating rotor possesses the advantages of the reciprocating machine and additionally the influence of pulsations resonance. Owing to this fact such machine makes it possible to achieve the highest slip velocity (emulsion acceleration) and the smallest homogenization ratio.

## 5 Conclusions

The conducted researches result in determining that the main factor of dispersion can be emulsion flow acceleration on the example of milk homogenization in a reciprocating pulsation, pulsation machine with a vibrating rotor, and valve homogenizer. Calculation of this parameter is not difficult in contrast to the slip velocity of a fat globule. The factor which connects emulsion flow acceleration to dispersion is the homogenization ratio. This ratio shows the concentration efficiency of energy on the breakup of fat globules. Homogenization ratio declines at applying vibration on the emulsion, increasing multiplicity of treatment and resonance effects.

Potentially acceleration factor and homogenization ratio can be common for such homogenizers as rotor–pulsation, spinneret, ultrasound, electrohydraulic, and mixers. This factor can be used as universal for making mathematical theory and comparing operating efficiency for such types of homogenizers.

## References


1. Wilbey, R.A.: Homogenization of milk: principles and mechanism of homogenization, effects and assessment of efficiency: valve homogenizers. In: Encyclopedia of Dairy Sciences, pp. 750–754 (2011)
2. Huppertz, T.: Homogenization of milk|Other types of homogenizer (High-Speed Mixing, Ultrasonics, Microfluidizers, Membrane Emulsification). In: Encyclopedia of Dairy Sciences, 2nd edn, pp. 761–764 (2011)
3. Samoichuk, K.O.: Vyznachennia universalnoho faktora dysperhuvannia zhyrovoi fazy moloka. Innovative Aspects in Food and Hospitality Industry Equipment Development Under Present Conditions, pp. 17–19 (2017)
4. Samoichuk, K.O.: Acceleration of emulsion as the basic factor of hydrodynamic dispergating of milk fat phase (in Russian). In: Breakthrough technics and technologies of food production in restaurant and trade fields, vol. 2, no. 26, pp. 210–220 (2017)
5. Nuzhin, E.V., Gladushnyak, A.K.: Gomogenizatsiya i gomogenizatoryi. Odessa, Pechatnyiy dom (2007)
6. Samoichuk, K., Palyanichka, N., Verholantseva, V., Levchenko, L.: Perspektyvy vykorystannia vibratsiinykh homohenizatoriv moloka. Vibratsii v tekhnitsi ta tekhnolohiiakh, pp. 36–38 (2017)
7. Samoichuk, K., Kiurchev, S., Oleksiienko, V., Palyanichka, N., Verholantseva, V.: Investigation of homogenization of milk in a pulsation machine with a vibrating rotor. East Europe J. **6/11**(84), 16–21 (2016)
8. Deinychenko, G.V., Samoichuk, K.O., Levchenko, L.V.: Analytical determination of dispersion at processing of milk in pulsation homogenizer (in Russian). Bull. Kharkiv Nat. Tech. Univ. Agric. Petro Vasilenko **179**, 32–39 (2016)

# Setting Ground Dimension-Type Series-Tillage Fertilizing, Sowing Complexes for Growing Grain Crops

**Vitaliy Serbiy, Volodymyr Diuzhaiev, Halyna Antonova and Olena Mykhailenko**

## 1 Introduction

The most important direction of increasing the efficiency of agricultural production and its competitiveness is the development and implementation of the latest technologies based on resource economy. This purpose can be reached due to the introduction of machinery of new generation to the process of technology, the distinguishing feature of which is multifunctionality.

In addition, the contradiction between the agrotechnical need for tillage and its negative effect on fertility, which manifests itself in the strengthening of erosion processes, and re-compaction of soil, led to a reassessment of the feasibility of repeated tillage in the cultivation of crops. This circumstance also led to the introduction of new technologies with a significant reduction of the number of operations of tillage and the unification of operations of different purposes into a single technological complex with the use of combined machines and operating elements.

One of the promising directions of the development of integrated mechanization of agricultural production is the creation of combined machines that allow to perform simultaneously several operations in one process: tillage, sowing, fertilization, and addition of herbicides [1]. The use of such machines reduces the number of passes of machine-tractor aggregates in the field, reduces the loss of time at idle passages, increases productivity, and reduces monetary and labor costs.

Such a combination is also appropriate agronomically, as favorable conditions for the growth of plants are created by accelerating the course of fieldwork, better

V. Serbiy (✉)
National Science Center "Institute of Mechanization and Electrification of Agriculture", Kiev, Ukraine
e-mail: svk_84@ukr.net

V. Diuzhaiev · H. Antonova · O. Mykhailenko
Tavria State Agrotechnological University, B. Khmelnitsky ave. 18, Melitopol 72310, Ukraine

© Springer Nature Switzerland AG 2019 
V. Nadykto (ed.), *Modern Development Paths of Agricultural Production*,
https://doi.org/10.1007/978-3-030-14918-5_22

supplying of water, and supporting of thermal regimes, and the consolidation of the soil and the destruction of its structure are eliminated.

Combined machines also correspond most fully to the current trends in tractor construction, which consists in the creation of energy-intensive tractors that cannot be downloaded by single-operator machines, especially in small-sized fields, on slopes and irrigated areas.

## 2 Review of Literary Sources

Mikhalieva et al. [2] state that the technological modernization of the agro-industrial complex allows to achieve savings in fuels and lubricants in 2–2.5 times, labor costs—up to three times, to ensure an increase in the productivity of agricultural crops by 30%. The concept proposed by these authors focuses on the formation of a new type of machine-tractor park (ICC), where single-operational units should be replaced by multifunctional, universally combined ones, able to adapt to those conditions that change due to the rapid change of working bodies and modular aggregation.

Such researches as M. V. Krasnoschokov, M. K. Mazitov, G. G. Maslov studied the optimal width of the multifunctional complex.

Maslov [3] substantiates the optimum width with the minimum of reduced costs and losses and constructs an econometric mathematical model in which the speed of an aggregate is given by a constant, which substantially restricts the field of search for finding the optimal parameters of the soil–cultivating complex, which ensure the minimum costs and losses. In addition, crop rotation is not taken into account for the estimation of crop sowing costs of all crops in this model.

Therefore, it was decided to improve the model significantly and to ensure the search for optimum width of the unit without setting of discrete parameters and constants, taking into account the appropriate cost of the tractor for all variants of machine-tractor units. It was also determined to take into account the risks of losses of agricultural crops depending on the delay of the sowing terms and parameters of the running gear system and its mass.

## 3 Results

The optimum of the target function of the tillage–fertilizing–sowing complex is the minimum amount of expenses for the implementation of combined technological operations for soil cultivation, sowing and addition of the main doses of mineral fertilizers and losses of the future harvest of agricultural crops, associated with violation of the optimal terms of sowing, delaying the duration of peeling of grain steels during sowing of intermediate crops and consolidation of soil by running gear systems.

As is well known [3–5], the delay in the processing of grain stubblefield after harvesting for 2–3 days reduces the yield of agricultural crops next year by 1.5–2 centners per 1 ha (Fig. 1). This circumstance imposes certain requirements for determining the operational parameters of the unit to reduce the impact on the profitability of production.

It is known that delays with the terms of sowing of grain crops according to the information of Nasonov [6] lead, respectively, to losses of the future yield on nonlinear dependence (Figs. 2, 3, and 4).

And the effect of the sealing action of the running gear system on the reduction of agricultural crops on the example of spring wheat is given in Fig. 5.

The methodology was compiled in order to calculate the pressure on the soil by the running system of the machine of the corresponding power. It was compiled which is essentially the following.

The model range of energy products is selected from the catalog of production of the Minsk tractor plant in 2016 [11] with capacities from 6.6 to 261 kW (Table 1).

Using the well-known expression for determining the maximum pressure of a single propeller on the support surface [12–14] and the data from Tables 1 and 2, the corresponding calculations for the presented model line were carried out and the

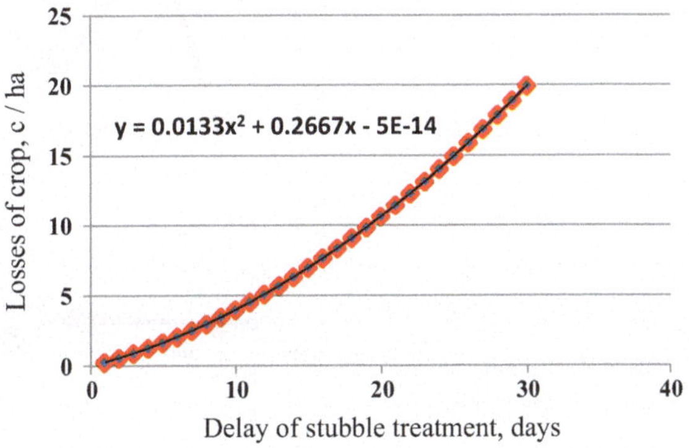


**Fig. 1** Loss of grain crops of future harvest due to delays of previous processing of stubble, c/ha

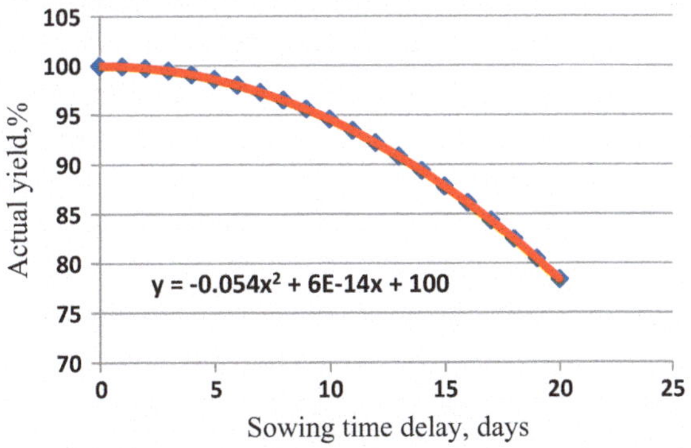


**Fig. 2** Schedule of actual yield of winter wheat from tightening of sowing terms, % [6]

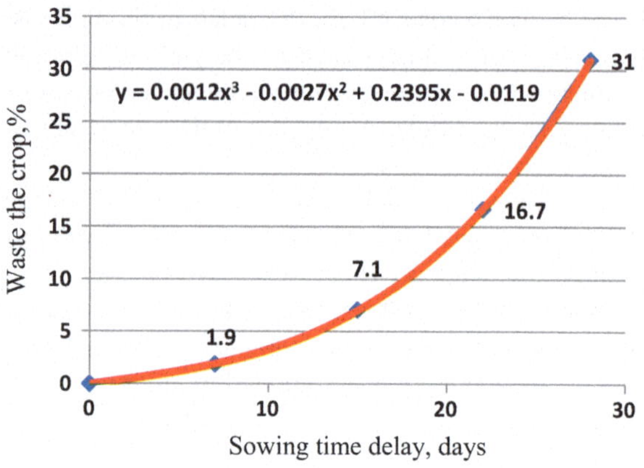


**Fig. 3** Diagram of reducing the yield of corn due to the delay in the terms of sowing, % [7]

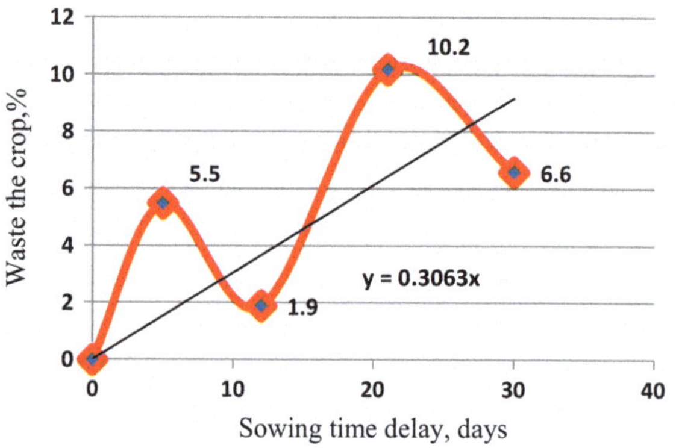


**Fig. 4** Diagram of reducing the yield soybean to the delay in the terms of sowing, % [8, 9]

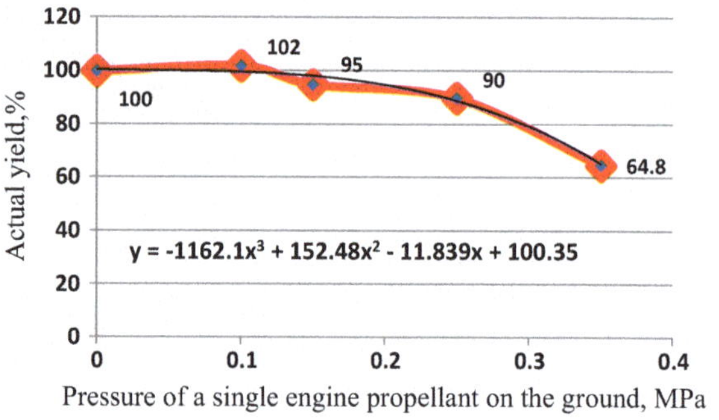


**Fig. 5** Diagram of the change of spring wheat yield depending on the sealing effect of the running gear system of the machine [10]

**Table 1** Technical characteristics of the model range of energy resources of the Belarusian Motor Vehicle Manufacturers Association from the 2016 product catalog

| Power tool | Brand | Model | Nomine flax power (kW) | Operating weight (kg) | Tires | |
|---|---|---|---|---|---|---|
| | | | | | Front | Back |
| 1 | 2 | 3 | 4 | 5 | 6 | 7 |
| Moto-block | Belarus | 09H | 6.6 | 176 | 6L-12 | |
| Mini tractor | Belarus | 132H | 9.6 | 532 | 6.5L-12 | 6.5L-12 |
| Mini tractor | Belarus | 152 | 9.6 | 650 | 210/75R13 | 210/75R13 |
| Tractor | Belarus | 311 | 24.3 | 1450 | 7.50L-16 | 12.4L-16 |
| Tractor | Belarus | 422 | 36.6 | 2025 | 210/80R16 | 11.2–20 |
| Tractor | Belarus | 510 | 41.9 | 3430 | 7.50-20/9.00-20 | 15.5R38/18.4 |
| Tractor | Belarus | 622 | 46 | 2410 | 12.4L-16 | 360/70R24 |
| Tractor | Belarus | 80.1 | 60 | 3770 | 9.0–20 | 15.5R38 |
| Tractor | Belarus | 922 | 70 | 4400 | 13.06.202 | 15.5R38 |
| Tractor | Belarus | 1021 | 77 | 4295 | 360/70R24 | 15.5R38 |
| Tractor | Belarus | 1220.3 | 90 | 5500 | 14.9R24 | 16.9R38 |
| Tractor | Belarus | 1221.2 | 90.4 | 5730 | 420/70R24 | 16.9R38 |
| Tractor | Belarus | 1523 | 109 | 6250 | 420/70R24 | 520/70R38 |
| Tractor | Belarus | 2022.3 | 156 | 7220 | 420/70R24 | 580/70R42 |
| Tractor | Belarus | 3022.2 | 223 | 11,500 | 540/65R30 | 580/70R42 |
| Tractor | Belarus | 3522 | 261 | 20,000 | 600/65R34 | 710/70R42 |
| Tractor | Belarus | 3525.6 | 261 | 12,000 | 600/65R34 | 710/70R42 |

schedule of dependence of the pressure of the propellant engine propulsion of the factory equipment on its power was constructed (Fig. 6).

Further, we will reveal the essence of the economic-mathematical model in the calculation of the standard series of tillage–fertilizing–sowing implement, and its structurally functional scheme is given in Fig. 7.

The target function of the given costs and losses is as follows:

$$C = Ц_{Т} \cdot \left(\frac{1}{A_{Т} \cdot F} + \frac{TPiTO_{Т}}{100 \cdot T_{г.з.т.} \cdot W_{зм}}\right) + Ц_{М} \cdot \left(\frac{1}{A_{М} \cdot F} + \frac{TPiTO_{М}}{100 \cdot T_{г.з.м.} \cdot W_{зм}}\right) + Z + U + Y + X, \; C \rightarrow min, \quad (1)$$

where

| | |
|---|---|
| $C$ | is given costs and losses per hectare, UAH/ha; |
| $Ц_M$ | is market value of the energy resource, UAH; |
| $Ц_M$ | is market value of the tillage–fertilizing–sowing complex, UAH; |
| $A_T$ | is lifetime of the energy resource, years; |

**Table 2** Specifications of the tires, which are completed with the basic factory models of power plants of the Belarusian Motor Vehicle Company

| Tire brand | Tire width (m) | The radius is not loaded wheels (m) | Intersection radius tires (m) | Air pressure in the tire (kPa) | К1 | К2 |
|---|---|---|---|---|---|---|
| 1 | 2 | 3 | 4 | 5 | 6 | 7 |
| 6L-12 | 0.155 | 0.285 | 0.0775 | 100 | 1.6 | 1.5 |
| 6,5L-12 | 0.168 | 0.285 | 0.084 | 100 | 1.6 | 1.5 |
| 210/75R13 | 0.21 | 0.32 | 0.105 | 100 | 1.4 | 1.5 |
| 12.4L-16 | 0.327 | 0.465 | 0.1635 | 220 | 1.3 | 1.5 |
| 11.2–20 | 0.274 | 0.4925 | 0.137 | 210 | 1.3 | 1.5 |
| 15.5R38 | 0.394 | 0.785 | 0.197 | 160 | 1.1 | 1.5 |
| 360/70R24 | 0.36 | 0.577 | 0.18 | 160 | 1.2 | 1.5 |
| 15.5R38 | 0.394 | 0.785 | 0.197 | 160 | 1.1 | 1.5 |
| 16.9R38 | 0.429 | 0.8375 | 0.2145 | 160 | 1.1 | 1.5 |
| 520/70R38 | 0.52 | 0.875 | 0.26 | 160 | 1.1 | 1.5 |
| 580/70R42 | 0.585 | 0.95 | 0.2925 | 160 | 1.1 | 1.5 |
| 710/70R42 | 0.716 | 1.03 | 0.358 | 100 | 1.1 | 1.5 |
| 16.9R30 | 0.429 | 0.735 | 0.2145 | 160 | 1.1 | 1.5 |

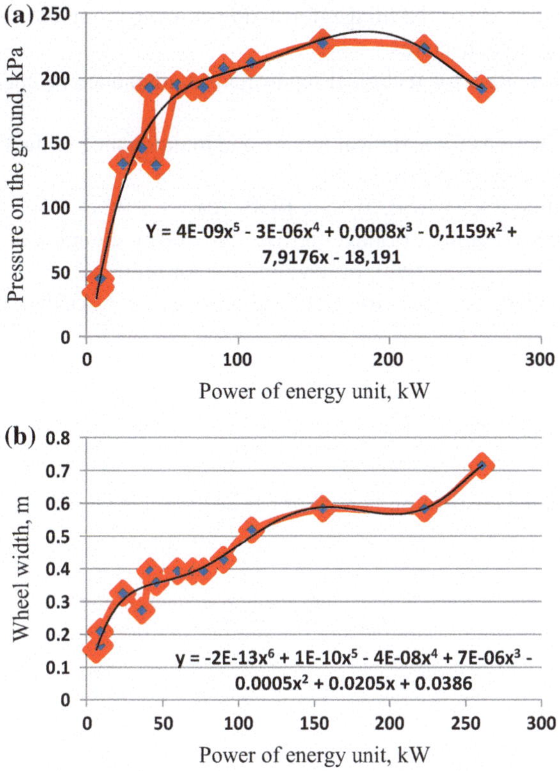


**Fig. 6** Charts of the change of pressure of a single engine on the ground (**a**) and the width of the wheel (**b**) depending on the power of energy unit of the factory equipment (dependencies for the range of tractors of the Belarusian Motor Vehicle Company "Belarus")

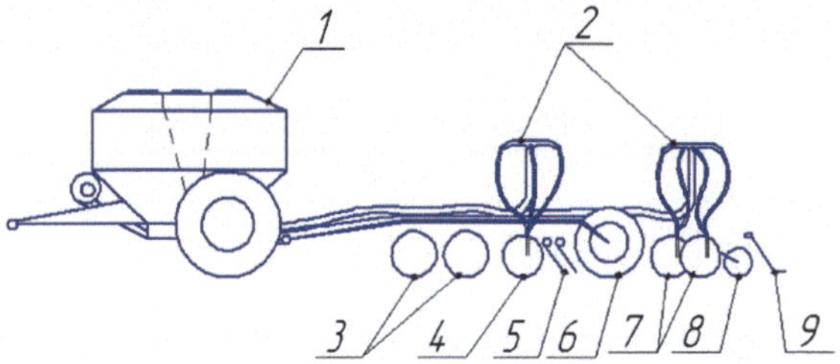


**Fig. 7** Structurally functional scheme of combined tillage fertilizing sowing implement: 1—sowing module with three-section bunker, dispensers, and fan; 2—two-current pneumatransmitting and distributing system; 3—disk tillage movable operating element; 4—two-disk coulter for adding main fertilizers; 5—spring harrows; 6—Bearing wheels; 7—two-disk coulter for adding seeds and starter fertilizers; 8—rolls; 9—stubble harrow

$A_M$ is service life of the tillage–fertilizing–sowing complex, years;
$F$ is field area, hectares;
$\text{TPiTO}_\text{T}$ is deductions for technical repair and maintenance of energy resources, %;
$\text{TPiTO}_\text{M}$ is deductions for technical repairs and maintenance of the tillage–fertilizing–sowing complex, %;
$T_{\text{Г.З.Т.}}$ is annual tractor load, $T_{\text{Г.З.}} = 1350$, year;
$T_{\text{Г.З.М.}}$ is annual loading of tillage–fertilizing–sowing complex of year;
$W_{\text{ЗМ}}$ is variable productivity of machine-tractor unit, ha/h;
$Z$ is the given expenses for fuel and human labor, UAH/ha;

$$Z = P\text{mm} + L_\text{p}, \tag{2}$$

where

$Lp$ is the estimated costs of human labor per hectare, UAH/ ha;

$$Lp = \frac{\text{К}_{\text{люд}} \cdot \text{n} \cdot \text{t}_{\text{зм}} \cdot \text{C}_{\text{тр}}}{\text{F}}, \tag{3}$$

where

$K_{\text{люд}}$ is the number of people employed at work, *n*.;
$n$ is number of working days, units;
$t_{\text{зм}}$ is time of change, year;
$C_{\text{тр}}$ is normal value of man-hours, UAH/ person·h, $C_{\text{тр}} = 17.4$ UAH/person·h;
$P\text{mm}$ is the indicated costs of fuel per hectare, UAH/ha;

$$P\text{mm} = \frac{B \cdot \left(P_{\text{уд}} + \varepsilon \cdot \rho \cdot S \cdot \left(\vartheta_\text{p}^2 - \vartheta_0^2\right)\right) \cdot v_\text{p} \cdot q \cdot n \cdot k \cdot t_{\text{зм}}}{F}, \tag{4}$$

where

$\vartheta_0$ is the initial adopted speed of the machine-tractor unit, which does not cause the increase of the traction resistance of ag. cars, m/s, (0.56 m/s);
ε is a coefficient that takes into account the discreteness of soil cultivation in the cross section of the layer and the transfer of kinetic energy to the soil parts;
$\text{P}_{\text{уд}}$ is specific tractive resistance of tillage–fertilizing–sowing complex, N/m, we will accept for calculations $\text{P}_{\text{уд}} = 3100$ N/m;
ρ is density of soil, kg/m$^3$;
$S$ is area of the lateral intersection of the soils layer, which is processed by the working body on the linear meter of the working width of the complex, m$^2$;
$q$ is the given expenses of fuel, g/kW, we will accept for calculations $q = 240$ g/W;

$K$ is number of changes per day, unit;
$U$ is losses of the future harvest associated with delaying the terms of peeling stubble (Fig. 1), UAH/ha;

$$U = \sum_{i=1}^{n} \Delta u(n_i) \cdot \acute{f}, \tag{5}$$

where

$n$ is number of working days, days;
$\Delta u$ is the function of future harvest losses, c/ha (Fig. 1);

$$\Delta u = 0.0133 \cdot n^2 + 0.2667 \cdot n, \tag{6}$$

where

$f$ is the daily volume of work, ha/day;
$Y$ is losses of the future yield from the delay of the optimal terms of sowing, c/ha;
$X$ is losses of the future harvest due to soil compaction cause by energy vehicles, c/ha.

Let us express the variable productivity of the tillage–fertilizing–sowing complex through the processing area $F$ and the number of days $n$:

$$\begin{cases} \acute{f} = \frac{F}{n} \\ W_{3M} = \frac{\acute{f}}{K \cdot t_{3M}} \end{cases} \tag{7}$$

where

$K$ is number of changes per day, units

We get,

$$W_{3M} = \frac{F}{n \cdot K \cdot t_{CM}}, \tag{8}$$

This yields

$$C = Ц_T \cdot \left(\frac{1}{A_T \cdot F} + \frac{TPiTO_T \cdot n \cdot K \cdot t_{CM}}{100 \cdot T_{г.з.т.} \cdot F}\right) + Ц_M \cdot \left(\frac{1}{A_M \cdot F} + \frac{TPiTO_M \cdot n \cdot K \cdot t_{CM}}{100 \cdot T_{г.з.м.} \cdot F}\right) + Z + U + Y + X, \tag{9}$$

Let us express the price of the tillage–fertilizing–sowing complex and energy resource:

$$Ц_T = f(N), Ц_M = f(B), \tag{10}$$

$$Ц_т = N \cdot \varepsilon_т \,, Ц_м = B \cdot \varepsilon_м, \tag{11}$$

where

$B$ is the working width of the tillage–fertilizing–sowing complex, *m*;
$N$ is capacity of energy vehicle, kW;
$\varepsilon_T$ is specific cost of the working width's unit of the tillage–fertilizing–sowing complex, UAH/m;
$\varepsilon_M$ is specific cost of power unit, UAH/m.

From Eq. (11), we obtain:

$$\varepsilon_т = \frac{Ц_т}{N}, \varepsilon_м = \frac{Ц_м}{B}, \tag{12}$$

Let us assume based on the results of the marketing analysis for mini-till tillage–fertilizing–sowing complex $\varepsilon_M = 493{,}000\,\text{UAH/m}$ and for energy means $\varepsilon_T = 3500\,\text{UAH/kW}$.

Since $B = \frac{W_{3M}}{0.1 \cdot \vartheta \cdot \tau}$, then substituting $Ц_м$(11), we obtain:

$$Ц_м = \frac{W_{зм}}{0{,}1 \cdot \vartheta \cdot \tau} \cdot \varepsilon_м\,, \tag{13}$$

We place formula (8) into Eq. (13):

$$Ц_м = \frac{F}{n \cdot K \cdot t_{см} \cdot 0{,}1 \cdot \vartheta \cdot \tau} \cdot \varepsilon_м\,, \tag{14}$$

We combine Eqs. (1) and (14):

$$\begin{aligned} C = {} & \varepsilon_T \cdot N \cdot \left( \frac{1}{A_T \cdot F} + \frac{\text{TPiTO}_T \cdot n \cdot K \cdot t_{CM}}{100 \cdot T_{Г.3.T.} \cdot F} \right) \\ & + \frac{F}{n \cdot K \cdot t_{CM} \cdot 0.1 \cdot \vartheta \cdot \tau} \cdot \varepsilon_M \cdot \left( \frac{1}{A_M \cdot F} + \frac{\text{TPiTO}_M \cdot n \cdot K \cdot t_{CM}}{100 \cdot T_{Г.3.M.} \cdot F} \right) \\ & + Z + U + Y + X, \end{aligned} \tag{15}$$

Let us shorten $F$,

$$\begin{aligned} C = {} & \frac{\varepsilon_T \cdot N}{F} \cdot \left( \frac{1}{A_T} + \frac{\text{TPiTO}_T \cdot n \cdot K \cdot t_{CM}}{100 \cdot T_{Г.3.T.}} \right) \\ & + \frac{\varepsilon_M}{n \cdot K \cdot t_{CM} \cdot 0.1 \cdot \vartheta \cdot \tau} \cdot \left( \frac{1}{A_M} + \frac{\text{TPiTO}_M \cdot n \cdot K \cdot t_{CM}}{100 \cdot T_{Г.3.M.} \cdot F} \right) \\ & + Z + U + Y + X, \end{aligned} \tag{16}$$

We will be keeping on reducing,

$$C = \frac{\varepsilon_{T} \cdot N}{F} \cdot \left( \frac{1}{A_{T}} + \frac{TPiTO_{T} \cdot n \cdot K \cdot t_{CM}}{100 \cdot T_{\Gamma.3.T.}} \right) + \frac{\varepsilon_{M}}{0,1 \cdot \vartheta \cdot \tau} \cdot \left( \frac{1}{n \cdot K \cdot t_{CM} \cdot A_{M}} + \frac{TPiTO_{M}}{100 \cdot T_{\Gamma.3.M.}} \right) + Z + U + Y + X, \quad (17)$$

where

τ is the coefficient of use of working time of change.

$$\tau = f(B, F) \rightarrow f(n, f)' = \frac{F}{n \cdot K \cdot t_{CM} \cdot 0.1 \cdot B \cdot \vartheta}, \quad (18)$$

Let us substitute τ,

$$C = \frac{\varepsilon_{T} \cdot N}{F} \cdot \left( \frac{1}{A_{T}} + \frac{TPiTO_{T} \cdot n \cdot K \cdot t_{CM}}{100 \cdot T_{\Gamma.3.T.}} \right) + \frac{\varepsilon_{M} \cdot B}{F} \cdot \left( \frac{1}{A_{M}} + \frac{TPiTO_{M} \cdot n \cdot K \cdot t_{CM}}{100 \cdot T_{\Gamma.3.M.}} \right) + Z + U + Y + X. \quad (19)$$

Let us make an algorithm for calculating the optimal width of the tillage–fertilizing–sowing complex. Facilities are divided into six groups by size (F):—up to 15 g (1 group);

- *15–150 ha (2 groups);*
- *151–1500 ha (group 3);*
- *1501–6000 ha (4 groups);*
- *6001–15,000 ha (group 5);*
- *more than 15,000 ha (group 6).*

We will introduce in the initial conditions for simulation of the machine-tractor unit of crop rotation with three agricultural crops: winter wheat, corn for grain and soy (Fig. 8). Let us give a plan of crop rotation and dates of seeding [13].

Thus, let us simulate the work of the tillage–fertilizing–sowing complex in an aggregate with a power mean for different values of width and high-speed regimes for operations of soil cultivation, fertilization; and sowing of grain corn, soya, winter wheat and seed crop.

1. Let us assume that the amount of each crop is 33.3% of the total area of fields, respectively, 5, 50, 250, 2000, 5000, 10,000 ha. Let us substitute an array of values of the fields' sizes in Eq. (1),

$$F_i = F[5, 50, 250, 2000, 5000, 10{,}000] \quad (20)$$

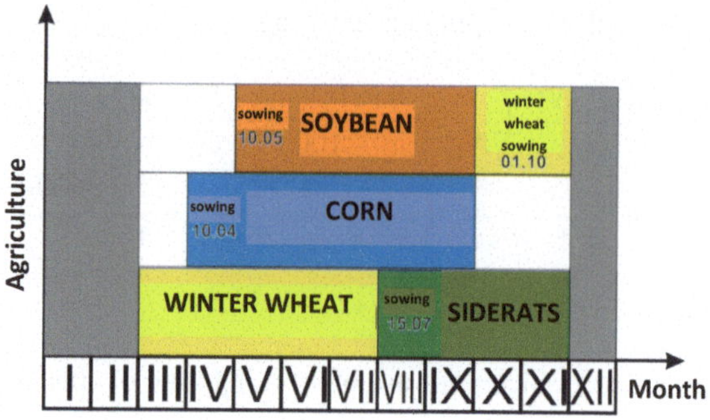


**Fig. 8** Plan of the crop rotation with the dates of sowing operations

2. Iteration of the working width of the seizure of the complex is carried out with an initial width $B_0$ in the range $B_{\max}$ and a sampling step $\Delta B$.
   We assign the initial and boundary conditions for modeling of width of the complex:

$$B_0 = 0.1\,\text{M},\ B_{\max}[1, \ldots, 20]\text{M},\ \Delta B = 0.5\,\text{M} \tag{21}$$

   We substitute the indexed parameter of the width of the complex Bi in Eq. (1)

$$B_{i+1} = B_i + \Delta B \tag{22}$$

3. Further, we iterate the days of operation of the complex with the initial delay of work $n_0$ in the range $n_{\max}$ with the step $\Delta n$.
   We will create the initial and boundary conditions for modeling the term of processing of operations and sowing by the complex:

$$n_0 = 0,\ n_{\max}[0, \ldots, 30],\ \Delta n = 1 \tag{23}$$

   We substitute the indexed parameter of the term of operation $n_i$ in Eq. (1):

$$n_i = n_i + \Delta n \tag{24}$$

4. We determine the losses of the future yield due to the delay in the processing of cereal grain stubble using Eqs. (5) and (6) and taking into account the cost of the crop:

$$\acute{U}_i = \sum_{i=1}^{n} \Delta u(n_i) \cdot \acute{f} \cdot C_{\text{yp}} \tag{25}$$

   where $C_{\text{yp}}$ is the cost of the harvest, UAH/ percent, wheat $C_{\text{yp}} = 500\,\text{UAH/c}$.
5. And we consider the costs to the unit of the area treated, UAH/ha:

$$U_i = \frac{\acute{U}_i}{F} \tag{26}$$

6. We determine the losses of the future harvest of all crops taking into account the delay of the optimal terms of sowing on dependences pic. 2–4.
7. We calculate the loss of the future harvest caused by soil compaction under the work of the machine's movable operating element, depending on the figure on dependence pic. 6.
8. The calculated fuel costs and wages are determined by Eqs. (2–4).
9. Further, we calculate the given costs and losses due to Eq. (3).
10. We enter the settlement data into the database for the corresponding values *B_i*, *n_i* and other regime and energy parameters of the complex.
11. Repeat p. 3–10 until *n*_max is reached.
12. Repeat step 2–10 until *B*_max is reached.
13. Repeat item 1–10 until the end of the array is reached $F = 10{,}000$ ha.
14. We find the variable productivity with each iteration of *n_i* according to the dependence $W_{3M} = \frac{F}{n \cdot K \cdot t_{3M}}$ and the speed $\vartheta = \frac{W_{3M_i}}{0.1 \cdot B_i \cdot \tau}$ and cut off the values above $\vartheta_{max}$ (4 м/c). We will accept for the variants of the complexes calculated $\tau = 0.78$, however, in the future calculations will be specified by the value $\tau$, which should correspond to the width of the unit and the corresponding field area with the assumption that the kinematic length of the unit is constant.
15. In the obtained region, we find *B* and *n* which correspond to the minimal given expenses *C*.

Let us make a block diagram of the algorithm and a program in the programming language VBA in calculating the optimal width of the soil–fertilizer–sowing complex (Figs. 9 and 10).

According to the results of the simulation, the calculated data are obtained for substantiation of the optimum width of the soil–fertilizing and fertilizing–seeding complexes for different areas of farms.

The values of the optimal parameters of the complexes for soil cultivation and sowing are given in Table 3.

**Fig. 9** Block diagram of the algorithm from the calculation the optimum width of soil–fertilizer–sowing complex

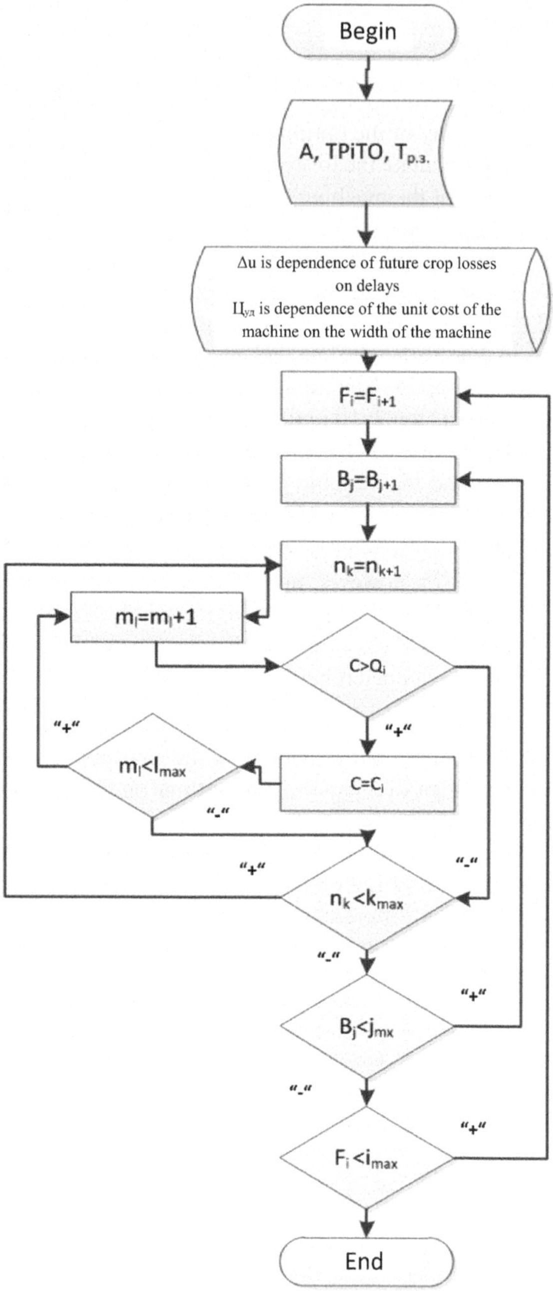

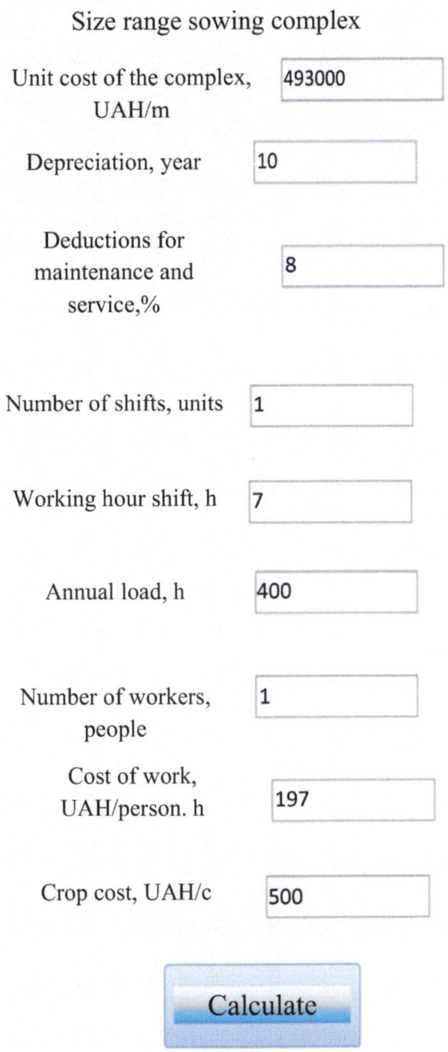


**Fig. 10** Screenshot of the program window to calculate the optimal width of the soil–cultivating complex

## 4 Conclusions

According to the results of modeling, optimal regime and energy parameters and a typical size of the fertilizing–sowing complex for different areas of farms during the growing of grain crops are determined. So far, it will be optimal for processing operations for a period of 3 and 7 days for farms with an area of cultivation of cereals 15 and 150 ha, respectively. Moreover, such requirements meet the working widths of the seizure of complexes of 0.3 and 1.2 m with a nominal power of the power means of 5 and 23 kW.

**Table 3** Rational characteristics of soil–cultivating and fertilizing–sowing complexes

| House area, $F$, ha | | Estimated working width of the seizure of the complex, $B$, m | Estimated number of units, units | Term of work, $n$, day | Mass, $t$ | Volume of bunker, $m^3$ | Speed of movement, $V$, m/s | Estimated rated power of the energy source, $N$, kW | The costs and losses are given, $C$, $10^3$·UAH/ha |
|---|---|---|---|---|---|---|---|---|---|
| 1 | | 2 | 3 | 4 | 5 | 6 | 7 | 8 | 9 |
| I | <15 | 0.3 | 1 | 3 | 0.4 | 0.28 | 2.83 | 5.0 | 3.27 |
| II | 75–150 | 1.2 | 1 | 4–7 | 1.3 | 1.10 | 3.0 | 23 | 1.51 |
| III | 300 | 2.8 | 1 | 5 | 2.9 | 2.58 | 3.63 | 76 | 1.45 |
| | 400 | 3.9 | 1 | 5 | 4.0 | 3.59 | 3.48 | 97 | 1.37 |
| | 500 | 5.0 | 1 | 5 | 5.1 | 4.60 | 3.39 | 118 | 1.34 |
| | 600–750 | 8.0 | 1–3 | 3–4 | 8.2 | 7.36 | 3.97 | 264 | 1.19 |
| IV | 1500–6000 | 9.6 | 1–5 | 5–7 | 9.8 | 8.84 | 3.79 | 266 | 0.99 |
| V | 10,000–15,000 | 10.6 | 7–8 | 7–8 | 10.9 | 9.81 | 3.73 | 266 | 0.97 |
| VI | >30,000 | 13.4 | 11 | 11 | 13.7 | 12.33 | 3.13 | 271 | 0.98 |

The optimal period of functioning of fertilizing–sowing complex for farms with an area of 300, 400, 500, and 750 ha is 5, 5, 5, and 4 days, respectively. As for the rational working width of capture, there are the calculated values (in Table 3) for 300 ha—2.8 m; 400 ha—3.9 m; 500 ha—5.0 m, and 750 ha—8.0 m due to which the maximum ecological effect will be obtained from the use of the corresponding machine and the maximum profitability for farms on the criterion of the minimum of reduced costs and losses of crop from soil compaction.

The optimal term for soil cultivating soil and sowing for farms with an area of 1500 and 15,000 ha is 5 and 8 days for a complex with a working width of 9.6 and 10.6 m, respectively, with a rated power of 266 kW.

For a farm with an area of over 30,000 ha, the optimal duration of work will be 11 days with the use of 11 units with a working width of 13.4 m and a rated power of 271 kW.

## References


1. Kryuchkov, M.M., Potapova, L.V., Lukyanova, O.V.: Application of soil-cultivating and sowing combined aggregates in the conditions of the Ryazan region. Ryazan, 158 p (2013)
2. Mikhalev, A.A., Yezhevsky, A.A., Krasnoshkekov, N.V.: On technological modernization of agricultural production in Russia. Machinery and equipment for the village, no. 4 (2005)
3. Maslov, G.G., Kheyfets, A.B.: The function of costs and losses in studies of multifunctional aggregates. Sci. J. Kuban State Univ. Agric. **115**, 51–65 (2016)
4. Agricultural systems in the Krasnodar Territory for 1990–1995 and for the period up to 2000. Recommendations, Krasnodar (1990)
5. Zhalnin, E.V., Savchenko, A.N.: Technologies of grain harvesting by combine harvesters. Rosselkhozizdat, 207 p (1985)
6. Khomenko, M.S., Zyryanov, V.A., Nasonov, V.A.: Mechanization of sowing of cereals and herbs. Directory, Harvest, 168 p (1989)
7. Sowing of corn: Crop timing [Electronic resource]. Journal access mode: http://www.agromage.com/stat_id.php?id=427-Titleofthescreen (2016)
8. Naumov, A.U.: Features of plant development and soybean yield, depending on the timing of its sowing. [Electronic resource]. Mode of access to page: http://cyberleninka.ru/article/n/osobennosti-razvitiya-rasteniy-i-urozhaynost-soi-v-zavisimosti-ot-srokov-eyo-poseva-Namefromthescreen (2016)
9. Correa, U.T.: Features of the formation of the crop and quality of the seeds of the soon-expanded varieties of soya lira depending on the time of sowing [Electronic resource] (2016). Access to the article: file: /// C: / Users / COMFY/ Downloads/ osobennosti-formirovaniya-urozhaya-i-kachestva-semyan-skorospelogo-sorta-soi-lira-v-zavisimosti-ot-sroka-poseva.pdf-Namefromthescreen
10. Ksenevich, I.P., Skotnikov, V.A., Lyasko, M.I.: Chain system-soil-harvest. Agropromizdat, 304 p (1985)
11. Products of the MTP [Electronic resource]. Access mode to page: http://www.belarus-tractor.com/catalog/-Nameofthescreen (2016)
12. Zyryanov, A.P., Pyatayev, M.V., Kuznetsov, N.A.: Reduction of impact of tractor wheels on soil. Vestnik KrasGaU (4), 223–227 (2014)
13. Mobile agricultural technology. Methods of determining of the impact of propulsion on the soil: GOST 26953-86. Introduction. 01/01/87. Publishing of standards, 1987.-11

14. Kalinin, E.I.: Estimation of the influence of the installation of dual tires on the towing of tractor engines in the composition of the arable aggregate on the agroforest of high humidity [Text]: a collection of scientific works. Kalinin, Y.I.: Visn. of Kharkov National tech Un of agriculture named after P. Vasilenko/ KHNTUA. X., 2010. Edition. 96: Resource-saving technologies, materials and equipment in a repair-out, pp. 382–391. ISBN 5-7987-0176 X. Bibliographer: pp. 390–391

# Consulting Services in Agriculture

**Nadiia Serskykh and Igor Britchenko**

## 1 Relevance of the Work

In our opinion, global processes in the modern economy, first of all affect the integration of markets, which leads to increased interconnectedness and mutual influence of national economies of different countries. This applies to almost all types of markets, including service markets. Service providers in Ukraine, both large and small, have been affected in recent years by a number of significant changes directly related to globalization: accession to the WTO, European integration processes, consumer demands, scientific and technological progress, revolutionary changes in corporate information society, rethinking approaches before doing business, developing new sales technologies, etc. In the context of these changes, the task of producers is to reduce costs, increase their own competitiveness by acquiring professional consulting services.

Analysis of recent research and publications. The growth of volumes of production and sales of services, both in the world and in Ukraine, is forcing many scholars to pay attention to the theoretical and practical aspects of the development of this market. The study of the market of services is devoted to the work of K. Antoniuk, V. Zaitseva, O. Morgulets, A. Rumyantsev, A. Starostina, M. Sagaidak, and F. Kotler.

Selection of previously unsettled parts of the general problem. The mentioned authors do not adequately cover the problems of the development of services in the field of informatization and professional advisory services in agriculture. Particular attention is also needed to the issue of outsourcing professional services, which to date are little studied.

N. Serskykh (✉)
Tavria State Agritechnological University, Melitopol 72300, Ukraine
e-mail: nadiia.serskykh@tsatu.edu.ua

I. Britchenko
Uzhgorod Trade and Economic Institute, Kyiv National University of Trade and Economics, Uzhgorod 88020, Ukraine

© Springer Nature Switzerland AG 2019 
V. Nadykto (ed.), *Modern Development Paths of Agricultural Production*,
https://doi.org/10.1007/978-3-030-14918-5_23

Setting objectives. The purpose of the article is to study and analyze the current state of the Ukrainian market of services, its structure and dynamics of development, the definition of the dominant sectors of the market of services, as well as substantiation of the possibilities of using professional advisory services in agriculture.

Presentation of the main research material. Today, there are many interpretations of the definition of "service," the marketing classic Philip Kotler stated that "a service is any measure or benefit that one party can offer to another and which is basically invisible and does not entail any control" [1]. K. McConnell and S. Bry believe that a service is something that is elusive (invisible) and in exchange for that consumer, firm, or government are ready to provide something of value [2]. In the business dictionary, the term "service" is interpreted as "a value action, an act or an effort made to meet needs or requirements" [3], K. V. Antonyuk proposes to consider the service as "a process that combines the simultaneous provision and receipt of other subjects of goods to meet their needs, mainly in intangible form and on a commercial basis" [4].

We believe that a service—a tangible or intangible type of non-stored economic activity—is not assigned and is usually consumed at a point of sale.

## 2 Research and Discussion of the Results

The service sector is a complex mechanism on which the economic development of the countries of the world depends—in developed countries services occupy about 70% of GDP. The branches of this sphere include transport, food, tourism, education, health care, telecommunications and communications, consulting, etc. Along with the traditional branches of the services market, new ones connected with the development of telecommunication networks, banking, political reforms, informatization, and globalization are emerging and developing.

The growth of the services sector is often associated with the growth of economic welfare within the country, and with the beginning of the twenty-first century, when the development of the Internet allowed to "erase" the borders between countries, this market in industrialized countries reached almost gigantic proportions.

In Ukraine, the services market in Ukraine has been rapidly developing. In 2016, the volume of goods sold through retail networks amounted to almost 556 billion hryvnias (of which 58% are goods produced in Ukraine). These figures exceed the corresponding sales volumes in 2010 by 69 and 53%, respectively. In the analyzed period, from 2010 to 2015, the share of services in the volume of goods sold amounted to 82–88%. If we compare the volumes of sales of services with the volumes of sales of goods manufactured in Ukraine, then in the above-mentioned period, they were higher by almost one and a half times. But in 2015 the situation on the market changed significantly, volumes of sales of services decreased by 65% compared to 2014, their share amounted to 123.3 billion hryvnias against 360.6 billion hryvnias (Fig. 1). In our opinion, this is due to the sharp reduction in the cost of the national currency, the

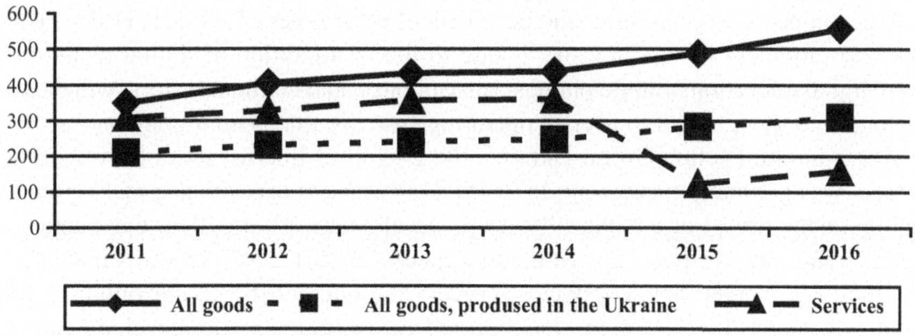


**Fig. 1** Sales volumes of goods and services in Ukraine, bln. UAH

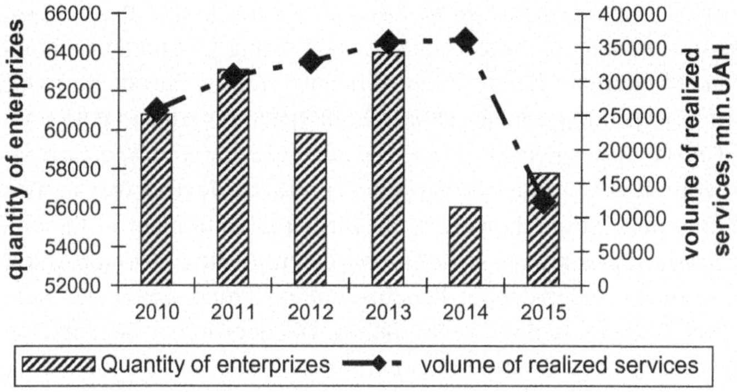


**Fig. 2** Dynamics of realization of services by Ukrainian enterprises

unstable political situation, the interruption of transport links with the Autonomous Republic of Crimea and military actions in the East [5].

Gradually, beginning in 2016, sales volumes of all types of services began to increase. Thus, according to the State Statistics Service of Ukraine, in 2017, the volume of sales of all types of non-financial services grew by UAH 23.4 million in comparison with 2016, and by 58.5 million UAH in comparison with 2016 year. In percentage terms, it was 1.14 and 1.5%, respectively.

In the period from 2010 to 2015, the number of enterprises of legal entities in the sphere of services in Ukraine decreased by 3454 units, the maximum number of enterprises reporting to the statistical agencies was recorded in 2013-63981 companies (Fig. 2).

In the structure of the Ukrainian market of services, the leading place is occupied by transport services and communications services—about 50%; transport services include: activity of railway, pipelines, water, aviation, and other ground transport; services for organization of trips, organization of cargo transportation, postal and courier services; warehousing and auxiliary transport activities. The second place in the volume of services rendered since 2013 includes information and telecommunication services: production of films, television and radio (12%); telecommunications

(69%); computer programming and provision of other types of services (19%). The rapid development of this industry is due to the proliferation of digital technologies: digital television, mobile phones, smartphones, and communicators, which are personalized devices with powerful operating systems, that is mini-computers.

According to Factum Group Ukraine, the growth of Internet users from mobile devices was gradually increasing. In 2015, 31% of users over the age of 15 used a mobile phone (26%) or a tablet (9%) to access the network. In 2016, these figures were 25 and 10%, respectively. In the first quarter of 2017, 57% of users indicated that they were using mobile devices to access the Internet [6]. In our opinion, the rapid growth of "mobile" users has become possible due to the development of communication technologies and an increase in the coverage area of the network in Ukraine.

The tendency to increase the volume of sales of services in the field of informatization remains unchanged. Since January, the leading Ukrainian mobile operators "Kyivstar" and "Vodafone" have redeemed licenses for the introduction of the fourth generation of mobile communication with high-speed broadband data transmission, the so-called 4G (fourth generation). The use of new technologies, in turn, should lead to increased volumes of production and sales of mobile devices that support the new standard of communication, as well as the further development of the e-commerce market in Ukraine. As a result of the improvement of the informatization process, commodity producers receive many benefits, among which: better and more efficient use of fixed assets, increase the level of managerial decisions made, and increase the economic efficiency of production in general. Gradually, the volume of sales of services in the field of professional, scientific and technical activities is increasing: technical and scientific research and development, consulting, engineering services, advertising services, etc.

Experts point out that in the context of globalization of the economy and the development of e-commerce, an important role will be played by outsourcing, as a way of transferring the secondary functions of enterprises to specialized companies. According to the Cambridge University Business Dictionary, outsourcing is a situation where a company uses a different organization to carry out its work and not its own employees [7]. Currently, many enterprises in Ukraine use the services of professionals in IT sphere, marketing, audit, finance, etc. Outsourcing allows companies to reduce their training costs and concentrate on key business functions. A wide popularity of outsourcing is used by agricultural enterprises. This is due to the low level of informatization in the village, lack of skilled personnel, etc. The main sources of information for modern producers are internal data: results of activity in previous years, production and sales volumes, cost structure, income level, past sales prices, etc. As for information from external sources, then, as a rule, only the secondary data from desk surveys, that is, those that manufacturers can get free of charge, reach entrepreneurs. Of course, there is not enough information to effectively manage the production and marketing activities of such information. Manufacturers need the freshest data from external sources, information from which is often inaccessible. The level of informatization of agro-industrial production depends directly on the level of development of the system of advisory services in the country [8].

Especially for meeting the needs of information resources of small commodity producers. In the process of activity, advisory services should ensure the provision of timely, constantly updated information and meet the requirements and needs of business entities. An objective need is to study demand and supply, to analyze markets for sales and place of sales, to carry out relevant marketing research, and so on. The main functions of the advisory services are:

- Integration, which consists in the introduction of scientifically substantiated research results directly in the production. This function is carried out at the expense of the development of databases of higher educational establishments, special research, and production units or laboratories, which can also provide services for the provision of seed material, offer up-to-date information of accounting and analytical character, develop advertising and effective promotional measures. Also, in the above-mentioned subdivisions there may be the implementation on a contractual basis of new business projects, provision of services for the research of agrarian markets with the subsequent help in optimizing the structure of production, etc.;
- The advisory function is directly related to the integration, that is, exclusively from all the results of scientific developments, experts can advise the counselors at each stage of the production and marketing of agricultural products;
- Information—is, first of all, in the formation of an effective information environment and promotion of scientific and technological progress in agrarian production. The implementation of this function is due to the obligatory carrying out of the following measures: holding of field days, seminars, and schools for demonstration and development of new technologies, exhibitions, and conferences with the distribution of analyzes and forecasts of conditions of production and marketing of agricultural products, mandatory submission of operational information on prices fluctuations, market conditions in the form of headlines in the media, price lists or postal materials.

Obligatory in the activity of the advisory service is the establishment and maintenance of feedback between production structures and scientific and educational institutions through the development of plans for conducting research works and dissemination of their effective results.

Due to the fact that advisory services have a huge amount of scientific and production information, the most important role for its systematization and distribution belongs to computer technology. The development and dissemination of the World Wide Web help to bring the necessary knowledge and materials to the addressee in full and in the shortest possible time. Consequently, the process of computerization of agricultural enterprises is one of the problems that need immediate resolution, because it is precisely at the expense of the remoteness of some industries from research centers, the lack of modern technology makes it almost impossible to obtain the most up-to-date information by manufacturers in a timely manner.

Professional services that will provide advisory services to agricultural producers will help:

- To identify and formulate problems, analyze them and evaluate the options for their solution;
- To increase motivation for independent acceptance and implementation of management decisions, evaluation of their results, etc..;
- To analyze and apply the latest developments and researches at all stages of production and management processes;
- Receive consultations on effective planning and management of the economy, financial analysis, optimization of the use of resources;
- Provide feedback to public administration and research institutions;
- To integrate knowledge from a variety of sources, including international ones, to conduct applied research;
- To make possible changes in the conduct of economic activity and production process;
- To disseminate innovations among producers of a certain region.

## 3 Conclusions

Dynamic development of the services market in Ukraine was observed until 2014, but with the start of hostilities in the Donbass and the occupation of the Autonomous Republic of Crimea, it slowed somewhat. The number of enterprises in all industries has decreased; however, since 2016 the volume of sales of services is gradually increasing. In the structure of the market of services, the leading sectors of transport, informatization, and communications. Gradually, the volume of sales of professional services, including information and consulting (advisory) services, in particular, in agriculture, is growing. The main problems of the development of the field of consulting services in agriculture are the low level of awareness of commodity producers and the lack of computerization of rural areas.

## References


1. Kotler, F.: Osnovy marketinga (1994)
2. Makkonnell, K.R., Bryu, S.L.: Ekonomiks: principy, problemy i politika (1993)
3. What is service? Definition and meaning http://www.businessdictionary.com/definition/service.html
4. Antonyuk, K.V.: Viznachalni faktori rozvitku suchasnoyi mizhnarodnoyi torgivli poslugami pokupcyami (2015)
5. Diyalnist pidpriyemstv sferi poslug [Elektronnij resurs]. Statistichnij byuleten. 2010–2017
6. Doslidzhennya internet-auditoriyi—[Elektronnij resurs]—Rezhim dostupu: http://inau.ua/proekty/doslidzhennya-internet-audytoriyi
7. Meaning of "outsourcing" in the English Dictionary. https://dictionary.cambridge.org/dictionary/english/outsourcing
8. Tarabrin, O.: Informacijno-analitichna sistema ekonomichnogo monitoringu gospodaryuyuchih sub'yektiv APK Ukrayini (2008)

# Development of Technology for the Hemp Stalks Preparation

**Viktor Sheichenko, Igor Marynchenko, Vitaliy Shevchuk and Natalia Zadosnaia**

## 1 Introduction

In modern production, industrial hemp is considered to be a highly profitable crop, provided by the absence of waste from its processing. All components of the plant are seeds, fiber (long, short), hemp (a by-product of fiber production) have a wide range of applications in many industries and national economy. Due to the unique natural properties of hemp, scientists constantly find new areas of non-traditional use for the plant that makes it attractive for production.

A wide range of uses (textile and food industries, medicine, energy, auto and aircraft construction, construction, etc.)—convincingly demonstrates the promise of using technical hemp.

In the process of hemp stalks preparation, the main parameters are formed that determine the qualitative and quantitative characteristics of the fiber. The technology of hemp stalks preparation is a sequence of complex biological and technological processes that begin with the wetting of hemp straw and continue with a number of

V. Sheichenko (✉)
Department of Machinery Repair and Construction Materials Technology, Poltava State Agrarian Academy, Skovoroda Str., 1/3, Poltava 36003, Ukraine
e-mail: vsheychenko@ukr.net

I. Marynchenko
Institute of Bast Crops, National Academy of Agrarian Sciences of Ukraine, Tereshchenkiv Street, 45, Hlukhiv, Sumy Region 41400, Ukraine

V. Shevchuk
Department of Processes, Machines and Equipment for Agricultural Production, Uman National University of Horticulture, Institutskaya Str., 1, Uman, Cherkassy Region 20300, Ukraine

N. Zadosnaia
Tavria State Agrotechnological University, B. Khmelnitsky ave, 18, Melitopol, Zaporizhia Region 72310, Ukraine

© Springer Nature Switzerland AG 2019

V. Nadykto (ed.), *Modern Development Paths of Agricultural Production*,
https://doi.org/10.1007/978-3-030-14918-5_24

technological operations: drying, stirring, baling the finished hemp stalks into rolls, and transporting it to processing plants.

The main factors that determine the output of fiber during the hemp stalks preparation include the uneven lying of the stalks along the length and thickness of the tape. This indicator, in turn, depends on the uniformity of the layer of hemp in ribbons. A significant influence on the quality of the fibers has lengthiness of stems, the timing of spreading and preparing of hemp stalks.

The main imperfections of the known methods of obtaining stalks of technical hemp include the dependence of the process of forming its quality indicators on weather conditions; the unevenness of the cross-linking and the strength of the fiber along the entire mass of the dispatched layer; a long duration of the technological process, which leads to a quantitative loss of production.

A resource-saving technology for obtaining hemp stalks has been developed, which allows optimizing the process of wetting at the initial stages of preparation for primary processing and intensifies the cleaning process of the stalks due to additional pressing with the help of improved mechanical influences. Due to this, there was a significant reduction in the time spent on obtaining and collecting trusts, the efficiency of the stages of its primary processing.

## *1.1 Analysis of Recent Studies and Publications*

One of the important processes of primary processing of hemp stems, which further affects the quality of the final product (fiber), is the process of preparing stalks. By mechanisms of destruction of bonds between fibers and tissues in the stem, the existing methods of preparing trusts can be divided into three groups: biological, chemical, and physicochemical [1–3].

By a biological method, the breaking of bonds between fiber and wood in the stem is achieved through the use of the vital activity of various microorganisms, in particular bacteria and mold fungi.

To biological method of hemp stalks preparation, it is possible to carry an aqueous watering (immersion of stalks in the water environment). Another method of biological wetting is dewed wetting, when the hemp stems, after collecting them, are spread on the field by a thin layer for the preparation of the stalk. Unlike water, where bacteria are the active substances, mold stimuli are the main pathogens in the case of dewed wetting. In the process of their vital activity, fungal spores produce substances (enzymes) that cause destruction of pectin substances and disrupt the bonds between tissues [4–6].

Technology of dewed wetting can be used in winter and spring. Depending on the climatic zone, the process of dewed wetting during the autumn period can last 25–40 days or more [7, 8].

The main disadvantage of this method of hemp stalks preparation is a large dependence on weather conditions. In the case of unfavorable autumn weather (no precipitation, or vice versa, a large number, a decrease in the average daily temperature), the process of stalks preparing is delayed, as a result of which the stalks do not have time to prepare for the appearance of the first snow.

Moisturizing with dew can be attributed to the method of preparing the stems, when the stems after cutting the seeds in the fall are left to stand on the root until spring. With this method, the mechanism of tissue destruction is the same as when the stems spread to the field [9–12].

The disadvantage of this method of hemp stalks preparation is that the field is not released in the autumn, and in the spring, the finished hemp stalks should be collected in a short time, which requires the use of additional equipment.

In the chemical method of hemp stalks preparation, the breaking of bonds between fiber and wood in the stem is achieved through the use of chemical preparations in which the decomposition of pectin and other substances into simpler and, ultimately, their destruction [7, 8]. However, the use of this method of hemp stalks preparation requires additional costs (financial, human, hourly), because the wetting liquid cannot be used many times.

With a physicochemical method, the breaking of bonds between fiber and wood in the stem is achieved through thermochemical hydrolysis when using steam and high temperature [13–16].

The best technological and operational properties are fiber obtained from stalks, prepared by water treatment and by steaming. However, as already noted above, the hemp stalks preparation by these methods provides for additional financial expenses and a large expenditure of manual labor. Therefore, the alternative is dewed wetting.

Note that each of the known methods of hemp stalks preparation has its own characteristics, advantages, and disadvantages. A simpler and cheaper method of dewed wetting, since it does not require the use of the additional preparation, eliminates the need for construction of treatment plants, and the number of necessary workers can be reduced to a minimum [17, 18].

Thus, we note that arose the problem of the need for the industry to switch to new energy and resource-saving technologies for the integrated preparation and processing of hemp raw materials, with the production of an undirected short fiber with specified final quality characteristics on its basis.

This will significantly expand the scope of application of hemp fibers in various sectors of the national economy of Ukraine and beyond. This will significantly expand the scope of application of hemp fibers in various sectors of the national economy of Ukraine and beyond. Therefore, the creation of new resource-saving technologies and equipment that would solve the problem of optimizing the hemp stalks preparation is an urgent task.

## 1.2 Statement of the Objective and Tasks of the Study

The objective of the study is to increase the efficiency of the hemp industry by developing a new resource-saving technology for the production of hemp stalks.

To achieve the set objective, the following tasks were solved:

- Summarize the information on existing technological schemes, machines and mechanisms for preparing the tape of stems for collection and further processing and develop a resource-saving technology for obtaining hemp stalks;
- Develop a technological scheme and perform a kinematics calculation of the working units of equipment for mechanical preparation of hemp stalks at separate stages;
- To carry out approbation of resource-saving technology for obtaining trusts on experimental equipment in the field;
- To investigate the quality of the fiber isolated after the processing of trusts obtained for resource-saving technology.

## 2 The Basic Part of the Study

Optimize the complex processes of the assembly cycle, and especially the process of hemp stalks preparation, is possible by combining them (combination). Due to this, it becomes possible to improve biological processes and to save significant production resources. The most promising scientific direction of the radical solution of the problem of increasing the efficiency of technological processes for the hemp stalks preparation is the transition to methods that allow the intensification of the cleaning process of the stalks due to the additional action by improved corrugated water-filled roller and finger wheels. This, in aggregate, provides optimal conditions for obtaining final products with predicted indicators.

According to the working process of the aggregate (Fig. 1), the windrow 4, previously formed by rotary rakes, is subjected to an additional narrowing by the finger wheel hay rakes 1. After this, depending on the raw material, are performed the rolling operation by a set of grooved rollers 2 (formed from one, two or three modernized rollers of different corrugation) and loosening operation with the help of finger wheels 3.

In the technological process of forming a tape of hemp stems, it is proposed to use the following technique: a modernized corrugated water-filled roller KVG-1.4, hinged finger wheel hay rakes GVK-6, and round baler PRP-1,6 for picking up of the received mass.

In the process of modernization, the surface of the KVG-1,4 water-filled roller was equipped with 12-angle corrugations in 185.3 mm increments (Fig. 2).

The most common round baler PRP-1,6 in the farms can ensure stable performance of the technological process if the mass of stalks in the windrow does not exceed 2.5–2.7 kg per meter and the width of it is 1.3 m, and stems have a certain elasticity. Such windrow parameters can be provided only if the number of stems does not exceed 850–900 thousand pieces per ha. This can be when harvesting stems from seed (wide-row) crops, and the width of the field with which rake is performed in one pass does not exceed 6 m.

On continuous sowing, the density of stalk can reach values of 2000–2200 thousand pieces per ha. That is, the baler of the aforementioned brand at the harvesting of continuous sowing will not always be able to ensure a stable process, and the

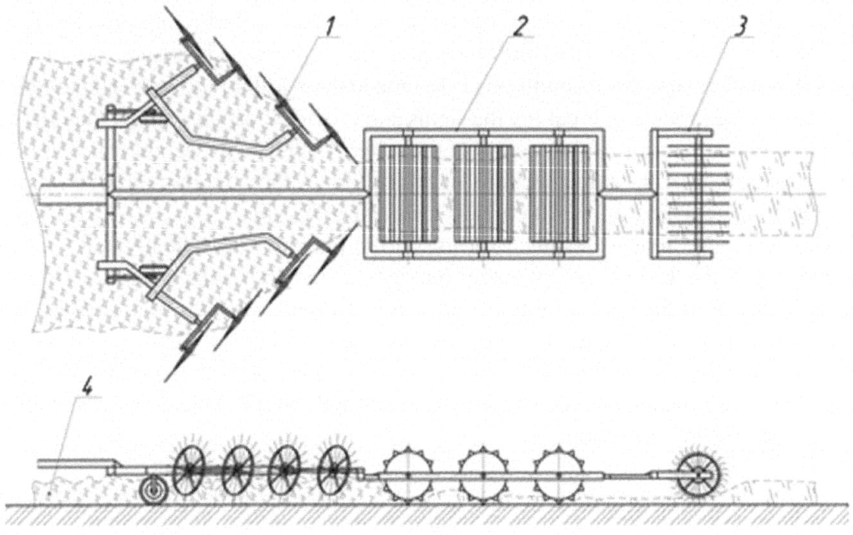


**Fig. 1** Technological scheme of the unit for the formation of tape and the hemp stalks preparation

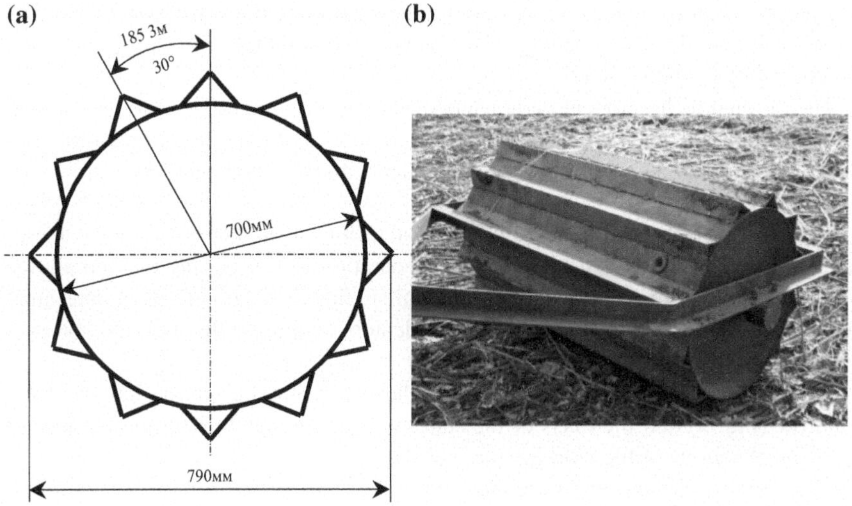


**Fig. 2** Modernized water-filled roller KVG-1,4: **a**—the scheme of placing of riffles on a surface of a roller; **b**—roller assembly

hemp stalks harvesting technology must also be used to collect hemp crops with a stalk density of more than 850–900 thousand pieces per ha. Therefore, one of the ways that will ensure these conditions is to reduce the width of the roll, increase the intensity of the process of breaking the stems and provide elasticity of raw materials. After the modernization of the section of the roller KVG-1.4, its weight was 350 kg. The total weight of the three sections of the rollers in the coupling is 1050 kg, in contrast to the existing ones, which have a weight of 977 kg.

The developed unit is designed for use in the preparation of hemp stalks tape. According to the technology of hemp harvesting, the following operations were used: threshing of the tops, cutting off the stems, forming a windrow, preparing the hemp stalks by laying it under the snow in winter and further rolling the hemp stalks in rolls. It should be noted that stems of hemp, in comparison with the stems of other crops, have significantly greater rigidity and strength, so the balers can not stably perform the rolling operation.

## 3 Results and Discussion

The proposed unit provides the execution of operations raking stems with reduced width of the windrow and rolling of the stems with simultaneous tearing them off the ground by rippled rollers.

The rational technological parameters of the experimental unit were determined in order to find the most effective technological modes of operation of the proposed unit, depending on the input raw materials factors and the final qualitative indices of short hemp fibers.

Based on the results of the research, rational technical characteristics of the aggregate for tape formation and hemp stalks preparation in roll coiling (shown above) were established and additional experimental studies were conducted to determine the effect of machining on the qualitative indicators of hemp stalks and bast obtained by the new resource-saving technology.

Figures 3 and 4 show the results of studies of the change in the height of the windrow (Fig. 3) and the mass of the stalks (Fig. 4), depending on the number of mechanical actions of the working parts of the unit.

According to the results of the analysis of the studies shown in Figs. 3 and 4, it is established that an increase in the number of mechanical influences on the hemp stems tape (onefold, twofold, and threefold rolling) leads to a corresponding decrease in its height and mass. So with a single rolling, the height was 29.0 cm at a weight of 3.4 kg per linear meter. When double rolling, these values decreased to 19.0 cm and 2.9 kg per linear meter, respectively. And with triple rolling, they were 17.0 cm and 2.4 kg per linear meter. The loosening process reduces the weight of the tape to 2.1 kg per linear meter, but under such conditions the height of the roll increases to 35.0 cm.

Dynamics of changes in the number of broken stems and the number of breaks in the hemp stalk depending on the variant of the unit action are shown in Fig. 5.

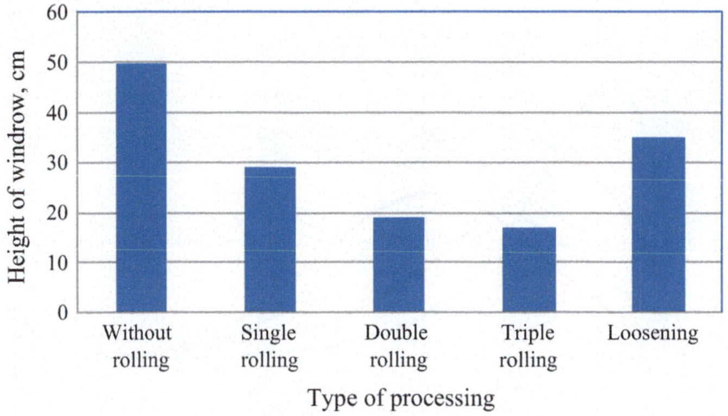


**Fig. 3** Dynamics of the change in the height of the windrow, depending on the number of mechanical actions of the working parts of the unit

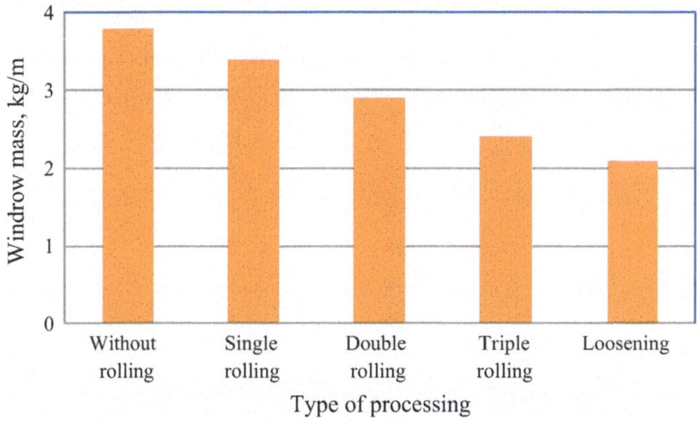


**Fig. 4** Dynamics of changes in the mass of stems, depending on the number of mechanical actions of the working parts of the unit

Analysis of graphical dependencies (Fig. 5) shows that as the mechanical actions increase, damage to hemp stems grows.

It is noted that the parameters of the hemp stalk tape are within the normative values for the further formation of rolls. However, with an increase in the number of mechanical actions the damage to the stems also increases, so it is necessary to investigate these processes in more detail.

For this purpose, the influence of the operations of mechanical preparation of the hemp stalks tape on the physical and geometric characteristics of the tape and the change in the qualitative indices of the short hemp fiber obtained were studied (Tables 1 and 2).

The qualitative characteristics of the bast obtained were evaluated according to the procedure given in Sects. 2.3.3, 2.3.4 [18, 19]. Comparative analysis of the data

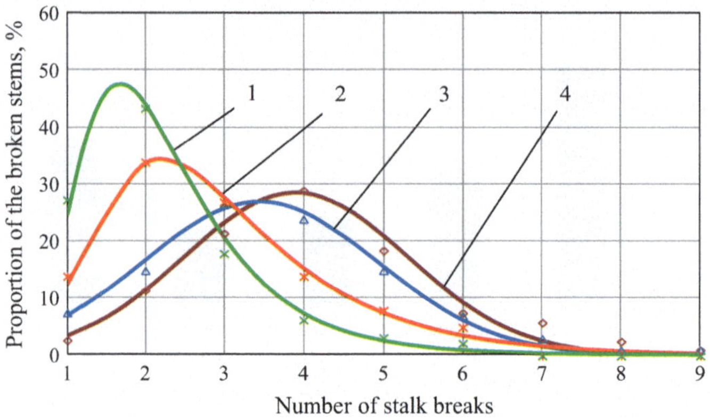


**Fig. 5** Dependence of the change in the number of breaks of stems depending on the action of the working parts of the unit: 1—without rolling; 2—single rolling; 3—double rolling; 4—triple rolling

**Table 1** Influence of preparatory operations on changing the hemp stalks tape characteristics

| Type of processing | Height of windrow | | Windrow mass | |
|---|---|---|---|---|
| | cm | % | kg per meter | % |
| Without rolling | 50.0 | 100 | 3.8 | 100 |
| Single rolling | 29.0 | 58 | 3.4 | 89.5 |
| Double rolling | 19.0 | 38 | 2.9 | 76.3 |
| Triple rolling | 17.0 | 34 | 2.4 | 63.2 |
| Loosening | 35.0 | 70 | 2.1 | 55.3 |

**Table 2** Influence of preparatory operations on the quality of short hemp

| Type of processing | Qualitative indicators of the hemp | | | |
|---|---|---|---|---|
| | Mass fraction of awn (%) | Mass fraction of lublike strands (%) | Breaking load (kg) | Deviation (kg) |
| Without rolling | 3.9 | 1.5 | 39.8 | ±7.86 |
| Single rolling | 4.9 | 0.8 | 39.0 | ±7.98 |
| Double rolling | 6.3 | 0.5 | 40.5 | ±8.29 |
| Triple rolling | 2.1 | 0.3 | 39.5 | ±5.64 |

presented in Table 2 with the existing standards allows us to conclude that a short fiber obtained from the hemp stalks through a new resource-saving technology fully meets the existing requirements.

## 4 Conclusions

A resource-saving technology for obtaining the hemp stalks has been developed based on the results of the studies. The technology is carried out due to the developed unit, which includes the modernized corrugated water-filled roller KVG-1.4, the finger wheel hay rakes GVK-6, and the round baler PRP-1,6.

The basic design features of the unit and the following indicators are established. The speed of movement of the reef when entering the stalks can be deviated from the perpendicular to the surface of the stem by an angle, the value of which varies between 15° and 18°. Due to the modernization of the section of the rollerKVG-1.4, the total weight of the three sections of the rollers in the coupling is increased to 1050 kg, which is 11% more of the original ones.

An increase in the number of mechanical influences on the hemp stems tape leads to a corresponding decrease in its height and mass. So with a single rolling, the height was 29.0 cm at a weight of 3.4 kg per linear meter. When double rolling, these values decreased to 19.0 cm and 2.9 kg per linear meter, respectively. And with triple rolling, they were 17 0 cm and 2.4 kg per linear meter. The loosening process reduces the weight of the tape to 2.1 kg per linear meter, but under such conditions, the height of the roll increases to 35.0 cm.

It has been established that a short fiber obtained from hemp stalks through a new resource-saving technology fully meets the existing requirements of normative and technical documentation.

## References

1. Migal, M.D.: Biologiya lubyanyh volokon konopel. Papirus, Sumy (2011)
2. Goloborodko, P.A., Sytnyk, V.P., Barannyk, V.G.: Selektsiya, tehnologiya vyrobnytstva ta pervynnoyi pererobky lonu i konopel, pp. 3–15 (2000)
3. Vyrovec, V.G., Sytnyk, V.P., Orlov, N.M.: Selektsiya, tehnologiya vyrobnytstva ta pervynnoyi pererobky lonu i konopel, pp. 25–33 (2000)
4. Chursina, L.A., Klevcov, K.N., Reshetej, A.A., Ageev, S.M.: Sovremennyj uroven ispolzovanija othodov na lno- i penkozavodah. UkrINTJeI, Kyiv (1993)
5. Zhuplatova, L.M.: Biologiya, vyroschuvannya, zbyrannya ta pervynna pererobka lonu i konopel, vol. 4, pp. 83–94 (2007)
6. Zhuplatova, L.M.: Problemy i perspektyvi v selektsiyi, genetytsi, tehnologiyi zbirannya, pererobtsi ta standartizatsiyi lubyanyh kultur, pp. 124–129 (2006)
7. Sheychenko, V.O., Marinchenko, I.O.: Silskogospodarski mashyny **29–30**, 168–174 (2014)
8. Venturi, P., Amaducci, S., Amaducci, M.T., Venturi, G.: J. Nat. Fibers 83–97 (2007)
9. Kolodinsky, J.: Natural fibers. Special Edition, pp. 58–63 (1997)

10. Gilyazetdinov, R.N., Barannik, V.G.: Selektsiya, tehnologiya vyrobnytstva ta pervynnoyi pererobky lonu i konopel, pp. 136–140 (2000)
11. Makaev, V.I., Lukyanenko, P.V., Gilyazetdinov, R.N.: Silskogospodarski mashyny **14**, 130–137 (2006)
12. Koropchenko, S.P., Gilyazetdinov, R.N., Lukyanenko, P.V.: Legka promislovist **2**, 46–47 (2006)
13. Gilyazetdinov, R.N.: Rozvytok naukovyh osnov stvorennya innovatsiynih tehnologiy pervinnoyi pererobki lubyanih kultur. Thesis by Ph.D. in Engineering, Herson (2009)
14. Hajlis, G.A., Kovalev, M.M.: Issledovanie selskohozjajstvennoj tehniki i obrabotka opytnyh danyh. Kolos, Moscow (1994)
15. Koropchenko, S.P.: Problemy legkoj i tekstilnoj promyshlennosti Ukrainy **1**(8), 313–315 (2004)
16. Wasko, J., Kozlovwski, R., Markovski, J.: Euroflax **1**, 6–10 (2004)
17. Ehrensing, D.T.: Feasibility of industrial hemp production in the United States Pacific Northwest (1998)
18. Penka korotkaja, Izdatel'stvo standartov, Moscow (1974)
19. Penka trepanaja, Izdatel'stvo standartov, Moscow (1976)

# Theoretical Studies of Stable Exploitation Conditions of a Three-Wheeled Tractor on the Field Slopes

**Viktor Sheichenko, Gedal Hailis, Igor Dudnikov and Tetiana Chorna**

## 1 Introduction

A known condition for ensuring the stability of an object on a support surface is the presence of at least three points of the support, and they are not located on a single straight line. Therefore, the stability of the machine on two wheels cannot be ensured. It is necessary to have a third reference point. For an agricultural implement, this support may be a tractor trailer. In the absence of a tractor, a rolling or dropping of the gun is observed. That is, you need three pillars for this machine. The article considers two layouts of a three-wheeled tractor. In the first scheme, one controllable wheel is presented in front and two behind. The second scheme features two wheels in front and one steering wheel in the rear.

### *1.1 Analysis of Recent Studies and Publications*

The work of tractors and agricultural machines depends on their design, the size of the fields where they operate, and the incline of these fields. In this regard, it is important to investigate the effect of the field slope angle to the horizon on the tractor stability during operation.

The paper [1] presents the results of a study of the stability of a four-wheel tractor, which is located on a field with a slope in either direction. The conditions of non-

V. Sheichenko · I. Dudnikov
Poltava State Agrarian Academy, Poltava, Ukraine

G. Hailis
Uman National University of Horticulture, Uman, Ukraine

T. Chorna (✉)
Tavria State Agrotechnological University, Melitopol, Ukraine
e-mail: tetiana.chorna@tsatu.edu.ua

© Springer Nature Switzerland AG 2019 
V. Nadykto (ed.), *Modern Development Paths of Agricultural Production*,
https://doi.org/10.1007/978-3-030-14918-5_25

displacement of the front and rear wheels of the tractor in the direction of the slope have been established. According to these conditions, it is necessary that the sum of the maximum frictional forces of both resting pairs of tractor wheels on the ground was greater than the force of the weight that falls on the axle.

The theoretical substantiation of the condition of transverse stability of a combine harvester during operation on slopes is given in [2]. According to the results of the research, the analysis of the position of the static and dynamic stability of the combine on transverse slopes has been carried out and the stability conditions for combines have been established both with and without the frame leveling system. The obtained dependencies are recommended to be used in the design of self-propelled bunker combine harvesters in order to determine its parameters under steady-state conditions on the slopes.

Numerous investigations have been made concerning the work of agricultural machines on the field slopes [1, 3–8], but the issue of a tricycle tractor stability on the slope fields is not sufficiently developed.

### *1.2 Statement of the Objective and Tasks of the Study*

The objective of the research is to increase the efficiency of machines functioning by establishing the conditions under which their steady work on the field slopes is ensured.

To achieve this goal, the following tasks were solved:

- to conduct theoretical studies of tricycle tractors under conditions of their operation on the slopes of fields and to determine the influence of the slope angle of the field on the conditions of stable operation of the machine;
- to establish the theoretical dependence of the tractor's incline on the parameters (distance between the middle of the rear wheels, the width of the wheel rim, the height of the tractor gravity center).

## 2 The Basic Part of the Study

Let's consider the work of a tractor that rests on three wheels under conditions when one of them is a front wheel, and two are the rear wheels. Scheme of this machine is shown in Fig. 1. The front wheel of this tractor is a steering wheel.

This tractor has such a peculiarity. Its weight, which falls on the front wheel, should be heavy enough to ensure the pressure of this wheel against the ground when the machine is moving on the road with a slope upward. Under these conditions, due to the transfer of tractor weight to the rear wheels, the load falls on the front wheel.

Under such a location of the tractor wheels, shown in Fig. 1, it can be assumed that the *ABDE* rectangular quadrilateral is located below the tractor, which includes

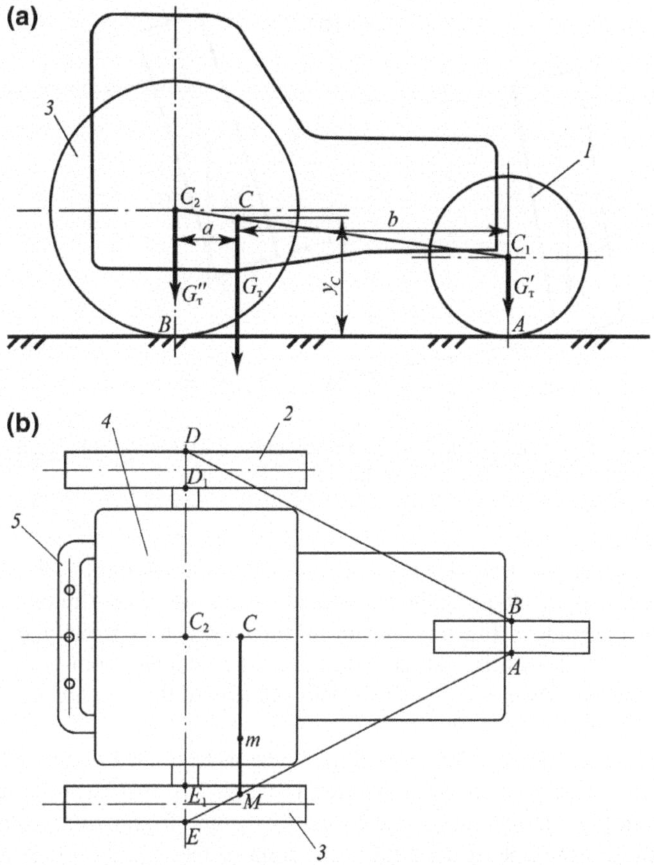


**Fig. 1** Scheme of the three-wheel tractor: 1—front wheel (steering wheel); 2—left rear wheel; 3—right rear wheel; 4—cabin; 5—hitch; A, B, D, $D_1$, E, and $E_1$ are the lower supporting points of the wheel rims

the lower support line *AB* of the front wheel and the lower lines $DD_1$ and $EE_1$ of the rear wheels (on the straight line *DE*).

Inside of this quadrilateral under the condition of a stable condition of a tractor there is a trace of its gravity center *C*, that is, the point of intersection of the vertical axis, passing from the tractor gravity center *C* down to the intersection with the horizontal surface of the ground. If the gravity center *C* is located on the ground inside the quadrilateral *ABDE*, the tractor is in a stable condition; if the trace does not lie inside the quadrilateral, it means that there is no stability.

In Fig. 1, it is shown that the weight of the tractor front element, which is indicated as $G'_T$, is transmitted to the front wheel. Under the machine operation on the slope of the field, the tractor wheels will turn, and the tractor will also turn. The diagrams of the tractor front wheel rotation on the slope are shown in Fig. 2. If the field is

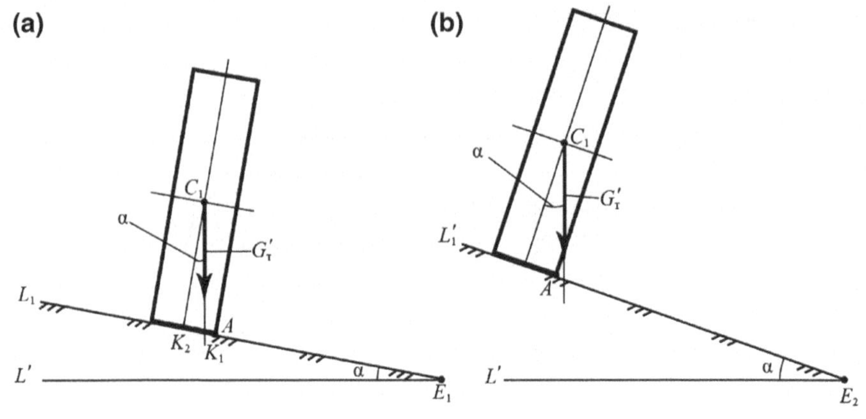


**Fig. 2** Scheme of the axle load of the sole front wheel of the tractor by $G_T'$ force on the field slope at a small angle $\alpha$ (**a**) and at a large angle $\alpha$ (**b**) of the field incline

inclined horizontally to a small angle $\alpha$ (Fig. 2a), then the force $G_T'$ will be deviating from the median plane dividing the cylindrical part of the wheel rim into two equal parts, also on the angle $\alpha$ (Fig. 2a). The line of force $G_T'$ action begins at the point $C_1$ and acts vertically downwards to the left of the extreme lower point $A$ of the wheel rim; this means that the force $G_T'$ action does not lead to the forward wheel shift to the right.

If the angle $\alpha$ of the field deviation $E_2L_1'$ upwards from the horizontal line $E_2L_1'$ is larger than the angle $\alpha$ in the previous case (Fig. 2a), that is, if the angle $\alpha$ is the same as in Fig. 2b, then the vertical force $G_T'$, going from point $C_1$ downwards (Fig. 2b), crosses the line of the field to the right of the extreme lower point $A$ of the wheel rim. In this case, it would be possible to roll over the point $A$ of the right wheel, but this is impossible due to the fact that the wheel is connected to the other part of the tractor, which has not yet been brought to such a position as to roll over (for the rollover of the tractor it is necessary, as already being noted, that the trace of the center $C$ of its weight on the ground appeared outside the quadrilateral $ABDE$ in Fig. 1).

Under the circumstances where the force $G_T'$ of a point $A$ is safe, it does not mean that such a force action is always safe; it is advisable to check this state of force $G_T'$ taking into account the above recommendations. Thus, working with such a tractor on the field slopes is better than the scheme in Fig. 2, as shows. It is even better if the force $G_T'$ acts on the $C_1K_1$ or $C_1K_2$ line on the same Fig. 2a. The angle $\alpha$ of a field slope incline, under which the vertical line of the force $G_T'$ action (Fig. 2) passes through point $A$, is called the limiting angle of the wheel $\alpha_{Kcr}$; it is determined by following dependence:

$$\alpha_{Kcr} = \mathrm{arc}tg \frac{b}{2 \cdot r_K} \qquad (1)$$

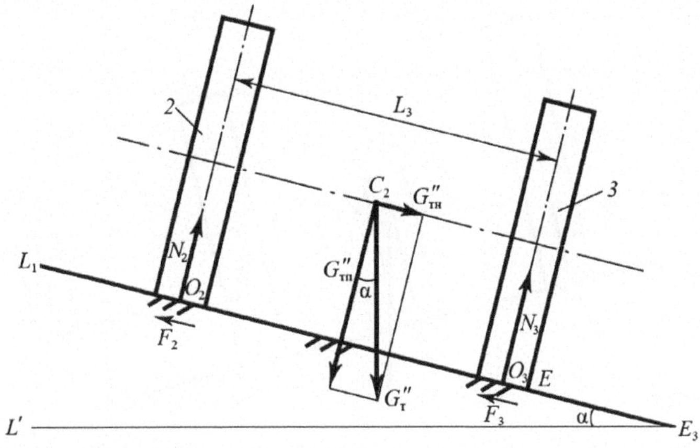


**Fig. 3** Rear view of a turned three-wheel tractor and forces acting on a transversal vertical plane where the center $C_2$ is located

where $b$ is the width of the wheel rim, $r_K$ is the wheel radius.

The $G_{\mathrm{T}}$ force of tractor weight can be divided into two components parallel to it, namely $G'_{\mathrm{T}}$ and $G''_{\mathrm{T}}$ (Fig. 1a). The force $G'_{\mathrm{T}}$ has already been mentioned above. The force $G''_{\mathrm{T}}$ is applied at the point $C_2$, representing the projection of the $C_2$ rear wheels axis on a vertical plane passing through the center of gravity $C_2$ of the rear element of the tractor. Let's assume that if the position of the machine changes, the position of the center of gravity $C_2$ does not change, that is, we assume that the moving of materials inside the machine during its movement does not occur, and the distance from the weight center $C_2$ to its reference plane and the wheels does not change. At such a turn of the tractor on the angle $\alpha$, the force $G''_{\mathrm{T}}$ will act on the surface of the field perpendicular to this surface with force $G''_{\mathrm{TP}}$ (Fig. 3) and parallel to the field surface with a sloping force $G''_{\mathrm{TH}}$, that is, the force $G''_{2}$ can be expanded into two such components as $G''_{\mathrm{TP}}$ and $G''_{\mathrm{TH}}$ that equal to:

$$\left.\begin{aligned} G''_{\mathrm{TP}} &= G''_{\mathrm{T}} \cos\alpha; \\ G''_{\mathrm{TH}} &= G''_{\mathrm{T}} \sin\alpha. \end{aligned}\right\} \tag{2}$$

As a result of action of the $G''_{\mathrm{TP}}$ component, the rear part of the machine is pressed against the soil surface (Fig. 3), and under the action of the constituent $G''_{\mathrm{TH}}$ the machine tends to move to the right, overcoming the resistance to the friction created by the soil in the zones of its contact with the lower wheel surface, where the points $O_2$ and $O_3$ situated (Fig. 3). All friction force of the soil and the rear wheels we indicated as $F$. This force consists of two components $F_2$ and $F_3$. The component $F_2$ operates in the zone of point $O_2$, and component $F_3$ operates in the zone of point $O_3$. The component $F_2$ equals to $f_{\mathrm{P}} \cdot N_2$, and the component $F_3$ equals to $f_{\mathrm{P}} \cdot N_3$, where $f_{\Pi}$ is the friction coefficient at rest, and $N_2$ and $N_3$ are the normal forces of

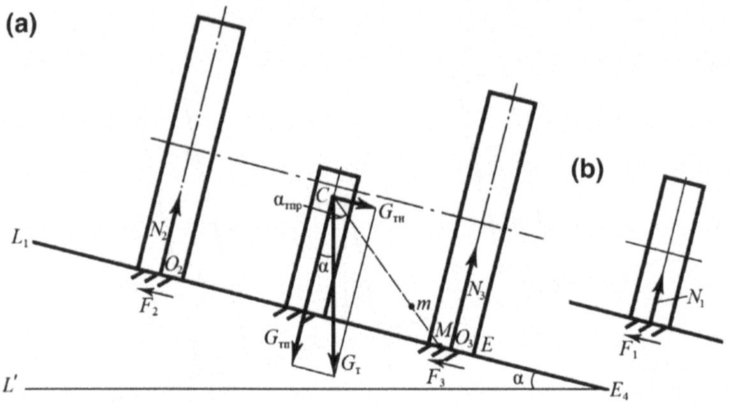


**Fig. 4** Rear view of a turned three-wheeled tractor and the forces acting on it in a transverse vertical plane, where its gravity center is located

the soil reaction in the zone of points $O_2$ and $O_3$. Then the condition of the absence of wheel slip on the soil will look like:

$$F_{2\max} + F_{3\max} > G''_{\mathrm{T}}, \tag{3}$$

or

$$f_{P\max} \cdot (N_1 + N_2) > G''_{\mathrm{T}}, \tag{4}$$

where $f_{P\max}$ is the maximum value of the static friction coefficient of the wheel on the soil.

## 3 Results and Discussion

Figures 1, 2, 3, and 4 show the weight force $G_{\mathrm{T}}$ of the tricycle tractor which is applied to the center $C$ of its weight, as well as the components of the force $G'_{\mathrm{T}}$ and $G''_{\mathrm{T}}$ acting on the point $C_1$ of the front wheel axis 1 and at the point $C_2$ on the rear axle 2 and 3. Dependence of forces $G'_{\mathrm{T}}$ and $G''_{\mathrm{T}}$ on the force $G_{\mathrm{T}}$ and distance $a$ and $b$ between the axes (Fig. 1a) are given in the equation

$$\left.\begin{aligned} G_{\mathrm{T}} &= G'_{\mathrm{T}} + G''_{\mathrm{T}}; \\ G''_{\mathrm{T}} \cdot a &= G'_{\mathrm{T}} \cdot b. \end{aligned}\right\} \tag{5}$$

Taking into account these two equations, we find:

$$G'_T = G\frac{a}{a+b}, \quad (6)$$

$$G''_T = G\frac{b}{a+b}. \quad (7)$$

If the three-wheeled tractor is in such position on the field slope shown in Fig. 4 (rear view), then the tractor will be influenced by such forces, which will increase the pressure of the right wheel to the ground, and the pressure of the left wheel on the ground will decrease. Under such conditions, a soil pressure force on the lower part of the front wheel also changes a bit (Fig. 4b).

As can be seen from Fig. 4 on the field slope, the gravity force $G_T$, acting on the whole tractor, is divided into two components $G_{TH}$ and $G_{TP}$. The component $G_{TP}$ presses the tractor to the inclined surface of the field, and the component $G_{TH}$ seeks to move the tractor to the right and down. This is prevented by friction forces $F_1$, $F_2$, and $F_3$, acting from the side of the soil on all tractor wheels. The force $F_1 = f_P{\cdot}N_1$, $F_2 = f_P{\cdot}N_2$, and force $F_3 = f_P{\cdot}N_3$, where $N_1$, $N_2$, and $N_3$ are the normal forces of the soil reaction on the rim of wheels 1, 2, and 3, and $f_P$ is the coefficient of soil friction at rest on the wheel rim. Force $G_T$ under these conditions acts vertically down.

The forces $G_T$, $G'_T$, and $G''_T$ are in such conditions in a transverse vertical plane.

Let's consider the scheme of a tricycle tractor with two wheels in front and one steering wheel in the rear (Fig. 5). The weight of the gun is applied at point C (center of gravity) and is equal.

$$G_M = G_1 + G_3,$$

where

$G_1$ is the pressure force of the front wheels 1 and 2 on the ground;
$G_3$ the force of the rear wheel pressure on the ground.

This scheme of a three-wheeled tractor has a special feature. The load on the front wheels $G_1$ is increased and can provide the pressing of the two front wheels to the ground when the machine moves along the field with a downward inclination. But the load decreases sharply on the rear wheel (Fig. 6).

The analysis of the schemes in Figs. 1 and 5 showed that the ABDE reference quadrilateral in both tractor configuration variants (first variant—Fig. 1, second variant—Fig. 5) coincides. The common conclusion will be for the two layouts of the tricycle tractor as follows. The tractor is in a steady-state (equilibrium) if the projection of the center of gravity is on the support surface inside the ABDE quadrilateral. If this condition is not met, then there is no stable equilibrium.

The weight of the front of the tractor $G_1$ is transmitted to the front two wheels (Figs. 5 and 6). The tractor's front wheels (Fig. 6) turn out to be lower than the rear wheel when operating on a downhill field and the tractor's gravity $G_M$ line will pass through points K (contact points of the rear wheels with the support surface). When the tractor's gravity $G_M$ line is to the right of the line connecting two points K, its stable equilibrium ends and the three-wheeled tractor can now tip over to the right.

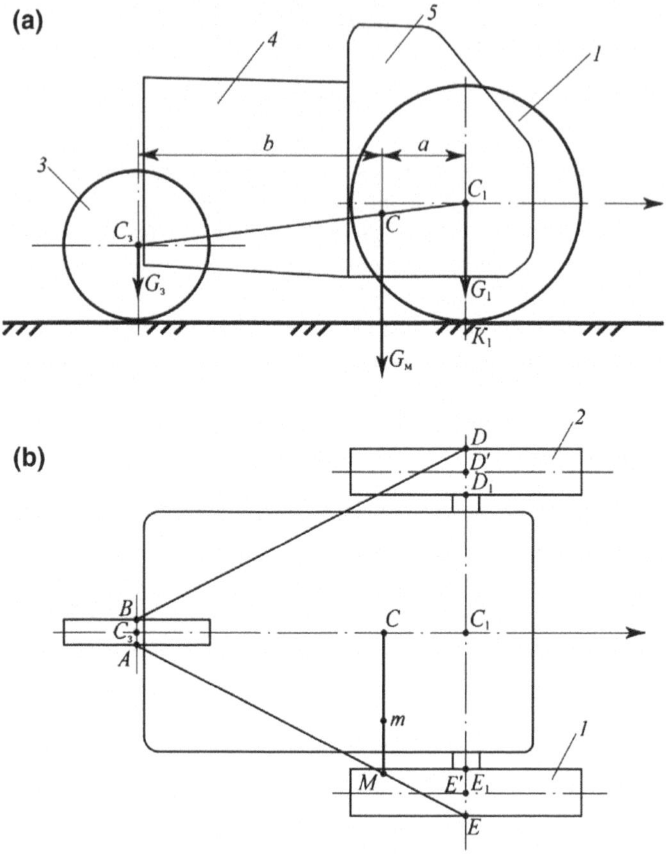


**Fig. 5** Scheme of a three-wheeled tractor: 1—front right wheel; 2—front left wheel; 3—rear wheel (controlled); 4—body; 5—cabin; A, B, D, D1, E1, are E are the reference lower rims of the wheels

Only the resistance of the air that the tractor must overcome at this turn when moving downwards can interfere with this turning. It is dangerous to drive along the road or the field (with a big downward slope).

Denote by βm the angle of inclination of the road or the field by which the tractor descends with a downward inclination (Fig. 6). Then we have:

$$\mathrm{tg}\beta_{\mathrm{M}} = \frac{C_3' C_3}{C_3' A_{\mathrm{M}}}, \tag{8}$$

where

$C_3' C_3$ the height at which the center $C_3$ of the rear wheel of the tractor appeared after its turning and lifting;

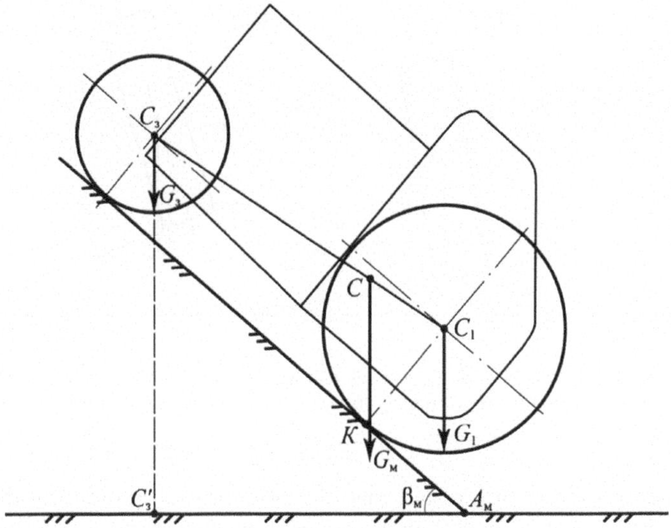


**Fig. 6** Scheme illustrating the movement of the machine to the right of the field with a downward slope (side view)

$C_3'A_M$ horizontal distance from a point $C_3'$ to a point $A_M$ after turning the tractor through an angle $\beta_M$.

As already mentioned, the angle $\beta_M$ by the formula (8) should be significantly less than the angle $\beta_M$, which is shown in Fig. 6. When working on fields with slopes, the tractor can move not only up and down along the slope, but also across. When moving across a field with a bias to the right or left, the working conditions of a single rear wheel deteriorate. It can begin to work with a bias or bend in some direction. Consider the possible load on this wheel. Scheme of the rear wheel turn of a three-wheeled tractor on a slope is shown in Fig. 7. If the inclination of the field is absent, then the angle $\alpha$ of its inclination to the horizon is zero, and the pressure force on ground $G_3$ wheels acts vertically and its action does not displace the wheel to the left or right. If the field is inclined to the right by the angle $\alpha$, then (Fig. 7b) the middle plane dividing the cylindrical part of the wheel rim into two equal parts will be inclined to the right across the field by the angle $\alpha$ (Fig. 7b). The line of action of the force $G_3$ begins at the point $C_3$ of the wheel and acts vertically downward to the left of the extreme lower point A of the wheel rim; this means that the force $G_3$ does not cause the rear wheel to roll over to the right.

If the angle $\alpha$, the deflection of the field $E_1E_1'$ upwards from the horizontal is greater than the angle $\alpha$ in the previous case (Fig. 7c), i.e., if the angle $\alpha$ is the same as in Fig. 7b, then the vertical force going down from $C_z$ down (Fig. 7c) will cross the field line to the right of the extreme lower point A of the rear wheel rim. In this case, it is possible to roll the wheel around point A to the right, but this is impossible due to the fact that the wheel is connected with the rest of the machine, which has

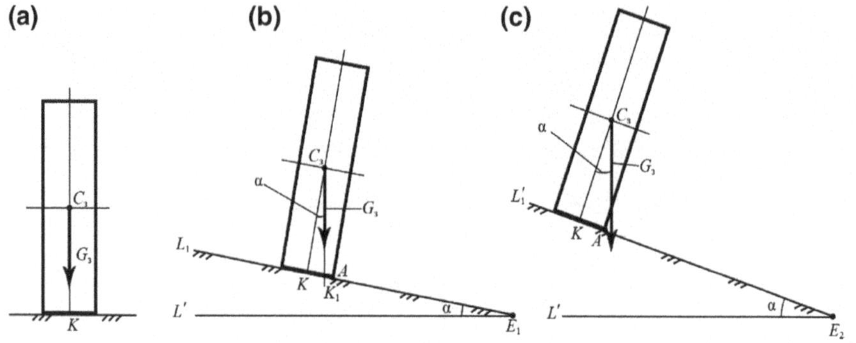


**Fig. 7** Scheme of action on the axis of a separate rear wheel of a tractor of force $G_3$ on a horizontal field (**a**) and on a field slope at a small angle $\alpha$ (**b**) and at a greater angle $\alpha$ (**c**) of the field inclination

not yet been brought to such a state that it is necessary to tip over (for the tractor to overturn, the center of gravity on the ground turned out to be outside the ABDE quadrilateral in Fig. 5).

The fact that in this case the action of a force $G_3$ to the right of point A turned out to be non-dangerous does not mean that such an action of force is always not dangerous; such an action of force $G_3$ must be verified with regard to the above instructions. Thus, it is better to work with such a machine on the slope of the field according to the scheme in Fig. 7b, and it would be even better if the force $G_3$ acts along the line $C_3K$ or $C_3K_1$ in the same Fig. 7. The angle of inclination $\alpha$ of the slope of the field, at which the vertical line of the force $G_3$ (Fig. 7) passes through point A, is the maximum allowable angle of inclination of the wheel $\alpha_{\text{кпр}}$; it is determined by the formula:

$$\alpha_{\text{кпр}} = \operatorname{arctg}\frac{b}{2r_{\text{к}}}, \tag{9}$$

where $b$—the width of the rim of the wheel, $r_{\text{к}}$—the radius of the wheel.

To improve the stability of the rear wheel of the tractor, it is necessary that the rim of the wheel be larger than shown in Fig. 7c, in order to achieve such a direction of action of force $G_3$, as shown in Fig. 7a, b.

Above, the force of gravity tractors $G_{\text{M}}$ was represented by two components parallel to it $G_1$ and $G_3$ (Fig. 5a). The force $G_1$ is applied at point $C_1$, representing the projection of the ED axis of the front wheels on the vertical plane. We consider that when the tractor's position changes, the position of the center of gravity $C_1$ does not change, i.e., we believe that the movement of materials inside the tractor does not occur while changing its position, and the distance from the center of gravity $C_1$ to the support plane and wheels does not change.

Let us denote the pressure of the front right wheel on the horizontal surface of the soil $G_{1\text{п}}$, and the force of pressure of the front left wheel on the horizontal surface of the soil $G_{1л}$ (Fig. 5). Then

$$\left.\begin{array}{l} G_1 = G_{1п} + G_{1л}; \\ G_{1п}(C_1E') = G_{1л}(C_1D'), \end{array}\right\} \tag{10}$$

where

$C_1E'$ the distance from the center $C_1$ of the front axle of the tractor to the middle plane of the cylindrical part of the rim of the right wheel;

$C_1D'$ distance from center $C_1$ of the tractor's front axle to the middle plane of the cylindrical part of the rim of the left wheel.

From Fig. 5, it follows that

$$\left.\begin{array}{l} G_M = G_1 + G_3; \\ G_1a = G_3b. \end{array}\right\} \tag{11}$$

Then

$$\left.\begin{array}{l} G_1 = G_M\frac{b}{a+b}; \\ G_3 = G_M\frac{a}{a+b}. \end{array}\right\} \tag{12}$$

Let's now consider turning the front of the tractor on the slope of the field to the right. With such a rotation in the transverse vertical plane at the angle $\alpha$ to the right, the force $G_1$ will act on the surface of the field $E_1L_1$ perpendicular to this surface with the force $G_1'$ (Fig. 8) and parallel to the surface of the field with the inclined force $G_1''$. We consider the force $G_1$ divided into two components $G_1'$ and $G_1''$ equal:

$$G_1' = G_1 \cos\alpha = G_M\frac{b}{a+b}\cos\alpha;$$

$$G_1'' = G_1 \sin\alpha = G_M\frac{b}{a+b}\sin\alpha. \tag{13}$$

The forces $G_1$, $G_1'$, $G_1''$ are located in a transverse vertical plane.

Under the influence of the component $G'$, the front of the tractor is pressed against the inclined surface of the field (Fig. 8), and under the effect of the component $G_1''$, the machine tends to move to the right, overcoming the resistance to friction that is created by the soil in the zones of its contact with the lower surfaces of the wheels, where the points $O_1$ and $O_2$ (rice 8). The whole force of friction of the soil on the lower surfaces of the wheels is denoted by Ft, this force consists of two components $F_1$ and $F_2$, of which the component $F_1$ acts in the zone of finding of the point $O_1$ of the wheel 1, and the component $F_2$ acts in the zone of finding of the point $O_2$ of the wheel 2. The component $F_1$ is equal $f_nN_1$, and the component $F_2$ is equal to $f_nN_2$, where $f_n$ is the coefficient of resting of the rest of the soil about the wheel, and $N_1$

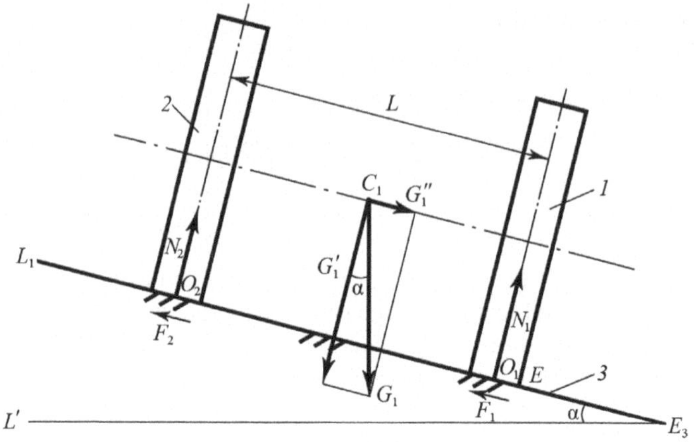


**Fig. 8** Rear view of the front of the three-wheeled tractor rotated to the right and the forces acting in the transverse plane in which the center of gravity $C_1$ of the front of the tractor is located: 1—the right wheel, 2—the left wheel, 3—the surface of the field

and $N_2$ are the normal forces of the soil reaction in the zone of finding points $O_1$ and $O_2$.

Then the condition that the wheels do not slip along the soil of the front of the tractor will be expressed as follows:

$$F_1\max + F_2\max > G_1'' \quad (14)$$

or

$$f_{\text{п}}\max(N_1 + N_2) > G_1'', \quad (15)$$

where $f_{\text{пmax}}$ is the maximum value of the static friction coefficient of the wheel on the soil.

Such are the peculiarities of the action of the forces on separate organs and units of a three-wheeled tractor while moving along the slopes of the fields. Now let's consider the actions of loads on the whole three-wheeled tractor with one rear wheel while working on the slopes of the fields.

Figures 5, 6, and 9 show the gravity of a three-wheeled tractor, applied at the center C of its gravity, as well as the components of the force acting on the point C1 on the axis of the front wheels 1 and 2 and at the point $C_z$ on the axle of the rear wheel.

If the entire tractor is in such a position on the slope of the field, as shown in Fig. 7 (rear view), then the forces at which the right wheel pressure will increase on the soil will increase, and the pressure of the left wheel on the ground will decrease. At the same time, the force of pressure on the ground of the rear wheel will change slightly (Fig. 9b).

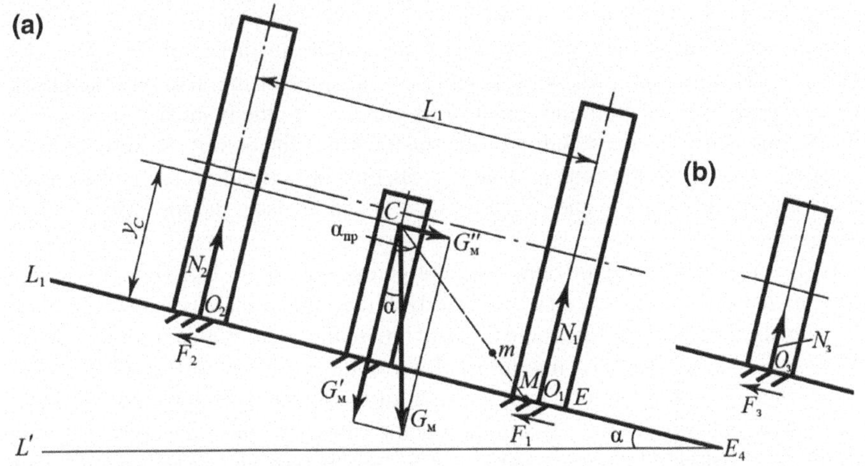


**Fig. 9** Rear view of the tractor on the slope of the tractor and the forces acting on it in the transverse vertical plane in which the center of gravity (**a**) is located, and the rear view of the separate rear wheel in this situation (**b**)

Such are the peculiarities of the action of the forces on separate organs and units of a three-wheeled tractor while moving along the slopes of the fields. Now let's consider the actions of loads on the whole three-wheeled tractor with one rear wheel while working on the slopes of the fields.

Figures 5, 6, and 9 show the gravity $G_M$ of a three-wheeled tractor, applied at the center C of its gravity, as well as the components $G_1$, $G_2$ of the force acting on the point $C_1$ on the axis of the front wheels 1 and 2 and at the point $C_3$ on the axle of the rear wheel.

If the entire tractor is in such a position on the slope of the field, as shown in Fig. 7 (rear view), then the forces at which the right wheel pressure will increase on the soil will increase, and the pressure of the left wheel on the ground will decrease. At the same time, the force $C_3$ of pressure on the ground of the rear wheel will change slightly (Fig. 9b).

## 4 Results and Discussion

As can be seen from Figs. 5, 6, 8, and 9, on the slope of the field, the gravity of the tractor $G_{\mathrm{M}}$ acting on it and decomposed into three components, and, providing the tractor is pressed against the field and the ability to work on this field.

As already indicated, in order for the tractor to be on the slope of the field in a steady state, it is necessary that the vertical line lowered from the center of its gravity C downwards crosses the surface of the soil (ground) within the ABDE quadrilateral in Fig. 5b. In our case, the force vertical $G_M$ applied at point C in Fig. 5a must intersect any point of the CmM line inside the ABDE quadrilateral (Fig. 5b). The tractor at the same time will work steadily on the slope of the field, but the closer the end of the vertical from the center of its gravity C to the boundary zone mC in Fig. 5b, the better for the stability of the tractor, but if the end of the vertical from the point C will be in the zone mM, in Fig. 5b, then because of vibration the tractor can work unstable and the end of the vector $G_M$ can intersect the line AE, which is unacceptable.

If the angle $\alpha$ is insignificant, then the component force $G''_M$ (Fig. 9) is also insignificant and the tractor can work steadily. If the angle $\alpha$ increases, then the force $G''_M$ will also increase, and the force $G_M$ will turn to the right. With a further increase in the angle $\alpha$, the force $G_M$ line turns, closer to the right wheel (Fig. 9), and may occur when the force line $G_M$ is to the right of the lower right points m and M of the right wheel (Figs. 5 and 9) and the tractor will overturn to the right. But this overthrow should not be in any case. The effect of the rear wheel of the tractor does not significantly affect this.

The angle $\alpha$, in which the power line of the tractor $G_M$ passes through the point M (Fig. 9), is called the limiting angle of the tractor and is denoted by $\alpha_{\text{пр}}$. It is shown in Fig. 9 and is determined from the equality:

$$tg\alpha_{\text{пр}} \approx \frac{L_1 - b}{2y_C}, \tag{16}$$

where

$L_1$ the distance between the middle of the cylindrical planes of the tractor's front wheels (Fig. 9);
$b$ the rim width of the wheel;
$y_C$ height of the center with the weight of the tractor (Fig. 5).

Then

$$\alpha_{\text{пр}} = \operatorname{arc}tg\frac{L_1 - b}{2y_C}. \tag{17}$$

As already mentioned above, the growth of the angle $\alpha$ to the value of $\alpha_{\text{пр}}$ in no case should be allowed. In order to make the work of a three-wheeled tractor on the slopes possible, it is necessary to properly calculate and test its capabilities so that it can carry out field work on such lands without danger.

## 5 Conclusions

1. For selected tractor layout schemes, steady-state conditions have been established according to which the vertical line lowered from its center of gravity downwards crossed the soil surface inside the support quadrilateral formed as a result of the connection of the supporting outer points of the wheel rims.
2. The dependence of the marginal angle of inclination of a tractor with one front wheel operated on the width of the rim and wheel radius, as well as the value of the marginal angle of inclination of a three-wheeled tractor, which has two forward non-steering wheels, is established, depending on the distance between the middle of the cylindrical planes of the front wheels, the width of the rim wheels, height of the center of gravity of the tractor, which made it possible to theoretically determine the conditions of safe operation of tractors on the slopes of the field.
3. The conditions of stability of a three-wheeled tractor depending on its design, the angle of the field to the horizon are substantiated, thereby creating the preconditions for increasing the accuracy of determining the safe conditions of operation of a three-wheeled tractor.

## References

1. Hailis, G.A., Sheichenko, V.A., Dudnykov, I.A., Murovanyi, I.S., Tolstushko, N.N., Shevchuk, V.V.: Silskogospodarski mashyny. In: Proceedings of LNTU, vol. 32, pp. 121–129, (2017)
2. Guskov, V.V., Velev, N.N., Atamanov, YuE: Tractors: Theory. Mashinostroenie, Moscow (1988)
3. Pryshliak, V.M.: Sci. Herald NULES Ukraine **144**(5), 134–142 (2010)
4. Kirienko, M.N., Polyanskiy, A.S., Zadorozhnyiy, V.V.: Bull. KhNTUA (Kharkiv) **107**(2), 22–24 (2011)
5. Usenko, M.V.: Modern technologies in mechanical engineering and transport, vol. 1, pp. 121–128 (2014)
6. Mohov, S.P., Ariko, S.E.: Proceedings of BSTU **2**(140), 45–48 (2011)
7. Raichman, D.B.: Proc. VNIIEMK **20**, 158–163 (1989)
8. Smolinskyi, S.V.: Mechanization and Electrification of Agriculture. Collected papers, vol. 98, no. 1, pp. 306–313 (2013)

# Research of the Cereal Materials Micronizer for Fodder Components Preparation in Animal Husbandry

**Alexander Skliar, Boris Boltyanskyi, Natalia Boltyanska and Denis Demyanenko**

## 1 Introduction

The problems of creation of the technical devices for grain material micronization, which effectively fits in the technological lines of fodder production, are the relatively high energy intensity of the process and the lack of a qualitative single-layered distribution of grain material along the surface of the heat treatment. Therefore, studies aimed at reducing the energy intensity of the process and improving the quality of single-layer grain material distribution with the simultaneous turning of the grains are relevant.

To date, a number of micronizers have been created, where the material moving process is carried out with the help of vibration transport or with the use of mechanical devices for the moving and turning of grain material [1] and still requires technological improvement. Past earlier theoretical studies were aimed mainly at determining the permissible deviations of the energy flow for grain material processing [2, 3] and the parameters of dosing devices [3]. At the same time, the issue of qualitative distribution during the movement of the processed material on the surface of the treatment was not considered in combination with the problems of energy intensity and the quality of the micronization process implementation.

To substantiate the structural and kinematic parameters of the micronizer, a mathematical model of grain material single-layer moving process on a sloping circular surface was created and laboratory researches of grain material physical and mechanical properties changes during the heat treatment process were carried out. To con-

A. Skliar · B. Boltyanskyi · N. Boltyanska (✉) · D. Demyanenko
Tavria State Agrotechnological University, B. Khmelnitsky ave. 18, Melitopol 72310, Ukraine
e-mail: nataliia.boltianska@tsatu.edu.ua

A. Skliar
e-mail: oleksandr.sklyar@tsatu.edu.ua

B. Boltyanskyi
e-mail: borys.boltianskyi@tsatu.edu.ua

© Springer Nature Switzerland AG 2019 
V. Nadykto (ed.), *Modern Development Paths of Agricultural Production*,
https://doi.org/10.1007/978-3-030-14918-5_26

firm the adequacy of the aforementioned mathematical model, experimental studies of grain material micronizer were conducted.

The purpose of this work is to conduct a comparative analysis of theoretical and experimental studies of grain material micronizer, as well as determine the optimal parameters of the micronization process.

## 2 Theoretical Researches

The quality of particles distribution on the treatment surface depends both on the shape of this surface, its kinematic parameters, physical and mechanical properties of the grain material [4, 5].

The shape of a circular inclined surface is described by a system of equations (1), where, successively, starting from $\alpha_o$, the angle of the surface inclination decreases by $\Delta\alpha h_i$ with step $h_i$.

The shape of the surface is determined on the basis of the dependence of grain material movement velocity along the inclined surface and the angle of this surface inclination at each step $h_i$.

$$
\begin{aligned}
x_{h_i} &= 0.5d + \sum_{i=1}^{h} \left\{ d_{h_i} \cos\left[ \alpha_o - \sum_{i=1}^{h_i} \Delta\alpha(h_i - 1) \right] \right\}; \\
y_{h_i} &= \sqrt{R_{h_i}^2 - x_{hi}^2}; \\
z_{h_i} &= 0.5d + \sum_{i=1}^{h} \left\{ d_{h_i} \sin\left[ \alpha_o - \sum_{i=1}^{h_i} \Delta\alpha(h_i - 1) \right] \right\},
\end{aligned}
\tag{1}
$$

where

$d$ working surface diameter;
$d_{hi}$ elementary layer diameter;
$\alpha_o$ initial inclination angle;
$\Delta\alpha h_i$ elementary layer inclination angle changing value;
$h_i$ step of working surface height changing.

The sum of the forces $\Sigma P_{ihi}$ acting on the contacts of elementary layers of particles located on a circular inclined surface (Fig. 1) is determined by expression

$$
\begin{aligned}
\sum p_{ih_i} &= \sum_{i=1}^{n_{hi}} m_ч \cos\alpha_i \cos\left( \frac{\alpha_i + \alpha_{i+1}}{2} - \alpha_i \right) + \\
&+ \sum_{i=n_н/6}^{h_{i-1}} \left\{ \sum_{i=1}^{n_{h-1}} \left[ p_{ih_i} \cos\left( \frac{\alpha_{i-} - \alpha_{i+1}}{2} \right) - mg \sin\alpha_i f_t - p_{ih_{i-1}} \sin\left( \frac{\alpha_i + \alpha_{i+1}}{2} - \alpha_i \right) f_t \right] \right\},
\end{aligned}
\tag{2}
$$

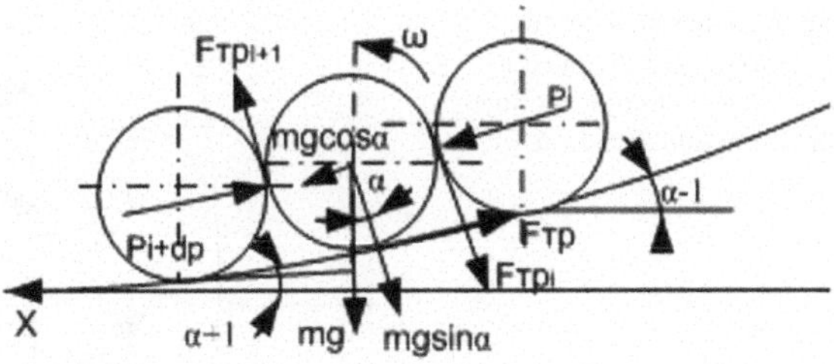


**Fig. 1** Scheme of forces acting on a particle of the *i*-th layer in the radial direction

where

| | |
|---|---|
| $\Sigma P_{ihi}$ | pressure of the material mass located above the $h_i$ layer; |
| $p_i$ | specific pressure per particle; |
| $nh_i$ | the number of particles on the layer $h_i$. |

The composition of different fraction particles arrangement (see Fig. 1) is determined from the condition (assumption) of even arrangement of particles in the zone of mutual intersection of adjacent elementary layers [6–8].

The pressure on each particle extends in proportion to the area occupied by the *i*-th particle in the pressure zone. The area occupied by a particle in the zone of intersection of particles is defined as $0.25\pi d_{av}^2$.

The condition of the particles motion in the radial direction is determined by an inequality that reflects the excess of the gravity forces components over the frictional forces between the particles and the plane of the inclined circular surface [9, 10].

$$mg\cos\alpha\sin\alpha(1-f_t)+p_i\cos\left(\pi-\frac{\alpha_{i-1}+\alpha_i}{2}\right)\geq(p_i+dp)\times\cos\left(\pi-\frac{\alpha_i+\alpha_{i+1}}{2}\right)f_t+F_{ot}\sin\psi \quad (3)$$

where

| | |
|---|---|
| $p_i$ | the specific pressure per particle; |
| $m$ | the mass of the particle; |
| $\alpha_i, \alpha_{i-1}$ | the angles of the surface incline on $x_i$ and $x_{i-1}$; |
| $F_{oti}$ | the repulsive force that occurs when particles collide; |
| $\psi$ | the angle between the direction of frictional force during the interaction of particles with the *y*-axis. |

To determine the conditions for moving particles along the circles of the inclined surface, we consider the scheme of forces acting on a particle of bulk material during circular horizontal oscillations of the inclined surface in one direction.

The main driving force in this case is the friction force of the particle, which should exceed the forces from the interaction of the contacting particles [11, 12] (see Fig. 2).

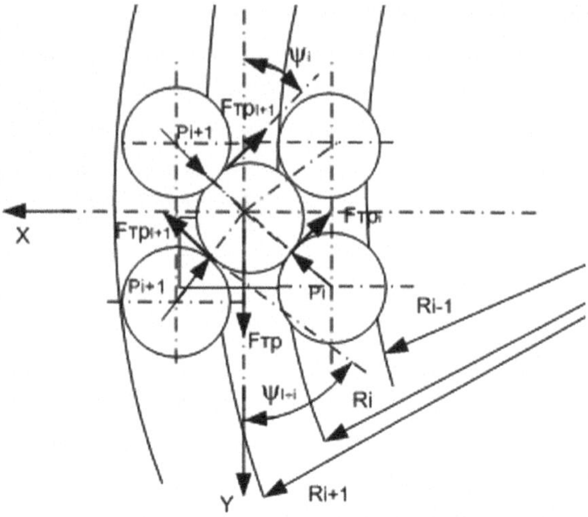


**Fig. 2** Scheme of forces acting on a particle during motion on a circular inclined surface in one direction

Obtained model of grain material single-layer movement process along the inclined circular surface allows us to determine the motion, inverting and collision properties of elastic particles, which simulating a discrete environment and parameters of micronizer with an inclined circular surface.

## 3 Micronizer Design

In order to carry out experimental studies with considering of necessary requirements for cereal materials micronization process, a grain materials micronizer with a sloping circular vibration surface was created.

The experimental micronizer sample (see Fig. 3) contains a hopper 1 and a cone 2 that are fixedly mounted on a vertical threaded axis 3, a sloping circular working surface 4, which is located on the bearing case 5. The drive of the working surface 4 is carried out from the shaft 6, which connected to the eccentric 7, which is driven to the rotational motion through the belting 8 from the electric motor 9.

The main tasks of the research are: to improve the micronizer by means of modernization of the constructive-technological scheme, based on a new set of structural elements, their mutual arrangement and the existence of links between them, provide high quality of the finished feed products, improve the quality of grain materials processing, simplify the design and reduce the energy intensity of the process.

The problem is solved by the fact that the micronizer drive is executed in the form of a motor with an eccentric that creates circular horizontal oscillations, and the working surface has a dynamic angle of inclination.

Using such type of surface and oscillations allows creating an even single-layered movement of particles on the surface of the treatment with simultaneous turning,

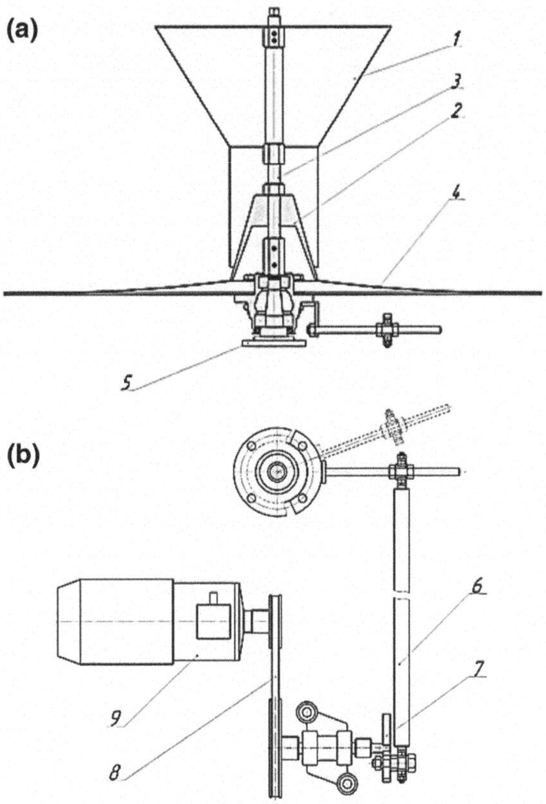


**Fig. 3** **a** Scheme of the micronizer working surface; **b** the micronizer drive diagram. 1—hopper; 2—cone; 3—vertical stationary axle; 4—working surface; 5—case with bearings; 6—shaft; 7—eccentric with a through groove; 8—belting; 9—electric motor

which greatly improves the quality of the micronization process and reduces energy costs. Creating a complex movement of grain material elements on a circular surface allows significantly reducing the size and metal capacity of the sample. Creating the possibility of smoother regulation of the frequency and amplitude of vibration allows achieving high quality of the micronization process when processing materials with different physical and mechanical properties.

Micronizer working process consists of following steps. From the hopper 1, located in the center above the working surface 4, the grain material under its own weight is evenly distributed in a circle through the gap between the wall of the hopper 1 and the cone located inside of the hopper. The creation of oscillations is carried out by an electric motor 9, which is connected to the eccentric 7 through the belting 8 and intermediate shaft.

**Table 1** Levels and intervals variation during conducting of experimental researches

| No. | Level of factors | | | |
|---|---|---|---|---|
| | Coded | | Decoded | |
| | $X_1$ | $X_2$ | $A$ | $n$ |
| 1 | 1 | 1 | 13 | 160 |
| 2 | −1 | 1 | 7 | 160 |
| 3 | −1 | −1 | 7 | 80 |
| 4 | 1 | 0 | 13 | 120 |
| 5 | 0 | 0 | 10 | 120 |
| 6 | 0 | −1 | 10 | 80 |
| 7 | 1 | −1 | 13 | 80 |
| 8 | 0 | 1 | 10 | 160 |
| 9 | −1 | 0 | 7 | 120 |

Feed regulation is carried out by raising or lowering the hopper latch. The regulation of the oscillations amplitude is carried out by eccentricity change on the eccentric cross-sectional groove. The regulation of the oscillation frequency is carried out by changing the voltage of the transformer, which is fed to the DC motor. Measurement of the exact number of vibrations is carried out by a meter connected to the magnetic sensor; the angle of working surface rotation is displayed on a scale.

Under the action of horizontal circular oscillations, the particles perform a complex movement with a turning over and gradually move to the periphery of the circular surface. During the movement, which is 60–90 s long and is regulated by changing the frequency and amplitude of oscillation, the grain material is irradiated with infrared rays. After that, under the action of its own weight, the grain material falls on a circular sloping conveyor, which moves grain to further processing.

## 4 Experimental Researches

The purpose of the research is to obtain experimental dependencies of grain material micronizer structural and kinematic parameters influence on the uniformity of feed distribution, energy intensity and the ability of grain particles to turn over.

To solve the problem, the method of multifactorial experiment planning was used and a three-level second order Box-Behnken design [13] was implemented.

As the result of previous studies, it was found that the frequency $n$ and the amplitude $A$ of the working surface oscillations are generally effects on the quality and energy intensity of the grain material feed process on an inclined circular vibration surface.

Levels and intervals of variation of factors are presented in Table 1.

The process of grain material (soybeans) moving on a sloping circular vibration surface was video recorded, after which the number of grains on each of the four sections of the surface was calculated in the stop-frame mode.

The results were compared with the reference values. The ability of grains to turn over was recorded visually using a control group of grains, which was partially colored in a contrasting color.

The studies were carried out in triple repetitions.

In order to obtain a regression model, the application package of "Statistica" software was used, which also automatically calculates and statistically estimates the significance of the regression coefficients for the Student criterion, its adequacy according to the Fisher criterion and the efficiency by the determination coefficient.

As a result of calculations, the equations of regression of the energy intensity, the uniformity of distribution and the ability of the particles of grain to turn over were obtained

$$T = -339.7778 + 13.7222x + 4.875y - 0.6111x^2 + 0.0167xy - 0.0175y^2 \quad (4)$$

$$U = -63.2963 + 16.9815x + 2.1542y - 0.9074x^2 - 0.0333xy - 0.0101y^2 \quad (5)$$

$$E = -8.0296 + 1.1259x + 0.0921y + 0.0315x^2 + 0.0021xy - 0.0003y^2 \quad (6)$$

During micronization process studies, a statistically significant positive and moderate correlation of the energy intensity and the ability of the particles to flip over with the frequency and amplitude of oscillations was obtained. In turn, the uniformity of the feed process reaches the maximum values at the basic levels of oscillations frequency and amplitude. Response surfaces of the Eqs. (4, 5 and 6) are presented in Fig. 4.

During the implementation of regression analysis, the zero-hypothesis is checked for the absence of independent factors interconnection. For this purpose, the verification of independent factors is used in the absence of strong linear correlation with the accepted level of significance $\alpha = 0.05$.

Considering the obtained regression equations, we can conclude that the optimum values of uniformity 92% and the ability of the grains to turn over 91% are achieved at an amplitude of oscillation $A = 10°$ and a vibration frequency $n = 120$. Under these conditions, the energy intensity of the process $E$ is 15.8 kJ/kg.

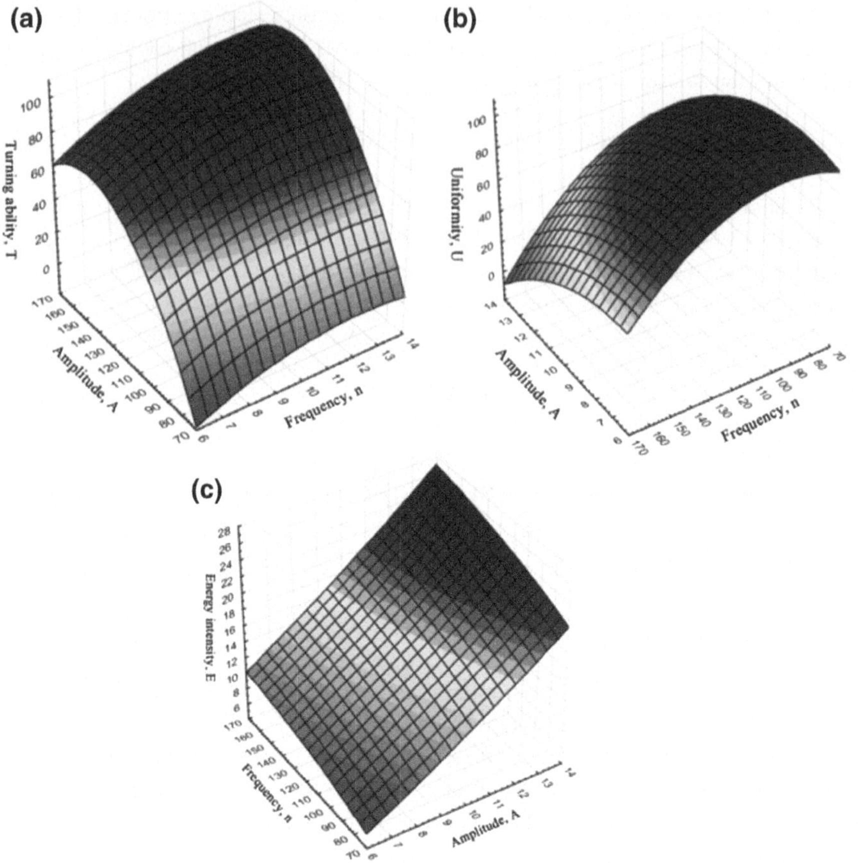


**Fig. 4** **a** Response surface that shows dependence of the ability of particles to turn $T$ on amplitude $A$ and the frequency of vibration $n$; **b** response surface that shows the dependence of the supply flow uniformity $U$ on amplitude $A$ and the frequency of vibration $n$; **c** response surface that shows the dependence of the energy intensity of the process $E$ on the amplitude $A$ and the frequency of vibration $n$

## 5 Comparative Analysis

As a result of the theoretical studies and calculations, the velocities of the particles moving on an inclined circular vibrating surface were obtained. The values obtained in the course of theoretical studies were compared with the experimental data, which is shown in the graph (Fig. 5).

The curve of velocity change of the particles moving on an inclined circular vibrational surface, obtained in the course of theoretical studies, is described by the equation $y = -1.632 \ln(x) + 5.7717$ and has the value of the approximation reliability $R^2 = 0.9934$. The curve of velocity change of the particles moving on an

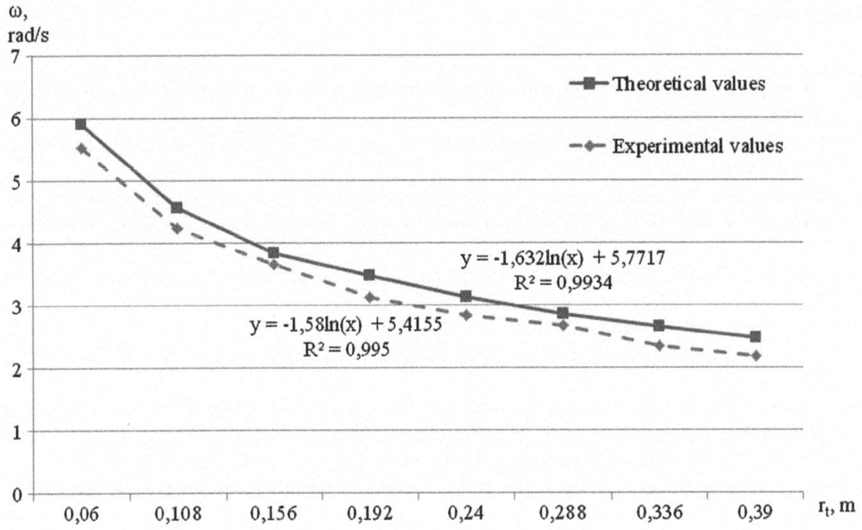


**Fig. 5** Comparison of the theoretical and experimental values of the speed of movement of grains as a function of remoteness from the center of the treatment surface

inclined circular vibrational surface, obtained in the course of experimental studies, is described by the equation $y = -1.58\ \ln(x) + 5.4155$ and has the value of the approximation reliability $R^2 = 0.995$.

The comparative analysis shows that the mathematical model of particles motion on an inclined circular vibrating surface is adequate, and the parameters of the micronization process are advisable to use during processing of grain material with various physical and mechanical properties [14, 15].

## 6 Conclusion

The purpose of the article was to carry out a comparative analysis of theoretical and experimental studies of a micronizer with an inclined circular vibrating surface. The result of this analysis is the experimental confirmation of the obtained mathematical models of the grain material movement and distribution during the heat treatment. Thanks to the universality, low metal intensity and energy costs, the obtained micronizer sample is able to increase the efficiency of feed material processing and to reduce energy consumption for the micronization process on livestock farms [16–18].

## References


1. Shatsky, V.: Modeling of feeds preparation mechanized processes. IZhA UAAN. Zaporizhzhya, X-PRESS, 140. (1998)
2. Braginets, N.: Micronization of grain for fodder purposes. In: Mechanization and Electrification of Agriculture, vol. 1, pp. 29–31 (1989)
3. Khudonogov, A.: Technology of wild-growing and agricultural raw materials processing with highly concentrated infrared heating. Dissertation of Doctor of Technical Sciences. Irkutsk, 392 (1988)
4. Kalnins, I., Pankov, Y.: Treatment of forage grains by infrared radiation. Feed Production, vol. 10, pp. 16–18 (1983)
5. Lukyanenko, V.: Mathematical model of interacting seeds motion on an inclined vibrating surface and numerical methods for solving systems of kinematic equations. Motrology **15**, 135–142 (2013)
6. Mishurov, N.: Substantiation of the optimal parameters and operating modes of the unit for processing grains by IR radiation. Dissertation of Candidate of Technical Sciences (1990)
7. Shcherbakov, V.: Chemistry and biochemistry of bulk grains. Food Industry, 163 (1977)
8. Maksakov, V.: Effect of temperature and other actions on the quality of protein feed. Bull. Agric. Sci. **11**, 95–98 (1983)
9. Revenko, I., Braginets, M., Rebenko, V.: Machines and equipment for animal husbandry. In: A Textbook. K: Condor, pp. 46–57 (2009)
10. Arsent'ev, V.A.: Methods of particles dynamics and discrete elements as a tool for research and optimization of processes for processing natural materials. Enrichment of Ores, vol. 1, pp. 30–35 (2010)
11. Nikolaev, S.: Molecular interaction. Samizdat **5**, 1–3 (2011)
12. Khimka, S.: The substantiation of parameters of the vibrating batcher for loose forages. Dissertation of Candidate of Technical Sciences, 23 (2012)
13. Karlin, S.: Bases of the theory of random processes. M. Mir. 61 (1971)
14. Zavaliy, A., Yanovich, I.: System of infra-red isothermal surface heating. Motrol **11B**, 172–175 (2009)
15. Zavaliy, A.: Optical-geometric models for designing reflecting surfaces of infrared drying devices. Motrol. **13**, 150–158 (2011)
16. Dale, N., Araba, M., Whittle, E.: Protein solubility as an indicator of optimum processing of soy-burn meal. In: Proceedings, Yeorgia Nutrition Conference, for The Feed Industry, Atlanta, LA, vol. 11, pp. 18–20 (1987)
17. Kalnins, I.: Preliminary processing of feed grain. Flour-and-Elevator Industry 5 (1984)
18. Nikitin, L., Babich, A., Drogobychsky, T.: Control of the infrared radiation device. Bull. Agrarian Sci. 66–67 (1997)

# Grains Dynamic Strength Determination and the Optimal Combination of Components of a Diamondiferous Layer of Grinding Wheels

**Olga Sushko, Serhii Kiurchev, Oleksandr Kolodii and Roman Bakardzhyiev**

## 1 Introduction

The process of blade treatment of hardened steels tool with SSM due to high quality is a competitive abrasive processing. Numerous advantages are due to less technological time, a significant reduction in the cycle of parts processing, while increasing the processing speed, reducing the roughness of the surface, and a significant increase in wear resistance of the tool [1]. The methodology for choosing the optimal combination of diamond strength grains properties and metal bond in relation to the processing of a particular material was not presented until this time. Recommendations on the use of certain diamond grains in sources are general and have very large ranges. Such recommendations, taking into account the high cost of diamond grains, lead to low efficiency of their use and, consequently, high cost of the diamond grinding process, which significantly impedes its application in the processing [2]. The traditional concentration of diamond grains (25, 50, 100, 150, 200%), which is used in existing diamond circles, requires a lot of clarification. In this case, the problem of optimal combination of metal bond strength properties and diamond grains should be solved in terms of preserving their integrity in the process of sintering diamond circles [3]. Therefore, it is necessary to identify the optimal combination of grades, grains, concentrations of diamond powders with the type of metal sheath, which provides a minimum defect of sintered grinding wheels (in terms of determining the conditions for maximizing the grains integrity) in order to increase the robustness of diamond circles.

O. Sushko (✉) · S. Kiurchev · O. Kolodii · R. Bakardzhyiev
Department of Metal Treatment Technology, Tavria State Agrotechnological University, B. Khmelnistskiy ave 18, Melitopol 72312, Ukraine
e-mail: olha.sushko@tsatu.edu.ua

© Springer Nature Switzerland AG 2019

V. Nadykto (ed.), *Modern Development Paths of Agricultural Production*,
https://doi.org/10.1007/978-3-030-14918-5_27

To solve this problem, a number of tasks were set:

- to determine the methodology of theoretical studies of system elements strength influence of the "diamond grain—the metal base of grinding wheel" on the intensity of their destruction;
- on the basis of the simulation of the stress–strain state of the sintered diamond-bearing layer of the circle to determine the effect of the metal bond strength properties, diamond grains and their concentration on the integrity of the grains in the abrasive tool and to provide recommendations on the optimal composition of the sintered diamond-metal compositions which would ensure a minimum defect of sintered grinding wheels.

## *1.1 Analysis of Recent Studies and Publications*

Currently, synthetic diamonds in accordance with DSTU 9206-80 produce grinding powders, micro-powders, and sub-micro-powders of 14 brands. This allows using diamonds differentially [3]. Most researchers tend to prefer to use for the study and description of the processes of destruction of fragile nonmetallic materials (in particular, abrasive processing) a kinetic theory of fracture. Moreover, at the final treatment (polishing, finishing) the prevailing fracture will be thermo-fluctuation mechanism of the bond breakage. In the case of force grinding (when the processing of construction materials is accompanied by the removal of a significant metal layer), the process of material destruction during formation can be conventionally considered as a process performed by the formation and development of cracks under the influence of mechanical forces (cutting forces) [4]. Formation in the treated material of the so-called defective layer is a consequence of the violation of material continuity during cutting of the development and intersection of micro-cracks. Proceeding from the basic provisions of the theory of destruction, the depth of propagation of these cracks will depend on the degree of stress-deformed state in the treated body and determined by the energy conditions of the processing process [5].

In determining the specific wear and the coefficient of using the potential cutting diamond grains properties of grinding wheels, the recuperation of diamond grains from the diamond-bearing layer of the disk was investigated. The quality and size of the grains were determined in order to further identification of the optimal combination of grades, granularity, and concentrations of diamond powders with the type of metal bond, which provides a minimum defect of sintered grinding wheels [6]. We have found that grinding wheels with characteristics that are significantly different from those mentioned in the marking fall into operation. That is, used circles with characteristics that are significantly different from those shown in the marking and cannot implement the necessary indicators of diamond-abrasive processing. It is revealed that diamond grinding wheels on metal bond have an initial defect in the form of diamond grains destroyed during the sintering process, which results in a decrease of 19 and 4% of the main and large fractions grains [7]. Consequently, to determine the conditions for the preparation of diamond circles with the maximum integrity of the grains, it is necessary to conduct a study of the dynamic strength of diamond-bearing layer grains.

### *1.2 Statement of the Objective and Tasks of the Study*

In the work, the method of grains dynamic strength determining of the diamond-bearing layer of grinding wheels is given in order to identify the conditions of their production with maximum grain integrity.

## 2 The Basic Part of the Study

An indicator of dynamic strength characterizes the ability of a diamond powder to withstand the shock, dynamic load experienced by diamonds when working with a diamond tool. This technological indicator is used predominantly to characterize the quality of AC 50 and more, grades of powders and granules 200/160 and above [1]. Dynamic strength can be determined by different methods, but the most common method is to test diamonds when vibrating load on devices such as "Fraitester" (Fig. 1), [2].

A powder sample of 2 karat weight is loaded into a narrow steel cylinder with a steel ball, which moves freely inside it. The cylinder is mounted on the shaft of the electric motor in a special device. When it is rotated, the reciprocating movement of the ball in the capsule along its axis is provided. The device records the number of cycles of sample loading with a ball in terms of the revolutions number. While testing, the grain is crushed (tightened). After a certain number of cycles, the sample is unloaded, sifted on to sieves, and determined the percentage of grains on the control sieves.

**Fig. 1** A device for determining the dynamic strength of diamond powders such as Fraitester

**Table 1** Changing range in the strength *Fi* index different powder grades, installed on the Fraitester device (given by the ISM of the National Academy of Sciences of Ukraine)

| Diamond brand | *Fi* index, relative to one | Powder granularity | | |
|---|---|---|---|---|
| | | 400/315 | 315/250 | 250/200 |
| AC50 | *Fi min* | 22 | 35 | 45 |
| | *Fi max* | 65 | 77 | 81 |
| | *Fi av.* | 35 | 47 | 54 |
| AC65 | *Fi min* | 26 | 45 | 59 |
| | *Fi max* | 48 | 76 | 97 |
| | *Fi av.* | 39 | 58 | 73 |
| AC80 | *Fi min* | 31 | 44 | 63 |
| | *Fi max* | 70 | 88 | 104 |
| | *Fi av.* | 50 | 64 | 81 |

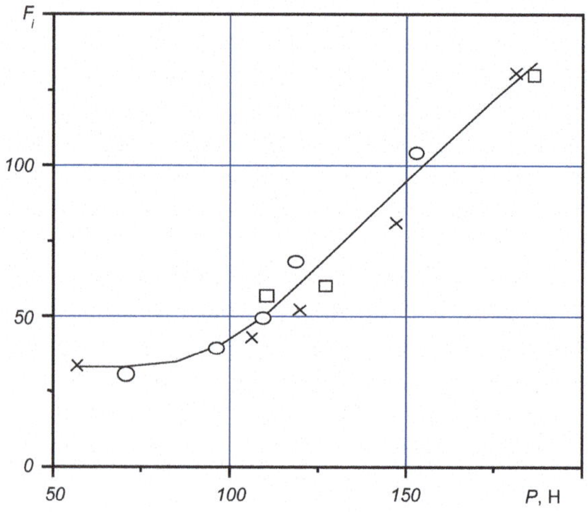


**Fig. 2** Dependence between indicators obtained on the devices Fraitester and DA-2M for grinding grain diamond 315/250: ○—AC50; ×—AC65; □—AC80

As a result of the tests, the number of bead impacts is about the grain is determined, which are necessary to destroy 50% of the grains of the powder sample. As a rule, there are usually at least two trials. The fixed number of cycles characterizes the dynamic strength of the powder (Fraitester *Fi* index). The results of the comparison correlation of *Fi* indexes and the strength of grain breaking of some domestic synthetic diamond brands are given in Table 1 and Fig. 2 [3].

In foreign practice, the dynamic strength of powder after a single shock loading in the Fraitester device for a given time is estimated as an index equals to the percentage of grains that have kept the original size. When carrying out the control on the installation of "Fraitester", you must take into account the following.

Since the application of the vibration frequency of the capsule leads to different results, the device is designed to ensure that the specified vibration frequency is consistent to within 0.1%. The weight of the ball and the shape of the capsule also affect the results of the control significantly. All of the above determines the complexity of the hardware design of the installation and its high cost. In view of the above, an express method for determining of powder grains dynamic strength was developed on the basis of another one, different from the described apparatus, in which the destruction of grains would occur with a greater number of destructive bodies. It can be expected that in this case, the stability of control will be higher. Therefore, a vortex device was chosen. It was designed in the ISM of the National Academy of Sciences of Ukraine in common with the research institute "Emalhimmash" for the special treatment of diamond powders.

The vortex machine is intended for the processes of dispersing multicomponent systems, mixing dry loose materials, selective crushing, form correction and oval shaping of powdered material grains, including dry powder particles or liquid. The material in the apparatus is processed by the influence of an alternating magnetic field on a mixture of treated material and ferromagnetic particles.

The operating chamber of the device is a container in the form of a cylinder made from a strong nonmagnetic material (Sital). The camera is installed in the device creating a rotating magnetic field with a magnetic induction value of 0.15 T. At the same time, it is loaded with non-uniform ferromagnetic particles and the portion of processed material in a ratio of no more than 1:10. Under the influence of the electromagnetic field, ferromagnetic particles acquire a rotary motion with a rotational speed of 3000 rpm and with frequency of oscillations of 1000 Hz relative to the vector of magnetic induction. A mixture of ferromagnetic particles and particles of the treated material is in a weighted state with the formation of a large number of counterflows. The processed particles collide with each other at high speed, as well as with ferromagnetic particles and the walls of the working chamber, which leads to an intensive shredding of the product.

For a uniform shredding of the material throughout the length of the working chamber, the design of the device provides for its reciprocating gradual movement in a direction parallel to its axis. The working bodies are pieces of steel wire with a diameter of 2–2.1 mm and length of 18–20 mm, which move chaotically under the action of a rotating magnetic field. The working chamber consists of five cylindrical sections of diameter 80 mm and length of 30 mm, each of which can be used for independent simultaneous tests.

The degree of powdered grains destruction is controlled by analysis of 120-mm sieves. The weight of 50 karat powder was also determined, which is optimal for sieves of this diameter. Reduction of the sample volume is possible only by reducing the cross section of screening cloth and increasing the stability of the vibration machine. When choosing the weight of working bodies, we are based on the fact that the intensity of powder grains processing must be high, in order to reduce the time for monitoring the dynamic strength.

**Table 2** Value of static $P_{st}$, dynamic $R_f$ strengths and coefficient $k$

| Grinding powders brand | $P_{st}$ (H) | $P_f$, number of cycles | $k \times 10^3$ (1/s) |
|---|---|---|---|
| AC100 | 11,110 | 70 | 1.56 |
| AC80 | 9300 | 60 | 2.44 |
| AC65 | 8360 | 50 | 3.17 |
| AC50 | 6460 | 40 | 3.00 |

On the other hand, the destruction should not go very intensively, in order to be able to study the kinetics of this process. Previous tests have shown that the most optimal for these purposes is the bond of working bodies weighing 15 g. The tests were carried out with five brands of diamond grinding powders (AC2–AC20) grains of 125/100 and seven marks (AC15–AC100) with grain size of 315/250.

The data processing by the least-squares method gives the following approximation of the relationship between the static strength $P_{st}$, the dynamic strength $P_f$ obtained on the Fraitester installation and the coefficient $k$, obtained on the vortex device of the ISM design:

$$P_{st} = 4069 + 1114/k \quad (1)$$

$$P_f = 23.05 + 75.14/k \quad (2)$$

Moreover, the correlation coefficient between $P_{st}$ and $k'^1$, $P_f$ and $k'^1$ is equal to 0.987 and 0.988, respectively. The degree of deviation of the experimental values from the calculated formula 2 for $P_{st}$ does not exceed 9% and for $P_f$–6%. The value of the static strength $R_{st}$, the dynamic strength $R_f$ obtained at the Fraitester, and the coefficient $k$ obtained on the vortex machine for diamond powders AC50–AC100 with grain size of 315/250 are given in Table 2.

Thus, it can be concluded from the conducted studies that the strength of diamond polishing powders can be controlled on the basis of the measurement of the coefficient k, which is determined by formula 2. As a device is used for machining powders, it can serve the vortex apparatus of the INM, as well as the type of ABC.

To obtain high-strength diamonds, the Ni-Mn-C system [2] was used. At optimal values of pressure and temperature, it creates correct, well-cut crystals the average content of which in the synthesis product reaches 25–30%, and maximum—40%. It is known that in diamond raw materials, which are removed from synthesis products, usually contain grains that differ in geometric parameters, as well as fused grains [4]. To select high-quality crystals, a technological scheme is used which includes selective (mechanical and ultrasonic) crushing of diamond raw materials, its separation according to the size and shape of the grains, as well as the separation of their surface roughness.

Mechanical crushing is carried out in a rotary crusher of continuous action, which provides a selective process of fractions and defective grains destruction.

**Table 3** Dynamic strength of new polishing powders

| Powder grain | AC50 | AC65 | AC80 |
|---|---|---|---|
| 500/400 | 15 | 20 | 25 |
| 400/315 | 25 | 30 | 35 |
| 315/250 | 40 | 45 | 70 |
| 250/200 | 45 | 60 | 80 |
| 200/160 | 50 | 70 | 90 |
| 160/125 | 60 | 80 | 100 |
| 125/100 | 70 | 90 | 110 |
| 100/80 | 80 | 100 | 120 |

To classify the size of grains, vibratory sieve design of ISM with a vibrating inclined deck, which is equipped with a special coating, is used. In this case, the content of particles of the main fraction in the powder of each grainy is at least 80%. The dynamic strength of new polishing powders is given in Table 3.

Taking into account the mentioned properties, diamond powders AC50 and AC80 received the highest quality category. These powders are intended for the manufacture of tools on metal bonds, which are used for the processing of building materials, ceramics, quartz glass, production of geological exploration crowns. Powders of the specified brands may be recommended for the manufacture of a corrective tool of various types, as well as for use in the manufacture of high-performance composite materials.

## 3 Results and Discussion

Thus, it can be concluded from the conducted researchers that the strength of diamond polishing powders can be controlled on the basis of the measurement of the coefficient $k$, which is determined by formula 2. As a device for processing powders can serve as a vortex apparatus INM, as well as apparatus of the type ABC.

## 4 Conclusions

The peculiarities of AC50 and AC80 powders (i.e., high strength, high isometric, and mostly smooth faces of crystals) cause the need for new connections that would ensure the safe containment of the grains in the instrument, as well as conducting research of the selection of optimal bonding structures for each of the specified brands.

Diamond grains manufactured by De Beers, according to the accepted classification in Ukraine can reach the strength of the order of AC200–AC250. Such powerful diamond grains together with the proposed method of forming a cutting sub-microrelief can open wide prospects for increasing the effectiveness of superhard diamond grinding materials.

## References

1. Semko, M.F., Hrabenko, A.I., Rab, A.F., Uzun'yan, M.D., Pyvovarov, M.S.: Osnovi almaznoho shlyfovanyya. Kyyiv, Tekhnika (1978)
2. Fedorovich, V.A.: Razrabotka nauchnyh osnov i sposobov prakticheskoy realizatsii upravleniya prisposablivaemostyu pri almaznom shlifovanii sverchtverdich materialov. Thesis by Dr. Sc. Eng., Kharkiv (2003)
3. Sushko, O.V.: Analiz vlastivostey almaznych zeren v abrazivnich instrumentah. Melitopol, TDATU, **17**(2):77–81 (2016)
4. Novikov, N.V., Androsov, I.M., Maystrenko, A.L.: Metodika opredeleniya prochnosti i treshinostoikosti polikristallicheskich sverchtverdyh materialov. Sverhtverdie materialy. **2**:33–37 (1982)
5. Tarasenko, V.V.: Analiz isnuyuchih teoriy ruynuvannya krishkich materialiv. Melitopol, TDATU, **16**(2):131–139 (2016)
6. Sushko, O.V.: Metodika rekuperatsii almaznogo poroshku z almazonosnogo sharu shlifuvalnich krugiv. Melitopol, TDATU, **17**(3):117–124 (2017)
7. Sushko, O.V.: Viznachennya pitomogo znosu ta koeffitsientu vikoristannya potentciynih rizalnih vlastivostey almaznih zeren shlifuvalnich krugiv. Materialy XI Miznarodnoyi naukovo-praktichnoyi konferenzii «Problemy konstrujuvannya, virobnitstva ta ekspluatatsii silskogospodarskoyi techniki» Kropivnitskiy: TsNTU (2017)

# Application of Phenoclimatographic Models in Stone Fruits Protecting from Spring Frosts

**Valentyna Odyntsova, Serhii Sushko, Larysa Bondarenko and Nina Shcherbakova**

## 1 Introduction

Providing the adaptation of stone fruits such as apricot, peach and cherry to the low air temperatures in winter and, especially, in the period after long thaws, as well as during spring frosts, it is necessary to take into account their biological characteristics of resistance to extreme temperature conditions for the climatic zone cultivation. When the stone fruit plantings are placed on the territory of the Southern Steppe of Ukraine, weather conditions of the winter–spring period can damage generative formations in varying degrees. The result depends on their anatomical, morphological and physiological features. In this regard, the study of the stone fruits needs to be heated during the period of biological rest and the beginning of intensive vegetation is a priority.

Rest is a necessary step in the life cycle of plants. To release biological rest, a number of fruit plants need the average daily air temperature in the range of 0–10 °C [1].

Due to the biological characteristics of stone fruits (short rest and early flowering), frequent spring frosts (once every 3–5 years) have a negative effect on the safety of generative buds. After blooming buds, during flowering, and especially during the formation of the ovary, their resistance to negative temperatures is almost completely lost. Spring frosts with an intensity of 1–3 °C below zero cause complete death or partial damage to the generative buds. That leads to significant crop losses.

To predict the date of completion of the period of biological dormancy in fruit crops, various methods have been developed: according to the date of intensive veg-

V. Odyntsova
Melitopol Experimental Gardening Station named after M.F. Sidorenko at Institute of Horticulture of the National Academy of Agrarian Sciences of Ukraine, Melitopol 72311, Ukraine

S. Sushko (✉) · L. Bondarenko · N. Shcherbakova
Tavria State Agrotechnological University, B. Khmelnitsky ave. 18, Melitopol 72310, Ukraine
e-mail: sls447@gmail.com

© Springer Nature Switzerland AG 2019

V. Nadykto (ed.), *Modern Development Paths of Agricultural Production*,
https://doi.org/10.1007/978-3-030-14918-5_28

etation of apricots using the transition date of average daily air temperature through 15 °C to lower values in autumn [2]; by cooling apple [3], peach [4], nectarine [5] at temperatures below 7.2 °C; on the accumulation of dynamic portions [6] and others.

Based on the experience of our many years of research [7, 8] and modern foreign researchers [9, 10], it has been found that the phenoclimatographic models developed by American scientists at University of Utah provide a fairly accurate forecast for the date of the onset of the deep dormancy end and the beginning of flowering. These original models are created on the basis of information on changes in temperature environmental conditions. To determine the date of deep dormancy completion, they proposed the chill unit (CU) model [11], and the dates of the onset of the phenological phases for the generative buds' development until the beginning of flowering are the accumulation of growing degree hour (GDH) with the application of the ASYMCUR model [12, 13]. It should be noted that these models are closely related. The basis of their development is on hourly maximum and minimum air temperatures. The date when the marginal accumulation of CU occurs is the starting point for the onset of GDH accumulation. It is necessary for the further development of the buds until the beginning of flowering. The boundary values of the chill units (CU) required for the completion of dormancy and the growing degree hours (GDHs), the accumulation of which is necessary for the plant to start flowering, vary depending on the crop and variety. When calculating the limit constant values of CU and GDH, not only the values of maximum and minimum air temperatures are used, but also data on the occurrence of the phenological phases for stone fruit development.

In conditions of irrigated gardening, an indirect method of protection is the most effective and reasonable when fine sprinkling is applied. The principle of the method is to wet the trees with irrigation water, followed by cooling of the buds due to evaporation of water from their surface (evaporative cooling of the buds). This leads to the flowering phase of fruit crops delaying for a later date, so that the generative buds in the least development-resistant phase do not fall under the influence of critical temperatures.

## 2 The Basic Part of the Study

Experimental data of phenophase onset of flowering trees were obtained in plantings of apricot varieties (*Prunus armeniaca L.*) Melitopol'skiy Luchistyy, peach (*Prunus persica* (*L.*) *Batsch*) *Ivan Tupitsyn* and cherries (*Prunus avium L.*) *Krupnoplodnaia*, which are placed on the experimental site of the Melitopol Experimental Horticulture Station named after M.F. Sidorenko IS NAAN in the city of Melitopol (46°50/N, 35°22/E). The height above sea level is 33 m. The climate is continental.

To calculate the constant boundary values of CU and GDH, we used the maximum and minimum air temperatures obtained at the meteorological station in Melitopol, located in close proximity to the plantations of the studied varieties of stone fruit. In the calculations, the perennial data (for at least ten years) of the phenophase of the onset of flowering of apricot, peach and sweet cherry in the garden were also used. These daily maximum and minimum temperatures were converted to hourly, synthesizing air temperature values as it was described in [14].

When using phenoclimatographic models, the following conditions should be taken into account: For accumulating one unit of CU per hour, the optimum temperature is 6 °C. At temperatures different from the optimum, the cooling process is less efficient. At temperatures below 1.4 °C and above 12.5 °C, CU accumulation does not occur, and temperatures above 16 °C introduce a negative effect to the accumulation process. One GDH unit is calculated as one hour at a temperature 1 °C higher than the baseline, equal to 4.5 °C for most fruit crops. At temperatures lower than the baseline, the growth and development of trees do not occur. The air temperature of 25 °C is the optimum when the greatest accumulation of GDH occurs (per one hour about 20.5 °C accumulates), and the temperature of 36 °C is critical, because there is a weak development of trees or its complete absence when the temperature is higher.

Predicting the release date of the studied crops from deep dormancy as reaching the CU limit value for each variety was calculated by summing up its hourly values for each day, starting from the date with a negative CU value, which is observed in autumn and corresponds to the period of the growing season end. The accumulation of GDH to the corresponding limit value is performed immediately after the maximum accumulation of CU values, up to the predicted start date of flowering for each stone fruit.

The total water content in the generative buds was determined in grams per gram of dry matter by drying the buds' samples at a temperature of 105 °C [15]. Mathematical processing of experimental data was provided through the methods of correlation and regression analysis [16]. To obtain analytical expressions for the velocity and acceleration of the process of hydration of the buds with the corresponding GDH value, the first and second derivatives of the functions obtained were calculated [17].

Anatomical and morphological changes of generative buds were studied according to the methodological guidelines [18].

## 3 Results and Discussion

Application of phenoclimatographic models and the statistical method of the smallest deviations [14] allowed to establish the limiting values of the CU index, which are necessary to be accumulated by apricot, peach and sweet cherry in order to release the period of deep dormancy. The limiting values for the GDH index have been defined as well. The accumulation of the GDH index is necessary to start flowering. It has been established that in order to release the period of biological rest *Melitopolskii Luchistyi* apricot needs to accumulate 940 °C CU, *Ivan Tupitsyn* peach requires 1200 and *Krupnoplodnaia* sweet cherry needs 1350 °C CU. To start flowering, the stone fruit crops of these varieties will need to be accumulated, respectively, at 3725 °C GDH, 4866 °C and 4839 °C GDH. A smaller value of the CU for apricot indicates an earlier period of its release from dormancy, which is consistent with the data of researchers [19] and indicates a weaker degree of frost resistance of this crop compared to peach or cherries. At the same time, the limiting value of GDH for apricot indicates the earlier terms of its flowering. After all, from these cultures, sweet cherry comes out of rest. However, in terms of the accumulation of GDH, the size of peach and cherry is almost the same. In most cases, that is confirmed by the timing of flowering of these crops in the garden.

Having determined the limiting values of phenoclimatographic indices, as a result of summing up their values for each day to the limiting values corresponding to 100% of CU and 100% of GDH, the dates for exiting the period of deep dormancy and the beginning of flowering for each culture of the corresponding variety for specific research years were predicted. A comparison of the estimated forecast dates of the onset of flowering and the actual dates of observations of the phenophase of the onset of flowering in the garden is presented in the table. The predicted dates of apricot emergence from dormancy do not contradict the findings of scientific studies about Melitopol conditions, where apricots leave dormancy in December to the second half of February depending on weather conditions and the varieties belonging to the ecologic and geographic group [20]. The later dates of dormancy breaking for peach and cherry are also consistent with their pomological characteristics on the basis of increased winter hardiness and cold resistance [21, 22]. Obtained by calculating the date of commencement of apricot flowering also does not contradict the fact that it blooms in earlier calendar periods than peach and cherry.

The results of validation of phenoclimatographic models by comparing the calculated and observed dates of the onset of flowering of apricot, peach, cherry in the field showed fairly high prediction accuracy (see Table 1). The discrepancy between the estimated and the actual dates of the start of flowering of stone fruit crops does not exceed three days. That, in turn, indicates the adequacy and representativeness of the phenoclimatographic models used in the climatic conditions of the Southern Steppe of Ukraine. It must be noted that the authors create models indicate that they are closely related. We also statistically proved that there is a functional nonlinear relationship between the values of accumulation of CU and GDH at $R^2 = 0.96$, $P = 0.01$.

Studies on the use and evaluation of phenoclimatographic models conducted in the climatic conditions of Iran [9], the mountainous region of Italy [10], Spain [23] and Japan [24] do not reject the possibility of their use and confirm rather high accuracy of prediction of the time needed for rest breaking and development of buds for fruit crops before flowering in comparison with other predictive models.

Over the years of research, we have established that the duration of deep dormancy (from beginning to completion) for apricot was 66 days on average, 93 days for peaches and 112 days for cherries. The period of accumulation to the GDH limit value, that is, the period after the dormancy breaking before the onset of flowering, was, on average, 114 days for apricot, 95 days for peach and 77 days for sweet cherries. It must be highlighted that the accumulation period of CU with the corresponding lower air temperatures is longer for sweet cherry rather than for apricot and peach. It means that the process of cooling generative buds for stone fruits in the autumn–winter period is more significant in their development regulation as well

**Table 1** Comparison of calculated and actual dates of the onset of flowering for apricot, peach and cherry

| Crop, variety | Test year | Estimated date of biological rest breaking | Date of flowering | | The difference between the dates of flowering, days |
|---|---|---|---|---|---|
| | | | Forecast | Actual | |
| *Melitopolskii Luchistyi* apricot | 2010 | 25.12.09 | 18.04 | 18.04 | 0 |
| | 2011 | 07.12.10 | 23.04 | 22.04 | +1 |
| | 2012 | 13.12.11 | 16.04 | 17.04 | −1 |
| | 2013 | 22.01.13 | 15.04 | 13.04 | +2 |
| | 2014 | 22.12.13 | 10.04 | 07.04 | +3 |
| | 2015 | 30.12.14 | 17.04 | 17.04 | 0 |
| *Ivan Tupitsyn* peach | 2010 | 05.02.10 | 24.04 | 22.04 | +2 |
| | 2011 | 28.12.10 | 27.04 | 28.04 | −1 |
| | 2012 | 31.12.11 | 20.04 | 23.04 | −3 |
| | 2013 | 10.02.13 | 21.04 | 19.04 | +2 |
| | 2014 | 12.01.14 | 17.04 | 17.04 | 0 |
| | 2015 | 28.01.15 | 24.04 | 21.04 | +3 |
| *Krupnoplodnaia* cherry | 2010 | 16.02.10 | 24.04 | 22.04 | +2 |
| | 2011 | 06.02.11 | 28.04 | 28.04 | 0 |
| | 2012 | 08.01.12 | 20.04 | 22.04 | −2 |
| | 2013 | 21.02.13 | 21.04 | 22.04 | −1 |
| | 2014 | 10.02.14 | 17.04 | 18.04 | −1 |
| | 2015 | 04.02.15 | 26.04 | 25.04 | +1 |

as degree of frost resistance control comparing to the need of plants for a thermal resource after dormancy breaking.

This means that due to the early release of apricot from a period of deep dormancy, when the probability of thawing in the second half of winter remains quite high, as well as with early periods of flowering under conditions of possible spring frosts, weather conditions can be the main cause of generative sphere damage. In other words, when placing apricot plantations on the territory of the Southern Steppe of Ukraine, it is necessary to take into account its anatomical, morphological and physiological features. Peach and cherry blossom later, but during this period the probability of spring frosts remains high. Therefore, it is necessary to take into account the demands of these stone fruits for heat when placing plantings in this soil-climatic zone.

After the trees leave the state of biological dormancy, the rate of GDH accumulation can be judged on the intensity of growth processes occurring in the buds of stone fruits up to the beginning of blossom. The scientists [24] found that in the winter–spring period, the water regime of the buds is closely related to extreme (minimum and maximum) air temperatures. At this time, the total water content

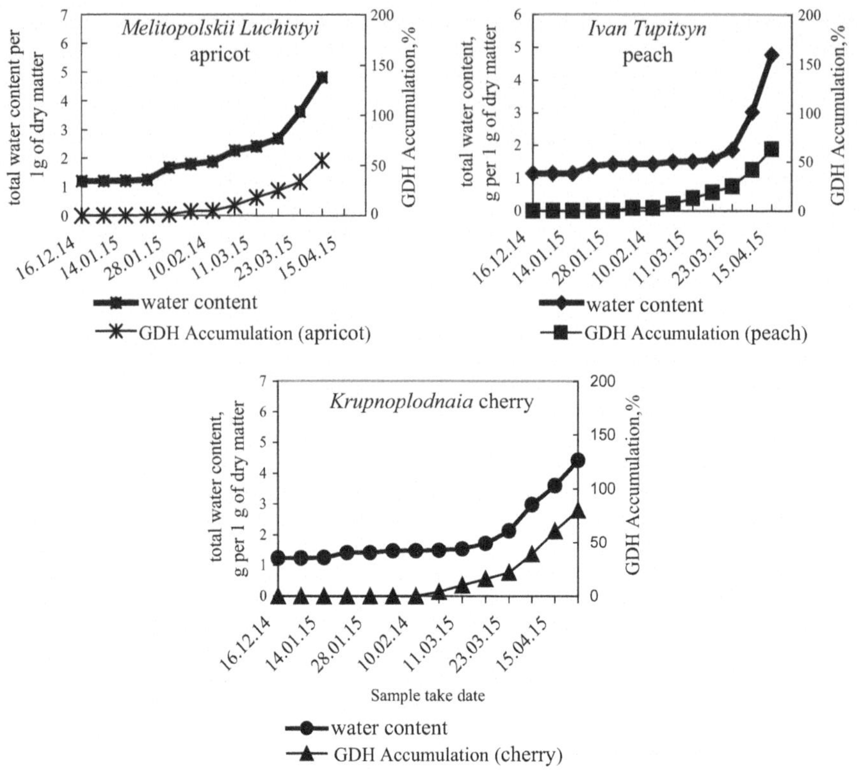


**Fig. 1** Dynamics of total water content in generative formations and GDH accumulation in the winter–spring period 2014–2015

increases in the generative buds of stone fruits. The intensity depends on the varietal characteristics and specific environmental conditions [25]. Based on the fact that the main input parameters of phenoclimatographic models are hourly maximum and minimum air temperatures, it can be assumed that there is a correlation between the water content of the generative buds of apricot, peach, cherry and GDH.

According to the presented dynamics of the total water content of the buds of the stone fruit crops under study in the winter–spring period, the rate of development of the generative organs and the intensity of GDH accumulation can be traced (see Fig. 1). Using the example of experimental data for the winter–spring period 2014–2015, the nature of changes in the total water content in the generative buds of apricot, peach and sweet cherry together with the accumulation of GDH of each stone fruit was established. Analyzing graphically presented data of water content of tissues of generative buds and the rate of accumulation of GDH, it was determined that both processes in their development tend to increase.

It has been established that in the winter period there is almost no increase in the total water content in the buds due to the fact that the trees have not released the state of deep dormancy yet. At this time, GDH accumulation is absent or does not exceed 1%. Then, in late winter to early spring, more intensive development of the buds begins, as evidenced by an increase in their water content and an increase in the rate of GDH. During the observation period, the total water content in the generative buds increased at different rates depending on the fruit species and weather conditions. At the beginning of the spring vegetation of stone fruit crops, more intensive growth of all indicators was observed. The highest degree of hydration of the reproductive formations of apricot, peach and cherry was noted before the trees blossom.

As a result of the regression analysis of multi-year data, a close nonlinear relationship was established between the indicator of generative buds' development in apricot, peach, cherry (by total water content), on the one hand, and the accumulation of GDH, on the other hand. Third-degree regression equations are obtained. Their reliability is confirmed by the coefficients of determination ($R^2 = 0.98$; 0.98; 0.95) that indicates the share of variations in the total water content in the buds of apricot, peach and cherry according to the action of the factor under study (GDH accumulation). The graphs (see Fig. 2a) clearly demonstrate the dependence of the water content in the generative buds of the studied crops on GDH accumulation.

The calculated first derivatives of the obtained functional dependencies, which characterized the intensity of generative formation development in apricot, peach and cherry at different air temperatures, were taken into account when determining GDH. It enabled to identify the patterns of this process (see Fig. 2b). A graphic representation of the values of the first derivatives showed that there were two stages of the growth rate (as for the total water content in the buds) in all stone fruits studied. The first was characterized as a stage with a slower pace of generative buds' development. Its duration lied in the range from 0 to 43% of GDH accumulation. In other words, at this stage, there was some inhibition of increasing the total water content up to the value of 43% GDH. The inflection points of the functions (the tops of the parabolas) correspond to the value of 43% GDH. After that, the second stage began. It had accelerated growth rate with a more rapid increase in the total water content in the generative formations of apricot, peach and cherry until the beginning of their blossom.

The calculations of the second derivatives characterized the rate of hydration of the generative buds of apricot, peach and cherry depending on the temperature conditions during the observation period. Graphs represent the nature of the change in acceleration (see Fig. 2c). The analysis of the presented graphs confirmed their general tendency that the point of intersection of the line in the acceleration function through the GDH axis, regardless of the stonecrops being studied, corresponded to 43% of the GDH.

That showed the possibility of predictions for the rate and acceleration of generative buds' development in winter and at the beginning of the growing season according to the phenoclimatographic indicator of GDH accumulation and the resulting functional dependencies.

Therefore, the water content in generative formations of apricot, peach and cherry increases with different intensities. It depends on the varietal characteristics and specific weather conditions (air temperature) during the winter–spring period of

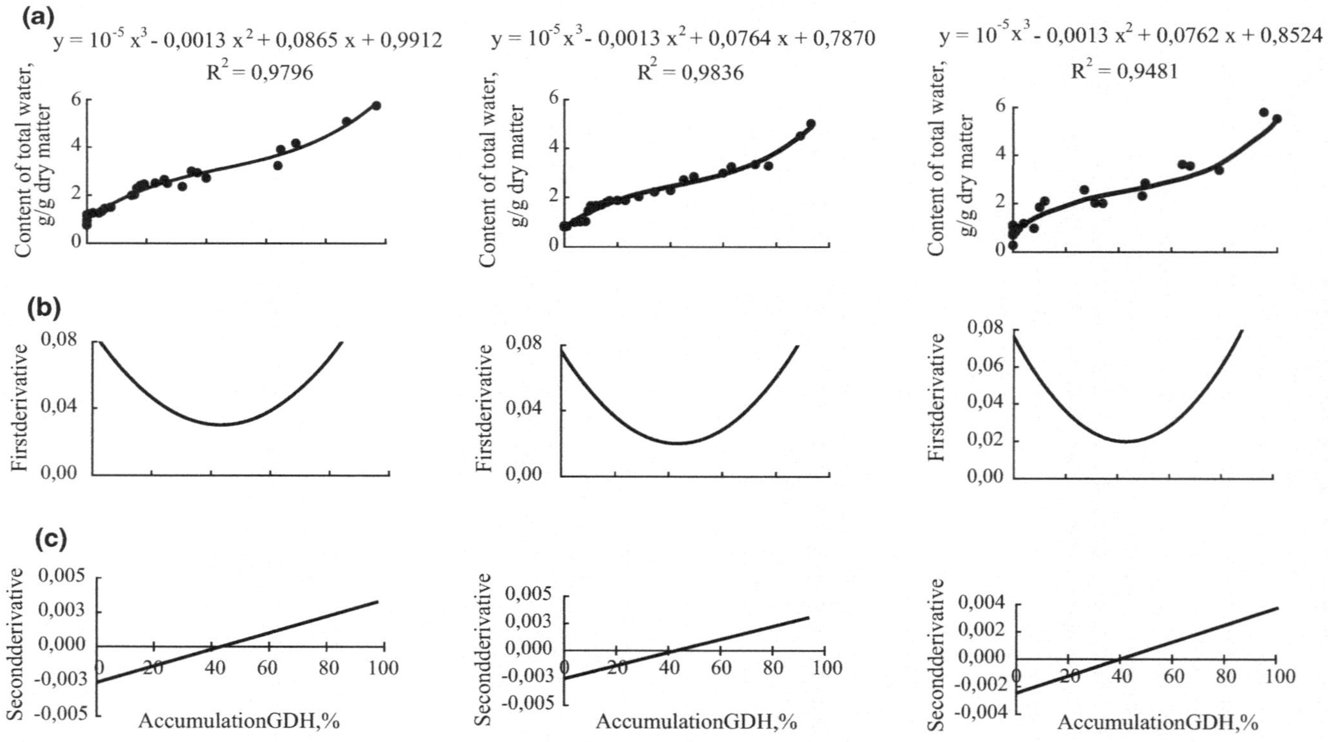


**Fig. 2** Dependence of water content in apricot (**A**) generative buds, peach (**B**), sweet cherry (**C**) on GDH changes (a) and graphs of their first (b) and second derivatives (c)

the year. As a result of the analysis of the obtained data, a general pattern of the relationship between the total amount of GDH and the level of water content of the generative buds was revealed. It was established that a significant increase in the intensity of the total water content occurred after 43% of GDH for all crops studied.

Studies of the anatomical and morphological features of the internal structure development in the anthers of apricot, peach and cherry blossoms showed that in the period from October to the end of January and the beginning of February, archesporial tissue was formed. That corresponded to a period of deep or biological rest. At this stage, generative buds had maximum resistance to negative air temperatures [26]. Application of phenoclimatographic model enabled to determine the date of biological rest breaking for studied crops. In 2014–2015, *Melitopolskii Luchistyi* apricot released from the state of biological rest on December 30, 2014, *Ivan Tupitsyn* peach on January 28, 2015, and sweet *Krupnoplodnaia* cherry on February 4, 2015 (see Table 1). These forecast dates were the starting point for the start of GDH accumulation. After the trees had left the state of biological rest, the generative buds were ready for their further development, but negative air temperatures hampered the onset of the subsequent stages of morphogenesis.

Therefore, the formation of maternal microspore cells proceeded somewhat slowly and lasted until the end of December for the apricot, until the end of January for the peach. For the cherries, this stage ended in early February. At the indicated time, GDH accumulation for all crops was up to 1%. The subsequent stage of reduction division with the formation of microspores was noted in apricot and peach on the first days of February, and in black cherry (at the beginning of the second decade of March). That led to a decrease in frost resistance of the buds. In this case, the accumulation of GDH was in the range from 5 to 10%. Subsequently, the tetrads decayed into microspores, forming separate pollen grains. When the stone fruit crops were passing that stage, they accumulated 11–42% of GDH. The weather conditions in the third decade of March to the second decade of April contributed to the further development of pollen. That period, there was a more active GDH accumulation, which reached 85% by the middle of April. By the time the generative buds dissolved in the pollen grains of all the studied crops, nuclear fission was observed with the formation of two-cell pollen with GDH of 90%. At that stage, frost resistance of generative formations was completely lost [26]. On apricot plantations, blossom was noted on April 17, 2015, peach—April 21, 2015, and sweet cherries—April 25, 2015.

Taking into account those results, we can conclude that the magnitude of the quantitative accumulation of GDH, along with the corresponding stages of the morphogenesis of generative buds (male gametophytes), gives an idea of the rate of their development from the moment they enter biological rest until buds open. It should be noted that more intensive accumulation of GDH occurred after achievement of that indicator of 42%, when microspore was formed in pollen grains and unicellular pollen began to form. The data obtained are consistent with the above material on a significant increase in the total water content in the buds after 43% GDH. As our studies have shown, accumulation of GDH for apricot, peach and cherry reached that magnitude during March. After that period, stable positive air temperatures were characteristics for the climatic zone.

That entails more intensive development of the generative organs of stone fruits. In such favorable weather conditions, the only exception is the probability of spring frosts and, as a consequence, the possibility of generative sphere damage for apricot, peach and cherry. In this regard, the largest accumulation of GDH can predict the physiological state of wintering and vegetative plants in extreme weather conditions and predict the magnitude of their potential productivity.

The establishment of the above-mentioned regularities made it possible to develop algorithms and devices for their implementation into controlling the system of fine sprinkling.

The main control functions are defined by local arithmetic logic procedures, which can be represented as follows:

$$U_{n,m} = \begin{cases} U_1, & \text{if } Y_1 < \text{con } X_{n,m} \le Y_2; \\ U_2, & \text{if } Y_2 < \text{con } X_{n,m} \le Y_3; \\ \dots & \dots \ \dots \ \dots \ \dots \ \dots \ \dots \\ U_L, & \text{if } Y_{L-1} < \text{con } X_{n,m} \le Y_L. \end{cases}$$

where

$X$ is the matrix of the internal conditions of the control algorithm,
$Y$ is the matrix of input signals from measuring sensors,
$U$ is the matrix of output signals to the executing devices of the irrigation system.

On the basis of the obtained CU limit values, the date of dormancy breaking was determined, and according to GDH, the dates of flowering for a particular year were predicted as well. According to the data obtained, the period of exposure of the plants to fine sprinkling was established, in other words, the onset and completion of evaporative cooling with the help of combined irrigation systems (with simultaneous activation of the above-crown and subcrown parts of the system). It has been established that evaporative cooling of the buds should begin on those dates when the accumulation of GDH reaches 30% of the maximum amount required to start flowering and end at 100%. Observations on the morphogenesis of apricot buds showed that at the time of switching on the irrigation system they were at the stage of 'microspore' development. When the system was turned off, it was 'pollen formation.'

In order to save irrigation water and the greatest effect of cooling the buds, it was determined to sprinkle during the daytime at an ambient temperature of ≥7 °C.

According to the indications of temperature changes in the generative buds of apricot and peach, the mode of operation of the irrigation system (irrigation pause) was determined. The duration of watering corresponded to the time when the buds of the trees were completely wetted with irrigation water. It amounted two minutes. The duration of the pause depended on the time of complete evaporation of water from the surface of the buds and ranged from 5 to 30 min depending on the influence of weather conditions of each particular day on the temperature of the buds. Automatic control of temperature changes in the generative buds was performed by means of a sensor (differential copper–constantan thermocouple with a self-recording potentiometer

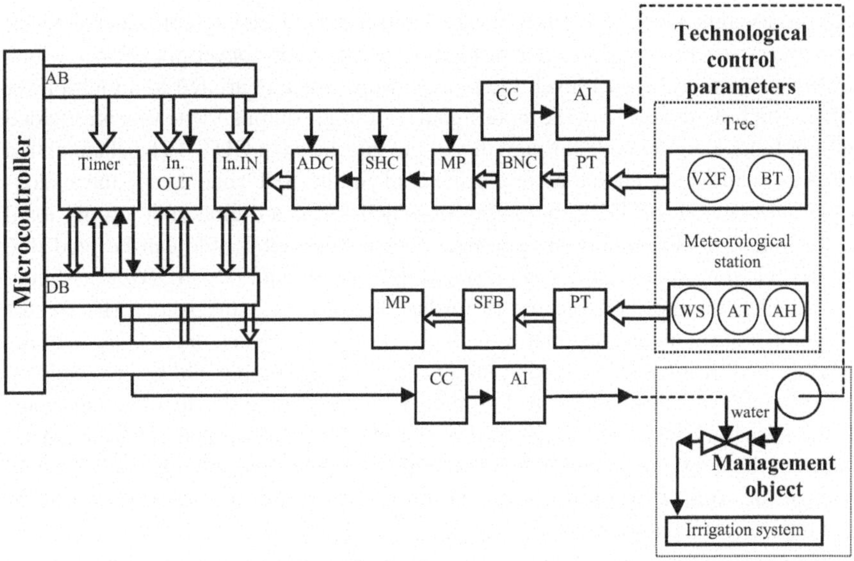


**Fig. 3** Functional block diagram of the irrigation control device: In.IN is an input interface; In.OUT is an output interface; PC is a primary converter interface; BNP is a normalizing converter block; MP is a multiplexer; SHC is sample and hold circuit; ADC is an analog-to-digital converter; DB is a data bus; AB is an address bus; CC is a control circuit; AI is an actuator interface; SFB is a signal formation block; Timer is a timer; VXF is a sensor of xylem flow velocity; BT is a sensor of a bud/leaf temperature; WS is a wind speed sensor; AT is an air temperature sensor; AH is an air humidity sensor

output). It was established that watering should resume when the temperature difference on the sensitive sensor elements reaches 2.8 °C. This principle was taken as the basis for the development of a bud cooling sensor, which consisted of four dry and four permanently wetted differential copper–constantan thermocouples, imitating dry and moistened surfaces of buds.

For the full automation of plant irrigation management, a special device has been developed. According to the established physiological parameters, the device provided such technological elements of irrigation control as 'start,' 'restore,' 'irrigation duration' signals and established the 'irrigation—pause' mode. The control system provided the automatic collection of information from the plant objects and meteorological changes in the environment as well as transmission of a control signal to the executive mechanisms of the irrigation system and the recording of incoming and outgoing information.

The functional–structural diagram of the irrigation control device is shown in Fig. 3.

The diagram provides the connection of the control object with the microcontroller data bus (MC) using the interface circuits of «In.IN». The technological parameters of the object (temperature and humidity, xylem flow velocity, etc.) in the interfaces of the primary transducers (PTs) are converted into electrical signals (constant voltage or frequency). After passing through the block of normalizing converters «BNP», which provides a standard signal level, the monitored parameters are fed to a multiplexer (MP), which switches one of the input signals to a single output. Switching is provided by supplying a digital code through the output interface (In.OUT.). The channel that has been switched is fed to the sample and hold circuit (SHC) and then to the analog-to-digital converter (ADC), the output of which forms a digital code proportional to the value of the monitored parameter. Then, the digital code can be read into the MP via the input interface (In.IN) and the system data bus (SDB). The digital code, which is read, is subjected to further digital processing in the MC for the irrigation control algorithms that were developed earlier. When the indication of the measurement result was necessary, the resulting information could be presented on a digital indicator through In.OUT. When, according to the results of calculations, it was necessary to give a signal to the control object (pump and electromagnetic valves of the irrigation system), In.OUT could be used to transmit the control signal through the control circuit (CC) and the interface of the actuator interface (AI) to turn on or off the pump or valve.

The design of the control circuit significantly depends on the type of the actuator [27]. The actuators, in our case, were contactless relay devices. To control them it was enough to send a signal to the input, which accepted only two states: low or high. The control circuit in that case had to perform the functions of a power amplifier operating in key mode. In the case when the monitored parameter was converted to a frequency, the procedure for introducing it into the processor was greatly simplified and after being formed in the signal formation block (SFB) and switching in the MC, it reduced to supplying a timer to the input. All other transformations associated with the calculation of the monitored parameter value were provided by the control program. In that case, the timer could also be used to form a control signal in a CC.

## 4 Conclusions

The obtained results showed that in the conditions of the Southern Steppe of Ukraine, it was possible to use phenoclimatographic models with a fairly high accuracy of forecasting the dates of the onset of blossom for apricot, peach and cherry. Phenoclimatographic model, which was based on the use of hourly maximum and minimum air temperatures, allowed to take into account the needs of stone fruits in certain temperature conditions necessary for their development and growth in the autumn–winter–spring period with various changes in weather conditions of a particular year.

According to the phenoclimatographic indicator of GDH accumulation and the resulting functional dependence, it was possible to predict the rate of physiological development of generative formations of stone fruits at different stages of morphogenesis in winter after trees break their biological rest and at the beginning of vegetation, i.e., during blossom.

The irrigation control device, which implements the specified diagram, could be connected to a wide range of sensors, as well as various types of standard actuators. Setting various matrices of the internal conditions of the control algorithm and changing the matrix of input signals from the sensors enabled easy and flexible adjustment of the output signal matrix to the executing devices of the irrigation system. That was the settings of the specified irrigation mode.

## References


1. Ivanov, V.F., Ivanova, A.S., Opanasenko i dr, N.Y.: Ekologiya plodovykh kul'tur. Kiyev: Agrarna nauka, 408 s (1998)
2. Vazhov, V.I., Sholokhov, A.M.: Agroklimaticheskaya otsenka usloviy perezimovki abrikosa v Krymu. Trudy Gosudarstvennogo Nikitskogo botanicheskogo sada. Pochvennyye i bioklimaticheskiye osobennosti sadovykh agrotsenozov Kryma. Yalta, T. LXXXIV, S. 78–87 (1981)
3. Eggert, F.P.: A study of rest in several varieties of apple and other fruit species grown in New York state. Proc. Am. Soc. Hort. Sci. **57**, 169–178 (1951)
4. Weinberger, J.: Chilling requirement of peach varieties. Proc. Am. Soc. Hort. Sci. **56**, 122–128 (1950)
5. Gelreath, P.R., Bushanan, D.W.: Rest prediction model for low chilling sungoldnectarine. Proc. Am. Soc. Hort. Sci. **116**, 270–273 (1981)
6. Erez, A., Fichman, S., Linsley-Noakes, G.C.: The dynamic model for rest completion in peach buds. Acta Hort. (276):165–174 (1990)
7. Odintsova, V.A.: Opredeleniye vykhoda plodovykh derev'yev iz perioda pokoya s pomoshch'yu fenoklimatograficheskikh modeley. Sadovodstvo i vinogradarstvo, №. 6, S. 10–12 (2005)
8. Odyntsova, V.A.: Zastosuvannya fenoklimatohrafichnykh modeley u rozvytku heneratyvnykh brun'ok abrykosa. Sadivnytstvo, Vyp.57, S. 87–91 (2005)
9. Razavi, F., Hajilou, J., Tabatabaei, S.J., Dadpour, M.R.: Comparison of chilling and heat requirement in some peach and apricot cultivars. Res. Plant Biol. **1**(2), 40–47 (2011)
10. Rea, R., Eccel, E.: Phenological models for blooming of apple in mountainous region. Int. J. Biometeorol. **51**, 1–16 (2006)
11. Richardson, E.A., Seeley, S.D., Walker, D.R.: A model for estimating the completion of rest for 'Redhaven' and 'Elberta' peach trees. Hort. Sci. **9**(4), 331–332 (1974)
12. Richardson, E.A., Anderson, J.L., Hatch, A.H., Seeley, S.D.: ASYMCUR, am asymmetric curvilinear fruit tree model. In: Abstracts XXI International Horticulture Congress, vol. 2, p. 2078 (1982)
13. Anderson, J.L., Richardson, E.A.: Validation of chill unit and flower bud phenology models for 'Montmorency' sour cherry. Acta Hort. **184**, 71–74 (1986)
14. Ashcroft, G.L., Richardson, E.A., Seeley, S.D.: A statistical method of determining chill unit and growing degree hour requirements for deciduous fruit buds. Hort. Sci. **12**, 347–348 (1977)
15. Yablonskiy, Ye.A.: Dinamika soderzhaniya vody v generativnykh pochkakh i odnoletnikh pobegakh razlichnykh po zimostoykosti sortov persika. Fiziologiya i biokhimiya kul'turnykh rasteniy, T. 13, №. 2, S. 200–205 (1981)
16. Dospekhov, B.A.: Metodika polevogo opyta (s osnovami statisticheskoy obrabotki rezul'tatov issledovaniy)– 5-ye izd., dop. i pererab. Moskva: Agropromyshlennoye izdatel'stvo, 351 s (1985)
17. El'sgol'ts, L.E.: Differentsial'nyye uravneniya i variatsionnoye ischisleniye. Moskva: Nauka, 424 s (1969)
18. Sholokhov, A.M.: Izucheniye morfogeneza tsvetkovykh pochek v svyazi s sortoispytaniyem i selektsiyey kostochkovykh na zimostoykost': Metodicheskiye ukazaniya. Yalta: Nikitskiy sad, pechatnyy tsekh nauchnoy informatsii, 14 s (1972)

19. Maksimova, I.N., Fedchenkova, G.A.: Zimneye-vesenneye razvitiye tsvetkovykh pochek abrikosa i ikh zimostoykost'. Biologicheskiye nauki, № 9, S. 75–78 (1974)
20. Fedchenkova, G.A.: Nekotoryye osobennosti prodolzhitel'nosti perioda pokoya u abrikosa v usloviyakh yuga stepnoy chasti Ukrainy. Yuzhnoye stepnoye sadovodstvo. Dnepropetrovsk: Promín', S. 218–222 (1973)
21. Turovtseva, M.I., Turovtsevoyi, V.O.: Rayonovani sorty plodovykh i yahidnykh kul'tur selektsiyi Instytutu zroshuvanoho sadivnytstva: dovidnyk/ za red. Kyev: Ahrarna nauka, S. 44 (2002)
22. Shestakov, I.I.: Persik. Ukazatel' zakonchennykh rabot/ sost. Tuz,A.S., Yakovlev, G.V., Dikiy, S.P. Maykop: Adygeyskoye otdeleniye Krasnodarskogoknizhnogoizdatel'stva, S. 66 (1973)
23. Alanco, J.M., Espiau, M.T., Sociasi Company, R.: Increase in the chill and heat requirements for blooming of the new almond cultivars. Options Méditerranéennes (94):65–69 (2010)
24. Pawasut, A., Fujishige, N., Kamane, Y., Yamaki, Y., Yonsjo, H.: Relationships between chilling and heat requirement for flowering in ornamental peaches. J. Japan. Soc. Hort. Sci. **73**(6), 519–523 (2004)
25. Yablonskiy, Y.A.: Tempy rosta plodovykh pochek i zimostoykost' sortov abrikosa, persika i mindalya. Trudy Gosudarstvennogo Nikitskogo botanicheskogo sada. Issledovaniya po fiziologii, embriologii i radiologii rasteniy, T. XLVI, S. 50–61 (1970)
26. Yelmanov, S.I., Yablonskiy, Y.A., Sholokhov, A.M.: Anatomo-morfologicheskiye i fiziologicheskiye issledovaniya tsvetkovykh pochek abrikosa v svyazi s ikh zimostoykost'yu. Trudy Gosudarstvennogo Nikitskogo botanicheskogo sada. Simferopol', T. XL, S. 65–79 (1969)
27. Olsson, G.: Piani Dzhanguido. Tsifrovyye sistemy avtomatizatsii i upravleniya. SPb.: Nevskiy Dialekt, 557 s (2001)

# Vacuum Cooling Technology for Pre-cooling of Cherry Fruits

**Oleksandr Lomeiko, Lilia Yefimenko and Vira Tarasenko**

## 1 Introduction

Formulation of the problem. Cherry—(Latin Prúnus ávium)—a tree plant, family Rosaceae. Cherries contain organic acids, sugars (fructose, glucose), vitamins C, A, B1, B2, E, PP, trace elements (iron, iodine), macroelements (potassium, calcium, magnesium, etc.), pectin substances, as well as large number of anthocyanins—substances from the group of flavonoids.

Cherry fruits are very useful for human health. They are in high demand through a pleasant sweet taste. The great demand for freshly harvested cherry fruits results in the growth of fruit production, the growth of available, and the emergence of new fruit and berry farms. Therefore, the problem of cooling and storing of cherry fruits are becoming actual [1].

Cherry fruits have the best flavors right after removing from the tree. Long-term storage in refrigerators dramatically worsens their taste. The primary objective of pre-cooling and storage is to create conditions in which the loss of nutrients would be minimal and the quality of products would remain the same as after harvesting [1].

Freshly harvested cherry fruits immediately suffer from high temperatures and loss of moisture, and therefore loss of vitamins, aromatic substances, and nutrients [2]. The product state of fruits without any additional measures is lost immediately, the sales period decreases, the price reduces. To reduce the speed of these processes, it is necessary to create certain collection, pre-cooling, sorting, and packaging conditions.

Cherry fruits are perishable products, and after 6 h after harvest, they lose half of their quality at temperature of 30 °C [3]. If, after harvesting, they are not cooled down as soon as possible, then the storage result will be unsatisfactory due to the lost quality of the fruits. Thus, to ensure that the harvested fruits are not lost, it is necessary to carefully observe the cooling and storage technologies. After harvesting,

O. Lomeiko (✉) · L. Yefimenko · V. Tarasenko
Department of Processing and Food Production Equipment named after professor F. Yalpachik, Tavria State Agrotechnological University, B.Khmelnytsky Ave 18, Melitopol 72312, Ukraine
e-mail: oleksandr.lomeiko@tsatu.edu.ua

© Springer Nature Switzerland AG 2019 
V. Nadykto (ed.), *Modern Development Paths of Agricultural Production*,
https://doi.org/10.1007/978-3-030-14918-5_29

it is necessary to provide pre-cooling of the cherry fruits as soon as possible. Fresh cherries of the most resistant varieties can be stored for 15 days, under to certain conditions. Optimal storage temperature of fresh cherries is 1–2 °C and relative humidity 80–90%.

Thus, the study of the processes of cooling and storing cherry fruit with the aim of extending shelf life and reducing weight loss is very important.

### *1.1 Analysis of Recent Research*

Analysis of the latest research and publications in the world indicates that technology of vacuum cooling of agricultural products is an extremely fast method of evaporative cooling, whose high efficiency achieves by reducing the time of the technological process [4]. Fruit and vegetable products that contain enough of water can be cool by the vacuum cooling process, which increases the efficiency of evaporation by reducing the pressure and boiling water temperature [5].

A high cooling rate is the main advantage of vacuum cooling. In addition, the loss of weight of fruits and vegetables cooled by vacuum can be reduced by adding a certain amount of water [6]. Vacuum cooling used as pre-cooling for products such as lettuce, mushrooms, broccoli, artichokes, carrots, cucumbers, peppermint, dill, arugula, green onions, and cut flowers in order to extend shelf life and improve quality [7].

## 2 Statement on the Purpose and Objectives of the Study

The purpose of this article is to investigate the effect of the vacuum cooling process on the quality and shelf life of cherry fruits.

To achieve the set objective, the following tasks were solved:

- to study patterns of influence of vacuum cooling parameters on the period of storage of cherry fruits;
- to study ways to reduce the weight loss of sweet cherry fruits in the process of vacuum cooling.

## 3 Solution

Because of theoretical studies, the regionalized varieties of sweet cherries of the late reaches were selected on the complex of economic and biological indicators: Krupnoplidna, Melitopolska Chorna, and Udivitelna, which were entered in the register of varieties of Ukraine [1]. Freshly picked cherry fruits were delivering in the experimental laboratory in the mornings. The temperature of the cherry fruits during this time was 25 °C. The weighing of the fruits before and after the cooling process carried out using electronic scales with an accuracy of ±0.01 g. Tests were carried out in the developed experimental installation for vacuum cooling of plant material [8].

The operating principle of the installation is as follows. The fruits of cherry are loaded into the chamber 6, placed on the shelf. The chamber closes with a hermetic lid. A vacuum pump and a compressor start to work. The pressure in the chamber reduced to the saturated pressure value. When the pressure in the vacuum chamber reaches the value of the initial working pressure, there is a flash point of the vacuum cooling process, and water begins to evaporate. After cooling of fruits to a given temperature the vacuum pump switches off, the vacuum fills. With hot air or water from the evaporator, it removes, and meltwater collected in the lower part of the chamber and due to the angle of the installation of the cylindrical chamber removes. After removing the meltwater, the chamber is ready for the next batch of fruits. The main advantage of vacuum cooling is the relatively high speed [5, 7] Studies have shown that in order to cool the cherry fruits from 25 to 2 °C it takes 40 min. In addition, the temperature on the surface and inside the fruit reduces evenly. While in conventional refrigeration, this process lasts much longer: for the surface of the cherry fruit—75 min, for cooling the middle of the fruit—200 min. Thus, the cooling is uneven. The results of the studies are presented (see Figs. 1 and 2).

For the purpose of studying the weight loss, vacuum cooling of the fruits of cherry carried out by three different methods at the most rational value of pressure 29 kPa [4].

In the first method, the thermocouple was installed at the center of the samples to accurately measure the temperature of the fruit center of cherry fruits after weighing and placement in a vacuum chamber.

The second thermocouple was freely suspended in the center of the camera. The external ambient temperature was measured using a third thermocouple, located around. The values of weight loss, temperature, time, and pressure were recording during the tests (Fig. 3).

In the second method, water was sprayed on the fruits of cherry after weighing. While in the third method, the product was covered with a polyethylene film after spraying water on the fruits.

Measurement of temperature in the second and third methods was carried out in the same way with the first method.

The period of cooling of cherry fruits from a temperature of 25 °C to a point where the product temperature reaches 2 °C is about 40 min with pressure in the vacuum chamber 29 kPa.

The test was finished at this point, because further reduction of pressure and an increase of cooling period lead to freezing of the product and consequently a decrease in its market value (Fig. 4).

Weight loss $G$, % is calculated by the formula:

$$G = \frac{G_{\text{in}} - G_i}{G_{\text{in}}}, \quad (1)$$

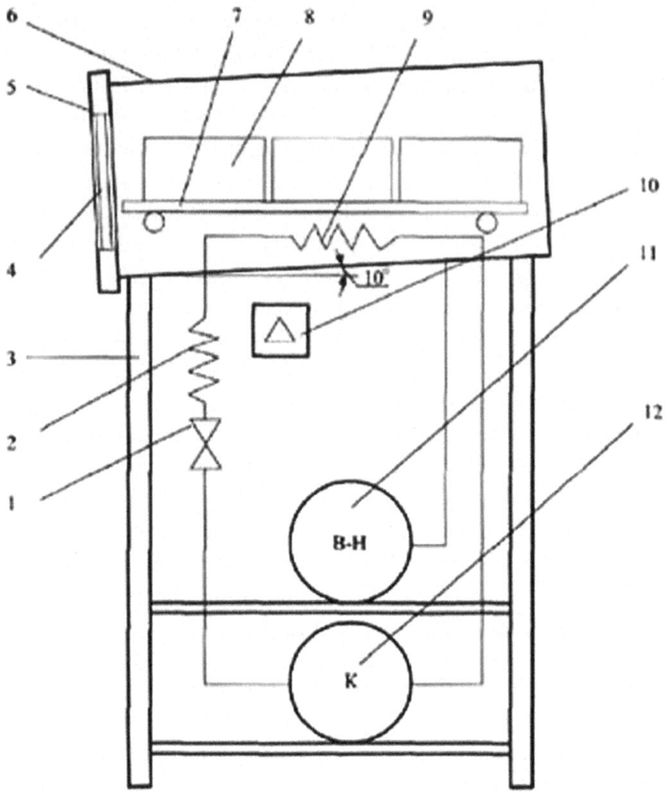


**Fig. 1** Scheme of an experimental installation for vacuum cooling of plant raw materials 1—thermostatic valve; 2—capacitor; 3—bed; 4—window; 5—lid; 6—vacuum chamber; 7—shelf; 8—product; 9—evaporator; 10—measuring devices; 11—vacuum pump; 12—compressor

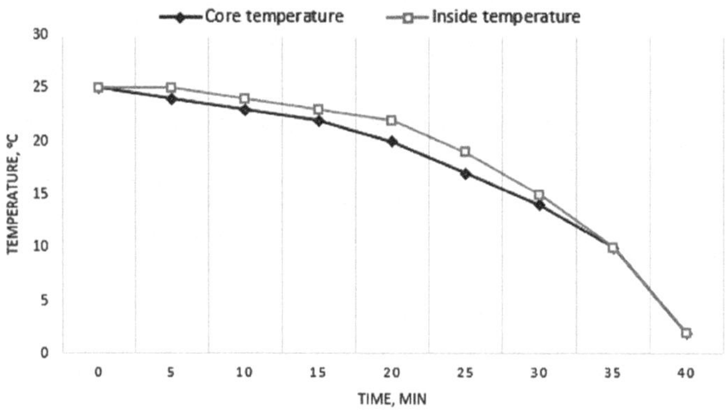


**Fig. 2** Reduction of core and inside temperature of cherry fruits in the process of vacuum cooling. The disadvantage of vacuum cooling is the loss of the weight of fruits and vegetables due to the evaporation of water [5, 9]

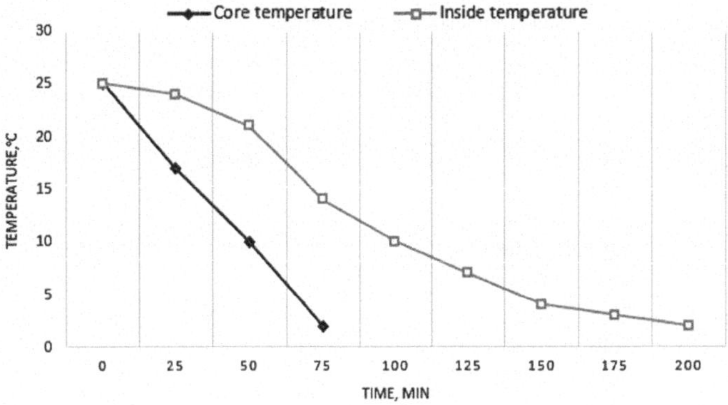


**Fig. 3** Reduction of core and inside temperature of cherry fruits in the process of refrigeration

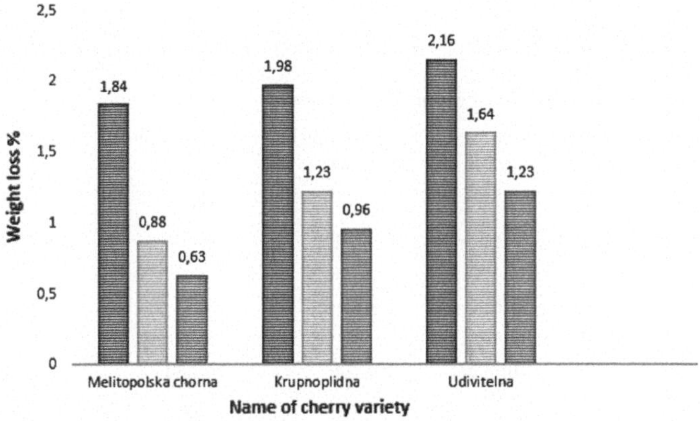


**Fig. 4** Weight loss of cherry fruits in the process of vacuum cooling

where

$G_{in}$ initial weight of cherry fruits, kg;
$G_i$ weight of cherry fruits at the test time, kg.

Analysis of the diagram shows that the highest value of mass loss in the ordinary vacuum cooling without the addition of water. These values for the varieties Melitopolska Chorna, Krupnoplidna, and Udivitelna are 1.84; 1.98; 2.16%, respectively. It is obvious that the weight loss value decreases when the product is vacuum cooled with sufficient moisture. The value of mass loss when adding water is 0.88% for Melitopolska Chorna's cherry fruits, 1.23% for the Krupnoplidna variety, 1.64% for the Udivitelna variety. The lowest values of loss of mass are when sprayed with water and enclosed with a polyethylene film: 0.63; 0.96; 1.23%, respectively, for Melitopolska Chorna, Krupnoplidna, Udivitelna (Table 1).

With the purpose of extending the shelf life and preserving the quality of cherry fruits, the regularities of the dynamics of biochemical substances and organoleptic

**Table 1** Parameters and regimes for storing of cherry fruits

| Storage method | Pretreatment | Storage temperature (°C) | Relative humidity (%) | Duration of storage |
|---|---|---|---|---|
| Refrigeration storage (control) | Fruits of cherries are collected, put in box pallets and sent for storage | 1–2 | 80–90 | 14 |
| Refrigeration storage with pre-vacuum cooling | The collected fruits of cherry are cooled in a vacuum chamber and sent to a refrigeration storage | 1–2 | 80–90 | 21 |

**Table 2** Weight loss of cherry fruits during vacuum cooling process

| Name of the cherry fruit's variety | Melitopolska Chorna | Krupnoplidna | Udivitelna |
|---|---|---|---|
| Initial weight (g) | 200 | 200 | 200 |
| Final weight (g) (without the addition of water) | 196.32 | 196.04 | 195.68 |
| Weight loss (%) (without the addition of water) | 1.84 | 1.98 | 2.16 |
| Final weight (g) (with the addition of water) | 198.24 | 197.54 | 196.72 |
| Weight loss (%) (with the addition of water) | 0.88 | 1.23 | 1.64 |
| Final weight (g) (with the addition of water and film coated) | 198.74 | 198.08 | 197.54 |
| Weight loss (%) (with the addition of water and film coated) | 0.88 | 1.23 | 1.64 |
| Initial cooling temperature (°C) | 25 | 25 | 25 |
| Final cooling temperature (°C) | 2 | 2 | 2 |
| Cooling time (h) | 0.66 | 0.67 | 0.67 |
| Initial pressure in vacuum chamber (kPa) | 101.3 | 101.3 | 101.3 |
| Working pressure in vacuum chamber (kPa) | 29 | 29 | 29 |

properties of cherry fruits under vacuum cooling and storage were studied. Table 2 shows comparative data of parameters and regimes of conventional cold storage (control method) and cold storage with the previous vacuum cooling of cherry fruits.

The results of the studies are presented in Tables 2 and 3.

**Table 3** Physico-chemical quality in dices of cherry fruits

| Indicator | Cherry fruits | | |
|---|---|---|---|
| | Immediately after harvesting | After refrigeration storage | After refrigeration storage with vacuum pre-cooling |
| Shelf life (days) | 0 | 14 | 21 |
| Dry matter content (%) | 17.3–19.8 | 16.37–18.79 | 17.05–19.56 |
| Sugar content (%) | 11.8–13.0 | 10.8–11.4 | 11.2–12.6 |
| Total acidity (%) | 0.53–0.68 | 0.48–0.62 | 0.5–0.65 |
| Vitamin C content (mg/100 g) | 11.2–12.4 | 4.5–5.1 | 6.9–8.1 |
| General sensory evaluation | 5 | 3.2 | 4.9 |

1. Vacuum cooling is a quick and effective method for cooling cherry fruits compared to conventional refrigeration.
2. The time of vacuum cooling of fruits of cherry fruits varieties Melitopolska Chorna, Krupnoplidna, and Udivitelna from a temperature of 25 to 2 °C is 40 min. Cooling on the surface and inside the fruit goes evenly.
3. The mode of vacuum cooling at a pressure of 29 kPa is rational to cool the fruits of cherry. Pressure reduction in the vacuum chamber from the atmospheric to the working takes about 5 min. The flash point in the vacuum cooling process occurs at a pressure of 29 kPa. Further lowering of pressure leads to freezing of products.
4. In the process of vacuum cooling of cherry fruits, the weight loss of varieties Melitopolska Chorna, Krupnoplidna and Udivitelna are 1.84; 1.98; 2.16%, respectively. Spraying water on the fruits of sweet cherry before vacuum cooling can reduce the weight loss values of 0.88, 1.23; 1.64%.

The lowest values of mass loss are when spraying water and coating with a polyethylene film: 0.63; 0.96; 1.23%, respectively, for Melitopolska Chorna, Krupnoplidna, and Udivitelna. It can be concluded that the spraying of water onto a product with subsequent coating with a polyethylene film before vacuum cooling is a factor that greatly reduces mass loss.

## 4 Conclusion

Due to vacuum cooling, the duration of short-term storage of cherry fruit is extended by 7 days. In addition, the fruits have a higher commercial quality, in particular, dry soluble substances by 3.88–3.9%, titrated acids by 1.69–3.07%, total sugars by 3.4–9.21%, vitamin C on 21, 43–25, 42% more than after the usual cold storage. The total organoleptic evaluation of the fruit with the previous vacuum cooling is 34% higher than in the control variant. Thus, it can be concluded that vacuum cooling is a fast and effective method for cooling cherry fruit as compared to conventional refrigeration.

## References

1. Turovtsev, M.I., Turovtsev, V.O.: Distributed varieties of fruits and berry crops of irrigated horticulture institute selection. Agrarian Science, Kyiv, 218 p (2002)
2. Lomeiko, O.P., Efimenko, L.V.: Theoretical study of the technology of vacuum cooling in the storage of crop production. In: Proceedings of the Tavria State Agrotechnological University, vol. 15, pp. 56–65 (2015)
3. Mishchenko V., Parfenova, N.: Biochemical and biological features of new varieties of cherries in the conditions of the Crimea. In: Gardening, no. 56, pp. 154–158 (2005)
4. Lomeiko O.P., Efimenko, L.V.: Using of vacuum cooled method for pre-cooling of cherry fruits. In: Current Problems of Energy and Ecology, pp. 276–279 (2016)
5. McDonald, K., Sun, D.W.: Vacuum cooling technology for the food processing industry: a review. J. Food Eng. **45**(2), 55–65 (2000)
6. Wang, L., Sun, D.W.: Rapid cooling of porous and moisture foods by using vacuum cooling. Trends Food Sci. Technol. **12**(5–6), 174–184 (2001)
7. Sun, D.W., Liyun, Z.: Vacuum cooling technology for the agri-food industry: past, present and future. J. Food Eng. **7**(2), 203–214 (2006)
8. Utility model patent UA 63714 MPK (2011/01) F25B 40/00 Installation for vacuum cooling of vegetable raw material. Stated. Jan 1/2010, published 25 Oct 2011, Bul. No. 20
9. Brosnan, T., Sun, D.W.: Compensation for water loss in vacuum pre-cooled lily flowers. J. Food Eng. **79**, 299–305 (2001)

# Sowing Units for Drilling Vegetable Crops

**Volodymyr Tarasenko, Larysa Bondarenko, Nina Shcherbakova and Natalia Horbova**

## 1 Introduction

One of the main tasks in the sowing vegetable crops is to ensure the best conditions for germination of seeds, as well as to obtain their optimum density with uniform distribution in the area of nutrition, which in turn depends on the sowing system quality of the drill. The use of commercially produced seeders with different sowing units and coulters of different designs for drilling small-seeded vegetable crops cannot ensure the required seeding rate and seeding depth, which leads to overspending of scarce costly seeds and a decrease in yield.

## 2 Materials and Methods

One of the most difficult and important issues in intensive technology for cultivating vegetable crops is obtaining a given number of plants per hectare, evenly distributed along the length of the row. The quality work of the seeder sowing units largely determines the variation of the intervals between the plants in a row and consequently the volume of the future yield. Pneumatic, hydraulic and other relatively complicated seeding devices are used to improve the uniformity of drilling [1, 2]. At the same time, the potential of constructively simple mechanical sowing units, in the direction of increasing the uniformity of drilling, has not been fully realized [3].

The general disadvantages of these seeders include seed damage when the seed disc leaves the filling zone, the seed leaks from the seed hopper when the sealing is broken and when the seed is transported to the ejection zone. Thus, the main reason

V. Tarasenko (✉) · L. Bondarenko · N. Shcherbakova · N. Horbova
Tavria State Agrotechnological University, B. Khmelnitsky ave. 18, Melitopol 72310, Ukraine
e-mail: mail@rosta.ua; volodymyr.tarasenko@tsatu.edu.ua

© Springer Nature Switzerland AG 2019

V. Nadykto (ed.), *Modern Development Paths of Agricultural Production*,
https://doi.org/10.1007/978-3-030-14918-5_30

for these drawbacks when sowing with drills mentioned above is the design of the sowing units and operating modes.

It should be noted that the sowing units for small-seeded crops have not been studied enough, and the domestic seeding machines commercially produced do not meet the agrotechnical requirements.

The purpose of the study is to increase the drilling quality of small-seeded crops in small rates by improving the design and technological schemes of mechanical sowing units.

The drilling occupies the leading place in the technological operation complex of vegetable crop cultivation. When drilling with seeder sowing units, seeds are placed in the longitudinal, transverse and vertical directions. Herewith, it is necessary to create sufficient conditions for the formation of optimal plant density and the obtaining of a programmed yield.

The sowing unit is one of the most important working tools of the seeder. It serves to select from the total mass of a certain number of seeds and to form the initial flow of them with specified parameters. Therefore, the merits and demerits of sowing, relative to the quality of seed distribution in the row and in general on the sown field, are mainly determined by the work of the sowing units.

Sowing units of seeders should meet the following basic agrotechnical requirements:

- to supply evenly seeds to the coulters;
- to ensure a stable sowing—to sow the same number of seeds per 1 m of the path, regardless of the thickness of their layer above the sowing units, the relief of the field, the inclination of the seeder, the change in the movement speed of the unit;
- not to damage seeds;
- to sow satisfactorily seeds of various crops, differing in shape, size, surface condition.

According to the technology of operation, the sowing units of the seeding machines can be divided into two groups: in the first group, seeds are taken in a continuous flow, and in the second one, they are taken one by one. The first units are used primarily in seeders, the latter—in seeders, potato planters and transplanters.

According to the principle of operation, the sowing units of the seeding machines can be mechanical and pneumatic. The first ones are the most common and are used in seed drills produced by SPC "ROSTA". Pneumatic units are applied in seeders for dotted sowing of calibrated and not calibrated seeds of row crops.

According to the design, the sowing units are divided into brush, drum (roller), disc (with vertical, inclined and horizontal axes of rotation) and belt. SPC "ROSTA" in its seeders uses mechanical brush and drum sowing units of various modifications.

The BBA brush-type seeder (see Fig. 1) is intended for sowing a wide range of crops: onion, cabbage, beet, dragee seeds, greens, feed and industrial crops. Seeds having close length, width and height are preferred for seeding with the BBA sowing unit.

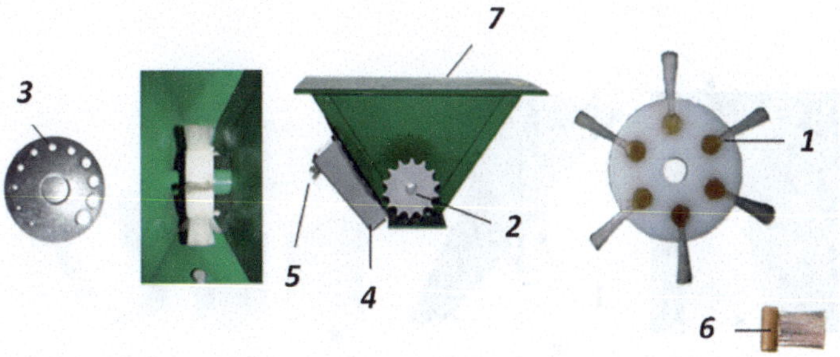


**Fig. 1** BBA sowing unit design: 1—sowing drum with brushes; 2—driven sprocket; 3—disc; 4—seed drill tube; 5—fixing screw; 6—brush; 7—hopper cover

SPC "ROSTA" produces a brush sowing unit of the basic modification BBA 96/6. 96—diameter of the drum on the brushes, mm; 6—number of brushes, pcs. (possibly 96/3; 96/2).

The BBA brush sowing unit (see Fig. 1) is a body of a hopper, on which such parts are mounted: a brush sowing drum 1 which rotates in sliding bearings, a sprocket 2 which is fixed on one shaft with a brush drum. At the rear of the hopper, there is a disc with calibrated holes of different diameters 3, which are selected depending on the seed size. The disc is pressed against the hopper with a guiding box, through which the seeds fall into the seed drill tube 4 and fixed with a screw 5. Six brushes 6 are mounted on the brush drum; it allows to regulate the seeding rate per one running metre by reducing the number of brushes on the drum, leaving three through one or two diametrically. In the upper part of the hopper, there is a cover of the hopper 7, which prevents the seed weathering and the ingress of moisture and various field impurities into the seeds.

In the "ROSTA" trading centres, there are special points where it is possible to select sowing units and to test them on special stands for the rate and uniformity of seeding. This is especially essentially for sowing expensive hybrids of vegetable crops. More information is available on our Website: www.rosta.ua.

A drum sowing unit with a passive ejector VPS is designed for sowing a wide range of seed crops: onion, cabbage, beet, dragee seeds, greens, feed and industrial crops. Seeds having close length, width and height are preferred for seeding with the VPS sowing unit.

SPC "ROSTA" produces VPS drum sowing units of two modifications: VPS 27/1-10/4 (possibly VPS 27/1-16/3, VPS 27/1-10/3, VPS 27/1-6/4) and VPS 52/1-24/3 (possibly 52/1-10/5; 52/1-10/3; 52/1-6/5; 52/1-6/10; 52/1-24/3; 52/1-24/5). Explanation of the abbreviation: V—drum; PS—passive ejector; 27 or 52—drum diameter in mm; 1—width of the groove for the ejector, mm; 10 or 24—number of bores in the drum, pcs; 3, 4 5, etc.—diameter of bores in the drum, mm (the depth of the bores is equal to their diameter). VPS 27/1 drum sowing units are installed on sections of SMK, SOR, SORL, SOML seeders.

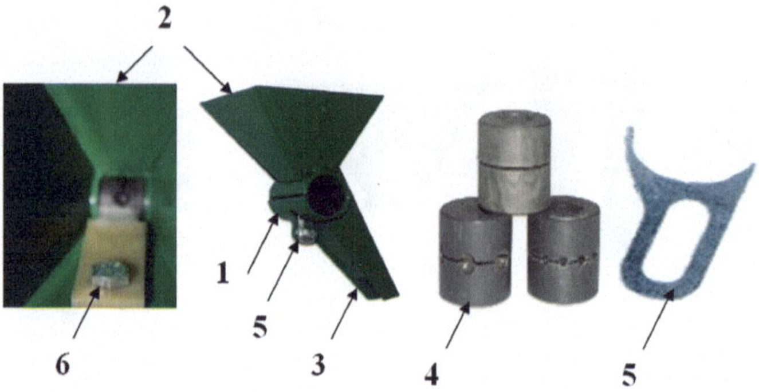


**Fig. 2** VPS 27/1 drum sowing unit: 1—body; 2—hopper; 3—seed drill tube; 4—drum; 5—ejector; 6—brush

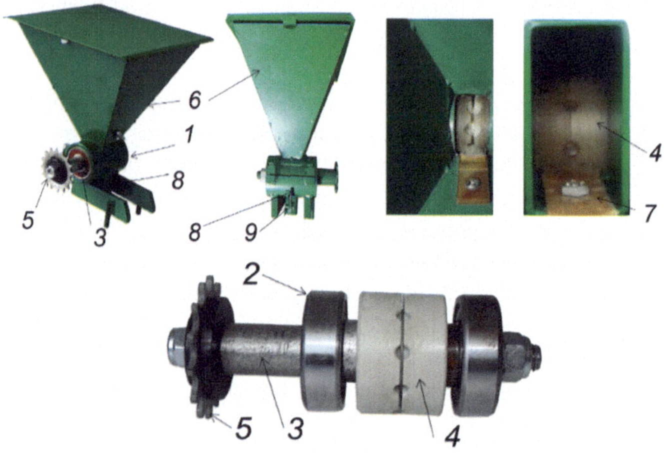


**Fig. 3** VPS 52/1 drum sowing unit: 1—body; 2—bearing; 3—driving shaft; 4—drum; 5—sprocket; 6—hopper; 7—brush; 8—ejector; 9—seed drill tube

The VPS 27/1 drum sowing unit (see Fig. 2) is a body 1, a hopper 2 and a seed drill tube 3 which are being mounted on it. A passive ejector 5 and a drum 4 with a diameter of 27 mm with bores according to the sown seeds are inserted into the body. A brush 6 for removing seed excess from the drum is fixed in the hopper.

The VPS 52/1 drum sowing unit (see Fig. 3) is a body, and the hopper of original construction is being mounted on it. The VPS 52/1 hopper is universal and allows to replace both units assembled and separately drums with different characteristics for various seeds.

The VPS 52/1 drum sowing units for precision seeding are installed on sections of SOR, SOM, SOT seeders instead of brush sowing units, which are included in the delivery of these seeders. All connecting dimensions of this sowing device correspond to the connecting dimensions on the platforms of these seeders. In order to replace

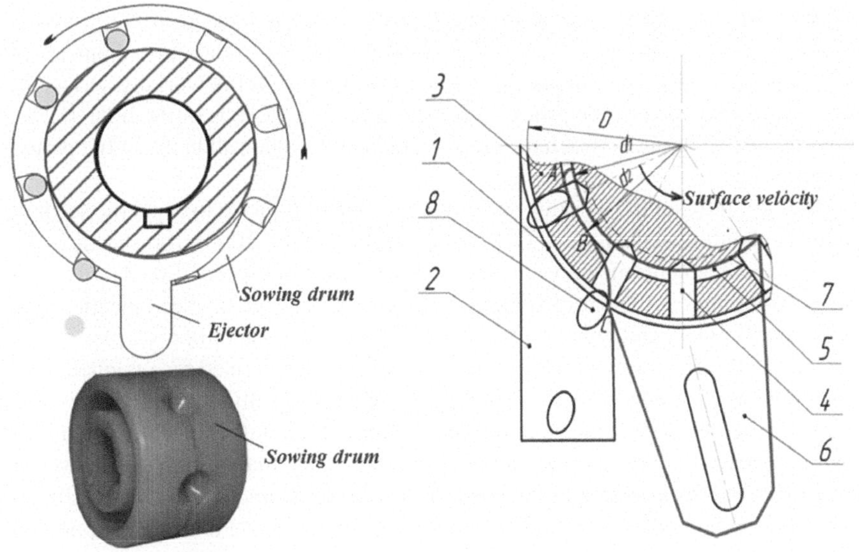


**Fig. 4** Principle of operation of the VPS 52/1 drum sowing unit with a passive ejector with brachistochronic curve BC: 1—body; 2—seed drill tube; 3—drum; 4—bore; 5—groove in the drum; 6—ejector; 7—support of the ejector; 8—seed

the brush sowing unit with a drum one, it is necessary to add additional chain links for a rear drive and remove them from the front one. For this purpose, three connecting links are installed on the chains.

The sowing unit consists of a steel cylindrical body 1 in which a driving shaft 3 with a sowing drum 4 and a drive sprocket 5 is mounted on bearings 2. The body has two milled grooves, a hopper 6 of the sowing unit is being mounted above the upper groove, and a seed drill tube 9 and a seed ejector 8 are being installed under the lower one. A brush 7 is fixed in the exit zone of the seed tube above the drum. The adjustment of the seeding rate is made by replacing the sowing drum 4 with the corresponding number of bores or by changing the drive gear ratio by installing changeable sprockets. For the convenience of disassembling the body and adjusting the seeding rate, the body is split design. By releasing the two screws in the rear of the casing, the driving shaft together with the bearings and the sowing drum can easily be removed for maintenance or replacement.

**The operating principle of the VPS 52/1 sowing unit (see Fig.** 4).

The sowing units receive a drive from the support-drive wheel of the sowing section via a chain drive. The seeding takes place when the drill moves and seeds 8 from the hopper (not shown in the drawing) are taken with bores 4 of the sowing drum 3 and are carried out separately from the bores to the seed drill tube. The seeds fall freely from bores 4 into the seed drill tube 2 under the influence of gravity, and the stuck seeds are thrown off by the passive ejector 6 forcefully.

Since the working surface of the ejector AB which is located in the groove 5 is formed along an arc of a circle with diameter d2 equal to the outer diameter of the groove, the velocity of the seeds 8 when bumping from the working surface of the ejector in the area of the groove 5 is directed along the axis of the bore 4. This prevents seed wedging between the ejector and the bore sidewall in the initial period of their ejecting.

The section BC of the working surface of the ejector 6 serves to expel the seeds at the final unloading of the bore 4. The working surface of the BC ejector is made in the form of a brachistochronic curve that ensures the maximum speed of seed movement with a minimum coefficient of friction. It ensures a high-quality seed ejection into the seed drill tube 2 and reduces the seed damage in the zone of bores 4 [4].

Next, seeds are laid in a furrow with a compacted bed formed by a coulter, after which the furrow is closed by a coverer and sealed with a rolling wheel. A new seed enters the vacated bore at the next turn of the drum. The number of seeds sown on a running metre will depend on the number of bores made in the drum and the speed of its rotation. These indicators depend on the drive parameters of the seeders on which the sowing unit is mounted and is determined according to the corresponding manuals.

The application of the VPS 52/1 drum sowing unit with a passive ejector with brachistochronic curve allows not only to sow small seeds of different crops, providing reliable one-grain dosing, and also to reduce the seed damage and increase the sowing quality.

It is possible to supply factory-made drums for a wide range of seeds: 10/5 (basic version); 10/3; 6/10; 6/5; 24/3; 24/5 (the first symbol is the number of bores, the second is their diameter). In the ROSTA trading centres, there are special points where it is possible to select sowing units and to test them on special stands for the rate and uniformity of seeding. This is especially essentially for sowing expensive hybrids of vegetable crops. More information is available on our Website: www.rosta.ua.

SPC "ROSTA" produces a wide range of seeders for sowing vegetable crops with BBA and VPS sowing units:

- hand seeders such as SMK, SOR and SORL (see Fig. 5);
- SOM and SOML seeders for motor blocks (see Fig. 6);
- SOT tractor seeders (see Fig. 7).

For row sowing of small-seeded vegetable crops in greenhouses, hotbeds and in the open ground, a whole range of SMK seeders is produced (see Fig. 5): SMK-1; SMK-2; SMK-3; SMK-4 and SMK-5 with VPS 27/1 sowing units with a dimensional row of drums for the corresponding crops: 27/1-10/4; 27/1-16/3; 27/1-6/4; 27/1-10/3 and a VPS 27/1 drum billet.

For dotted sowing of vegetable and melon crops, they offer a wide range of SORL seeders with VPS 27/1 sowing units (see Fig. 5) and SOR seeders (see Fig. 5): SOP 1/1; SOR 1/2 and SOR 2/1, which can be equipped with such sowing units as BBA 96/6; VPS 27/1 or VPS 52/1.

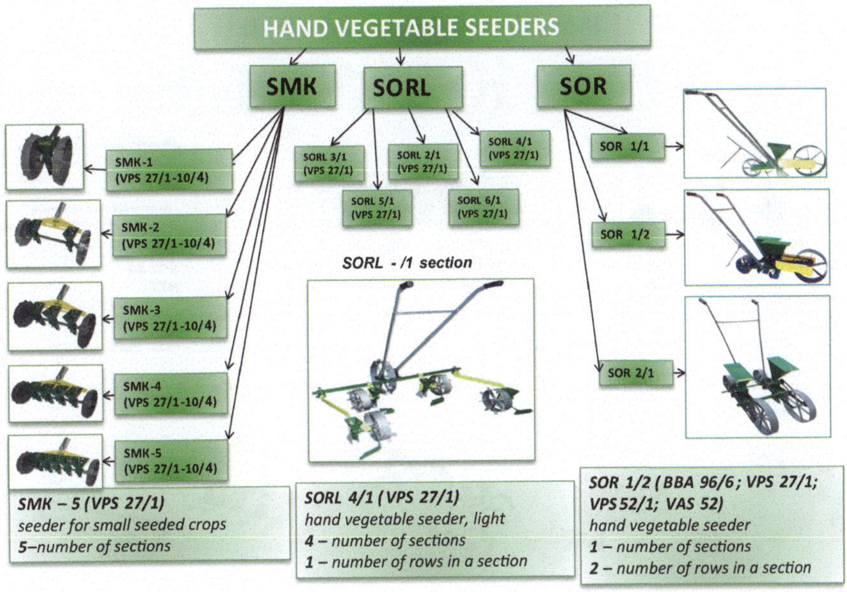


**Fig. 5** Classification of hand vegetable drills produced by SPC "ROSTA"

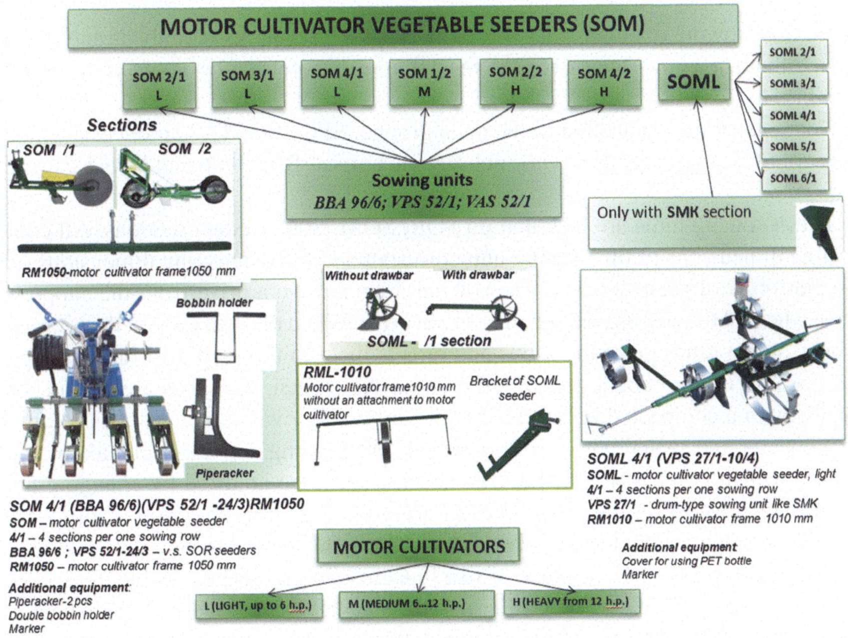


**Fig. 6** Classification of vegetable seeders for motor blocks produced by SPC "ROSTA"

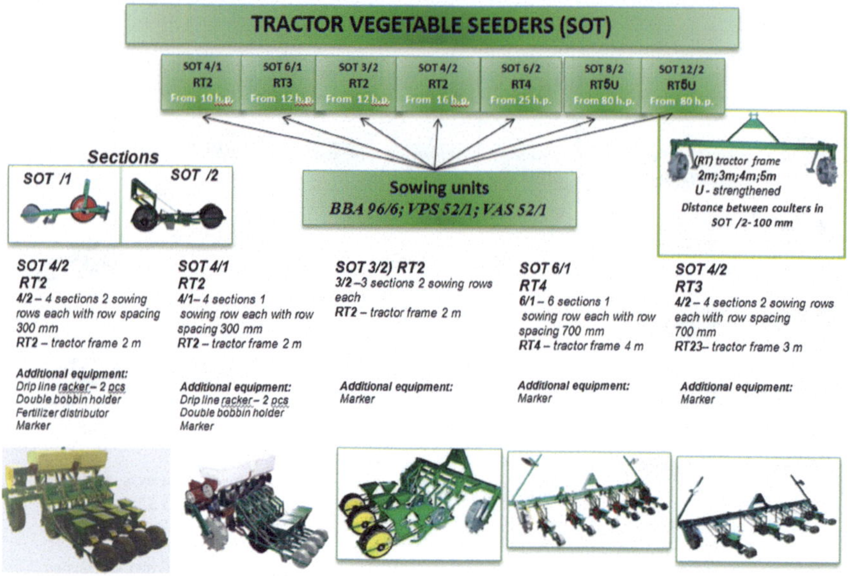


**Fig. 7** Classification of vegetable tractor seeders, produced by SPC "ROSTA"

For row sowing of small-seed vegetable crops in greenhouses, hotbeds and in the open ground, a whole range of SOM vegetable seeders for motor blocks is produced (see Fig. 6).

For vegetable crop drilling in the open ground, SPC "ROSTA" produces a whole range of SOT tractor seeders with such sowing units as BBA 96/6 and VPS 52/1 (see Fig. 7).

Brush sowing units are installed on SOR, SOM, SOT seeders sections. All connecting dimensions of this sowing unit correspond to the connecting dimensions on the platforms of these seeders. The sowing units receive a drive from the support-drive wheel of the sowing section via a chain drive. When the drill moves, the rotating brush sowing drum takes the seeds and directs them to the seed drill tube through the opening in the hopper and in the rotating disc. Then, the seeds are laid in the furrow with a compacted bed formed by a coulter, after which the groove is closed by a coverer and sealed with a rolling wheel. The seeding depth is regulated by the coulter, which is fixed with a nut. The seeding rates are adjusted by changing the number of brushes on the brush sowing drum and selecting the ratio of the teeth number of the drive and the driven sprockets of the unit drive. On the chain, there are two connecting links at a distance of five links from each other to enable the seeder to be equipped with the VPS sowing units. Thus, the chain for a forward wheel is shortened by removing a piece of a chain between locks, and for a back wheel, it is extended.

On Fig. 8, a general view of the SOR 1/1 vegetable single-row seeder with a brush-type sowing unit BBA 96/6 is shown.

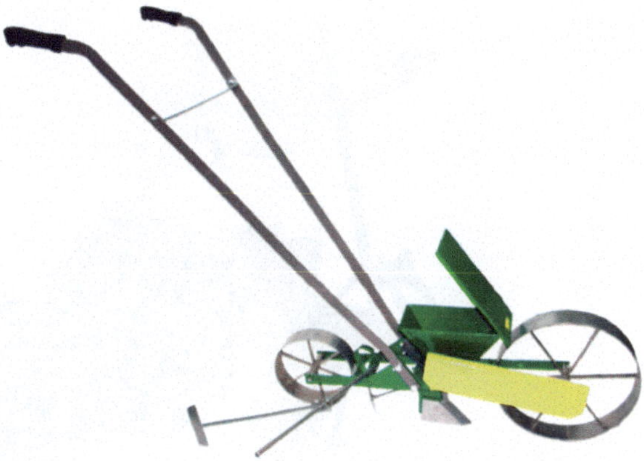

**Fig. 8** SOR 1/1 vegetable single-row seeder with a brush-type sowing unit BBA 96/6, general view

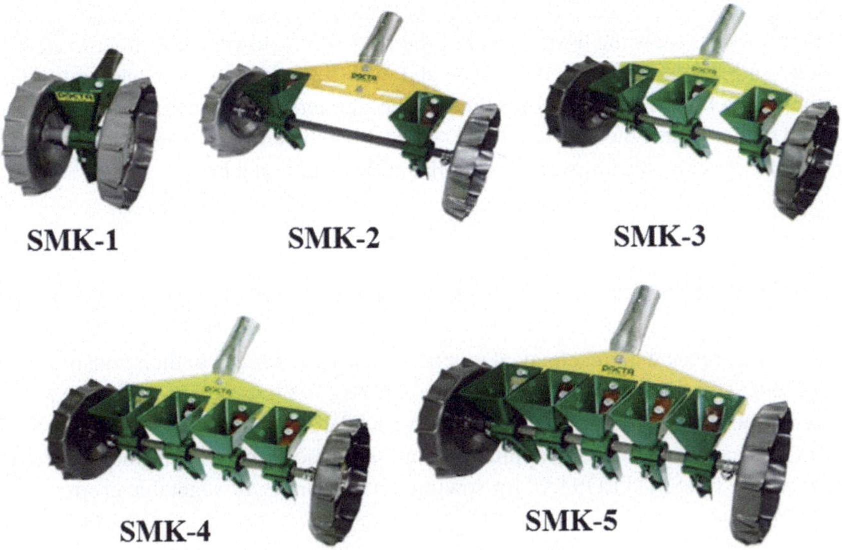


**Fig. 9** Seeders for small-seeded crops of the SMK-1 … 5 series

Seeders SMK-1 … 5 (VPS 27/1-10/4) (see Fig. 9) are made according to TU U30952138-001-2004 and are intended for row sowing of small-seeded vegetable crops in greenhouses, hotbeds and in open ground. Seeders from SMK-2 have a universal design, which is changed by adding sowing sections with a drum sowing unit with a passive ejector to a shaft with stamped wheels (VPS 27/1-10/4 is a base model).

Seeders SMK-1, SMK-2 (with adjustable row spacing from 6 to 24 cm) and SMK-5 (row spacing 6 cm) are mass-produced. Purchasing additional seeding sections of the SMK seeder, you can also get a 3- and 4-row drill.

**Fig. 10** Basic model of the SOT 4/2 × 30 vegetable tractor seeder

Figure 10 presents the basic model of the SOT 4/2 × 30 vegetable tractor seeder, which can be equipped with BBA 96/6 and VPS 52/1 sowing units and additional equipment: a bobbin holder, a drip line racker, markers and a fertilizer distributor.

The seeder is designed for row (SOT 4/1 × 30) and band twin-row planting (SOT 4/2 × 30) of vegetable crops, both on a flat surface and on a bed.

## 3 Conclusion

Planting machines with mechanical precision sowing units due to their constructive, operational simplicity, reliability and low cost have not lost their significance at the present time. The BBA 96/6, VPS 27/1 and VPS 52/1 sowing units developed by SPC "ROSTA" have found a wide application on hand, motor block and tractor seeders produced by the SPC "ROSTA" for sowing a wide variety of vegetable crops.

## References


1. Krasnoshchekov, N.V., Lipkovich, E.I.: The concept of developing a system of machine technologies in crop production. Tractors Farm Mach. **8**, 3–6 (2008)
2. Kem. A.A.: Substantiation of parameters and operating modes of the sowing units for drilling small-seeded crops: the author's abstract. Candidate of Engineering Science, Omsk, 16 p (1992)
3. Belodedov, V.A.: Optimization of single-grain sowing units in connection with the longitudinal uniformity of the seed distribution: author's abstract. Doctor of Engineering Science, M. 30 p (1985)
4. Tarasenko, V.V.: The sowing unit. Patent of Ukraine No. 117875. Bul. No. 13 dated 10 July 2017

# Examining the Creative Potential of Engineering Students

**Natalia Sosnytska, Olena Titova, Svitlana Symonenko and Olena Kravets**

## 1 Introduction

As creativity is one of the most desirable skills which potential employer expects from the engineering staff, university teachers are trying to develop their engineering students' creative potential. In this respect, there is a need to measure the creative potential level to assess the educational results. There are a great number of approaches and methods for creativity measuring. They are mainly represented by various tests and checklists which are aimed to diagnose both individual creativity in general and engineer's creativity as well. While training agricultural engineers, teachers need a special tool to control the level of their students' creative potential. The solution, which we were searching for, had to consider the specifics of agricultural engineers' professional activity and had to be suitable for application at any stage of studying by both teachers and students. An original tool has been designed. It was tested and applied to measure the creative potential of agricultural engineering students.

### *1.1 Analysis of Recent Studies and Publications*

Psychometric approaches to research of creativity provided the study of individual creative potential. First, batteries of tests were designed more than 60 years ago (Torrance Tests of creative Thinking—TTCT, Guilford's Alternative Uses Test, Wal-

N. Sosnytska
Department of Mathematics and Physics, Tavria Stat Agrotechnological University,
B. Khmelnytskyi ave., 18, Melitopol 72310, Ukraine

O. Titova (✉) · S. Symonenko · O. Kravets
Department of Foreign Languages, Tavria State Agrotechnological University,
B. Khmelnytskyi ave., 18, Melitopol 72310, Ukraine
e-mail: olena.titova@tsatu.edu.ua

© Springer Nature Switzerland AG 2019

V. Nadykto (ed.), *Modern Development Paths of Agricultural Production*,
https://doi.org/10.1007/978-3-030-14918-5_31

lach and Kogan's creative thinking tests). The tests [1–4] originally included tasks for divergent thinking assessment as well as problem-solving skills. Creativity tests allow to measure cognitive functions, individual traits, and motivation related to creativity [5]. Despite the fact that tests of creativity have been widely used, most of them are not able to assess creativity in a whole but only some aspects [6]. Creativity checklists are also criticized as they are usually either poorly designed or they are not used effectively [7]. Still, they are recommended to use [8, 9].

Scientists support one more efficient method to assess creativity. It relates to the review of the students' creative results which they store in their portfolios (projects, ideas, models, etc.) year in year out [6, 7]. According to consensual assessment technique (CAT) [10], which is another common creativity examining method, qualified experts assess the creative potential in specific domains. However, predictive validity of this judgment will depend on the level of the expertise [11]. On the other hand, it looks impossible to measure person's creativity by means of DT tests, IQ tests, or CAT. At least, those tools are able to assess some separate aspects of a creative potential [12]. Recent reviews confirm the need to design a concerted conception of creativity as a complicated construct. In this respect creativity assessment has to rely on that conception [13].

Because of mentioned findings, we hypothesized that an instrument for the assessment of the engineer's creative potential had to be designed considering the potential's complex structure. In this case, the tool might be efficiently used by teachers as well as students.

### *1.2 Statement of the Objective and Tasks of the Study*

As our study focuses on the assessment of agricultural engineer's creative potential, we believe that not much attention has been paid to this exact issue so far. Thus, the objective of the study was to assess a creative potential of agro-engineering students.

In this regard, the paper aims to achieve the objective through the following tasks:

- to define an engineer's creative potential as a complex structure;
- to analyze existing methods of creative potential assessment;
- to develop a special instrument for creative potential measuring;
- to apply the instrument for assessment of the creative potential of the engineering students at Ukrainian agricultural universities.

## 2 Basic Part of the Study

Since engineering schools are focusing their graduates on innovative activity in future job, it is highly important to control the level of students' creative potential. For this study, the term of *engineer's creative potential* has been defined as "an integrative quality of an expert based on the genetic (natural) faculties and inclinations of an individual [14]. It reflects one's abilities to perform an innovative engineering activity." Careful analysis of different approaches and opinions [4, 7, 10, 14–16] allowed us to identify the term as an integrative characteristic which relies on individual's inclination and gift for innovative engineering. *Inherited trait*, which is not changeable, makes up the base of the creative potential structure. A number of changeable components complete it. They are *intelligence and creativity, reflection, motivation and will,* as well as *productive activity*. The creative potential development can be described by the dynamic model. When the components (intelligence and creativity, reflection, motivation and will, productive activity) advance, the engineering student's creative potential progresses. The most relevant idea, which the dynamic model illustrates, is that all the components have to be developed simultaneously. This condition requires a specific pedagogical system which can enable good results. We consider that the definition as well as the components structure and interrelation inside the creative potential do not depend on the number and the content of the components which are seen by different researches. This means that the creative potential, which is based on the faculties, needs well-balanced systematic development of all its elements.

As those components can be developed, teachers are searching for effective ways to measure the level of creativity and innovation maturity. Scientists [1, 5, 6, 11–13] consider that the wide range of definitions, diverse approaches to understanding the phenomenon and evidently the complex structure of the creative potential provide a variety of approaches.

Among the great number of methods for evaluating creativity, there are some separate ones which can be effectively integrated into the engineering education curriculum. All the measuring instruments can be classified according to the approaches to creativity definition: process, product, person, and press [13]. The process approach is considered to be the most common. It is represented by the following tools. Torrance Tests of Creative Thinking in Voronin's interpretation is intended to reveal the creative abilities of the respondent through the application of unfinished drawings. The main research indicators (introduced by J. Gilford) are originality and flexibility. Although, the results of the test can be interpreted after the speed (performance) and complexity (elaboration) are assessed. Drawing completion test [17] is used to study the individual features of non-verbal components of creative imagination. The test is common while candidates (especially military) are applying for a job. The method of spontaneous description of unregulated activity is aimed at fixing and analyzing the free-time activities that the student performs voluntarily without reminding or coercing when he or she is not engaged in studies.

Personal approach for measuring creativity normally includes variable questionnaires. These are checklists for assessment of creative thinking and behavior. They were developed to identify the attraction of a student to complexity, flexibility in behavior, intuition, emotional stability, risk taking, as well as independence, responsibility, and tolerance. Those indicators are available to external observation in different situations (both in classes and during individual educational activity). The assessment can be performed by students themselves, their tutors, psychologists, parents, peers, etc. This group of methods includes such questionnaires as *How Do*

*You Think?* [18], the Creative Personality Scale [19], *How Creative Are You?* [20] the Creative Achievement Questionnaire [21] and others. In addition, nowadays, psychologists apply separate tests and "batteries" for diagnostics of various aspects of individual creativity: a method for studying personal creative abilities [22], a test of verbal creativity (remote associations) [23], *Creativity* test [24]. Those tools enable to reveal the individual creative potential as well.

Another widely recommended approach of creativity assessment is represented by The Consensual Assessment Technique (CAT) [10]. Its main idea is the product-based measuring of a creative potential. It is usually performed by recognized experts.

The idea to study environmental factors, which influence the creativity, is realized in a press approach. There is a set of tools for estimating creative learning climate (pedagogical activity, relationships between students as well as between students and teachers, physical environment, available materials and methods). The set is made up by such techniques as the College and University Classroom Environment Inventory [25], Assessing the Climate for Creativity [26], the Team Climate Inventory [27], and other instruments.

The current study has led to a special diagnostic tool which was made up of several tests as an integrated test. We analyzed the Intelligent Structure Test [28], the Mechanical Comprehension Test [29], Memory Tests, Tests for Logical and Conceptual Thinking [30], and the Test for Technical Abilities Estimation [15]. After a purposeful phased selection, a number of tasks were chosen and included in the integrated test. Its purpose was to examine memory, the ability to do mental arithmetic quickly, the ability to classify and analyze, spatial thinking, convergent thinking as well as mechanical intelligence.

As a result, the test consisted of 24 tasks, which were divided into three parts. The first part was aimed at checking memory. The second part of the test included nine adapted multiple-choice questions for testing mechanical aptitude, spatial visualization, skills for physics application as well as deduction of the way how things work. Finally, 14 tasks of the third part were selected to assess the level of logic and conceptual thinking, the ability to analyze and classify data, make inferences and express thoughts effectively.

The maximum score for the whole test was 25. The test had a 30-min time limit which had been defined experimentally after trials. Two options were available—paper-and-pencil version and computer-based one.

The diagnostic tool was tested in three stages. The first stage included a primary expert evaluation. During the second stage, the instrument was under the secondary expert assessment. Definition of test limit time was one of the issues. At this stage, a computer-based version was tested as well. The data on the test stability were analyzed. Test–retest reliability showed a correlation coefficient of 0.69. Coefficients of the test results coherence for paper-and-pencil and computer-based versions were also calculated. Correlation was 0.89. The limit time in both cases also coincided.

At the third stage, a pilot test was passed by the engineering students at Tavria State Agrotechnological University (Melitopol). Some experts (the lecturers, profession and practice teachers) were invited to examine the creative potential of the students which were participating in trial test. The experts used the Renzulli creativity questionnaire. The aim was to obtain statistically reliable results which could be used for assessment of the validity and reliability of the examining instrument.

To confirm the validity of the tool, the results of the test were compared with GPA, an indication of a student's academic achievement (correlation coefficient was 0.7) and Renzulli creativity indices (correlation coefficient was 0.58). The statistical analysis of the test results (matching results to normal distribution, comparing asymmetry and excess, calculating the index of complexity and the index of discrimination) made it possible to conclude that the diagnostic instrument, which was developed and tested, was valid, contained the optimal set of questions in terms of complexity. The details have already been presented to the scientific community [31].

## 3 Results and Discussion

In 2016 and 2017, engineering students, who were having conventional training at five Ukrainian Agrarian Universities [National University of Life and Environmental Sciences of Ukraine is marked as (1) on Fig. 1, Dnipro State Agrarian and Economic University (2), Kharkiv National Technical University of Agriculture after Petro Vasylenko (3), Uman National University of Horticulture (4), and Tavria State Agrotechnological University (5)], took part in the testing. The total number of participants was 411.

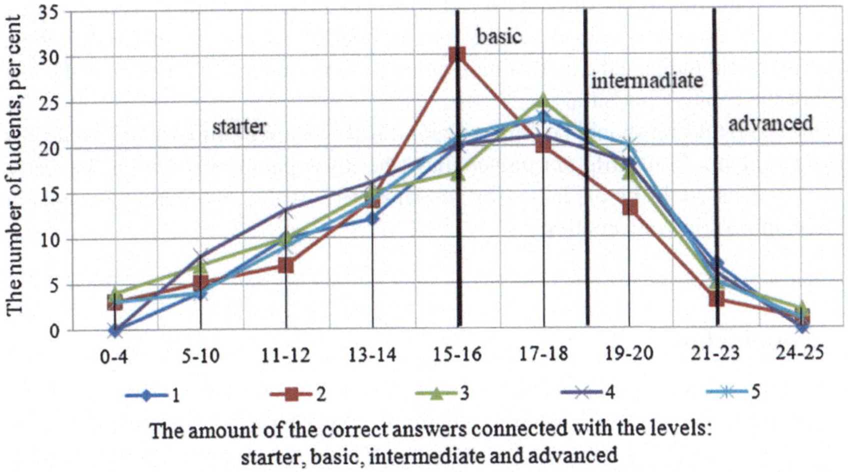


**Fig. 1** Distribution of students' results according to the levels of their creative potential at universities: National University of Life and Environmental Sciences of Ukraine (**1**), Dnipro State Agrarian and Economic University (**2**), Kharkiv National Technical University of Agriculture after Petro Vasylenko (**3**), Uman National University of Horticulture (**4**), Tavria State Agrotechnological University (**5**)

**Table 1** Assessment scale

| Creative potential level | Starter | Basic | Intermediate | Advanced |
|---|---|---|---|---|
| Test score | <15 | 15–18.5 | 19–22.5 | 23–25 |
| ECTS grades | F, FX | D, E | B, C | A |

It was predicted that the testing results would reveal four levels of creative potential development: starter, basic, intermediate, and advanced. The respondents, who scored less than 15 points out of 25, obtained a *starter* level. It equaled F and FX grades in the European Credit Transfer and Accumulation System (Table 1). A *basic* level score was 15–18.5 points (D and E grades in ECTS), an *intermediate* level with 19–22.5 points might equal to B and C grades as well as an *advanced* level which ranged 23–25 points and represented A grade.

The results of the testing are presented in Fig. 1.

Analysis of the test results showed that their distribution by the levels of development of the creative potential corresponded to the normal distribution. This confirmed the data reliability. A significant number of tested students (41–52%) demonstrated the basic level, when 25–35% showed the starter level. A lower percentage (from 13 to 23%) of the respondents could be considered as the ones who had the intermediate level. A small group of engineering students (just 6–7%) claimed the advanced level of the creative potential development. These results match the human resources issue which was elicited by the employers—the lack of young creative engineers [32].

Further analysis revealed another problem. The number of students, who coped one-fifth of the questions, did not exceed 10%. The time, which they needed to pass the test, was also under consideration. The respondents spent 3–8 min for the whole test. This could indicate the lack of motivation to work on the test rather than the low level of the creative potential.

One more finding should be emphasized. It is the relative similarity of the results among students from different universities (the discrepancy was within the range of 10%). This could be explained by the similarity of the systematic curricula at Ukrainian agrarian universities.

## 4 Conclusion

To guarantee the results of teaching creativity to engineering students, it is vital to examine the level of their creative potential development during studying. After the definition of the creative potential complicated structure and the analysis of the common methods for engineering creativity evaluation, a special diagnostic instrument was developed. It was represented as a paper-and-pencil and computer-based tool and included questions which allow to assess student's memory, mechanical intelligence, ability to classify and analyze, as well as spatial and convergent thinking. The instrument enabled to diagnose four levels of the creative potential (starter, basic, intermediate, and advanced) for 411 agricultural engineering students at five Ukrainian universities.

Further research on the issue should be directed toward the detailed analysis of the test results separately for each academic year to establish how student's creative potential changes with the time. The study of the features of domestic and foreign engineering education will enable to identify the factors that support and enhance creativity development for engineering students.

## References


1. Kim, K.H.: Can we trust creativity tests? a review of the torrance tests of creative thinking (TTCT). Creativity Res. J. **18**, 3–14 (2006). (in English)
2. Michael, W.B., Wright, C.R.: Psychometric issues in the assessment of creativity. J.A. Glover, R.R. J. **3**, 58–74 (1989). (in English)
3. Torrance, E.P.: The Torrance Tests of Creative Thinking: Norms-Technical Manual. Personal Press, Lexington, MA (1966). (in English)
4. Wallach, M.A., Kogan, N.: Modes of Thinking in Young Children: A Study of the Creativity Intelligence Distinction. Holt, Rinehart & Winston, New York (1965). (in English)
5. Cropley, D.H., Kaufman, J.C.: Measuring functional creativity: non-expert raters and the creative solution diagnosis scale. J. Creative Behav. **46**, 119–137 (2012). (University of Connecticut, USA, in English)
6. Adams, J.P., Turner, S.: Problem solving and creativity for undergraduate engineers: process or product? In: Engineering Education. Loughborough University (2008). (in English)
7. Piirto, J.: Understanding Creativity. Great Potential Press, Tucson (2004). (in English)
8. Gilliam, J.E., Carpenter, B.O., Christensen, J.R.: Gifted and Talented Evaluation Scales. Prufrock Press, Waco, TX (1996). (in English)
9. Renzulli, J.S., Smith, L.H., White, A.J., Callahan, C.M., Hartman, R.K., Westberg, K.L.: Scales for rating the behavioral characteristics of superior students. In: Mansfield Center, CT: Creative Learning (1997). (in English)
10. Amabile, T.M.: Social psychology of creativity: a consensual assessment technique. J. Pers. Soc. Psychol. **43**, 997–1013 (1982). (in English)
11. Kaufman, J.C., Baer, J.: Beyond new and appropriate: who decides what is creative? Creativity Res. J. Measuring Creativity. **24**(1), 83–91 (2012). (in English)
12. Piffer, D.: Can creativity be measured? an attempt to clarify the notion of creativity and general directions for future research. Thinking Skills Creativity **7**(3), 258–264 (2012). (in English)
13. Said-Metwaly, S., Van den Noortgate, W., Kyndt, E.: Approaches to measuring creativity: a systematic literature review. Creativity Theor. Res. Appl. **4**(2), 238–275 (2017). (in English)
14. Titova, O.A.: Struktura tvorchogo potetsialu inzhenera agrarnogo profiliu. Naukovyi visnyk NUBiP of Ukraine Pedagogika, Psychology & Philosophy Series. Kyiv, Milenium. **253**, 289–297 (2016). (in Ukrainian)
15. Koshuk, O.B.: Metodyka formuvannia tekhnichnykh zdibnostei maibutnikh inzheneriv-agrarnykiv u protsesi vyvchennia kursu 'Silckogospodarski mashyny'. Ph.D. Thesis in Education, Kyiv (2005). (in Ukrainian)
16. Popova, P.: Rozvytok tvorchogo potentsialu maibutniogo ingenera v protsesi profesiinoi pidgotovky u vyschomu tekhnichnomu navchalnomu zakladi. Ph.D. Thesis in Education, Zaporizhia (2006). (in Ukrainian)
17. Wartegg, E.: Gestaltung und Charakter. Ausdrucksdeutung zeichnerischer Gestaltung und Entwurf einer charakterologischen Typologie, Leipzig (1939). (in German)
18. Davis, G.A., Subkoviak, M.J.: Multidimensional analysis of a personality based test of creative potential. J. Educ. Meas. **12**, 37–43 (1975). (in English)

19. Gough, H.G.: A creative personality scale for the adjective check list. J. Pers. Soc. Psychol. **37**(8), 1398–1405 (1979). (in English)
20. Raudsepp, E.: How creative are you? Perigee, New York (1981). (in English)
21. Carson, S.H., Peterson, J B., Higgins, D.M.: Reliability, validity, and factor structure of the creative achievement questionnaire. Creat. Res. J. **17**, 37–50 (2005). http://dx.doi.org/10.1207/s15326934crj1701_4. (in English)
22. Tunik, Ye.Ye.: Psikhodiagnostika tvorcheskogo mishleniya. Kreativnie testi. Saint Petersburg (1997). (in Russian)
23. Mednick, S.A.: The Remote Associates Test. J. Creat. Behav. **2**, 213–214 (1968). http://dx.doi.org/10.1002/j.2162-6057.1968.tb00104.x. (in English)
24. Vyshniakova, N.F.: Kreativnaya psihopedagogika. Psihologiya tvorcheskogo obucheniya. Minsk (1995). (in Russian)
25. Fraser, B.J., Treagust, D.F., Dennis, N.C.: Development of an instrument for assessing classroom psychosocial environment at universities and colleges. Stud. High. Educ. **11**(1), 43–54 (2006). (in English)
26. Amabile, T.M., Gryskiewicz, N.D.: The creative environment scales: work environment inventory. Creat. Res. J. **2**(4), 231–253 (1989). (in English)
27. Anderson, N.R., West, M.A.: Measuring climate for work group innovation: development and validation of the team climate inventory. J. Organ. Behav. **19**(3), 235–258 (1998). (in English)
28. Amthauer, R.: Intelligenz-Struktur-Test 2000 R: I-S-T 2000 R Manual. Verl. für Psychologie 2. Göttingen : Hogrefe. 2 Auflage. DNB 965201929 (2001). (in German)
29. Bennett, G.K.: Manual for the Bennett Mechanical Comprehension Test, Forms S and T. The Psychological Corporation, San Antonio (1969). (in English)
30. Stolyarenko, L.D.: Osnovi psikhologii. Praktikum. 7 ed., Moscow (2006). (in Russian)
31. Titova, O.A.: Validyzatsiia instrumentariiu dlia diagnostyky tvorchogo potentsialu maibutniogo inzhenera-agrarnyka. Doslidzhenia riznykh napriamkiv rozvytku psykhologii ta pedagogiky. Odesa (2015). (in Ukrainian)
32. Sanghi, D.: The 21st Century Engineering Crisis. EDU TECH (2010). (in English)

# Development of Communicative Competence as a Precondition of Competitive Software Engineer Formation

**Svitlana Symonenko, Nataliia Zaitseva, Olena Titova and Margaryta Vynogradova**

## 1 Introduction

The European vector of Ukrainian education requires modifying priorities in the educational process, not only from the side of educational institutions and their direct representatives, lecturers, but also from the side of future graduates themselves as they are the prospective showpiece of higher educational institutions and a precondition for a certain university prominent reputation. Changing the worldview towards European society integration, joining the European education system and employing in the European labour market are rather confusing milestones for young man careers. Students do not know how to reach their most desired aim, namely, prosperity. Today students understand the importance of employee correspondence to international qualification frameworks. A potential employee should demonstrate a series of competences in the course of a job interview to be hired.

## 2 The Basic Part of the Study

The engineering profession is one of the most difficult ones, since it depends upon a successful combination of creative thinking with the manufacturing perfection. It is based on a thousand-year experience and requires a significant knowledge update at least every 10 years. The speed of the methodology updates, the constant addition of new aspects, the emergence of innovative technologies based on the profound knowledge of natural and technical disciplines forming the engineering speciality fundamentals.

S. Symonenko (✉) · N. Zaitseva · O. Titova · M. Vynogradova
Department of Foreign Languages, Tavria State Agrotechnological University,
B. Khmelnytskyi ave., 18, Melitopol 72312, Ukraine
e-mail: svitlana.symonenko@tsatu.edu.ua

© Springer Nature Switzerland AG 2019

V. Nadykto (ed.), *Modern Development Paths of Agricultural Production*,
https://doi.org/10.1007/978-3-030-14918-5_32

In Ukraine, because of the low level of social protection, the greatest stimulus from the side of an employer is the financial reward, and from the side of the worker it is effective knowledge application in the workplace for career prospects. Gradually, a manager or a team leader may face the communication problems between professionals who lack an ability to work in a team or build permanent relationships with clients.

European employers and HR experts foresee communication problems within the company and do their best to prevent them by means of thorough selection of employment procedures. According to the survey conducted by Joblift GmbH online job resource [1], the communicative competence belongs to the top 5 demands from employers. In 2017, Joblift meta job search engine analyzed more than 31 million vacancy advertisements on the European job market looking for employers' demands for one or more of 78 personality traits. Interpersonal behaviour as team spirit and communication skills (place 1 and place 4) is mentioned in the top five of more than 9 million vacancies in Germany. In the UK, communication skills rate second, and in France, social skills and team spirit rank high places, too—positions 3 and 5.

On closer examination of requirements to software engineers employing in Germany, the communicative competence should be emphasized, too: it ranks first in the top 11 soft skill list from computerwoche.de [2] as the ability to establish consensus and comprehension of how to realize one's goals through effective communicating with other people.

The situation is not going to change by 2020 according to the data from forbes.com [3]: leadership, communication and collaboration tend to rate first in the traditional soft skill list globally. More than 100 top HR managers, CEOs and recruiters (including the ones from Adobe, Prezi and SchooLinks) are willing to hire entry-level job seekers with no exact experience in the area they are being interviewed provided that they have a solid foundation of soft skills.

In Ukraine, an ability of presenting oneself and perceiving other people's perspective is considered by managers and generalists as teamwork essentials and preconditions for quick promotion obtaining. According to 21 senior managers opinion from the survey on the dou.ua [4], communicative skills imply Ukrainian soft developers' ability to produce and present the idea orally and in writing, to hold the team meetings, to report to product owners and to mentor newcomers both in Ukrainian and in English. The given examples prove the necessity of the presence and development of communicative skills for engineers.

The situation with forming communicative abilities and competences in the world and in Europe differs dramatically. The Framework for Qualifications of the European Higher Education Area (QF-EHEA), based on the Bologna Declaration (1999) and Dublin Descriptors (2005), was effectively adapted and taken into account for developing national qualification frameworks. The National Qualifications Framework of Ukraine was developed 12 years later and is still being introduced to educational institutions, governmental bodies and stakeholders.

Inherent-level descriptors of the National Qualifications Framework of Ukraine [5] include both a generalized ten-level description (integral competence) and the basic competence system (knowledge, skills, communication, autonomy and responsibility), reflecting the personal progress of learning achievements. They are:

- integral competence (ability to perform tasks, solve tasks and problems);
- knowledge (depth, character, range);
- skills (performing tasks of different types, problem-solving, activity evaluating, analysing, etc.);
- communication (interacting with people, working in a team, reporting information);
- autonomy and responsibility (task performance control, independence, responsibility for work).

The international experience of implementation of National Qualifications Frameworks worldwide and the long-term development of the National Qualifications Framework of Ukraine resulted in unique profound qualification gradation. The Cabinet of Ministries of Ukraine introduced level descriptors for integral competences that evolve from level to level: for example, level 0 includes an ability to act adequately in familiar simple situations under direct control and readiness for systematic training; level 9 provides an ability to identify and resolve socially significant systematic problems in a particular activity area that are crucial to sustainable development and tasks requiring new systemically important knowledge creation and development of transmission technologies. Similar descriptors are not available in the national frameworks of European countries except Germany and the UK of Great Britain and Northern Ireland, which nevertheless contain similar general competences for all levels.

The study shows that another distinct feature of the National Qualifications Framework of Ukraine is the inclusion of communicative competence descriptors, which are absent in the most qualifications frameworks of European countries with the exception of Bulgaria and Scotland, and is included in qualifications frameworks of Germany and Poland as components of other competences and skills. A special attention should be paid to communicative competence descriptors referring to higher education (levels 5 to 7) in the National Qualifications Framework of Ukraine. It can be seen that these three-level descriptors emphasize the necessity of communicative competence both in academic and professional environment:

- Level 5 necessitates interaction, collaboration with a wide social circle (colleagues, managers, customers) for professional or educational activity fulfilment.
- Level 6 outlines presenting information, ideas, problems, decisions and own experience in a professional activity field to specialists and non-specialists; ability to form a communicative strategy effectively.
- Level 7 foresees comprehensive and univocal reporting own conclusions, knowledge and definitions to specialists and non-specialists, particularly to learners; foreign language application in professional activity.

The higher education standard of Ukraine for IT-specialists and software engineers provides a series of competences necessary for successful career building. The most crucial ones are an ability to communicate in a foreign language both verbally and in writing, an ability to conduct the theoretical and applied research at the appropriate level, an ability to motivate people and move towards a common goal, to work in a team and an ability to communicate with representatives of other professional groups at different levels (with experts from other fields of knowledge or types of economic activities).

The concept of communicative competence is being the object of particular attention of scholars, teachers, employers, etc. There are hundreds of definitions of communicative competence, from the simple ones as the ability to interact well with others (Spitzberg 1988) to rather complicated such as personal mastering the communicative qualities associated with the need to interact with other people, with the objects of the surrounding world and its information flows, the ability to find, transform and transmit information, to perform various social roles in groups and teams [6].

To understand the necessity of communicative competence for software engineers is essential to elucidate the concept of communication. The term communication has more than two hundred definitions nowadays. Scholars commonly give definitions of communication which are related to their fields (pedagogical science, psychology, political science, etc.). Having analyzed numerous definitions, T. Goban-Klas outlines seven common definition types [7]: (1) communication as a transmission of information, ideas, emotions and skills; (2) communication as understanding of other people; (3) communication as an influence on people with the help of signs and symbols; (4) communication as an association using a language or symbols; (5) communication as an interaction with the help of symbols; (6) communication as an exchange of meanings between people; (7) communication as a component of a social process which expresses the group norms, provides public control, distributes roles, achieves coordination of efforts and so on. A Guide to the Project Management Body of Knowledge (PMBOK[R] Guide) [8] by Project Management Institute, Inc. gives the definition of communication as processes required to ensure the planning, creation, distribution, control and monitoring of project information.

Various definitions of the concept are not contrary to each other; they supplement each other, giving more details to the communication comprehension.

In the course of our research, it is reasonable to focus on different functions of communication. According to Moseeva, communication performs the following functions within combined teams [9]: (a) an informative function (information transfer, provision of information that facilitates decision-making); (b) a motivational function (encouraging employees to perform their duties better by persuasion, suggestion, orders, instructions through communication; (c) a control function (monitoring based on hierarchy and subordination behaviour of employees by various means of communication); (d) an expressive function (promoting the emotional expression of feelings, experiences, attitudes to what is happening, allowing people to meet social needs); (e) an integrative function means that communication helps to consolidate the organization, to join forces to achieve goals.

Moseeva considers communication both a phenomenon and a process. The researcher insists that as a phenomenon, communication includes interaction between structural units and participants on the basis of established regulations and rules. As a process, communication is the direct interaction of project participants, employees and concerned individuals. It is evident that incorrectly built communication poses the greatest risk to any project and its successful completion; sometimes uninterrupted communication can result in unstable social relations in the team, which can negatively affect its further work.

## 3 Results and Discussion

All the information mentioned above let us assume that communication has to be taught and improved constantly, especially at any level of teenage and adult education, in order to educate highly qualified in-demand employees overseas. It should be explained to potential specialists that in Europe, the communicative competence level is determined orally and in writing in the course of interviews and at assessment centres with the same complexity degree as a professional competence. A stubborn stereotype of an engineer at a drawing board or at a computer has become outdated. An engineer spends most of their time examining requirements, collecting data, exchanging ideas with colleagues, collaborating with customers and reporting to them, managing subordinates and controlling each project development. A software engineer has a wider range of responsibilities which include co-working with both experts and nonprofessionals, converting data from conventional for software developers form into user-friendly information and visualizing own ideas in graphical form and in writing. Hence, communicative competence is an important component of work routine, particularly for professional and business communication of a software engineer.

At Tavria State Agrotechnological University, software engineers start learning English the same way all the other students do: the course outcomes are set. The freshmen are explained that not vocabulary use or translation skills are to be proved at the end of the course. The goal of a Bachelor of Science in Software Engineering in the context of command of English is to establish effective communication and to fulfil their communication needs.

The experience shows that foreign language teachers are able to exert considerable influence on students' motivation to improve their communication competence as well as to master English.

In order to adjust the communicative competence development in software engineering education to a changeable and expansible list of communicative skills demanded from graduates by potential employers, a series of arrangements have been made at Tavria University:

- During panel discussions with manufacturers, entrepreneurs and stakeholders, the list of both professional and personal skills important for successful employment and career building has been drawn up.
- Communicative competence descriptors referring to higher education (levels 5 to 7) from the National Qualifications Framework of Ukraine have been taken into

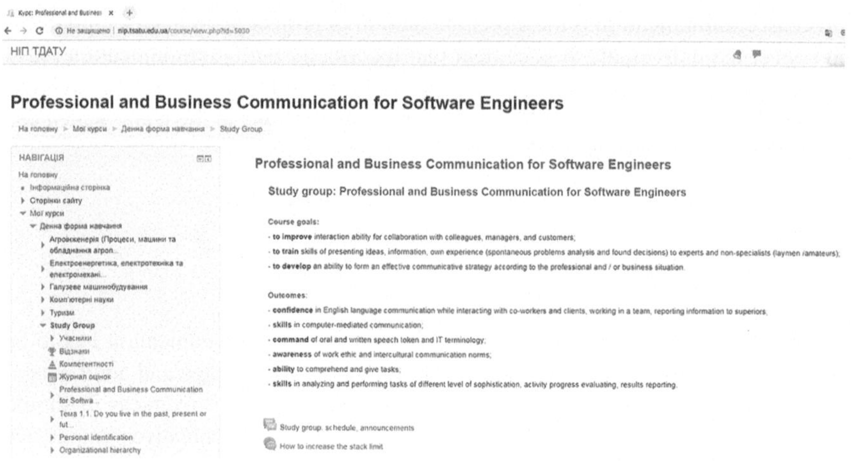


**Fig. 1** Study group "professional and business communication for software engineers" on the official Tavria State Agrotechnological University portal

account by the university teaching staff and corresponding activities have been included into educational programs in English.
- A study group for software engineering students has been formed in order to help young people to advance in communicative competence (Fig. 1).
- The survey among the most perspective employers, instructors and students has been conducted and output data has been analysed.

The mentioned above survey (Fig. 2) included the list of 10 prevalent communicative competence components which should have been ranked corresponding to their importance from the responders' point of view: work ethic, communication, positive attitude, self-motivation, team spirit, negotiation, ability to network, emotional intelligence, presentation skills and active listening skills. The responders have been also asked to add some items of their choice to complement the profile of workers of promise.

The first group of responders consisted of 31 proprietors and directors of regional enterprises. According to the survey results, experienced managers mostly assess candidates according to their behaviour towards the company and interpersonal relationship. Communication ranks first in their list. Approximately a quarter of employers put team spirit and work ethic high on a list, and almost 10% of responders mentioned active listening skills and negotiation (8%) in the top five (places 4 and 5). Among responders' own demands, loyalty, time management, initiative and critical thinking have been added.

The survey results indicate that teachers are more individual-oriented with regard to communicative competence development: most of 43 instructors believed that personal mode of operation and emotional sensitivity are more valuable to an employee than, for example, a team spirit: 18% of responders put emotional intelligence first on

**Survey for employers**
**Choose 5 of 10 the most important communicative competence components you consider to be crucial for the career building of a young specialist:**

Work ethic ☐
Communication ☐
Positive attitude ☐
Self-motivation ☐
Team spirit ☐
Negotiation ☐
Ability to network ☐
Emotional intelligence ☐
Presentation skills ☐
Active listening skills ☐

**Additional list:**
------------- -------------
------------- -------------
------------- -------------

**Survey for students**
**Choose 5 of 10 the most important communicative competence components you consider to be crucial for your career building:**

Work ethic ☐
Communication ☐
Positive attitude ☐
Self-motivation ☐
Team spirit ☐
Negotiation ☐
Ability to network ☐
Emotional intelligence ☐
Presentation skills ☐
Active listening skills ☐

**Additional list:**
------------- -------------
------------- -------------
------------- -------------

**Fig. 2** Survey forms for students and potential employers

a list, 16% ranked communication second and 13% placed presentation skills third. Teachers also would like holders of a bachelor's degree to demonstrate creativity, time management and initiative; therefore, these components were named them in the additional list.

Students mainly rely on the interdependence between the behaviour demonstrated by all the communication participants. Most of 129 responders expect the same attitude from their potential managers and co-workers they express themselves; hence, they try to reflect in their top-list interpersonal relationship aspects of communicative competence components which they would like to develop. Communication ranks first in the students' list (32%), 24% ranked team spirit second and 11% placed positive attitude third. The extra-list of software engineering students included flexibility, time management, intercultural understanding and leadership.

In general, the importance of communication is indicated by these data (Table 1).

## 4 Conclusion

On the basis of the data analysis, it is reasonable to take certain actions in order to accelerate and strengthen communicative competence development in the university study:

- to inform students about the necessity of being a competent communicator to become competitive software engineers;
- to support extracurricular study groups and students' clubs for communicative competence development;

**Table 1** Survey results

| Employers | | Teachers | | Students | |
|---|---|---|---|---|---|
| Communicative competence components | % | Communicative competence components | % | Communicative competence components | % |
| Communication | 25 | Emotional intelligence | 18 | Communication | 32 |
| Team spirit | 15 | Communication | 16 | Team spirit | 24 |
| Work ethic | 14 | Presentation skills | 13 | Positive attitude | 11 |
| Active listening skills | 9 | Team spirit | 11 | Work ethic | 9 |
| Negotiation | 8 | Positive attitude | 9 | Self-motivation | 6 |
| Positive attitude | 7 | Work ethic | 8 | Emotional intelligence | 6 |
| Self-motivation | 7 | Active listening skills | 8 | Ability to network | 5 |
| Emotional intelligence | 6 | Self-motivation | 7 | Presentation skills | 3 |
| Presentation skills | 5 | Negotiation | 6 | Active listening skills | 2 |
| Ability to network | 4 | Ability to network | 4 | Negotiation | 2 |
| *Additional list composed by responders* | | | | | |
| Loyalty | 17 | Creativity | 21 | Leadership | 23 |
| Time management | 10 | Time management | 7 | Flexibility | 12 |
| Initiative | 8 | Initiative | 5 | Time management | 8 |
| Critical thinking | 5 | | | Intercultural understanding | 6 |

- to extend available courses through communication-oriented tasks, activities and projects;
- to introduce courses for professional and business communication into educational programs;
- to provide mutual assistance of university instructors, employers and stakeholders with regard to providing practical application of communicative competence in the working place;
- to popularize the experience of Tavria State Agrotechnological University through workshops, seminars, webinars and panel discussions with leading institutions of higher education.

To conclude, it is necessary to emphasize that Ukrainian software engineers and developers are in great demand both in Ukraine and abroad among employers of different levels and statuses due to their working efficiency and learning capability. To function effectively in the working environment regardless of the company type and size, country, it is crucially important to have a high level of communicative competence.

## References

1. Von Teamgeist bis Selbstständigkeit – das sind die gefragtesten Soft Skills (2018). Available at: https://joblift.de/Presse/von-teamgeist-bis-selbststndigkeit-das-sind-die-gefragtesten-soft-skills (in German)
2. Köppler, S.: Die elf wichtigsten soft skills (2017). Available at: https://www.computerwoche.de/a/die-elf-wichtigsten-soft-skills,1902818 (in German)
3. Beaton, C.: Top employers say millennials need these 4 skills in 2017 (2017). Available at: https://www.forbes.com/sites/carolinebeaton/2017/01/06/top-employers-say-millennials-need-these-4-skills-in-2017/#381017ec7fe4
4. Donchenko, V.: Bez kakih soft skillov ne dorasti do Senior- i Lead-pozicij: vsjo o kommunikacijah (2017). Available at: https://dou.ua/lenta/articles/soft-skills-for-seniors-1/ (in Russian)
5. Postanova Kabinetu Ministriv Ukrai'ny "Pro zatverdzhennja Nacional'noi' ramky kvalifikacij" vid 23 lystopada 2011 r. № 1341 (2011). Available at: http://zakon1.rada.gov.ua/laws/show/1341-2011-%D0%BF (in Ukrainian)
6. Khutorskoy, A.: Technology of designing key and subject competences. MOTSSW, pp. 65–79 (2006)
7. Goban–Klas, T.: Media i komunikowanie masowe. Teorie i analizy prasy, radia, telewizji i Internetu, Warszawa (in Polish) (2002)
8. A Guide to the Project Management Body of Knowledge (PMBOK® Guide) – Sixth Edition (Project Management Institute, 2017) 756 p
9. Moseeva, G., Skopin, O.: Kommunikacija kak kljuchevaja kompetencija menedzhera proektov (2011). Available at: http://sovman.ru/article/1104/ (in Russian)

# Analysis of Main Process Characteristics of Infrared Drying in the Moving Layer of Grain Produce

Igor Palamarchuk, Serhii Kiurchev, Liudmyla Kiurcheva and Valentyna Verkholantseva

## 1 Introduction

Mechanism and intensity of moisture transfer in the material depend on interrelated complex of processes of moisture and material debonding, diffusion of the steam-gas medium through capillary-porous structure of the material, physical and mechanical methods of efficiency enhancement of heat mass exchange. Widespread occurrence among the last possesses realization of pulse power factors, thermal radiation processing methods. In addition to increasing thermal potential of processing, the combination of such methods of impact on the technological media in a process of drying enables providing required qualitative measures of the products via usage of pseudo-rare layer of the product, applying conveyor mechanisms realized in the developed vibrating conveyor wave system [1–12].

Thus, drying gives many benefits to society as it enables to store and process hundreds of thousands of tons of harvested grain every year, to improve its quality. It determines the objective and tasks of further improvement of the technology and techniques of grain drying.

### *1.1 Analysis of Recent Studies and Publications*

Intensity of the researched process of drying is determined with mass transfer coefficient $\beta$. Its value is greatly influenced by such parameters as specific mass of the

I. Palamarchuk
National University of Life and Environmental Sciences of Ukraine Kyiv, Heroes of Defense, 15, Kiev 03041, Ukraine

S. Kiurchev · L. Kiurcheva · V. Verkholantseva (✉)
Tavria State Agrotechnological University, B. Khmelnitsky ave., 18, Melitopol 72312, Ukraine
e-mail: valentyna.verkholantseva@gmail.com

© Springer Nature Switzerland AG 2019

V. Nadykto (ed.), *Modern Development Paths of Agricultural Production*,
https://doi.org/10.1007/978-3-030-14918-5_33

produce $P_S$, velocity of the conveyor belt $\upsilon_c$, particle sizes of raw products $d_з$, moisture density $\rho_B$, thermal diffusivity $a$, volume efficiency of the process on moisture $\Pi_v$, specific temperature of evaporation $r$. Besides, course of the researched process under the impact of microwave exposure is determined with exposure rate $N_{оп}$ in correlation to the energy $Q_B$ for evaporation and correspondingly with value $r$ [13, 2, 14].

Specific flow of radiant energy on the unit of area of exposure $S_{оп}$, mass of the processed produce and mass of evaporated moisture are calculated with the formulas:

$$N_s = \frac{N_B}{S}, \text{ W/m}^2 \quad (1)$$

$$N_m = \frac{N_{оп}}{m_{\mathcal{H}}}, \text{ W/kg} \quad (2)$$

$$N_B = \frac{N_{оп}}{m_B}, \text{ W/kg} \quad (3)$$

Mass of the evaporated moisture makes up

$$m_B = m_б + m_н - m_к$$

where $m_б$—mass of the weighing bottle; $m_н$—mass of the batch or produce to be dried; $m_к$—mass of the produce in the weighing bottle after drying.

Efficiency on the processed produce $\Pi_m$ and on the evaporated moisture $\Pi_в$ is calculated with the formulas:

$$\Pi_m = \frac{m_{пр}}{\tau} = \frac{m_{\mathcal{H}}}{\tau}, \text{ kg/h} \quad (4)$$

$$\Pi_B = \frac{m_B}{\tau}, \text{ kg/h} \quad (5)$$

where $\tau$—processing time of the produce.

Volume efficiency $\Pi_v$ makes up

$$\Pi_V = V_m \cdot \Pi_B, \text{ m}^3/\text{s} \quad (6)$$

where $V_m = 1/\rho_B = 0.849$ m$^3$/kg; $\rho_B = 1.178$ kgr/m$^3$ for moisture.

Thereafter specific flows of energy on the unit of mass of moisture $E_B$ and produce $E_m$ are determined with the formulas:

$$E_B = \frac{E_{оп}}{m_B}; \; E_m = \frac{E_{оп}}{m_{\mathcal{H}}}, \text{ J/kg} \quad (7)$$

The main specific values of the researched process are estimated on the basis of numerical values of heat-mass exchange parameters that were received via experi-

**Table 1** Initial data for estimating mass exchange parameters at thermal radiation drying of grain raw products in the moving layer of produce

| Time (s) | $П_V$ (m$^3$/s) | *t* (C) | *Xi* (kg/kg) | *Xp* | Δ*X* (kg/kg) | $S_З$ (m$^2$) | *β* (m/s) |
|---|---|---|---|---|---|---|---|
| | | 10.8 | 0.008 | | | 0.1413 | |
| 166.7 | 0.013990 | 39 | | 0.049 | 0.041 | 0.1413 | 2.415 |
| | | 11.7 | 0.0085 | | 0 | 0.1413 | |
| 83.3 | 0.034142 | 26.4 | | 0.0217 | 0.0132 | 0.1413 | 18.305 |
| | | 11.8 | 0.0086 | | 0 | 0.1413 | |
| 42.4 | 0.036566 | 24 | | 0.0196 | 0.011 | 0.1413 | 23.526 |
| | | 13.9 | 0.0093 | | 0 | 0.1413 | |
| 23.8 | 0.091901 | 19.5 | | 0.0135 | 0.0042 | 0.1413 | 154.856 |
| | | 14.2 | 0.0099 | | 0 | 0.1413 | |
| 8.8 | 0.251884 | 17 | | 0.0112 | 0.0013 | 0.1413 | 1371.244 |
| | | 11.1 | 0.0083 | | 0 | 0.1413 | |
| 83.3 | 0.033739 | 13.1 | | 0.0091 | 0.0008 | 0.1413 | 298.465 |
| | | 10.4 | 0.0078 | | 0 | 0.1413 | |
| 83.3 | 0.032748 | 15.2 | | 0.011 | 0.0032 | 0.1413 | 72.427 |
| | | 11.7 | 0.0086 | | 0 | 0.1413 | |
| 83.3 | 0.034142 | 26.4 | | 0.0205 | 0.0119 | 0.1413 | 20.305 |
| | | 10 | 0.0075 | | 0 | 0.1413 | |
| 833 | 0.031538 | 33.3 | | 0.0313 | 0.0238 | 0.1413 | 9.378 |
| | | 12.8 | 0.0089 | | 0 | 0.1413 | |
| 83.3 | 0.031978 | 38.4 | | 0.0464 | 0.0375 | 0.1413 | 6.035 |
| | | 14.1 | 0.0101 | | 0 | 0.1413 | |
| 83.3 | 0.046280 | 35.2 | | 0.0373 | 0.0272 | 0.1413 | 12.042 |
| | | 13.4 | 0.009 | | 0 | 0.1413 | |
| 83.3 | 0.044447 | 36.2 | | 0.0383 | 0.0293 | 0.1413 | 10.736 |
| | | 13.4 | 0.009 | | 0 | 0.1413 | |
| 83.3 | 0.048151 | 32.3 | | 0.0288 | 0.0198 | 0.1413 | 17.211 |

mental researches of infrared drying in the moving layer of grain produce (Tables 1 and 2), applying the second theorem of similarity and theory of "dimensions."

### 1.2 Statement of the Objective and Tasks of the Study

The objective of this study is to obtain an algorithm for designing of drying machines; infrared exposure of the granular agricultural produce under the conditions of its transportation through the working area in the pseudo-rare layer is used as intensi-

**Table 2** Specific energy and technological values of thermal radiation drying of grain raw products in the moving layer of produce

| Energy consumption on product processing $E_m$ (MJ/kg) | Energy consumption on evaporated moisture $E_в$ (MJ/kg) | Velocity of moisture evaporating $dW/dt$ (%/min) | $E_{оп}$ (J) |
|---|---|---|---|
| 9.86 | 65.53 | 0.31 | 50,000 |
| 4.23 | 26.85 | 0.28 | 25,000 |
| 2.16 | 25.07 | 0.26 | 12,711.86 |
| 1.30 | 9.98 | 0.21 | 7142.857 |
| 0.48 | 3.64 | 0.11 | 2631.579 |
| 1.46 | 9.06 | 0.05 | 8333.333 |
| 2.96 | 18.66 | 0.07 | 16,666.67 |
| 4.23 | 26.85 | 0.28 | 25,000 |
| 5.95 | 38.76 | 0.38 | 33,333.33 |
| 7.33 | 47.78 | 0.56 | 41,666.67 |
| 4.67 | 19.81 | 0.04 | 25,000 |
| 4.74 | 20.63 | 0.10 | 25,000 |
| 4.36 | 19.04 | 0.14 | 25,000 |

fying factors. In order to attain this objective, the following tasks are to be solved: analysis of the factor space of the developed oscillating system of heat mass exchange, deriving dependences between the given parameters with application of the "theory of dimensions", substitution of the attained equations with the correlations between similarity criteria, obtaining criterial equation of the mass exchange, composing required algorithm of calculation.

## 2 The Basic Part of the Study

It is necessary to include combination of the Pekle, Stanton, and Burdo criteria for further mathematical analysis, taking into consideration the peculiarities of the researched process.

The Pekle criterion is determined as

$$Pe = \frac{v_c \cdot d}{a} \tag{8}$$

where

| | |
|---|---|
| $a = 12.6 \times 10^{-8}$ m²/s | thermal diffusivity of the grain; |
| $d$ | typical diameter of the grain. |

The Stanton criterion is calculated with the formula:

**Table 3** Main parameters of mass exchange of infrared drying of the rape in the moving layer of produce

| $N_{оп}$ (W) | $N_s$ (W/m$^2$) | $\beta$ (m/s) | $\upsilon_c$ (m/s) | *St* | *Pe* | *Bu* |
|---|---|---|---|---|---|---|
| 300 | 6666.67 | 2.415 | 0.0015 | 1609.96 | 23.80952 | 0.015829 |
| 300 | 6666.67 | 18.305 | 0.003 | 6101.70 | 47.61905 | 0.006486 |
| 300 | 6666.67 | 23.526 | 0.0059 | 3987.40 | 93.65079 | 0.006056 |
| 300 | 6666.67 | 154.856 | 0.0105 | 14748.19 | 166.6667 | 0.00241 |
| 300 | 6666.67 | 1371.244 | 0.0285 | 48113.84 | 452.381 | 0.000879 |
| 100 | 2222.22 | 298.465 | 0.003 | 99488.50 | 47.61905 | 0.004376 |
| 200 | 4444.44 | 72.427 | 0.003 | 24142.18 | 47.61905 | 0.005635 |
| 300 | 6666.67 | 20.305 | 0.003 | 6768.27 | 47.61905 | 0.006486 |
| 400 | 8888.89 | 9.378 | 0.003 | 3126.05 | 47.61905 | 0.008192 |
| 500 | 11111.11 | 6.035 | 0.003 | 2011.69 | 47.61905 | 0.009233 |
| 300 | 6666.67 | 12.042 | 0.003 | 4013.89 | 47.61905 | 0.004785 |
| 300 | 6666.67 | 10.736 | 0.003 | 3578.57 | 47.61905 | 0.004982 |
| 300 | 6666.67 | 17.211 | 0.003 | 5736.86 | 47.61905 | 0.004599 |

$$St = \frac{\beta}{\upsilon_c} \quad (9)$$

In order to determine the Burdo criterion, the dependence is used

$$Bu = \frac{E_o}{Q_B} \quad (10)$$

$$E_o = N_o \cdot \tau;\ N_o = N_{on} + N_{np};$$

| | |
|---|---|
| $N_{np} = 300$ W | capability of the operating mechanism |
| $Q_B = m_B \cdot r$ | quantity of heat that is required for evaporating mass of moisture $m_B$; |
| $r$ | specific heat of evaporation, $r = 2.3 \times 10^6$ J/kg (m$^2$/s$^2$) |

Thus, $Bu = \frac{N_o \tau}{m_B r}$

Taking into account that

$$\Pi_V = \frac{m_B}{\rho_B \tau} \cdot \frac{N_O r}{N_O r} = \frac{N_O}{\rho_B r} \cdot \frac{1}{Bu} \quad (11)$$

$$Bu = \frac{N_o}{\Pi_V \rho_B r}$$

The calculation data are tabulated into Table 3 using obtained formulas.

Dependences of the represented parameters are found on the basis of the experimental data, as well as exponential rates are found with the help of methods of grapho-analytical analysis that enabled obtaining equation of the researched process of mass exchange in criterial representation:

$$\beta_{e\phi} = APe^{1.08}\, Bu^{1.2} \cdot \frac{1}{\upsilon}\left(\frac{\Pi_V r}{a^3}\right)^{1.2} \cdot \left(\frac{P_S d}{\rho}\right)^{1.56} \cdot \upsilon_c \tag{12}$$

Recommended row of operating mode parameters under conditions of thermal radiation drying of grain raw produce in the pseudo-rare layer of produce can be generated with the help of the obtained criterial equation.

## 3 Conclusion

Criterial equation of the mass exchange process in thermal radiation drying in the pseudo-rare layer of produce had been built with the help of experimental research data, method of "dimensional analysis" and Federman-Buckingham theorem. It is determined with the Pekle, Stanton, and Burdo criteria, the function which characterizes energy heat mass exchange parameters, efficiency of the researched process; and it enables generating recommended row at designing conveyor infrared dryers for the systems of storage of the granular agricultural produce.

## References

1. Aynshteyn, V.G., Baskakov, A.P., Berg, B.V., et al.: Psevdoozhizhenie, M, Himiya (1991)
2. Burdo, O.G., Kalinin, L.G.: Prikladnoe modelirovanie protsessov perenosa v tehnologicheskih sistemah. Druk, Odessa (2008)
3. Taganov, I.N.: Modelirovanie protsessov masso- i energoperenosa, L., Himiya (1979)
4. Burdo, O.G.: Energeticheskiy monitoring pischevyih proizvodstv. Poligraf, Odessa (2008)
5. Burdo, O.G.: Evolyutsiya sushilnyih ustanovok. Poligraf, Odessa (2010)
6. Palamarchuk, I.P., Bernik, P.S., Stotsko, Z.A., Yaskov, V.V., Zozulyak, I.A.: TeplomasoobmInnI protsesi ta obladnannya pererobnogo ta harchovogo virobnitstva, L., Beskid Bit (2006)
7. Bernik, P.S., Palamarchuk, I.P., Romanov, A.M.: Prospects of using of vibrodrive in processing of agricultural products. Agriculture: science and practice, Lviv, Ukrainian–Austrian (1996)
8. Baum, A.E., Rezchikov, V.A.: Sushka zerna, M., Kolos (1983)
9. Kotov, B.I., Kyfiak, V.V.: Identification of dynamic modes of heating and drying of grain products infrared exposure. Bull. NUBiP Technol. Energetics Agric. Sect. **194**(2), 165–170 (2014)
10. Filonenko, G.K., Grishin, M.A., Goldenberg, Y.A.M, Kossek, V.K.: Sushka pischevyih rastitelnyih materialov, M., Pisch. prom-st (1971)
11. Stankevich, G.M., Strahova, T.V., Atanazevich, V.I.: Sushka zerna. Lyibid, Kiev (1997)
12. Malezhyk, I.F., Dubkovetskyi, I.V., Veselovska, T.I.E.: Design of the dryer with thermal-radiation energy supply. Research work of NUFT (2008)
13. Samochetov, V.F., Dzhorogyan, G.A.: Zernosushenie, M, Kolos (1970)
14. Efremov, G.I.: Makrokinetika protsessov perenosa, M (2001)

# Geometrical Parameters for Distribution Systems of Hydraulic Machines

Angela Voloshina, Anatolii Panchenko, Igor Panchenko and Andrii Zasiadko

## 1 Introduction

Analysis of volumetric hydraulic machines showed the advantages of orbital hydraulic motor application on self-propelled machinery to drive active working tools [1–3]. Such hydraulic motors are characterized by high specific power with low specific metal consumption, good power characteristics, compactness, and the gearless use. The great positive side of these hydraulic motors is the possibility of their installation directly into the driving mechanism of conveyors, winches, tappers, wheel hubs, etc. The main reasons restraining the wide use of power full-flow hydraulic actuators for active working tools on self-propelled machinery are not only the limited range of existing hydraulic machines, but also the lack of comprehensive research in the design of hydraulic machines of rotary action [4, 5]. Therefore, there is an urgent need to carry out studies defining the laws of supplying fluid to the working chambers of a hydraulic machine, using the example of the distribution system of an orbital hydraulic machine in order to develop new and improve existing designs of hydraulic machines. Improvements in the output characteristics of orbital hydraulic machines can be achieved by solving an important problem which includes the study of the influence of changes in the geometric parameters of the distribution mechanism on the functional parameters of these hydraulic machines.

---

A. Voloshina (✉) · A. Panchenko · I. Panchenko
Tavria State Agrotechnological University, B. Khmelnitsky ave. 18, Melitopol 72310, Ukraine
e-mail: voloshinaa2012@gmail.com

A. Zasiadko
Berdyansk College of the Tavria State Agrotechnological University, Eastern ave. 23, Berdyansk 71108, Ukraine

© Springer Nature Switzerland AG 2019

V. Nadykto (ed.), *Modern Development Paths of Agricultural Production*,
https://doi.org/10.1007/978-3-030-14918-5_34

## 2 Review of Literary Sources

Simulation of the flow of working fluid moving through the channels of gerotor motors [6, 7] justifies the causes of cavitation in the distribution zone. Mathematical expressions for estimating the working fluid flow in gerotor hydraulic machines are given in [8]. The modeling of the working fluid supply by the distribution system to the working chambers of the orbital hydraulic motor has not been considered.

According to geometrical, mathematical, and hydrodynamic models [9, 10], theoretical studies of the influence of the geometrical parameters of the flow parts of the gerotor pump on its output characteristics were carried out. However, the work of the rotors of the gerotor pump is fundamentally different from the work of the rotors of the orbital hydraulic motor. For the operation of the gerotor pump, there is no need to create a rotating hydraulic field of the working fluid. To create a rotating hydraulic field in an orbital hydraulic motor, an end distribution system with segment windows is used [11, 12].

With many different designs, the orbital hydraulic machines can be combined in three main units [1, 2], which determine the effectiveness of these hydraulic machines. This is a gear pair, with a special cycloidal tooth profile [13], a mechanism that compensates the orbital movement of the rotor [12] and a distribution mechanism [11] that creates the hydraulic field necessary for the operation of a gear pair. The models describing the working processes occurring in the distribution systems of orbital hydraulic machines [12] have been proposed, and the influence of their design features on the output characteristics of these hydraulic machines [11] was investigated. The influence of the shape and geometrical parameters of distribution windows on the functional parameters of orbital hydraulic motors has not been studied.

The analysis of the studies related to the design of orbital hydraulic machines allowed us to conclude that they were carried out without considering a number of important factors, such as the work of a distribution system taking into account the geometrical parameters of rotating and fixed distributors, the shape of distribution windows, the number and overlap angle.

Thus, the study of the influence of changes in the geometric parameters of the distribution system, with windows made in the form of a groove, on the output characteristics of the orbital hydraulic machine is an urgent task aimed at improving the output characteristics of these hydraulic machines.

## 3 Research Methodology

To study the influence of geometrical parameters of the distribution system with windows made in the form of a groove on the output characteristics of an orbital hydraulic motor, it is necessary:

- to design diagrams, a mathematical apparatus, and a calculation algorithm which will allow to investigate the effect of changes in the geometric parameters of a distribution system with windows made in the form of a groove on the output characteristics of an orbital hydraulic motor;
- to develop the initial data and initial conditions for modeling the operation of the distribution system with windows made in the form of a groove;
- explore the relationship between geometric parameters of the distribution system and output characteristics of an orbital hydraulic motor.

## 4 Results

The main characteristic of a distribution system is its throughput (flow rate of the working fluid), given by the flow area of this system [12, 14]. The area of the passage is formed by the overlap areas of the windows of the fixed distributor with the windows of the rotating distributor. The window overlap areas depend on the number of the windows involved in the distribution of the working fluid, their shape, and geometrical parameters.

Figure 1 shows the elements of the distribution system for rotating (Fig. 1a) and fixed (Fig. 1b) distributors. On the end surfaces of rotating and stationary distributors, distribution windows are made in the form of a groove. The working fluid is supplied to the working chambers of the hydraulic motor or displaced out through the distribution windows. In further consideration, rotating and stationary distributors will mean their surfaces where distribution windows are located. During the operation, the end surfaces of the rotating and stationary distributors being in a contact form distribution zone together with the distribution windows [12].

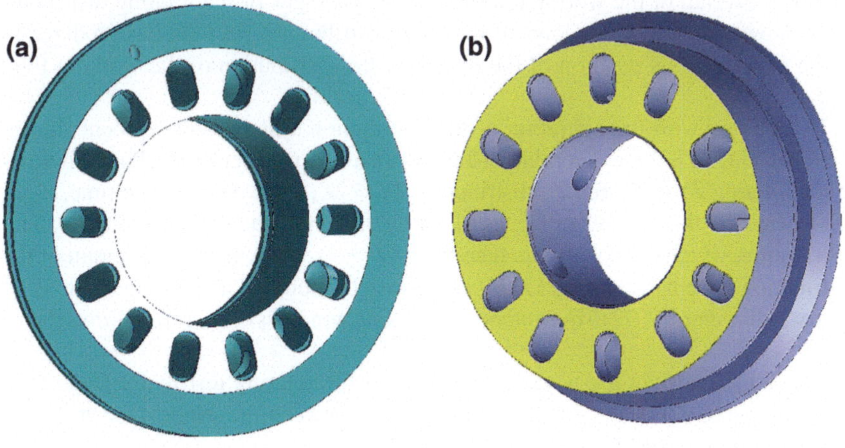


**Fig. 1** Contacting surfaces of fixed (**a**) and rotating (**b**) distributors with distribution windows

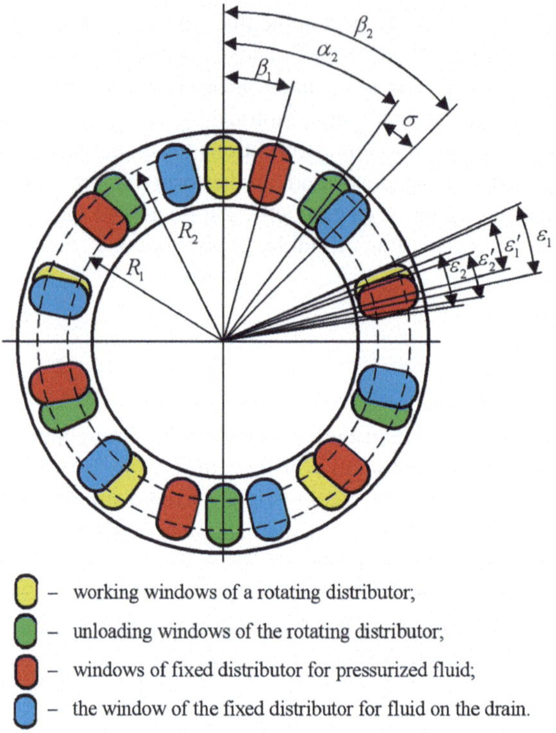


**Fig. 2** Design diagram of the distribution system with windows made in the form of a groove

The power and tank windows are made on the end surface of the fixed distributor. There are working and discharge passages on the end surface of the rotating distributor. The overlap of the working and discharge passages of the rotating distributor on the power and tank windows of the fixed distributor reflect the design diagram of the instantaneous position of the phases of the distribution of the working fluid (Fig. 2).

By analogy with the operation of the distribution system with segment-shaped windows [11, 12], there is a relationship between the number of working windows of the rotating $Z_1$ and fixed $Z_2$ distributors $Z_2 = 2Z_1 + 2$. At the same time, $Z_2 = 2Z$, where $Z$ is the number of cycles given by the kinematic diagram of the system distribution, which cannot be less than three ($Z \geq 3$). The kinematic diagram of the distribution system means [11, 12, 14] the ratio of the number of outgoing windows of the fixed distributor $Z_2/2$ to the number of working windows of the rotating distributor $Z_1/2$.

The angle between the working windows of the rotating distributor (Fig. 2) in the static position is $\alpha = 2\pi/Z_1$, and the angle between the windows of the fixed distributor is $\beta = 2\pi/Z_2$.

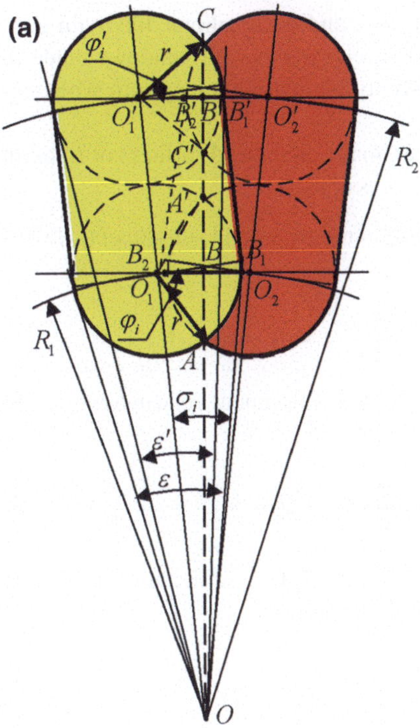


**Fig. 3** Calculation diagram for determining the area of the overlap: **a** $R_2 \cdot \sin\frac{\sigma_i'}{2} < r$; **b** $R_2 \cdot \sin\frac{\sigma_i'}{2} \geq r$

When numbering the windows, we agreed to consider the first window for the rotating distributor window located in the upper part of the circuit on the vertical axis (Fig. 2). The first window of the fixed distributor is the window located to the right of the first window of the rotating distributor. The subsequent numbering of the windows of the rotating and stationary distributors is carried out clockwise. We consider windows of the rotating distributor with odd numbers working, and with even ones—discharge. We consider windows of a stationary distributor with odd numbers as power windows, and with even ones—tank windows.

Let us assume that $O_1$ and $O_2$ are the centers of the lower semicircles of the windows (grooves) of the rotating and stationary distributors, respectively. $O'_1$ and $O'_2$ are the centers of the upper semicircles. $O$ is the center of the circles on which the centers of the semicircles are located: the lower ones with radius $R_1$ and the upper ones with radius $R_2$ (Fig. 3).

When determining the angles between a rotating distributor windows $\varepsilon_1$ and a fixed distributor $\varepsilon_2$, it is necessary to observe the condition $\varepsilon_1 + \varepsilon_2 = 2\pi/Z_2$. It is known [2, 11, 12] that the distribution system with equal geometrical parameters of the windows of rotating and fixed distributors ($\varepsilon_1 = \varepsilon_2 = \varepsilon$) has the greatest throughput.

Taking into account the accepted assumptions and designations, the main geometrical parameters characterizing the distribution system with windows made in the form of a groove are determined similarly to the distribution system with segment windows [2, 11, 12].

The angles of distribution windows (Fig. 2), made on the end surfaces of rotating and fixed distributors, are described by the expressions:

- the angles of the location of the first working window $\alpha_1$ and subsequent $\alpha_i$ windows of the rotating distributor

$$\alpha_1(t) = 2\pi \cdot n \cdot t,\ \alpha_2(t) = \alpha_1(t) + \alpha, \ldots, \alpha_i(t) = \alpha_{i-1}(t) + \alpha. \quad (1)$$

- the angles of the first discharge window $\beta_1$ and subsequent $\beta_i$ windows of the fixed distributor

$$\beta_1 = \frac{\pi}{Z_2},\ \beta_2 = \beta_1 + \beta, \ldots, \beta_i = \beta_{i-1} + \beta. \quad (2)$$

The angle between the following centers (Fig. 2) of the rotating and fixed valve windows that are in overlap is determined by the expression:

$$\sigma_i(t) = \sigma_i'(t) = |\beta_i - \alpha_i(t)|. \quad (3)$$

Moreover, to overlap the lower semicircles of the windows (grooves), must satisfy the condition $\sigma_i \le \varepsilon$, and for the upper windows—$\sigma_i' \le \varepsilon'$.

The angles (Fig. 3) between the lower $\varepsilon$ and upper $\varepsilon'$ of the semicircles of the windows (grooves) of the rotating and fixed distributors will be equal to:

$$\varepsilon = \frac{\pi}{Z_2} \text{ или } \varepsilon = 2\arcsin\frac{r}{R_1}, \quad (4)$$

$$\varepsilon' = 2\arcsin\frac{r}{R_2},$$

and the radii of the lower and upper semicircles of the windows (grooves) of the rotating and fixed distributors will be determined by the expression:

$$r = R_1 \cdot \sin\frac{\varepsilon}{2}. \quad (5)$$

Based on the accepted assumptions and notation, the area of overlap of the $S_i$ windows of the rotating and stationary distributors, which is equal to the area of the $AB_1B_1'CB_2'B_2$ figure (Fig. 3), will be equal to:

$$S_i = S_{1i} + S_{2i} + S_{3i}, \quad (6)$$

where $S_{1i}$ is the area of the $AB_1B_2$ figure; $S_{2i}$ is the area of the $CB_1'B_2'$ figure; $S_{3i}$ is the area of $B_1B_1'B_2'B_2$ figure.

The area $S_{1i}$ of the $AB_1B_2$ figure (Fig. 3a) is determined from the expression:

$$S_{1i} = \frac{1}{2} \cdot \left(S_{1i}' + S_{1i}''\right), \tag{7}$$

where $S_{1i}'$ is the area of the $AB_1A''$ segment; $S_{1i}''$ is the area of the $AB_2A'$ segment.

$$S_{1i}' = \frac{1}{2} \cdot r^2 \cdot (2\varphi_{1i} - \sin 2\varphi_{1i}),\ S_{1i}'' = \frac{1}{2} \cdot r^2 \cdot (2\varphi_{2i} - \sin 2\varphi_{2i}),$$

where $\varphi_{1i}$ is the angle limiting the size of the segment $AB_1A'$, glad; $\varphi_{2i}$ is the angle limiting the size of the segment $AB_2A'$. In case of equality of geometrical parameters ($\varepsilon_1 = \varepsilon_2 = \varepsilon$) of distribution windows of rotating and stationary distributors, $\varphi_{1i} = \varphi_{2i} = \varphi_i$.

Then, the area $S_{1i}$ of the $AB_1B_2$ figure will be equal to:

$$S_{1i} = \frac{1}{2} \cdot r^2 \cdot (2\varphi_i - \sin 2\varphi_i). \tag{8}$$

To determine the angle $\varphi_i$ limiting the size of the $AB_1A'$ segments, we consider the triangle $\Delta AO_1B$.

From the triangle $\Delta AO_1B$

$$O_1B = r \cdot \cos\varphi_i, \tag{9}$$

and from the triangle $\Delta OO_1B$:

$$O_1B = R_1 \cdot \sin\frac{\sigma_i}{2}. \tag{10}$$

Equating the expressions (9) and (10), we obtain the expression for determining the angle $\varphi_i$:

$$\varphi_i = \arccos\left(\frac{R_1}{r} \cdot \sin\frac{\sigma_i}{2}\right). \tag{11}$$

The area $S_{2i}$ of the $CB_1'B_2'$ shape and the angle $\varphi_i'$ (Fig. 3a) are determined similarly from the expressions:

$$S_{2i} = \frac{1}{2} \cdot r^2 \cdot \left(2\varphi_i' - \sin 2\varphi_i'\right);$$
$$\varphi_i' = \arccos\left(\frac{R_2}{r} \cdot \sin\frac{\sigma_i'}{2}\right). \tag{12}$$

The area of the shape $B_1B_1'B_2'B_2$ (Fig. 3a) is determined from the expression:

$$S_{3i} = \frac{1}{2}\left(B_1B_2 + B_1'B_2'\right) \cdot BB', \tag{13}$$

where $B_1B_2$ and $B_1'B_2'$ are the bases of the trapezoid $B_1B_1'B_2'B_2$; $BB'$ is the height of the trapezoid $B_1B_1'B_2'B_2$, $BB' = R_2 - R_1$.

$$\begin{aligned} B_1B_2 &= 2 \cdot (r - O_1B) = 2 \cdot \left(r - R_1 \cdot \sin\frac{\sigma_i}{2}\right); \\ B_1'B_2' &= 2 \cdot \left(r - O_1'B'\right) = 2 \cdot \left(r - R_2 \cdot \sin\frac{\sigma_i'}{2}\right). \end{aligned} \tag{14}$$

Then, the area $S_{3i}$ of the shape $B_1B_1'B_2'B_2$ will be equal to:

$$S_{3i} = (R_2 - R_1) \cdot \left(2r - R_1 \cdot \sin\frac{\sigma_i}{2} - R_2 \cdot \sin\frac{\sigma_i'}{2}\right). \tag{15}$$

Expression (15) is valid only if $R_2 \cdot \sin\frac{\sigma_i'}{2} < r$. Provided $R_2 \cdot \sin\frac{\sigma_i'}{2} \geq r$ the area $S_{3i}$ corresponds to the area of the triangle $\Delta B_1B_2C$ (Fig. 3b):

$$S_{3i} = \frac{1}{2} \cdot B_1B_2 \cdot BC, \tag{16}$$

where $B_1B_2$ is the base of the triangle $\Delta B_1B_2C$. It is determined from the expression (14); $BC$ is the height of the triangle $\Delta B_1B_2C$

$$BC = \frac{B_1B_2}{2} \cdot tg\left(\pi - \frac{\sigma_i}{2}\right) = \left(r - R_1 \cdot \sin\frac{\sigma_i}{2}\right) \cdot tg\left(\pi - \frac{\sigma_i}{2}\right).$$

Then, the area $S_{3i}$ of the triangle $\Delta B_1B_2C$ will be equal to:

$$S_{3i} = \left(r - R_1 \cdot \sin\frac{\sigma_i}{2}\right)^2 \cdot tg\left(\pi - \frac{\sigma_i}{2}\right). \tag{17}$$

Based on the accepted assumptions and notation, the overlap area $S_i$ of the rotating and fixed window distributors will be equal to:

– when $R_2 \cdot \sin\frac{\sigma_i'}{2} < r$

$$\begin{aligned} S_i = {} & \frac{1}{2} \cdot r^2 \cdot \left(2\varphi_{1i} - \sin 2\varphi_{1i} + 2\varphi_{1i}' - \sin 2\varphi_{1i}'\right) \\ & + (R_2 - R_1) \cdot \left(2r - R_1 \cdot \sin\frac{\sigma_i}{2} - R_2 \cdot \sin\frac{\sigma_i'}{2}\right); \end{aligned} \tag{18}$$

– when $R_2 \cdot \sin\frac{\sigma_i'}{2} \geq r$

$$S_i = \frac{1}{2} \cdot r^2 \cdot (2\varphi_{1i} - \sin 2\varphi_{1i}) + \left(r - R_1 \cdot \sin \frac{\sigma_i}{2}\right)^2 \cdot tg\left(\pi - \frac{\sigma_i}{2}\right). \quad (19)$$

As a result of the research, design diagrams, a mathematical tool, and a calculation algorithm have been developed that allow investigating the effect of changing the geometric parameters of a distribution system with windows made in the form of a groove for the output characteristics of an orbital hydraulic motor.

The relationship between the geometrical parameters of the distribution system with the windows made in the form of a groove and the output characteristics of the orbital hydraulic motor was investigated using the *VISSIM* simulation package.

When modeling the work of a distribution system with windows made in the form of a groove, we accepted the following initial data and initial conditions [11, 12]:

- Kinematic and dynamic viscosity coefficients are constant.
- The modulus of elasticity of the working fluid is constant.
- The density of the working fluid is $\rho = 0.89$ g/cm$^3$.
- The coefficient of dynamic viscosity is $\mu' = 0.267 \times 10^{-7}$ MPa s.
- The inlet pressure (discharge pressure) is taken to be $p_{ip} = 16$ MPa.
- The outlet pressure (discharge pressure) is assumed to be $p_{op} = 0$ MPa.
- The angular velocity of the motor shaft is equal to $\omega = 68$ s$^{-1}$.
- The geometrical parameters of windows of rotating and fixed distributors are the same.
- The radius of the lower semicircles is $R_1 = 33$ mm.
- The radius of the upper semicircles is $R_2 = R_1 + 2r$ mm.

Figure 4 shows the structural–functional diagram of the relationship of the geometric parameters of the distribution system and the output characteristics of the orbital hydraulic motor.

**Block A** sets the initial parameters for modeling the operation of a distribution system with windows made in the form of a groove.

The radii of the lower and upper semicircles of the windows of the distributor and the spool, described by expression (5), are determined in **Block B**.

The angles of distribution windows made on the end surfaces of the distributor and the spool described by expressions (1) and (2) are determined in **Block C**.

**Block D** allows to determine the opening angles of the distribution windows described by expression (3), and the angles between the centers of the windows described by expression (4) are determined in **Block E**.

The overlap areas of the lower and upper semicircles of the distribution windows, described by expressions (8) and (12), are determined in **Block F**, and the auxiliary overlap angles, described by expression (11), in **Block G**.

In **Block H**, the overlap area of the central part of the distribution windows is determined, described by expressions (15) and (17).

The total area of overlap of distribution windows, made in the form of a groove, is determined in **Block K**.

It is known [11–13] that the main influence on the change in throughput of distribution systems of orbital hydraulic motors is exerted by the shape of distribution

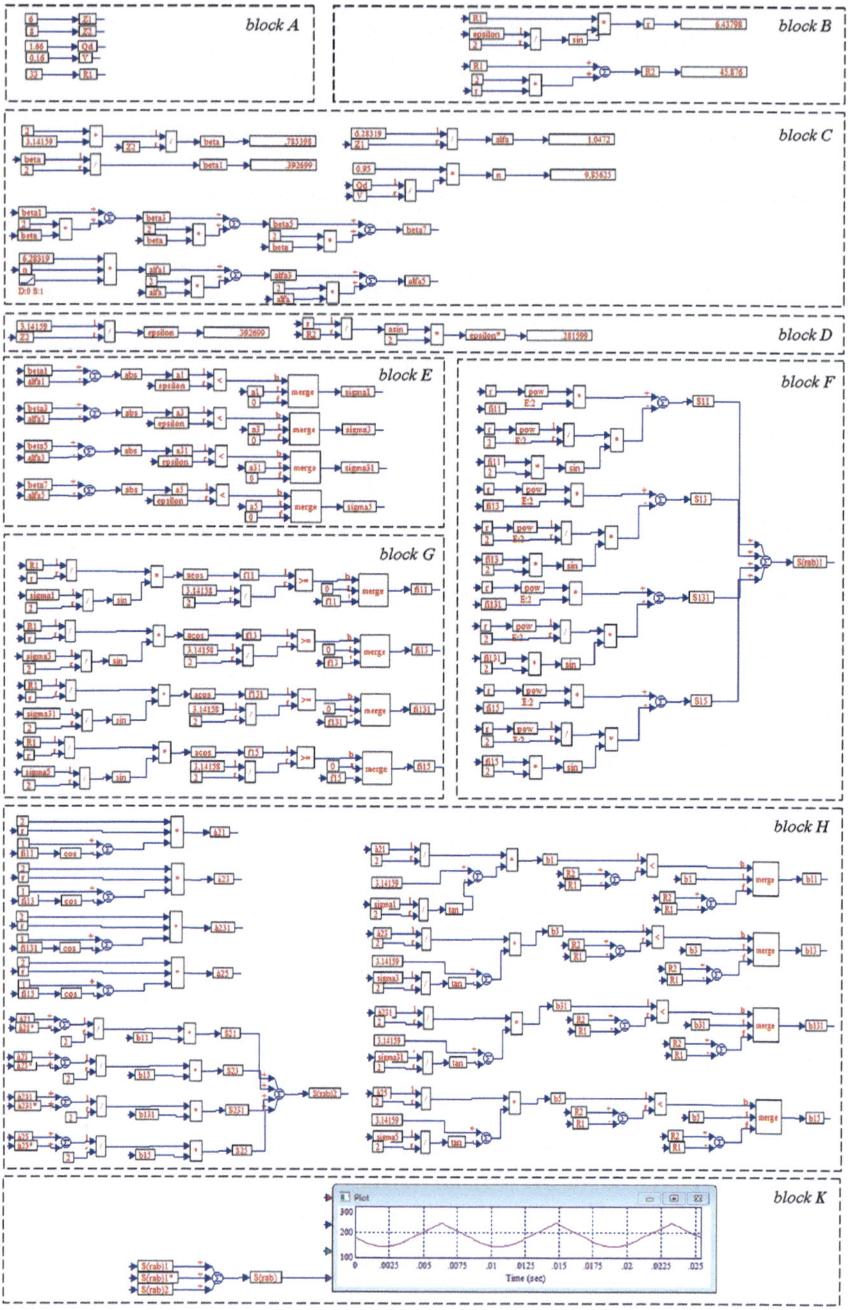


**Fig. 4** Structural and functional diagrams of the relationship between the geometric parameters of the distribution system and the output characteristics of the orbital hydraulic motor

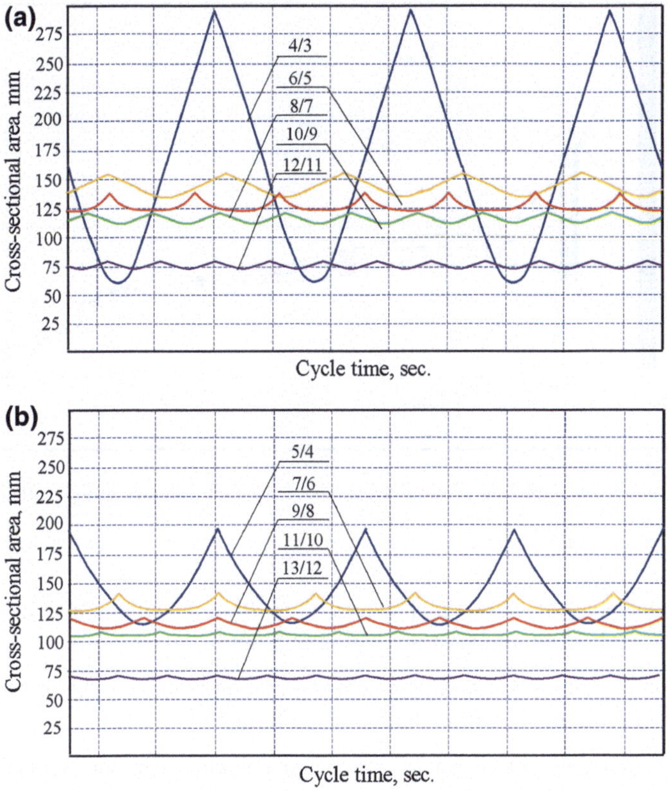


**Fig. 5** Change in the flow area of the distribution system with windows made in the form of a groove according to the kinematic diagrams: **a** even and **b** odd

windows and their number. The number of distribution windows is set by the kinematic diagram which is characterized by the ratio of the number of discharge windows of the rotating distributor to the number of working windows of the fixed distributor. Figure 5 presents the change in the cross-sectional area according to the kinematic diagram of the distribution system during the time corresponding to one cycle.

Analysis of changes in the flow area when using even kinematic diagrams (Fig. 5a) shows that the area varies within the following limits:

- for the kinematic diagram 4/3—from 60 to 300 mm$^2$ with an average value of 180 mm$^2$ and an amplitude of fluctuations of 240 mm$^2$;
- for the kinematic diagram 6/5—from 128 to 148 mm$^2$ with an average value of 138 mm$^2$ and an amplitude of oscillation of 20 mm$^2$;
- for the kinematic diagram 8/7—from 119 to 131 mm$^2$ with an average value of 125 mm$^2$ and an amplitude of oscillations of 12 mm$^2$;
- for the kinematic diagram 10/9—from 106 to 112 mm$^2$ with an average value of 109 mm$^2$ and an amplitude of oscillation of 6 mm$^2$;

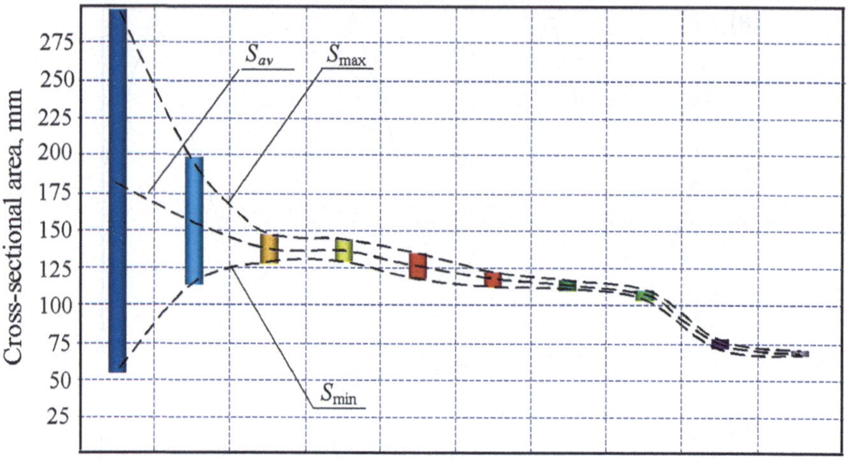


**Fig. 6** Change in the amplitude of the fluctuations of the flow area according to the kinematic diagrams

- for the kinematic diagram 12/11—from 73 to 78 mm$^2$ with an average value of 75.5 mm$^2$ and an amplitude of oscillation of 5 mm$^2$.

Analysis of changes in the flow area using an odd kinematic diagram (Fig. 5b) shows that the area varies within the following limits:

- for the kinematic diagram 5/4—from 120 to 190 mm$^2$ with an average value of 155 mm$^2$ and an amplitude of oscillations of 70 mm$^2$;
- for the kinematic diagram 7/6—from 126 to 139 mm$^2$ with an average value of 132.5 mm$^2$ and an amplitude of oscillations of 13 mm$^2$;
- for the kinematic diagram 9/8—from 113 to 121 mm$^2$ with an average value of 117 mm$^2$ and an amplitude of 8 oscillations of mm$^2$;
- for a kinematic diagram 11/10—from 101 to 106 mm$^2$ with an average value of 103.5 mm$^2$ and an amplitude of oscillation of 5 mm$^2$;
- for the kinematic diagram 13/12—from 68 to 72 mm$^2$ with an average value of 70 mm$^2$ and an amplitude of oscillations of 4 mm$^2$.

Figure 6 shows the dependences characterizing the changes in the amplitude of the oscillations of the flow area of the distribution system with windows made in the form of a groove depending on the kinematic schemes. The oscillation amplitude is understood as the difference between the maximum $S_{max}$ and the minimum $S_{min}$ values of the cross-sectional area.

The analysis of these dependences shows that for the kinematic diagram of 4/3, the amplitude of fluctuations is 1.3 times higher than the average value $S_{av}$ of the cross-sectional area. For the kinematic diagram 5/4, the amplitude of fluctuations is 45% of the average value $S_{av}$ of the cross-sectional area. Such fluctuations in the flow area of the distribution system cause significant fluctuations in the pressure of the working fluid, which negatively affects the operation of the hydraulic motor as a whole. Therefore, the application of the kinematic diagram 4/3 and 5/4, when designing orbital hydraulic motors, is impractical.

Analysis of the change in the flow area in distribution systems of orbital hydraulic motors with kinematic diagrams 12/11 and 13/12 shows that there is a significant (up to 25%) reduction of the cross-sectional area.

Thus, when designing orbital hydraulic motors, the most rational kinematic diagrams of distribution systems with windows made in the form of a groove can be considered diagrams from 6/5 to 11/10, ensuring the stability of the cross-sectional area.

## 5 Conclusions

As a result of the research, design diagrams, a mathematical tool, and a calculation algorithm have been developed, which allowed to investigate how the changes in geometric parameters of a hydraulic motor distribution system with windows made in the form of a groove influence on the motor output characteristics. The initial data and initial conditions for the operation of the distribution system modeling with windows made in the form of a groove were substantiated. The influence of the geometric parameters of the distribution system with windows made in the form of a groove on the output characteristics of the orbital hydraulic motor was investigated. The study revealed that the most rational kinematic diagrams can be considered the ones from 6/5 to 11/10, ensuring the stability of the transmission capacity of distribution systems with windows made in the form of a groove.

## References

1. Erasov, F.N.: New planetary hydraulic drive machines. In: Mechanical Engineering, Kiev (1969)
2. Panchenko, A.I., Voloshina, A.A.: Planetary rotary hydraulic motors. In: Calculation and Designing. Publishing and Printing Center "Lux", Melitopol (2016)
3. Panchenko, A., Voloshina, A., Kiurchev, S., et al.: Development of the universal model of mechatronic system with a hydraulic drive. Eastern Eur. J. Enterp. Technol. **4**(7, 94), 51–60 (2018). https://doi.org/10.15587/1729-4061.2018.139577
4. Biryukov, B.N.: Rotary piston hydraulic machines. In: Mechanical Engineering (1977)
5. Bashta, T.M.: Mechanical engineering hydraulics. In: Mechanical engineering (1971)
6. Velev, E.: Study cavitation gerotor motors, using computer simulation. In: XV International Scientific Conference: Renewable Energies and Innovative Technologies, pp. 64–66 (2016)
7. Marcu, I.L., Pop, I.I.: Interconnection possibilities for the working volumes of the alternating hydraulic motors. In: Scientific Bulletin of the Politehnica University of Timisoara, Special issue: Transactions on Mechanics: The 6th International Conference on Hydraulic Machinery and Hydrodynamics, pp. 365–370 (2004)
8. Yang, D., Yan, J., Tong, S.: Flowrate formulation of deviation function based gerotor pumps. J. Mech. Des. **132**(6), 064503–064505 (2010). https://doi.org/10.1115/1.4001595

9. Altare, G., Rundo, M.: Computational fluid dynamics analysis of gerotor lubricating pumps at high-speed: geometric features influencing the filling capability. J. Fluids Eng. **38**(11), FE-15-1757 (2016), https://doi.org/10.1115/1.4033675
10. Chiu-Fan, H.: Flow characteristics of gerotor pumps with novel variable clearance designs. J. Fluids Eng. **137**(4), FE-14-1137 (2015), https://doi.org/10.1115/1.4029274
11. Voloshina, A., Panchenko, A., Boltyansky, O., Panchenko, I., Titova, O.: Justification of the kinematic diagrams for the distribution system of a planetary hydraulic motor. Int. J. Eng. Technol. **7**(4.3), 6–11 (2018). https://doi.org/10.14419/ijet.v7i4.3.19544
12. Panchenko, A.,Voloshina, A., Boltyansky, O., et al.: Designing the flow-through parts of distribution systems for the PRG series planetary hydraulic motors. Eastern Eur. J. Enterp. Technol. **3**(1, 93), 67–77 (2018). https://doi.org/10.15587/1729-4061.2018.132504
13. Panchenko, A., Voloshina, A., Milaeva, I., Panchenko, I., Titova, O.: The influence of the form error after rotor manufacturing on the output characteristics of an orbital hydraulic motor. Int. J. Eng. Technol. **7**(4, 3), 1–5 (2018). https://doi.org/10.14419/ijet.v7i4.3.19542
14. Panchenko, A., Voloshina, A., Vereshaga, V., Zuev, A.: Mathematical model of the end distributive system with the slot-shaped windows. Proc. Tavria State Agrotechnological Univ. **11**(6), 322–331 (2011)

# Part II

# Mathematical Model Changing the Value of the Process of Leakage Current in 0.38 kV Networks

**Viacheslav Gerasymenko, Volodymyr Kozyrskyi, Natalia Maiborodina and Oleksandr Kovalov**

## 1 Introduction

It is known that the operational reliability of electric power equipment operating in the agro-industrial complex of Ukraine remains rather low.

The analysis of works in this direction shows that this issue is partially solved, for example, research on energy saving in an asynchronous electric engine from the position of constructive execution of the active parts of the latter [1], methods of optimization of electric energy losses in an asynchronous electric motor [2, 3], the construction of an asynchronous electric monitors with minimal losses [4].

However, unexplored operational modes of operation of the electric motor remain at other deviations from the nominal regime. It is known that a weak element of an electric motor is its insulating structure. In the process of exploitation of an electric motor in agriculture, its insulation is effected: an electric field, mechanical and thermal loads, moist air, chemically active reagents of the environment, and eventually as a result of these effects, the insulation gradually changes its properties [5, 6]. With the

The original version of the chapter has been revised: Reference 8 Ovcharov, S., & Strebkov, A. (2015) has been corrected. A correction to the chapter can be found at https://doi.org/10.1007/978-3-030-14918-5_80

V. Gerasymenko (✉) · N. Maiborodina
Nizhyn Agrotechnical Institute, National University of Life and Environmental Sciences of Ukraine, Shevchenko Street, 10, Nizhyn 16600, Ukraine
e-mail: syavagvp@gmail.com

V. Kozyrskyi
National University of Life and Environmental Sciences of Ukraine, Heroyiv Oborony Street, 15, Kiev 03041, Ukraine

O. Kovalov
Tavria State Agrotechnological University, B. Khmelnitskogo Avenue, 18, Melitopol 72310, Ukraine

© Springer Nature Switzerland AG 2019, corrected publication 2024

V. Nadykto (ed.), *Modern Development Paths of Agricultural Production*,
https://doi.org/10.1007/978-3-030-14918-5_35

loss of properties by insulation, the current of leakage also increases. Consequently, the study of the magnitude and dynamics of the current leakage increase is relevant.

## 2 Literature Review and Problem Statement

The current of leakage depends on the amount of overload of the engine, an increase in supply voltage, a decrease in insulation resistance, and other factors. Currently, the main method for controlling the state of isolation remains the method of measuring the resistance of insulation using a megohmmeter. The main drawback of this method is the periodicity of the insulation resistance control. It should also be noted that the value of the insulation resistance measured by the megohmmeter does not correspond to the actual value at the alternating voltage [7].

Extremely negatively affects the condition of insulation of the electric motor of moist air, since virtually all structural elements of the electric motor are in contact with ambient air, which contains moisture. As a result of this contact, there is a moisture exchange between the insulation and surrounding air, the insulation is moisturized and loses electrical resistance and electrical strength, faster aging with thermal loads.

Having analyzed the scientific literature, one can construct the following scheme of reasons for the emergence of leakage current in networks with a voltage of 0.38 kV for agricultural buildings (Fig. 1).

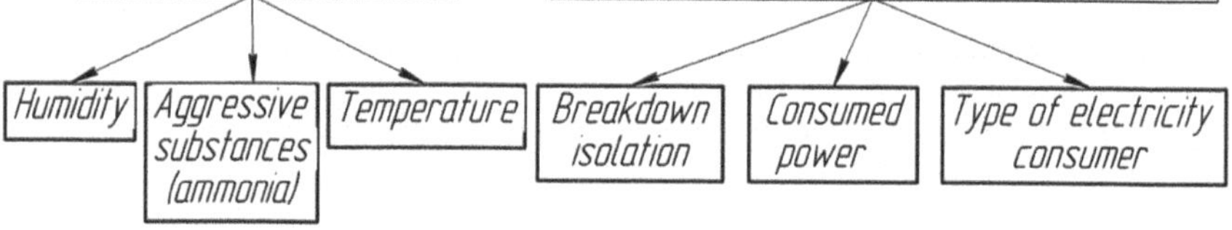


**Fig. 1** Causes of the leakage current

## 3 The Purpose and Objectives of the Research

The purpose of this study is to study the process of the emergence of dangerous leakage currents in electrical engineering complexes and systems of livestock buildings, the establishment of regularities and interconnections.

To achieve the goal formulated the following tasks:

- to conduct experimental research on the production facility. To determine the features of the functioning of the electrotechnical complex;
- to develop a mathematical model of the electric motor. On the basis of it, to analyze the process of formation of leakage currents in the feeding systems of livestock buildings.

## 4 Results and Discussion

The collection of information on the distribution of leakage current in electrical wiring is associated with significant material costs, so the task was to obtain the necessary information by planning an experiment and using the methods of mathematical statistics in the processing of data. The main here is the definition of the minimum required number of measurements of the investigated parameter and statistical verification of the hypothesis about the law of distribution of the random variable.

In the course of the experimental study, special measuring equipment was used—current jumper multifunctional UNI-T UT201, which allows to capture current 0.001 A and V&A VA18B multimeter, which can be connected to a laptop or computer via a USB connector, thus capturing the received data and the transformer current, which allows to record differential leakage current to ground. The scheme of measuring the leakage current is shown in Fig. 2.

The experimental results of measuring the value of leakage current in the form of histograms are shown in Fig. 3.

From this histogram, it is clear that the leakage current in electric motors reaches 30 mA. A well-known work in which the G-shaped scheme of replacing an asynchronous electric motor is used [8], the disadvantage of this work is the failure to take into account the phenomena of vortex currents and reverse magnetization. Therefore, for simulating leakage currents, a T-shaped asynchronous electric motor replacement scheme [9] (Fig. 4) is supplemented branches with a current of leakage and a capacitive displacement current through isolation.

In the substitution scheme, the following items, their parameters, and electrical quantities are indicated:

$R_1, X_1$ active and reactive resistance of the scattering of the stator winding;
$R_2', X_2'$ active and reactive scattering resistance of the rotor windings, which are brought to the stator winding;

$R_M$, $X_M$ active and reactive supports of the magnetizing circuit;
$R'_2\frac{(1-s)}{s}$ resistance, which takes into account the load of the engine and depends on sliding;
$R_B$ resistance to leakage of current through insulation;
$X_{C3}$ capacitive bias current resistance;
$U_1$ supply voltage;
$E_1$ EMF of the magnetizing circuit.

Complete complex supports of the branches of the circle:

$$\begin{aligned} i &= \sqrt{-1}; \\ Z_1 &= R_1 + i \cdot X_1; \\ Z_M &= R_M + i \cdot X_M; \\ Z_2 &= R_2 + i \cdot X_2 + R'_2 \cdot \frac{1-s}{s}; \\ Z_3 &= \frac{-i}{\omega \cdot X_{C3}}. \end{aligned} \tag{1}$$

For the above scheme, we write the system according to Kirchhoff's laws for unknown currents $I_1$, $I_2$, $I_0$, $I_B$, $I_3$, and EMF $E_1$ and solve it using symbolic mathematics in the environment of Mathcad.

Expressions for the required values are determined in the general form:

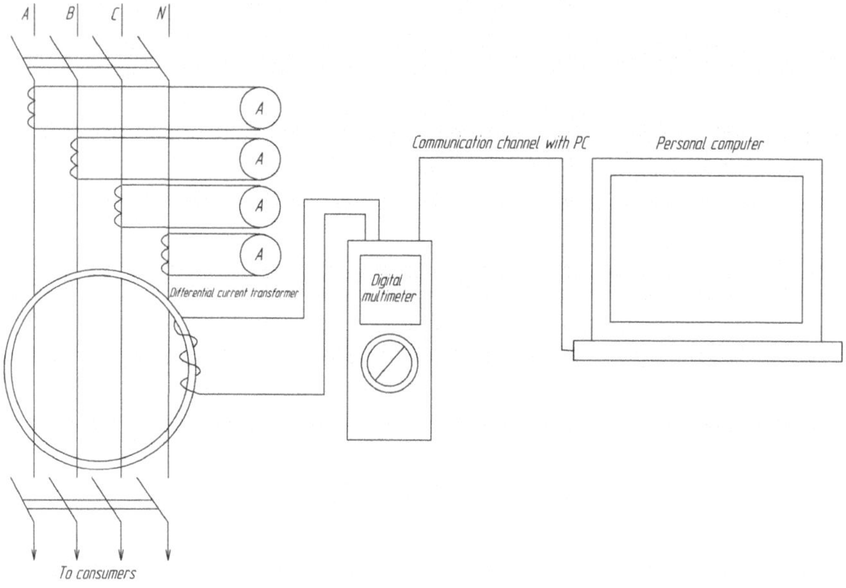


**Fig. 2** Scheme of measuring current leakage

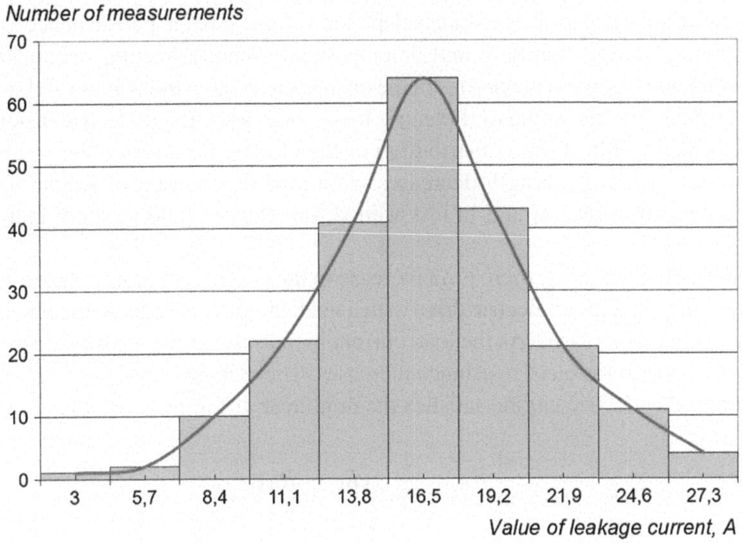


**Fig. 3** Histogram of current of electric engine leakage

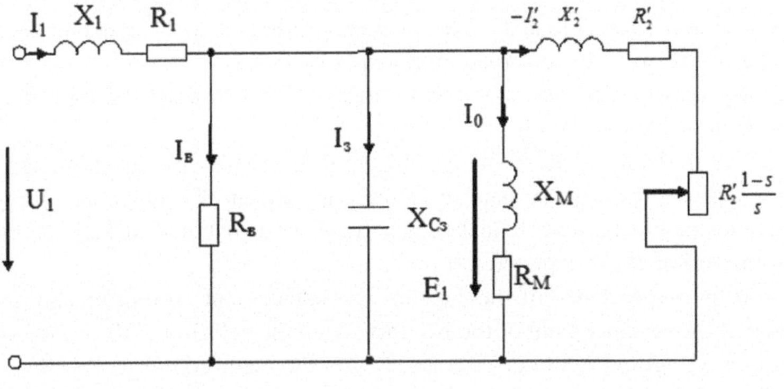


**Fig. 4** T-shaped scheme of replacement of an asynchronous electric engine

$$
\begin{aligned}
U_1 + E_1 &= I_1 + Z_1;\\
E_1 &= -I_0 \cdot Z_0;\\
E_1 &= I_2 \cdot k \cdot Z_2;\\
E_1 &= -I_3 \cdot Z_3;\\
I_1 &= I_0 + I_3 + I_{\mathrm{B}} - \frac{I_2}{k};\\
E_1 &= -I_{\mathrm{B}} \cdot R_{\mathrm{B}}.
\end{aligned}
\tag{2}
$$

The value of insulation resistance depends on the design performance of the electric motor, loading, humidity, and air temperature. When absorbing or condensing the moisture, the electrical properties of the insulation change: the volume and surface resistance decrease, the angle of dielectric losses increases, the dielectric instability increases. As a result of the redistribution of the electric field inside the insulating material, the electrical strength decreases. Saturated damp areas of isolation have a greater dielectric insight, and in less wetted, the electric field strength increases sharply.

At humidified isolation, there is an increase in the coefficient of impulse strength.

When working with an electric drive with a load, the current reduces the insulation and its resistance increases. As the load current increases, the drying time decreases, and the rate of restoration of resistance increases. The change in resistance $R_{\mathrm{B}}$ in the operation of the electric engine satisfies the nonlinear differential equation:

$$\frac{\mathrm{d}}{\mathrm{d}t} R_{\mathrm{B}} = \frac{1}{T_{\mathrm{C}}} \cdot (R_{\mathrm{C}} - R_{\mathrm{B}}), \tag{3}$$

where $T_{\mathrm{C}}$—constant time of change of resistance in the process of operation of an electric engine with a load, *s*.

Permanent time $T_{\mathrm{C}}$ depends on the load factor, humidity, and air temperature. For a 4 A series engine under load $I = I_{\mathrm{H}}$, the temperature of 20 °C depends $T_{\mathrm{C}}$ on the relative humidity, $w_n\%$ is described graphically by Fig. 5.

The graph of the dependence of the current of the leakage through $R_{\mathrm{B}}$ time-isolation from $t$ is shown in Fig. 6.

When the moisture content of the insulation decreases its dielectric insight $e$ and the reactive capacitive resistance of the bias current $X_3$ increases. The graph dependencies of the current of leakage $I_{\mathrm{B}}$ from the resistance of leakage current through insulation $R_{\mathrm{B}}$ is depicted in Fig. 7.

We find the dependence of the capacitive displacement current on the reactive resistance of the displacement of the isolation from the solution of the above system of equations, the graph $I_3$ of dependence $X_{\mathrm{C}3}$ on the pictured in Fig. 8.

Spatial surface development (dependence of active losses in isolation from current and resistance of leakage) $M_{m,n} = R_{\mathrm{B}}(|I_{\mathrm{B}}|, P_{\mathrm{B}m})$ is depicted in Fig. 9.

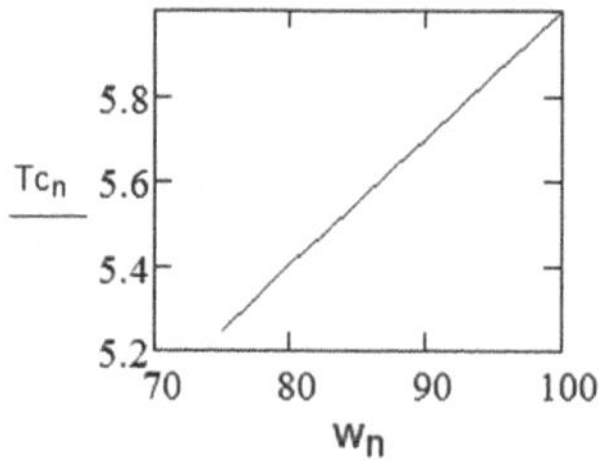


**Fig. 5** Dependence of the constant time of change of resistance $T_{\mathrm{C}}$ from relative humidity $w_n\%$

## 5 Conclusion

After analyzing the current-flow histogram of the electric motor, it can be argued that the distribution of leakage current is close to the normal law of the distribution of random variables, which allows using parametric statistical methods for constructing empirical dependencies.

From this histogram, it is seen that the largest magnitude of the current leakage of the working electric motor reaches 30 mA.

With the help of the developed mathematical model of the analysis of operating modes of the asynchronous motor, taking into account leakage currents and displacement through isolation, we can investigate:

1. Dependence of the leakage current due to the active leakage resistance and the resistance of the displacement of the insulation of the electric motor.

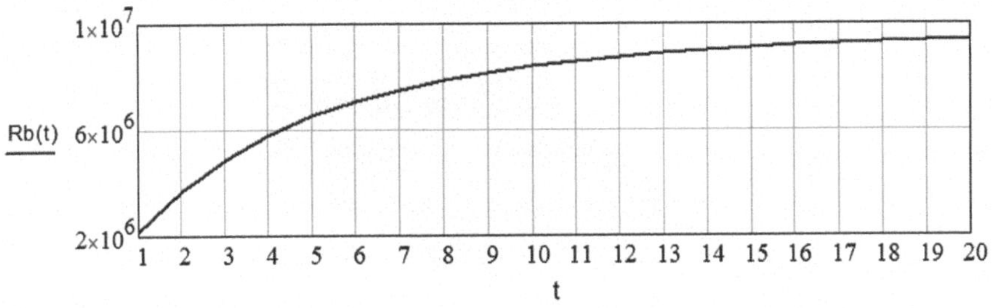


**Fig. 6** Change of insulation resistance of the engine during drying

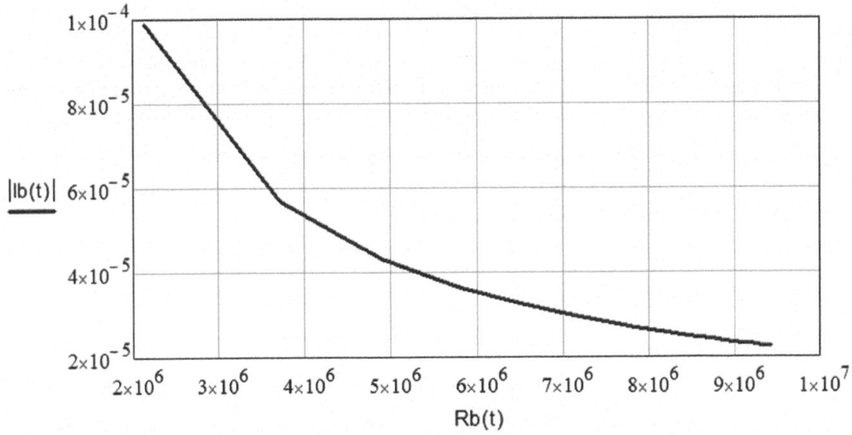


**Fig. 7** Dependence of the leakage current $I_B$ on the resistance of the leakage current $R_B$ through insulation

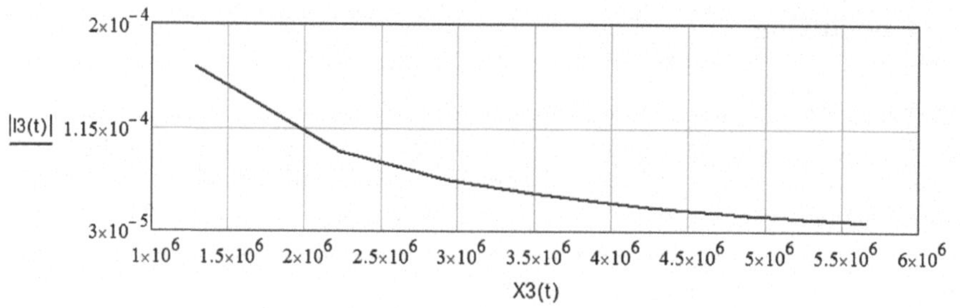


**Fig. 8** Dependence of capacitive displacement current $I_3$ on reactive resistance of displacement of insulation $X_3$

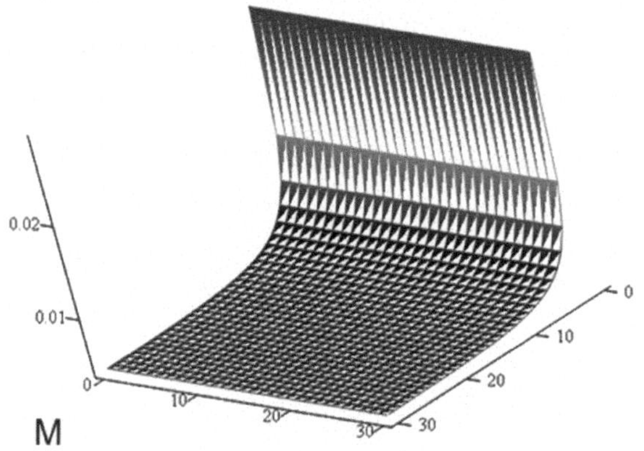


**Fig. 9** Dependence of active losses in isolation from leakage current $I_B$ and resistance of leakage current $R_B$ through isolation

2. Dependence of leakage current at simultaneous change of voltage and slip of the electric motor (fast process) or change of humidity and temperature of isolation together in different interconnections (slow process).
3. Dependence of leakage currents on separate factors of the operating mode of the electric motor and insulation parameters.

## References

1. Sun, D.S.: Research on voltage-chopping and energy-saving controlling technology for three-phase AC asynchronous motor. Adv. Mater. Res. **433–440**, 1033–1037 (2012). https://doi.org/10.4028/www.scientific.net/amr.433-440.1033
2. Hung, N.T., Thien, N.C., Nguyen, T.P., Le, V.S., Tuan, D.A.: Optimization of electric energy in three-phase induction motor by balancing of torque and flux dependent losses. In: Lecture Notes in Electrical Engineering, pp. 497–507 (2014). https://doi.org/10.1007/978-3-642-41968-3_50

3. Grouni, S., Ibtiouen, R., Kidouche, M., Touhami, O.: Novel loss optimization in induction machines with optimum rotor flux control. Int. J. Syst. Control. **1**(4), 163–169 (2010)
4. Kosmodamianskii, A.S., Vorob'ev, V.I., Pugachev, A.A.: Induction motor drives with minimal power losses. Russ. Electr. Eng. **83**(12), 667–671 (2012). https://doi.org/10.3103/s1068371212120073
5. Kozyrsky, V., Gerasimenko, V., Mayborodina, N.: Increased reliability of protection against leakage currents in networks of 0.38 kV. Eng. Technol. Agroindustrial Complex **2**(29), 22–24 (2012)
6. Masorsky, V.I.: Insulation control in networks with grounded neutral. In: Automation and Electrification of Enterprises. Energoatomizdat, Moscow (1983)
7. Manoylov, V.E.: Fundamentals of electrical safety, 5th edn. Energoatomizdat, Moscow (1991)
8. Ovcharov, S., & Strebkov, A.(2015). Active energy losses research in an asynchronous electric motor in operating terms. Eastern-European Journal of Enterprise Technologies, **2**(**8**(74), 22–28. https://doi.org/10.15587/1729-4061.2015.39026
9. Bruskin, D.E., Zorokhovich, A.E., Khvostov, V.S.: Electric machines and micromachines. In: Textbook for Electrotechnical Specialist Universities—3rd edn., Revised and Additional—M.: Higher Education, sc, 528 pp. ill (1990). ISBN:5-06-000-725-1

# Experimental Study of Positive Influence on Growth of Seeds of Electric Field a High Voltage

**Vadim Hulevskyi, Yurii Stopin, Yulia Postol and Mariia Dudina**

## 1 Introduction

The purpose of the study is to confirm experimentally by positive effect of the electric field of a high voltage, direct current on the rate and degree of plant cultivated seeds germination.

Problem setting is to increase the yield of crops by rational use of cultivated land and the production of environmentally friendly products. The solution of this problem can be found in the development of methods for influencing agricultural crops by various physical factors that stimulate the growth and development of plants and, ultimately, the productivity of the crops themselves.

Modern agriculture is impossible without developments and introduction of new technologies with the purpose of determination and use in practice of the most effective and economical productive processes.

The main stage in increasing the yield of agricultural crops is the preparation of seeds, for example, heat treatment of seeds before sowing, increasing germination, the appearance of rapid shoots, and the disinfection of seeds. To protect the phytosanitary state of the seed fund, the choice of such treatment is important, which, along with effective suppression in the seeds of pathogenic microflora of different etiologies, simultaneously ensure the activation of growth processes with minimal expenditure of energy, labor, and resources.

As a result of research having been conducted by many authors, it was established that electrical phenomena play an important role in the life of plants [1–5]. In the process of vital activity of any living organism, weak electric currents, the so-called biocurrents flow inside it. They arise, in particular, in response to external stimuli in plants, so it can be assumed that an external electric field can have a significant effect on the rate of development of any plant.

V. Hulevskyi (✉) · Y. Stopin · Y. Postol · M. Dudina
Tavria State Agrotechnological University, B. Khmelnitskogo Avenue, 18, Melitopol 72310, Ukraine
e-mail: vadym.hulevskyi@tsatu.edu.ua

© Springer Nature Switzerland AG 2019

V. Nadykto (ed.), *Modern Development Paths of Agricultural Production*,
https://doi.org/10.1007/978-3-030-14918-5_36

The most effective of the achieved results are the processes of the direct action of electric energy concentrated in the electric field on materials processed without intermediate energy transformations and, therefore, without additional losses [6].

Along with the traditional applications of electricity in agriculture, strong electric fields and electrical discharges are used.

This is due to the availability of a number of advantages of these technologies over traditional methods of influencing the material being processed. The fact is that practically all substances, both electrically conductive and dielectric, can be charged in one way or another and exposed to the force electric fields and discharges. This implies to universality of electrotechnology methods.

Methods of influence of electric fields and discharges on materials in many cases include some degree of activation of the material or its surface [7, 8].

For materials that have strict limitations on chemical and thermal effects, activation by the action of low-temperature none equilibrium electrical discharges, such as corona, barrier, flare and glow, in various gases, mainly in air, has great potential.

The effect produced by these processing methods depends on many factors, among which there is the quality of the processed seeds, the processing time before sowing into the soil and other external factors. The results of studies, conducted by scientists indicate, that the response of seeds to the same factor affecting during processing can have different effect [9, 10].

Processing in a high-voltage electric field of direct current is characterized by a high degree of energy efficiency, low energy losses for heating the processed object. The mechanism of action of the high-voltage electric field consists at the activation of the electronic complex of molecules, the ionization of these molecules, the formation of free radicals, that is, when molecules go into an excited state. This leads to changes in the electrical properties of seeds and increases water absorption.

The action of the high-voltage direct current electric field is characterized by a high degree of energy efficiency, low energy losses for heating the processing object [11, 12]. This method provides a working cycle in a few seconds or minutes and is one of the most energy-efficient. Although the fact that the molecules are in the excited state only a few seconds, this is enough to strengthen the work of the enzyme systems of seeds. Therefore, for the effective functioning of such an impact, it is necessary to determine the exact value of the electric field strength in the seed treatment zone. However, when processing seeds this way, it is difficult to determine the optimal field strength for seeds of different crops, so the criterion for assessing the effectiveness of pre-sowing processing should not only enable to evaluate the effectiveness of various effects on seeds, but also, if possible, to obtain these data in a short time. In such cases, either the yield of some known crop which may be taken as close substitute or the average of the yield of some group of crops is taken.

For the research was sunflower selected, as the most common technical crop in Ukraine. Since the requirements of sunflower to the climate, especially to the temperature, are high, it is necessary to know at what temperature those or other seeds germinate. F. Haberlandt studied the lowest (8–9 °C), optimal (28 °C), and maximum (35 °C) temperatures at which it is possible to observe the germination of sunflower seeds with the greatest rapidity. Thus, when processing seeds in an electric field of a high DC voltage, it is necessary to control the heating temperature of the seeds.

## 2 Materials and Methods

To prove the positive effect of a high-voltage electric field on the growth of seeds of cultivated plants, an experimental setup has been developed, and its scheme is shown in Fig. 1. The installation consists of a plane-parallel electrode system with a top electrode (1) and an electrode collector (2), on the surface of which the material, being processed is located. The collector is connected to a source with a high constant voltage (positive or negative). The power source contains an autotransformer *TV1*, a high-voltage transformer *TV2* with a transformation ratio $k = 45$, a rectifier *VD1-VD4*. The maximum electric field strength is $E_0 = 6.2$ kV/cm.

At the first stage of the research, identical seeds of sunflower were selected. The weight of the seeds was determined by the formula:

$$m = V \cdot \rho = \frac{\pi}{6} a \cdot b^2 \cdot \rho, \tag{1}$$

where $\rho$ is seed density, kg/m$^3$, *a*, *b*—width and length of seeds, *m*.

When placed between flat electrodes in the entire volume of the grain mass, which is a heterogeneous system (seeds and air), ionization processes begin to occur. The field strength inside the ellipsoidal particle will be:

$$E = \frac{E_o}{1 + \Phi_1(\varepsilon - 1)}, \tag{2}$$

where $E_0$ is the external field strength, *V/m*; $\Phi_1$ is the depolarization coefficient, which depends on the shape of the particle, that is, on the ratio of the minor axis of the ellipsoid to its larger axis.

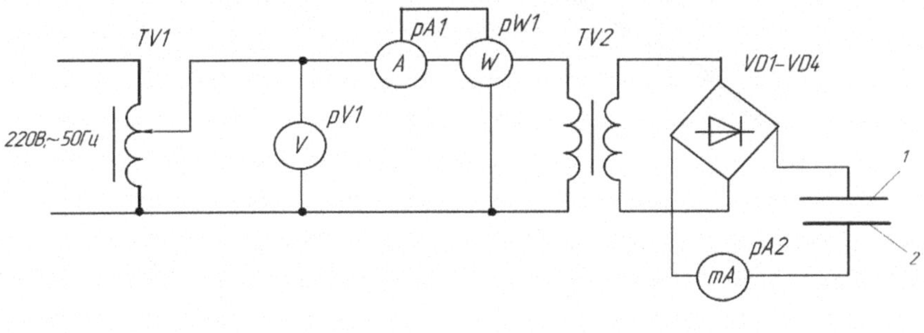


**Fig. 1** Experimental setup

Seeds prepared for the study were not subjected to chemical and thermal effects. The ambient temperature during seed germination was also not taken into account. Seed treatment was carried out at intervals of 4, 8, 12, 16 min. The grain temperature did not exceed the permissible values at which biological destruction of the cellular structure of the seeds could be observed.

## 3 Results and Analysis

As a result of processing, grain is the subject to the influence of both chemical ionization products (chemically active ozone and nitrogen oxides) and electrical discharges. When passing conduction currents and discharge currents in the seed mass, heat is released. The discharge current depends on the presence of free charges in the air gap and conductive dust on the surface of the seeds.

The dependence of the heating temperature of the sunflower seed treatment time at the maximum field strength is shown in Fig. 2.

For planting, the seeds were distributed among seven containers.

The groups were divided as follows:

- Seeds, being were exposed to the electric field of high-voltage DC;
- Seeds of the control group.

The results were recorded daily for 20 days (Fig. 3).

The total number of sprouted seeds depending on the treatment time is expressed as a percentage and is presented in Table 1.

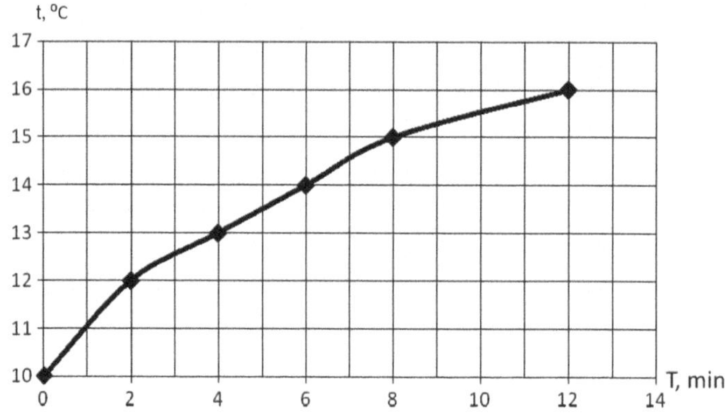


**Fig. 2** Addiction of the temperature of sunflower seeds heating from the time of processing by a constant electric field

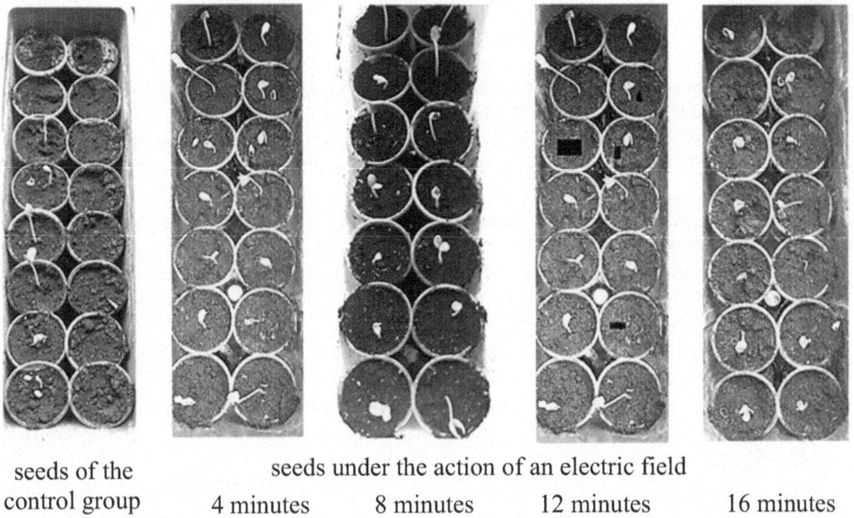


**Fig. 3** Research results

**Table 1** Amount of seeds germinated

| Duration of seed treatment | Germination of seeds (%) |
|---|---|
| 4 min | 60 |
| 8 min | 75 |
| 12 min | 80 |
| 16 min | 90 |
| Control | 50 |

# 4 Conclusions

This experimental research is of great practical importance, since the introduction of new technologies in the conditions Zaporizhia region allows to predict the yield and improve the quality of cultivated plants.

Studies have shown that processing sunflower seeds in an electric field of high voltage positively affects the rate and degree of germination. Such stimulation of the seed material at an exposure temperature of up to 35 °C makes it possible to increase the biological activity of the seeds without damaging the tissue and structure of the product. From this, we conclude that electrical stimulation accelerates the germination of seeds. Nevertheless, more research is needed in this area. Now the results will be evaluated during the growing season at the growth stage before the transition to the flowering phase.

## References


1. Miano, A.C., Forti, V.A., Abud, H.F., Gomes-Junior, F.G., Cicero, S.M., Augusto, P.E.D.: Seed Sci. Technol. **43**(2), 297–302 (2015). https://doi.org/10.15258/sst.2015.43.2.10
2. Radzevicius, A., Sakalauskiene, S., Dagys, M., Simniskis, R., Karkleliene, R., Bobinas, C., Duchovskis, P.: Zemdirbyste-Agric. **100**(2), 179–184 (2013). https://doi.org/10.13080/z-a.2013.100.023
3. Songnuan, W., Kirawanich, P.: J. Electrostat. **70**(5), 445–450 (2012). https://doi.org/10.1016/j.elstat.2012.06.004
4. Iwata, S., Okumura, T., Muramoto, Y., Shimizu, N.: In: Annual Report Conference on Electrical Insulation and Dielectric Phenomena (CEIDP), pp. 179–182 (2011) https://doi.org/10.1109/ceidp.2011.6232626
5. Gogo, E.O., Huyskens-Keil, S., Krimlowski, A., Ulrichs, C., Schmidt, U., Opiyo, A., Dannehl, D.: J. Appl. Bot. Food Qual. **89**, 60–67 (2016). https://doi.org/10.5073/jabfq.2016.089.007
6. Hulevskyi, V.B., Kuznietsov, I.O.: Naukovi pratsi pivdennoho filialu natsionalnoho universytetu bioresursiv i pryrodokorystuvannia Ukrainy «Krymskyi ahrotekhnolohichnyi universytet» **163**, 148–151 (2014)
7. Wang, S.M., Wang, J.F., Guo, Y.B.: Pol. J. Food Nutr. Sci. **68**(3), 195–205 (2018). https://doi.org/10.1515/pjfns-2017-0025
8. Hulevskyi, V.B., Yu, O., Stopin, N.P.: Pierova, Materialy VII Mizhnarodnoi naukovo-tekhnichnoi konferentsii «Enerhozabezpechennia tekhnolohichnykh protsesiv» (8–9 chervnia 2017 roku). 53–55 (2017)
9. Ichim, D., Creanga, D., Rapa, A.: J. Electrostat. **65**(7), 408–413 (2007). https://doi.org/10.1016/j.elstat.2006.10.009
10. Aguilar, J.O., Rivero, D.S., Puentes, A.E., Perilla, P.E.V., Navarro, A.M.S.: In: ICHEAP12: 12th International Conference on Chemical and Process Engineering, 43, pp. 169–174 (2015). https://doi.org/10.3303/cet1543029
11. Okumura, T., Muramoto, Y., Shimizu, N.: IEEE Trans. Dielectr. Electr. Insul. **21**(2), 913–917 (2014). https://doi.org/10.1109/tdei.2013.004085
12. Shi, M.F., Fan, J.J., Li, S.J., Yu, X.L., Liang, X.M.: Environmental technology and resource utilization II, 675–677, 1142–1145 (2014). https://doi.org/10.4028/www.scientific.net/AMM.675-677.1142

# The Usage of Electricity Charged Aerosol for Greenhouse Cooling: Problems and Prospects

**Anton Kashkarov, Volodymyr Diordiiev, Andrii Sabo, Gennadii Novikov and Olexandr Diordiiev**

## 1 Introduction

Greenhouses of Ukraine and developing countries require energy consumption reduction and improving the efficiency of the technological processes [1]. The fogging in greenhouses is one of these technologies for such purposes as:

- to water plants and seedlings;
- to cool greenhouses, that allows distributing of the production cycle for more favorable conditions to the sale the products in the hot period;
- to implement the moisture management for the vegetative stage of the plant;
- to manage the thermal inertia of the greenhouse for the reduction of the effects of short-term temperature differences.

Some short review for the equipment for the fog formation in greenhouses, its technical characteristics, and cost can be found on the Web sites of companies that implement ready-made technological solutions and are widely represented on the Internet (for Ukraine: www.poliv-tuman.com.ua; www.fogfresh.com.ua [1]).

## 2 Literature Review and Problem Statement

Stanciu [2] and Chen [3] consider cooling greenhouses by evaporation or with artificial and natural air-conditioning systems. With the help of specialized software, the effect of orientation of greenhouses on the required heating or cooling power [2] was studied, taking into account the design features of the greenhouse frame [3].

A. Kashkarov (✉) · V. Diordiiev · A. Sabo · O. Diordiiev
Tavria State Agrotechnological University, B. Khmelnitskogo Avenue, 18, Melitopol 72310, Ukraine
e-mail: anton.kashkarov@tsatu.edu.ua

G. Novikov
ASCON, Lenin Street, 68, Akimovka Settlement, Zaporozhye Region 72500, Ukraine

© Springer Nature Switzerland AG 2019

V. Nadykto (ed.), *Modern Development Paths of Agricultural Production*,
https://doi.org/10.1007/978-3-030-14918-5_37

The works of Davies and Lychnos [4, 5] consider the operation of the "humidifier + dehumidifier" system, which makes it possible to realize a closed air exchange system without access to external air. For small and family farms in Ukraine and developing countries, the usage of their work results in complicated by the low production of farms, limited financial resources, and the complexity of implementation and operation for this technology.

Problems with substantiation of the modes for natural and artificial ventilation are studied on the basis of hydrodynamic equations [6], regulation of air exchange and temperature control [7] by artificial and natural methods [8]. At the same time, issues of the charge dynamics, humidity, and temperature are considered indirectly.

The basic greenhouse fogging kit includes [1]: fog machine, pump with engine and timer, stainless steel brass nozzles with the anti-drop system, plastic fittings, brass, pipe, multi-function timer, supply hose.

The process of water filtering is very important for the fog formation in the greenhouse. Water quality affects the operation of pumps efficiency. For fog machine reliable operation, the choice of a water filtration system is very important as it increases its cost.

In the proceedings of Lekomtsev [9], the use of electro-aerosol technologies in livestock buildings was studied, which resulted in the reduction of water pressure and the requirements for water cleaning, simplification for the compressor equipment requirements. In addition, the issue of humidification and the reduction of dust in the air at different modes of charging drops of water aerosol were considered in the mentioned works. The authors concluded that the effect of the aerosol charges on the evaporation efficiency.

It is well known that the role of private vegetable growing in greenhouses is very important for the economies of developing countries like Ukraine. Therefore, it is relevantly to study the prospects of electrical aerosol usage for greenhouse cooling based on evaporation for the further sustainable greenhouse business development.

## 3 Statement of the Objective and Tasks of the Study

The development of a new method is to intensify the aerosol evaporation for greenhouse cooling with the electrical technology usage.

## 4 Materials and Methods

It is usual for the majority of cooling complexes that the fogging system is based on the liquid evaporation (water for this case), i.e., during evaporation water changes from a liquid to a gaseous state, while the surrounding air takes the heat energy required for the water heating [4.18 kJ/(kg × K)] and evaporation (2.256 kJ/kg). It is the reason why the ambient temperature decreases. As water has one of the highest values in nature for the heat capacity coefficient of the phase transition, it is recommended for optimal system operation to grind water into droplets with the size of 3–7 microns [1] for a rapid evaporation effect and increasing of the system efficiency. The water evaporation, which lies in the principle of fogging system operation, strongly depends on the ambient temperature, relative air humidity, ambient pressure; i.e., the drier and hotter, the higher the system efficiency. Mist-forming systems moisturize an overdried hot air and reduce its temperature and dustiness. Thus, the fogging system should not be considered as an ordinary cooling system, but rather as a system for creating and maintaining a comfortable microclimate.

The moisture evaporation limit in the greenhouse is abridged with a dew point at the current temperature and humidity [9]. The required mass of the sprayed liquid could be determined by the expression

$$m_1 = \Delta d \cdot V_a \frac{346}{273 + t_a} \cdot \frac{p}{99.3}, \tag{1}$$

where

$\Delta d$ difference in moisture content of air, kg/kg;
$p$ atmospheric pressure, kPa;
$t_a$ air temperature, °C;
$V_a$ air volume, $s^3$.

In the simplest version, the rate of evaporation of a spherical drop [9], fixed with reference to an infinite homogeneous medium, was considered by Maxwell.

$$I = 4\pi \cdot r \cdot D \frac{M}{RT} (\rho_{sv} - \rho_\infty), \tag{2}$$

where

$D$ coefficient of mutual diffusion of water vapor and air molecules, $0.259 \times 10^{-4}$ $m^2/s$;
$M$ molar mass of water in gaseous state, $18 \times 10^{-3}$ kg/mol;
$r$ radius of a drop, M;
$R$ universal gas constant, 831 J/(mol × K);
$T$ diffusible gas temperature, K;
$\rho_{sv}$ density of saturated vapor above the surface of the drop at a temperature $T$, $kg/m^3$;
$\rho_\infty$ vapor density at an infinite distance from the drop, $kg/m^3$.

Vapor density $\rho_{sv}$ and $\rho_\infty$ related through relative humidity $\varphi$ [%] air is given as

$$\rho_\infty = \frac{\rho_{sv} \cdot \varphi}{100}$$

Let us assume the coefficient $I_\varphi$, which characterizes the ability of the medium, where evaporation occurs, to retain moisture without condensing [$kg/m^3$]:

**Table 1** Values of the coefficient, characterizing the ability of the medium to retain moisture without condensation

| $T$, K (°C) | $\rho_{sv}$ ($10^{-3}$ kg/m$^3$) | $\varphi$ (%) | $I_\varphi$ ($10^{-3}$ kg/m$^3$) | $T$, K (°C) | $\rho_{sv}$ ($10^{-3}$ kg/m$^3$) | $\varphi$ (%) | $I_\varphi$ ($10^{-3}$ kg/m$^3$) |
|---|---|---|---|---|---|---|---|
| 1 | 2 | 3 | 4 | 1 | 2 | 3 | 4 |
| 293 (20) | 17.3 | 20 | 13.84 | 313 (40) | 51.2 | 20 | 40.96 |
| | | 40 | 10.38 | | | 40 | 30.72 |
| | | 60 | 6.92 | | | 60 | 20.48 |
| | | 80 | 3.46 | | | 80 | 10.24 |
| 303 (30) | 30.4 | 20 | 24.32 | 323 (50) | 82.8 | 20 | 66.24 |
| | | 40 | 18.24 | | | 40 | 49.68 |
| | | 60 | 12.16 | | | 60 | 33.12 |
| | | 80 | 6.08 | | | 80 | 16.56 |

$$I_\varphi = \rho_{sv} - \rho_\infty = \rho_{sv}\left(1 - \frac{\varphi}{100}\right)$$

Taking into account that the density of saturated steam depends on the ambient temperature, the coefficient $I_\varphi$ could be represented as tabular values (Table 1).

The reduction of the single-drop radius during the evaporation is described by the Maxwell formula, used by Lekomtsev [9], and it would be

$$r^2 = r_0^2 - \frac{2D}{\rho_l} \cdot I_\varphi \cdot t_e, \tag{3}$$

where

$r_0$ initial radius of a drop, m;
$t_e$ evaporation time, c;
$\rho_l$ liquid density, kg/m$^3$.

The aerosol, introduced into the room, is precipitated due to the gravity and electrostatic scattering. For complete evaporation of electro-aerosol droplets, the evaporation time should be less than the time of deposition, $t_e < t_d$. Conversely, from (Eq. 3), it is possible to determine the evaporation time $t_e$:

$$t_e = \frac{\left(r_0^2 - r^2\right) \cdot \rho_d}{2 \cdot D \cdot I_\varphi} \tag{4}$$

where

$r$ minimal drop radius ($r = 10^{-6}$ m), m.

Uncharged aerosol is precipitated by gravity. The deposition time $t_d$ is determined from the expression

**Table 2** Air dynamic viscosity values in dependence of the temperature

| T, K (°C) | μ ($10^{-6}$ N × s/m$^2$) | T, K (°C) | μ ($10^{-6}$ N × s/m$^2$) | T, K (°C) | μ ($10^{-6}$ N × s/m$^2$) |
|---|---|---|---|---|---|
| 1 | 2 | 1 | 2 | 1 | 2 |
| 273 (0) | 17.2 | 293 (20) | 18.1 | 313 (40) | 19.1 |
| 283 (10) | 17.6 | 303 (30) | 18.6 | 323 (50) | 19.6 |

$$t_d = \frac{9 \cdot h \cdot \mu}{2 \cdot r_0^2 \cdot g \cdot \rho_l} \tag{5}$$

where

$h$ deposition height, m;
$\mu$ dynamic air viscosity (Table 2), N × s/m$^2$.

The charged aerosol is additionally affected by electrostatic dispersion forces. The rate of electrostatic scattering for droplets with a diameter of less than 50 μm exceeds the rate of gravitational deposition. Then, the time of electrical aerosol deposition on the surface is determined by the expression

$$t_{\mathrm{d}} = \frac{6 \cdot \pi \cdot h \cdot \mu \cdot r_0}{E \cdot q} \tag{6}$$

where

$E$ electric field strength, V/m;
$q$ electric charge of aerosol particles, C.

$$E = \frac{U}{l} \tag{7}$$

where

$U$ voltage on the spray nozzle plates, V;
$l$ nozzle spacing, m.

$$q = \frac{2}{3}\pi^3 \cdot \varepsilon_0 \cdot E \cdot r_0^2 \tag{8}$$

Then, taking into account (Eqs. 7 and 8), the deposition time of the charged aerosol (6) can be represented as

$$t_{\mathrm{d}} = \frac{9 \cdot h \cdot \mu \cdot l^2}{(U \cdot \pi)^2 \cdot \varepsilon_0 \cdot r_0} \tag{9}$$

where

$\varepsilon_0$ dielectric constant, $\varepsilon_0 = 8.85 \times 10^{-12}$ $C^2/(N\ m^2)$.

To satisfy the evaporation condition $t_e < t_d$, it is possible to calculate the maximum allowable aerosol droplet radii $r_0$:

- for uncharged aerosol (Eq. 4) и (Eq. 5)

$$r_0^2 \cdot \left(r_0^2 - r^2\right) \leq \frac{9 \cdot h \cdot \mu \cdot D \cdot I_\varphi}{g \cdot \rho_l^2} \tag{10}$$

- for charged aerosol (Eq. 4) и (Eq. 6)

$$r_0\left(r_0^2 - r\right) \leq \frac{18 \cdot h \cdot \mu \cdot l^2 \cdot D \cdot I_\varphi}{(U \cdot \pi)^2 \cdot \varepsilon_0 \cdot \rho_l} \tag{11}$$

Let us solve the indicated equations with respect to $r_0$, taking as condition that $r_0 > 0$. Equation (10) is solved as quadratic (Eq. 11) by Cardano's formulas, reducing the three-term cubic equation to quadratic. To determine $r_0 = f(h)$, it gives the following decisions:

- for uncharged aerosol

$$r_0 = \sqrt{\frac{1}{2}\left[r^2 + \sqrt{r^4 + \frac{36 \cdot h \cdot \mu \cdot D \cdot I_\varphi}{g \cdot \rho_1^2}}\right]}, \tag{12}$$

- for charged aerosol

$$r_0 = z + \frac{r^2}{3 \cdot z} \tag{13}$$

where

$z$ additional variable.

$$z = \sqrt[3]{\frac{9 \cdot h \cdot \mu \cdot l^2 \cdot D \cdot I_\varphi}{(U \cdot \pi)^2 \cdot \varepsilon_0 \cdot \rho_1} + \sqrt{\left(\frac{9 \cdot h \cdot \mu \cdot l^2 \cdot D \cdot I_\varphi}{(U \cdot \pi)^2 \cdot \varepsilon_0 \cdot \rho_1}\right)^2 - \frac{r^6}{27}}} \tag{14}$$

## 5 Result and Discussion

The dependence of the evaporation time (Eq. 5) and the deposition of electric aerosol droplets (Eqs. 6 and 9) at the distance 2 m from the generator (Fig. 1) show that the charged aerosol is deposited faster, and it was confirmed in the works of Lekomtsev [9]. The calculations were performed under the following conditions: Deposition height is 2 m; the distance between the electrodes of the device charging aerosol is 5 mm; air temperatures are $t_a = 30$ °C, $t_a = 20$ °C; humidity is 40%; respectively, the difference in deposition time is less for large droplet radii—over 90 μm.

The graphical plots (Eqs. 12 and 13) for dependence from the condition of complete evaporation and the deposition distance (Fig. 2) show higher requirements for droplet grinding and a higher deposition rate. This leads to more strict requirements for nozzles and water purity, which affects the overall cost for the humidification system. The calculations were performed under the following conditions: voltage on the electrodes of the aerosol charging device—$U = 1$ kV; the distance between the electrodes of the device charging aerosol—5 mm; air temperatures are $t_a = 30$ °C, $t_a = 20$ °C; humidity is 40%.

From the plots of the conditions for aerosol complete evaporation (Fig. 2), it could be seen that as the temperature rises and the precipitation height increases, the permissible sizes of aerosol droplets increase. In the area of the calculation, the

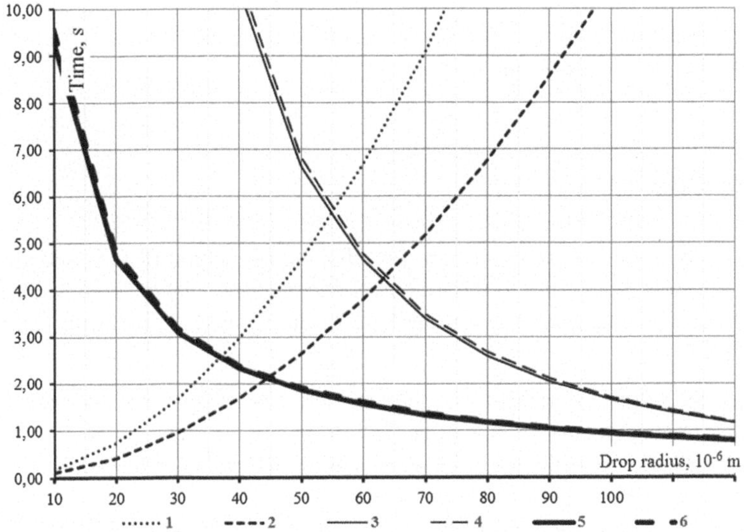


**Fig. 1** Dependence of the time for the deposition and evaporation of electrical aerosol droplets $t$ on radius $r$: 1—evaporation at $t_B = 20$ °C; 2—evaporation $t_a = 30$ °C; 3—gravitational sedimentation at $t_a = 20$ °C; 4—gravitational sedimentation at $t_a = 30$ °C and $U = 0$ V; 5—aerosol charge $U = 1$ kV; 6—aerosol charge $U = 1.5$ kV

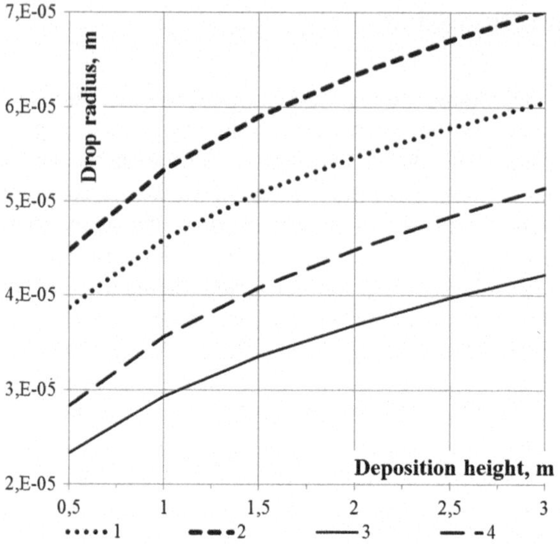


**Fig. 2** Condition of complete evaporation for a droplet radius $r_0$ depending on the distance to the evaporation surface (height of deposition): 1—gravitational deposition at $t_a = 20$ °C; 2—gravitational sedimentation at $t_a = 30$ °C; 3—deposition of electrical aerosol at $t_a = 20$ °C; 4—deposition of electrical aerosol at $t_a = 30$ °C

graphs of charged and uncharged aerosol with the same parameters of temperature and humidity have not extreme points.

## 6 Conclusion

The analysis results of the deposition time for the uncharged (Eq. 5) and charged (Eq. 6) aerosols (according to the graphical dependences of Fig. 1) as well as the conditions for its complete evaporation (Eqs. 12 and 13; Fig. 2) give us the following conclusions:

- the use of a charged aerosol as compared to an uncharged one for greenhouse cooling requires a smaller size of aerosol droplets (a decrease of up to 67% and up to 71% of the radius of uncharged aerosol droplets for humidity 40%, air temperature 20 and 30 °C, respectively);
- use aerosol charging to intensify air cooling in the greenhouse by humidifying is not rational without additional devices as it implies a finer grinding of the aerosol and increased demands for the solution purity;
- the obtained solutions (Eqs. 12 and 13) of equations satisfying the inequalities of the conditions for complete evaporation (Eqs. 10 and 11) allow us to study the

mutual influence of the environmental parameters, the solution conditions, and the aerosol generator features.

In the presented formulas, the excess of the drop maximum charge, when the drop is divided, is not taken into account. The maximum charge of an aerosol drop could be used as limitation in the future mathematical models of the operation of electrical complex for greenhouse cooling and its interaction with an object.

Based on the obtained results and mathematical expressions, it is planned to focus on creating an electrical cooling complex for greenhouses based on electrostatic slowing down of aerosol particles and to study the behavior of particles with a small charge for the further scientific speculations. It is determined by the facts that droplet deposition (Fig. 1) and the requirements for the minimum particle radius for evaporation (Fig. 2) will tend to the parameters of the gravitational deposition and evaporation, but to preserve the property of expansion of the aerosol cloud due to the same charges of aerosol particles.

## References


1. Rain&Fog: http://www.poliv-tuman.com.ua/904-2/
2. Stanciu, C., Stanciu, D., Dobrovicescum, A.: Effect of greenhouse orientation with respect to E-W axis on its required heating and cooling loads. Energy Procedia **85**, 498–504 (2016). https://doi.org/10.1016/j.egypro.2015.12.234
3. Chen, J., Cai, Y., Xu, F., Hu, H., Ai, Q.: Analysis and optimization of the fan-pad evaporative cooling system for greenhouse based on CFD. Adv. Mech. Eng. **2014**, 712740, 8 p (2014). http://dx.doi.org/10.1155/2014/712740
4. Lychnos, G., Davies, P.A.: A solar powered liquid-desiccant cooling system for greenhouses. Acta Horti. **797**, 95–101 (2008); https://doi.org/10.17660/actahortic.2008.797.11
5. Elsarrag, E., Igobo, O.N., Alhorr, Y., Davies, P.A.: Solar pond powered liquid desiccant evaporative cooling. Renew. Sustain. Energy Rev. **58**, 124–140 (2016). https://doi.org/10.1016/j.rser.2015.12.053
6. da Silva, R.C., Cordeiro, J.J.F., Pandorfi, H., Vigoderis, R.B., Guiselini, C.: Simulation of ventilation systems in a protected environment using computational fluid dynamics. Engenharia Agricola **37**(3), 414–425 (2017). https://doi.org/10.1590/1809-4430-eng.agric.v37n3p414-425/201
7. Flores-Velazquez, J., Montero, J.I., Baeza, E.J., Lopez, J.C.: Mechanical and natural ventilation systems in a greenhouse designed using computational fluid dynamics. Int. J. Agric. Biol. Eng. **7**(1), 1–16 (2014). https://doi.org/10.3965/j.ijabe.20140701.001
8. Nicolosi, G., Volpe, R., Messineo, A.: An innovative adaptive control system to regulate microclimatic conditions in a greenhouse. Energies **10**, 722 (2017). https://doi.org/10.3390/en10050722
9. Lekomtsev, P.L.: Electro aerosol technology in agricultural production, 314 p. (Лекомцев П. Л. Электроаэрозольные технологии в сельском хозяйственном производстве: дис. … доктора техн. наук: 05.20.02/Лекомцев Петр Леонидович. М., 2006, 314 с) (2006)

# Development of a Motor Speed Observer for a Electrified Soil-Cultivating Motoblock

**Oleksandr Kovalov, Sergey Kvitka, Oleksandr Solomakha and Viacheslav Gerasymenko**

## 1 Introduction

Modern agricultural production in Ukraine and abroad is characterized by massive use of mobile energy means of small mechanization in the form of small-sized motoblocks, mini tractors, and various specialized units with internal combustion engines [1]. At the same time, conducted comparative tests [2] found that the electrified motor blocks with traction motors DC and AC have a number of advantages compared with motoblock with the internal combustion engine: ease of control, simplicity of launch and stop, reliability and efficiency in the work, absence of gas pollution [3].

## 2 Literature Review and Problem Statement

The work by M. Korchemnyi deserves a great deal of attention [1], which shows the results of the study of motoblocks with traction asynchronous motors with squirrel cage rotor and phase motors and direct current motors with independent and consecutive excitation. It is found that the moment of speed on the engine shaft under the tillage is unstable and changes in time by 40–50%. In addition, there is a oscillation of the moment of resistance with the amplitude of ±20% of the average value. The effectiveness of motoblocks with electric motors is confirmed in the works [4, 5], which is devoted to the comparative technical and energy assessment of technical

O. Kovalov (✉) · S. Kvitka · O. Solomakha
Tavria State Agrotechnological University, B. Khmelnitskogo Avenue, 18, Melitopol 72310, Ukraine
e-mail: oleksandr.kovalov@tsatu.edu.ua

V. Gerasymenko
Nizhyn Agrotechnical Institute, National University of Life and Environmental Sciences of Ukraine, Shevchenko Street, 10, Nizhyn 16600, Ukraine

© Springer Nature Switzerland AG 2019

V. Nadykto (ed.), *Modern Development Paths of Agricultural Production*,
https://doi.org/10.1007/978-3-030-14918-5_38

level of motoblocks according to the passport data for motoblock with DIC and electric drive. In the work by Kusov [6], there are results of systemic research of multi-purpose motoblocks, which got the most widespread, including the electric drive. We discuss the information about the appropriate range of operating moving speed of motoblock during tillage: 0.3, 0.6, 0.9, and 1.2 mps. In the works [7, 8], the results of the investigation of speed observers in electric drives with constant and alternating current engines are given, which are rather complex in structure and realization.

## 3 Statement of the Objective and Tasks of the Study

The aim of the research is to increase the operational reliability of DC motor in the drive of small-sized soil-cultivating motoblock by introducing a speed observation. To achieve this goal, the following tasks were put:

- substantiation and development of the structural and functional scheme of motoblock;
- synthesis of the motoblock speed observer;
- research of efficiency and basic properties of the observer by modeling in the program MATLAB/Simulink.

## 4 Results and Discussion

DC Motoblock receives power from the thyristor converter. The thyristor converter has a control system with a classic subordinate principle of regulation-the external contour of speed regulation, internal contour of current regulation. Since the speed sensor in the designed unit is absent, we use the estimate of the speed that is calculated by the observant. Excitation of the motor is unregulated and we assume that the current excitation does not change. Structural and functional scheme of the motoblocks is shown in Fig. 1.

The object of system control is the direct current motor (DCM) of independent excitation, which can be described by the following equations

$$M = C_{\mathrm{M}} F I_{\mathrm{a}}; \tag{1}$$

$$E = C_{\mathrm{e}} F \omega; \tag{2}$$

$$I_{\mathrm{a}} = \frac{1}{R_{\mathrm{a}}(T_{\mathrm{e}} p + 1)}(U_{\mathrm{a}} - E); \tag{3}$$

$$\omega = \frac{1}{Jp}(M - M_{\mathrm{c}}), \tag{4}$$

where

| | |
|---|---|
| $M$ | electromagnetic moment DCM; |
| $C_{\text{M}}, C_{\text{e}}$ | constructive coefficients of DCM; |
| $F$ | magnetic flux, (since the excitation current is invariable, then $f$ = const); |
| $I_{\text{a}}$ | rectified current (anchor current of DCM); |
| $\omega$ | angular velocity of the shaft; |
| $p$ | operator of differentiation; |
| $R_{\text{a}}, T_{\text{e}}$ | active resistance and constant time of the circumference of the DCM; |
| $U_{\text{a}}$ | rectified voltage TP, (voltage is applied to the anchor of DCM); |
| $J$ | moment of inertia of DCM and mechanism, given to the shaft of DCM; |
| $M_{\text{c}}$ | moment of resistance (load), driven to the shaft DCM. |

Thermistor converter (TP) can be described by the following expression [9]

$$U_{\text{a}} = \frac{k}{T_{\mu} p + 1} u, \tag{5}$$

where

| | |
|---|---|
| $k$ | is the coefficient of amplification TP; |
| $T_{\mu}$ | constant time TP; |
| $u$ | control action, output of the current regulator. |

Due to the small value of $T_{\mu}$ (on 2–3 orders), in relatively expected constant time of the speed regulation process, let's neglect $T_{\mu}$. Then expression (5) is written in the type of

$$U_{\text{a}} = ku. \tag{6}$$

We make reasonable assumption that the moment of loading, attached to the engine shaft is slow variable at the time of the speed observer, then it can be written

$$\dot{M}_{\text{c}} = 0. \tag{7}$$

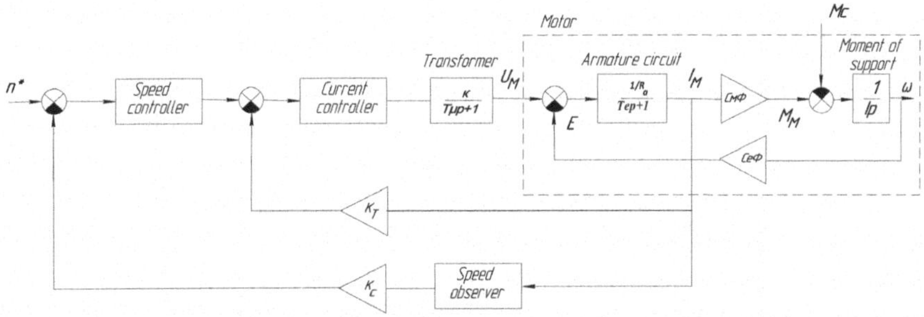


**Fig. 1** Structural and functional scheme of electric drive control to the motoblock

For the synthesis of the speed observer, there are expressions (1)–(4), (6) in the form of Koshi, taking into account: $T_e = \frac{L_a}{R_a}$

$$\dot{I}_a = \frac{1}{L_a}(-R_a I_a - C_e \Phi \omega + ku), \tag{8}$$

$$\dot{\omega} = \frac{1}{J}(C_M \Phi I_a - M_c). \tag{9}$$

Let's introduce Eqs. (7)–(9) in a matrix form

$$\dot{x} = Ax + Bu, \tag{10}$$

$$\dot{y} = Cx, \tag{11}$$

where

$x$ a vector of the state, $x = \begin{bmatrix} I_a \\ \omega \\ M_c \end{bmatrix}$

$y$ a measurement vector.

$$A = \begin{bmatrix} -\frac{1}{T_e} & -\frac{C_e \Phi}{L_a} & 0 \\ \frac{C_M \Phi}{J} & 0 & -\frac{1}{J} \\ 0 & 0 & 0 \end{bmatrix}, \quad B = \begin{bmatrix} \frac{k}{L_a} \\ 0 \\ 0 \end{bmatrix}, \quad C = \begin{bmatrix} k_i \; 0 \; 0 \end{bmatrix}.$$

Check the condition of the system's observation (10) and (11). To ensure that the system was fully observant and enough to have zero input and output signals $x = 0$. The system (10) and (11) corresponds this criterium.

Write the observer

$$\dot{\hat{x}} = A\hat{x} + Bu + Q_y(\hat{y} - y), \tag{12}$$

where

$\hat{x}, \hat{y}$ estimations of the vector of state and measurement vector.

Equation (12) can be converted with regard to (11) and $Q_x = -Q_y C$

$$\dot{\hat{x}} = A\hat{x} + Bu - Q_x(\hat{x} - x), \tag{13}$$

where $Q_x = \begin{bmatrix} q_1 \; 0 \; 0 \\ q_2 \; 0 \; 0 \\ q_3 \; 0 \; 0 \end{bmatrix}$.

According to [9], the characteristic equation for (13) can be written as

$$|Ep - A + Q_x| = 0, \tag{14}$$

where

$E$ is the identity matrix.

After transformations (14), there is the equation

$$p^3 + p^2\left(\frac{1}{T_e} + q_1\right) + p\frac{C_e\Phi}{L_a}\left(\frac{C_M\Phi}{J} - q_2\right) + q_3\frac{C_e\Phi}{L_a J} = 0. \tag{15}$$

Using the modal control let's distribute the polynomial root according to Batervort

$$p^3 + 2\omega_0 p^2 + 2\omega_0^2 p + \omega_0^3 = 0. \tag{16}$$

To do this, compare the coefficients with the operator relevant powers of differentiation

$$2\omega_0 = \frac{1}{T_e} + q_1. \tag{17}$$

$$2\omega_0^2 = \frac{C_e\Phi}{L_a}\left(\frac{C_M\Phi}{J} - q_2\right). \tag{18}$$

$$\omega_0^3 = q_3\frac{C_e\Phi}{L_a J} \tag{19}$$

Using (17)–(19) let's define coefficients of the observer

$$q_1 = 2\omega_0 - \frac{1}{T_e};\ q_2 = \frac{C_M\Phi}{J} - \frac{2\omega_0^2 L_a}{C_e\Phi};\ q_3 = \frac{\omega_0^3 L_a J}{C_e\Phi}. \tag{20}$$

Considering that the persistence level $\eta = 0.707\omega_0$, and the course of the transitional phase $t = 3\eta^{-1}$, $\omega_0$ can be defined

$$\omega_0 = \frac{4.24}{t}. \tag{21}$$

Let's write the equation of the observer in a scalar form

$$\dot{\hat{I}}_d = -\frac{1}{T_e}\hat{I}_d - \frac{C_e\Phi}{L_a}\hat{\omega} + \frac{k}{L_a}u - q_1(\hat{I}_d - I_d). \tag{22}$$

$$\dot{\hat{\omega}} = \frac{C_M\Phi}{J}\hat{I}_d - \frac{\hat{M}_c}{J} - q_2(\hat{I}_d - I_d). \tag{23}$$

$$\dot{\hat{M}}_{\mathrm{c}} = -q_3(\hat{I}_{\mathrm{d}} - I_{\mathrm{d}}). \tag{24}$$

Thus, the observer was received (22)–(24) with coefficients (20) with the distribution of the roots of the characteristic equation according to Batervort.

We will draw a study of the received observer by modeling in the program MATLAB/Simulink. Parameters of speed regulators and current can be defined using the built-in tuner in the PID controller. In the regulators, restrictions on the minimum and maximum value of output were used. Modeling of the system work was conducted and shown in Fig. 1.

At the time of 1 s, the task was submitted to the speed equal to 50% of the nominal (400 rpm). Engine acceleration occurred and at the time of 4 s, the nominal moment of loading was applied to the engine. During the period from 12 to 18 s, the imitation of spilling was conducted (the value of the moment decreased to the value of friction), then the moment of loading restored. At the time of 25 s, the assignments for the speed were changed by 100%, at the presence of the moment of resistance. In the period from 32 to 38 s, the imitation of slippage was conducted restoring the loading moment.

It should be noted that all the influences—change of speed and moment—are carried out by a jump. It was necessary to investigate the quality of the obtained system. Usually, in real conditions, these influences occur smoothly.

## 5 Conclusion

The structural and functional scheme of management is substantiated and the synthesis of the speed of the electric motoblock is performed. The research of the received observer was done by modeling in the program MATLAB/Simulink. When analyzing the dependencies depicted in Figs. 2, 3, 4 and 5, it is possible to conclude that the simulation showed high dynamic characteristics of the developed system. The use of the observer allows to exclude the unreliable element of the speed sensor, which will positively affect the performance, reliability, and accuracy of the system. In this case, all the properties of the electromechanical system of tillage of soil on the base of compact electrified soil-cultivating motoblock are kept.

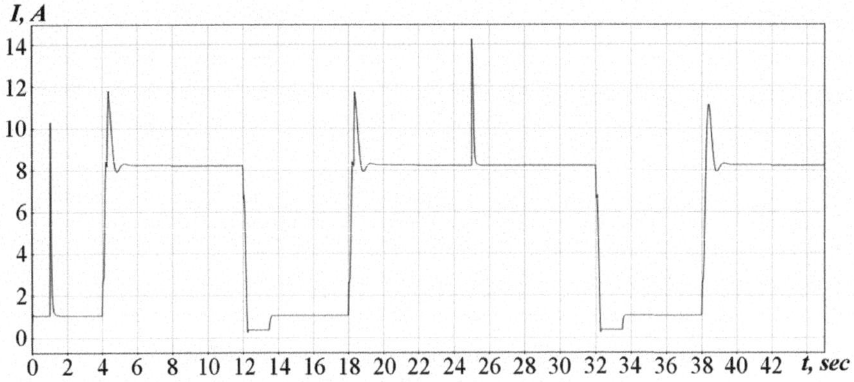


**Fig. 2** Actual motor current dependency of time

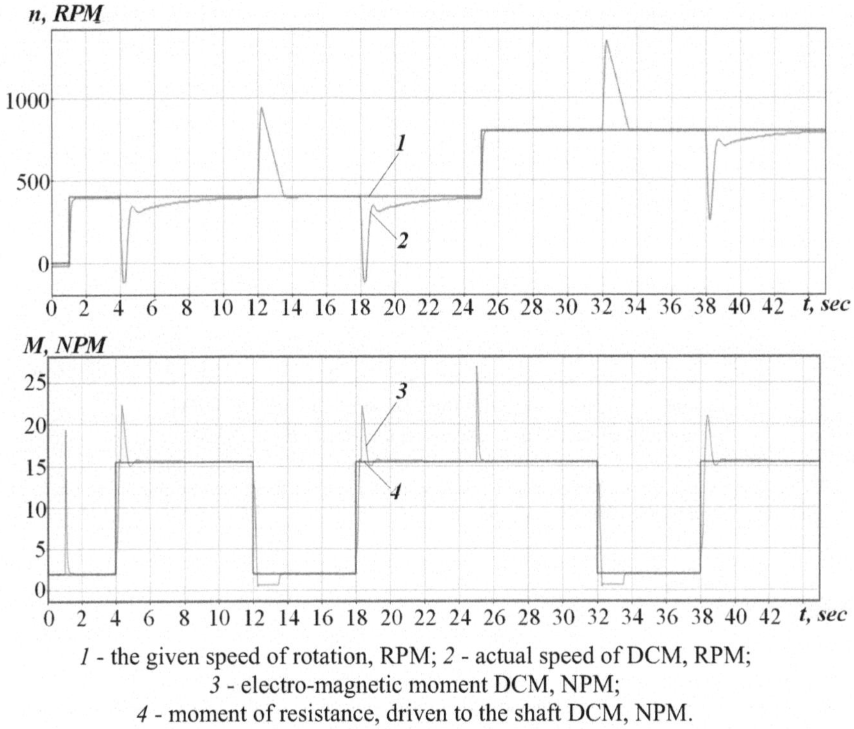


**Fig. 3** Dependence of the given speed of rotation, actual speed, electromagnetic moment and the moment of resistance, evaluated by the observer of time

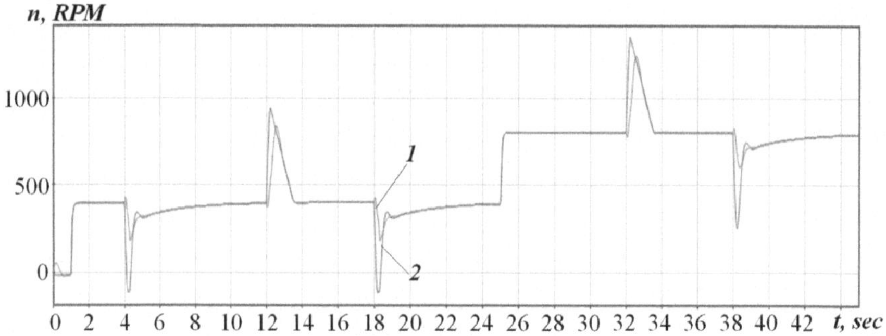


*1* - speed evaluated with the help of the observer, RPM;
*2*-actual speed of DCM, RPM

**Fig. 4** Dependence of the speed evaluated with the help of the observer and actual speed of DCM, evaluated by the observer of time

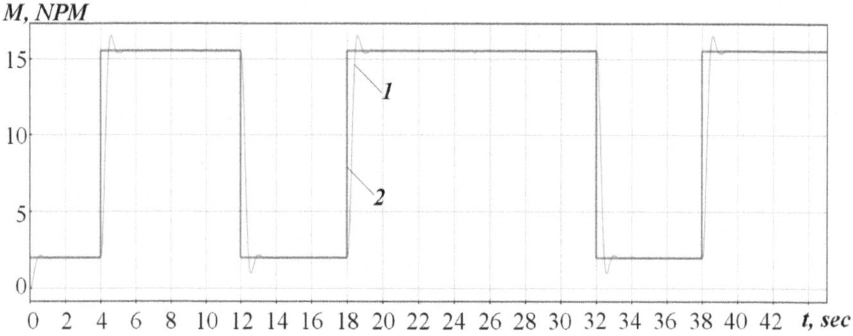


*1* - moment of resistance, with the help of the observer, NPM;
*2* - moment of resistance, driven to the shaft DCM, NPM.

**Fig. 5** Dependence of the moment of resistance, with the help of the observer and the moment of resistance, evaluated by the observer of time

# References


1. Korchemnyi, M., Savchenko, I., Yusupov, N., Husakov, S.: Elektryfikatsiia **8**, 30–31 (1997)
2. Kovalov, O.V., Kutsenko, Y.M., Rubtsov, M.O., Nazarian, H.N.: Pratsi Tavriiskoho derzhavnoho ahrotekhnolohichnoho universytetu **11**(4), 187–200 (2011)
3. Kovalov, O.V.: Visnyk Dnipropetrovskoho derzhavnoho ahrarnoho universytetu **2**, 80–84 (2010)
4. Kovalov, O.V.: Pratsi Tavriiskoho derzhavnoho ahrotekhnolohichnoho universytetu **11**(3), 155–164 (2011)
5. Kovalov, O.V., Katiukha, A.A., Nazarian, H.N.: Pratsi TDATA **7**(3), 93–99 (2007)
6. Kusov, T.T.: Traktoryi i selskohozyaystvennyie mashinyi **10**, 12–15 (1988)
7. Sorokin, V.A., Grekov, E.L., Bessonov, V.G.: Electrotechnical Syst. Complexes **4**(37), 4–10 (2017). https://doi.org/10.18503/2311-8318

8. Ya, V., Frolov, R.I.: Zhiligotov. Zapiski Gornogo instituta **229**, 92–97 (2018). https://doi.org/10.25515/pmi.2018.1.92
9. Potapenko, E.M., Kazurova, A.E.: Osnovyi teorii i metodyi avtomaticheskogo upravleniya, Zaporozhye (2013)

# Experimental Investigations of Functional Properties of Biofuel Processed in the Electrotechnological Complex

**Kushlyk Ruslan, Nazarenko Igor and Kushlyk Roman**

## 1 Introduction

Rising oil prices due to reduced oil reserves in oilfields affects the growth of light petroleum products that causing the increase in agricultural production. Due to the growing demand for oil in the world, its cost and deficit will increase steadily, reaching 16 million barrels by 2025 [1]. This necessitates the search for the replacement of oil products or the mixing of petroleum products with other inorganic substances, such as vegetable origin.

The analysis of different types of alternative fuels showed that the most promising item for Ukraine is the use of mixed fuel which consists of rapeseed oil (MERO) and diesel fuel (BP) methyl ester [2–4].

According to the state standards of Ukraine, it is possible to add a volumetric fraction of methyl/ethyl esters of fatty acids (MEZHK) in diesel fuel of not more than 7%. As practice shows, the biofilm viscosity increases significantly with an increase in the IEO in the DP of more than 7%. As a result, there is a coking of fuel equipment, a decrease in the power of a diesel engine, and an increase in fuel consumption [5, 6].

In view of this, the development of devices that would reduce the viscosity of biofuels, which includes more than 7% of the IEO, does not lose its relevance nowadays.

K. Ruslan · N. Igor · K. Roman (✉)
Tavria State Agrotechnological University, 18 B. Khmelnytsky Ave, Melitopol, Zaporizhzhia obl 72312, Ukraine
e-mail: kushlykroman@ukr.net

© Springer Nature Switzerland AG 2019

V. Nadykto (ed.), *Modern Development Paths of Agricultural Production*,
https://doi.org/10.1007/978-3-030-14918-5_39

## 2 Analysis of Literary Data and Problem Statement

The European Union plans to translate about a quarter (23%) of all European automotive fleet by 2020 to alternative fuels: natural gas—10% (23.5 million cars), biogas—8% (18.8 million cars), and hydrogen (fuel elements)—5% (11.7 million cars) [7].

The most promising alternative fuels are fuels derived from plant raw materials [8]. Thus, in Europe, the use of rapeseed oil from petroleum and products of its chemical processing is expanding [9–11]. However, the most widespread was the methyl ether of rapeseed oil [12, 13].

The analysis of the results of the research work [14] showed that the effective power of the engine when operating on the MERO in nominal mode decreases by 0.8–3.1%. Hourly (l/h) and specific (g/kWhh) fuel consumption increases by 1.9–4.2% and 2.5–7.5%, respectively. At the same time, the presence of hydrocarbons in the exhaust gases is reduced by 1.8–8.3% in relation to work on mineral diesel fuel.

Significant improvements in diesel performance occur on mixed fuels. It is established that when diesel engine is operating on a blended fuel, the effective power of the diesel engine is reduced by nominal mode to 1%. The growth of hourly expenses does not exceed 2.6%. The reduced abdominal discomfort reaches 40% [6].

The production of high-quality biofuel mix requires its processing. From the choice of the appropriate equipment or devices for the specified stage of the process, in essence, the efficiency of the technology used depends.

Hydrodynamic mixers are effective devices that perform two functions: processing and pumping liquids [15]. The disadvantages of a hydrodynamic mixer are the lack of high-quality mixing of the mixture and the inability to control the intensity of the mixing process in the mixer.

To activate the methanol molecules of vegetable oil, it is proposed to use a bioreactor whose reaction volume is filled with ferromagnetic particles [6]. The disadvantage of this technological line is that after the separator, a small part of the raw glycerin remains in the MERO, which leads to stratification of the fuel.

For efficient operation of diesels working on biofuel developed mechanical devices that provide support for the required composition of the mixture when refuel a tractor. An example of such devices is rotary-pulsating apparatus. The disadvantage of these devices is the non-compliance with a given percentage of mixed liquids [16].

One of the ways to increase the efficiency of using biofuels is to treat it with ultrasound (UZ) directly in the engine power supply [17]. The disadvantage of a device for ultrasonic treatment of biofuels is a slight decrease in the viscosity of fuel due to the low throughput of an ultrasonic emitter. The device only works at positive air temperatures.

Mixer filters of biofuel have been used in the power supply of diesel engines of autotractor technology [18]. The disadvantage of the filter mixer is the lack of high-quality mixing of mineral fuel and vegetable oil.

Most researchers, using mixtures of diesel fuel and methyl ether, have experimentally established that it is not enough to intensify the process of obtaining a qualitative biofuel only by mechanical, cavitation, or thermal treatment. Further search for resource-saving technologies for obtaining such fuel is required. Undoubted interest in this direction is the technical solutions aimed at the creation of electrotechnological systems with simultaneous cavitation and ultrahigh-frequency (UHF) electromagnetic effects on its physical and chemical properties. It is the fact that studies of such processes have not been conducted, due to their relevance and high practical significance.

## 3 Purpose and Tasks of the Research

The purpose of experimental research is to detect the influence of ultrasonic and ultrahigh-frequency (UHF) electromagnetic processing of the electrotechnological complex on the viscosity and density of the mixed biofuel. Simultaneous treatment of a mixed biofuel with a high concentration of MEPO in the DP can improve its functional properties.

To achieve this goal, you need to solve the following tasks:

- to determine the joint effect of ultrasound and microwave electromagnetic field of the electrotechnological complex on the biofilm viscosity, depending on the processing time;
- to determine the joint effect of ultrasound and microwave electromagnetic field of the electrotechnological complex on the biofuel density, depending on the concentration of the MERO in diesel fuel;
- to determine the time of stabilization of functional properties of biofuel after treatment in the electrotechnical complex.

## 4 Materials and Methods of Research on the Influence of Ultrasonic and Microwave Electromagnetic Treatment on the Viscosity and Density of the Mixed Biofuel

### *4.1 Materials, Equipment, and Devices Used in Experimental Studies*

In order to improve the functional properties of the mixed biofilm by ultrasonic and microwave electromagnetic treatment, an electrotechnical complex was developed and produced. The functional scheme of the complex is presented in Fig. 1. The general view of the complex is presented in Fig. 2. The research was carried out using commercial mineral diesel fuel L-0.2-62 and rapeseed oil methyl ester. The METRO

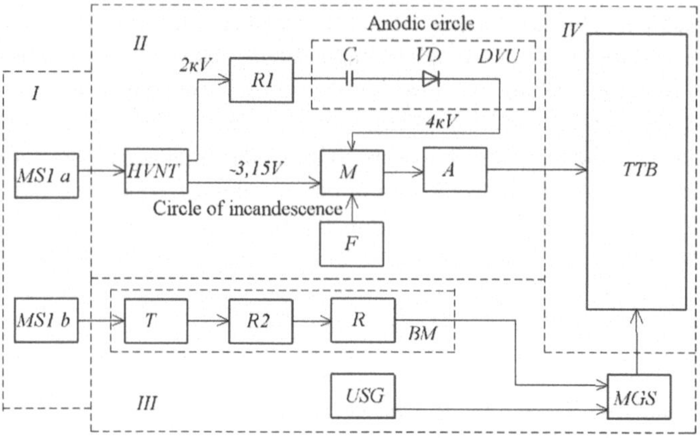


**Fig. 1** Scheme of electric function of the electrotechnological complex: I—block of operating mode switches; II—microwave unit; III—ultrasonic unit; IV—microwave chamber for processing biofuel; MS1a, b—mode switch; HVNT—high-voltage network transformer; R1—rectifier; C—high-voltage capacitor; VD—high-voltage diode; DVU—doubling voltage unit; M—magnetron; A—an antenna; F—fan; T—transformer; R2—rectifier; R—reactor; BM—block of magnetization; USG—ultrasonic generator; MGS—magnetostrictor; TTB—tank for treating of biofuel

**Fig. 2** General view of the electrotechnological complex for the treatment of mixed biofuels

was manufactured at the enterprise "Bionafta Ukraine" LLC (city of Pavlograd, Ukraine).

Experimental samples of the mixtures were prepared from diesel fuel and MERO in percentage terms: 90% AP + 10% MERO (mixture 1), 80% AP + 20% MERO (mixture 2), and 70% AP + 30% MERO (mixture 3).

At the experimental installation [19], it was established that the optimal time of treatment with a mixed biofuel ultrasound and microwave electromagnetic field is

5 min. Therefore, the processing of mixtures in the electrotechnical complex was also carried out for 5 min. The UH-8 water thermostat (Germany) was used to thermostat the treated samples of fuel mixtures. The measurement of the viscosity of the treated mixtures was carried out with viscosimeter VTL-4 (VTL—viscosity of transparent liquids) in Ukraine. The measurement of the density was carried out using the hydrometer (Ukraine).

## 4.2 *Method for Determining the Viscosity and Density of Diesel Fuel, MERO, and Their Mixtures*

The method for determining the viscosity and density of diesel fuel, MERO, and raw mixtures was in the next. Before the start of operation, the capacity was washed with solvent and dried in air. Each of three containers was poured into 100 L of mixed biofuel. This fuel consisted of 10 L MERO and 90 L of DF; 20 L MERO and 80 L DF; 30 L MERO and 70 L DF.

The mixture was stirred with an electric stirrer for 5 min. With viscosimeter VTL-4 with a fivefold repetition at a temperature of 20 °C, the viscosity of the DF and MERO and their mixtures was determined. It was determined that the average viscosity of the L-0.2-62 grade was 4301 $mm^2/s$, and MERO—11,630 $mm^2/s$. The average viscosity and density of the blended fuel are given in Table 1.

The method of joint exposure to ultrasound and microwave electromagnetic field was as follows. A tank of 100 L was filled with a mixture number 1 and supplied biofuels to the electrotechnical complex. Processing mode is flowing. The main condition—a blended biofuel—should be in the chamber for processing 5 min from the moment it is applied to the chamber until its leakage. After processing 100 L of biofuel mixture number 1, samples were taken at three different tank points. Then, the viscosity and the density of the treated sample were determined. By the same method, mixtures 2 and 3 were processed and their viscosity and density were determined.

A sample of the treated sample was taken in a test tube, cooled to room temperature. The specimen was then thermostated at 20 °C for 15 min. Then, the viscosity and density of the test specimen were measured.

**Table 1** Average viscosity and density of mixed fuels

| Index | Concentration of the MERO in diesel mineral fuel (%) | | |
|---|---|---|---|
| | 10 | 20 | 30 |
| Viscosity ($mm^2/s$) | 4.66 | 5.04 | 5.58 |
| Density ($kg/m^3$) | 841 | 846 | 852 |

## 5 Results of Studies on Viscosity and Density of Mixed Biofuel Processed in the Electrotechnical Complex

In Fig. 3, the received data on the temperature of heating of mixtures of 1, 2, and 3 (10, 20, and 30% of the MERO in DP) in the process of their processing in the electrotechnological complex is presented.

The treatment of mixtures by ultrasound was constantly carried out. The treatment of mixtures of microwave electromagnetic fields was carried out with a duty cycle of 35%.

It is established that the higher the proportion of MERO in diesel fuel samples and the greater time it is processed, the higher the temperature of the samples heated. Thus, during the processing of mixtures of biofuel for 1 min, the sample temperature was in the range of 31–35 °C, for the duration of the dues 3 min—46–50 °C and 5 min—58–63 °C (Fig. 3).

The results of the determination of the viscosity of the mixed biofuel, processed in the electrotechnological complex for 5 min, are presented in Fig. 4.

Data analysis (Fig. 4) shows that with the observation time, the viscosity of all fuel mixtures is initially reduced and then set at a certain level. Moreover, with a decrease in the proportion of the MERO the period of the permanent viscosity of the mixture is set later. For example, in mixture 3 (curve 4, Fig. 4), the period of stable viscosity occurs after the fourth day. For mixture 1 (curve 2, Fig. 4), this period takes place not later than the seventh day of observation.

Interestingly enough is the fact that in mixtures with a lower proportion of the MEO, the intensity of reduction of its viscosity is higher. Thus, the viscosity of mixtures 1 and 2 (curves 2 and 3 in Fig. 4) after 1 and 2.5 days of observation becomes lower than the net diesel fuel (line 1, Fig. 4).

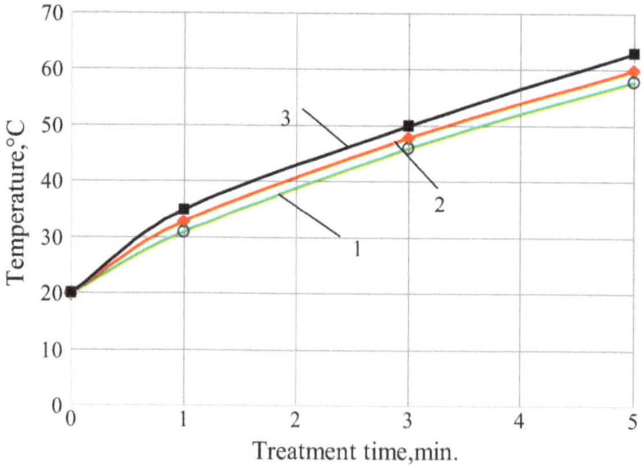


**Fig. 3** Dependence of the temperature of the biofuel from the time of its treatment: 1, 2, and 3—mixture 1, 2, and 3, respectively

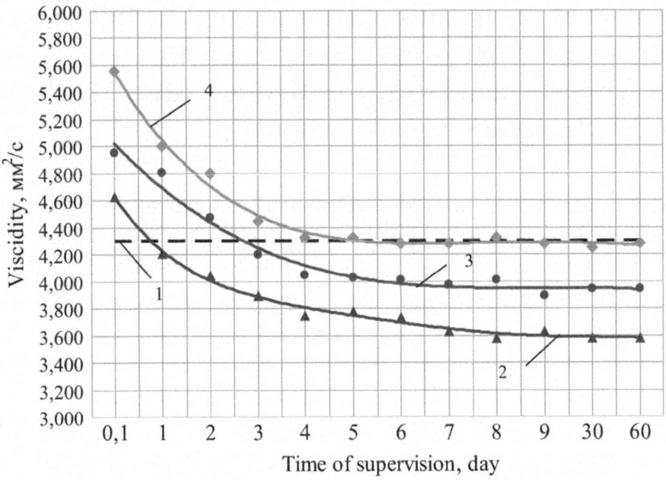


**Fig. 4** Dependence of the viscosity of the mixed biofilm on time of observation after processing of fuel by ultrasound and microwave waves for 5 min: 1—DF; 2, 3, 4—mixture of 10, 20, and 30% of MERO in DF, respectively

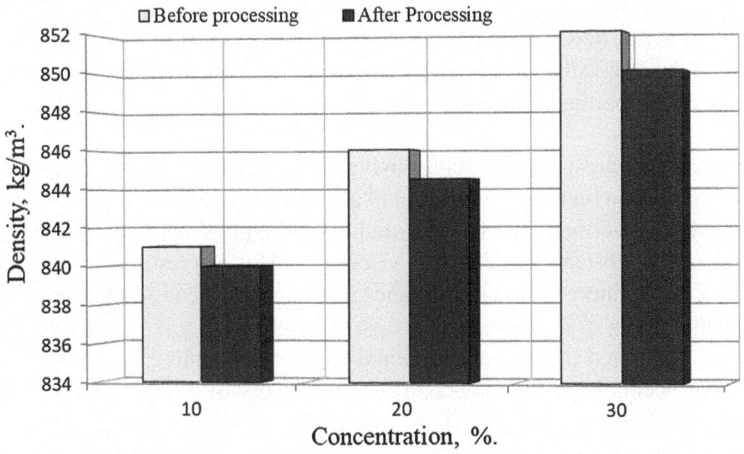


**Fig. 5** Dependence of biofuel density on the concentration of MERO in diesel fuel before processing and after processing in the electrotechnical complex after 0.1 day of observation

Figure 5 shows the dependence of the biofilter density on the concentration of MERO in the DP before processing and after processing in the electrotechnical complex in 0.1 day of observation.

Analysis of studies (Fig. 5) shows that after processing of biofuels in 0,1 day of observation the density of all mixtures decreased by an average of 1–2 kg/m$^3$. However, after 7 days, the density of all mixtures was the same as it was before processing.

## 6 Discussion of the Results of Research on the Influence of Ultrasound, Microwave Electromagnetic Field on the Viscosity and Density of Biofuels After Treatment in the Electrotechnological Complex

At present, researchers are offered various methods and technical means, which allow to change physicochemical parameters of oil fuel products. The influence on the kinetics of processes that occur in petroleum products can be chemicals (catalysts, surfactants, additives) and physical fields (thermal, ultrasound, electromagnetic).

As a result of such an intervention, the radius of the nucleus and the thickness of the adsorption–solvent shell of a complex structural unit, which is an element of the oil disperse system, change. However, the action on petroleum products with the help of chemicals leads to an accelerated depreciation of technological equipment.

Nowadays, many researchers pay special attention to the methods of intensive processing of petroleum products. The basis of such technologies lies in the method of pulsed energy influence due to acoustic waves and cavitation. Cavitation destroys the bond between individual molecules and affects the change in the structural viscosity of the petroleum product. Under the influence of the cavitation of high intensity over a long period of time, C–C bonds in paraffin molecules are violated. As a result, changes in the physicochemical composition of oil products occur. This leads to a decrease in the molecular weight, the temperature of crystallization, and also to the change in the properties of the oil product (viscosity, density, flash point, etc.) [20, 21].

Of particular interest in terms of improving the functional properties of the blended fuel is a joint action on biosolar ultrasound and microwave electromagnetic field. In particular, [22] describes experimental studies on the physical modeling of heating processes of the substances of different viscosities. This can result in a more intense increase in temperature. This circumstance in some cases leads to an improvement in its quality [23].

It should be noted that the experimental studies of blending biofuel processing in the electrotechnology complex confirmed the theory of the destruction of bonds between individual molecules. Obviously, such a mechanism of the influence of ultrasound and microwave electromagnetic field on biofuels is the factor which improves the functional properties of the mixed fuel.

Taking into account the given circumstances, experimental studies of the joint effect of ultrasound and microwave electromagnetic fields on bioheat for 5 min were conducted. It has been established that the combined processing of mixtures 1 and 2 UZ and microwave electromagnetic fields for 5 min in the electrotechnological complex has reduced the viscosity of biofuels by 22.8 and 22.69%, respectively, in relation to untreated biofuels. This circumstance justifies the expediency of consistent processing.

However, the assertion that the possibility of introducing the results on an industrial plant for the treatment of a blended biosolar ultrasound and microwave electromagnetic field for 5 min will be premature. So, there are still a number of unresolved issues concerning the influence of ultrasound and microwave fields on the viscosity of the liquid that moves in the object of control. It would be highly advisable to carry out appropriate research. This would give an opportunity to broaden the understanding of the processes that occur in the dynamic mode of biofuel processing.

## 7 Conclusions

1. An analysis of world and domestic scientific and practical experience has shown that the use of methyl esters in diesel fuel more than 7% negatively affects the quality of the mixed biofuels. As a result, there is a coking of fuel equipment, a reduction in diesel engine power, and an increase in fuel consumption. One of the ways to eliminate these shortcomings is to use ultrasound and microwave electromagnetic fields to process biofuels.
2. It was established that the combined treatment of mixtures 1 and 2 (10 and 20% of MERO in DP) with ultrasound and microwave electromagnetic field in the electrotechnical complex for 5 min allowed to decrease the viscosity of biofuel by 22.8 and 22.69%, respectively, in relation to unfinished biofuel. This substantiates the expediency of consistent processing.
3. It was established that the indices of the density of the mixed biofuel after treatment in 7 days became the ones that were before processing.
4. It was established that stabilization of functional properties of biofilm (viscosity and density) occurs 7 days after ultrasound and microwave electromagnetic field processing. This causes the need for the appropriate time elapses before using biofuels.

## References

1. Marchuk, O., Khokman, K.H., Lissitsa, A.: Discussion paper no. 111 (2010)
2. Perederiy, N.O.: Sci. Periodical Nat. Agric. Univ. **131**, 300–304 (2008)
3. Kuc, T.V.: Sci. Periodical Nat. Univ. Bioresour. Using Nat. Ukraine, Ed. **141**, 196–203 (2009)
4. Semenov, V.G.: Tractors Farm Mach. **5**, 7–8 (2007)
5. Gromakov, A.V.: Mech. Electrification Agric. **8**, 24–25 (2011)
6. Nagornov, S.A., Fokin, R.V., Malakhov, K.S.: Announcer of Moscow State Technical University. N.E. Bauman—Ser. "Engineering", **4**, 122 (2009)
7. Khachyian, A.S.: Engine Constr. **1**, 28–31 (2004)
8. Shkalykova, V.N., Patrakhaltsev, N.N.: Publishing house of Peoples' Friendship University of Russia, 64 (1993)
9. Georgogianni, K.G., Kontominas, M.G., Pomonis, P.J., Avlonitis, D., Gergis, V.: Energy Fuels. **22**(3), 2110–2115 (2008). https://doi.org/10.1021/ef700784j
10. Georgogianni, K.G., Kontominas, M.G., Pomonis, P.J., Avlonitis, D., Gergis, V.: Fuel Process. Technol. **89**(5), 503–509 (2008). https://doi.org/10.1016/j.fuproc.2007.10.004
11. Candeia, R.A., Silva, M.C.D., Carvalho Filho, J.R., Brasilino, M.G.A., Bicudo, T.C., Santos, I.M.G., Souza, A.C.: Fuel **88**(4), 738–743 (2009). https://doi.org/10.1016/j.fuel.2008.10.015

12. Lin, B.F., Huang, J.H., Huang, D.Y.: Fuel **88**(9), 1779–1785 (2009). https://doi.org/10.1016/j.fuel.2009.04.006
13. Markov, V.A., Shuster, A.Y., Devyanin, S.N.: Transp. Altern. Fuel. **4**, 33–37 (2009)
14. Nagornov, S.A., Fokin, R.V.: Actual questions of agrarian science and education. In: Materials of International Scientific-Practical Conference Dedicated to the 65th Anniversary of Ulyanovsk State Agricultural Academy (20–22 May 2008). V. VI. Technical Providing AIC—The Ulyanovsk State Agricultural Academy, pp. 123–128 (2008)
15. Bogdanovich, V.P., Gromakov, A.V.: The state and prospects of the development of agricultural machinery. In: Materials of the 4th International Scientific-Practical Conferences Within the Framework of the 14th International Agroindustrial Exhibition "Interagromash-2011" (Rostov-on-Don, DonExpo Exhibition Center) March 2–3, Rostov, pp. 70–72 (2011)
16. Nagornov, S.A., Matvieiev, O.V., Fokin, R.V.: Digest of scientific works of the Federal State Budget Scientific Institution All-Russian Scientific Research Institute for the Use of Machinery and Petroleum Products in Agriculture. Iss. 14. Federal State Budget Scientific Institution All-Russian Scientific Research Institute for the Use of Machinery and Petroleum Products in Agriculture, Tambov, pp. 33–39 (2008)
17. Fadeev, S.A.: Abstract of thesis of dissertation of candidate of engineering sciences, 05.20.03, Saratov. 14 p (2011)
18. Ukhanov, A.P., Ukhanov, D.A., Kriukov, V.V., Sidorov, E.A., Godina, E.D.: Pat. 2486949 RF, IPC B01F5/06. The mixer is a filter of mineral fuel and vegetable oil. Applicant and patent holder Federal State Budget Educational Institution of Higher Professional Education "Penza State Agricultural Academy" 06.04.2012, Bul. 11 (2013)
19. Kushlyk, R., Nazarenko, I., Kushlyk, R., Nadykto, V.: Eastern Eur. J. Enterp. Technol. **2/1**(86), 34–41 (2017). https://doi.org/10.15587/1729-4061.2017.95985
20. Promtov, M.A.: Chem. Pet. Eng. **2**, 6–8 (2008)
21. Nemchin, A.F., Myhaylyk, V.A., Todorashko, G.T., Schepin, E.V.: Ind. Heat Eng. **24**(6), 60–63 (2002)
22. Kashif, M., Yahya, N., Zaid, H.M.: J. Appl. Sci. **8**, 366–370 (2011)
23. Komarov, V.V.: Phys. Wave Process. Radio Syst. V **13**(4), 57–63 (2010)

# Hybrid Power System Stochastic Optimization

**Olga Lysenko, Mykola Kuznetsov, Andriy Chebanov and Svitlana Adamova**

## 1 Introduction

Energy is the basis of social development. Since the industrial revolution, we have experienced steam age and electrical age. Every change is accompanied by a change in the pattern of energy and marks great progress of human civilization. In the last centuries, electricity is the basis of global social and economic development accompanied by the improvement of industrialization and energy application. During this period, mining non-renewable resources and consuming fossil energy continuously lead to resource tension, air pollution, greenhouse effect, climate change, and so on.

Energy application model is changing which is aimed at reducing the impact of fossil energy on human beings. Clean energy substitution and electric energy substitution are the core of energy consumption patterns transformation [1].

The optimum ratio of individual elements in renewable energy-based hybrid power systems (HPSs) is determined with taking a number of important factors into account, namely: the availability of traditional energy sources; climatic (meteorological) conditions; structure of energy supply systems and energy consumption; requirements for the quality of electric and thermal energy; requirements for the schedule of energy supply; environmental and economic factors. The need to optimize the structure of generating capacities is a peculiarity of local power systems. Such optimization should take into account peculiarities of energy consumption, reliability requirements, available renewable energy sources (solar, wind), cost indicators. The combination of photovoltaic panels and wind generators boosts overall output energy. However, the system of energy accumulation should provide continuous power sup-

O. Lysenko (✉) · A. Chebanov · S. Adamova
Department of Power Engineering and Automation, Faculty of Energy, Tavria State Agrotechnological University, Melitopol 72312, Ukraine
e-mail: olga.lysenko@tsatu.edu.ua

M. Kuznetsov
Department of Integrated Power Systems, Institute of Renewable Energy, National Academy of Sciences of Ukraine, Kiev 02000, Ukraine

© Springer Nature Switzerland AG 2019

V. Nadykto (ed.), *Modern Development Paths of Agricultural Production*,
https://doi.org/10.1007/978-3-030-14918-5_40

ply and cover possible shortage of electricity from renewable sources. Such systems may include an accumulator storage battery, hydrogen containers and fuel cells, etc.

## 2 Problem Description

On condition of increase in the share of HPS in the energy sector, a significant number of publications on methods of computational optimization have appeared recently to achieve technical and economic reliability in hybrid systems. The use of deterministic methods has provided this optimization, with economic indicators being generally uses, such as net present value of the system, cost of energy. The research [2–5] explores the HOMER software. A promising way to increase the technical and economic characteristics of a standalone power supply is shown in [6]. In publication [7], we proposed an automatic programming method for autonomous grids in isolated islands.

Thus, different optimization methods may be used to achieve the technical and economic optimality of hybrid systems [8], but, on the other hand, the availability of power generation by wind and solar power plants depends on the state of the weather, and it introduces uncertainty in the operation of the power grid. The level of energy consumption is random as a rule. Therefore, the optimization of this hybrid power system should take into account its stochasticity.

Probabilistic techniques can be further classified into two types. They include the analytical and Monte Carlo simulation (MCS) methods [9].

In [10], a demand model is used to synthesize high-resolution domestic load profiles, and these are used within Monte Carlo analysis to determine how much peak shaving could be achieved with storage. Fixed network topologies are examined, or a Monte Carlo approach is used to quantify the investment needs at voltage level in [11]. The non-intrusive probabilistic approach adopted in [12] study combines a finite number of deterministic thermodynamic process simulations with an artificial neural network approximation model and Monte Carlo simulations to assess the effect of system uncertainties on the predicted performance of the NMT system.

In this paper, the authors investigate the solution of hybrid power system optimization which is based on probabilistic statistical analysis and Monte Carlo simulation (MCS) methods.

## 3 Materials and Methods

Several reliability indices are introduced in published sources. Loss of load expected (LOLE), loss of energy expected (LOEE) or expected energy not supplied (EENS), loss of power supply probability (LPSP), and equivalent loss factor (ELF) are some of the most common indices used in the reliability evaluation of generating systems [9].

The use of HPS that does not have a guaranteed source of primary energy (primarily, this applies to wind and solar energy) has led to new problems in ensuring reliability in managing the development of the UES. The fact that, along with the power shortage potential in the UES, there is the possibility of an excess capacity is an important problem if, at a high value of the current capacity at the WPP and SPP, the level of consumption decreases. Fluctuations in the energy balance need to be offset by changing the power of other traditional sources, which requires an appropriate regulatory range to ensure the regulatory quality of the electricity. The power of the WPP and SPP can vary within a certain range from zero to the maximum potentially available capacity; therefore, reliable assurance of the needs of consumers is possible only with certain probability. Consequently, it is necessary to introduce criteria that take into account the occurrence of a deficit or excess capacity and the probability of occurrence of such situations.

It is precisely this type of criteria that includes the EXC index; it is also possible to introduce indexes similar to LOLE or LOEE, where the corresponding events relate to excess power or energy. Example:

$$\mathrm{LOGE} = \sum_{i \in S_1} p_i T_i, \tag{1}$$

where $S_1$ shall mean the set of all states of the system associated with the loss of generated energy, $p_i$ shall mean the probability thereof, and $T_i$ shall mean duration;

$$\mathrm{LOGP} = \mathrm{LOGE}/T, \text{ or } \mathrm{LOGP} = \sum_{k \in S_1} p_k G_k, \tag{2}$$

where $G_k$ shall mean excessive generated energy.

In the absence of accumulation, there should be LOLP = 1 − LOGP, and when choosing the capacity of batteries, the condition of optimization can be considered as min (LOLP + LOGP).

When optimizing the number of generating objects, the target function is deterministic, and restrictions are stochastic due to the physical nature of the energy carrier. Additional optimization criteria also have a stochastic nature and can relate to the amount of lost or underexposed energy (M-R criterion), permissible deviations of voltage (D-model), etc. [13].

Let max $P_\Delta$ and min $P_\Delta$ denote, respectively, the maximum and minimum values of the imbalance of power at a given time horizon. If the density distribution of the power imbalance $f_\Delta P(p) = F'_\Delta P(p)$, is known, then for the proportion of excess or, more precisely, unused energy, the LOGP estimate can be modified:

$$\mathrm{WE}(T) = T \int_0^{\max P_\Delta} p \cdot f_{\Delta P}(p)\mathrm{d}p, \tag{3}$$

and the likelihood of excess capacity

$$\mathrm{EXC} = \int_{0}^{\max P_\Delta} f_{\Delta P}(p)\mathrm{d}p = 1 - F_{\Delta P}(0). \tag{4}$$

For lack of energy, that is, loss of load, it is similar:

$$\mathrm{LPS}(T) = T \int_{\min P_\Delta}^{0} p \cdot f_{\Delta P}(p)\mathrm{d}p, \tag{5}$$

$$\mathrm{LOLP} = \int_{\min P_\Delta}^{0} f_{\Delta P}(p)\mathrm{d}p = F_{\Delta P}(0). \tag{6}$$

With continuous change of generating and consumption capacities, their exact coincidence (zero balance) is considered as a point event with zero probability. The pHPSence of buffer power (battery) can provide a zero balance with a certain probability, and the function of distributing power deviations from the needs will have a noticeable peak at zero point.

In the previous probability formulas (4)–(6), the time interval $T$ does not formally belong to the right parts; here, it implies that the probability distribution is valid for this time horizon, that is, for the set of possible realizations of the random process of a certain duration. For stationary processes, the length of an interval can be arbitrary. As a rule, in order to get the stationary position, it is necessary to allocate a purely stochastic component from the actual process of energy balancing, for which it is necessary to determine separately the systematic components of generation and consumption—seasonal dependence, average daily turnover, etc. [14, 15].

In the problems of stochastic optimization, it is often necessary to observe the changes in mathematical expectation and dispersion at the same time, to avoid exceeding their prescribed values or admitting this excess with a certain probability. To solve this problem, models with mixed conditions (two-criterion and multi-criteria problems) [16] are considered, which will require the establishment of a hierarchy of criteria in terms of Pareto-optimality.

To use the criteria defined by the formulas (3)–(6), it is necessary to evaluate the behavior of the random process that describes the balance (or the imbalance) of the generated and consumed power:

$$P_\Delta(t) = P_\mathrm{G}(t) - P_\mathrm{L}(t) \tag{7}$$

$$P_G = P_R + P_K \tag{8}$$

where $P_\mathrm{R} = P_\mathrm{W} + P_\mathrm{PV}$ shall mean the capacity of renewable energy sources that are of occasional nature (the wind $P_\mathrm{W}$ and the sun $P_\mathrm{PV}$);

$P_K$ shall mean controlled power (traditional sources);
$P_L$ shall mean power load (consumption).

In order to determine the possibility of energy accumulation and making sharp changes in the balance of generation and consumption smooth, it is necessary to take into account not only the total number of different implementations, but also their consistency. The subject of the study should be the duration of the periods of excess or lack of power, the accumulated amount of unrealized energy, the average and marginal need for compensating capacities (accumulation, auxiliary or reserve generators, etc.). Parametric description does not provide this possibility, as opposed to simulation.

For simulation of HPS, the representation of instantaneous power in the form of averaged value $\omega(t)$ for a given season (trend curve), average daily value as a random variable and current short-term changes $U(t)$, concerning deviations from average power as a random process are proposed. Consequently, the instantaneous power function in the general case will look [15]:

$$P_i(t) = \omega_i(t) + \sigma_i\varepsilon + U_i(t), \tag{9}$$

where $\sigma_i$ shall mean standard deviation of average daily values for the $i$th participant; $\varepsilon$ shall mean standard Gaussian distributed random variable.

Such a representation shows good convergence with the actual nature of the generation of wind power plants. A similar view is proposed for photovoltaic installations [17], although as an option for simulating a random process, evenly distributed random variables can be used.

Normal wind and solar energy consumption mode are badly cohered with traditional electricity needs, so some assumptions about the consumption regime need to be made to calculate the adequacy of the power grid. It is natural to assume that the system contains controlled sources of energy (electricity, diesel generators, small hydroelectric power stations, etc.), whose scheduled work corresponds to the average consumption level and additional generations of HPS. This will allow us to assess the impact of the stochastic component, which is a major contributor to wind and solar energy. Thus, we accept:

$$P_K(t) = \omega_L(t) - \omega_R(t), \tag{10}$$

$$P_\Delta(t) = \sigma_W\varepsilon_1 + \sigma_{PV}\varepsilon_2 - \sigma_L\varepsilon_3 + U_W(t) + U_{PV}(t) - U_L(t), \tag{11}$$

where $\varepsilon_i$ shall mean independent standard random variables;
$M[P_\Delta(t)] = 0$ shall mean mathematical expectation of the imbalance equal to zero.

## 4 Result and Discussion

We shall compare different approaches to the power balance modeling with actual data on wind speed, solar radiation and electricity consumption. The data of January and July 2016 for the region of Zaporizhzhya, have been selected. A separate settlement in the region of Zaporizhzhya is considered to be a consumer (one point of connection T1, Mordvinivka Village), a group of settlements (four points of connection T1–T4); the initial data have been partially reflected in the research [18, 19]. Probability that the balance exceeds the level $p_1$ (kW) shall be determined with the formula:

$$\mathrm{EXC} = \int_{p_1}^{\infty} f_{\Delta P}(p)\mathrm{d}p = 1 - F_{\Delta P}(p_1) = 1 - \Phi\left(\frac{p_1}{\sigma_\Delta}\right). \tag{12}$$

The inverse problem suggests the maximum power of the imbalance with a given confidence probability.

We shall compare calculated and measured probabilities for several variants of construction of power complexes when the power of WPP (W) and SPP (S) varies from 0 to 200 kW. It is possible to estimate the average power or the absence of such an estimate. Consider for example the consumption of T1 in January, the confidence probability of 0.8, and the level of imbalance over 8% of the average power Tables 1 and 2. The estimation of the unbalance modulus shall mean the symmetric levels of excess and insufficient energy (in this case, EXC = |LOLP| = 0.1).

Comparison of calculated and measured values indicates a satisfactory accuracy of the representation of the distribution as a normal one. It is noteworthy that the predic-

**Table 1** Probability of limiting values of consumer T1 imbalance, estimate

| Criterion | Method of determination | $W = 0$, $S = 0$ | $W = 200$, $S = 200$ | $W = 200$, $S = 0$ | $W = 0$, $S = 200$ | $W = 100$, $S = 100$ |
|---|---|---|---|---|---|---|
| Probability 0.8 (kW) | Estimation | 45 | 86 | 81 | 55 | 60 |
| | Measurements | 45 | 85 | 83 | 56 | 62 |
| $\lvert P\Delta\rvert > 80$ kW (probability) | Estimation | 0.011 | 0.116 | 0.104 | 0.030 | 0.044 |
| | Measurements | 0.011 | 0.117 | 0.109 | 0.024 | 0.042 |

**Table 2** Probability of limiting values of consumer T1 imbalance, without the estimate

| Criterion | Method of determination | $W = 0$, $S = 0$ | $W = 200$, $S = 200$ | $W = 200$, $S = 0$ | $W = 0$, $S = 200$ | $W = 100$, $S = 100$ |
|---|---|---|---|---|---|---|
| Probability 0.8 (kW) | Estimation | 115 | 141 | 139 | 119 | 121 |
| | Measurements | 123 | 143 | 144 | 126 | 125 |
| $\lvert P\Delta\rvert > 80$ kW (probability) | Estimation | 0.187 | 0.234 | 0.231 | 0.195 | 0.199 |
| | Measurements | 0.200 | 0.242 | 0.237 | 0.203 | 0.210 |

**Table 3** Potential to reach the limiting values of the imbalance for group T1–T4, July

| Criterion | Method of determination | $W = 0$, $S = 0$ | $W = 400$, $S = 400$ | $W = 400$, $S = 0$ | $W = 0$, $S = 400$ | $W = 200$, $S = 200$ | $W = 100$, $S = 100$ |
|---|---|---|---|---|---|---|---|
| Probability 0.8 (kW) | Estimation | 377 | 412 | 408 | 381 | 388 | 381 |
| | Measurements | 360 | 405 | 400 | 380 | 382 | 374 |
| $\|P\Delta\| >$ 300 kW (probability) | Estimation | 0.154 | 0.176 | 0.173 | 0.157 | 0.161 | 0.157 |
| | Measurements | 0.125 | 0.145 | 0.140 | 0.130 | 0.130 | 0.125 |

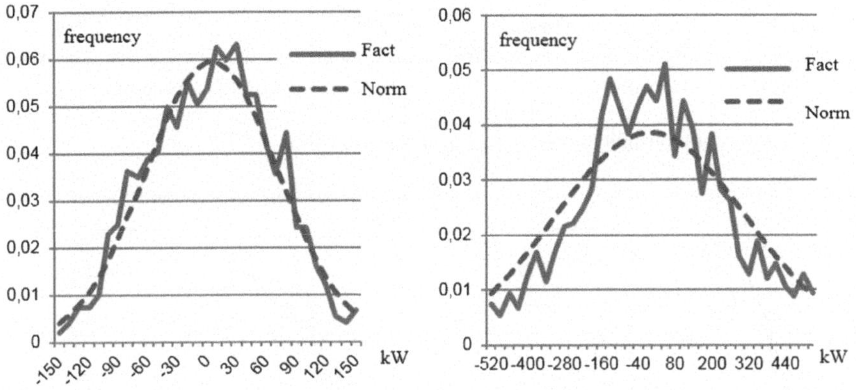


(a) Consumer T1, January, Estimate (b) Consumers T1-T4, July, without Estimate

**Fig. 1** Examples of distribution of power imbalances on condition of HPS availability

tion of average daily power can significantly reduce the probability of exceeding the specified levels of imbalance, thereby reducing the need for additional compensating capacities (standby or accumulative).

Similarly, consider the unpredicted T1–T4 consumption in July (Table 3). As measured values, the mean indications of positive and negative values of the imbalance are given, since in the summer there was a certain asymmetry, as well as the deviation from normal distribution by a parameter of excess (Fig. 1b).

The impact of the SPP on the indicators of imbalance (Table 3) is still less than the effect of WPP with equal capacity, regardless of the summer season. Obviously, the factor of use of the installed capacity is more impressive, as the time of solar energy is limited. Note that the average current power of SPP in July was 90 kW, with WPP being 110 kW, on condition of a nominal capacity of 400 kW. In January, the respective values were 14 kW (SPP) and 80 kW (WPP) on condition of nominal capacities of 200 kW.

Consider how close the modeling results and the actual (measured) values are, for example, small consumers (variants T1 and T1–T4). Figure 2 shows the distribution of power imbalances for “clean” (without HPS) consumption.

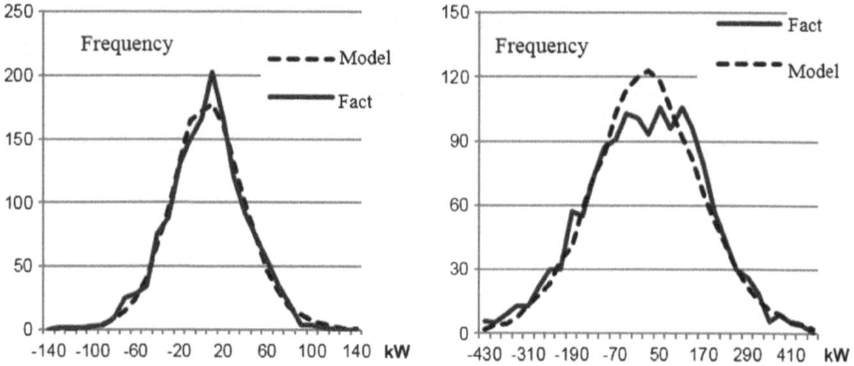


(a) for One Consumer T1 (b) for the Group of Consumers T1-T4

**Fig. 2** Examples of probability of imbalance functions distribution

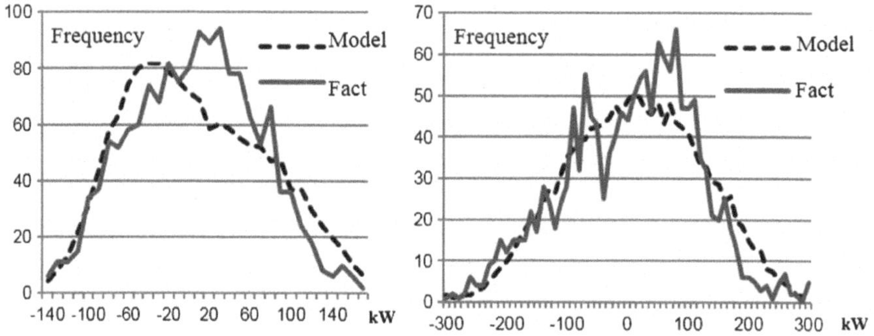


(a) Taking Estimating into Account (b) No Estimating

**Fig. 3** Examples of the probability of imbalance distribution upon the availability of HPS 200 kW

In the absence of HPS, the actual and simulated distributions coincide in almost the entire range, except for the zero-value range.

In the presence of WPP and SPP, certain model features may arise. Thus, for Consumer T1, in the presence of WPP, the simulated values of the stochastic component in the considered example had distribution differences compared with actual ones, taking into account the estimating capabilities per day (Fig. 3a) and without estimating (Fig. 3b). It should be taken into account that the power distribution of WPP differs from normal (rather lognormal); therefore, the wind speed was modeled by the Weibull distribution. The actual wind was somewhat different from this distribution due to the limited actual sample size. Differences of fact and models are noticeable in the area of small values of imbalance (up to 100 kW, or up to 10% of average consumption), but significant imbalances have the same distribution.

Replacing WPP with SPP in this example, we get a somewhat different nature of the differences between the facts and the model (Fig. 4a). In case of consistent use,

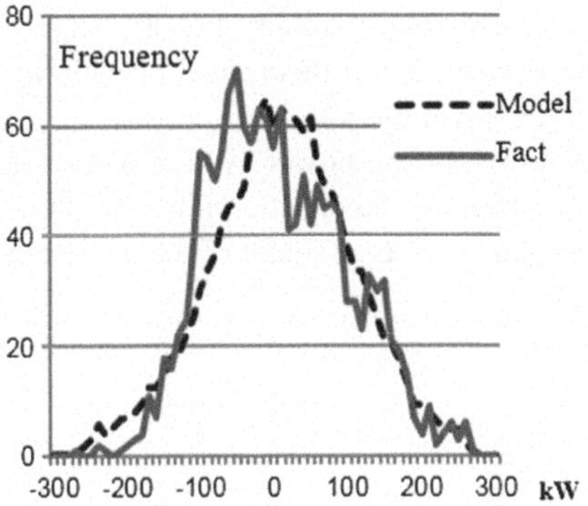


(a) up on the Availability of SPP 200 kW

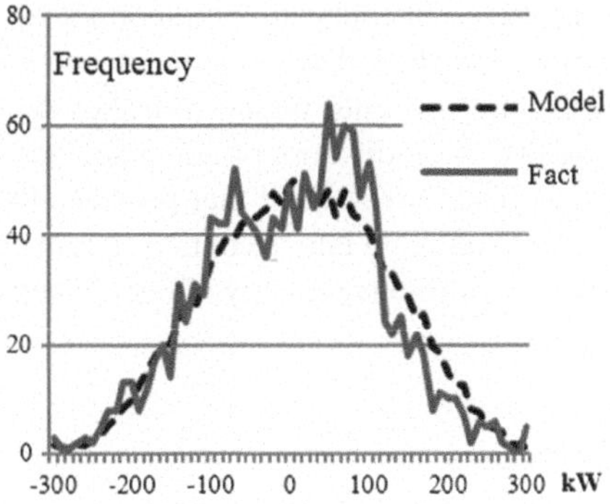


(b) up on the Availability of WPP and SPP at the Same Time

**Fig. 4** Examples of the probability of imbalance distribution without estimating

the effect of WPP is more noticeable (Fig. 4b), since in this example, winter actual data are considered when the wind has higher energy and the solar radiation is lower.

As we see, the presence of solar generation in the considered example increases the number of negative values of the balance of power; the presence of wind power plants increases the positive values. The combination of WPP and SPP aligns the distribution of negative and positive values. In general, differences in model and measured values are noticeable in the area of small imbalances that have little effect on the adequacy of the power grid. Instead, significant imbalances are modeled quite reliably.

## 5 Conclusion

In general, the parametric description allows calculating the integral values of the deviation of the generated power from the consumption needs with satisfactory accuracy. This allows us to apply the indices of the adequacy of the power system to optimize the volumes of renewable energy that has a random nature (wind, solar). In the case of parallel operation with the power grid, such an assessment would be sufficient. However, an important option for local power systems is the ability to accumulate energy to smooth the sharp changes in the balance of generation and consumption. Parametric description does not provide this possibility, as opposed to simulation, whereas it is necessary to determine the statistical characteristics of the actual set of measurements for the correct model. The assumption of the independence of the successive elementary events, which in this example is represented by a 30-min implementation of the generation/consumption of electricity, allows us to calculate the probability of events such as repeating several consecutive values of the unbalance of a single sign (this refers to the calculation of indices like LPSP or LOEE). This indicator is important when choosing the optimum size of compensating capacities. The correspondence of the measured and theoretical probability distribution depends on the choice of the mathematical model.

Thus, the indicators of adequacy and reliability of the power system receive a probabilistic assessment that provides the possibility of stochastic optimization of the system taking into account different criteria and risk assessment of the deviation from the normative quality of energy supply.

## References


1. Hongxia, Z., Yuqing, T., Deying, C.: In: MATEC Web of Conferences, 108, 14006. EDP Sciences (2017). https://doi.org/10.1051/matecconf/201710814006
2. Lopez, N., Espiritu, J.F.: Procedia Comput. Sci. **6**, 463–468 (2011). https://doi.org/10.1016/j.procs.2011.08.086
3. Krishan, O.: Procedia Comput. Sci. **125**, 321–328 (2018). https://doi.org/10.1016/j.procs.2017.12.043
4. Nayar, C., Tang, M., Suponthana, W.: Sustainable energy technologies. In: IEEE International Conference, pp. 1076–1080 (2008). https://doi.org/10.1109/icset.2008.4747166
5. Zailan, R., Zaini, S.N., Rashid, M.I.M., Razak, A.A.: In: MATEC Web of Conferences, 131, 02001. EDP Sciences (2017). https://doi.org/10.1051/matecconf/201713102001
6. Muravyev, D.I., Lukutin, B.V.: In: MATEC Web of Conferences, 110, 01052. EDP Sciences (2017). https://doi.org/10.1051/matecconf/201711001052
7. Fan, L., Wang, K., Shi, W.: In: MATEC Web of Conferences, 42. EDP Sciences (2016). https://doi.org/10.1051/matecconf/20164207001
8. Okinda, V.O., Odero, N.A.: Int. J. Res. Eng. Technol. **4**(11), 153–163 (2015)
9. Wu, Y.K., Chang, L.T., Chang, S.M., Hsieh, T.Y., Lee, T.C.: In: MATEC Web of Conferences, 55, 06001 (2015). https://doi.org/10.1051/matecconf/20165506001
10. Pimm, A.J., Cockerill, T.T., Taylor, P.G.: J. Energy Storage. **16**, 231–242 (2018). https://doi.org/10.1016/j.est.2018.02.002
11. Marwitz, S., Olk, C.: J. Phys. Conf. Ser. **977**(1), 012006. IOP Publishing (2018)
12. Kolios, A., Jiang, Y., Somorin, T., Sowale, A., Anastasopoulou, A., Anthony, E.J., Collins, M.: Energy Convers. Manag. **163**, 74–85 (2018). https://doi.org/10.1016/j.enconman.2018.02.046
13. Gitman, M.B.: Vvedenie v stohasticheskuyu optimizatsiyu, Permian (2014)
14. Kuznietsov, M.P., Uzheiko, S.O.: Vidnovliuvana enerhetyka **3**, 6–12 (2016)
15. Kuznetsov, N.P.: Alternativnaya energetika i ekologiya **23**, 44–48 (2014)
16. Niknam, T.: Appl. Energy **88**, 4817–4830 (2011). https://doi.org/10.1016/j.apenergy.2011.06.023
17. Kuznietsov, M.M.: Vidnovliuvana enerhetyka. **1**, 12–16 (2016)
18. Kuznietsov, M.P., Lysenko, O.V.: Naukovyi visnyk Tavriiskoho derzhavnoho ahrotekhnolohichnoho universytetu [Elektronnyi resurs]. **7**(1), 3–10 (2017). Rezhym dostupu: http://nauka.tsatu.edu.ua/e-journals-tdatu/V7T1.html
19. Kuznetsov, N.P., Lyisenko, O.V.: Problemyi regionalnoy energetiki. **2**(34), 139–147 (2017)

# Energy Saving in the Technological Process of the Grain Grinding

**Marina Postnikova, Evgeniy Mikhailov, Dina Nesterchuk and Olga Rechina**

## 1 Introduction

The complexity of the technological processes of producing all-mash determines high demands to the control over the equipment operation. It is worth reducing the specific energy consumption for the all-mash production in order to reduce the cost of production. Therefore, the correct choice of operating modes of a grinder is of great importance for the all-mash production. The above-mentioned fact specifies the relevance of the research aimed at solving this applies scientific task.

### *1.1 Analysis of Recent Studies and Publications*

Since new methods of husbandry had been implemented, the backgrounds for the transformation of such an important aspect of economic life, as energy saving began to appear from the abstract to the real factor of the economy. It is found that the most informative indicator for determining energy saving operating modes is the specific electric energy consumption for processing the crop production [1, 2].

The study identified the main factors: the grinder's productivity, the power, and the load factor of the electrical equipment that influence the amount of specific energy consumption in the grinding process. The majority of the previous researches were devoted to the technological issues of grinding with a hammer grinder [3, 4], as well as the operating modes of the machine [5]. Based on the analytical review of the literature on theory and practice of the grinding process in modern feed rooms, a mathematical model of the research object was developed to identify minimum

M. Postnikova (✉) · E. Mikhailov · D. Nesterchuk · O. Rechina
Tavria State Agrotechnological University, B. Khmelnitskogo Avenue, 18, Melitopol 72310, Ukraine
e-mail: marina.postnikova@tsatu.edu.ua

© Springer Nature Switzerland AG 2019

V. Nadykto (ed.), *Modern Development Paths of Agricultural Production*,
https://doi.org/10.1007/978-3-030-14918-5_41

energy consumption [6, 7]. There are not so many researches on the energy intensity of the grain milling processes and its connection with technological factors [6, 7].

The analysis of existing feed rooms on livestock farms shows that the actual productivity of feed lines is lower than the project one, and the quality of feed processing does not meet the modern requirements. The following data is given in the grain grinder's certificate: specified engine power is $P = 32.2$ kW; productivity of the feed grain grinding $Q = 4$–$6$ t/h; specific energy consumption $W = 5.4$–$8.0$ kW h/t; [4] that can be reduced by optimizing the technological process of grinding the grain. That is why the enhancement of the energy efficiency of feed lines is an urgent problem nowadays.

### *1.2 Statement of the Objective and Tasks of the Study*

The aim of the research was reduction of electric energy consumption in electromechanical grinding systems by using energy saving operating modes based on the regularities of the specific electric energy consumption.

To achieve the above-mentioned aim, the following tasks were solved:

- to identify the main factors influencing the amount of energy per one unit of the finished product;
- to find a specific energy consumption and optimal operating modes of the electric equipment when grinding by the grain mill DB-5-1.

## 2 Results and Discussion

When solving the problem of a grain mill optimization, it is necessary to choose an optimization parameter, enabling to evaluate the investigated object, and which, in the form of mathematical relation, is determined by means of independent factors. Correctly chosen optimization parameter determines the task of investigation, and to solve it, it is necessary to obtain a mathematical model convenient for searching the minimum of the optimization parameter having in mind corresponding factors that determine the grain mill properties. Such mathematical model can be successfully obtained by rearranging the initial model of a grain mill by the method of mathematical experiment planning (MEP). In earlier theoretical researches [6], the regression equation of the optimization parameter in the function of factors was obtained in the coded units

$$\begin{aligned}\hat{y} = 3.68 + 0.816x_1 - 1.28x_2 + 1.36x_3 - 0.349x_1x_2 \\ + 0.292x_1x_3 - 0.583x_2x_3 + 0.738x_1^2 - 0.524x_2^2 + 0.74x_3^2,\end{aligned} \tag{1}$$

and physical units

$$W = 16.2 - 1.313P + 2.42Q - 15.41K - 0.035PQ - 0.29PK + 1.46QK$$
$$+ 0.0295P^2 - 0.13Q^2 + 18.5K^2. \quad (2)$$

In formulas (1) and (2), the following is accepted: $\hat{y}$, $W$—specific energy consumption; $x_1$, $P$—power; $x_2$, $Q$—productivity; and $x_3$, K—coefficient of electrical equipment loading, in coded and normalized factor values correspondingly.

To investigate the grain mill optimum area on the basis of the second-order regression Eq. (1), let us use one of the effective methods for searching optimum, called the canonical transformation of the grain mill mathematical model. The main step of the canonical transformation is the transfer of the origin of coordinates to a new point S of the factor space and the rotation of the old axes at a certain angle to align with the main axes of the geometric surface. As a result, linear terms of the equation disappear and the value of the absolute term in the original equation changes. To find coordinates of the new center of the yield surface of $S_z$, it is necessary to differentiate Eq. (1) for each variable $x_i$ and to equal partial derivatives to zero

$$\frac{\partial \hat{y}}{\partial x_1} = 0.816 - 0.349x_2 + 0.292x_3 + 1.476x_1 = 0;$$
$$\frac{\partial \hat{y}}{\partial x_2} = -1.28 - 0.349x_1 - 0.583x_3 - 1.048x_2 = 0;$$
$$\frac{\partial \hat{y}}{\partial x_3} = 1.36 + 0.292x_1 - 0.583x_2 + 1.48x_3 = 0. \quad (3)$$

Solving the system of Eqs. (3), we obtain the value of the coordinates of yield surface center $S_Z$.

$$x_{1S} = -0.47; \quad x_{2S} = -0.5; \quad x_{3S} = -1.02. \quad (4)$$

When putting the coordinate values to the original Eq. (1), we can determine the optimization parameter value $Y_S = 3.11$.

The following values of factors and objective function correspond to the coordinates $x_{1S}$, $x_{2S}$ and $x_{3S}$, as well as the values of $Y_S$, in physical units (6)

$$P = 22.65\ \text{kW}; \quad Q = 4.0\ \text{t/h}; \quad K = 0.4; \quad Y_S = W = 3.11\ \text{kW h/t}. \quad (5)$$

To obtain analytical and graphical interpretation of optimization parameter optimum described by Eq. (1), and to determine the values of $P$, $Q$, and $K$ factors, close to optimal ones, let us make use of the method of bidimensional sections of the yield surface [7].

Regarding the adequate regression Eq. (1) we obtained, let us have a close look at three variants of bidimensional sections of the yield surface, with the construction and analysis of the curves family.

**Variant 1**. Let us accept $x_1 = 0$ in (1) that corresponds to the value of the factor $P = 25$ kW. In this case, bidimensional sections of the yield surface will characterize the optimization parameter depending on the productivity $Q$ ($x_2$) and the loading coefficient $K$ ($x_3$). Equation (1) then will look like

$$\hat{y} = 3.68 - 1.28x_2 + 1.36x_3 - 0.583x_2x_3 - 0.524x_2^2 + 0.74x_3^2. \quad (6)$$

Now let us define the coordinates of the surface center $S_Z$ by means of the system of equations in partial derivatives

$$\frac{\partial \hat{y}}{\partial x_2} = -1.28 - 0.583x_3 - 1.048x_2 = 0;$$
$$\frac{\partial \hat{y}}{\partial x_3} = 1.36 - 0.583x_2 + 1.48x_3 = 0.$$
$$x_{2S} = -0.58; \quad x_{3S} = -1.15; \quad Y_S = 3.27, \quad (7)$$

that satisfies $Q = 3.84$ t/h and K $= 0.37$.

The value of $Y_s$ is determined after placing the values of $x_{2S}$ and $x_{3S}$ in (6). For the canonical transformation (6) by (7), the characteristic determinant is formed

$$f(B) = \begin{vmatrix} -1.048 - B & 0.5 \cdot (-0.583) \\ 0.5 \cdot (-0.583) & 1.48 - B \end{vmatrix} = (-1.048 - B)(1.48 - B) - 0.085 = 0.$$

The eigenvalue of this characteristic equation will be $B_{22} = 1.51$; $B_{33} = -1.08$. Form that, the canonical equation has the following form

$$Y - 3.27 = 1.51X_2^2 - 1.08X_3^2. \quad (8)$$

Since the coefficients in the canonical equation $B_{22}$ and $B_{33}$ have different signs, in accordance with [7], the level lines at bidimensional sections of the yield surface under the received values of the optimization parameter $Y_S$ will be hyperbolae with an elongated saddle along the $X_3$ axis. Contour curves for different $Y_S$ values are shown in Fig. 1. The $Y_S$ value in the center denotes the minimum optimization parameter value $W$.

According to Fig. 1, the optimum area near the center $S_Z$ is characterized by the values of $x_2$ and $x_3$ factors in the range: $x_2$ from −4.1 to −7.5 and $x_3$ from −0.25 to −1.5, which corresponds to the factors $Q = 3.2$–3.5 t/h and $K = 0.45$–0.3. In this case, the minimum value of the optimization parameter will be

$$W = Y_S = 3.27\,\text{kW h/t and} \quad P = 25\,\text{kW}.$$

**Variant 2.** Let us accept $x_2 = 0$ in Eq. (1) that corresponds to the value of the factor $Q = 5$ t/h. In this case, bidimensional sections of the yield surface will characterize

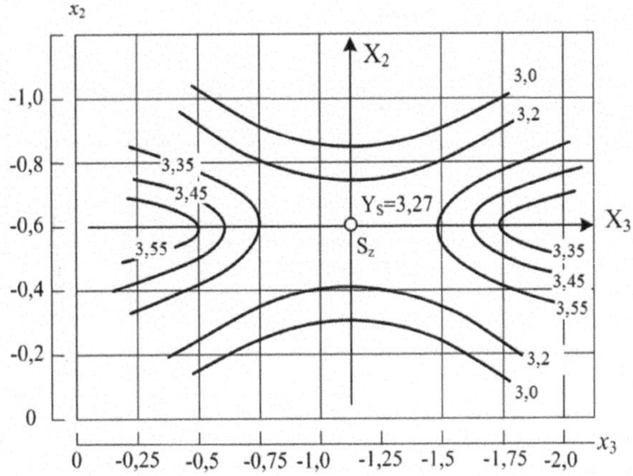


**Fig. 1** Contour curves of bidimensional sections of the yield surface if $x_1 = 0$ ($P = 25$ kW) and $Y_S = 3.27$; 3.35; 3.45; 3.55

the optimization parameter depending on the power $P$ ($x_1$) and the loading coefficient $K$ ($x_3$) of the grain mill. Then, Eq. (1) will have the following look

$$\hat{y} = 3.68 + 0.816x_1 + 1.36x_3 + 0.292x_1x_3 + 0.738x_1^2 + 0.74x_3^2. \quad (9)$$

Let us find the coordinates of the surface center $S_Z$ and the value of the optimization parameter $Y_S$ solving the system of equations in partial derivatives

$$\frac{\partial \hat{y}}{\partial x_1} = 0.816 + 0.292x_3 + 1.476x_1 = 0;$$
$$\frac{\partial \hat{y}}{\partial x_3} = 1.36 + 0.292x_1 + 1.48x_3 = 0. \quad (10)$$

As follows from the solution of equation system (10), we obtain the value of the coordinates, and by solving (9), the parameter $Y_S$ will be derived: $x_{1S} = -0.39$; $x_{3S} = -0.84$; $Y_S = 2.95$, that corresponds to $P = 23.05$ kW; $K = 0.44$. For the canonical transformation (9), the characteristic determinant is formed by Eq. (10)

$$f(B) = \begin{vmatrix} 1.476 - B & 0.5 \cdot 0.292 \\ 0.5 \cdot 0.292 & 1.48 - B \end{vmatrix} = (1.476 - B)(1.48 - B) - 0.021 = 0.$$

The eigenvalues of this characteristic equation will be: $B_{11} = 1.62$; $B_{33} = 1.33$. In this case, the canonical equation looks like

$$Y - 2.95 = 1.62X_2^2 + 1.33X_3^2. \quad (11)$$

Since the coefficients $B_{11}$ and $B_{33}$ have the same signs $B_{11} > 0$ and $B_{33} > 0$, the contour lines in the bidimensional sections of the yield surface will be ellipses according to [7], and the yield surface center will be the minimum of the W parameter.

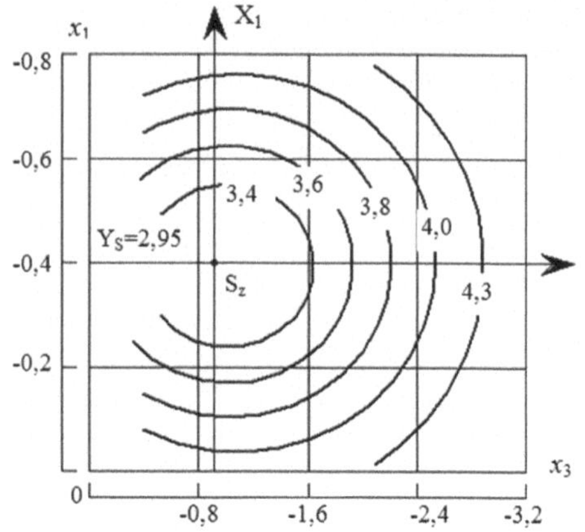


**Fig. 2** Contour curves of the bidimensional sections of the yield surface where $x_2 = 0$ ($Q = 5$ t/h) $Y_S = 2.95$; 3.4; 3.6; 3.8; 4.0; 4.3

The contour curves for various $Y_S$ values are shown in Fig. 2. The Figure demonstrates that the optimum area near $S_Z$ center is characterized by the values of $x_1$ and $x_3$ factors where $x_1$ from $-0.56$ to $-0.24$ and $x_3$ from $-0.2$ to $-1.6$ that corresponds to the factors in physical units

$$\text{P} = 25.31\text{–}23.8 \ \text{kW and K} = 0.56 - 0.28.$$

The minimum value of the optimization parameter in center of this case is

$$W = Y_S = 2.95\,\text{kW h/t} \quad \text{and } Q = 5\,\text{t/h}.$$

**Variant 3**. Let us accept $x_3 = 0$ in Eq. (1) that corresponds to the value of the factor $K = 0.6$. In this case, bidimensional sections of the yield surface will characterize the optimization parameter depending on the power $P$ ($x_1$) and productivity $Q$ ($x_2$). The regression Eq. (1) will then look like

$$\hat{y} = 3.68 + 0.816x_1 - 1.28x_2 - 0.349x_1x_2 + 0.738x_1^2 - 0.524x_2^2. \quad (12)$$

The yield surface center coordinates and $Y_S$ value are also determined here by the solution of the following equation system in partial derivatives

$$\frac{\partial \hat{y}}{\partial x_1} = 0.816 - 0.349x_2 + 1.476x_1 = 0;$$
$$\frac{\partial \hat{y}}{\partial x_2} = -1.28 - 0.349x_1 - 1.048x_2 = 0. \quad (13)$$

After solving the system above (13), we obtain the value of the center $S_Z$ coordinates, and after their substitution in (12), the parameter $Y_S$ will be specified:

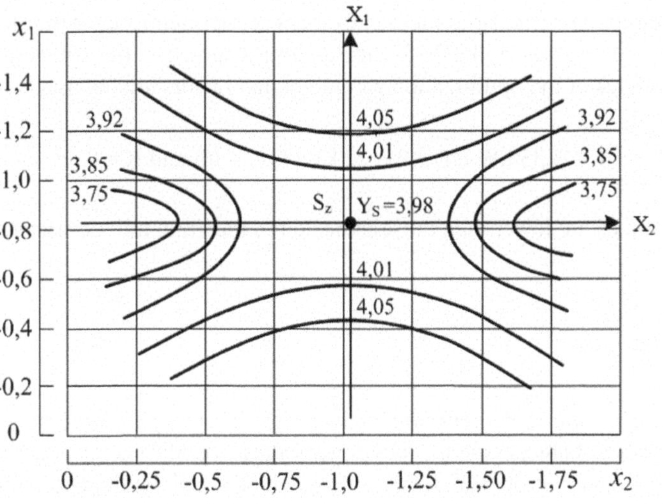


**Fig. 3** Contour curves of the bidimensional sections of the yield surface where $x_3 = 0$ $(K = 0.6)$, $Y_S = 3.98$; 3.92; 3.85; 3.75

$x_{1S} = -0.78$; $x_{2S} = -0.96$; $Y_S = 3.98$, that corresponds to: $P = 21.1$ kW; $Q = 3.08$ t/h; $W = Y_S = 3.98$ kW h/t.

The characteristic determinant for system (13) is the following

$$f(B) = \begin{vmatrix} 1.476 - B & 0.5 \cdot (-0.349) \\ 0.5 \cdot (-0.349) & -1.048 - B \end{vmatrix} = (1.476 - B)(-1.048 - B) - 0.03 = 0.$$

The eigenvalues of this characteristic equation will be: $B_{11} = 1.49$; $B_{22} = -1.06$. From that the canonical equation looks like

$$Y - 3.98 = 1.49X_1^2 - 1.06X_2^2. \tag{14}$$

In this case, the coefficients in the canonical equation $B_{11}$ and $B_{22}$ have different signs. That is why contour lines in the bidimensional sections of the yield surface will be hyperbolic curves according to [6], where $Y_S$ has various values. Contour curves for $x_3 = 0$ for various $Y_S$ values are shown in Fig. 3.

According to Fig. 3, the optimum area of the optimization parameter near the $S_Z$ center is characterized by the values of $x_1$ and $x_2$ factors in the range: $x_1$ from $-0.55$ to $-1.0$ and $x_2$ from $-0.65$ to $-1.35$ that corresponds to the factors in physical units

$$P = 22.25 - 20.0\,\text{kW} \quad \text{and}\ Q = 6.3 - 2.3\,\text{t/h}.$$

The minimum value of the optimization parameter here is

W = $Y_S$ = 3.98 kW h/t and the loading coefficient K = 0.6.

Based on analytical and graphical analyses of the optimum area of the optimization parameter of the grain mill DB-5-1 technological process, the following average optimum values of the optimization parameter and factors are obtained:

$$W = 3.33 \text{ kW h/t}; \quad P = 23 \text{ kW}; \; Q = 4 \text{ t/h and K} = 0.45.$$

These values are lower than those indicated in the grain mill certificate.

## 3 Conclusion

1. For the first time, as the criterion for assessing the energy saving operating modes of electromechanical systems for grain grinding, specific energy consumption per unit of output is proposed, taking into account actual equipment loading which ensures optimization of the grain mill productivity at a minimum specific energy consumption.
2. The study identified the main factors that affect the amount of energy consumption per unit of finished product, namely grain mill productivity, power, and loading coefficient of electrical equipment.
3. A theoretical method for reconstruction of the mathematical model of a grain mill was developed first for the purpose of determining the optimal value of specific energy consumption using the method of mathematical experiment planning.
4. On the basis of analytical and graphical analysis of the optimum area of the optimization parameter and factors, the following average values are obtained: $W = 3.33$ kW h/t; $P = 23$ kW; $Q = 4$ t/h; $\text{K} = 0.45$, which are lower than those indicated in the grain mill certificate.
5. The results of the investigations carried out are the basis for the development of scientifically substantiated norms of specific energy consumption in the technological process of the grain grinding.

## References


1. Postnikova, M.V., Ye, V., Mykhailov, O.P.: Karpova, Naukovyi visnyk Tavriiskoho derzhavnoho ahrotekhnolohichnoho universytetu (Elektronnyi resurs) 2(1): 169–173 (2012). Rezhym dostupu: nbuv.gov.ua/e-journals/nvtdau/index.html
2. Postnikova, M.V., Nazarian, H.N., Ye, V., Mykhailov, O.P.: Karpova, Pratsi Tavriiskoho derzhavnoho ahrotekhnolohichnoho universytetu **12**(2), 111–116 (2012)
3. Bocharova, T.A.: Tehnologiya hraneniya i pererabotki produkcii rastenievodstva s osnovami standartizacii/(Barnaul, 2008)
4. Tekhnolohiia vyrobnytstva kombikormiv (Elektronnyi resurs). Rezhym dostupu: http://www.kurcha.com.ua/index.php
5. Yu, M.: Kutsenko, Problemy enerhozabezpechennia ta enerhozberezhennia v APK Ukrainy **153**, 98–100 (2014)

6. Nazarian, H.N., Postnikova, M.V., Karpova, O.P.: Rishennia zadach optymizatsii obiektiv doslidzhennia metodom planuvannia matematychnoho eksperymentu (Melitopol 2012)
7. Melnikov, S.V., Alyoshin, V.R., Roschin, P.M.: Planirovanie eksperimenta v issledovaniyah selskohozyaystvennyih protsessov. Kolos, Lenynhrad (1980)

# Determination of the Duration of Spherical-Shaped Berries Freezing Under the Conditions Stationary Heat Flow

**Nikolai Struchaiev, Yulia Postol, Yurii Stopin and Ivan Borokhov**

## 1 Introduction

Currant, gooseberry, grapes—one of the popular cultures with rich chemical composition, pleasant aroma, and harmonious taste—find wide consumption not only as a table food, but is also often used in dietary and therapeutic nutrition. However, the conditions of growth determine the seasonality of its consumption. Therefore, the search for ways to prolong the period of consumption of grapes by determining the optimal treatment regimes plays an important role, since in the frozen grapes for a long time, almost all vitamins and trace elements contained in berries are retained.

### *1.1 Analysis of Recent Studies and Publications*

In the modern world practice of food storage, an increasing distribution is found by low-temperature freezing [1–3]. The essence of the process of freezing is the lowering of the temperature below the cryoscopic until complete or partial conversion into ice of the free and bound moisture in the product. The most suitable for freezing the temperature from −18 to −23 °C—at this temperature, freezing allows you to conserve berries in 10 … 12 months with minimal loss of food and organoleptic qualities [10–12].

The main technological goal for freezing is the production of fine-crystalline ice structures in products without a significant redistribution of moisture and minimal damage to the structure. This is achieved only at high freezing rates [3], which depend on a number of characteristics: thermophysical properties, dimensional-mass characteristics, structural-mechanical (rheological) properties.

N. Struchaiev (✉) · Y. Postol · Y. Stopin · I. Borokhov
Tavria State Agrotechnological University,
B. Khmelnitskogo Avenue, 18, Melitopol 72310, Ukraine
e-mail: mykola.struchaiev@tsatu.edu.ua

© Springer Nature Switzerland AG 2019

V. Nadykto (ed.), *Modern Development Paths of Agricultural Production*,
https://doi.org/10.1007/978-3-030-14918-5_42

### 1.2 Statement of the Objective and Tasks of the Study

The aim of the study was to establish the possibility of increasing the efficiency of freezing spherical berries by determining the optimum duration of the process under conditions of a stationary heat flux.

To achieve this goal, the following tasks were set:

1. To propose a technique for calculating the duration of the process of freezing berries of spherical shape;
2. To check the correspondence of theoretical calculations to the actual time of freezing of berries of spherical shape.

## 2 The Basic Part of the Study

### 2.1 Model Representation of the Problem and Its Solution

To calculate the optimal freezing temperature of grapes, we choose as a physical model of the problem a simplified representation of the berries in the form of a sphere. The calculation of the duration of the freezing process will be carried out under the following assumptions: before the freezing, the berry is cooled to cryoscopic temperature; the berries surface is cooled during convective heat exchange with the environment, the temperature of which remains constant; the coefficient of heat transfer from the external surface of the berries to the environment remains constant ($\alpha = \text{const}$) over the entire temperature range; the thermal conductivity coefficient $\lambda$ of the over-frozen berries is assumed to be averaged for each design temperature layer.

Then, the mathematical problem of non-stationary heat conductivity in a homogeneous and isotropic body for a one-dimensional heat flux and the absence of internal sources of heat can be represented in spherical coordinates by a linear homogeneous second-order partial differential equation [4] with the initial condition

$$\frac{\partial t(r,\tau)}{\partial \tau} = a\left(\frac{\partial^2 \tau(r,\tau)}{\partial r^2} + \frac{2}{r}\frac{\partial t(r,\tau)}{\partial r}\right), \quad \begin{array}{l} t = t_0 \ \text{ at } \tau = 0; \\ t < \infty \text{ at } \tau > 0, 0 < r < R \end{array} \tag{1}$$

$$t(r_1 0) = t_M. \tag{2}$$

the boundary condition of the third kind for convective heat transfer between the surface of the body and the environment at a constant heat flux and the symmetry condition

$$\lambda\frac{\partial t}{\partial r} + \alpha[t(R,\tau) - t_c] = 0 \quad \text{at } t > 0, r = 0 \tag{3}$$

$$\frac{\partial t}{\partial r} = 0, \quad (t(0, \tau) \neq \infty) \tag{4}$$

where $r$ is the current radius, $m$; $\tau$—time, $s$; $a = \lambda / c_p \rho$—coefficient of temperature conductivity, $m^2/s$; $\lambda$—coefficient of thermal conductivity, W/(m K); $c_p$—isobar specific heat, J/kg K; $\rho$—density, $kg/m^3$; $t$—current temperature, $K$; $\alpha$—coefficient of heat transfer to the environment; $R$—characteristic size—radius of the berries, $m$; $t_c$—temperature of the cooling medium, $K$.

The problem represented by Eqs. (1) … (4) can be formulated in a dimensionless form. To do this, we introduce a dimensionless excess temperature $\theta$, the relative coordinate $\zeta$, and the relative time (Fourier number) $Fo$.

$$\theta = \frac{t(r, \tau) - t_н}{t_c - t_н}, y = \frac{r}{R}, Fo = \frac{a \cdot \tau}{R^2}$$

where $t_H$ is the initial temperature of the berries; $Fo$ is the criterion of thermal homochromy for unsteady heat exchange processes, characterizing the relationship between the rate of change in thermal conditions in the environment, and the rate of adjustment of the temperature field inside the body.

Then, the temperature will be a function of the relative variables $y$ and $Fo$, i.e., $\theta = \theta(y, Fo)$, and the process of freezing grapes can be described by a mathematical model

$$\frac{\partial \theta(y, Fo)}{\partial Fo} = \frac{\partial^2 \theta(y, Fo)}{\partial y^2} + \frac{2}{y} \frac{\partial \theta(y, Fo)}{\partial y} \tag{5}$$

and we have

$$\begin{aligned} &\theta = 1 \quad \text{at } Fo = 0; \\ &\theta < \infty \text{ at } Fo > 0, 0 < y \leq 1; \end{aligned}$$

At the initial time, $\theta(y, 0) = 0$. Then, the cooling problem (5) can be represented in the form

$$\begin{aligned} &\frac{\partial \theta}{\partial Fo} - Bi[1 - \theta(1, Fo)] = 0, \\ &\frac{\partial \theta}{\partial Fo} = 0 \quad \text{at} \quad \theta(0, Fo) \neq \infty \end{aligned} \tag{6}$$

where $Bi = \alpha R / \lambda$ is the similarity criterion $Bio$, which characterizes the ratio of conductive $R/\lambda$ and convective $1/\alpha$ thermal resistances at the body–environment boundary.

### 2.2 Materials and Methods of Research

The research method is based on an improved method for studying the process of freezing fruits and berries [3, 8, 9, 12]. The grapes were frozen in the air in a stationary refrigerating chamber, the temperature of which was maintained with the help of two refrigerating machines FAL-56. A two-stage freeze was applied with a stop at an intermediate cooling temperature, which prevented possible damage to the berries when the internal moisture was transferred to the crystal structure. The temperature in the center of the berry was measured by a calibrated chromel–alumel thermocouple TXA, emf. which is 8.3 mV/100 °C. Before freezing, the berries were initially kept in the pre-cooling chamber at a temperature of 5 … 2 °C. Then, the cooled berries [10, 11] were frozen in a freezing chamber with forced circulation of air at a speed of 1.0 … 2.5 m/s at a relative humidity of $88 < \varphi,\% < 93$. To freeze faster, the compartment for freezing was pre-cooled to a temperature of minus $40 \pm 2$ °C. The freezing was considered complete when the berries freezed to the full depth and reached the temperature of $-20 \pm 2$ °C in the center of the berry. At this temperature, protein denaturation in grapes decreases substantially, which creates optimal conditions for its long-term storage.

## 3 Results and Discussion

### 3.1 The Theoretical Solution of the Problem of Freezing Berries of Spherical Shape

We solve the problem of cooling and freezing spherical berries under isobaric convective heat transfer conditions by a numerical method of approximate solution of the heat equation—the finite difference method, or, as it is also called, the grid method [5]. It based on the replacement of the derivatives entering into the differential equation by the difference relations in separate discrete points—grid nodes. We solve the problem of cooling and freezing spherical berries under isobaric convective heat transfer conditions by a numerical method of approximate solution of the heat equation—the finite difference method, or, as it is also called, the grid method [5]. It based on the replacement of the derivatives entering into the differential equation by the difference relations in separate discrete points—grid nodes. In this case, the problem of approximate numerical integration of Eq. (6) by the grid method is reduced to finding the approximate value of the relative temperature $\theta$ function at internal nodes (*ih*, *kl*), where $i = 0, 1, 2 \ldots$, $h = 1/n$, $k = 0, 1, 2 \ldots l$.

Since the function $T(x, \tau)$ depends on two variables $x$ and $\tau$, we use a grid of rectangular type. The final result of the solution is given by the expression according to which the temperature value at a given point (node) is a function of time.

We replace the partial derivatives $\frac{\partial\theta}{\partial\tau}$ and $\frac{\partial^2\theta}{\partial\tau^2}$ at the node $(ih, kl)$ by their difference relations

$$\frac{\theta_{i,k+1}-\theta_{i,k}}{l}+\varepsilon_1=\frac{\theta_{i+1,\kappa}-2\theta_{i,k}+\theta_{i-1,\kappa}}{h^2}+\varepsilon_2+\frac{2}{ih}\left(\frac{\theta_{i+1,\kappa}-\theta_{i-1,\kappa}}{2h}+\varepsilon_3\right), \tag{7}$$

where $\varepsilon_1$, $\varepsilon_2$, $\varepsilon_3$ are the remainder terms tending to zero as $l$ tends to zero, and $h$; $i$, $h$, $k$, $l$ are nodes of numerical integration; $n$ is the number of divisions of the conditional segment. From Eq. (7), we obtain the explicit formula for calculating the temperature at the internal points of the grape for any $i \neq 0$.

$$\theta_{i,k+1}=(1-\frac{2l}{h^2})\cdot\theta_{i,k}+\frac{1}{h^2}(1+\frac{1}{i})\cdot\theta_{i+1,k}+\frac{1}{h^2}(1-\frac{1}{i})\theta_{i-1,\kappa}+lR_{i1k},$$

$$R_{i,k}=\varepsilon_2+\frac{2}{ih}\varepsilon_3-\varepsilon_1 \tag{8}$$

Let us denote $T_i$ the approximate value of the function $\theta_{ik}$ at the nodes $(ih, kl)$ and let us leave the remainder term $R_{i,k}$ in Eq. (8), then

$$T_{i,k+1}=(1-2n^2l)T_{i,k}+n_2l(1+\frac{1}{i})T_{i+1,k}+n^2l(1-\frac{1}{i})T_{i-1,k} \tag{9}$$

Equation (9) allows us to calculate the temperature of the layer $k + 1$ from the values of the temperature of the previous layer.

To determine the temperature at the center of the berry, we use Eq. (6) for $y = 0$ in the form

$$\left(\frac{\partial\theta}{\partial Fo}\right)_{y=0}=3\left(\frac{\partial^2\theta}{\partial y^2}\right)_{y=0} \tag{10}$$

Substitution of derivatives

$$\left(\frac{d\theta}{dF}\right)_{y=0}=\frac{\theta_{o,k+1}-\theta_{o,k}}{l}+\varepsilon' \text{ и}$$

$$\left(\frac{d^2\theta}{dy^2}\right)_{y=0}=\frac{\theta_{1,k}-2\theta_{0,k}+\theta_{1,k}}{h^2}+\varepsilon''=\frac{2(\vartheta_{1,k}-\theta_{0,x})}{h^2}+\varepsilon''$$

gives

$$\frac{\theta_{0,k+1}-\theta_{0,k}}{l}+\varepsilon'=\frac{6}{h^2}(\theta_{1,k}-\theta_{0,k})+3\varepsilon'' \tag{11}$$

From the Eq. (11), we find the temperature at the center of the berry

$$\theta_{0,k+1} = \left(1 - \frac{6l}{h^2}\right)\theta_{0,k} + \frac{6l}{h^2}\theta_{1,k} + R_{0,k}l, R_{o,\kappa} = 3\varepsilon'' - \varepsilon'$$

Dropping the term $R_{o,k}$, we obtain in an explicit form the formula for the temperature in the center

$$T_{0,k+1} = (1 - 6n^2 l)T_{0,k} + 6n^2 l T_{l,k}. \tag{12}$$

To determine the temperature of the grape skin, we can use Eq. (9) for $y = 1$ at the time $Fo = (k + 1) \cdot l$ in the form

$$T_{n,k+1} = (1 - 2n^2 l)T_{n,k} + n(n + 1)lT_{n+1,k} + n(n + 1)lT_{n-1,k}$$

Taking into account the temperature in the additional node $[(n + 1)h, kl]$

$$T_{n+1,k} = \frac{2Bi}{n}(1 - Tn, k) + T_{n-1,k}$$

we obtain the final formula for determining the temperature of the grape skin

$$T_{n,k+1} = \left[1 - 2n^2 l - 2(n + 1)lBi\right]T_{n,k} + 2n^2 l T_{n-1,k} + 2(n + 1)lBi \tag{13}$$

The optimal dimensionless freezing time of berries can be determined by iteration the equations obtained with respect to the dimensionless time *Fo*. We using the cooling ball model of the Lykov [6] applied to the Planck formula [7]

$$\begin{aligned} Fo = {} & \frac{q \cdot \rho_2 \cdot \omega}{c_1\rho_1(t_{\kappa p} - t_c)}\left[\frac{Z}{Bi} + Z^2\left(\frac{1}{2} - \frac{1}{Bi}\right) + \frac{1}{3}Z^3\left(\frac{1}{Bi} - 1\right)\right] \\ & + \frac{8c_2\rho_2\left(t_{\text{ц}н} - t_{\kappa p}\right)}{15c_1\rho_1\left(t_{\kappa p} - t_e\right)}\left[\frac{Z}{Bi} + Z^2\left(\frac{1}{2} - \frac{3}{2Bi}\right) + Z^3\left(\frac{1}{Bi} - \frac{2}{3}\right) - \frac{1}{4}\left(\frac{1}{Bi} - 1\right)Z^4\right] \\ & + \frac{1}{3}\left(\frac{1}{Bi} + \frac{1}{2}\right)\ln\left(1 + \frac{Bi \cdot Z}{1 - Z}\right) - \frac{1}{6\left(\frac{1}{Bi} - 1\right)^2}\ln\left[1 + Z(Bi - 1)\right] - \frac{2 - Bi}{6(1 - Bi)}Z + \frac{1}{6}Z^2 \end{aligned} \tag{14}$$

where $q$—is the specific heat of ice formation, J/(kg K); $\rho_1$, $\rho_2$—density of the previous frozen and subsequent unfrozen layer of pulp, kg/m$^3$; $\omega$—moisture content; $c_1$ and $c_2$ are the specific heat capacities of the frozen and unfrozen layers J/(kg deg); $Z$—is the relative thickness of the frozen layer; $t_{ct}$—is the cryoscopic temperature of grapes, *K*; $t_c$—temperature in the center of the berry, *K*.

In calculating the relative freezing time of grapes according to (14), the number of *Bio* was calculated by finding the coefficient of convective heat transfer to the external medium $\alpha$ from the criterion dependence $Nu = f(Pr, Re)$ determined by the Nusselt number *Nu*, Reynolds *Re* and Prandtl *Pr* under any mode of blowing a berry (as a ball) according to the Rantz–Marshall equation [6]

$$Nu = 2 + 0.6 \cdot Pr^{1/3} \cdot Re^{1/2},\ Nu = \alpha \cdot R/\lambda_\infty,\ Re = u \cdot R/\nu_\infty,\ Pr = c_\infty \rho_\infty/\lambda_\infty$$

Taking into account the temperature dependence of the thermophysical properties of the medium and the body

$$Bi = \frac{\lambda_\infty(t)}{\lambda(t)} \cdot \frac{\left(2 + 0.6 \cdot Pr^{1/3} Re^{1/2}\right)}{2} \quad (15)$$

where $u$—is the velocity of the blowing air, m/s: $R$—is the characteristic size (berry radius), *m*; $c_\infty$, $\lambda_\infty$, $\nu_\infty$ and $\rho_\infty$—specific heat, J/(kg K); thermal conductivity, W/m K; kinematic viscosity $m^2$/s and air density kg/$m^3$, respectively; for air $Pr(-40\ °C) = 0.72$.

The Bio numbers ($0.05 < Bi < 0.15$) calculated for the free-convective mode of blowing berries of spherical shape, for example, grapes or forced circulation of the air ($10 < Nu < 30$), allow to consider the freezing process as an external task, whose internal thermal resistance $R/\lambda$ is less than the external $1/\alpha$, due to which the temperature difference inside the body is less than the difference in temperature between the surface and the environment. In this case, the freezing of berries can be referred to as the cooling of a thermally conditionally thin body (thickness $R$ is small, and the thermal conductivity of grapes $\lambda$ is significant), so the cooling rate of the entire volume is determined primarily by the size of the body.

The transition from the calculated dimensionless time *Fo* to the real-time $\tau$ of freezing of the grapes was carried out on the basis of the definition of the Fourier number

$$Fo = a\tau/R^2, \quad (16)$$

then

$$\tau = \frac{R^2}{a} \cdot Fo \quad (17)$$

## 3.2 *Experimental Verification of the Results of Calculating the Time of Freezing of Berries of Spherical Shape*

In order to establish the correspondence of theoretical calculations to the actual time of freezing grapes, an experimental verification of this characteristic was carried out. During the experiments, the error in determining the operating parameters and calculating the similarity numbers was: dimensions $\varepsilon$ ($R$) = 3.5%, speed $\varepsilon$ ($u$) =

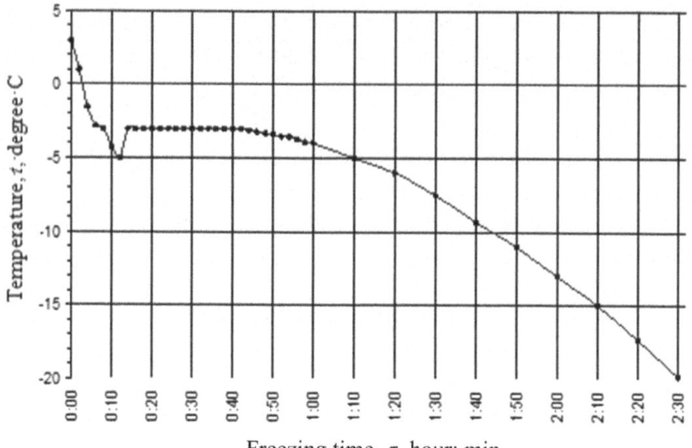


**Fig. 1** Typical temperature change of grapes when frozen in a refrigerated chamber under controlled stationary (adiabatic) conditions

8.8%, temperature $\varepsilon$ ($t$) = 2.5%, time $\varepsilon$ ($\tau$) = 0.5%, Reynolds number $\varepsilon$ ($Re$) = 9.5%, Nusselt number $\varepsilon$ ($Nu$) = 4.8%, and Bio number $\varepsilon$ ($Bi$) = 4.8%. The value of the thermophysical characteristics of the medium and grapes was adjusted, depending on their temperature.

When freezing grapes, the following stages are usually distinguished: cooling to the point of freezing, freezing in the area of maximum ice crystallization, and further cooling (freezing) of products to storage temperature. These stages are manifested in the peculiarities of the behavior of the temperature field, which are indicated in Fig. 1 by the freezing curve of grapes.

Here it should be noted that the tissue juice of grapes is a dissociated colloidal solution of mineral and organic substances. Unlike single-component aqueous solutions of eutectic (cryohydrate) concentration, its freezing takes place not at a constant temperature, but at a variable temperature.

It can be seen from the graph that at the cooling stage (minutes) ($0 < \tau < 8$), a rapid drop in the temperature of the berry occurs at first: the grapes are undercooled, and for some time, ice crystals have not yet formed. The appearance of nuclei of the solid phase (crystals) from the tissue fluid upon reaching the freezing point (−5.0 °C) in the region of maximum crystal formation (minutes) ($8 < \tau < 15$) leads to a rapid transformation into ice of the water contained in the cells that is in the bound condition. Occurring with the release of latent heat leads to an increase in the temperature of the tissues to the cryoscopic (−2.8 °C … −3.6 °C), which is held at this level for some time. At this stage ($12 < \tau$, min < 15), about 72% of water passes into the ice. Then, the gradual cooling of the frozen berries continues ($15 < \tau$, min < 150)—their temperature drops to minus 20 °C, and then, they are stored at the freezing temperature.

Since at this temperature, a part of the water retained by the macromolecular matrix of plant berry tissues still remains in the liquid phase, and the process of complete freezing could be continued to cryohydrate temperature (−55 … −65 °C). However, this is impractical for a number of reasons. Obviously, as a result of the achieved temperature in frozen berries, there is practically no metabolism, oxidation-reduction processes are blocked, and microbiological activity is suppressed, which creates the necessary conditions for long-term storage. At the same time, the freezing of berries to cryohydrate temperature, as shown by the estimates taking into account the cooling rate determined from the freeze schedule, will take at least 1.4 h. The overexpenditure required for such a freeze is economically unjustified and inefficient.

For example, we calculate the duration of freezing of grape berries with a diameter $D = 2.5 \times 10^{-2}$ m or $R = (2.5/2) \times 10^{-2} = 1.25 \times 10^{-2}$ m and a temperature in the refrigerator minus 20 °C at $Bi < 0.15$ and $Fo = 3.65$. We assume that the average coefficient of thermal diffusivity of grapes was calculated from thermophysical characteristics taking into account their contribution to the individual stages of the cooling and freezing process,

$$a = 13.12 \times 10^{-8}\ \mathrm{m}^2/\mathrm{s}.$$

Then by formula (17)

$$\tau = \frac{R^2}{a} \cdot Fo = \frac{(0.0125м)^2 \cdot 7.3}{13.12 \times 10^{-8} м^2/c} = 8694\mathrm{s} \approx 2.41\mathrm{h}$$

Divergence of the estimated time freezing of grapes with experimentally measured $\tau_{exp}$. (see Fig. 1) does not exceed 5%, which can be considered a completely acceptable result, taking into account the assumptions made in the formulation of problem (1) … (4) and the approximate nature of numerical integration. Comparison of the freezing time of a berry of spherical shape ($\Phi = 1/3$), calculated under the same conditions according to the classical Planck formula [7], recommended by the International Institute of Refrigeration,

$$\tau' = \Phi \cdot \frac{q\rho\omega D}{2 \cdot (t_{кр} - t_{с})} \left( \frac{D}{4\lambda} + \frac{1}{\alpha} \right) \qquad (18)$$

indicates the discrepancy of $\tau'$ both with the experimental expand with the calculated time $\tau$. Without additional corrections, formula (18) can only be used as an estimate.

## 4 Conclusions

1. A procedure is proposed for calculating the duration of the process of freezing berries of spherical shape in conditions of a stationary heat flux.
2. The conformity of theoretical calculations to the actual time of freezing of berries of spherical shape is verified.

The calculations of the freezing time of grape berries under steady-state heat flow conditions are in satisfactory agreement with the experimentally established time of two-stage freezing in air with compulsory circulation, accurately reflect real processes, and can be used to select optimal operating conditions and process parameters for industrial applications.

It should be noted here that the results of calculating the optimal duration of freezing grapes and the temperature fields of its berries when using the averaged values of thermophysical characteristics in the proposed model, with the nonlinear character of their temperature dependences noted at [7] at temperatures below 0 °C, can be completely corrected when performing calculations in boundary nonlinear boundary value problems of heat conduction taking into account the two-step nature of such a process.

## References

1. Ajzenberg, V.Y.: Hranenie i pererabotka kartofelya, ovoshchej, plodov i vinograda. Kolos, Moscow (1979)
2. Hayakawa, K.I.: Science et Technique du Froid (1977)
3. Sahagian, M.E., Goff, H.D.: Effect of freezing rate on the thermal, mechanical and physical aging properties. Thermochim. Acta. **246**(2):271–283 (1994). https://doi.org/10.1016/0040-6031(94)80095-2
4. Tihonov, A.I., Samarskij, A.A.: Uravneniya matematicheskoj fiziki. Nauka, Moscow (1972)
5. Forsajt, Dzh, Vazov, V.: Raznostnye metody resheniya differencial'nyh uravnenij v chastnyh proizvodnyh. Inostrannaya literatura, Moscow (1963)
6. Teoriya teplo-massoobmena Vysshaya shkola, Moscow (1979)
7. Ginzburg, A.S., Gromov, A.I., Krasovskaya, G.I.: Teplofizicheskie harakteristiki pishchevyh produktov. Pishchevaya promyshlennost', Moscow (1980)
8. Kolbe, E., Kramer, D.: Planning for seafood freezing. Fairbanks, Alaska (2007)
9. Skrupskis, I.A., Gross, U., Rucins, M., Aboltins, A.: Foodbalt-Proceedings-2008, 79–83 (2008)
10. Kampuse, S, Kampuss, K.: Acta Horticulturae **599**, 711–717. https://doi.org/10.17660/actahortic.2003.599.93
11. Aboltins, A., Boca, S., Galoburda, R., Gross, U., Skrupskis, I.: Heat-mass transfer in layer of Berries during freezing process. Refrigeration for Sustainable Development. Book of Abstracts. Prague (2011)
12. Yalpachik, V., Struchaev, M, Tarasenko, V.: Experimental determination of the coefficient of thermal conductivity during freezing. In: Proceedings of the Taurian State Agrotechnological University: Scientific Specialized Edition/TDATU—Melitopol, 2017, vol. 1(17), pp. 113–118

# Energy-Saving Control of Asynchronous Electric Motors for Driving Working Machines

**Oleksandr Vovk, Sergey Kvitka, Serhii Halko and Oleksandr Strebkov**

## 1 Introduction

Electromechanical energy conversion in all areas of human activity is carried out with the help of an electric driver, which accounts for consumption from 50 to 70% of all electricity, which is produced in different countries of Europe [1–4]. Most of the all-electric drives are AC drives—according to different data from 68 to 85% (mostly, asynchronous electric drives) [5, 6].

During a considerable operating time, asynchronous electric motors, which are power drive units of asynchronous electric drives, operate with not the maximum efficiency due to not optimal loading. In different European countries, the number of asynchronous electric drives with such electric motors varies from 80 to 90% [7]. Using devices to regulate the operation of these asynchronous electric motors will let to save up to 40–50% of the energy they consume [7, 8], and the saving 1 kW of power consumption costs 4–5 times cheaper than the cost of 1 kW of the power of the newly commissioned power-generating unit [5]. Therefore, the saving of electricity during the operation of three-phase asynchronous electric motors of the drive of working machines of the process lines is an actual national economic problem, the solution of which was devoted to many scientific and practical works.

## 2 Literature Review and Formulation of the Problem

At the present stage, to solve this problem worldwide, adjustable asynchronous electric drives are used, which have a significant effect both in saving electricity and in other indicators of technological process. However, these automatic control systems

O. Vovk (✉) · S. Kvitka · S. Halko · O. Strebkov
Faculty of Energy, Tavria State Agrotechnological University,
Zaporozhye region, Melitopol 72310, Ukraine
e-mail: oleksandr.vovk@tsatu.edu.ua

© Springer Nature Switzerland AG 2019

V. Nadykto (ed.), *Modern Development Paths of Agricultural Production*,
https://doi.org/10.1007/978-3-030-14918-5_43

for electric drives have their advantages and disadvantages, which are described in detail in [9–13]. In addition to the shortcomings listed in mentioned publications, the regulated asynchronous drive has one more drawback—all existing algorithms of its operation and the means for their implementation do not take into account the type of mechanical characteristic of the working machine which is driven by the asynchronous electric motor of the drive.

## 3 Purpose and Objectives of the Research

Therefore, the article proposes to consider the control of an asynchronous electric motor for the drive of working machines, taking into account the type of their mechanical characteristics for minimizing the losses of active power in the electric motor. The control is proposed to be carried out with the help of the applied voltage taking into account the load of the electric motor. The purpose of the article is to establish the dependence between the voltage at the terminals of the electric motor and the coefficient of its load, taking into account the type of mechanical characteristic of the working machine with a minimum of losses of active power in the electric motor.

## 4 Materials and Methods of the Research

In order to establish the relationship between the losses of active power in an asynchronous electric motor, the applied voltage and the load factor of an asynchronous electric motor, taking into account the type of mechanical characteristic of the working machine, we consider the process of electromechanical energy conversion in an electric motor using its replacement circuit (Fig. 1).

The scheme (Fig. 1) indicates: $\dot{U}_1$—the complex of the applied value of the applied phase voltage; $\dot{I}_1$—complex of effective value of consumed phase current; $\dot{I}_2''$—complex of the current value of the load current of the load branch of the circuit;

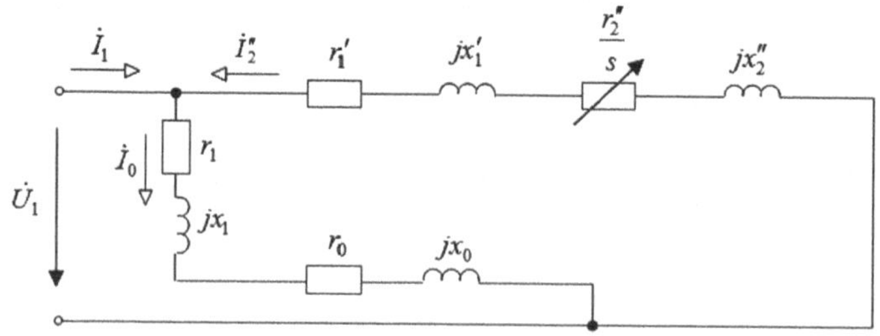


**Fig. 1** Circuit for replacing one phase of a three-phase asynchronous electric motor for complexes of effective values

$\dot{I}_0$—complex of the current value of the current intensity of the magnetization branch of the circuit; $s$—electric motor slip; and $r_1'$, $x_1'$, $r_2''$, $x_2''$, $r_1$, $x_1$, $r_0$, $x_0$—parameters of the circuit.

The loss of active power in the magnetizing branch of the scheme (constant losses) is:

$$\Delta P_{\text{const}} = (r_1 + r_0) \cdot I_0^2 = \frac{(r_1 + r_0)}{(r_1 + r_0)^2 + (x_1 + x_0)^2} \cdot U_1^2. \tag{1}$$

At the nominal applied voltage $U_{1n}$, expression (1) will look like:

$$\Delta P_{\text{const}.n} = \frac{(r_1 + r_0)}{(r_1 + r_0)^2 + (x_1 + x_0)^2} \cdot U_{1n}^2, \tag{2}$$

Dividing (1) by (2), we obtain:

$$\Delta P_{\text{const}} = \Delta P_{\text{const}.n} \cdot \left(\frac{U_1}{U_{1n}}\right)^2 = \Delta P_{\text{const}.n} \cdot k_u^2. \tag{3}$$

The loss of active power in the load branch of the scheme (variable losses) is:

$$\Delta P_{\text{var}} = (r_1' + r_2'') \cdot (I_2'')^2 = \frac{(r_1' + r_2'')}{\left(r_1' + \frac{r_2''}{s}\right)^2 + (x_1' + x_2'')^2} \cdot U_1^2. \tag{4}$$

At nominal applied voltage and rated load (i.e., with nominal slip $s_{\text{n}}$), expression (4) will look like:

$$\Delta P_{\text{var}.n} = \frac{(r_1' + r_2'')}{\left(r_1' + \frac{r_2''}{s_n}\right)^2 + (x_1' + x_2'')^2} \cdot U_{1n}^2. \tag{5}$$

Dividing (4) by (5), we obtain:

$$\Delta P_{\text{var}} = \Delta P_{\text{var}.n} \cdot \frac{\left(r_1' + \frac{r_2''}{s_n}\right)^2 + (x_1' + x_2'')^2}{\left(r_1' + \frac{r_2''}{s}\right)^2 + (x_1' + x_2'')^2} \cdot k_u^2. \tag{6}$$

In turn, the total losses of active power in the asynchronous motor are:

$$\Delta P = \Delta P_{\text{const}.n} \cdot k_u^2 + \Delta P_{\text{var}.n} \cdot \frac{\left(r_1' + \frac{r_2''}{s_n}\right)^2 + (x_1' + x_2'')^2}{\left(r_1' + \frac{r_2''}{s}\right)^2 + (x_1' + x_2'')^2} \cdot k_u^2. \tag{7}$$

The paper [14] shows how the sliding of an asynchronous motor depends on the applied voltage, load, and the type of mechanical characteristic of the working machine. Therefore, for the electric motor of a drive of a working machine with a mechanical characteristic independent of speed, expression (7) takes the form:

$$\Delta P = \Delta P_{\text{const}.n} \cdot k_u^2 + \Delta P_{\text{var}.n} \cdot \frac{\left(r_1' + \frac{r_2''}{s_n}\right)^2 + \left(x_1' + x_2''\right)^2}{\left(r_1' + \frac{r_2'' \cdot k_u^2}{s_n \cdot k_l}\right)^2 + \left(x_1' + x_2''\right)^2} \cdot k_u^2, \tag{8}$$

where $k_l$ is the electric motor load factor, equal to the ratio of the current torque on the electric motor shaft to the rated torque.

For the electric motor of a drive of a working machine with a linearly increasing mechanical characteristic, expression (7) takes the form:

$$\Delta P = \Delta P_{\text{const}.n} \cdot k_u^2 + \Delta P_{\text{var}.n} \cdot \frac{\left(r_1' + \frac{r_2''}{s_n}\right)^2 + \left(x_1' + x_2''\right)^2}{\left(r_1' + \frac{r_2'' \cdot \left(\frac{k_u^2}{k_l \cdot s_n} + \frac{1-m_0}{1-s_n}\right)}{m_0 + \frac{1-m_0}{1-s_n}}\right)^2 + \left(x_1' + x_2''\right)^2} \cdot k_u^2, \tag{9}$$

where $m_0$ is the relative value of the starting torque of the working machine.

For the electric motor of a drive of a working machine with a nonlinearly increasing mechanical characteristic, expression (7) takes the form:

$$\Delta P = \Delta P_{\text{const}.n} \cdot k_u^2 + \Delta P_{\text{var}.n} \times \frac{\left(\left(r_1' + \frac{r_2''}{s_n}\right)^2 + \left(x_1' + x_2''\right)^2\right) \cdot k_u^2}{\left(r_1' + \frac{r_2''}{1 + \frac{(1-s_n)}{k_l \cdot 2 s_n}\left(k_u^2 - \sqrt{k_u^4 (1-s_n) + k_l \cdot 4 s_n (1-m_0)\left(k_u^2 + k_l \cdot s_n m_0\right)}\right)}\right)^2 + \left(x_1' + x_2''\right)^2}. \tag{10}$$

For the electric motor of a drive of a working machine with a nonlinearly decreasing mechanical characteristic, expression (7) takes the form:

$$\Delta P = \Delta P_{\text{const}.n} \cdot k_u^2 + \Delta P_{\text{var}.n} \times \frac{\left(\left(r_1' + \frac{r_2''}{s_n}\right)^2 + \left(x_1' + x_2''\right)^2\right) \cdot k_u^2}{\left(r_1' + \frac{r_2''}{\frac{1}{2}\left(\left(1 + \frac{k_l}{k_u^2} s_n m_0\right) - \sqrt{\left(1 - \frac{k_l}{k_u^2} s_n m_0\right)^2 - 4 \frac{k_l}{k_u^2}(1-m_0)(1-s_n)}\right)}\right)^2 + \left(x_1' + x_2''\right)^2}. \tag{11}$$

## 5 Results and Discussion

We establish dependence $k_l = f\ (k_l)$ for asynchronous electric motors of a drive of various working machines, in which the active power losses will be minimal. That is, we define, using Eqs. (8)–(11), what should be the coefficient $k_u$ with a certain value of the coefficient $k_l$, so that the power losses in the electric motor would be minimal if the electric motor drives the working machines with different types of mechanical characteristics.

When we disclose the parentheses in (8)–(11), we obtain the equation of the eighth and higher order relative $k_u$. Therefore, it is not possible to analytically derive equations for the calculation of $k_u$ depending on $k_l$. This dependence can only be obtained numerically for a specific asynchronous electric motor, using a program that allows us to find the minimum of the function. The results of calculation and approximation of the curve for an asynchronous electric motor 4AM90L4 (rated power 2.2 kW) in the case of driving a working machine with a speed-independent mechanical characteristic are shown in Fig. 2.

The result of the approximation of the cited dependence is the equation in which the power losses in said asynchronous electric motor will be minimal:

$$k_u = -0.895 \cdot k_l^2 + 1.784 \cdot k_l + 0.089. \tag{12}$$

In the case of a drive, which is indicated by an electric motor of working machines with other types of mechanical characteristics, the dependencies $k_u = f(k_l)$ will be the same as in Fig. 2. The equation at which the power losses in this asynchronous

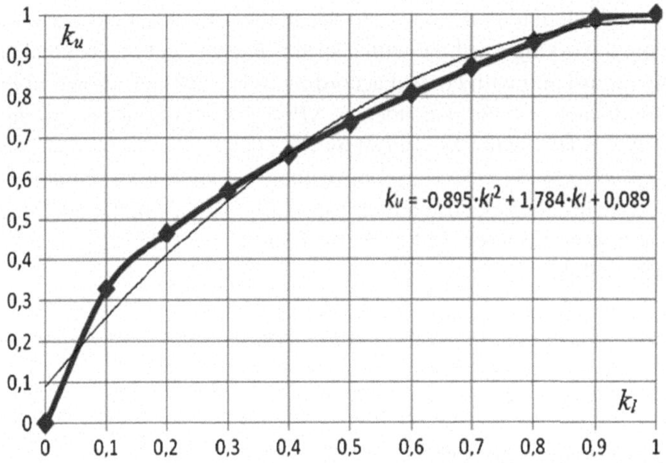


**Fig. 2** Dependence $k_u = f(k_l)$ for the 4AM90L4 asynchronous motor in the case of driving a working machine with a speed-independent mechanical characteristic

electric motor will be minimal in the case of driving a working machine with a linearly increasing mechanical characteristic is:

$$k_u = -0.83 \cdot k_l^2 + 1.709 \cdot k_l + 0.09. \quad (13)$$

In the case of driving a working machine with a nonlinearly increasing mechanical characteristic:

$$k_u = -0.872 \cdot k_l^2 + 1.758 \cdot k_l + 0.09. \quad (14)$$

In the case of driving a working machine with a nonlinearly decreasing mechanical characteristic:

$$k_u = -0.783 \cdot k_l^2 + 1.653 \cdot k_l + 0.091. \quad (15)$$

As can be seen from the obtained Eqs. (12)–(15), the coefficients in them are different; they characterize the type of mechanical characteristics of the working machine and the design parameters of the electric motor. Consequently, the control over the asynchronous electric motor with the minimum of active power losses, with the help of the applied voltage, depending on its load, should be carried out according to the previously obtained dependences $k_u = f(k_l)$, which take into account the design parameters of the electric motor and the type of mechanical characteristic of the working machine. The calculation should be carried out in a program that allows you to find the minimum of the function, specifying the load of the electric motor and using expressions (8)–(11). After the calculation, the obtained dependences are approximated and equations for controlling the electric motor are obtained.

The results of calculating the power losses in the 4AM90L4 electric motor in the case of driving a working machine with a speed-independent mechanical characteristic without control and with control according to Eq. (12) are shown in Fig. 3.

Figure 3 indicates: 1—the dependence $\Delta P = f(k_l)$ without control and 2—the dependence $\Delta P = f(k_l)$ with the control by Eq. (12).

As can be seen from Fig. 3, during control, there is a significant reduction in power losses in an asynchronous electric motor, especially with a small load. For example, if the electric motor will work with 50% load during the year [15], then the annual energy savings will be:

$$\Delta W = \Delta P' \cdot t \cdot N, \quad (16)$$

where

$\Delta W$ energy saved per year, kWh;
$\Delta P'$ power loss difference without control and during control, kW;
$t$ the running time of the electric motor per day, $h$;
$N$ number of working days per year.

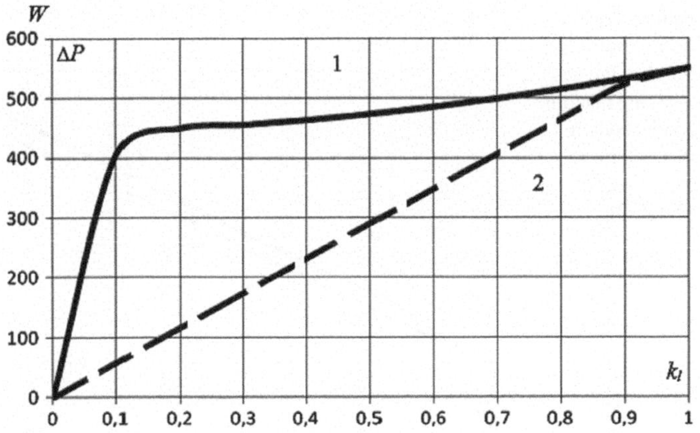


**Fig. 3** Dependence $\Delta P = f(k_l)$ for the asynchronous 4AM90L4 motor in the case of driving a working machine with a mechanical characteristic, which is independent of speed

$$\Delta W = 0.18325 \cdot 8 \cdot 250 = 366.5\,\text{kWh}.$$

In the case of a drive with the indicated electric motor of working machines with other types of mechanical characteristics, in the absence of control and with the control according to the corresponding equation, the dependences of $\Delta P = f(k_l)$ will be similar to those shown in Fig. 3. The annual energy savings during the operation of the electric motor with 50% load will be:

- In the case of driving a working machine with a linearly increasing mechanical characteristic:

$$\Delta W = 0.19434 \cdot 8 \cdot 250 = 388.68\,\text{kWh};$$

- In the case of driving a working machine with a nonlinearly increasing mechanical characteristic:

$$\Delta W = 0.18707 \cdot 8 \cdot 250 = 374.14\,\text{kWh};$$

- In the case of driving a working machine with a nonlinearly decreasing mechanical characteristic:

$$\Delta W = 0.20022 \cdot 8 \cdot 250 = 400.44\,\text{kWh}.$$

The obtained Eqs. (12)–(15) were checked in laboratory conditions on an asynchronous electric motor 4AM90L4. The results of the experimental test for a working

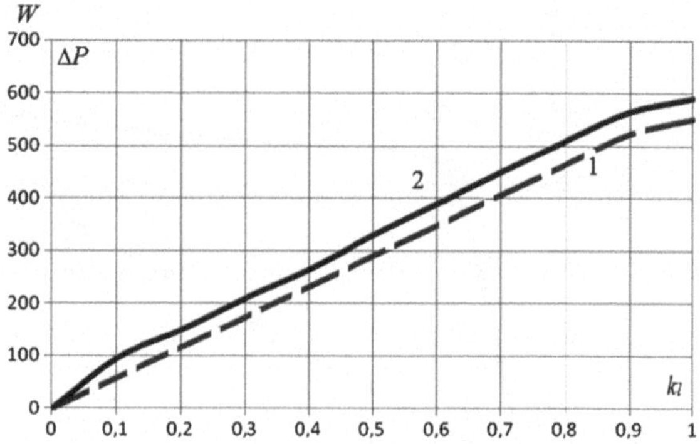


**Fig. 4** Dependence $\Delta P = f(k_l)$ for the asynchronous motor 4AM90L4 in the case of driving a working machine with a speed-independent mechanical characteristic (conveyor) under control according to Eq. (12)

machine with a speed-independent mechanical characteristic (for a conveyor) are shown in Fig. 4.

Figure 4 denotes: 1—the calculated dependence $\Delta P = f(k_l)$ and 2—the experimental dependence $\Delta P = f(k_l)$. As can be seen from the obtained results, the convergence of the calculated values and the experimental values is within the permissible limits, since the standard deviation does not exceed 10%. The deviation is caused by the following:

(1) The asynchronous electric motor was in operation for some time; hence, the values of its parameters differ from the calculated ones due to insignificant operational defects.
(2) During obtaining Eq. (12), the saturation of the steel of the asynchronous electric motor and the change in its parameters due to heating were not taken into account.

The results of experimental checks for working machines with other types of mechanical characteristics are of a similar nature to that shown in Fig. 4. The root-mean-square deviations are the same; their causes are similar to those shown above.

## 6 Conclusions

As a result of the research, it was suggested to control the asynchronous motor by minimizing the losses of active power with the help of the applied voltage, depending on its load. The control is carried out according to the previously obtained equations of the applied voltage as a function of the motor load, which take into account the design parameters of the motor and the type of mechanical characteristic of the driven working machine. Calculation $k_u = f(k_l)$ at $\Delta P \to$ min is proposed in a program that allows finding the minimum of the function, given by the load of the electric motor and using expressions (8)–(11). After calculation, the obtained dependences are approximated and equations for controlling the motor are obtained.

## References

1. Barsky, V.A., Beshta, A.S., Gorbachev, N.V., Zagirnyak, M.V. and others.: Energy saving. Power Eng. Energy Audit **9**(115), 2–11 (2013)
2. Mirchevski, S., Bartłomiejczyk, M., Hamacek, Š., Hamacková, H.: Result and solutions of young r & s for innovations and progress, 5–9 (2013)
3. Hussein, S.: Energy Power Eng. **3**, 107–112 (2011)
4. Gavrila, H., Manescu, V., (Paltanea), Paltanea, G., Scutaru, G., Peter, I.: Procedia. Eng. **181**:568–574 (2017)
5. Mirchevski, S.: Electronics **16**, 46–49 (2012)
6. Gulyaev, I.V., Tutaev, G.M.: Vector control systems of an electric drive based on an asynchronous valve motor: monograph (2010)
7. Koschnick, G., Sattler, B.: ZVEI—German Electrical and Electronic (2015)
8. Borisevich, A.V.: Energy-Saving Vector Control of Asynchronous Electric Motors: A Review of the State and New Results (2015)
9. Kozyaruk, A.E., Vasiliev, B.Y.: Bulletin of the south ural state university. Ser. Power Eng. **15**, 47–53 (2015)
10. Ushakov, A.S., Kolganov, A.R.: Bull. Ivanovo State Power Univ. **2**, 1–8 (2012)
11. Nazipova, L.R.: Science, Education, Society: Materials of the VI International Scientific and Practical Conference, vol. 2, 326–329 (2017)
12. Jannati, M., Anbaran, SA, Asgari, SH., Goh, WY., others.: Renew. Sustain. Energy Rev. **75**, 1306–1319 (2017)
13. Kohlrusz, G., Fodor, D.: Hung. J. Ind. Chem. Veszprém **39**, 265–270 (2011)
14. Vovk, O. Yu., Kvitka, S.O., Kvitka, O.S.: Visitor of KhNTUsG, Petra Vasilenka, 164, 121–123 (2015)
15. Kostić, M.: Elec. Energ. **23**, 333–344 (2010)

# Part III

# The Role of Social Capital in Development of Agricultural Entrepreneurship

**Mykola Hrytsaienko, Halyna Hrytsaienko, Larysa Andrieieva and Larysa Boltianska**

## 1 Problem Statement

The main purpose of the development of the agrarian sector is to directly improve the well-being of people living in rural areas, and indirectly, to lay the foundations for strengthening regional and national development in general. Public trust, solidarity, and sense of belonging are the social capital that can increase the effectiveness of population participation in rural development programs. That is why the study of the essence and features of the social capital of the agrarian sphere becomes an actuality as a starting point for defining the directions of institutional transformation in the countryside and the development of agrarian entrepreneurship.

## 2 Analysis of Recent Research and Publications

### *2.1 Entrepreneurship*

The content of the concept of "entrepreneurship" was first disclosed by R. Cantyleon (1680–1734) in his essay Essay Sur La Nature Du Commerce En Général, 1730. He considered entrepreneurship as an economic function of a special kind and emphasized the always present element of risk [1]. J.-B. This (1767–1832) in The Treatise of Political Economy, or a Simple Presentation of a Method for the Formation, Distribution, and Consumption of Wealth ("Traité d'économie politique ou Simple Exposé de la manière dont se forment, se distribuent et se conomment les richesses," 1803) formulated his own view on entrepreneurship as a rational combination of factors

M. Hrytsaienko (✉) · H. Hrytsaienko · L. Andrieieva · L. Boltianska
Department of Entrepreneurship, Trades and Exchange Activities, Tavria State Agrotechnological University, B. Khmelnytsky Ave., 18, Melitopol, Ukraine
e-mail: nick.gric@gmail.com

© Springer Nature Switzerland AG 2019

V. Nadykto (ed.), *Modern Development Paths of Agricultural Production*,
https://doi.org/10.1007/978-3-030-14918-5_44

of production at this point in the market space [2]. J. Schumpeter (1883–1950) in his book Theory of Economic Development (Theorie der wirtschaftlichen Entwicklung, 1911) reinterpreted the notion of "entrepreneurship" as a complex phenomenon characterized by innovation, a new combination factor of production through the introduction of new types of techniques and technologies, the creation of new products, the development of new markets, sources of raw materials; application of new developments, new technological solutions; risk and uncertainty, focus on receiving profits [3, pp. 131–141]. F. Hayek (1899–1992) saw the essence of entrepreneurship in the search and study of new economic opportunities, behavioral characteristics, and not specific activities. Due to a number of functions such as decisionmaking and responsibility, he linked entrepreneurship with managerial activity, although he distinguished between the concept of "entrepreneur" and "manager" because of their different functional essence [4, pp. 51–62].

At present, the views of scholars on the essence of the concept of "entrepreneurship" can be divided as a set of relations (Azizov [5, p. 84]), activity (Law of Ukraine "On Entrepreneurship" [6], Berezin [7, p. 358]), the style of behavior (Lantuch [8, p. 145], Rudkovskaya [9, p. 89]), the institute (Lyakhovets [10, p. 30], Tsapovskaya [11, p. 49]), the creation of value (Nesterenko [12, p. 40], Timmons [13]), and the factor of development (Lysenko [14, p. 218], Ryabokon et al. [15, p. 86]).

Undoubtedly, the authors of all the statements presented are correct, but their definitions are somewhat one-sided, since entrepreneurship, as the main institute of market economics, can be considered as a value creation activity that requires the formation of a coherent system of the relevant relations, requires an entrepreneur of a certain type of behavior, is a factor development of economy and society as a whole. In our opinion, the most precise and concrete definition of the essence of the concept of "entrepreneurship" is given in the Law of Ukraine "On Entrepreneurship": "Entrepreneurship is a direct independent, systematic, at its own risk, activity on the production of products, the execution of works, provision of services for the purpose of profit, which is carried out by natural and legal persons registered as subjects of entrepreneurial activity in accordance with the procedure established by the legislation" [6].

## 2.2 Agrarian Enterprise

Features of the agrarian sector of the economy caused the emergence of the concept of "agrarian entrepreneurship." Artiyam [16, p. 5], Kormyshkin [17, p. 54], and Mayovets [18, p. 13] offer a view on agrarian entrepreneurship as an activity; Ivanyuta [19, p. 62]—as a set of relationships; Malik et al. [20, p. 35]—as an institution; Prutskaya [21, p. 28]—as a set of subjects of entrepreneurial activity. The author's theoretical views on the concept of "agrarian entrepreneurship" coincide with the opinion of Z. Artistyak, who believes that "agrarian entrepreneurship is an initiative, independent activity of citizens and their associations, aimed at obtaining income at their own risk and for property liability in within defined organizational and legal forms operating in the agricultural sector of the economy" [16, p. 5].

## *2.3 Social Capital*

For the first time, the term "social capital" was used in the early twentieth century by the American researcher L. J. Hanifan, who, based on the study of the activities of rural schools as community centers, published an article in 1916, in which he concluded that the success of educational activities is possible only on the basis of social capital, that is, numerous social connections between participants in educational activities, local community, and government agencies [22]. The founder of the theory of socioeconomic crises, the prominent domestic economist M. Tugan-Baranovsky wrote the fundamental work "Social Foundations of Cooperation" [23] in the same year, laying the foundations for the future theory of social capital in it.

The direct formation of the theory of social capital can be attributed to the 60 years of the last century—the recognition of the scientific community of the works of American economists Schultz [24] and Becker [25] to create a concept of human capital, which revealed the economic role of education, the relationship of investment into human capital with an increase in the level of labor productivity.

Semantic analysis of the phenomenon of social capital held by French sociologist P. Bourdieu, social capital is defined as the amount of resources and substantiate the possibility of realization of social capital through its conversion into other forms of capital [26]. In the works of American economist and sociologist J. Coleman, social capital is a resource that contributes to the development of the economy and the improvement of society as a whole [27]. In the definition of the essence of social capital, an American philosopher, political scientist, political economist, and writer of Japanese origin, F. Fukuyama, emphasized that "these are norms, informal norms, or values that make collective actions possible in groups of people" [28].

An institutional approach to the interpretation of the essence of the notion of social capital is presented in the works of American political scientist R. Patmona, who revealed the main sources of its formation—networks, norms, and trust [29], and American economist D. North, who disclosed social capital as an element of the institutional environment of human behavior [30]. In the institutional context of social capital, the leading Ukrainian scientist O. Shpiculyak emphasized [31]. In his view, "the institutional dimension of social capital should be seen as a set of tangible and intangible assets to ascertain the formation mechanism of social and economic security" [31, p. 64].

The problem of social capital in recent years has steadfastly become one of the main interests of the scientific community, going far beyond purely economic discourse. Thus, K. Jennings, analyzing the impact of the external conflict on the formation of social capital, concluded that the presence of an external threat contributes to increasing the cohesion of the population and also causes a higher level of social capital [32]. Researchers O. A. Taabba and S. Ankrah investigated the place of social capital in cooperation between universities and industry in the field of technology transfer [33].

In the recent literature, these common approaches to the definition of "social capital" as the aggregate resources of society (Coleman [27, p. 35], Bourdieu [26, p. 66], Bogush [34, p. 18], Shpylulyak and Mazur [31, p. 64]), as a combination of social ties (Bert [35, p. 9], Peldem [36, p. 629], Roskolotko [37, p. 17]), and as membership in a particular social network (Portes [38, p. 6], Fukuyama [39], Pryatelchuk [40, p. 51]). In our view, the definition that identifies the concept of "capital" and "resources" or sees it as a set of social relations, membership in a particular social network does not reflect significant capital characteristics—the ability to generate income. We have formulated its own definition of social capital as a set of relationships can generate income and socioeconomic relations that occur in a particular social network based on existing standards and trust. We believe that trust in the fundamental component of social capital.

## 2.4 Communication of Social Capital and Entrepreneurship

Scientists from different countries are interested in various aspects of the interconnection of entrepreneurship and social capital. Thus, Fam and Talaver [41] investigated the relationship between gender, social capital, and access to financing of micro, small, and medium enterprises in the manufacturing sector in Vietnam. Rodrigo-Alarcón et al. [42] have developed a configuration model to explain how the combination of social capital and internal capabilities can contribute to the relationship between entrepreneurial orientation and efficiency. Hoogendorn et al. [43] put forward the hypothesis that stable entrepreneurs, working in conditions of market imperfection and unfavorable institutional contexts with broad requirements to the knowledge base, suffer from greater institutional barriers and face with different types of risks when creating a business than ordinary entrepreneurs.

Social capital is formed on the basis of relationships within groups of people, which are united by common interests and support informal contacts for the purpose of mutual benefit and assistance. Social capital is inextricably linked with institutional formations, which is literally shaped by them. We believe that social capital plays a leading role in the development of entrepreneurship.

There are scholars who study various aspects of social capital in relation to agribusiness. Yes, Ruiz-Ortega et al. [44], on the example of 292 Spanish agribusiness firms, analyzed the relationship between social capital, entrepreneurship and strategic behavior, and productivity of companies. Having determined the essence of a completely closed network, scientists have, on the basis of empirical studies, concluded that closed interorganizational relations are the most controversial aspect of social capital, since they create a network paradox, and although they provide the opportunity to obtain implicit knowledge, valuable ideas, and new possibilities, they also create restrictions in generation and access to new ideas due to excessive information and short-sightedness, inertia, and blockage, thus hindering the development of entrepreneurial orientation.

We consider the logical opinion of V. Ryabokon, O. Shpikuliak, and V. Pehov that the specifics of the formation and use of social capital in the countryside require the introduction of the concept of "agrarian social capital" in the scientific circulation, which they propose to define as "… a set of relationships, relations between people, property, production and labor potential of agrarian enterprises operating in agriculture and performing a specific role in the rural economy—production of agricultural products for food security by holding you" [15, p. 87]. Based on the institutional determinant of the formation of social capital, O. Shpiculak and G. Mazur emphasize that in the village, the problems of social capital formation must be considered from the standpoint of relations between individuals and economic agents [31, p. 64]. Agree with their opinion that formal features of social capital in the village are organizations and structures that use individual-collective links between people to meet socioeconomic needs.

## 3 Materials and Methods of Research

The theoretical and methodological basis of the research is the publication of Ukrainian and foreign scholars on the development of the agrarian complex and the institutional foundations of agrarian entrepreneurial activity. In solving the research tasks, the determining role is played by methods: monographic—for studying the peculiarities of entrepreneurial activity in the agrarian sector; economic-statistical, scientific comparison—to assess the impact of the elements of social capital on the development of agrarian business; comparative analysis and other methods of information processing—for the empirical confirmation of scientific conclusions, as well as substantiation of the logic of approaches to improvement of institutional mechanisms of regulation of agrarian entrepreneurship.

The information base of the research is the results of a scientific search, obtained in the process of studying the essence and functional characteristics of the development of entrepreneurship, agrarian entrepreneurship, and social capital, data from the World Bank, the European Social Survey, the State Statistics Service of Ukraine, the Institute of Sociology of the National Academy of Sciences of Ukraine, scientific articles and monographs, own authors' work.

# 4 Main Results of the Study

## *4.1 Trust and Entrepreneurship*

During the study of data from 23 European countries for 2016, a direct proportional dependence of the gross domestic product (GDP) on purchasing power parity per capita was found [45] ($y$) from the Business Performance Index, which is concluded by the World Bank [46] ($x$):

$$y = 1520x - 75{,}887(R = 0.44) \quad (1)$$

According to calculations, an increase in the index of ease of doing business per unit increases the size of GDP by purchasing power parity per capita by $1520 USA, and the density of the connection between the signs is average (the correlation coefficient is 0.44).

Similar calculations were made on the basis of Ukraine data for 2010–2017 Years:

$$y = 12.4x + 7713.9(R = 0.34) \quad (2)$$

The relation of GDP per purchasing power parity per capita with the index of ease of doing business in Ukraine also turned out to be directly proportional: an increase in the factor of one unit is accompanied by an appropriate change in the resultant attribute by 12.4 units, and the density ratio is average (the correlation coefficient is 0.34).

The correlation-regression analysis of the connection between the Index of Business Etiquette ($y$) and the level of public trust ($x$), which was determined according to the European Social Survey [47], revealed a directly proportional relationship between them:

$$y = 2.44x + 63.77(R = 0.57) \quad (3)$$

According to calculations, increasing the confidence level by one unit (on a scale from 1 to 10) contributes to a proportional change in the Business Risk Index, respectively, by 2.4 units.

GDP per purchasing power parity per capita ($y$) also has a positive correlation with the level of trust in society ($x$):

$$y = 10{,}660x - 15{,}965(R = 0.72) \quad (4)$$

Changing the level of confidence by one unit entails a corresponding change in the resultant attribute of $10,660 USA.

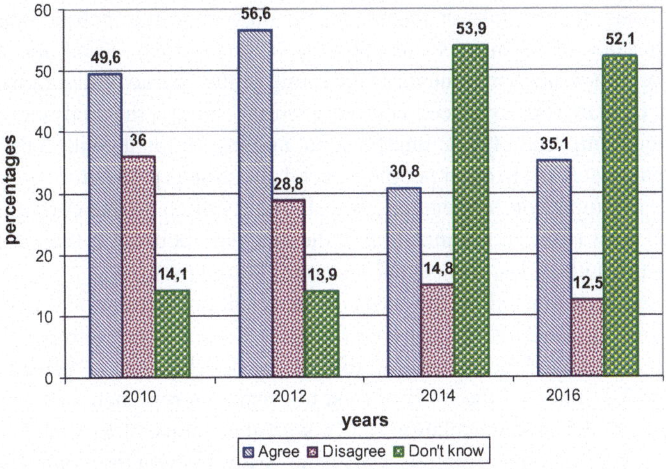


**Fig. 1** Dynamics of distribution of opinion of respondents of Ukraine regarding the statement "Do not trust anyone—the safest," %. *Source* Built according to [48]

That is, the link between GDP—an important indicator of the country's economic strength, as well as business conditions and trust—is one of the factors that affects them.

Similar calculations were made according to the data of Ukraine for 2010–2017 years. At the same time, the level of trust in society was determined by the percentage of respondents who disagree with the expression "Do not trust anyone—the safest" [48]. Due to the significant decrease in the level of trust in society in recent years ($x$), its dependence on the index of ease of doing business ($y_1$) and GDP on purchasing power parity per capita ($y_2$) is inversely proportional:

$$y_1 = -1.024x + 75.08(R = 0.98) \quad (5)$$

$$y_2 = -17.03x + 8741.8(R = 0.45) \quad (6)$$

In Fig. 1, data on the level of trust/distrust in Ukrainian society are given.

For the period of 2010–2016, the level of trust (the percentage of respondents who disagree with the statement "Do not trust the person is the safest") ranged from a maximum of 36% in 2010 to an appropriate minimum of 12.5% in 2016. At the same time, the maximum level of distrust was observed in 2012—56.6%. Despite the fact that in 2014 and 2016 the level of apparent mistrust decreased significantly (to 30.8% and 35.1% respectively), the level of total trust in these years is also the smallest (14.8% and 12.5% respectively). At the same time, the number of respondents who failed to answer this question, respectively, by years was 53.9 and 52.1%.

On the basis of the study of the Doing Business research methodology, it was assumed that one of its main socioeconomic factors—trust—is influenced by the formation of individual components of the rating of the degree of favorable business conditions. Considering its place in entrepreneurial activity, the analysis focuses on two of its most important types: interpersonal and institutional. Within the limits of interpersonal trust (trust to other people) special attention is paid to the trust of family and relatives, compatriots, colleagues, as well as directly to private entrepreneurs. In institutional trust (trust in organizations), the analysis focuses on the trust of local authorities, tax authorities, banks, and insurance companies.

The research conducted on the basis of the results of the national annual monitoring surveys of 1992–2016 [48] confirmed the existence of a direct relationship between the attitude of the population toward the development of private business (business) in Ukraine and the level of trust to private entrepreneurs (the coefficient of regression is 0.5, and correlation 0.42), with their consent to work (regression coefficient is 0.275, and correlation 0.6) or the desire to open their case (coefficient of regression is 0.8, and correlation 0.87). At the same time, the attitude toward the development of private entrepreneurship and the level of trust to the heads of state enterprises has a reverse relationship (the coefficient of regression is −0.25, and the correlation is 0.34).

As a result of the research, the relationship between the respondents' desire to open their own case (the effective feature) and the following factors: dependence of the population of Ukraine on the development of private enterprise (correlation coefficient 0.87, indicating a strong link between the signs) was revealed, with the consent to work personally a private entrepreneur (a correlation coefficient of 0.8, which also indicates a strong link between the signs), confidence level for private entrepreneurs (a correlation coefficient of 0.39, indicating a moderate connection between features).

In the framework of the study, a model of the dependence of the desire to open their own business ($Y$) from the levels of trust of family and relatives ($x_1$), compatriots ($x_2$), and colleagues ($x_3$) was constructed:

$$Y = -2.20 + 1.42x_1 + 0.24x_2 - 0.58x_3 (R = 0.97) \tag{7}$$

The coefficient of multiple correlations shows the presence of a dense bond (close to the functional one) between the factors under consideration. The coefficients of regression indicate the most positive influence on the desire to open their own business, namely the trust of family and relatives ($a_1 = 1.42$) among the factors included in the model. At the same time, the level of trust in colleagues reduces the desire to open their own business (the coefficient of regression is −0.58).

The dependence of the desire to open its own business ($Y$) from the levels of trust of the tax inspectorate ($x_1$), local authorities ($x_2$), banks ($x_3$), and insurance companies ($x_4$) is described by the equation:

$$Y = 0.92 + 0.26x_1 + 0.17x_2 - 0.89x_3 + 1.48x_4 (R = 0.53) \tag{8}$$

In the presence of an average link between the factors included in the model, coefficient of multiple correlation is $R = 0.53$. The growth of trust in the tax inspectorate, local authorities, and insurance companies increases the desire to open their own business (regression coefficients $a_1 = 0.26$, $a_2 = 0.17$, and $a_4 = 1.48$). On the contrary, increasing the level of trust to banks prevents the desire to open their own business (regression coefficient $a_3 = -0.89$). This fact may be related to the more significant (in comparison with bank deposits) levels of business risks.

## 4.2 *Social Capital in Agrarian Entrepreneurship*

Formation and use of social capital in the agrarian sector will promote the development of entrepreneurial activity and reduce the transaction costs of its subjects, which will occur as a result of overcoming the costs associated with access to markets, resources, information, investments, innovations, etc. One of the main directions of the solution of these issues is the development of institutions of agrarian entrepreneurship, and, first of all, agricultural cooperation.

In Table 1, data on the availability of agricultural servicing cooperatives, private peasant, and farm enterprises in Ukraine for the years 2010–2016 are presented.

For 2010–2016, the number of agricultural service cooperatives increased by 21.4%, and the average annual growth rate was 3.3%. However, in absolute terms, their number is extremely small and does not satisfy the real needs of the agrarian sector of the economy. 27.6% of the existing agricultural service cooperatives are engaged in milk production, 27.1%—land cultivation and harvesting, 3.0%—meat production, 11.8%—growing of fruits and vegetables, 6.4%—grain production, 23.9%—providing other services.

Despite the fact that for 2010–2016, the number of private peasant farms decreased by 10.2% (on average, by 1.8% reduction per year), farms by 18.9% (on average per year by 3.4%), and their number and contribution to the country's economy remain significant. Thus, in 2016, farms of the population produced 43% of the total volume of agricultural products, including 38.7% of crop production, 54.4% of livestock products. The share of farms in agricultural production in 2016 amounted to 8.7%, including 11.2% in crop production and 2% in livestock production.

Given the self-sufficiency of the rural population by food, the leading role of small businesses (private farms, as well as farms, including 81% of which have cultivated land with an area of up to 500 ha) in support of food security of the country is undeniable. Therefore, we believe that the issue of cooperation in the agrarian sector of the economy is relevant in the first place precisely for them.

The inclusion of man in strong collectivist ties, the establishment and fulfillment of mutual obligations among citizens, the strengthening of their responsibilities of responsibility and cooperation between them always used by villagers to improve the quality of their own lives and the life of the entire community. The larger the size of the social capital of the rural community, the greater the additional resources owned by its carriers. Unfortunately, the current state of rural society is characterized by

**Table 1** Dynamics of availability of agricultural service cooperatives in Ukraine

| Indexes | 2010 | 2011 | 2012 | 2013 | 2014 | 2015 | 2016 | 2016% by 2010 |
|---|---|---|---|---|---|---|---|---|
| Number of agricultural servicing cooperatives, units | 838 | 920 | 947 | 1035 | 928 | 949 | 1017 | 121.4 |
| Private peasant farms | | | | | | | | |
| Quantity, thousand units | 4540.4 | 4359 | 4301.8 | 4241.6 | 4136.8 | 4108.4 | 4075.2 | 89.8 |
| Area of land for commercial agricultural production, thousand hectares | 2841.8 | 2819.9 | 2884.1 | 2858 | 2820.2 | 2837.1 | 2818.9 | 99.2 |
| Number of cattle, heads | 2454.2 | 2383.7 | 2566.8 | 2519.7 | 2177.3 | 2479.8 | 2468.4 | 100.6 |
| Number of pigs, heads | 3319.3 | 3045.4 | 3031.3 | 3049 | 2779.8 | 3375 | 3103.2 | 93.5 |
| Farms | | | | | | | | |
| Quantity, units | 41,524 | 40,965 | 40,676 | 40,752 | 32,133 | 32,303 | 33,682 | 81.1 |
| The area of agricultural land, total, thousand hectares | 4290.8 | 4345.9 | 4389.4 | 4451.7 | 4707.7 | 4343.7 | 4437.9 | 103.4 |
| Number of cattle, thousand heads | 95.3 | 103 | 109.9 | 111.8 | 103.7 | 106.3 | 105.7 | 110.9 |
| Number of pigs, thousand heads | 294.8 | 260.4 | 265.2 | 272.9 | 249.9 | 276.1 | 273 | 92.6 |

*Source* Compiled according to the State Statistics Service of Ukraine

the destruction of social capital, which consists of the collapse of traditional ties and relations between people, a crisis of mutual trust. Many factors contributed to this, including the breakdown of cultural traditions in the countryside, the devaluation of spiritual values as a result of the chronic poverty of the majority of the rural population, the consolidation of the philosophy of individualism in the mass consciousness, the lack of adequate dialog between the authorities and the population, and so on.

In these conditions, the question of the possibility of increasing the social capital of the village based on the consistent improvement of living conditions and the revival of the peasant's positive traditions, collective participation in solving force Majeure's life problems, ensuring participation of the population in the assessment of the efficiency and safety of the state and municipal government, and the adoption of based on such assessment of organizational decisions, as well as real anti-corruption measures under public control. Particular attention should be paid to the formation of positive-oriented social networks, integrated with the process of socioeconomic

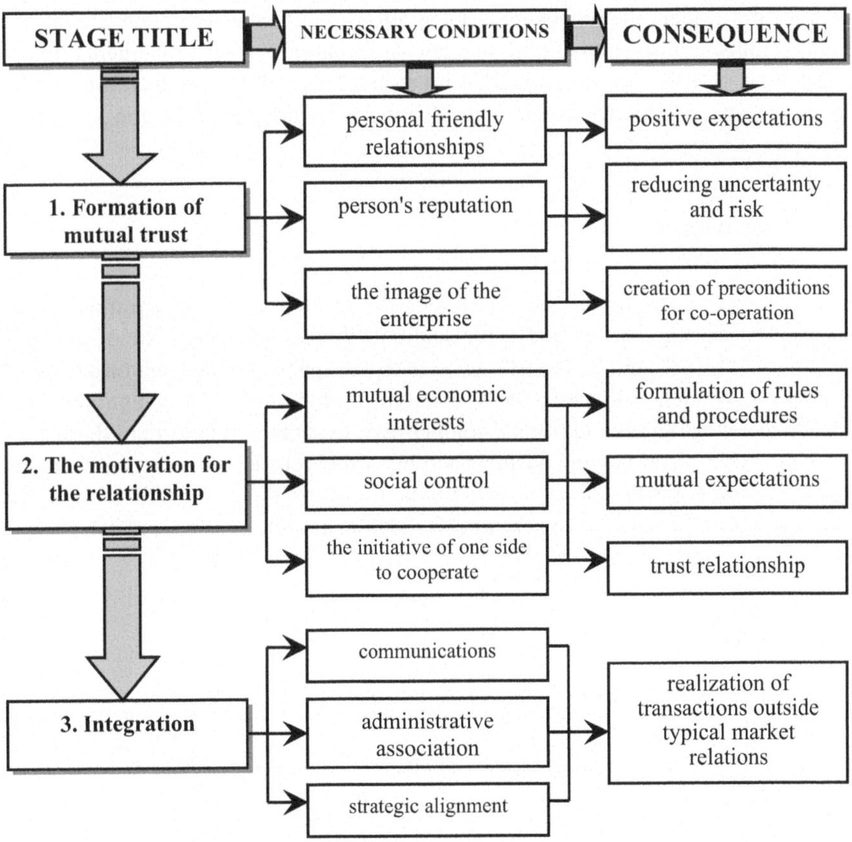


**Fig. 2** General scheme of formation of preconditions for the development of cooperation in the countryside. *Source* Composed on the basis of working out literary sources

development of the village, as well as the development of intellectual potential on the basis of effective reform of the existing education system.

Fig. 2 shows the general scheme of formation of preconditions for the development of cooperation in the countryside. The scheme includes the main stages, the necessary conditions, as well as the consequences of each of them.

So, to create mutual trust among potential subjects of cooperation (first stage) must have personal friendships, positive personal reputation management, and appropriate image of the company, leading to the emergence of some positive expectations, reducing uncertainty and risk, and facilitate the creation of the past conditions for cooperation.

The impulse to relations between the subjects of cooperation (second stage) is the mutual economic interests of the parties, the availability, and the possibility of social control and initiative of one party to cooperation—a special form of organization of economic activity to achieve common goals. As a result, closer confidential relations are formed, mutual expectations are improved, and rules and procedures of interaction are formulated.

In the third stage, the processes of integration—the formation of stable communication links, administrative and strategic association for the implementation of transactions outside the typical market relations. This will help reduce transaction costs and create additional competitive advantages for cooperation actors.

## 5 Conclusions

In the context of institutional transformations, the development of the agrarian sphere of the economy requires the search for qualitatively new resources, one of which is social capital, which simultaneously serves as a prerequisite for the creation and operation of cooperation in this area of the economy. Subjects of agrarian entrepreneurship, cooperating, receive additional competitive advantages as a result of the shared use of their own resources and business activity. Prospects for further research in this area are defining the directions of institutional transformations that will promote the development of the agrarian sphere of cooperation, as well as social capital.

## References

1. Kantil'jon, R.: Ocherk o prirode torgovli voobshhe (Essay on the nature of trade in general). Delo, Moscow (1955)
2. Sjej, Zh-B: Traktat politicheskoj jekonomii (Treatise of political economy). Delo, Moscow, Russia (2000)
3. Shumpeter, J.: Teorija jekonomicheskogo razvitija (Theory of Economic Development). Direkmedia, Moscow (2007)
4. Hajek, F.: Poznanie, konkurencija i svoboda (Cognition, Competition and Freedom). Jeksmo, Moscow (1999)

5. Azizov, S.: Evolution of scientific views on the concept of "entrepreneurship" and its current characteristics. Ekon. i menedzhment kul'tury **1**, 84–90 (2015)
6. The Verkhovna Rada of Ukraine.: The Law of Ukraine "About entrepreneurship", available at: http://zakon5.rada.gov.ua/laws/show/698-12 (1991). Accessed 03 Oct 2018
7. Berezina, L.: Entrepreneurship: epistemology and role in the development of relations between the subjects of agrarian and industrial complex. Naukovyj visnyk NLTU Ukr. **23**(14), 356–362 (2013)
8. Lantukh, I.: Entrepreneurship as an economic definition. Naukovyj visnyk Poltavs'koho universytetu ekonomiky i torhivli **1**(69), **2**, 141–147 (2015)
9. Rudkovs'ka, V.: Entrepreneurship as a special type of economic thinking. Visnyk Kyivs'koho natsional'noho universytetu imeni Tarasa Shevchenka. Ekonomika **124**, **125**, 89–91 (2011)
10. Liakhovets', O.: Entrepreneurship as a component of the institutional environment of modernization. Naukovyj visnyk Mizhnarodnoho humanitarnoho universytetu. Seriia: Ekon. i menedzhment **22**, 30–33 (2016)
11. Tsapovs'ka, M.: Classical and alternative approaches to the analysis of entrepreneurship in the Ukrainian economic thought of the second half of the nineteenth and early twentieth centuries. Materialy kruhloho stolu (Round table materials) Pidpryiemnytstvo iak rushijna syla suspil'noho prohresu (Conference Entrepreneurship as the driving force of social progress), National Academy of Sciences of Ukraine, Kiev, Ukraine, 1 December, 49–51 (2016)
12. Nesterenko, O.: Austrian business theory. Materialy kruhloho stolu [Round table materials] Pidpryiemnytstvo iak rushijna syla suspil'noho prohresu (Conference Entrepreneurship as the driving force of social progress). National Academy of Sciences of Ukraine, Kiev, Ukraine, 1 December, 40 (2016)
13. Timmons, J.A., Spinelli, S.: New Venture Creation: Entrepreneurship for the 21st Century. McGraw-Hill/Irwin, USA (2004)
14. Lysenko, V.: Theoretical aspects of the definition of the essence of entrepreneurship. Visnyk Kharkivs'koho natsional'noho tekhnichnoho universytetu sil's'koho hospodarstva imeni Petra Vasylenka **150**, 218–224 (2014)
15. Riabokon', V. Shpykuliak, O., Pekhov, V.: Social capital as an institutional determinant of rural development and agrarian entrepreneurship. Ekon. Finansy. Menedzhment: aktual'ni pytannia nauky i praktyky **1**, 83–93 (2015)
16. Artymiak, Z.: Development of forms of entrepreneurship in the agrarian sector and their efficiency. Abstract of Ph.D. dissertation, Economics and Enterprise Management, State Agrarian University of Lviv, Lviv, Ukraine (2008)
17. Kormyshkin, Yu.: Genesis of scientific views on the interpretation of the concept of "agrarian entrepreneurship" at the present stage of development. Visnyk Berdians'koho universytetu menedzhmentu i biznesu **4**(36), 51–55 (2016)
18. Majovets', Ye.: Ahrarne pidpryiemnytstvo v Ukraini (Agrarian Entrepreneurship in Ukraine), Vydav. tsentr LNU im. Ivana Franka, Lviv, Ukraine (2006)
19. Ivaniuta, V.: The role of agrarian entrepreneurship in the development of rural areas in the context of European integration. Ekonomichnyj visnyk **1**, 61–67 (2016)
20. Malik, M. Shpykuliak, O., Suprun, O.: Development of Entrepreneurship in the Agrarian Sector: Macroeconomic Aspect. Naukovyj visnyk Uzhhorods'koho universytetu. Seriia Ekon **1**(49), **2**, 32–37 (2017)
21. Pruts'ka, T.: Institutional support for the development of entrepreneurial activity in the agrarian sector. Ph.D. Thesis, Economics and management of the national economy, National Agrarian University of Vinnitsa, Vinnitsa, Ukraine (2015)
22. Hanifan, L.J.: The rural school community center. Ann. Am. Acad. Polit. Soc. Sci. **67**, 130–138 (1916)
23. Tugan-Baranovskij, M.: Social'nye osnovy kooperacii (Social foundations of cooperation), 4th edn. Jekonomika, Moscow (1989)
24. Schultz, T.W.: The Economic Value of Education. Columbia University Press, New York and London (1963)

25. Becker, G.S.: Investment in human capital: a theoretical analysis. J. Polit. Econ. **70**(5), 2: Investment in Human Beings, 9–49 (1962)
26. Burd'e, P.: Forms of capital. Jekonomicheskaja sociologija **3**(5), 60–74 (2002)
27. Coleman, J.S.: Foundations of Social Theory. Cambridge (MA), UK (1990)
28. Fukuyama, F.: What is social capital? Kyiv lecture by Francis Fukuyama", Day, (Online), vol. 177, available at: http://day.kyiv.ua/uk/article/podrobici/shcho-take-socialniy-kapital (2006). Accessed 03 Oct 2018
29. Putnem, R.: Who Killed Civic America? Prospect, 66–72 (1996)
30. Nort, D.: Instituty, institucional'nye izmenenija i funkcionirovanie jekonomiki (Institutions, institutional changes and the functioning of the economy). Fond jekonomicheskoj knigi "Nachala", Moscow, Russia (1997)
31. Shpykuliak, O., Mazur, H.: Institutional principles of the formation of social capital in the mechanism of development of rural areas. Ekon. APK **8**, 63–68 (2014)
32. Jennings, C.: Social capital, conflict and welfare. J. Dev. Econ. **124**, 157–167 (2017)
33. Al-Tabbaa, O., Ankrah, S.: Social capital to facilitate 'engineered' university–industry collaboration for technology transfer: a dynamic perspective. Technol. Forecast. Soc. Chang. **104**, 1–15 (2016)
34. Bohush, L.: Social capital and socio-humanitarian space: methodology of definition, interconnections, development vectors. Ekon. ta derzhava **4**, 15–20 (2010)
35. Burt, R.: Structural Holes the Social Structure of Competition. Cambridge, UK (1992)
36. Paldam, M.: Social capital: one or many? Definition and measurement. J. Econ. Surv. **14**(5), 629–654 (2000)
37. Roskolot'ko, I.: Social capital as a productive resource of social ties. Aktual'ni problemy derzhavnoho upravlinnia **2**, 17–20 (2013)
38. Portes, A.: Social capital: its origins and application in modern sociology. Ann. Rev. Sociol. **24**, 1–24 (1998)
39. Fukujama, F.: Doverie: social'nye dobrodeteli i put' k procvetaniju (Trust: social virtues and the path to prosperity), OOO "Izdatel'stvo ACT", ZAO NPP "Ermak". Russia, Moscow (2004)
40. Pryiatel'chuk, A.: Property and capital: essence, structure and main functions. Visnyk Zhytomyrs'koho derzhavnoho universytetu imeni Ivana Franka **63**, 48–52 (2012)
41. Pham, T., Talavera, O.: Discrimination, social capital, and financial constraints: the case of Viet Nam. World Dev. **102**, 228–242 (2018)
42. Rodrigo-Alarcón, J., Parra-Requena, G., García-Villaverde, P.M.: Effectiveness of entrepreneurial orientation: The role of social capital and capabilities. Inv. Europeas de Direccion y Economia de la Empresa **20**(3), 131–139 (2014)
43. Hoogendoorn, B., van der Zwan, P., Thurik, R.: Sustainable entrepreneurship: the role of perceived barriers and risk. J. Bus. Ethics (Online), available at: https://link.springer.com/article/10.1007/s10551-017-3646-8 (2017). Accessed 4 Oct 2018
44. Ruiz-Ortega, M.J., Parra-Requena, G., García-Villaverde, P.M., Rodrigo-Alarcon, J.: How does the closure of interorganizational relationships affect entrepreneurial orientation? Bus. Res. Q. **20**, 178–191 (2017)
45. Gross domestic product per capita based on purchasing-power-parity in current prices: World Data Atlas (2018) available at: https://knoema.com/atlas/ranks/GDP-per-capita-based-on-PPP. Accessed 03 Oct. 2018
46. Doing Business: Measuring Business Regulations (2018) available at: http://www.doingbusiness.org/. Accessed 03 Oct. 2018
47. European Social Survey (2018) available at: http://www.europeansocialsurvey.org/. Accessed 03 Oct 2018
48. Results of national annual monitoring surveys 1992–2016, "Ukrainian society: monitoring social change", available at: http://i-soc.com.ua/assets/files/monitoring/dodatki2016.pdf (2016). Accessed 03 Oct 2018

# Methodological Aspects of Forming Mathematic Models of Management of Socio-economic Systems Development

**Oleksii Hudzynskyi, Yulia Hudzynska, Svitlana Sudomyr and Mariia Sudomyr**

## 1 Introduction

In modern conditions of computerization, technologization of information systems formation, one problem of the important ones that needs to be solved is the development of methodological aspects of forming up adaptive models for ensuring the socio-economic systems development of various types of their development, behavior and civilized value orientations.

Such an aspect of the problem of ensuring the unity of developed models in their organic integrity is not considered in the theoretical and methodological terms. The emphasis is placed on the formalization and development of mathematical models of purely technical and technological orientation as a methodological toolkit. With this approach, the qualitative characteristics (parameters) of the system are dropped: Their target orientation, target installations and the level of formation of information potential decrease. Empirical studies confirm our conclusion. A mathematical reflection of the solution of the development of socio-economic systems under such an inappropriate approach does not fulfill its purposeful mission.

O. Hudzynskyi · Y. Hudzynska
Department of Management Department Named After Prof. Y. S. Zavadskyi, National University of Bioresources and Natural Resources of Ukraine, 11 Heroiv Oborony Str., Kiev 3127, Ukraine

S. Sudomyr (✉)
Department of Enterprise Economics Department, Separated Unit of the National University of Bioresources and Natural Resources of Ukraine "Berezhany Agricultural Institute", 20 Academichna Str., Berezhany 47502, Ukraine
e-mail: sudomyr_s@ukr.net

M. Sudomyr
Candidate of Economic Sciences, Raivska Str., Berezhany 47502, Ukraine
e-mail: sudomyr@ukr.net

© Springer Nature Switzerland AG 2019

V. Nadykto (ed.), *Modern Development Paths of Agricultural Production*,
https://doi.org/10.1007/978-3-030-14918-5_45

### *1.1 Analysis of Recent Studies and Publications*

Romashko [1], Nesterenko [2], Savras [1], Seliverstov [2] et al. made a considerable contribution in solving issues of modeling, forecasting and creating systems for socio-economic development analysis. However, the existing semantic and mathematic modeling systems for forecasting possible directions of economic development of socio-economic systems at macro- and microeconomic levels do not provide a response to the number of outstanding issues: What criteria basis is required to assess efficiency of the system of measures for mitigation and elimination of weaknesses and threats; what is the focus of development system under conditions of instability and crisis; what is the main criterium for development system focus assessment under conditions of instability and crisis; and what management system is required under conditions of instability and crisis?

All this requires a new scientific approach to forming semantic and mathematic models of socio-economic systems development under conditions of instability and crisis on the basis of synergetic methodology.

## 2 Statement on the Purpose and Objectives of the Study

To develop methodological aspects of forming adaptive models for ensuring socio-economic systems development for different types of their development, behavior and civilized value orientations. To substantiate the scientific approach to forming semantic and mathematic models of socio-economic systems development under conditions of instability and crisis on the basis of synergetic methodology. To propose a technology that is built on compliance with the requirements for solving problems with the highest efficiency to ensure the achievement of goals of socio-economic systems development.

## 3 Solution

In this regard, in order to increase the effectiveness of mathematical modeling as an organizational and interacting factor, we propose a technology that is built in compliance with the requirements for solving problems with the highest efficiency (social, environmental, organizational, intellectual, economic effects) to ensure the achievement of the goals of socio-economic systems development.

At a fundamental level of the logic of this methodological approach, we set up requirements of the main system-forming principles, which are represented in Fig. 1.

We propose to carry out a general assessment of development management efficiency of complex socio-economic systems with synergetic orientation using the following formula (1):

| | | |
|---|---|---|
| Systemacity | Target focus | Adaptability |
| Formalization | **Principles of formation of development models of socio-economic systems** | Complexity |
| Semanticism | | Creativity |
| Priority | | Algorithmization |
| Informativeness | | Strategic importance |
| Efficiency | | Emergence |
| Organic unity | Capacity for activation | Synergy |

**Fig. 1** Basic principles of the methodology of forming a general model of effective development of enterprises as system integrity (semantic algorithmic formalized approach)

$$PУP_{cec} = C_{p} \times K_{оц}, \tag{1}$$

where

$PУP_{cec}$ is development management efficiency of complex socio-economic systems with synergetic orientation;
$C_p$ is strategical development;
$K_{оц}$ is goal reasonability coefficient.

Harmonious balance of constituent elements as subsystems in a unity with organizational factor is the basis for ensuring the synergetic effect. At the same time, we focus attention on taking into account balance in all directions as systems characterizing development levels of basic elements for ensuring the synergetic effect of activities at the enterprises. We propose to determine the synergetic effect using the following formula [3]:

$$C_{еф} = S_{еф} - (P_{вмг} + P_{дсжп} + P_{з} + P_{д} + P_{p} + P_{cc} + P_{p}), \tag{2}$$

where

$C_{еф}$ is the synergetic effect;
$S_{еф}$ is the sum of effects from the simultaneous introduction of all components;
$P_{вмг}$ is a level of measure and harmony relations;
$P_{дсжп}$ is an access level of living labor subjects to use of capitals—materialized labor and biological, created by nature;
$P_{з}$ is a balance level in management system triad: functions, duties and responsibilities—resources, rights and power;
$P_{д}$ is an access level to the distribution of living labor results from use in the unity of the specified types of capital;
$P_{pn}$ is a level of rationally necessary combination of ownership and management;

$P_{cc}$ is a level of justice perception in the system of ensuring unity of corporate interests through the development system of ownership and power relations of individual and collective orientation;

$P_p$ is a level of development and dynamic balance of formal and informal institutions in the system of ownership and power and social and labor relations, etc. The specified levels will be determined using the formulas (3–9):

$$P_{вмг} = \frac{ФК_{P_{вмг}}}{НК_{P_{вмг}}}, \quad (3)$$

$$P_{дсжсп} = \frac{ФК_{P_{дсжсп}}}{НК_{P_{дсжсп}}}, \quad (4)$$

$$P_{з} = \frac{ФК_{P_{з}}}{НК_{P_{з}}}, \quad (5)$$

$$P_{д} = \frac{ФК_{P_{д}}}{НК_{P_{д}}}, \quad (6)$$

$$P_{p} = \frac{ФК_{P_{p}}}{НК_{P_{p}}}, \quad (7)$$

$$P_{cc} = \frac{ФК_{P_{cc}}}{НК_{P_{cc}}}, \quad (8)$$

$$P_{рд} = \frac{ФК_{P_{рд}}}{НК_{P_{рд}}}, \quad (9)$$

where

*ФК* is an actual coefficient;

*НК* is a standard coefficient.

The estimated absolute indicator value is determined on the basis of correlation dependence model of resulting indicator in a combination of factors influencing it.

The formula for determining the standard integrated indicator is as follows:

$$У = \pm К \pm К_1 Р_{вмг} \pm К_2 Р_{дсжсп} \pm К_3 Р_з \pm К_4 Р_д \pm К_5 Р_р \pm К_6 Р_{сс} \pm К_7 Р_{рд}, \quad (10)$$

where

| | |
|---|---|
| $У$ | is an integrated indicator; |
| $К$ | is an absolute term of equation (constant model value); |
| $К_1 ... К_7$ | are regression coefficients upon factor indicators; |
| $Р_{гз}, Р_{вмг}, Р_{дсжсп}, Р_з, Р_д, Р_р, Р_{сс}, Р_{рд}$ | are factors influencing the value of an integrated indicator. |

The specified methodologies provide for the use of weighting coefficients. However, we believe that the synergetic effect can be ensured only by the formed system of impact factors and justified systemic criteria for their assessment. Such conclusion is grounded by almost equal significance of each factor in the system, and our understanding is that each factor when not taken into account may create chaos in the system integrity (according to chaos and system resonant excitation theories).

Management of organizational formations development has its own features and depends on the following: development types and directions, organization models, types of their behavior and response, reorganization and restructuring, integration and diversification processes, organization culture models and organizational culture.

The main features of management of organizational formations development were partially disclosed by the scientists in theoretical researches of the following organization models: Fayol's mechanistic model, Parson's natural model, North's institutional model, Hall's conflict model, Bogdanov's process model, Franchuk's problem model, Burns and Stalker's organic model. The stage of theories development includes organization models of competitive, innovative, strategical and anti-crisis orientation. Organization culture models are of concern and require in-depth study and determining features in justification of the socio-economical systems development directions. The latter are determined by the following: expansion of interrelations both inside the socio-economical systems and between the systems upon their interaction; increase of the system informative capacity; increasing pace of the system progressive development; complicating interaction of social systems, economics and nature; available and optimally possible system potential; global communicating, organizational and interacting processes of associative, cluster and corporate orientation; improvement, differentiation and integration of the system elements; and expansion of the range of real opportunities for further development of the socio-economic systems [4].

Our position lies in forming the assessment methodology in terms of models of enterprise operation as socio-economic systems proposed by the authors: survival model; survival model on the basis of simple reproduction; model of moderate growth with signs of socio-economic development; model of organization development on the basis of expanded reproduction; model of organization development on the basis of innovative approach; model of organization development on the basis of cluster approach; model of organization development on the basis of export-oriented approach; model of organization development of leadership orientation; model of organization development on the basis of interaction of ecological, social, intellectual and economic factors in their unity; model of organization development with orientation on dynamically changing components in the external environment; model of organization development of varying behavioral orientation; model of organization with orientation on expanding geographical market segments; model of organization with developed system of anti-crisis activities; and model of organization with developed system of competitive orientation.

We propose to assess the specified models by the following criteria: objectives achievement level of agricultural enterprises; performance of functionally supporting subsystems; effectiveness of management decisions; and efficiency of management by integrated indicators.

Sustainable development model is based on the principles of equal opportunities for justified satisfaction of aesthetic, spiritual and economic needs of a person implemented via humanitarization, ecologization, biologization and socialization of agrarian production, preservation and transfer of the generations code, harmonization of coexistence of human and nature.

Therefore, we believe that sustainable development should be assessed using the index of ($I_{sr}$): ecological ($I_e$), economic ($I_{ek}$), social ($I_c$) and intellectual ($I_{il}^2$) development [formula (11)]:

$$C_p = \sqrt[4]{I_e^2 + I_{ec}^2 + I_c^2 + I_i^2}, \tag{11}$$

where

$C_p$ is a sustainable development;
$I_e$ is an ecological development;
$I_{ec}$ is an economic development;
$I_c$ is a social development;
$I_i$ is an intellectual development.

In socio-economic systems, the main system-forming element of mathematical, cybernetic and other models is a semantic (meaning—semantic) approach, and more complete semantic models are based on target installations and target programs of actions.

The last ones are based on value orientations, which are the basic, fundamental foundation for choosing the organization's models.

Mathematical and cybernetic models as mechanisms are a connecting element between semantic models of target installations and techno-technological means of information display of content-oriented programs of the dynamically strategic developmental orientation of socio-economic systems. Only in this approach, it is possible to ensure an optimal balance of all the component elements of the system target and functionally providing direction of their synergistic performance.

This approach is an axiom in the formation of effective management systems.

The second direction of action is the functional-providing components (as subsystems) development of the systemic integrity of the semantic direction of the synergetic direction development of socio-economic systems. With this approach, a new essence and direction of the mathematical and cybernetic orientation are being developed. At the same time, we place great importance on the philosophical approach to the formation of such systems in their organic unity, interconnection and development: value orientations, target settings, organizational interaction, organizational behavior, organizational culture, organizational processes, organizational climate.

In this aspect, in our opinion, one should approach to the development of theoretical and methodological principles for the formation of socio-economic systems of synergistic direction. Therefore, we support the position of scientists who propose to develop the logic of actions in the scheme "system component of the system" and not vice versa [4]. In this aspect, all components of the system integrity should be explored in a more profound version in conjunction with the initial basis of research, which is the "system" and its system properties.

At the same time, we make an emphasis on following the logic of actions such as: "values behavior model of organization-general system strategy of development strategy as means of implementation of target-setting strategies for development" of creative, innovative, investment, incremental, strategic, competitive, anti-crisis, social, ecological, organizational, economic, activating, technical and technological, biological, intellectual, export-oriented and other directions of development.

Schematically, this relationship is represented in Fig. 2.

## 4 Conclusion

From a methodological point of view, a mathematical model acts as the main system-linking component of organizational interaction (subsystem) between semantic models in the context of all components of a functional-providing direction with technical information support, which forms the basis for the formation of effective systems of socio-economic entities management and provides their successful functioning with rather high synergistic effects.

However, at the same time, we emphasize on the fact that the system-synergistic effect in managing the activities of enterprises as socio-economic systems can only be ensured in the following conditions: changes in value orientations; ethical components in hierarchical systems; reasonable models of organizations; constructed on the unity of interests in an optimally balanced system of social inequality on the criterion of providing sufficiently necessary level and quality of life of the population of the first group of the three-level system by the level of profitability and the formed labor capital.

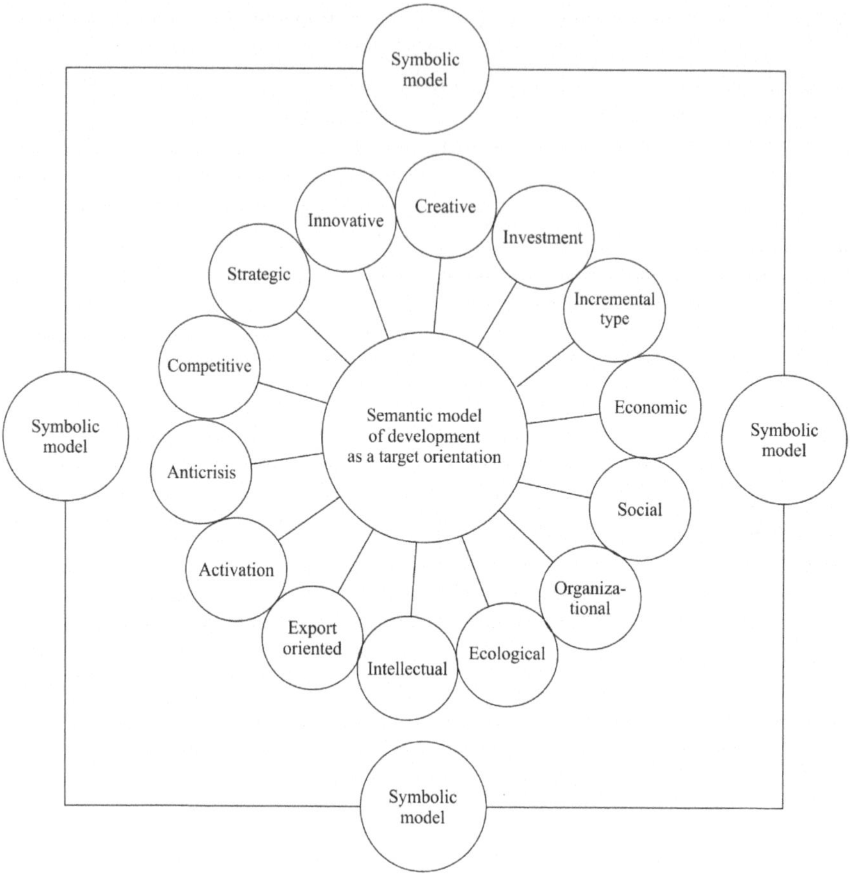


**Fig. 2** Interconnection of mathematical and semantic models of socio-economic systems development as target guideposts and installations

This should be supported by a well-founded system of accounting and analytical support for managing of organizations' activities, which is being developed by us in the context of the proposed methodological approach to formation of models of implementation of socio-economic programs according to the directions of development, types of behavior and reaction of enterprises as system integrity in carrying out systemic changes, taking into account the life cycles of their development.

## References

1. Romashko, S.M., Savras, I.Z., Seliverstov, R.H., Yurynets, R.V.: Modeling of socio-economic development of territories: scientific and methodological development: Kyiv, NAPA, 44 (2013)
2. Nesterenko, S.A.: Multifactor model of agricultural enterprises competitiveness assessment: economics, management. Monthly Theor. J. Res. Pract. **12**, 79–83 (2013)
3. Hudzynskyi, O.D., Sudomyr, S.M., Hurenko, T.O.: Theoretical and methodological principles of effective management of enterprise development. Kyiv, K.: "Komprynt", 411, (2017)
4. Sudomyr, S.M.: Formation of a management system for the development of agricultural enterprises: theory, methodology. Kyiv, K.: "Komprynt", 483 (2015)

# Modeling Innovative Economic Activity of Peasant Farm

**Sergey Kalchenko, Tetiana Popova, Denis Eremenko and Larisa Eremenko**

## 1 Introduction

One of the features of the world economic development is strengthening of the role of small business whose activity is not just limited to ensuring maximum profits. Small-scaled production forms also contribute to solving the problem of employment as well as reducing social tension. Entrepreneurial units that function in the agrarian sector play an important role in this process. Keeping land in family ownership in agricultural production contributes to solving environmental problems in the process of land utilization. At the same time, it should be noted that there are some specific problems in the area of peasant farms functioning in agriculture due to the peculiarities of modern national economy.

The peculiarity of the current state of the functioning of Ukrainian agrarian sector is the factual dominance of small-scale forms of farming such as farmers and private peasant farms in the production of agricultural raw materials. It should be noted that the share thereof in the total volume of fruit and berry production is more than 80%, with over 97% in potato production and about 75% in milk production [1]. However, this segment of agrarian production is not considered at present as a subject of entrepreneurial activity in the legal sense, although, for the representatives of local government, these farms play an important role in improving the overall situation in

S. Kalchenko (✉) · T. Popova · D. Eremenko · L. Eremenko
Tavria State Agrotechnological University,
B. Khmelnitsky ave. 18, Melitopol 72310, Ukraine
e-mail: serhii.kalchenko@tsatu.edu.ua; przixl136@gmail.com

T. Popova
e-mail: tetiana.popova@tsatu.edu.ua

D. Eremenko
e-mail: denis.eremenko@tsatu.edu.ua; eremenko.denis.mobile@gmail.com

L. Eremenko
e-mail: larisa.eremenko@tsatu.edu.ua

© Springer Nature Switzerland AG 2019

V. Nadykto (ed.), *Modern Development Paths of Agricultural Production*,
https://doi.org/10.1007/978-3-030-14918-5_46

agriculture. One of the government initiatives in that direction includes a complex of programs aimed at peasant farm creation [2]. The purpose of such actions is to stimulate the transformation of the most economically active part of farm households into small farm enterprises, although the actual implementation of this program is moving rather slowly at present. In our opinion, one of the problems hindering the economic development of small forms of agricultural production is the lack of a wide range of patterns of their economic activity, aimed at maximizing efficiency.

*The purpose of this study* is to develop scientifically based forms of peasant farm activities in the agrarian sector of Ukraine.

## 2 Review of Literary Sources

In our opinion, private peasant farms are a form of agrarian production, where their members are interested in preserving and improving the quality of natural resources, as well as ensuring sustainable development of rural areas. However, their status as well as the prospects for further activity, as an essential part of the agrarian sector of the economy, requires further consideration. A labor-balance sheet concept developed by O. Chayanov justifies the tendency of peasant farms to active manual labor exploitation since a preindustrial village; the average farm was not able to use high-tech equipment due to the lack of financial or land resources [3]. Nowadays we need some reforms to change the role of peasant farms in domestic agrarian market development. It will give possibilities to successively balance the domestic agrarian economy and successfully negotiate the challenges for rural and urban economies transformation.

A. Chelintsev, defining economic features of the peasant economy, considered it as a labor force, pointed out that "… the average level of the family farming is maintained as a result of two indicators, biological and family-oriented. According to them the growth of a family is based on the number of small (minor) children who subsequently become adults, thus increasing the numbers of workers. At the same time, the separation of young people from their families at a certain stage of life, leads to a decrease in the overall productivity of the family in which they grew" [4].

Therefore, in order to support the development of rural areas, it is necessary to increase the number of small-scaled agrarian production. One of the ways to ensure it is to choose scientifically justified patterns of their activity based on the assessing of a complex of indicators.

## 3 Research Methodology

The activities of most peasant farms cannot be adequately evaluated using traditional indicators of economic efficiency (profit, profitability, etc.). Their functioning has to be oriented at exploring biological and social laws rather than economic issues and determining the customer needs and their changes.

According to the approach, developed by T. Yavorskaya, the economic efficiency of the peasant farm production can be measured not only taking into account the production volume, but also the production costs, in particular, labor costs, including wages and taxes. Due to estimated effect of crop production in a particular household, it was concluded that the peasant farm is able to function, even if it pays a fixed tax [5].

In our opinion, in order to determine the level of efficiency of peasant farm agricultural production it is necessary to use a combined approach that is based on using conditional indicators of the production, sale and agricultural product consumption. It is also necessary to take into consideration the influence of production on the welfare of the peasant family through the use of the measure of the revenue share in the total budget of the family. It is considered as an indicator of social efficiency.

Scientists of NSC "Institute of Agrarian Economics" of NAASU developed an approach of production efficiency assessment in the small-scale agricultural sector, based on the index of conditional gross income, "which is defined as the difference between the gross proceeds of sales and the value of consumed products and costs without a wage" [6]. Taking the given methodology as a basis, we suggest that the following indicators to estimate the level of economic efficiency of a peasant farm activity should be used:

$$\mathrm{GI} = \mathrm{B} - \mathrm{PC} \tag{1}$$

$$\mathrm{CNI} = \mathrm{GI} - \mathrm{Wc}, \tag{2}$$

where GI shall mean gross income, PC shall mean production costs, CNI shall mean conditional net income, and Wc shall mean conditional wages.

The "price gain" indicator was used to assess the social effect of sale optimizing of goods. The gain is received by a peasant farm being a producer. A consumer benefits from this fact too, because he buys farm products without using the services of a mediator. The total sum of the price gain of both a producer and a consumer is calculated according to the following formulas:

$$\mathrm{Ppg} = \sum_{i=1}^{n} (\mathrm{P\kappa}_i - \mathrm{Po}_i)\mathrm{Q}_i; \tag{3}$$

$$\mathrm{Cpg} = \sum_{i=1}^{n} (\mathrm{P\kappa}_i - \mathrm{Pp}_i)\mathrm{Q}_i; \tag{4}$$

where Ppg shall mean the price gain of a producer; Cpg shall mean the price gain of a consumer; Pк shall mean the compromise price; Po shall mean the wholesale price; Pp shall mean the retail price; Q shall mean the volume of production; and *i* shall mean the product type.

## 4 Results

The results of data analysis (Fig. 1) convincingly testify the dominant role of peasant farms of Ukraine in some areas of production. This tendency is especially relevant for crop production. In particular, in 2000–2016 the average share of peasant farms in the production of potatoes, vegetables, fruits and berries together, grain and sunflower was 87, 77, 73, 19 and 14%, respectively. Crop production is stable enough because of low indicator fluctuations.

A slightly different trend is observed in animal husbandry. During the analyzed period, the share of peasant farms in the production of meat and eggs decreased significantly by 38 and 20 points, respectively. The volumes of milk production were stable. The average value of its share in total production was 68%. This situation is caused by the number of factors. Firstly, agricultural producers are more likely to participate in cattle-breeding, which require a lot of investments, and crop production that is considered as profitable. As a result, the most labor-intensive industries (fruit, vegetable and potato production) are not quite attractive for producers. The share of agricultural enterprises in the total milk production is relatively insignificant as well. This is due to the high needs for pastures and livestock buildings [7]. The circumstances determined the preference of the recommended forms and areas of farm activities.

The basic principles of creating models of peasant farm functioning:

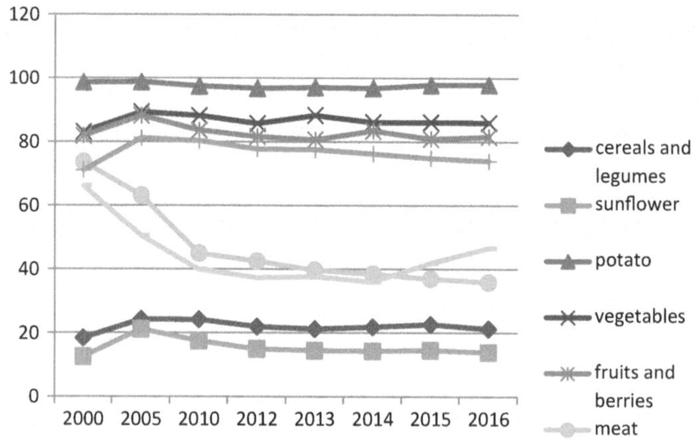


**Fig. 1** Dynamics of the peasant farm share in the total production of agricultural products

**Table 1** Recommended area of land, hectares[a]

| Cultures | Year | | | | | |
|---|---|---|---|---|---|---|
| | 1 | 2 | 3 | 4 | 5 | 6 |
| Potato | 5.92 | 5.92 | 5.33 | 4.74 | 4.14 | 2.96 |
| Carrot | 0.59 | 0.59 | 0.59 | 0.59 | 0.59 | 0.59 |
| Onion | 0.59 | 0.59 | 0.59 | 0.59 | 0.59 | 0.59 |
| Merry | 0.83 | 0.83 | 0.95 | 1.18 | 1.30 | 1.89 |
| Apricot | 0.71 | 0.71 | 0.83 | 1.07 | 1.18 | 1.54 |
| Peach | 0.59 | 0.59 | 0.71 | 0.83 | 1.07 | 1.18 |
| Apples | 0.59 | 0.59 | 0.71 | 0.71 | 0.83 | 0.95 |
| Pear | 0.47 | 0.47 | 0.47 | 0.47 | 0.47 | 0.47 |
| Plum | 0.36 | 0.36 | 0.36 | 0.36 | 0.36 | 0.36 |
| Hazelnut | 0.59 | 0.59 | 0.59 | 0.59 | 0.59 | 0.59 |
| Currant | 0.59 | 0.59 | 0.59 | 0.59 | 0.59 | 0.59 |
| Blackberry | 2.40 | 2.40 | 2.40 | 2.40 | 2.40 | 2.40 |

[a]According to the results of the authors' researches

1. Taking into account the regional specificity of the market situation of agricultural raw materials. The lack of competition among agricultural enterprises led to the development of fruit and vegetable production.
2. Stimulation of the consolidation of peasant farm efforts for the implementation of economic activity on the basis of cooperation.
3. Justification of mobile or immobile machine use for decreasing of manual labor.

The object of our research is the service cooperative "Holmodmash-1", which is located in the district of Semenivka of the Region of Zaporizhzhia. The cooperative includes 120 farms, and its total land area is 16.01 ha. In order to increase the efficiency of peasant farm functioning, it is necessary to reduce the amount of intensive labor activities by producing both fruits and vegetables. The amount of recommended land use structure is given in Table 1.

Its feature is dynamism, gradual reduction of areas under vegetable crops and the increase of lands under orchard for the period of six years. This time interval is chosen, taking into account the fact that the vast majority of orchard trees are in the stage of fruiting in the sixth year. Part of the area is allocated to crops, such as potatoes, carrots and onions, because they are constantly in demand in the regional market.

The diverse structure of fruit and berry produce reduces the peak load of peasant households. It also allows the business unit to be protected from possible price fluctuations for certain types of goods, to reduce the negative effect of natural and climatic phenomena in the risk-prone area. For the period of six years, due to the change of activities, the costs of physical labor reduced by 11 thousand man-years and other costs increased by 1.7 thousand euros or doubled (Table 2). However, fruit and berries productions increase the conditional net income from 164 thousand euros

**Table 2** Cost-effectiveness of the proposed measures[a]

| Ser. No. | Indicator | Year | | | | | |
|---|---|---|---|---|---|---|---|
| | | 1 | 2 | 3 | 4 | 5 | 6 |
| 1 | Labor costs, thousand man-hours | 50.4 | 49.6 | 47.2 | 45.6 | 43.2 | 41.6 |
| 2 | Production costs, thousand euros | 1.5 | 1.7 | 1.8 | 1.9 | 2.4 | 3.2 |
| 3 | Cost of manufactured goods, thousand euros | 373.3 | 386.7 | 415.5 | 523.2 | 570.7 | 600.0 |
| 4 | Revenue, thousand euros | 276.5 | 288.5 | 307.7 | 358.4 | 422.7 | 444.5 |
| 5 | Conditional wage, thousand euros | 110.9 | 108.3 | 104.3 | 99.7 | 94.9 | 91.2 |
| 6 | Revenue, thousand euros | 165.6 | 180.3 | 203.5 | 258.7 | 327.7 | 353.3 |
| 7 | Conditional net income, thousand euros | 164.1 | 178.5 | 201.7 | 256.8 | 325.3 | 350.1 |
| 8 | Conditional level of profitability, % | 145.9 | 162.3 | 190.2 | 252.7 | 334.2 | 370.9 |
| 9 | Level of marketability, % | 74.1 | 74.6 | 74.1 | 68.5 | 74.1 | 74.1 |
| 10 | Total income of families, thousand euros | 2583.5 | 2596.8 | 2625.6 | 2733.3 | 2780.8 | 2810.1 |
| 11 | Percentage of family income spent on manufactured goods in the total income of families, % | 14.5 | 14.9 | 15.8 | 19.1 | 20.5 | 21.4 |

[a]According to the results of the authors' researches

to 350 thousand euros and the level of profitability goes up by 3.7 times. It is offered to remain the same level of marketability that equals to 74%. It enables to increase sales of these products because of existent opportunities to select vegetables, fruits and berries in accordance with market requirements.

One of the purposes of our study is to determine the structure of a crop area in order to get the maximum price effect. The object of our research is the service cooperative "Yagidka-2", which is located in the district of Vesele of the Region of Zaporizhzhia. Its total land area is 25.8 ha. About 20 ha are used for agricultural production, and the rest of the space is under residential or commercial buildings.

Furthermore, the task was to determine the optimum structure of land in accordance with the chosen criterion when introducing a new technology of growing vegetables and fruit crops. It involves the active use of means of small mechanization, increasing the area under orchards. The typical vegetable and orchard crops grown in the peasant farms of the region were selected. To calculate farmers' working time, we take into consideration some characteristics of peasant farms. The total duration of fieldwork is nine effective working months (from March to November).

In winter, peasant households do not produce any crops. According to anonymous surveys, the number of working days in a month is 25 and farmers work about 4 h per day. The average number of household members is 2.8. Thus, the total working time of the peasant farm is 604.8 thousand man-hours.

Legend:
$x_1$ shall mean the land area of tomatoes;
$x_2$ shall mean the land area of potatoes;
$x_3$ shall mean the land area of carrots;
$x_4$ shall mean the land area of cherries;
$x_5$ shall mean the land area of apples;
$x_6$ shall mean the land area of peaches.

A system of limitations of inequalities for active technology is drawn

(1) in accordance with the land area:

$$x_1 + x_2 + x_3 + x_4 + x_5 + x_6 \leq 2000$$

(2) in accordance with the area of individual crops:

$$100 \leq x_1 \leq 500;$$

$$300 \leq x_2 \leq 1000;$$

$$40 \leq x_3 \leq 400;$$

$$120 \leq x_4 \leq 1000;$$

$$60 \leq x_5 \leq 400;$$

$$120 \leq x_6 \leq 400; \; x \geq 0$$

(3) in accordance with production factors:

$$43x_1 + 52x_2 + 27x_3 + 23x_4 + 35x_5 + 29x_6 \leq 604{,}800$$

The value of the objective function determines the structure of the land areas for the specified crops to ensure the maximum price effect:

$$f(x) = 41x_1 + 190x_2 + 36x_3 + 500x_4 + 78x_5 + 574x_6$$

An optimum solution was obtained by realization of a MathCAD simplex method. According to the results for the maximum price effect, the following land areas are recommended: 1 ha shall be under tomatoes land area; 4 ha shall be under potatoes land area; 0.4 ha shall be under carrots land area; 10 ha shall be under cherries land area; 0.6 ha shall be under apples land area; and 4 ha shall be under peaches land area. With this distribution, the total amount of the price effect shall be 146,000€.

The system of limitations for the designed technological pattern includes the set of inequalities:

(1) in access to the land area:

$$x_1 + x_2 + x_3 + x_4 + x_5 + x_6 \leq 2000$$

$$100 \leq x_1 \leq 500;$$

$$300 \leq x_2 \leq 1000;$$

$$40 \leq x_3 \leq 400;$$

$$120 \leq x_4 \leq 1000;$$

$$60 \leq x_5 \leq 400;$$

$$x_6 \leq 400; \; x \geq 0$$

(2) in access to production resources:

$$31x_1 + 35x_2 + 20x_3 + 18x_4 + 18x_5 + 12x_6 \leq 604{,}800$$

The maximum price effect shall be achieved when the objective function, which determines the structure of the land areas for specified crops, is described as:

$$f(x) = 62x_1 + 274x_2 + 55x_3 + 764x_4 + 112x_5 + 858x_6$$

According to the optimum solution of an optimization problem to get the maximum price effect, the following land areas are recommended: 1 ha shall be under

**Table 3** Efficiency of land use optimization

| No. | Indicator | Existing structure of land use[a] | Project structure of land use | | Deviation of a project structure from an existing land use structure, +, − | |
|---|---|---|---|---|---|---|
| | | | Existing technology | Designed technology | Existing technology | Designed technology |
| 1 | Labor costs, thousand man-hours | 35.37 | 47.16 | 31.74 | 11.79 | −3.63 |
| 2 | Production costs, thousand euro | 23.33 | 25.33 | 40.00 | 2.00 | 16.67 |
| 3 | Cost of manufactured goods, thousand euro | 53.97 | 101.35 | 141.89 | 47.39 | 87.93 |
| 4 | Revenue, thousand UAH | 18.89 | 70.95 | 99.33 | 52.06 | 80.44 |
| 5 | Conditional wage, thousand euro | 10.38 | 13.83 | 9.31 | 3.46 | −1.07 |
| 6 | Revenue, thousand euro | −4.44 | 45.61 | 59.33 | 50.06 | 63.77 |
| 7 | Conditional net income, thousand euro | −14.82 | 31.78 | 50.02 | 46.60 | 64.84 |
| 8 | Conditional level of profitability, % | −43.96 | 81.14 | 101.44 | 125.11 | 145.40 |
| 9 | Level of marketability, % | 35.00 | 70.00 | 70.00 | 35.00 | 35.00 |
| 10 | Total income of families, thousand euro | 433.17 | 480.55 | 521.09 | 47.39 | 87.93 |
| 11 | Percentage of family income spent on manufactured goods in the total income of families, % | 12.46 | 21.09 | 27.23 | 8.63 | 14.77 |

[a]According to the survey

tomatoes land area; 4.3 ha shall be under potatoes land area; 0.2 ha shall be under carrots land area; 9.8 ha shall be under cherries land area; 0.8 ha shall be under apples land area; and 3.9 ha shall be under peaches land area. This land use structure could give the price effect which is equal to 43,500€ (Table 3).

In the case of optimization of a land use structure, it is predicted that the conditional level of profitability will increase to 145%, and the share of production in the peasant farm budget will rise by almost 10 points. This is an evidence of increasing social impact of the practical implementation of the project.

However, it should be noted that the effective implementation of these measures will require significant efforts from the farmers. They should be aimed at the consolidation of small farms in order to reconcile production and economic activity. According to V. Diesperov consolidated development of small-scale agricultural producers is one of the trends for the Ukrainian agrarian sector [8]. Representatives of small agribusiness should be familiarized with the peculiarities of inter-branch association setting up, as well as actively use modern technical and technological patterns.

Zbarsky V. proves, that one of the problems of peasant farm effective development in Ukraine is the lack of proper information support. The result of this process is in sufficient scientific developments in the area of peasant farm production activities [9]. Mostly, there are no unified technological maps of growing crops for farms with an area of up to 10–50 ha. Also, there are almost no models of typical livestock and poultry complexes that would be able to support a normal peasant family with minimal financial resources. Chelintsev A. stressed the need of appropriate infrastructure for the peasant farm development [10]. The problem is also quite relevant nowadays. Today the researches are considering social and economic aspects of peasant farm functioning. Therefore, Lupenko Y., Malik M. and Kisil M. claim that the organization of small-scale production in Ukraine has to be based on scientifically justified approaches [11].

## 5 Conclusions

The article is devoted to the research of the problem of effective economic activity organization of peasant farms in Ukraine. Their importance in the agricultural production was indicated here. The research is focused on the substantiation of the transformation necessity of methodical approaches to the evaluation of the results of peasant farm activities. The authors define the indicators that reflect the economic and social nature of peasant farm functioning. The forms of organization of peasant farm agricultural production are determined. The calculations prove the positive impact of the proposed recommendations on the level of peasant farm efficiency. Further research results will be implemented in the development of regional programs for the small-scaled agricultural business.

## References


1. Agriculture of Ukraine 2016: Statistical Collection (2017)
2. Lupenko, Y., Malik, M.: Program for the development of private peasant farms in Ukraine up to 2020 (project) (2013)
3. Chayanov, A.V.: Peasant Farm: Selected Works (1989)
4. Chelyntsev, A.N.: Theoretical foundations of the organization of the farm (1919)
5. Yavorska T.I.: Small Business in Agriculture: Theory and Practice (2012)
6. Orlaty, M.: Social sphere of the village (1997)
7. Goryov, V.P., Zbarsky, V.K.: Small entrepreneurship: progression, problems (2011)
8. Diesperov, V.S.: Trends in rural household development—nomy of agroindustrial complex **5**, 70–72 (in Ukrainian) (2015)
9. Zbarsky, V.K.: Formation and development of small forms of farming in the countryside (2006)
10. Chelyntsev, A.N.: Russian agriculture before the revolution (1928)
11. Lupenko, Y., Malik, M., Kisil, M.: Methodical recommendations on the organization of family farms in Ukraine (on the example of milk production) (2014)

# Prospects of Ukraine on the World Market of Dairy Desserts

**Tatiana Krasnoded, Tetiana Popova, Tetiana Bakina and Olena Vasylchenko**

## 1 Introduction

The production of dairy desserts in Ukraine is the prospective way to enter the external market of dairy products. This happens due to certain circumstances. Firstly, Ukrainian dairy desserts are competitive in terms of price and taste. Secondly, with the low rate of dairy production, the domestic producers should combine production using less milk and more its substitutes, as well as jam, berries, vegetable oils, animal fats, etc. Therefore, Ukrainian producers are more or less competitive in the category of dairy desserts. For this reason, in order to increase the level of stability in the foreign market, certain economic mechanisms of functioning of the dairy dessert industry should be developed, from theoretical concepts, which are undeveloped in this area, to practical measures in the international context.

## 2 Review of Literary Sources

The problems of milk and dairy market functioning are researched by many scholars and practitioners, which includes the global level due to the relevance of the topic. The tendencies of the world milk and dairy product market development are broadly

T. Krasnoded (✉) · T. Popova · T. Bakina · O. Vasylchenko
Tavria State Agrotechnological University, B. Khmelnitsky ave. 18, Melitopol 72310, Ukraine
e-mail: tetiana.krasnodied@tsatu.edu.ua

T. Popova
e-mail: tetianapopova@tsatu.edu.ua

T. Bakina
e-mail: tatiana.bakina@tsatu.edu.ua

O. Vasylchenko
e-mail: olena.vasylchenko@tsatu.edu.ua

© Springer Nature Switzerland AG 2019

V. Nadykto (ed.), *Modern Development Paths of Agricultural Production*,
https://doi.org/10.1007/978-3-030-14918-5_47

analyzed by the researches. Gawaz Ye.G., Galushko V.P., Ladyka V., Novak I.M. analyzed the current state and prospects of dairy market product development. They revealed the problems of its functioning at all levels, ranging from the production of raw materials to the product export. It was emphasized that export-oriented producers should use innovative approaches to running their business based on the investments [1–5]. Tivonchuk C.V. stated that the functioning of the world dairy product and milk market is under the influence of many factors, so each country needs to keep pace with global trends and use all the potential opportunities to strengthen its position in order to maintain the market [6]. Baboshko O.V., Verhovets O.A. examined the problems and prospects of the Russian market of dairy products facing economic sanctions. They emphasized the importance of enterprise crediting and the lack of the state support. The main issue is the organization of such a state structure that would protect the interests of the industry and could respond quickly to all kinds of changes [7, 8]. Rozanova T.P., Magomedov A.D. support the investigations of other researches that structural changes in the dairy industry should take place with the financial and other kinds of support from the government in order to ensure export-oriented production. Moreover, the complexity of tasks require a new quality of management at all levels of enterprise functioning [9].

However, the problems of the functioning of a sectoral milk dessert market have not been investigated enough by researches. According to the studies, the growth of specialized markets depends on the development of every single industry. Therefore, it is appropriate to elaborate development strategies for any national market within the framework of the global from the position of specialized markets. For instance, the dairy dessert market might exemplify the dairy market or the canned meat market may be considered within the general market of meat products. These issues should be explored at the global level because it is important to determine the place of a particular country in the foreign market of a specific industry.

The purpose of the study is to highlight the peculiarities of the domestic dairy dessert production and to identify competitive strategies of Ukrainian producers at different stages and levels of their operation on the foreign market.

## 3 Research Methodology

This study is based on the concepts of economic theory, economics of industrial markets, trade and industrial organizations, scientific developments of researches, and economists on the dairy production and trade issues. A dialectical method and systematic approach to study the situation on the world market of dairy products have been used.

The author used abstract-logical and economic-statistical methods to determine the theoretical concepts of the enterprise functioning on the milk dessert market. The main tendencies in the market of dairy desserts in Ukraine were identified by using the method of average and relative quantities, and dynamics. An index analysis was applied to summarize the characteristics of changes in the domestic market of

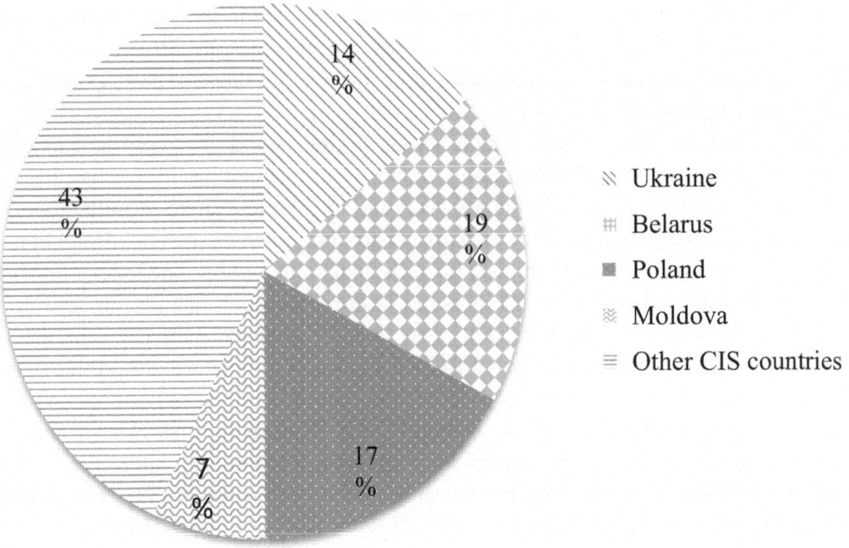


**Fig. 1** Share of Ukrainian products on the dairy dessert market of the CIS. *Source* State Statistics Committee of Ukraine

dairy dessert. Forecast of indicators was developed by means of extrapolation. The connection between certain groups of indicators was described using correlation and regression analysis. A graphic method allowed providing a graphic representation of economic phenomena and processes in time.

## 4 Results

A multi-stage social survey, conducted by the Kyiv Scientific and Research Institute of Food Hygiene of the Ministry of Health, shows that 27% of the Ukrainian population does not consume milk in the pure form as well as other dairy products such as sour cream, kefir, and butter. However, they consume dairy desserts, the basis of production of which is milk, cream, or wet cheese [10]. Such products are in demand both in Ukraine and in the foreign market (Fig. 1).

Milk desserts both in the domestic and in the world market of dairy products are presented by ice cream, condensed milk and cream, flavored milk and milk drinks, cheeses, cream, and sour tomatoes with fillers, yoghurts, glazed cheeses, etc. Today, more than 95 of the large dairy processing plants in Ukraine are engaged in the production of such types of products [11]. Some of enterprises are oriented toward the foreign trade. The most powerful of them have a high market share in Ukraine and have high potential on the global level (Table 1). Thus, an overview of the indicators has shown that the production capacity of holdings and companies such as Danone-Unimilk, Dairy Alliance, Terra Food, and Lustdorf exceeds 100 thousand tons per annum. The average indicator for these enterprises in 2017 was equal to 115.5 thousand tons per year. At the same time, the average share of dairy production in the total volume was 21.25%.

The dairy processing plants, which belong to the second group on the capacity with volumes from 50 to 99 thousand tons per year, reached the following ratings. They produced 71.3 thousand tons per year, and the average share of milk desserts in this volume was equal to 20.57%. This group of companies includes Halychyna, PepsiCo (Wimm-Bill-Dann), Almira, Lactalis, and some others. The latter group of the most powerful dairies that are involved in the production of milk desserts and foreign trade have production capacity less than 50 thousand tons per year. Their average production capacity is equal to 23.7 thousand tons per year and the share of milk dessert products averages 20.22%. An example of such enterprises is the Pervomaisky Dairy Canning Plant, Ternopil Dairy Plant, Kupyansk Dairy Canning Plant, Lubny Dairy Plant, BershadMoloko (Roshen Corporation), and others. The high share of the domestic dairy market of the most powerful dairy factories shows that their products are in demand on the domestic market (7.02, 4.36, and 1.47%, respectively, in groups).

At the same time, most enterprises maintain trade and export relations with the nearest neighbors of Ukraine and have access to the markets of Poland, Russia, Belarus, Azerbaijan, Moldova, Georgia, and other CIS countries. The demand for such types of products as condensed milk, condensed cream, milk mixes for cocktails, ice cream, chocolate and flavored butter, and certain types of yoghurts is very high because of relatively long-term storage.

For over twenty companies, leading in 2017, the average indicators of their activities were calculated. Thus, the total production capacity for dairy products was 58.7 thousand tons. The share of milk desserts in the total milk production capacity and the share of milk desserts exported averaged a number of 20.55 and 6.95%, respectively. Using similar data for dairy producers in the previous years, it is possible to predict the dynamics of some indicators for a group of enterprises using the method of extrapolation based on the average level values.

In the practice of socioeconomic forecasting, the linear-extrapolation approach prevails. On the basis of statistical analysis, tendencies and parameters of the system's development in the past are revealed and transmitted to the future.

To assess the predictive values, the confidence limits (intervals) for the average indicator are calculated:

$$\bar{y} - t_\alpha \cdot S_{\bar{y}} \le y_{i+L} \le \bar{y} + t_\alpha \cdot S_{\bar{y}}, \quad (1)$$

$\bar{y}$ is the mean value of the data series, and $t_\alpha$ is the table value $t$ according to Student's criterion at the given level of significance $\alpha$, $\alpha = \left(\frac{2}{n+1}\right)$, and the given number of degrees of freedom, which is compared $(n-1)$, $t_\alpha = 0.217$;

$S_{\bar{y}}$ is a standard error of an average parameter: $S_{\bar{y}} = \frac{S}{\sqrt{n}}$, where $S = \frac{\sqrt{\sum (y-\bar{y})^2}}{n-1}$

**Table 1** The most powerful dairy processing enterprises of Ukraine, engaged in the production of milk desserts[a]

| Number | Enterprise | Share of the domestic milk market products, % | Production capacity in 2017 | | |
|---|---|---|---|---|---|
| | | | Total for dairy products, thous. tons | Including milk desserts, % of total production capacity | Share of export of milk desserts, % |
| 1 | Danone-Unimilkholding | 8.1 | 133 | 22 | 6 |
| 2 | DairyAllianceHolding | 7.1 | 117 | 24 | 4 |
| 3 | LustdorfCompany | 6.7 | 110 | 19 | 8 |
| 4 | TerraFoodHolding | 6.2 | 102 | 20 | 11 |
| 5 | GaliciaCompany | 5.7 | 93 | 21 | 4 |
| 6 | PepsiCo (Wimm-Bill-Dunn) Company | 5.5 | 91 | 19 | 8 |
| 7 | Company “Almira” | 5.0 | 82 | 13 | 9 |
| 8 | Company “Lactalis” | 3.8 | 62 | 18 | 6 |
| 9 | CJSC “Hercules” | 3.7 | 60 | 28 | 12 |
| 10 | Company “Milkland” | 3.4 | 56 | 30 | 7 |
| 11 | “Pridneprovsky” plant | 3.4 | 55 | 15 | 7 |
| 12 | PervomaiskyDairyCanningPlant | 2.9 | 47 | 23 | 3 |
| 13 | TernopilDairyPlant | 2.2 | 36 | 18 | 8 |
| 14 | Kupyanskydairycanning plant | 1.7 | 27 | 24 | 3 |

(continued)

**Table 1** (continued)

| Number | Enterprise | Share of the domestic milk market products, % | Production capacity in 2017 | | |
|---|---|---|---|---|---|
| | | | Total for dairy products, thous. tons | Including milk desserts, % of total production capacity | Share of export of milk desserts, % |
| 15 | LubnyDairyPlant | 1.6 | 26 | 11 | 11 |
| 16 | Bershadmoloko (Roshen Corporation) | 1.0 | 16 | 24 | 7 |
| 17 | Andrushevskybuttercheesefactory | 1.0 | 16 | 12 | 8 |
| 18 | MolvestHolding | 1.0 | 16 | 16 | 6 |
| 19 | Proviative Company (Proviaut LLC, PJSC “Ichnyansky Milk and Tinned Combine,” agricultural enterprise “Mayak”) | 0.9 | 15 | 23 | 5 |
| 20 | CompanyRud | 0.9 | 14 | 31 | 6 |

[a] *Source* State Statistics Committee of Ukraine
[a] The share of exports of milk desserts is indicated without taking into account the information on some individual markets

**Table 2** Forecast of production capacity of dairy processing plants in Ukraine, the share of dairy desserts in their overall production and the share of dairy dessert products exports[a]

| Year | Average production capacity of the group of leading dairy processing plants of Ukraine, ths. tons | Average share of milk desserts in total production, % | Average share of milk dessert exports, % |
|---|---|---|---|
| 2010 | 69.9 | 16.32 | 5.14 |
| 2011 | 69.2 | 17.65 | 5.96 |
| 2012 | 67.6 | 18.90 | 4.29 |
| 2013 | 68.0 | 19.79 | 6.08 |
| 2014 | 64.5 | 21.25 | 7.11 |
| 2015 | 60.3 | 20.03 | 6.12 |
| 2016 | 58.2 | 20.40 | 6.94 |
| 2017 | 58.7 | 20.55 | 6.95 |
| Forecast dynamics of indicators | | | |
| 2018 | [$64.40 \leq y \leq 64.69$] | [$19.31 \leq y \leq 19.40$] | [$6.04 \leq y \leq 6.10$] |
| 2019 | [$63.75 \leq y \leq 64.00$] | [$19.70 \leq y \leq 19.77$] | [$6.16 \leq y \leq 6.21$] |
| 2020 | [$63.09 \leq y \leq 63.33$] | [$19.97 \leq y \leq 20.03$] | [$6.19 \leq y \leq 6.24$] |

[a] Authors' calculations

Data of Table 2 show that a decrease in average capacity of dairy processing plants causes an increase of market share and the average share of dairy dessert exports. It means the existence of strong correlation between these two factors, and it is reflected by the correlation coefficients that are equal to 0.73 and 0.66, respectively. According to the determination indicators, an increase of 46.7% in the share of dairy desserts in the dairy market depends not only on the growth of dairy production capacities, but also on other factors. The influence of other elements causes the raise of the dairy export share for 56.4%. So, it is possible to capture some parts of the market not only due to the increase in production volumes, but through the implementation of an effective policy of behavior of these enterprises operating means of marketing, price formation according to the quality of products, elimination of competitors, etc.

All enterprises entering the foreign market are certified to meet the requirements of ISO 9001. This is a quality management system, being certified with which proves that the company can produce goods at a stable level of quality and constantly increase it. ISO 22000 guarantees consumers the safety of raw materials, impurities, and components used in production, as well as the safety of the finished product, as all its key factors, including microbiological, chemical, and physical, are under the full control of the enterprise, the standard combines the standard requirements of ISO 9001 and HACCP principles. ISO 14001 is one of environmental management standards. Its implementation at the enterprise means that the manufacturer seeks to minimize its own negative impact on the environment.

Most Ukrainian enterprises provide their employees with the opportunities to upgrade the skills. Also, they collaborate and share their experience with leading producers from Germany, Austria, Bulgaria, Russia, Denmark, and Switzerland. This allows them to diversify high-quality products, which are accepted by consumers in Ukraine and in the CIS countries. Certain types of their products were awarded with diploma at various international exhibitions [12].

In order to identify a strategy to strengthen foreign trade positions and increase the profitability of milk and dessert product exports, it is necessary to take into account the peculiarities of the domestic producer operation on the relevant world market, including advantages and weaknesses of this situation. The following features are noted. Firstly, the problematic nature of the dairy industry development does not provide sufficient amount of milk raw to the producers of milk desserts, as well as other dairy products. Dairy processing enterprises do not work at full capacity. Secondly, only large milk processing plants, which produce a lot of goods for export, have the necessary international certification, modern modernized equipment, skilled and experienced employees can take part in foreign trade. Thirdly, the Ukrainian producers of dairy desserts are competitive enough in terms of price and quality, but are not sufficiently competitive on the assortment and volume of production. The only loyal price policy cannot provide stability in the foreign market. Fourth, excellent flavor properties of domestic dairy desserts might not guarantee high-quality standards. Entrepreneurs may continue to use too low-quality milk and fat substitutes, dyes, and nutritional supplements, taking into account minor regulatory constraints on this issue. Fifth, domestic producers of dairy desserts passively implement international marketing strategies. The dairy and dessert sector of the international market is not sufficiently researched by specialists.

Taking into account analytical data and features of functioning of Ukrainian producers in the domestic and foreign markets of dairy desserts, several strategic directions for their further development and promotion to the foreign market should be implemented. The producers have to:

- respond adequately to the consumer needs, widening the product range;
- involve productive diversification in light of foreign experience;
- adhere to the requirements of international production standards and maintain a high product quality;
- examine the world market of dairy desserts, finding additional sales channels and establishing partnerships with foreign consumers;
- require and consume high-quality products and support of domestic producers by customers;
- improve of the regulatory framework for the milk processing enterprises by the state. For example, it is necessary to enact a law prohibiting of the usage of palm oil or certain types of food additives that are not allowed in EU countries, but are still used in Ukraine,
- strengthen the quality control on the domestic market of dairy products and introduction of penalties and other administrative sanctions, etc.

## 5 Conclusions

In the view of the aforesaid, Ukrainian producers of dairy desserts have currently taken an ambivalent position on the world market. On the one hand, some domestic dairy processing plants have significant prospects for further development and entry to new segments of this market. On the other hand, the other group of enterprises is struggling to keep their shares on the domestic market. This is caused by the crisis in the livestock sector and some problems in the development the dairy product subcomplex. Such a principle difference is demonstrated by a number of functioning peculiarities of the dairy dessert domestic producers. Implementation of the strategies for the development of enterprises that are or are planning to become "players" of the world market of dairy products, including desserts, should be done on both consumer and state levels. At the same time, the history of each company should begin with an ambitious dream of creating a new and modernized dairy business that meets the needs of both the current market and the policy of European integration adopted by Ukraine.

## References


1. Gawaza Ye, G.: The world market for milk and dairy products: trends and prospects for Ukraine. Econ. AIC **7**, 106–113 (2015)
2. Galushko, V.P.: Foreign systems for ensuring the quality of milk in the production chain and the possibility of their use in Ukraine. Econ. Agroind. Complex **3**, 137–142 (2011)
3. Ladyka, V.: World dairy market: state and prospects. Propozytsiya **3**, 42–43; **4**, 31–33 (2009)
4. The World Dairy Market, 2011–2013.: Pisheprom of Ukraine **18**(226), 13–17 (2013)
5. Novak, I.M.: Trends in the development of the global milk market and the experience of international trade in foreign markets of domestic dairy product manufacturers. Innov. Econ. **10**, 98–102 (2012)
6. Tivonchuk, C.V.: The world market for milk and dairy products: features of formation and development trends. Bull. Agrarian Sci. Black Sea Reg. **2**, 57–64 (2014)
7. FAO: World Milk Market Grows [Virtual Resource]. Available at: http://www.agronews.ru/NewsP.php?NId=58713
8. Baboshko, O.V., Veroevets, O.A.: Problems and prospects of the russian market of dairy products in the conditions of economic sanctions. Siberian Trade Econ. J. **2**(20), 98–101 (2015)
9. Rozanova, T.P.: The market of milk and dairy products in Russia: trends and prospects. Vestn. REU **2**, 45–54 (2013)
10. AGRO NEWS (Virtual Resource). Available at: http://www.agronews.us/node/24090
11. OECD-FAO Agricultural Outlook 2014–2023. OECD Publishing, Paris, 329 p (2014)
12. The official site of Food and agriculture organization of the United Nations, statistics division (Virtual Resource). Available at: http://faostat3.fao.org. Accessed 28 Apr 2015

# Cognitive Modeling in the Regional Strategic Management

**Olha Nazarova, Elena Shevchuk, Svitlana Plotnichenko and Nonna Surzhenko**

## 1 Introduction

Despite the significant research results achieved by scientists, important aspects of the content and features of the development of a regional development strategy require special analysis and study. Basically, planning is based on the forecast of regional development programs, individual standards and indicators. Undiscovered completely in research remain questions of organizational and methodological formation of a system of indicative planning at the regional level. Solving these problems will allow complementing the system of strategic management and planning with regional economies. Cognitive modeling is intended for structuring, analyzing, and making management decisions in complex and uncertain situations (geopolitical, domestic political, military, etc.), in the absence of quantitative or statistical information about the processes taking place in such situations. Cognitive modeling for regional development will provide an opportunity to identify key factors: target, leverage and indicators, analysis and forecasting that will allow to manage the system—regional development.

### *1.1 Analysis of Recent Research*

Problems of regional planning and management are explored in the works of T. Avdeeva, I. Aleksandrov, N. Baransky, O. Golosov, N. Ketova, Y. Kolesnikova, A.

O. Nazarova (✉)
Department Higher Mathematics and Physics, Tavria State Agrotechnological University, B. Khmelnistskiy ave 18, Melitopol 72312, Ukraine
e-mail: nazarova777o@gmail.com

E. Shevchuk · S. Plotnichenko · N. Surzhenko
Department of Management, Tavria State Agrotechnological University, B. Khmelnistskiy ave 18, Melitopol 72312, Ukraine

© Springer Nature Switzerland AG 2019

V. Nadykto (ed.), *Modern Development Paths of Agricultural Production*,
https://doi.org/10.1007/978-3-030-14918-5_48

Morozova, V. Nemchinova, N. Nekrasova, V. Ovchinnikov, R. Popova, A. Probst, J. Feigin, V. Filipov, and R. Schnipper. The papers indicate the specifics of socioeconomic system management at the regional level, analyze and present possible regional development strategies, and characterize the main factors of development. There are theoretical and methodological approaches to indicative planning in modern conditions.

The problems of indicative planning are considered in the works of foreign economists L. Volras, J. Hicks, G. Debre, and others. Particularly noteworthy are the theoretical approaches of Ye. A. Yasin, who conceptually substantiated the essence of indicative planning at the state level.

The most advanced in management technology is currently considered cognitive modeling, the essence of which boils down to the fact that the definition and solution of problems are based on internal and external experience.

The technology of cognitive modeling is based on determining possible and rational ways of managing the situation, acting proactively and not bringing potentially dangerous situations in conflict situations, and in case of occurrence making a rational decision in the interests of economic entities [2, 3].

## 2 Statement on the Purpose and Objectives of the Study

The goal is to develop a cognitive map for factors of regional management, their analysis, scripting and interpreting results building a curvilinear relationship to predict results in the region of the Zaporozhye.

To achieve the goal, the following tasks were set and solved in the robot:

- identification of key factors, building a cognitive map;
- on the basis of the braking and acceleration matrices, the target factors, levers, and indicators of the regional strategic development system are determined;
- built scenarios for the specified plan of experiments;
- analysis of situations (scenarios) was made, and the optimal scenarios for the development of the region were highlighted;
- for optimal scenarios, curvilinear dependencies were obtained for predicting target factors, and surfaces (response functions) for prediction were constructed;
- given recommendations for the strategic development of the region.

## 3 Solution

In regional management in various areas: economic, social and environmental, there are changes in the parameters of the system. This changes can be quantitative and qualitative. Quantitative changes indicate, as a rule, economic growth. Qualitative ones lead to changes in the structural characteristics and content of the development itself [10, 11].

However, one should not exaggerate the significance of only the economic factors of the development of the region. Sociocultural, demographic, environmental and other factors shape the environment where human life takes place—the most valuable capital of any organization and democratic society.

Thus, the socioeconomic development of the region has many aspects. The most important aspects, in our opinion, for the areas are the following:

- the growth of production and income and hence the investment attractiveness;
- increasing incomes of the population, improving the quality of medical services, increasing the level of education—and this means positive changes in the demographic situation and creating comfortable living conditions.

The strategic development of region is a complex and multi-purpose process. Its content is due to the individual characteristics of region. Therefore, it is very important to analyze the internal and external environment of the object of study. Elements of such an analysis are contained in the following groups of factors: natural, demographic, economic-geographical, and economic [7, 8].

In today's conditions, objectively related to market relations, the role and significance of the regional economy and management have increased. Most of the problems raised during the reform process are interconnected with the influence of various regional factors and conditions.

Regional management as a kind of special management is a set of principles, methods, forms, and means of influencing the economic activity of the region. Taking into account the current conditions, regional management is the management of socioeconomic processes in the region in the conditions of transition of its economy to market relations.

Regional development refers to the change in the socioeconomic structure of the region, which ensures economical use of resources and maximum needs of the territory. In this approach, significant changes are made in the main management methods and its assessment [9].

There are nine main factors of the regional management system:

Analysis (see Fig. 1) allows us to identify the main factors necessary for constructing a cognitive model and describing the system of regional economic development [1, 4, 5]. In the obtained cognitive map of factors, the sign "+" means a direct change of two related factors and the sign "–", respectively, the inverse change of two related factors.

The matrices of the interconnection of the factors are represented by the adjacency matrix:

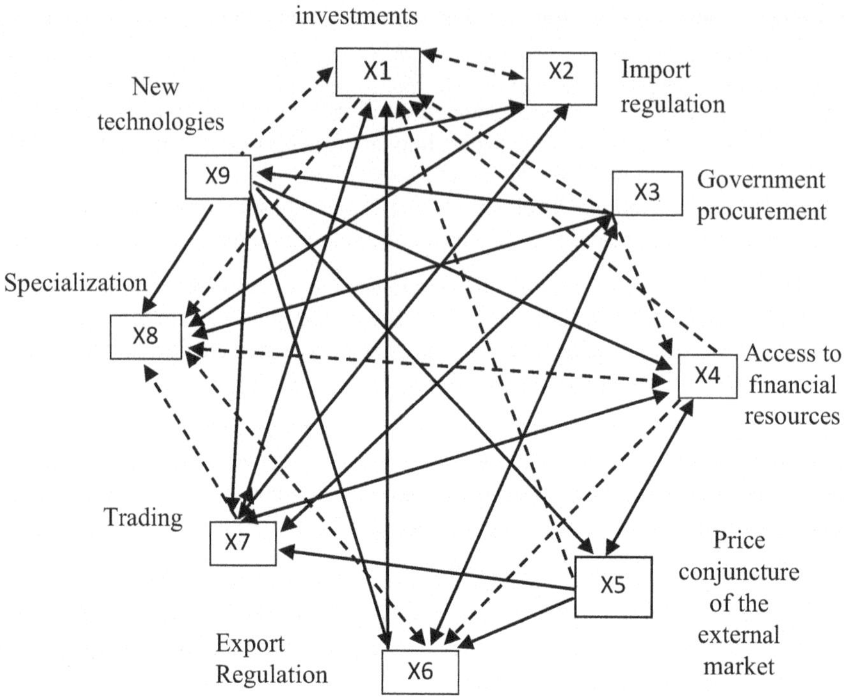


**Fig. 1** Cognitive model of interaction of factors of the system development of the region

$$A := \begin{pmatrix} 0 & -1 & 0 & 0 & 0 & 0 & 0 & -1 & -1 \\ -1 & 0 & 0 & 0 & 0 & 0 & 1 & 1 & 1 \\ -1 & 0 & 0 & -1 & 0 & 1 & 1 & 1 & 1 \\ -1 & 0 & 0 & 0 & 1 & -1 & 1 & -1 & 1 \\ -1 & 0 & 1 & 1 & 0 & 1 & 1 & -1 & 0 \\ 1 & 0 & 1 & 0 & 0 & 0 & 0 & -1 & 1 \\ 1 & 1 & 1 & 1 & 0 & 0 & 0 & -1 & 0 \\ 0 & 0 & 0 & -1 & 0 & 0 & 0 & 0 & 1 \\ 0 & 1 & 0 & 1 & 1 & 1 & 1 & 1 & 0 \end{pmatrix}$$

where:

1 increase (fall) of factor $X_i$, leads to an increase (fall) $X_j$;
$-1$ growth (fall) of factor $X_i$, entails a fall (growth) of $X_j$;
0 the relationship between the factors $X_i$ and $X_j$ is missing or weak

From the basic factors on the basis of the acceleration and deceleration matrices, the control factors through which control actions are fed into the system are selected. Thus, the factors can be divided into groups (see Fig. 2):

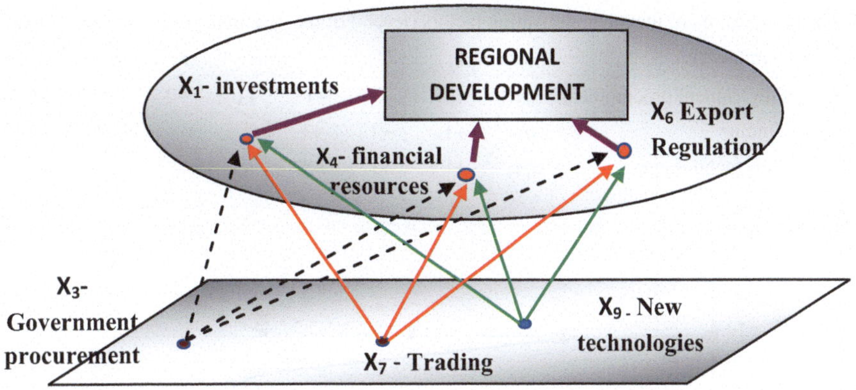


**Fig. 2** Diagram of the influence of the main factors of the development of the region [2, 6]

- **Target factors**—the change of which is the purpose of system management: $X_1$—investments, $X_4$—financial resources, $X_6$ Export Regulation.
- **Factors—levers** (controllers)—potentially possible levers of action on the situation: $X_3$—Government procurement, $X_7$—Trading, $X_9$—New technologies.

**Factors—indicators** that display and explain the development and explain the development of processes in the problem situation: $X_2$—Import regulation, $X_5$—Price conjuncture of the external market, $X_8$—Specialization.

The scenario for the development of a region is proposed to specify a set of external and internal conditions that influence and largely determine the parameters of the regional economy. The study of these conditions, the determination of their influence on the general course of development of regional systems, is an important element in the process of strategic choice. It is especially important to take into account the interrelations of the regional economy with the rest of the economy of Ukraine. All this means that for such calculations a special tool is needed—cognitive modeling and scenario analysis.

Scenarios make it possible to analyze and plan non-standard situations, determine the conditions under which a favorable or unfavorable situation may arise, and help to assess and determine key factors that are intensively affecting the system.

To obtain scenarios, the processes of disturbance propagation on graph $G$ along a certain route $M$ are considered. Modeling impulse processes (Fig. 1) will provide an opportunity to identify key factors and build scenarios for various conditions.

The value of impulse in vertices xi at the moment t is described by the function:

$$U_i(t+1) = U_i(t) + \sum_{j=1}^{n} f(V_j, V_i) p_j(t) \tag{1}$$

where $p_j\ (t)$ is the vector of change in the values of the parameters of the vertices of a weighted non-oriented graph with the corresponding simulation factors.

**Table 1** Indicators of the main motivating factors (levers) of the system the development region

| Levers of influence | | Impul. | 1 | 2 | 3 | 4 | 5 | 6 | 7 |
|---|---|---|---|---|---|---|---|---|---|
| $V_3$ | Government procurement | $q_3$ | | −1 | +1 | +1 | −1 | | +1 |
| $V_7$ | Tradin | $q_7$ | +1 | −1 | +1 | | | −1 | +1 |
| $V_9$ | New technologies | $q_9$ | +1 | +1 | −1 | −1 | +1 | +1 | +1 |

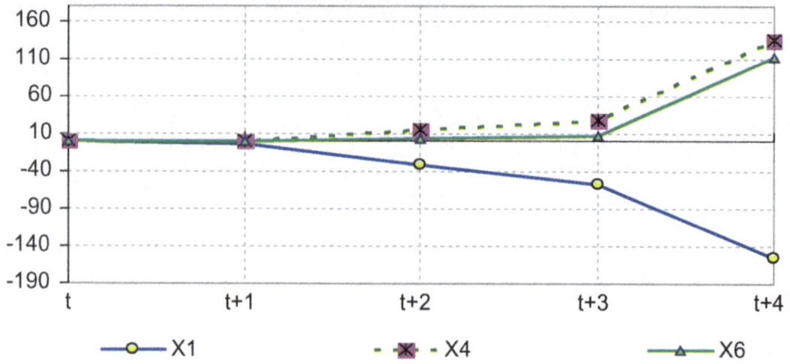


**Fig. 3** Графік внесення збурень $q_3 = +1, q_7 = +1, q_9 = +1$ ($X_3$—Government procurement, $X_7$—Tradin, $X_9$—New technologies)

$t$ is the take of modeling $t = 0, 1, 2, 3,\ldots n$, reflecting the sequence of changes in the state of the regional development system;

$U_i\ (t + 1)$, $U_i\ (t)$ are the values of the factor $i$ on the simulation takt $t + 1$ and $t$ accordingly;

$f(V_j, V_i)$, weight factor influence $x_j$ on $x_i$, $j \in n$;

$N$ is the number of factors directly influencing factor $x_i$.

When implementing strategic planning, an experiment plan is selected (Table 1).

From the scenarios discussed, the following is noted (see Fig. 3), on which the influence of factors-levers significantly changes the indicators of the target factors of the system development the region.

Reflection on the graphs is the result of the scenario, where the axis of abscissa is the takts, the ordinate axis is the value of the factors in the conventional units.

Scenario, where the mutual influence of the factors leads to the best indicators factors of the system, involves further study of the factors with the construction of the predicted dependence (linear or nonlinear).

The viewed scenarios allow us to conclude the main factors that influence the changes in the system are the development of the region are: $X_3$—Government procurement, $X_7$—Tradin, $X_9$—New technologies.

In this scenario, it can be noted that the improvement of indicators on these factors leads to an increase in exports by almost 5 times after 3 takt, an increase in financial resources by 5–8 times, while investment is significantly reduced. This can be explained by the fact that the region is able to independently carry out its projects and financing. With this combination of factors, the region develops independently.

**Table 2** Indicators of the main factors (levers) of the development, the Zaporizhia region, million UAH

| Year | Export regulation | Access to financial resources | Investments | New technologies |
|---|---|---|---|---|
| 2010 | 1156.6 | 2458.9 | 7963.8 | 8107.1 |
| 2011 | 4,151,226 | 5344.9 | 6676.5 | 8513.4 |
| 2012 | 4,004,800 | 3658.1 | 7204.4 | 9419.9 |
| 2013 | 3,678,507 | 6048.3 | 6838.8 | 10,248.5 |
| 2014 | 3,730,224 | −9079.9 | 7034.5 | 9487.5 |
| 2015 | 2,931,027 | −2724.3 | 7794.3 | 11,003.6 |
| 2016 | 2,292,805 | 13,799.3 | 11,039.7 | 11,530.7 |
| 2017 | 2,980,861 | 20,951.6 | 8549.9 | 13,379.3 |

It is obvious that technical progress in science and technology will lead to an increase in indicators for the Zaporizhia region (Table 2).

As a result of the scenario analysis, three main motivating factors were identified that influence the level of development of the region and require consideration and analysis.

With the help of correlation regression analysis, a correlation matrix was obtained, which confirms a strong relationship between the factors and the result.

The second-order polynomial for factors $Y_1$- Access to financial resources (2) and $Y_2$ - Export regulation (3) has the form:

$$y1 = -0.00001 + 55.283x_1 - 18.671x_2 + 0.0057x_2^2 - 0.014x_1x_2 + 0.006x_2^2 \quad (2)$$

$$y2 = -0.0000004 + 14,634.369x_1 - 625.335x_2 + 0.753x_2^2 - 2.523x_1x_2 + 0.922x_2^2 \quad (3)$$

where $x_1$—Government procurement, $x_2$—New technologies.

The obtained regression equation allows us to make a reliable forecast for 92–96% of data ($R^2 = 0.92, R^2 = 0.96$). Fisher kriteriy, the model is adequate to the data under investigation. The considered scenarios allow us to make a forecast for the Zaporizhia region (data for analysis 2010–2018) for the obtained dependence (2) for the factor.

As a result of the scenario analysis, three main motivating factors. Access to financial resources (see Fig. 4a) and Export Regulation (see Fig. 4b) were identified that influence the level of development of the region and require consideration and analysis.

The viewed scenarios allow us to conclude that the main factors that influence the changes in the system are the development of the region are: $X_3$—Government procurement, $X_7$—Tradin, $X_9$—New technologies.

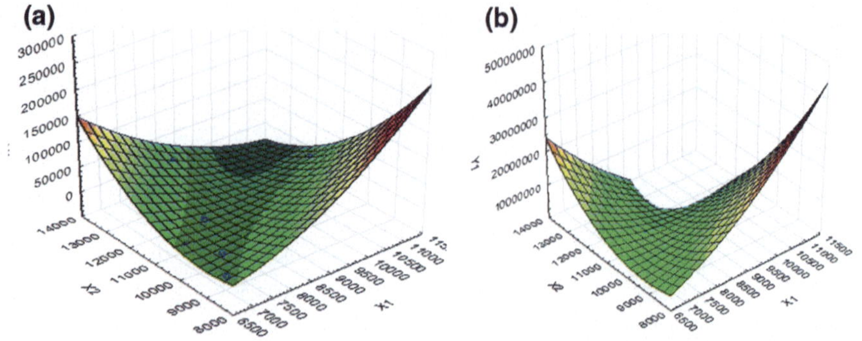


**Fig. 4** Impulse schedule $q_3 = +1, q_7 = +1, q_9 = +1$ ($X_3$—Government procurement, $X_7$—Tradin, $X_9$—New technologies)

For effective socioeconomic development of the region, the formation and development of new industries, and industries at the high-tech level, it is necessary to increase the technical level of industries and sectors of the economy.

On the basis of the obtained results, we can conclude that when managing a region, the manager's strategy should be aimed at modernizing the regional economy, that is, developing industries with high technologies: the sphere of information technologies, nanotechnology, and others, as well as staff development.

## 4 Conclusion

A cognitive model for managing regional development factors is developed; the main factors are determined. On the basis of scenario analysis, strategies for regional management were obtained. For the target factors, dependencies were obtained for forecasting the regional socioeconomic system.

Successful implementation of strategies in the regions is always characterized by the skillful application of a number of these approaches, often in a variety of combinations. However, successful implementation is characterized by two peculiarities: Managers use these approaches, taking into account their advantages and disadvantages, and realistically assess the situation.

## References

1. Axelrod, R.: The Structure of Decision: Cognitive Maps of Political Elites. Princeton University Press, Princeton (1976)
2. Kornoushenko, Y.K.: Upravleniye protsessami v slaboformalizovannykh sredakh pri stabilizatsii grafovykh modeley sredy.// Trudy Instituta problem upravleniya. Tom 2, pp. 82–95 (1999)

3. Kulinich, A.A.: Metodologiya kognitivnogo modelirovaniya slozhnykh plokho opredelennykh situatsiy. Vtoraya mezhdunarodnaya konferentsiya po problemam upravleniya (Moskva, Institut problem upravleniya, 17–19 iyunya 2003 g.). Izbrannyye trudy, pp. 219–227
4. Nazarova, O.P.: Kognitivnoe modelirovanie regional'noj jekonomiki IV Mizhnarodna naukovo-praktichna konferencija «Derzhava i pravo: problemi stanovlennja i strategija rozvitku», Sumi, pp 549–552 (2011)
5. Nazarova, O.P.: Kognitivnyj podhod k formirovaniju sistemy rychagov regional'nogo razvitija. Mizhnarodna nau-kova-praktichna konferencii "Osnovni naprjami innovacijnogo rozvitku virobnictva ta pererobki produkcii APK", L'viv—Ivano-Frankivs'k, pp 118–122 (2011)
6. Nazarova, O.P.: Teoretiko—metodichni osnovi riziku menedzhmenta v upravlins'kih rishennjah. Zbirnik naukovih prac' Tavrijs'kogo derzhavnogo agrotehnologichnogo universitetu (ekonomichni nauki) №1(29)—Melitopol', pp. 105–109 (2015)
7. Plotnichenko, S.R.: Operational management in the enterprise management system/S.R. Plotnichenko. Collection of scientific works of TDATU (economic sciences). In: Sinyaya, L.V. (eds) Melitopol: View Melitopol typography "Lux", №1 (30) pp. 59–64 (2016)
8. Plotnichenko, S.R.: Analysis of transport systems modeling methods. O.P. Nazarova, SR Plotnichenko, NV Surzhenko. Collection of scientific works of TDATU (economic sciences). In Sinyaeva, L.V. (ed) Melitopol: View Melitopol typography "Lux", № 1 (30), pp. 53–55 (2017)
9. Surzhenko, N.V.: Obosnovaniye sovremennykh aspektov strategicheskogo planirovaniya predpriyatiy. Sbornik nauchnykh trudov Tavricheskogo gosudarstvennogo agrotekhnologicheskogo universiteta (ekonomicheskiye nauki). Pod red. L.V. Sinyayevoi.—Melitopol': Izd-vo Melitopol'skaya tipografiya«Lyuks», №1 (30), pp. 73–76 (2016)
10. Shevchuk, O.Y.: Konkurentna stratehiya yak instrument marketynhovoho menedzhmentu. Zbirnyk naukovykh prats′ TDATU seriya «Ekonomichni nauky, №3 (32), pp. 139–146 (2016)
11. Shevchuk, O.Y.: Stratehichne upravlinnya sotsial′no-ekonomichnoho rozvytku malykh ta serednikh mist: tendentsiyi ta problemy. Zbirnyk naukovykh prats′ TDATU seriya «Ekonomichni nauky» **3**(35), 168–174 (2017)

# Managing Competitiveness of the Enterprise: Theoretical-Methodological Aspect

**Svitlana Nesterenko, Svitlana Rozumenko, Oleg Kravets and Liudmyla Redko**

## 1 Introduction

Formulation of the problem. There is a process of development, the promotion of competition among commodity producers in various fields, under different economic conditions during the period of the establishment of market relations in Ukraine. The transformation of organizational and legal mechanisms of business management of enterprises is impossible without the active and well-founded policy of regulation of property relations.

The current state of production-economic relations in Ukraine is characterized by a large number of contradictions. The complicated processes of reforming the economic relations that are taking place in our country and the development of new market conditions require new approaches to the organization of production and economic activity, principles and methods of production management, as the fundamental basis for ensuring the competitiveness and the relative competitiveness.

Management of enterprise competitiveness is a certain aspect of enterprise management, aimed at the formation, development and implementation of competitive advantages and ensuring the viability of the enterprise as a subject of economic competition. These circumstances determine the choice and relevance of the research topic.

### 1.1 Analysis of Recent Research

The foreign scholars, Smith [1], Ricardo [2] and Mill [3], have devoted many works to the study of the theoretical foundations of the competitiveness of enterprises.

S. Nesterenko (✉) · S. Rozumenko · O. Kravets · L. Redko
Department of Management, Tavria State Agrotechnological University, B. Khmelnistskiy ave 18, Melitopol 72312, Ukraine
e-mail: svitlana.nesterenko@tsatu.edu.ua

© Springer Nature Switzerland AG 2019

V. Nadykto (ed.), *Modern Development Paths of Agricultural Production*,
https://doi.org/10.1007/978-3-030-14918-5_49

They considered the competition mainly in terms of price competition. The study of various forms of enterprise competition was reflected in the works of Schumpeter [4], Keynes [5], Hayek [6], Fathundinov [7, 8] Chamberlin [9] and Erhard [10]. To the issue of the competition of enterprises among representatives of managerial sciences has been given great attention in the writings of Ansoff [11], Drucker [12], Kotler [13], Porter [14, 15] Gudzinsky [16], Sudomir [17] and Ushachev [18].

## 2 Statement on the Purpose and Objectives of the Study

The purpose of the article is to substantiate theoretical and methodological aspects of enterprise competitiveness management.

The achieving of this purpose influences on solving the following tasks:

- substantiation of system nature and conceptual control apparatus;
- disclosure of managerial and economic foundations of the functional integrity model of a competitive enterprise; and
- substantiation of factors and construction of the mechanism for managing the competitiveness of enterprises.

## 3 Solution

Presentation of the main material. The achieving of the competitiveness, that is, the main purpose of economic operators with a total participation, can be accomplished through consistency in the implementation of the plans. The use of the term "management" in relation to the development of enterprises seems obvious and ordinary, but it is difficult to give a general and accurate definition. Therefore, the concept of "management" from the point of view of the economy, sociology and philosophy is multilateral and multi-valued interpretation. According to well-known economists B. A. Raizberg and R. A. Fatkhutdinova, tangentially to the economic processes of "Management is a conscious action of a person on various objects and processes in the environment, carried out with the aim of receiving a certain orientation and achieving the desired results" [19, p. 20]. With the view of the above mentioned authors one can agree, because they accurately and clearly reveal the essence of management, due to the harmonious unity that affects the solution of practical problems.

Management is a common function of the organized systems inherent in any society, and an objective, natural process, that is, inextricably linked with the development of productive forces of society and social change. Systemic nature, social character of labor, human relations in the process of production and the exchange of products of material activity cause the need for action of subjects of management to objects for the reduction of the structure, qualitative specification, the implementation of the program and the activities of economic entities.

It is accepted to distinguish internal and external management in the management theory. Internal management is a set of elements of a controlled system, which is formed within the enterprise itself. External management is a set of elements that are not subject to the internal control system of the enterprise, since they are formed outside.

There were always two components in the management: the subject that manages and the managed object. The first is an active part of the entire managerial system, which carries out an action and in a such way manages the passive part of the system.

Therefore, the managerial and economic basis of the model of functional integrity of a competitive enterprise is a set of basic laws and principles that underlies the choice of functions in accordance with the levels of management and integrity of enterprises that enable the most effective achievement of the purposes of the enterprise. They are considered in management, economic theory, organization theory, sociology and other sciences.

The functional integrity of the enterprise is complex-organized communications and management in the economic subjects of the market economy that are organized in their essence, quality, purpose and multifunctional because of the complexity and ambiguity of organizational systems. The processes of the relationship between the levels of management, as well as the division and cooperation of labor in the enterprise as a system of functions, are conditioned by objective laws and are based on certain principles of the integrity of the enterprise. Under the management of the enterprise, in authors opinion is the allocation of functional areas and the implementation of the functions of the enterprise, which provide conditions for effective work of employees on the enterprise. In the management of an economic object, in our opinion, it should be more correct to speak not about a single one, but about a prevailing management method or a combination of methods that are aimed at achieving the purpose [20].

The main system-forming criterion is the purposefulness that is associated with an aim. The purpose is one of the basic concepts in the systematic approach to identifying the relationship between the theory of management and the category. Purposes can be different. They depend on the external and internal factors that influence the object of management. The theoretical significance for the analysis and improvement of the functional integrity of the enterprise as an important part of the science of management in market conditions are the knowledge and detailed consideration of the constituent parts of its main components (principles, functions and other elements) and the development of a methodology for its quantitative and qualitative basis. Optimization of functional integrity is explained by:

the effectiveness of technical, technological, social, economic, information and other processes of operation in the enterprise; and

the nature of managerial work and its quality: the knowledge of the leaders of causal relationships and relations within and outside the enterprise, the orderly knowledge of management of economic entities in marketing high-tech and highly informative conditions, as well as practical skills and skills in the division of labor and controlling sphere over specialized lines.

The system of functional integrity of the enterprise is a set of elements that are in communication with each other, form a certain unity as a result of the coordination and management of the enterprise. The functional integrity of a business entity in a market environment "grows" for a certain purpose and principles, functions and norms, methods and tools for delegation of authority and responsibility between employees of the enterprise. It acts as a mean of transforming the theory of management into a practical mean of harmonizing the development of a staff with the conditions of reproduction, where purposes, values and norms of management reveal the essence of employee activity. The functional integrity of an enterprise forms two aspects: structural and managerial.

Both aspects are considered sufficiently in the scientific literature, but the question remains to assess their qualitative characteristics, that is, the methodical approach to determining the effectiveness of their composition and application in the situation of orientation of enterprises to the external environment, in particular, on the consumer.

In this case, the process character of the functional integrity of the enterprise is highlighted when professionally trained managers form an enterprise and manage it through the setting of purposes and the development of methods and techniques that lead to logical relationships between managers and employees.

Many scholars [20] consider the systematic approach to the functional integrity of the enterprise in the complex the empirical indicators of enterprise management that takes into account the development of technologies, information flows, as well as relationships as links between people and groups of people, which take different part in the economic, spiritual, political life of the enterprise. Changes in the nature of relationships are conditioned by changes in the social situation and the social form of the interacting. In contrast to the process approach that scientifically explains the nature of managerial work and causal relationships are presented in the form of the development of theories and mechanisms, we suggest using the system approach to functional optimization. Yes, this is a complicated process, that is, based on scientific knowledge, it takes into account the system of complex relationships, and it is influenced by numerous and varied factors of the external and internal environment.

The main purposes of object management of economic activity should be called the most typical: the aim of sustaining system; the aim of leaving the undesirable state of the system; and the aims of development of the system (Table 1) [21].

The purpose of the system is implemented through the functions of management, which represents the essence of the management process. Building an empirical model of enterprise functional integrity will give the possibility, with scientific substantiation, to judge the real nature and content of changes, to make practical recommendations for adjusting enterprise management in accordance with the main objectives and goals of society. Comparison of the changes which are taking place in the practice of functional structures of enterprises construction will allow to manage the economic entities, plan and predict the dynamics of their development [22].

Development is the process of turning from one state to another, in-depth moving from the old qualitative state to a new quality state, from simple to complex, and from lower to higher [20].

**Table 1** Managing the competitiveness of enterprises as a system

| Subsystem | General characterization |
|---|---|
| 1 | 2 |
| Justification of the target orientation of the enterprise as the main system of creating a criterion | Ensure effective functioning of enterprises, their strategic development; real and potential competitiveness |
| Target focus on managing current and strategic competitiveness | 1. Formation of systematic principles for the implementation of the target orientation of enterprises by correspondent goals and objectives of their strategic development of competitive development<br>2. Formation of management systems for adapted goals that should be achieved and tasks that should be solved in the current and strategic terms |
| Objects of competitiveness management | – Goals of functioning and development of socioeconomic systems<br>– Tasks to strengthen the competitiveness of enterprises<br>– Factors for providing the enterprise competitiveness<br>– Strategies of enterprise development<br>– Competitive advantages, priorities<br>– Stages, techniques<br>– Potentials<br>– Marking communications<br>– Consumers<br>– Functionally protecting subsystems<br>– Competitors<br>– Products, goods<br>– Human, technical, financial and other capita<br>– Components of organizational development of enterprises (values, organizational culture, culture of organization, management profile, organizational behavior)<br>– Employees of enterprises, staff<br>– Centers of responsibility |
| Subjects of competitiveness management | – Linear and functional management control<br>– Staff of enterprises<br>– Formal and informal institutions |

(continued)

**Table 1** (continued)

| Subsystem | General characterization |
|---|---|
| 1 | 2 |
| Methodological tools for managing competitiveness of enterprises | Theories: competition, competitiveness, organizational systems, flexibility, marketing communications, management and others<br>Means of socioeconomic systems and principles of competitive potential formation, systems of competitiveness, methodological tools for research and evaluation of enterprises activity |
| The mechanisms of enterprise competitiveness management | The current and operating system of organizational, economic, socio-psychological and legal influence target objects when the providing enterprise competitiveness is achieved |
| Providing of resources | Informational, financial, biological, technical, personnel, mental, organizational, innovative |

In economic literature, the main factors that cause instability of the Ukrainian economy are quite well developed (Fig. 1).

From the above mentioned, it turns out that the transition to market relations leads to the emergence of a number of problems in the management which have not been investigated by the scientists or practitioners in our country.

The functional integrity of empirical indicators system grows from the signs of socioeconomic sphere, needs and behavior of the population, which is based on the norms and values inherent in a particular society. Obviously in this case, it becomes the means of enterprise economic development and the maintenance of the harmony of its development on the conditions of reconstitution, where the social values and norms do not conflict with the economic goals of the enterprise.

The following methodological approaches can be founded in the basis of allocation and systematization of functional indicators:

- clear allocation in the sphere of certain function of optimizing the enterprise structural integrity as a system object, in which the economic and social indicators should "work", ensuring an adequate reflection of the statics and dynamics of the manifestations of the sphere constituent; the essence of the function;
- the allocation of the system of economic and social indicators, their differentiation according to the nature of the objects measured by them;
- the allocation of indicators system-forming basis, the definition of a methodological, instrumental approach to the measurement of management functions;
- accounting for the dynamics and measurement of precisely economic and social indicators; and
- the definition of the main and additional methods of obtaining data.

The economic model of managing the functional integrity of an enterprise with the allocation of four components (production, social, marketing, cultural and spiritual) takes into account not only the features of a market economy and the purpose of enterprise management; its peculiarity lies in the fact that the primary importance is given to the function of activating human potential in the management of the enterprise.

Obviously, it is difficult to find absolutely empirical equivalents of functioning economic object model. Such model is intended to integrate a system set of indicators, and this set is represented in the model isomorphically by its most important characteristics: both the actual object of research and the theoretical substantive ideas about it. The model is designed to implement the principle of integrity of management functions. Only this factor will ensure the consistency of scientific purpose and method. Functional models are the organizations that are organized, regulated, dynamic, all components of which interact and seek to achieve in general results, useful for the system.

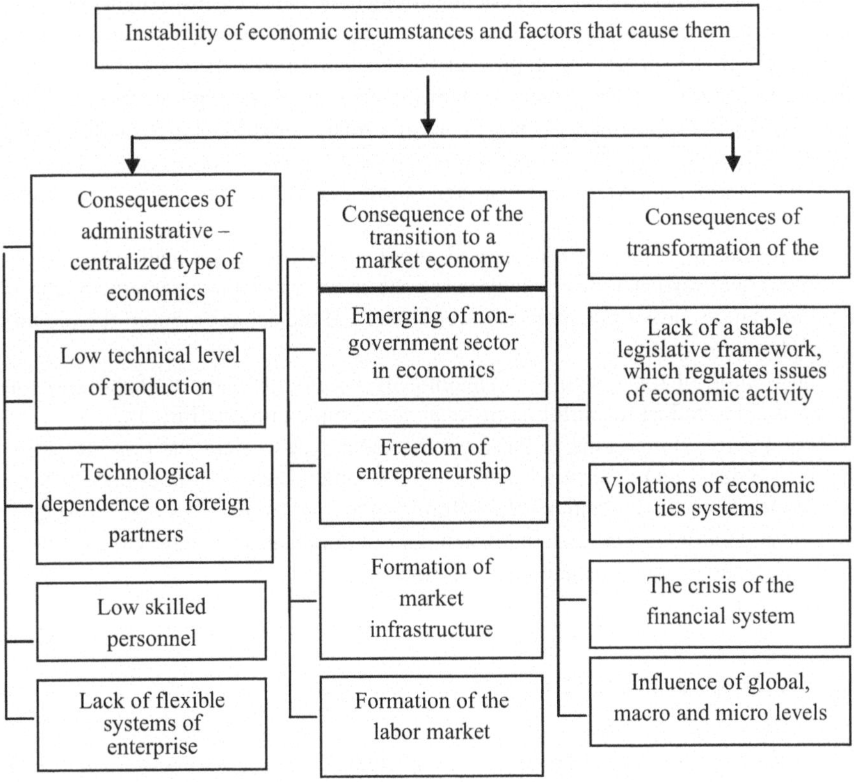


**Fig. 1** Instability of economic circumstances and factors that cause them

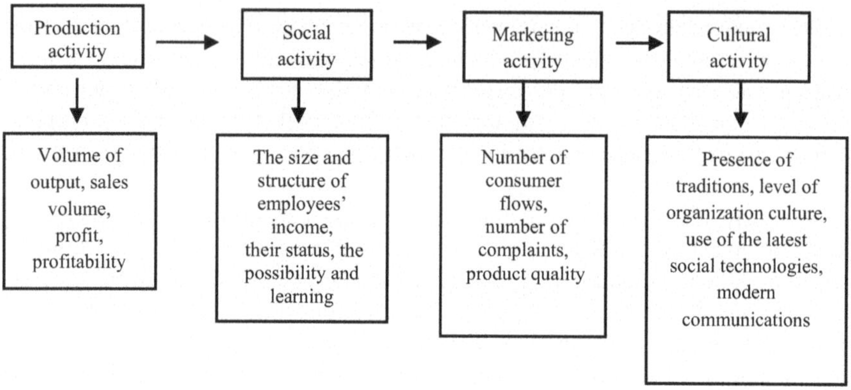


**Fig. 2** System of indicators of the management model of the competitive enterprise functional integrity

Management mechanism as a system is divided into subsystems—organizational, economic, socio-psychological, legal provision. Each of them has, in addition to general ones, its own specific principles, tasks, goals and methods of regulation (self-regulation). Statistic management system includes organizational structures, management structures, managerial personnel and management techniques. The management process connects two main management blocks. This process drives the system in motion, maintaining it in a state of order and providing dynamic development (Fig. 2).

The management process is a targeted activity of the subjects of ownership, power and functional responsibility, which using the methods of management functions implementation, provides solutions to problems and the achievement of the organizational system goals.

Functioning of the system of managing by the competitiveness factors of the enterprise is to reflect its multidimensional and complex nature (Fig. 3).

The effective functioning of these management system elements provides competitive products and enterprise stability in a competitive environment. Achieving management objectives requires the defining of resources, the allocation of stages, elements, functions, organizational structures, communication with the external environment, feedback and management methods.

## 4 Conclusion

Enterprise competitiveness management represents the process of systematic use of management functions in flexible organizational and production structures. On the basis of modern science and technique achievements, the regulation of production processes should be carried out on the principles of a proactive approach to solv-

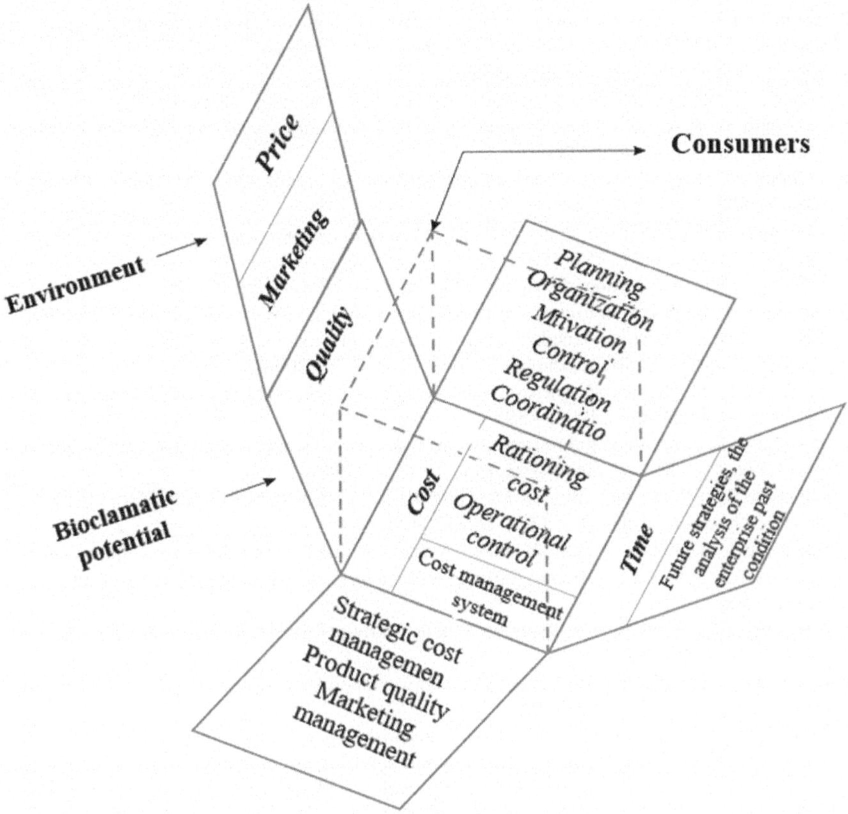


**Fig. 3** System of management enterprise competitiveness factors

ing problematic situations in the integrity of harmonious unity and interaction of factors: managerial, organizational, biological, technological, technical, personnel, financial and economic, innovation and investment, marketing, information and communication and psychological direction, which ensures the effective functioning and development of enterprises in a competitive environment.

## References


1. Smit, A.: Issledovaniya o prirode i prichinah bogatstva narodov. – M.: Ekonomika 237s. 292 (1993)
2. Rikardo, D.: Nachala politicheskoy ekonomii. Antologiya ekonomicheskoy klassiki : v 2-h t. – M.: MP «Ekonom», – T. 1. – S. 397–473 (1993)
3. Mill Dzh, S.: Osnovyi politicheskoy ekonomii. V 3-h tomah.—M.: Progress, T.1. – 495s.; T.2. – 480s.; T.3. – 447s. (Seriya«Ekonomicheskaya myisl Zapada» (1980)

4. Shumpeter Y. Teoriya ekonomicheskogo razvitiya: (Issledovanie predprinimatelskoy pribyili, kapitala, kredita, protsenta i tsikla kon'yunturyi).—M.: Progress. – 456s (1982)
5. Keyns, Dzh, M.: Izbrannyie proizvedeniya
6. Hayek, F.A.: Poznanie, konkurentsiya, svoboda. – SPb: Pnevma.—395s (1999)
7. Fatxundinov, R.A., Osovs'ka Pidruchny'k, G.V.: Upravlinnya konkurentozdatnistyu organizaciyi. K,: «Kondor», 470s (2009)
8. Fathutdinov, R.A.: Upravlenie konkurentosposobnost'yu organizacii. – M. : Izd-vo Eksmo, – 544 s (2004)
9. Chemberlin, E.: Teoriya monopolisticheskoy konkurentsii: per. s angl. pod red. Yu. Ya. Olevicha. – M.:Ekonomika, 351s (1996)
10. Erhard, L.: Blagosostoyanie dlya vseh.—M.: Nachala – Press, 108s (1991)
11. Ansoff, I.: Novaya korporativnaya strategiya: per. s angl. I. Ansoff. – M.: Piter 457s (1999)
12. Druker Piter, F.: Praktika menedzhmenta. M.: Izd. dom Vilyams, 397s (2007)
13. Kotler, F.: Marketing. Menedzhment. —SPb.: Piter, .752s (2000)
14. Porter M. E. Mezhdunarodnaya konkurentsiya: per. s angl.; pod red. V. D. Schetinina. – M.: Mezhdunar. otnosheniya, 1993. – 896 s
15. Porter, M.: Konkurentnoe preimuschestvo. Kak dostich' vysokogo rezul'tata i obespechit' ego ustoychivost'. – M. : Al'pina Biznes Buks, – 715 s (2006)
16. Gudzy'ns'ky'j O.D., Sudomy'r S. M., Gurenko T. O. Menedzhment pidpry'yemny'cz'koyi diyal'nosti: navch.posibny'k K.: IPK DSZU, 2011. S
17. Gudzy'ns'ky'j, O.D., Gajdamak, N.V., Sudomy'r, S.M.: Upravlinnya rezul'taty'vnistyu diyal'nosti pidpry'yemstv (teorety'ko-metodologichny'j aspekt): monografiya., T.O. – K.:, 2011. 175 s
18. Ushachev, I.G.: Nuzhnyi novyie podhodyi k agrarnoy politike. APK: ekonomika, upravlenie. – # 1. – S. 4 – 12 (2006)
19. Rayzenberg, B.A., Fathutdinov, R.A.: Upravlenie ekonomikoy: uchebnik. – M.: ZAO "Biznes-shkola Intel-Sintez". – 784 s (1999)
20. Nesterenko, S.A.: Upravlinnya konkurentospromozhnistyu pidpry'yemstv yak sy'stemna cilisnist' Zbirny'k naukovy'x pracz' Tavrijs'kogo derzhavnogo agrotexnichnogo universy'tetu (ekonomichni nauky'). Za red. M.F. Kropy'vka. – Melitopol': Vy'd-vo Melitopol's'ka ty'pografiya Lyuks» , – # 3 (27) S. 88–94 (2014)
21. Nesterenko, S.A.: Upravlinnya konkurentospromozhnistyu gospodaryuyuchy'x sub'yektiv v agrarnij sferi ekonomiky'. Visny'k Xarkivs'kogo nacional'nogo agrarnogo universy'tetu: Zb. nauk. pracz'. Seriya: Ekonomika APK i pry'rodokory'stuvannya. – 2009. – Vy'p. 12. – S. 96–104
22. Nesterenko, S.A.: Zabezpechennya ekvivalentnosti integrovany'x formuvan' rozpodil'chy'x vidnosy'n yak mexanizm pidvy'shhennya yix konkurentospromozhnosti. Zb. nauk. pracz' Tavrijs'kogo derzhavnogo agrotexnologichnogo universy'tetu. Seriya: Ekonomichni nauky'. – 2011. – # 2 (14). – S. 209–222

# International Ranking and Clustering Systems in Complex Evaluation of Demographic and Migration Processes

**Svitlana Nesterenko, Dmytro Vasylkivskyi, Raisa Kvasnytska and Ihor Lapshyn**

## 1 Introduction

The basis of a reasonable approach to decision-making in the demographic sphere is a systematic statistical analysis of the current demographic situation of the regions and the country, which is based on the statistical and mathematical tools of scientific research: socioeconomic infrastructure, the service sector and the specifics of production, people's living standards, labor market trends, their specifics, and the dynamics of migration processes.

The differentiation of demographic processes in Ukrainian regions is a consequence of historical, economic, political, and other factors, which have been primarily reflected on the level of demographic development, the regions of the country were divided into three groups: regions with complex, mediated, and favorable demographic situation.

The generalized reflection of the socioeconomic welfare of any country, which is shaped by both past and current social processes, is the demographic situation in this country. It is the demographic state that is one of the determining factors for

S. Nesterenko (✉)
Department of Management, Tavria State Agrotechnological University,
B. Khmelnistskiy ave 18, Melitopol 72312, Ukraine
e-mail: svitlana.nesterenko@tsatu.edu.ua

D. Vasylkivskyi · I. Lapshyn
International Economic Relations Department,
Khmelnytskyi National University, Khmelnytskyi, Ukraine
e-mail: vasilkivsky@gmail.com

I. Lapshyn
e-mail: igorlaps8@gmail.com

R. Kvasnytska
Finance, Banking and Insurance Department,
Khmelnytskyi National University, Khmelnytskyi, Ukraine
e-mail: rstepanivna@gmail.com

© Springer Nature Switzerland AG 2019

V. Nadykto (ed.), *Modern Development Paths of Agricultural Production*,
https://doi.org/10.1007/978-3-030-14918-5_50

ensuring a stable and secure development of the country, and optimal demographic development is a priority interest of the state, which is dualistic both as a factor and as a result of its functioning.

Thus, in the years of independence, Ukraine has experienced a demographic situation characterized by a decline in population, a rise in mortality, a decline in fertility, aging of the population and significant migration processes. These negative demographic processes occur in times of complex crisis in the domestic economy, which, in turn, contribute to the deepening and demographic crisis in the country.

The deepening of the same demographic crisis in the country today is not only one of the factors of insufficiently effective socioeconomic development, but also threatens its future and national security of the state. Due to an inadequate analytical definition of the demographic situation in Ukraine, the level of realization of the priority directions of the state social policy is distorted, and the main threats to national security in this area remain unresolved.

Therefore, today the issue of improving the approaches to assessing the demographic situation is extremely relevant. After all, it is the analysis of the demographic situation in the country and its individual regions that allows negative trends forecasting and identifying problems with the qualitative composition of the population and contributes to the formation of an effective regional policy on the socio-demographic development of a particular territory.

Creation of favorable opportunities for demographic improvement and formation of preconditions for the gradual stabilization of the population and ensuring a stable tendency to improve the quality of Ukraine's demographic potential requires the study of the problematic aspects of carrying out a statistical analysis of the demographic situation in the country and in its separate regions.

## *1.1 Analysis of Recent Research*

The research of issues and problems of the demographic situation development is highlighted in the writings of many scholars. Thus, the nature, constituents, and the state of the demographic situation in Ukraine are considered by such Ukrainian scientists as: N. Vlasenko, G. Gerasymenko, V. Goryovyi, E. Libanova, O. Mykolyuk, V. Novikov, S. Pyrozhkov, L. Stelmakh, O. Shekera, and others. Their works give a general description of the demographic situation in the country, analyze the changes in the population structure, identify the main factors of demographic changes, describe their models, and also highlight the interrelationships. However, methodological nuances, comparative analysis using computer systems and technologies, and statistical tools with the ability to evaluate and compare the demographic situation of the regions need further development and improvement.

For today, there are few research approaches to the analysis of the demographic situation in the country based on such methods of statistical analysis as the ranking method and the clusterization method. However, it is precisely for the construction of a coherent picture of the demographic situation in the regions of Ukraine that the transition from the study of individual values of demographic indicators to their distribution and grouping is necessary in order to comprehensively assess the dynamics of the demographic situation of the regions.

## 1.2 The Current Demographic Situation

Today, the issues of the implementation effectiveness of national social and demographic policies, programs of socioeconomic development at the regional and local levels need more attention. At the same time, the primary promising goal of demographic development is the creation of favorable opportunities for demographic improvement and the formation of prerequisites for the gradual stabilization of the population and ensuring a steady trend toward improving the quality of demographic potential [1].

This is explained by the fact that in modern Ukraine, the demographic situation, which is determined by the structure of the population and the nature of its movement, types and mode of reproduction, fertility and mortality rates, and the number of marriages and divorces, is characterized by great severity and intensity [2].

So, more than 100 years ago, demographic processes attracted the attention of specialists only, but now statistics on the demographic situation on the planet—the focus of attention of both international organizations and governments of countries and the general public. Therefore, the basis of a reasoned approach to decision-making in the demographic field is a systematic statistical analysis of the current demographic state of individual regions and the country as a whole, the results of which determine the trends and patterns of demographic processes and phenomena, assesses the inter-regional differences in demographic processes [3].

It should be noted that the population of Ukraine as of January 1, 2017, according to statistics, amounted to 42,584.5 thousand people (excluding the temporarily occupied territory of the Autonomous Republic of Crimea and the city of Sevastopol). To provide more characteristics, we will consider the main demographic indicators of the population from 2005 to 2017 in Table 1 [4–7].

Data in Table 1 indicate an annual decrease in the population of Ukraine, which in total amounted to 4696.3 thousand people (9.93%) from 2005 to 2017. At the same time, the natural movement of the population is marked by a significant excess of the number of deceased persons over the number of births, which for the analyzed period amounted to 145,743 people (40.95%).

The reduction in the population is accompanied by a virtually steady increase in the number (and, accordingly, of the share) of persons over 60 years of age. If the population census in 1959 recorded 4387.0 thousand people (10.5%), over 60 years old, and the 1970 census—6563.8 thousand people (13.9%), then in 1979 there were 7764.8 thousand people in Ukraine (15.7%) more than 60 years old, in 1989–9256 thousand people (18%), and in 2001–10,323 thousand people (21.4%).

The reasons for this have historical roots and accumulated over a long period of time. The First and Second World Wars, three famines (1921, 1932–1933, 1947),

**Table 1** Dynamics of the basic demographic indicators of the population of Ukraine

| Indicators | Population, thousand people | Number of births, people | Number of deceased, people | Natural movement of population (increase /reduce), people | Migration balance, people |
|---|---|---|---|---|---|
| 2005 | 47,280.8 | 426,085 | 781,964 | −355,879 | 4583 |
| 2015 | 42,928.9 | 411,783 | 594,795 | −183,012 | 14,233 |
| 2016 | 42,760.5 | 397,039 | 583,631 | −186,592 | 10,620 |
| 2017 | 42,584.5 | 3,639,897 | 574,123 | −210,136 | 11,997 |
| Deviation of 2017 from 2005 | | | | | |
| Absolute | −4696.3 | −62.098 | −207.841 | −145.743 | 7414 |
| Relative, % | −9.93 | −14.57 | −26.58 | −40.95 | 161.77 |

mass repressions of the 1930s and 1950s, the Chernobyl disaster, the protracted systemic crisis of the 1990s, had a very negative effect on the natural process of demographic evolution, led to a noticeable deformation of the age structure of the population, premature death of about 16 million people. In conjunction with a low birth rate over the past 40 years, this led to depopulation [5–9].

A correlation analysis was conducted to find a mutual relationship between economic developments. A correlation index is a numerical measure of some type of correlation, meaning a statistical relationship between two variables. It assumes values in the range from −1 to +1, where +1 indicates the strongest possible agreement and −1 the strongest possible disagreement. In our case, such indicators as GDP per capita and other demographic indicators demonstrate the development of the country.

Results showed that during slow economic development, lack of stability, the population receives less financial resource to achieve desired way of living. That affects the birth rate, (Correlation index: 0.65), mortality rate (C.I.: −0.58), and life expectancy of the population (C.I.: 0.65). This result confirms that the economic factor strongly affects both the indicator of the average life expectancy and the demographic situation in the country in general.

Lack of stability and the decline of economy affect migration rate. The number of Ukrainian migrants is lower than in the early 1990s in general. By 2005, the number of Ukrainian migrants abroad has decreased to 6.17 million people; in 2016, it has reached almost 5 million people according to the UN data.

Most of the migrants come from the west regions of the country. The main destination countries for migrant workers are Poland, the Russian Federation, the Czech Republic and Italy. Ukrainian migrants make significant the impact on the development of above mentioned countries. Economic and demographic situation in Ukraine affects on other countries as well and influence on labor forces in European region in some cases it affects on international affairs. Therefore, the situation in Ukraine is one of international importance.

Thus, the differentiation of demographic processes in the regions is a consequence of historical, economic, political, and other factors, which is primarily reflected in the degree of development of socioeconomic infrastructure, the sphere of services and production specifics, people's living standards, labor market trends and their specifics, and the dynamics of migration processes. By the level of demographic development, the regions of the country are divided into three groups: regions with complex, mediated, and favorable demographic situation.

## 2 Ranking Analysis

Since the indicators of mechanical and natural movement of the population describe the demographic situation of the country and the region, it is expedient to use the method of comparing these indicators based on the general—integrated assessment, which is determined on the basis of statistical data characterizing the demographic situation. To build a coherent picture of the demographic situation in the regions of Ukraine, it is necessary to shift from the values of demographic indicators to their distribution using the ranking method according to the nominal scale.

Thus, according to the results of the integrated assessment of the relevant indicators, the regions are assigned ranks, the comparison of which simplifies the process of comparative analysis of the demographic situation in the regions. This approach allows us to form a region's ranking based on the demographic situation based on the rankings that consist of a total score assessment.

In addition, it is possible to analyze the dynamics of the demographic situation of the regions according to their rank and rating, which results in the trends of demographic development of individual regions and the country as a whole.

Consequently, the basis for integrated assessment is the summing up of all demographic indicators for each region to the total rank, that is, to the nominal scale. Based on the analysis of the positions of scientists on the specification of the indicators characterizing the demographic situation, we consider it appropriate to carry out the distribution of demographic indicators into two groups:

1. indicators, the growth of which shows the improvement of the demographic situation
   (indicators-stimulants): the total fertility rate, the total marital ratio, the burden of individuals from 0-15 years, the number of women of reproductive age and the average life expectancy at birth;
2. indicators, the decrease of which demonstrates the deterioration of the demographic
   situation (indicators-de-stimulators): the total mortality rate, the mortality rate of children under 1 year of 1000 live births, the total divorce rate and the demographic burden of persons aged 65 and over.

Thus, according to the calculations of the first group of indicators, the demographic situation will be better in those regions where the values of the indicators will be higher. That is, the higher the value of the indicator or rank is, the better is the situation.

Conversely, according to the calculations of the second group of indicators, the demographic state will be better in those regions where the corresponding values will be lower. The calculation and analysis of the above-identified groups of demographic indicators, the formation of ranks of regions on their basis, is based on statistical data for 2005, 2016 [10] (Tables 2 and 3).

Table 2 shows the ranks and ratings of the regions of the country, with the best indicator—there is a higher nominal value of the rank. Thus, the best demographic status in 2016 is in the Chernihiv region, which ranked first, while the last place in the ranking of 23 regions of Ukraine was taken by Rivne region.

As for the Kyiv region, in 2005 the complex assessment (rating) of the demographic indicators-stimulants was 13, in 2016–15. That is, there is a positive movement of the region in the ranking, with no significant impact on the increase in the fertility and marriage rate.

By comparing comprehensive assessments, that is, the ranking of areas by the development of the demographic situation, we get three groups of regions with the corresponding rating: high, medium, and low. It should be noted that the higher the rating, the better the demographic situation in the region.

Thus, in 2005, the following areas, such as Volyn, Zakarpattya, Ivano-Frankivsk, Lviv, Rivne, Ternopil, Khmelnytsky, and Chernivtsi, are among the regions with a high ranking. The low-ranking group includes the following areas: Kirovograd, Mykolayiv, Sumy, Kherson, and Chernihiv oblasts. All other areas belong to the middle group. In 2016, the high-ranking group included the following areas: Ivano-Frankivsk, Lvivska, Odeska, Rivnenska, Chernivetska, and Kyiv city. Among the low rating areas were Zaporizka, Kirovogradska, Mykolaivska, Poltavska, Sumska, Khersonska, Kharkivska, Cherkaska, and Chernihivska. All other areas belong to the middle group. In 2005, the following situation was observed according to indicators-destimulators, with the following high-ranking groups including: Volynska, Zakarpatska, Ivano-Frankivska, Lvivska, Ternopilska, Chernivetska regions, and Kyiv city. The low-ranking group includes the following areas: Dnipropetrovska, Kirovogradska, Cherkaska, and Chernihivska. All other areas belong to the middle group. In 2016, the situation has almost not changed; however, the demographic situation became even worse, where it was disappointing.

Comparing the comprehensive assessments, i.e., the ranking of the regions with the development of the demographic situation (Fig. 1), we receive three groups of regions with the corresponding ratings: high, medium, and low. Note that in Fig. 1, the higher the rating, the better the demographic situation. Thus, in 2005, the following areas fall into the high-ranking group: Volyn, Zakarpattia, Ivano-Frankivsk, Lviv, Rivne, Ternopil, Khmelnytsky, and Chernivtsi. The regions with low ratings include Kirovograd, Mykolaiv, Sumy, Kherson, and Chernihiv oblasts. All other areas belong to the middle group.

In 2016, the high-ranking group included the following regions: Ivano-Frankivsk, Lviv, Odesa, Rivne, Chernivtsi, and Kyiv. The regions with low rating include

**Table 2** Ratings and ranks of regions by the demographic indicators-stimulators group

| Indicators | Regions | | | | | | | | | | | | | |
|---|---|---|---|---|---|---|---|---|---|---|---|---|---|---|
| | Total fertility rate | | Total marriage rate | | Demographic load by individuals from 0 to 15 years | | Number of women of reproductive age per thousand people | | Average life expectancy at birth | | General Rating | Rank | General Rating | Rank |
| | 2005 | 2016 | 2005 | 2016 | 2005 | 2016 | 2005 | 2016 | 2005 | 2016 | 2005 | 2005 | 2016 | 2016 |
| Vinnytska | 13 | 11 | 6 | 5 | 14 | 11 | 13 | 11 | 16 | 17 | 62 | 13 | 55 | 12 |
| Volynska | 21 | 21 | 15 | 13 | 22 | 22 | 21 | 22 | 12 | 15 | 91 | 19 | 93 | 22 |
| Dnipropetrovska | 8 | 9 | 20 | 7 | 4 | 8 | 7 | 10 | 4 | 2 | 43 | 9 | 36 | 6 |
| Zhytomyrska | 19 | 17 | 11 | 7 | 17 | 18 | 18 | 15 | 1 | 1 | 66 | 15 | 58 | 13 |
| Zakarpatska | 22 | 22 | 15 | 17 | 21 | 21 | 22 | 21 | 12 | 9 | 92 | 20 | 90 | 19 |
| Zaporizka | 7 | 7 | 11 | 7 | 3 | 6 | 6 | 6 | 9 | 10 | 36 | 5 | 36 | 6 |
| Ivano-Frankivska | 20 | 12 | 15 | 14 | 20 | 19 | 19 | 14 | 19 | 20 | 93 | 21 | 79 | 15 |
| Kyivska | 12 | 19 | 21 | 22 | 8 | 16 | 12 | 18 | 9 | 4 | 62 | 13 | 79 | 15 |
| Kirovogradska | 11 | 10 | 3 | 1 | 12 | 8 | 8 | 9 | 5 | 6 | 39 | 8 | 34 | 5 |
| Lvivska | 17 | 13 | 8 | 19 | 16 | 15 | 16 | 16 | 21 | 21 | 78 | 17 | 84 | 17 |
| Mykolaivska | 9 | 8 | 8 | 14 | 9 | 10 | 9 | 8 | 3 | 7 | 38 | 7 | 47 | 9 |
| Odeska | 14 | 20 | 11 | 21 | 9 | 17 | 15 | 19 | 2 | 8 | 51 | 11 | 85 | 18 |

(continued)

**Table 2** (continued)

| Indicators | Regions | | | | | | | | | | | | | |
|---|---|---|---|---|---|---|---|---|---|---|---|---|---|---|
| | Total fertility rate | | Total marriage rate | | Demographic load by individuals from 0 to 15 years | | Number of women of reproductive age per thousand people | | Average life expectancy at birth | | General Rating | Rank | General Rating | Rank |
| | 2005 | 2016 | 2005 | 2016 | 2005 | 2016 | 2005 | 2016 | 2005 | 2016 | 2005 | 2005 | 2016 | 2016 |
| Poltavska | 3 | 6 | 6 | 5 | 6 | 4 | 3 | 4 | 11 | 12 | 29 | 3 | 31 | 3 |
| Rivnenska | 23 | 23 | 15 | 14 | 23 | 23 | 23 | 23 | 16 | 14 | 100 | 23 | 97 | 23 |
| Sumska | 1 | 1 | 2 | 1 | 5 | 1 | 1 | 1 | 8 | 13 | 17 | 1 | 17 | 2 |
| Ternopilska | 15 | 5 | 15 | 7 | 18 | 13 | 16 | 7 | 22 | 22 | 86 | 18 | 54 | 11 |
| Kharkivska | 2 | 2 | 8 | 18 | 2 | 2 | 2 | 5 | 15 | 11 | 29 | 3 | 38 | 8 |
| Khersonska | 10 | 15 | 4 | 7 | 13 | 12 | 11 | 13 | 6 | 3 | 44 | 10 | 50 | 10 |
| Khmelnytska | 16 | 14 | 14 | 7 | 14 | 13 | 14 | 12 | 18 | 16 | 76 | 16 | 62 | 14 |
| Cherkaska | 5 | 4 | 4 | 4 | 11 | 5 | 4 | 2 | 12 | 18 | 36 | 5 | 33 | 4 |
| Chernivetska | 18 | 16 | 22 | 19 | 19 | 19 | 20 | 17 | 19 | 19 | 98 | 22 | 90 | 19 |
| Chernihivska | 6 | 3 | 1 | 1 | 6 | 3 | 5 | 3 | 6 | 5 | 24 | 2 | 15 | 1 |
| Kyiv city | 4 | 18 | 23 | 23 | 1 | 7 | 9 | 20 | 23 | 23 | 60 | 12 | 91 | 21 |

**Table 3** Ratings and ranks of regions by the demographic indicators-destimulators group

| Indicators | Regions | | | | | | | | | | | |
|---|---|---|---|---|---|---|---|---|---|---|---|---|
| | Total mortality rate | | Mortality rate of children under 1 year of age per 1000 live births | | Total divorce rate | | Demographic load by individuals aged 65 years | | General Rating | Rank | General Rating | Rank |
| | 2005 | 2016 | 2005 | 2016 | 2005 | 2016 | 2005 | 2016 | 2005 | 2005 | 2016 | 2016 |
| Vinnytska | 16 | 11 | 9 | 6 | 8 | 12 | 22 | 20 | 55 | 16 | 49 | 10 |
| Volynska | 8 | 7 | 4 | 11 | 3 | 4 | 10 | 3 | 25 | 3 | 25 | 4 |
| Dnipropetrovska | 15 | 18 | 11 | 15 | 23 | 22 | 13 | 14 | 62 | 18 | 69 | 20 |
| Zhytomyrska | 18 | 16 | 10 | 13 | 8 | 7 | 17 | 15 | 53 | 15 | 51 | 12 |
| Zakarpatska | 2 | 2 | 20 | 23 | 1 | 1 | 2 | 1 | 25 | 3 | 27 | 7 |
| Zaporizka | 10 | 15 | 16 | 6 | 19 | 20 | 12 | 16 | 57 | 17 | 57 | 18 |
| Ivano-Frankivska | 3 | 4 | 23 | 9 | 5 | 6 | 9 | 5 | 40 | 7 | 24 | 1 |
| Kyivska | 17 | 17 | 1 | 1 | 13 | 23 | 14 | 10 | 45 | 11 | 51 | 12 |
| Kirovogradska | 20 | 21 | 16 | 21 | 17 | 12 | 16 | 21 | 69 | 20 | 75 | 22 |
| Lvivska | 4 | 6 | 4 | 9 | 2 | 2 | 7 | 7 | 17 | 2 | 24 | 1 |
| Mykolaivska | 11 | 10 | 3 | 16 | 22 | 21 | 5 | 11 | 41 | 8 | 58 | 19 |
| Odeska | 13 | 9 | 15 | 13 | 11 | 12 | 4 | 8 | 43 | 9 | 42 | 9 |

(continued)

**Table 3** (continued)

| Indicators | Regions | | | | | | | | | | | |
|---|---|---|---|---|---|---|---|---|---|---|---|---|
| | Total mortality rate | | Mortality rate of children under 1 year of age per 1000 live births | | Total divorce rate | | Demographic load by individuals aged 65 years | | General Rating | Rank | General Rating | Rank |
| | 2005 | 2016 | 2005 | 2016 | 2005 | 2016 | 2005 | 2016 | 2005 | 2005 | 2016 | 2016 |
| Poltavska | 21 | 22 | 6 | 4 | 19 | 10 | 20 | 19 | 66 | 19 | 55 | 16 |
| Rivnenska | 5 | 3 | 14 | 17 | 3 | 2 | 3 | 2 | 25 | 3 | 24 | 1 |
| Sumska | 22 | 20 | 19 | 5 | 13 | 7 | 19 | 18 | 73 | 22 | 50 | 11 |
| Ternopilska | 7 | 8 | 8 | 2 | 6 | 5 | 15 | 11 | 36 | 6 | 26 | 6 |
| Kharkivska | 12 | 14 | 11 | 12 | 16 | 17 | 11 | 13 | 50 | 13 | 56 | 17 |
| Khersonska | 9 | 13 | 18 | 22 | 17 | 10 | 5 | 9 | 49 | 12 | 54 | 15 |
| Khmelnytska | 14 | 12 | 7 | 6 | 13 | 17 | 18 | 17 | 52 | 14 | 52 | 14 |
| Cherkaska | 19 | 19 | 11 | 18 | 19 | 12 | 21 | 22 | 70 | 21 | 71 | 21 |
| Chernivetska | 6 | 5 | 22 | 19 | 8 | 12 | 8 | 5 | 44 | 10 | 41 | 8 |
| Chernihivska | 23 | 23 | 21 | 20 | 7 | 9 | 23 | 23 | 74 | 23 | 75 | 22 |
| Kyiv city | 1 | 1 | 1 | 3 | 11 | 17 | 1 | 4 | 14 | 1 | 25 | 4 |

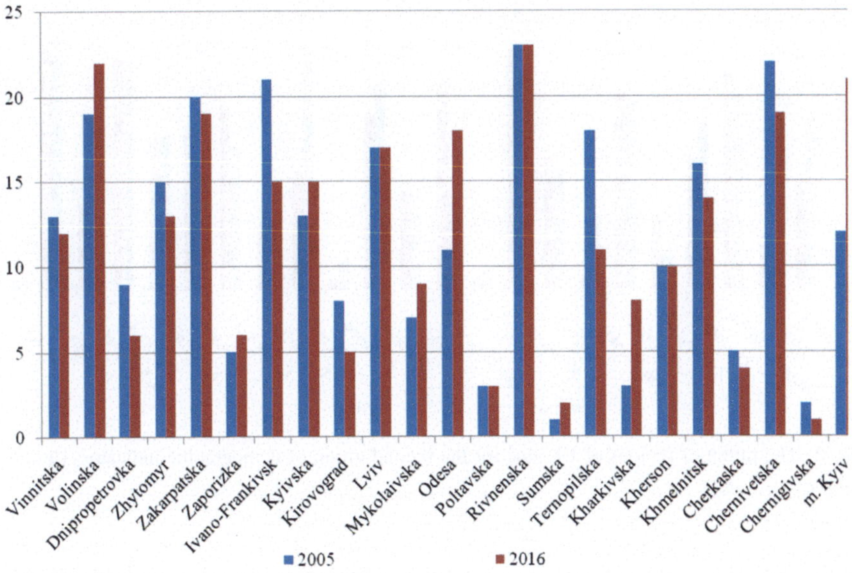


**Fig. 1** The ranking of regions of Ukraine from the first group of demographic indicators (more is better)

Zaporozhye, Kirovograd, Nikolaev, Poltava, Sumy, Kherson, Kharkiv, Cherkasy, and Chernihiv. All other areas belong to the middle group.

As for the indicators of the second group (Fig. 2), in 2005 the following situation was observed, the following areas were included in the high-ranking group: Volyn, Zakarpattia, Ivano-Frankivsk, Lviv, Ternopil, Chernivtsi regions, and Kyiv. The regions with low ratings include Dnipropetrovsk, Kirovograd, Cherkasy, and Chernihiv. All other areas belong to the middle group. In 2016, the situation was almost unchanged; mainly where, often, the demographic situation was disappointing, it became even worse.

Thus, the most depressed region from the demographic point of view is the north and east of the country, preceding almost all criteria of the west, and partly the south of the country. Since 2005, the situation didn't change dramatically; in general, no significant changes took a place. However, these changes become apparent if they are viewed within specific indicators and areas. In general, the situation in the country has not become better and remains disappointing.

## 3 Cluster Analysis

On the basis of the algorithm for full connection, we have built a dendrogram for 23 observations (regions); Fig. 3 reflects the clusterization of the regions of Ukraine

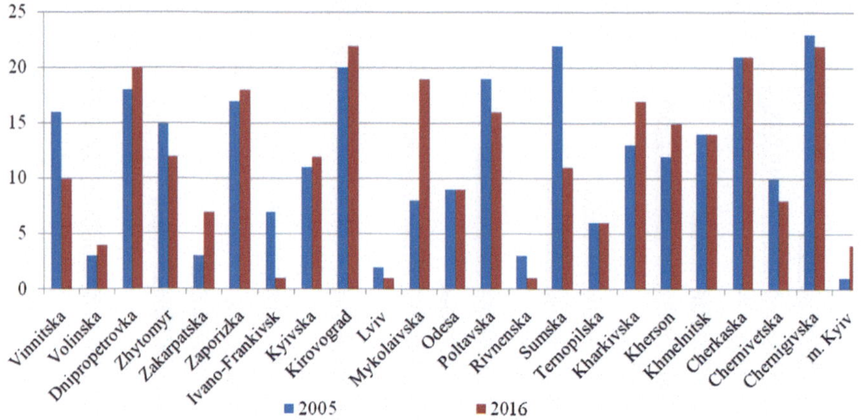


**Fig. 2** The rating of regions of Ukraine for the second group of demographic indicators (less is better)

by the total rating of their indicators-stimulators and indicators-destimulators. Thus, by indicators-stimulators, all regions were ranked in four clusters. Below the left side, there are areas characterized by the highest demographic status—1-th cluster (5 oblasts and the city of Kyiv). The 2-nd cluster includes five oblasts and is characterized by a higher average demographic development. The situation in 3-rd cluster, which includes eight regions, is worse. The worst demographic situation is in 4-th clusters, which includes three oblasts.

It is clear from the left side of the figure that the east, north and southeast of the country are depressed. In the west and the northwest, the situation is better. According to the indicators-destimulator, the dendrogram of cluster analysis also consists of 4-th clusters (four oblasts and the city of Kyiv), which are characterized by the highest demographic state located on top of the right-hand side of the figure and belong to 1-st cluster. Accordingly, the bottom 3 areas are composed of four clusters, which are characterized by the worst demographic situation. 2-nd and 3-rd clusters contain, respectively, eight and seven regions. The situation in these clusters is mediocre.

## 4 Conclusion

The overall demographic and economic situation in Ukraine is constantly getting worse. Migration processes have been affected by negative consequences of economic development. Labor forces in the form of migrants are looking for ways to better conditions of living, so the young population migrates to neighboring countries and contributes to their development at a time when this is not happening in Ukraine.

In general, Ukraine and its regions have been characterized by a sharp difference in demographic indicators. As already mentioned, the best demographic situation in

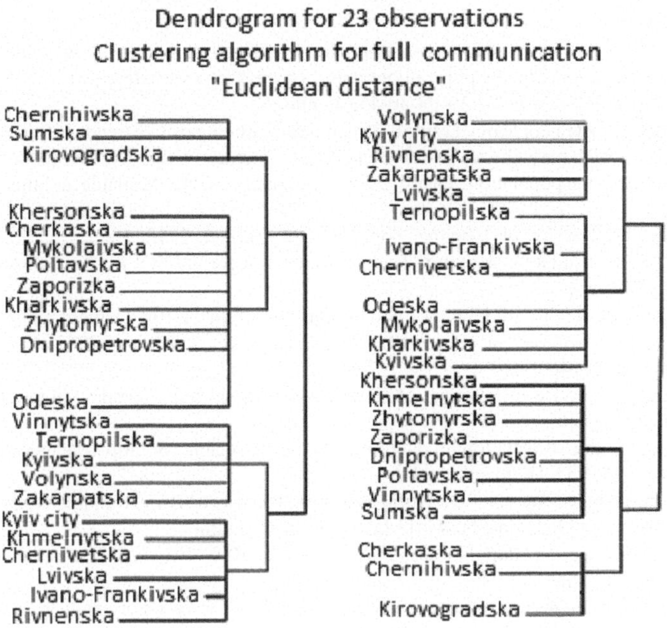


**Fig. 3** Results of cluster analysis in two groups of demographic indicators

the west and the south is worse in the east and north. The demographic and economic problems in Ukraine influence on other countries and in general on a region by the migrants.

The use of ranking analysis allowed obtaining detailed understanding of the situation of what is happening in Ukraine. The use of the ranking and cluster analysis method for grouping the regions of the country into two groups of demographic indicators shows almost identical results to those obtained by grouping the regions by rank and ratings derived from comprehensive estimates of the demographic situation in the country.

This fact suggests that the use of these methods is expedient and can be used to carry out a comprehensive assessment of the demographic situation in the country. However, the methods and methodology for the application of such statistical methods of analysis require further improvement of their practical application.

## References

1. Novikov, V.M.: Social infrastructure as a factor in preserving the demographic potential of Ukraine. Demography and Social Economy (2004)
2. Pyrozhkov, S.I.: The demographic factor in the global development strategy of Ukraine. Demografija ta social'na ekonomika, (Demography and social economy), 2004-1-2, 5-21 (in

Ukr.) (2004)
3. Shekera, O.H.: Demographic situation in the world and in Ukraine. Available at: http://healthysociety.com.ua/index.php?option=com_content&view=article&id=121:2011-07-03-09-43-30&catid=36:2011-04-19-08-30-45&Itemid=58
4. Mykolyuk, O.: Are people in Ukraine the main value? About positive trends in the demography of Ukraine, which have already positive tendencies recorded by the statistics and how the state is preparing for the population census. “Den” Newspaper (2011). Available at: https://day.kyiv.ua/uk
5. State statistics service of Ukraine, population of Ukraine 2010, Demographic Yearbook, Kyiv (2011). Available at: http://database.ukrcensus.gov.ua/PXWEB2007/ukr/publ_new1/2011/publ2011.asp
6. State statistics service of Ukraine, population of Ukraine 2014, Demographic Yearbook, Kyiv (2015). Available at: http://database.ukrcensus.gov.ua/PXWEB2007/ukr/publ_new1/2015/naselen_2014pdf.pdf
7. State statistics service of Ukraine. Metadata. Available at: http://database.ukrcensus.gov.ua/MULT/Dialog/statfile_c.asp
8. The World Bank. Population in Ukraine (2015). Available at: http://data.worldbank.org/indicator/SP.POP.TOTL?locations=UA
9. Central Intelligence Agency. The World Factbook, Ukraine (2016). Available at: https://www.cia.gov/library/publications/the-world-factbook/geos/up.html
10. Statistical Information “Population and Migration” for 2005–2016 (2017). Available at: http://www.ukrstat.gov.ua/
11. UN data base migration statistics. Available at: http://www.un.org/en/development/desa/population/migration/data/estimates2/estimates17.shtml

# Ukraine in the Context of the World Organic Production of Agricultural Products

Roman Oleksenko, Iryna Kolokolchykova and Olena Syzonenko

## 1 Introduction

The Earth, namely agricultural lands, is the main component of the ecological system of the planet. Soil fertility and its ability to reproduce provide the existence of all living things. Negligent use of land resources leads to a decrease in the fertile of soil layer. Agrarians around the world are looking for new production technologies, sometimes increasing the amount of poison and pesticides. This further harms the land and leads to irreversible consequences.

According to Karl Marks, "even a whole society, nation, and even all existing societies, taken together, are not owners of the land." They are only the owners who do not use it. And as "boni patres familias" (good parents of the family), they should leave it better for next generations. Therefore, this issue is extremely relevant both in terms of agrarian ecological production and business [1].

As a result, people began to understand their own mistakes and tried to correct it by applying organic production methods in agricultural. Today the organic production is the most intensively growing sector of the world's economy.

R. Oleksenko (✉) · I. Kolokolchykova · O. Syzonenko
Tavria State Agrotechnological University, Melitopol, Ukraine
e-mail: roman.xdsl@ukr.net

I. Kolokolchykova
e-mail: kolokolchikovairina@gmail.com

O. Syzonenko
e-mail: syzonenko.lena@gmail.com

© Springer Nature Switzerland AG 2019

V. Nadykto (ed.), *Modern Development Paths of Agricultural Production*,
https://doi.org/10.1007/978-3-030-14918-5_51

## 2 Organic Production in the World: Industry Analysis

"Today organic agricultural farming is developing successfully. The global demand for organic food increases with the growing awareness of consumers of the importance of the environmental aspect of agricultural production and the impact of such products on human health. It also depends on the growth of household incomes and the convergence of organic products directly with the consumer" [2].

Leading countries in organic production are Australia, Argentina, China, the USA, and Spain. These are countries with the largest areas of certified organic land. The organic sector in Asia is developing very rapidly. China and India have the highest rates here. There are large quantities of agricultural lands certified according to organic standards in these countries. Ukraine occupies 20th place in this ranking with an area of organic land 381,173 ha [3] (Table 1).

In 2016, there were identified 178 countries with organic agricultural in the world [4]. The total area of agricultural land occupied by organic production amounted to 57.8 million hectares (46.9% are in Australia). The largest share of agricultural land under organic production from their total area is in Liechtenstein (37.7%). The current legislation on the organic production of organic products has 87 countries in the world (Table 2).

The organic production has steady development trends in developed countries and EU countries. The manufacturers understand the need and social orientation of this process (Fig. 1). Entrepreneurs of the agrarian sector want to achieve not only the food security of mankind, but also the preservation of natural resources of the Earth and improve the population of the planet [5].

Organic farming is perceived as a holistic business management system with the best of business practices, the using of advanced technologies, and marketing approaches in the formation and functioning of the organic products market. Retail

**Table 1** World ranking of the total area of organic agricultural lands, 2016

| No | Countries | Hectares | No | Countries | Hectares |
|---|---|---|---|---|---|
| 1 | Australia | 27145021 | 11 | Canada | 1099014 |
| 2 | Argentina | 3011794 | 12 | Brazil | 750000 |
| 3 | China | 2281215 | 13 | Mexico | 673968 |
| 4 | United States | 2031318 | 14 | Austria | 571385 |
| 5 | Spain | 2018802 | 15 | Sweden | 552695 |
| 6 | Italy | 1796363 | 16 | Poland | 536579 |
| 7 | Uruguay | 1656952 | 17 | Turkey | 5237767 |
| 8 | France | 1538047 | 18 | United Kingdom | 490205 |
| 9 | India | 1490000 | 19 | Czech Republic | 488591 |
| 10 | Germany | 1251320 | 20 | Ukraine | 381173 |

**Table 2** Organic agricultural: key indicators and leadership, 2016

| Indicators | The world | Leader countries |
|---|---|---|
| Countries with certified organic farming data | 178 countries | – |
| The total area of agricultural land occupied by organic production | 57.8 million hectares | Australia (27.1 million hectares), Argentina (3.0 million hectares), China (2.3 million hectares) |
| The share of agricultural land under organic production from the total area of agricultural land | 1.2% | Liechtenstein (37.7%), French Polynesia (31.1%), Samoa (22.4%) |
| Area under wilderness and other non-agricultural land | 39.9 million hectares | Finland (11.6 million hectares), Zambia (6.7 million hectares), India (4.2 million hectares) |
| Manufacturers | 2.7 million manufacturers | India (835 000), Uganda (210 352), Mexico (210 000) |
| Organic market | 89.7 billion USD | The USA ($43.1 billion—equivalent to € 38.9 billion); Germany ($10.5 billion—equivalent to € 9.5 billion); France ($7.5 billion—equivalent to € 6.7 billion) |
| Consumption per capita | 2.1 USD (11.3 Euro) | Switzerland ($304—equivalent to € 274); Denmark ($252—equivalent to € 227); Sweden ($218—equivalent to € 197) |
| Number of countries with legislation on organic production and circulation of organic products | 87 countries[a] | |

*Note* [a] data for 2017 year

sales of organic products from the leading countries of the world are presented dynamically in Fig. 2. The stimulation of producers of organic products in the world takes place through national programs for the development of this area of agricultural activity.

Ukraine is a country that only beginning to develop in this direction and has great potential. Being 20th in the ranking of the total area of organic land (381,173 ha), Ukraine is promising to make new progress. Now Ukraine has 15 certification struc-

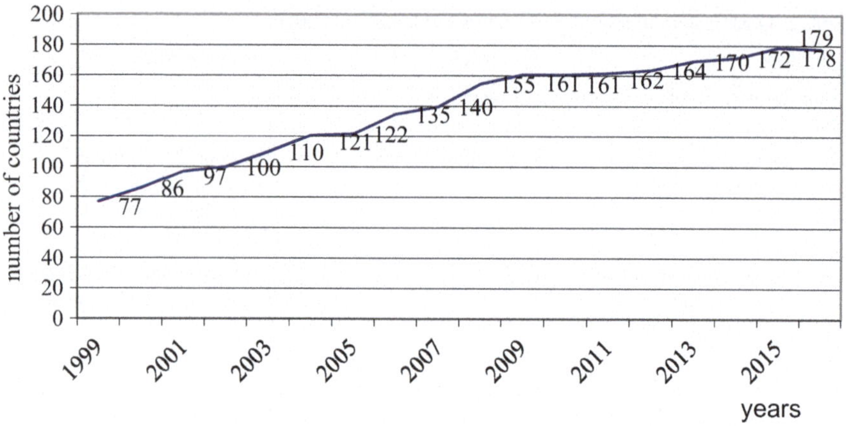


**Fig. 1** Dynamics of the countries with data on organic production, 1999–2016. *Source* Plant Growing of Ukraine. Statistical Collection [5]

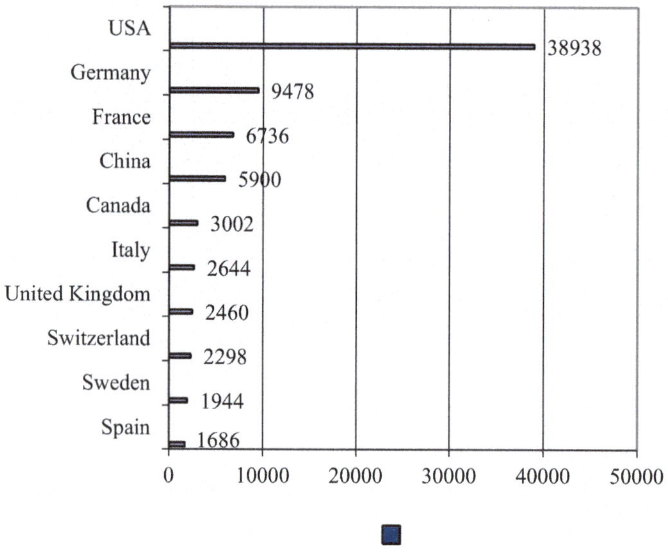


**Fig. 2** Retail sales of the world's largest markets for organic food, 2018. *Source* Plant Growing of Ukraine. Statistical Collection [5]

tures for organic production, which operate in the country and are recognized by the European Commission [5].

## 3 Current State of Organic Sector in Ukraine

Ukrainian organic certified agricultural market with an emphasis on exports began its development in the 1990s due to the demand for organic products on the international market. However, organic production originated long before, thanks to the work of local scientists of organic production in Ukraine.

There were passed the Law of Ukraine "On the production and circulation of organic agricultural products and raw materials", 2013 [6]; The Law of Ukraine "On basic principles and requirements for organic production, turning and marking of organic products", 2018 [7]. In 2015, Ukraine has adopted the National Organic Logo. Such use of the logo is voluntary and can be used for organic products of export, processing products, and raw materials. It can be used for organic products of export, processed products, and raw materials.

Unfortunately, today in Ukrainian, organic agricultural, which is implemented mainly by small and medium enterprises, is not a priority for the country. However, the relevant legislative framework has been already formed in the country [5].

Most of the organic for the domestic market are certified according to the EU organic agricultural legislation. There are no private standards and labels in Ukraine. The two main associations representing the private organic sector of Ukraine are Organic Federation of Ukraine and Certified Organic Producers Union "Organic Ukraine" [6].

Official IFOAM statistical surveys confirm that Ukrainian enterprises are now actively involved in organic farming. The total area of organic agricultural lands is increasing. There is also an increase in the number of enterprises that deal with organic matter. Already, the domestic market of organic food is actively forming [5, 8].

In the dynamics of recent years, the area under organic agricultural lands has increased significantly. In quantitative terms, this figure is not large, but the trend increase is clearly shown in Table 3.

In 2017, the share of organic agricultural lands was 1.52%. And this indicator is significant. But, if we look at the dynamics relative to the base year (2005), the growth rate is 73.57% in the reported year.

The sizes of Ukrainian-certified organic farms have different sizes: from several hectares to several thousand hectares of arable lands [8]. Most Ukrainian organic farms are located in Kyiv, Odessa, Kherson, Kharkiv, Zhytomyr, and Lviv regions (Table 4).

Domestic market of Ukrainian organic products is quite young. The main channels for selling organic products are supermarket chains in Ukraine (Good Wine, Silpo, Auchan, MegaMarket, etc.). In addition, there are small specialized stores of healthy food and online stores.

**Table 3** Agricultural lands in Ukraine, including organic ones, 2005–2017

| Years | Total area of agricultural lands, thousand hectares | Organic agricultural lands, thousand hectares | The share of organic agricultural lands to total (%) | Basic rates of growth of organic agricultural land (%) |
|---|---|---|---|---|
| 2005 | 26043.6 | 241980 | 0.93 | – |
| … | … | … | … | … |
| 2010 | 26951.5 | 270226 | 1.00 | 111.67 |
| 2011 | 27670.5 | 270320 | 0.97 | 11.71 |
| 2012 | 27801.3 | 272850 | 9.8 | 112.76 |
| 2013 | 27573.1 | 393400 | 1.43 | 162.57 |
| 2014 | 27239.1 | 400764 | 1.47 | 165.62 |
| 2015 | 26901.8 | 410550 | 1.53 | 169.66 |
| 2016 | 27026.0 | 381173 | 1.41 | 157.52 |
| 2017 | 27585.2 | 420000 | 1.52 | 173.57 |

**Table 4** Number of enterprises engaged in organic production by region, 31.12.2016

| No | Region | Number of enterprises | No | Region | Number of enterprises |
|---|---|---|---|---|---|
| 1 | Kiev | 58 | 13 | Kirovograd | 14 |
| 2 | Odessa | 38 | 14 | Mykolayiv | 13 |
| 3 | Kherson | 38 | 15 | Zaporizhzhia | 12 |
| 4 | Kharkiv | 29 | 16 | Zacarpattia | 12 |
| 5 | Zytomyr | 27 | 17 | Volyn | 10 |
| 6 | Lviv | 26 | 18 | Ivano-Frankivsk | 9 |
| 7 | Dnipropetrovsk | 23 | 19 | Rivne | 8 |
| 8 | Khmelnytskyi | 21 | 20 | Ternopil | 7 |
| 9 | Vinnytsia | 19 | 21 | Summy | 7 |
| 10 | Poltava | 18 | 22 | Chernivtsi | 4 |
| 11 | Chernigov | 16 | 23 | Donetsk | 2 |
| 12 | Cherkasy | 14 | 24 | Lugansk | 1 |
| Total number of enterprises engaged in organic production—426 | | | | | |

*Source* 7

Certified organic agricultural and processing products are produced in Ukraine: various vegetables, seasonal fruits and berries, pumpkin, melon, watermelon, eggs, mushrooms, greens, nuts, honey, cereals, flour, flakes, jam, syrups, juices, butter, spices, bakery, dairy, and meat products. Imported organic products in Ukraine include baby food, tea, coffee, sugar, spices, fruits, vegetables, pasta, chocolate, butter, wine, and beer.

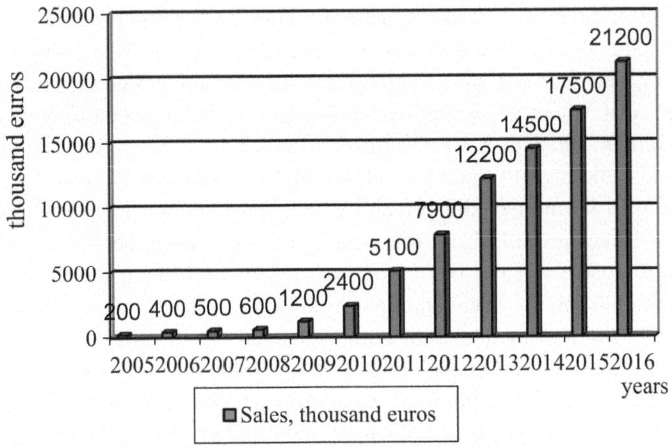


**Fig. 3** The volumes of retail trade in organic products on the domestic market in Ukraine, 2005–2016. *Source* Federation of Organic Movement of Ukraine [9]

Federation and Organic Movement's studies in Ukraine show that the modern domestic consumer market for organic products in the country is growing and has the corresponding sizes.

Studies of the Federation of Organic Movement of Ukraine show that the modern domestic consumer market for organic products in the country is growing and has the corresponding sizes (Fig. 3).

In 2017, the main organic products that Ukraine supplied to its foreign markets to its customers were corn, wheat, soybeans, barley, sunflower, millet, blueberries (frozen), oats, millet, apples (fresh), buckwheat, birch juice, flax, rye, sea buckthorn (frozen), blackberry (frozen), hips (frozen), coriander, peas, elderberry flower (frozen), strawberry (frozen), cranberry (frozen), chamomile (dried), hemp, raspberry (frozen), and sunflower oil (The Law of Ukraine "On the production and circulation of organic agricultural products and raw materials") [6].

The main importing countries for Ukrainian organic products are the Netherlands, Germany, Great Britain, Italy, Austria, Poland, Switzerland, Belgium, Czech Republic, Bulgaria, and Hungry. In addition, Ukrainian products are bought by the USA, Canada, Australia, and some Asian countries (The Law of Ukraine "On the production and circulation of organic agricultural products and raw materials") [6].

## 4 Conclusion

Ukraine is a state that increases the pace of production of organic products. The country holds the 20th place in the ranking of the countries-producers of this product and has basic developments in this direction, namely the leading organ of certification "Organic Standard", 15 certification bodies on organic production, adjusted production sales through the network of supermarkets Good Wine, Silpo, Auchan, Megamarket. The main organic products (in volume) that Ukraine supplies to foreign markets are corn, wheat, soybeans, barley, sunflower, millet, oats, millet, buckwheat, apples. Ukrainian legislation, which already has the Law on Organic Production, Distribution and Labeling of Organic Products, and the National Organic Logo should direct efforts to harmonize the principles of environmental legal regulation with the order in force in the EU. At the same time, work must be carried out to create the consciousness of the buyers of such products through the involvement of non-governmental public organizations through the introduction of environmental education. On the part of the producers, they need optimization of tax legislation in order to ensure economic development and stimulate ecologically safe behavior. Thus, the application of a comprehensive approach to the implementation of the proposed measures will bring positive changes and rapid growth of the market capacity of organic products both internally and internationally.

## References


1. Marx, K., Engels, F.: Soch. 2nd ed. T. 25.Economics, Moscow 437 p (1984)
2. The World of Organic Agriculture 2018
3. Willer, H., Lernaud, J.: The world of organic agriculture. Statistics and new trends 2018 (2018)
4. The World of Organic Farming 2018
5. Plant Growing of Ukraine 2017. Statistical Collection
6. The Law of Ukraine "On the production and circulation of organic agricultural products and raw materials", 2013
7. The Law of Ukraine "On basic principles and requirements for organic production, turning and marking of organic products", 2018
8. Organic map of Ukraine, 2017
9. Federation of Organic Movement of Ukraine, 2016

# Development and Incipience Decentralization of Authority in Ukraine and Formation of Its Impact on Local Budgets' Financial Capacity

**Svitlana Osypenko, Inna Kohut, Olena Iatsukh and Elvina Abliazova**

## 1 Introduction

As a result of the aggravation of the political situation and the intensification of the economic crisis in Ukraine, the question of the formation of a law-governed state with a socially oriented market economy is extremely urgent. The key direction among the complex of reforms initiated in Ukraine is the decentralization of power, as it is the realization of a decentralization policy that will allow building a more efficient system of public administration, fundamentally updating the content of activities of both executive and local self-government bodies.

According to the provisions of the Strategy for Sustainable Development "Ukraine 2020" [1], the goal of decentralization policy is to move away from the centralized model of governance in the state, to ensure the capacity of local self-government and to build an effective system of territorial organization of power in Ukraine, full implementation of the provisions of the charter on local self-government, the principles of subsidiarity, universality and financial self-sufficiency of territories.

## 2 Problem Statement

At the present stage of its development, Ukraine faced a number of problems in the system of state and local authorities, namely:

- excessive fragmentation of territorial communities and an increase in the depopulation of the country's rural population. So, since 1991, the number of rural populations has decreased by 2.5 million people, and the number of rural settlements—by 348 units. At the same time, the number of village councils increased

S. Osypenko (✉) · I. Kohut · O. Iatsukh · E. Abliazova
Tavria State Agrotechnological University, B. Khmelnitsky ave., 18, Melitopol 72312, Ukraine
e-mail: sa_osypenko@i.ua

© Springer Nature Switzerland AG 2019

V. Nadykto (ed.), *Modern Development Paths of Agricultural Production*,
https://doi.org/10.1007/978-3-030-14918-5_52

by 1067 units. In Ukraine, more than 12 thousand territorial communities have been formed; more than 6 thousand communities have a population of less than 3 thousand people, of which 4809 communities have fewer than 1000 people and less than 500 people in 1129 communities [1];

- disorderly system of the administrative–territorial system of the country and low level of development of the institution of local self-government, which is due to: lack of a unified classification of administrative–territorial units; legislative unregulated legal status of administrative–territorial units, the absence of clearly defined criteria at the legislative level for the formation of district and districts in cities; significant disproportion within the financial and economic potential of the administrative–territorial units of the same level, etc.;
- low level of financial capacity of local self-government bodies. It should be noted that the local government of village communities had virtually no financial resources to carry out its own provided by this law powers. According to the All-Ukrainian Association of Village and Village Councils, only 5% of rural communities can be considered self-sufficient, capable of providing socioeconomic development of controlled territories [2];
- declaring the principle of subsidiarity within the framework of income formation and implementation of the expenditures of local budgets, as well as the absence of a clear division of powers between local governments and local state administrations;
- high degree of deterioration or the absence of communal infrastructure in general, as well as low qualifications of local government officials. In particular, in more than half of rural communities, local self-government bodies are represented only by the chairman and secretary, and as a result, virtually no services from local governments receive no residents [3];
- legislative disregard for issues of ownership and disposition of territorial communities to land outside the settlement, as well as the transfer of these land plots from state ownership to communal.

All these problems arose not immediately, but, firstly, as a result of the Soviet principle of the formation and functioning of local self-government and territorial organization, which did not work in the conditions of the newest Ukraine [4]. That is, in an attempt to build a decentralized model of state power, the administrative–territorial system of the country was not reformed, which completely does not correspond to the conditions of decentralization. After all, as a result of the implementation of the policy of decentralization of power, there is a change in the management system from the monopoly–hierarchical to the partner-horizontal, and the key part of the powers and financial resources goes from the state authorities to the bodies of local self-government [5]. Secondly, the lack of an integrated approach to the implementation of the policy of decentralization of power, as well as the fragmentation of reforms in this direction, caused such a difficult situation.

## 3 Literature Review

In theoretical and practical research, the notion of "decentralization" is multidimensional and is considered in various ways. At the semantic level, decentralization (from the Latin *de*—"opposition" and *centralis*—"central") is interpreted as the destruction, weakening or abolition of centralization [6, p. 8].

As American Scientist M. Bell notes, "decentralization is a way of measuring the size of the transition of the national economy from the administrative–command to the market" [7, p. 237].

Scientists have identified several types of decentralization that have certain characteristics, implications and conditions of the implementation, which can be implemented, both independently and in a relationship with each other, in different forms and in various combinations, both within the country and in certain branches of national economy [8, p. 149].

So, V. Zaychikova distinguishes three types of decentralization, namely: economic, administrative and political. At the same time we note, that fiscal decentralization is a kind of economic decentralization [9, p. 58].

In turn, S. Volokhova observes that the following forms (types) of decentralization are distinguished in science: political (democratic), administrative (bureaucratic), economic (market, financial), territorial–spatial [10, p. 282]. G. Sach, P. Pavlichenko, M. Polonsky, A. Ursu [11, p. 8] and N. Bikadorova [4, p. 146] distinguish administrative, political and financial decentralization.

Also, along with administrative and political decentralization, domestic scholars allocate budget decentralization and decentralization of financial resources to local budgets.

Taking into account the existence of certain differences in view of domestic scientists regarding the definition of the types of decentralization, in their studies they unanimously point out that the key role in implementing the policy of decentralization of the country's power belongs to the process of separating functions and powers between central and local levels of government within the framework of provision citizens of the country, public goods and services, as well as the separation of financial resources to the extent sufficient to finance the local authorities' own powers about self-government. As a result, local self-government bodies reach a sufficient level of financial autonomy in solving their own self-governing powers.

In foreign countries, fiscal decentralization is realized through self-financing by collecting fees for services rendered to consumers, the application of a scheme of joint financing of services through cash and labor contributions of consumers, an increase in local revenues through the introduction of property tax or sales tax, the use of intergovernmental transfers, the approval of municipal borrowings and mobilizing resources of central or local authorities through loan guarantees. In addition, the practical realization of fiscal decentralization is based on the principle of subsidiarity. In accordance with this principle, it is expedient to retain the authority of the center only on condition of their more effective implementation, which requires clear legislative consolidation and financial and economic substantiation of the vertical distribution of powers between levels of government in the state [9, p. 59].

The effectiveness of the implementation of these provisions was substantiated in the scientific works of a number of foreign scientists–financiers.

So, the founder of the theory of fiscal decentralization is the English Economist Ch. Tibu, who stated that "it is fiscal decentralization that increases the level of competition between local authorities, which significantly limits the size of the public sector. In addition, decentralization increases efficiency, because, unlike central government, the local has more complete information regarding the needs of representatives of the territorial community" [12, p. 93].

In our opinion, the most comprehensive and thorough approach to the definition of the essence of fiscal decentralization was defined by R. Masgrave and P. Masgraviev, who from the viewpoint of the allocation function characterize fiscal decentralization as the most effective system for allocating financial resources. The policy of local authorities involves the differentiation of decisions in the financial sphere in order to best reflect the preferences of their residents [13, p. 60–98].

In turn, the British Economist O. Outs in his studies defines fiscal decentralization as the right of local governments to make decisions on financial matters and territorial management issues on their own. In addition, the so-called decentralization theorem, based on the following assertions: the benefits associated with the decentralization of public services, is spatially constrained; providing consumers with benefits to private or public goods, as well as their attitude to various public goods is not the same; the reaction of consumers to "factors of budget localization" is mobile. This is the difference between services provided at the local level and tax rates. In other words, each consumer, changing his place of residence, can pick up the most favorable services and pay for them (local tax rates). If these conditions are not fulfilled, then, according to O. Outs, there are no grounds for decentralization in terms of economic efficiency [9, p. 63].

Consequently, as a result of the analysis of foreign experience in implementing the policy of decentralization of power, we note that the term "fiscal decentralization" used by foreign scholars is used to define the process of empowering the rights and competencies of local authorities in the field of local finance.

Definition of this category has not been reflected in the norms of the current domestic legislation. However, in their research domestic scientists usually defining the same process, use in parallel concepts such as "fiscal decentralization" or "financial decentralization" and "budget decentralization."

## 4 Methodology, Variables and Data

The following indices are used to assess the level of fiscal decentralization in foreign practice: the ratio of local government incomes to total incomes of power and the ratio of the expenditures of local authorities to total expenditures of power [14].

The first indicator shows the degree of involvement of local authorities in the processes of mobilizing public resources through the tax system. However, this index does not consider the responsibility of local authorities for providing public goods at the expense of funds received from external sources. This kind of social activity can be better assessed as a share of the expenditures carried out by subnational authorities. However, in this case, there are several limiting factors when comparing the expenditures of the state and local authorities. In particular, in the case where the local government acts as a central government agent (e.g., state administrations), it is not financially autonomous. Therefore, this indicator does not reflect the decentralization of the expenditures, and centrally collected taxes do not show genuine decentralization of incomes. One more limitation of the index is that it does not consider the number of local authorities involved in the distribution of the expenditures. Under the same conditions, more local governments will have a higher degree of fiscal decentralization [14].

One of the most commonly used methods in European countries for evaluating financial decentralization is the methodology developed by the Organisation for Economic Cooperation and Development (OECD).

The method for assessing the degree of fiscal decentralization of the OECD is based on an analysis of the autonomy of regional and local authorities in the fiscal and expenditure policy, the distribution order and the type of intergovernmental transfers received from higher level authorities, as well as the degree of independence within the framework of debt policy. However, the characteristic of this system is the use of only qualitative indicators, which to some extent complicates the definition of the degree of fiscal decentralization.

Cherniavsky A.V. notes that the criteria for the budgetary autonomy of local authorities include the right of subnational authorities to set tax rates and tax base, or only tax rates, or tax base. In other cases, fiscal independence is lacking or significantly limited [10, p. 137].

As for the research of domestic scientists, the most widespread approach to determining the degree of financial decentralization in general, as A. Luchka notes, is the application of such macroeconomic indicators: the share of revenues and expenditures of local budgets in GDP and in the consolidated budget of the country [15, p. 375].

Also, in the domestic financial literature, there is a somewhat different approach, on the basis of which one can determine the level of financial decentralization for each link of local budgets:

Coefficient of financial decentralization on incomes of the consolidated budget of Ukraine is calculated as the ratio between the revenues of local budgets of Ukraine and incomes of the consolidated budget of Ukraine.
Coefficient of financial decentralization for consolidated budget expenditures is calculated as the ratio between expenditures of local budgets of Ukraine and expenditures of the consolidated budget of Ukraine.

Coefficient of decentralization of parts of local budgets is calculated as the ratio of incomes of a certain level of local budgets (revenues of the budgets of the ARC, regions, district, city, settlement, village) with local budget revenues.
Generalized coefficient of financial decentralization can be calculated in two ways: Firstly, it is calculated as the sum of the coefficient of financial decentralization for the consolidated budget revenues and the coefficient of financial decentralization in the consolidated budget expenditures is divided into two; secondly, it is calculated as the ratio of the coefficient of decentralization of parts of local budgets to the number of levels of local budgets [4, p. 150].

In our view, the use of these methods of macroeconomic indicators is entirely justified, since decentralization of financial resources is a generic concept and covers the whole system of local finance.

## 5 Empirical Analysis

The practical realization of the reform of local self-government was started in mid-2015 through the implementation of administrative–territorial reform, namely the association of territorial communities in order to strengthen their capacity. The key parameters of the rating of the assessment of the capacity of the united territorial communities (UTCs) are as follows: the number of UTC; their area; the number of united communities; the number of UTCs from the population of less than 5 thousand people.

Thus, taking into account a certain legal inconsistency in the implementation of this reform, legislative changes have, however, become the impetus for the intensification of the process of building a new system of local self-government. In 2015, the regional state administrations formed the prospective plans for the association of territorial communities (except for the Transcarpathian and Ternopil regions). It should be noted that not all promising plans were agreed upon by the Cabinet of Ministers of Ukraine. In general, almost 30% of the perspective plans for the association of territorial communities did not find support from the Cabinet of Ministers of Ukraine and were transferred to the regional state administrations for further elaboration. Besides, the approval of prospective plans for the association of territorial communities, in our opinion, contradicts the principle of voluntariness. In addition, according to the data [16], almost all regions have territorial communities that are not included in the promising plan of the association; that is, within these territories there is no alternative to the association. In the future, this may slow down the implementation of the administrative–territorial reform, as well as weaken the financial capacity of territorial communities that have not gone through the association procedure.

In addition, a significant part of the promising plans was formed without following the main requirements of the current methodology, as evidenced by significant differences between the plans of long-term plans approved by the regional councils and approved by the government.

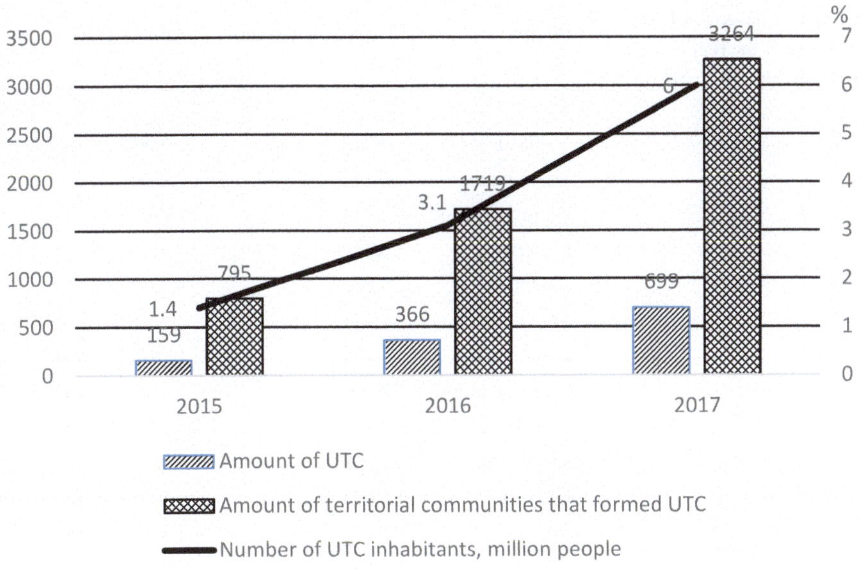


**Fig. 1** Formation dynamics of UTC in Ukraine (*based on data* [18])

It should be noted that the changes to the long-term plans continue to this day, despite that in fact for 2015–2017 3264 settlements voluntarily united in 699 united territorial communities, while this is a quarter of the planned volume (Fig. 1). In addition, the amount of UTC residents during this period increased from 1.4 million to 6 million; that is, 1.4% of the population of Ukraine are residents of UTC.

The next indicator included by the experts on decentralization to the characteristics of the capacity of the united territorial communities and to determine their ranking is the area of UTC. So, characterizing this indicator, it should be noted that the volume of UTC area is also unevenly distributed among regions (Table 1). The regions-leaders are Zhytomyr region, in which the area of UTC is 56.4% of the total area of the oblast, Zaporizhzhya region—54.9%, Dnipropetrovsk region—49.1%, Volyn region—45.2%, Ternopil region—39.2%, respectively. It should be noted that the Dnipropetrovsk, Zhytomyr and Zaporizhzhya regions are among the largest regions in Ukraine.

In addition, Ukraine already has 5 UTC the area which coincides with the territory of respective administrative districts: Narodytska (Zhytomyr region), Starosynyavska, Letychivska (Khmelnytsky region), Snovska (Chernihiv region) and Apostolivska (Dnipropetrovsk region).

On the one hand, large UTCs have the opportunity to significantly reduce management costs, as well as have a higher level of solvency due to the increase of taxation objects by local taxes and fees, which, accordingly, will ensure the growth of local budget revenues . On the other hand, large-scale UTCs, hypothetical social

**Table 1** Analysis of UTC coverage of the region's area in the period 2015–2017

| Region | Area of the region, sq.km. | Area of UTC | |
|---|---|---|---|
| | | sq.km. | % |
| 1 | 2 | 3 | 4 |
| Vinnitsa | 26,513 | 3896 | 14.7 |
| Volyn region | 20,144 | 9095 | 45.2 |
| Dnipropetrovsk | 31,914 | 15,669 | 49.1 |
| Donetsk | 26,517 | 5293 | 20 |
| Zhytomyr | 29,832 | 16,811 | 56.4 |
| Transcarpathian | 12,777 | 616 | 4.8 |
| Zaporizhzhya | 27,180 | 14,922 | 54.9 |
| Ivano-Frankivsk | 13,900 | 2342 | 16.8 |
| Kievskaya | 28,131 | 2231 | 7.9 |
| Kirovograd | 24,588 | 252 | 10.3 |
| Lugansk | 26,684 | 5331 | 20 |
| Lviv | 21,833 | 3467 | 15.9 |
| Nikolaev | 24,598 | 8917 | 36.3 |
| Odessa | 33,310 | 7378 | 22.2 |
| Poltava | 28,748 | 7750 | 27.0 |
| Rivne | 20,047 | 4777 | 23.8 |
| Sumy | 23,834 | 7990 | 33.5 |
| Ternopil | 13,823 | 5422 | 39.2 |
| Kharkiv | 31,415 | 3676 | 11.7 |
| Kherson | 28,461 | 7209 | 25.3 |
| Khmelnitsky | 20,645 | 11,149 | 5.4 |
| Cherkassy | 20,900 | 3730 | 17.8 |
| Chernivtsi | 8097 | 2656 | 32.8 |
| Chernihiv | 31,865 | 16,396 | 51.5 |

*Source* Compiled from [18]

and public services, which are concentrated within the administrative center of UTC, are becoming far removed from the community members who became the part of UTC.

In addition, the outlined trends indicate that today the issue of reforming the district and regional levels of government is already pressing.

Also, one of the key indicators for determining the level of UTC capacity is to determine the number of UTCs from the population of less than 5 thousand people.

According to the method, capable territorial communities are territorial communities of villages (settlements, cities), which, as a result of a voluntary association, can independently or through appropriate local self-government authorities provide the corresponding level of services, in particular in the field of education, culture, health care, social protection, housing and communal services, taking into account human resources, financial support and infrastructure development of the respective administrative–territorial unit [17]. However, based on the experience of European countries, it should be noted that communities with a population of more than 5 thousand inhabitants are able to achieve a capable status.

For example, for Ukraine, the formation of territorial communities with a population of less than 5 thousand inhabitants is characteristic. Accordingly, in 85% of Ukraine's regions, the number of people in OTs ranges from 3 to 5 thousand people. Only in Transcarpathian and Kharkiv regions, there is no OTG with a population of less than 5 thousand people.

However, today important are not quantitative but qualitative parameters of decentralization of local self-government authorities. To ensure the sustainable development of the territory and the provision of quality services to the population, the community must be financially capable.

Overall, the result of the reform was an increase in the interest of local governments in increasing revenues to local budgets, finding reserves for their filling and improving the efficiency of administering taxes and fees. The dynamics of revenues to the budgets of UTC is ahead of the growth rate of income at other levels of local budgets. Receipt of own revenues of local budgets 366 UTC in 2017 grew by 87% and amounted to UAH 9.3 billion (+4.3 billion UAH). And the growth rate of own revenues of 159 OTGs in 2016 is +34.2% (+1.1 billion UAH), which is almost 3% more than the average in Ukraine [18].

Given the expansion of the income base of local budgets, an increase in the share of local budgets in the total consolidated budget of Ukraine has taken place. According to the results of 2017, the share of local budgets has exceeded fifty percent mark. The share of local budgets in the consolidated budget incomes of Ukraine amounted to 51.2%, which is almost 6% more than the same indicator for 2015 (Table 2).

Simultaneously with an increase in the share of local budgets in the financial resources of the state, their share in the volume of gross domestic product is increasing. In 2014–2015, this figure was 6.4 and 6.1%, respectively, and in 2016—7.1%; by the end of 2017, the share of local budgets in the country's GDP is 7.7%.

Using the approach of domestic scientists in the area of determining the level of financial decentralization, it can be noted that in fact, during 2014–2017, the calculated coefficients increase (Table 3).

Thus, the generalized financial decentralization ratio in 2017 was 0.51, correspondingly increasing by 1 point compared to 2014 and by 4 points compared to 2015 and 2016, indicating that the local authorities are still starting to form a significant financial resource to be able to effectively manage and direct them to community development.

**Table 2** Share of incomes of local budgets of Ukraine in incomes of the consolidated budget of Ukraine and GDP, %

| Year | Share in the consolidated budget incomes of Ukraine | | Share in the redistribution of GDP | |
|---|---|---|---|---|
| | Without transfers | Taking into account transfers | Without transfers | Taking into account transfers |
| 2008 | 24.8 | 46.1 | 8.3 | 14.5 |
| 2009 | 26.0 | 49.3 | 7.9 | 14.7 |
| 2010 | 25.9 | 50.6 | 7.2 | 14.2 |
| 2011 | 21.7 | 45.5 | 6.4 | 13.4 |
| 2012 | 22.6 | 50.5 | 6.9 | 15.4 |
| 2013 | 22.7 | 49.9 | 6.9 | 14.5 |
| 2014[a] | 22.1 | 50.8 | 6.4 | 14.7 |
| 2015[b] | 18.5 | 45.6 | 6.1 | 14.8 |
| 2016[b] | 21.8 | 46.7 | 7.1 | 15.3 |
| 2017[b] | 22.6 | 51.2 | 7.7 | 16.8 |

[a]The data of budget reporting of the Autonomous Republic of Crimea and the city of Sevastopol for the first quarter of 2014 are taken into account
[b]Data are given without taking into account the Autonomous Republic of Crimea, the city of Sevastopol and parts of the zone of ATO

**Table 3** Level of financial decentralization in Ukraine

| Indexes | Years | | | |
|---|---|---|---|---|
| | 2014 | 2015 | 2016 | 2017 |
| Coefficient of financial decentralization on incomes of the consolidated budget of Ukraine | 0.5 | 0.45 | 0.46 | 0.51 |
| Coefficient of financial decentralization for consolidated budget expenditures | 0.43 | 0.41 | 0.41 | 0.41 |
| Generalized coefficient of financial decentralization | 0.46 | 0.43 | 0.43 | 0.47 |

## 6 Conclusion

For today, the peculiarity of the decentralization implementation in Ukraine is that the reform of the institution of local self-government, the administrative–territorial organization and the state regional policy is simultaneously taking place. It should be noted that the administrative and territorial reform plays the decisive role in the process of strengthening decentralization in Ukraine, since this reform is intended to ensure the formation of financially capable communities that can not only meet the minimum social needs of the population, but also provide integrated and systemic development of the territories, increase the level financial autonomy of local authorities within the framework of ensuring their own powers, promote the balance of national interests with the interests of rural areas, stimulate entrepreneurial activity, etc.

Despite the strengthening of the positive trend regarding the implementation of the administrative–territorial reform in Ukraine in the period which was set aside by the concept of reforming local self-government and territorial organization of the authorities for realization of the reform, namely in the period from 2015 to 2017, only a quarter of the planned volume united territorial communities was formed.

The adopted changes to the budget law can be considered as the first step toward a decentralized government reform, as evidenced by the share of revenues and expenditures of local budgets in the consolidated budget and GDP. However, it is too early to give an overall assessment of the reform of local self-government and the territorial organization of authority in Ukraine, since the process is not yet complete.

## References


1. Pro Stratehiiu staloho rozvytku «Ukraina – 2020»: Ukaz Prezydenta Ukrainy vid 12.01.2015 r. № 5/2015 [Online]. Retrieved from: http://zakon2.rada.gov.ua/laws/show/5/2015 [in Ukrainian]
2. Mistsevi finansy: pidruchnyk: [avt. kol.; za red. d-ra. ekonom. nauk, prof. O. B. Zhykhor, d-ra ekonom. nauk, prof. O. P. Kyrylenko]. *K.: UBS NBU*, 579 p (2015) [in Ukrainian]
3. Tkachuk, A.: Miliardy, vytracheni na ATR, pidut na rozvytok vsiiei Ukrainy. *Informatsiino-analitychnyi resurs «Dialoh UA»*. [Online]. Retrieved from: http://www.dialogs.org.ua/ru/dialog/page22-57.html [in Ukrainian]
4. Bykadorova, N.O.: Finansova detsentralizatsiia mistsevoho samovriaduvannia. *Ekonomichnyi visnyk Donbasu: nauk. zhurn./ Luhan. nats. un-t im. T. Shevchenka, In-t ekon. promyslovosti Nats. akad. nauk Ukrainy*, **2**(20), 145–151 (2010) [in Ukrainian]
5. Osypenko, S.O.: Mistsevi biudzhety v umovakh detsentralizatsii finansovykh resursiv: dys. kand. ekonom. nauk: 08.00.08. *Melitopol*, 234 p (2016) [in Ukrainian]
6. Drozdovska, O.S.: Finansova detsentralizatsiia: obiektyvna neobkhidnist ta osnovni tsili. *Naukvi pratsi NDFI*, **2**(17):8–14 (2002) [in Ukrainian]
7. Bell, M., Adams, C.: Fiscal decentralization indicators: local democratic governance. In: Proceedings of 92' annual conference, pp. 236–241. National Tax Association, Atlanta (1999) [in English]
8. Stukalo, N.V., Derkach, M.I., Stukalo, M.V.: Teoretyko-metodolohichni zasady fiskalnoi detsentralizatsii. *Ekonomichnyi prostir: zb. nauk. pr. Prydniprov. derzh. akad. bud-va ta arkhit*, **40**, 147–154 (2010) [in Ukrainian]
9. Zaichykova, V.V.: Mistsevi finansy Ukrainy ta yevropeiskykh krain. *K.: NDFI*, 299 p (2007) [in Ukrainian]
10. Cherniavskyi, A., Vartapetov, K.: Biudzhetnaia detsentralyzatsyia v stranakh s perekhodnoi ekonomykoj. *Voprosy ekonomyky*, **11**, 126–141 (2004) [in Russian]
11. Detsentralizatsiia ta efektyvne mistseve samovriaduvannia: navch. posib. dlia posadovtsiv mistsevykh ta rehionalnykh orhaniv vlady ta fakhivtsiv z rozvytku mistsevoho samovriaduvannia (2007). *K.: PROON MPVSRT*, 269 p [in Ukrainian]
12. Tiebout, C.: Economic Theory of Fiscak Decentralzation. *Public Finances: Sources, and Utilization*. Princeton, New Jersey, pp. 76–96 (1961) [in English]
13. Masgrave, R.A., Masgrave, P.B.: Public finance in theory and practice, 5th edn. McGraw-Hill, Inc., 627 p (1989) [in English]

14. Pylypiv, V.V., Makarenko, L.M.: Detsentralizatsiia upravlinnia sotsialno-ekonomichnym rozvytkom Ukrainy. *Aktualni problemy rozvytku ekonomichnoho rehionu: zb. nauk. Pr.*, Vyp. 4(2) (2008) [Online] Retrieved from: http://www.nbuv.gov.ua/portal/Soc_gum/aprer/2008_4_2/103.pdf [in Ukrainian]
15. Luchka, A.V.: Teoretyko-metodolohichni osnovy otsiniuvannia finansovoi ta biudzhetnoi samostiinosti orhaniv mistsevoho samovriaduvannia. *Ekonomichnyi analiz: zb. nauk. pr. Ternopilskoho nats. ekon. un-t,* Vyp. 9. Chastyna 3, pp. 375–381 (2011) [in Ukrainian].
16. Butska, O.I., Tymoshenko, O.V.: Komunalnyi bank yak instytut spryiannia sotsialno-ekonomichnoho rozvytku terytorialnykh hromad. *Investytsii: praktyka ta dosvid*, 20, 44–46 (2013) [in Ukrainian]
17. Bondaruk, T.H.: Zarubizhnyi dosvid mistsevoho opodatkuvannia ta mozhlyvosti yoho zastosuvannia v Ukraini. *Zb. nauk. prats Nats-ho un-tu derzh. podatkovoi sluzhby Ukrainy. Irpin*, 1:56–66 (2011) [in Ukrainian]
18. Monitorynh protsesu detsentralizatsii vlady ta reformuvannia mistsevoho samovriaduvannia. [Online]. Retrieved from: http://decentralization.gov.ua/monitoring2018_01_ua#main-info [in Ukrainian]

# Analyses of Personnel Usage at Agricultural Enterprises

**Yurii Prus, Tetiana Yavorska, Olena Voronianska and Olexiy Petryha**

## 1 Introduction

The main reason to investigate personnel usage is that economic science explores the ways of the most effective usage of resources to meet the needs of society. At the same time, considerable attention belongs to labor itself as an essential factor of production. The growth of labor productivity is a key factor for ensuring sustainable and strong economic development of the country and raising the people's standards of living. Modern development of enterprises is characterized by growing role of the human factor because financial result is more and more determined by the quality and motivation of its employees. Indicators that analyze the labor force efficiency and labor costs are significantly important. They provide essential information about the state of the workforce, and the results of the analysis can be vital for making important managerial decisions related to the goals and strategies of enterprises and industry as well. Therefore, an analysis of the current situation of personnel usage is indispensable.

Y. Prus (✉) · T. Yavorska · O. Voronianska
Tavria State Agrotechnological University, Melitopol 72310, Ukraine
e-mail: yurii.prus@tsatu.edu.ua

T. Yavorska
e-mail: super_yavort@ukr.net

O. Voronianska
e-mail: elenavoron1955@gmail.com

O. Petryha
Dnipro's Branch of Open International University of Human Development "Ukraine", Dnipro 49112, Ukraine

© Springer Nature Switzerland AG 2019

V. Nadykto (ed.), *Modern Development Paths of Agricultural Production*,
https://doi.org/10.1007/978-3-030-14918-5_53

# 2 Methods, Information, and Results

## 2.1 Methods and Information

There are a large number of publications in the economic literature that highlights the problems of efficient usage of personnel at the enterprise.

In particular, attention is paid to the management of human resources at farms [1–3]. The attraction, motivation, and retention of sufficient and skilled labor force remain an important problem for a long time [4, 5]. Based on the current system of staff motivation which aim is to increase in labor productivity, the main measures for improving the usage of personnel at agricultural enterprises are substantiated [6]. Labor productivity at agrarian enterprises is the subject of analysis both at the global level [7–9] and at the level of independent states [10–15]. Scientists have determined various factors of labor productivity and ways to increase it. The dynamics of wages and the factors affecting them are also an important part of the scientific study of personnel [16]. Forecasting models of the labor market and wages [17–20] are developed based on these factors.

However, despite the wide range of existing scientific outcomes, the problem of staff usage at agricultural enterprises requires further researches. The main indicators that characterize the effectiveness of the personnel usage are the number of employees, the number of manufactured products or volume of performed work, the cost of working time, as well as value of the personnel movement.

Gross production at Ukrainian agricultural enterprises is an initial result of the interaction of manufacturing factors, the material, and cost outcome of other outputs of production activity. It acts in kind (all the products produced in a given year in a particular industry) and in value form (natural products estimated at comparable prices of the given year). The latest comparative prices were determined by the State Statistics Service of Ukraine in 2010 and based on them—average market prices. Determination of gross production in value form is necessary to ensure comparability in time and space parameters of gross production, productivity, and other cost factors of production efficiency within the industry and individual enterprises.

In order to determine labor productivity in the current year prices, the total incomes of enterprises, profit, or profit from sales of agricultural products as well as gross added value may be used. In our study, labor productivity was determined by dividing the above-mentioned cost indicators into the number of employees in a certain period. Relative labor productivity ($LP_R$) is determined by the formula:

$$LP_R = \frac{GP_0}{AEN_1}, \quad (1)$$

where

$GP_0$ gross agricultural production in the base year;
$AEN_1$ average number of employees in the reported year.

Absolute change in labor productivity based on gross production ($\Delta LP_{GP}$) is determined by the formula:

$$\Delta LP_{GP} = LP_1 - LP_R, \quad (2)$$

where

$LP_1$ labor productivity in the reported year.

Absolute change in labor productivity based on the number of employees is determined by the formula:

$$\Delta LP_{AEN} = LP_R - LP_0, \quad (3)$$

where

$LP_0$ labor productivity in the base year.

The determination of real wages in purchasing power in 2017 was calculated by multiplying the nominal wage by the corresponding annual consumer price indices:

- up to 2010—$i_{2011} \cdot i_{2012} \cdot i_{2013} \cdot i_{2014} \cdot i_{2015} \cdot i_{2016}$;
- up to 2011—$i_{2012} \cdot i_{2013} \cdot i_{2014} \cdot i_{2015} \cdot i_{2016}$;
- up to 2012—$i_{2013} \cdot i_{2014} \cdot i_{2015} \cdot i_{2016}$;
- up to 2013—$i_{2014} \cdot i_{2015} \cdot i_{2016}$;
- up to 2014—$i_{2015} \cdot i_{2016}$;
- up to 2015—$i_{2016}$.

The change of nominal wages is determined by the formula:

$$\Delta W_{nom} = W_{nom}^{n} \cdot W_{nom}^{n-1}, \quad (4)$$

where

$W_{nom}^{n}$ nominal wage in period $n$;
$W_{nom}^{n-1}$ nominal wage in the period which precedes $n$.

The change in the minimum wage is determined by the formula:

$$\Delta W_{min} = W_{min}^{n} \cdot W_{min}^{n-1}, \quad (5)$$

where

$W_{min}^{n}$ minimum wage in period $n$;
$W_{min}^{n-1}$ minimum wage during the period which precedes $n$.

The nominal wage change via the company's decision is determined by the formula:

$$\Delta W_{nom}^{c} = \Delta W_{nom} - \Delta W_{min}, \quad (6)$$

The minimum required wage growth in comparison with the previous year which takes into consideration the inflationary growth of consumer prices is determined by the formula:

$$\Delta W_{\text{infl}} = W_{\text{nom}}^{n-1} \cdot i_{\text{infl}}^{n} - W_{\text{nom}}^{n-1}, \tag{7}$$

where

$i_{\text{infl}}^{n}$ index of inflation during period $n$.

Compensation of the inflation with the help of the growth of the minimum wage is determined by the formula:

$$\Delta W_{\min} - \Delta W_{\text{infl}}, \tag{8}$$

Compensation of the inflation with the help of the growth of nominal wages is determined by the formula:

$$\Delta W_{\text{nom}} - \Delta W_{\text{infl}}, \tag{9}$$

To predict nominal wages paying attention to the growth of real wages, we propose to use the following formula:

$$W_{\text{nom}}^{k} = W_{\text{nom}}^{n} \cdot i_{\text{infl}}^{k} \cdot i_{rw}^{k}, \tag{10}$$

where

$W_{\text{nom}}^{k}$ projected nominal wage during period $k$;
$W_{\text{nom}}^{n}$ actual nominal wage during period $n$;
$k$ periods of forecasting (1, 2 years, etc. ; in 2017 $n = 0$, in 2020 $k = 3$);
$i_{\text{infl}}^{k}$ inflation index for the period of forecasting;
$i_{rw}^{k}$ the real wage index for the period of forecasting.

To predict gross added value, we propose to use the following formula:

$$\text{GAV}^{k} = (\text{LP}_{\text{GAV}}^{k} \cdot \text{ANE}^{k})/1000, \tag{11}$$

where

$\text{GAV}^{k}$ projected gross added value during period $k$, billion UAH;
$\text{LP}_{\text{GAV}}^{k}$ projected labor productivity per employee in gross added value during period $k$, thousand UAH;
$k$ periods of forecasting (for 2020 $k = 3$);
$\text{ANE}^{k}$ estimated number of employees during period $k$, thousand people (determined by formulas 12 and 13).

Our researches were based on the materials of the State Statistic Service of Ukraine for 2010–2017—Statistical Yearbook "Agriculture of Ukraine" and Statistical Yearbook "Labor of Ukraine"—the data of which was used directly or to calculate the required indicators.

## 2.2 *Labor Productivity*

Our researches show that labor productivity at agrarian enterprises of Ukraine declined during 1990–2000 (Fig. 1). The main reasons for such a negative trend were over-employment, de-industrialization of agricultural production, declining crop yields, and livestock productivity.

Since 2000, the annual labor productivity of the agrarian sector of the country has steadily increased. As it is shown in Fig. 1, changes in labor productivity in agriculture were based on changes in labor productivity in crop production, while the cost of livestock production in the value of gross agricultural production was relatively lower; moreover, this share declined from 29 to 23% during 2010–2017.

It was determined that the basis for increasing labor productivity was mostly because of the increase in production volumes rather than decrease in labor force in the process of reforming agrarian enterprises (Table 1).

Thus, changes in labor productivity were based mainly on changes in gross production in 2011, 2012, 2013, 2015, and 2017. In particular, its decline in 2012, 2015, and 2017 led to decrease in labor productivity comparing to 2011, 2014, and 2016.

In agriculture, the staff contribution is also characterized by an indicator of the amount of agricultural land per employee in crop production and number of animals per employee in livestock (Fig. 2). The amount of agricultural land and the rela-

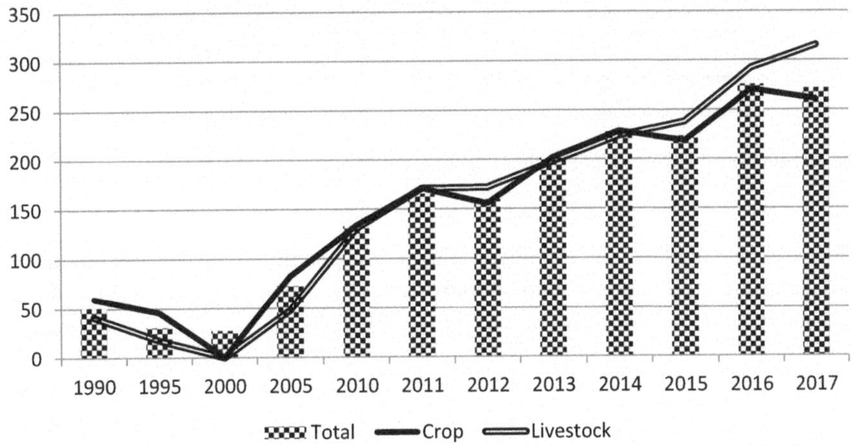


**Fig. 1** Labor productivity of Ukrainian agricultural enterprises, thousand UAH/person

**Table 1** Influence of factors on change of labor productivity at Ukrainian agricultural enterprises

| Indicators | Years | | | | | | | |
|---|---|---|---|---|---|---|---|---|
| | 2010 | 2011 | 2012 | 2013 | 2014 | 2015 | 2016 | 2017 |
| Cost of gross production in agriculture, UAH billions | 94.1 | 121.0 | 113.1 | 136.6 | 139.1 | 131.9 | 145.1 | 140.5 |
| Average number of employees in agricultural production, thousand persons | 709.1 | 732.6 | 708.2 | 678.8 | 610.6 | 590.7 | 527.1 | 517.6 |
| Labor productivity, thousands of UAH per person | | | | | | | | |
| – Real | 132.7 | 165.2 | 159.7 | 201.2 | 227.8 | 223.3 | 275.3 | 271.5 |
| – Relative | x | 128.4 | 170.9 | 166.6 | 223.7 | 235.5 | 250.2 | 280.4 |
| Changes of labor productivity, total, including via: | x | 32.5 | −5.5 | 41.5 | 26.6 | −4.5 | 52.0 | −3.8 |
| – Gross production | x | 36.8 | −11.2 | 34.6 | 4.1 | −12.2 | 25.1 | −8.9 |
| – Number of employees | x | −4.3 | 5.7 | 6.9 | 22.5 | 7.7 | 26.9 | 5.1 |

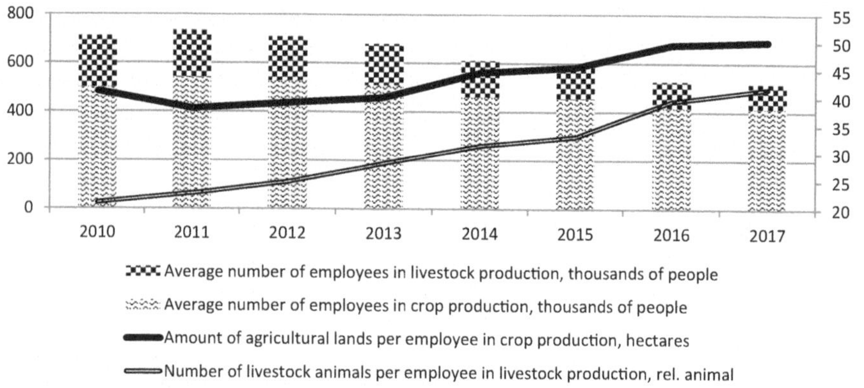


**Fig. 2** Number of employees at Ukrainian agrarian enterprises

tive livestock calculated per employee at the corresponding branches was gradually increasing due to reduction in workers and practically unchanged areas of agricultural land and quantity of livestock animals during 2010–2017.

## 2.3 Number of Employees

The number of employees is a factor that has permanently influenced the growth of labor productivity during the whole period of research except 2011. Usage of new technologies and modern high-productivity agricultural machinery lead to reducing of labor need. In general, from 1990 to 2010, the number of agribusiness workers decreased by 3.2 million people, including in animal husbandry—by 1.8 million people, and from 2010 to 2017—by 191.5 and 107.9 thousand people consequently (Fig. 2).

Data analysis shows that the total number of employees decreased by 27.0% during 2010–2017, and it was more rapidly, by 51.6%, in livestock production in comparison with crop growing.

It was determined that the most precise dynamics of the number of workers at agricultural enterprises in the Zaporizhzhya region is described by the equations:

$$\tilde{y}_t = 556.24 - 17.28t, \text{(Crop production)} \tag{12}$$

$$\tilde{y}_t = 227.51 - 15.91t, \text{(Livestock)} \tag{13}$$

These equations show that the number of workers in crop production decreased an average annually by 17.3 thousand people and in animal husbandry—by 15.9 thousand of people during 2010–2017.

Using these equations for prediction, we have found that the number of workers in crop production will be with a probability of 95% on average 383.4 thousand of people, ranging from 331.1 to 435.7 thousand of people, and in livestock production—68.5 thousand of people ranging from 58.8 to 78.1 thousand of people in 2018–2020. Thus, based on existing tendencies at agricultural enterprises on average for 2018–2020 comparing with 2010–2016, the number of employees is expected to decrease from 19 to 39%, including in crop production—from 8 to 31%, and in livestock production—from 50 to 62%.

During the years between 2010 and 2017, decrease in crop production at agricultural enterprises was determined by reduction in the share of areas under vegetable and fodder crops from 7.7 to 3.7% while the share of cereals and technical crops increased from 95.3 to 96.3%. The process of replacement of labor-intensive crops by less labor-intensive is still ongoing at agrarian enterprises although the share of less labor-intensive crops is still significant. In the near future, reduction in crop production will be affected as it was mentioned above by more intensive usage of foreign and domestic high-tech equipment—tractors, combines, and agricultural machines. For instance, only 1–3% of purchased combines were domestic in Zaporizhzhya region during 2011–2016 (thus, most of the purchased combines are foreign and most of them are high yielding). Meanwhile, the share of purchased tractors with power more than 100 kW increased from 24 to 35%. It should be emphasized that the share of domestic tractors among them increased from 14 to 35%.

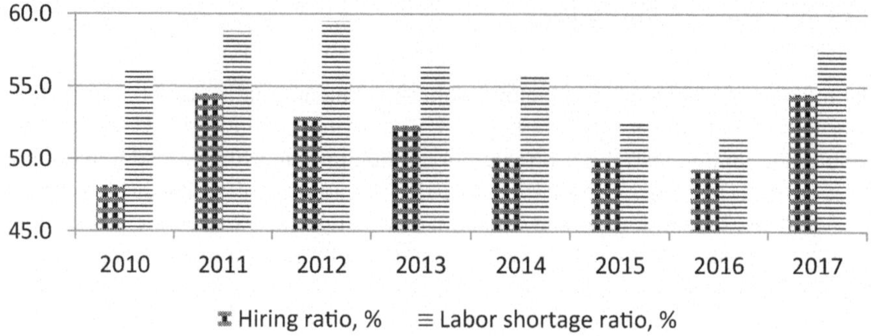


**Fig. 3** Movement of personnel at Ukrainian agricultural enterprises

Reduction in livestock workers at Ukrainian agricultural enterprises was, first of all, due to a significant change in the structure of the industry during 2010–2017. For instance, the livestock population decreased by 359.8 thousand of heads and its share in the structure of the total population—from 45.1 to 21.3%. At the same time, the number of pigs and poultry increased by 26.9 and 68.9%, respectively, and their shares increased from 19.1 to 23.8% and from 32.6 to 54.0%, respectively. The level of mechanization and automation of production in pig breeding and poultry farming has always been higher comparing to other branches of animal husbandry. It is still growing via the current development of technological progress and leads to reduction in demand of workers.

In general, it was possible to talk about the absolute redundancy of labor at agricultural enterprises during the early 1990s. Nowadays, in our opinion, the redundancy of the workforce is relative. The main reason for this is that it relativity occurs in animal husbandry as a result of almost threefold reduction in a livestock population. For example, 61 relative heads were on each 100 hectares of agricultural lands in 1990 while there are only 20 heads in 2017. Thus, development of the livestock sector, increase in livestock, and growth of investments will contribute, at least, to the slowdown in the ratio of labor shortage at the market that has recently been significant (Fig. 3).

The high level of staff movement at agricultural enterprises depends on, first of all, seasonal production in crop growing as the main branch of agriculture in Ukraine. In order to save money that spent on staff retention in the absence of fieldwork, agricultural enterprises release their employees with the consent of the parties. During this time, employees register at employment centers and receive unemployment payment. At the end of the specified period, employees are hired again to the agricultural enterprises. Consequently, the cost of staff retention in agricultural enterprises except the enterprises itself is actually carried out by taxpayers.

## 2.4 Costs of Staff Retention

The efficiency of staffing is characterized not only by labor productivity and the loading of lands or livestock per employee but by the enterprise expenses on the staff. Types of staff retention expenses are different: the first group should include labor costs related to the wage fund (direct wages, pay for non-working time, bonuses, and irregular payments, subsistence wages with deductions for social purposes) and to the second one which includes expenses for the retention of labor which do not belong to the wage fund (privileges, services, money and subsistence assistance, housing payment, social security, expenses on vocational training, cultural and household services, other labor costs).

Domestic statistics up to 2015 did not investigate the agricultural enterprises in case of the costs of staff retention which do not belong to the wage fund. For a long time, the payroll was only displayed in the reports along with deductions for social purposes which are presented in our calculations (Table 2).

Thus, the expenses of staff retention per worker increased by 3 times during 2010–2017. However, they were lagging behind the growth of current expenses and revenues of enterprises. Therefore, personnel costs represented by a small share of current expenditures and a small proportion of revenues were spent on their financing. The enterprises' expenditures for the staff retention, which do not belong to the wage fund, constituted only 2% of the total amount of labor costs in 2015–2017.

The efficiency of personnel usage which was calculated based on income and profit increased considerably during 2010–2017. Of course, inflation and exchange rate influence significantly on these values but the analysis of the transferred profit by the aggregate index of agricultural production costs, which allows estimating the purchasing power of profit, still shows the increase in labor productivity by 2.1 times (see Table 2).

It was proved that the ratio of the personnel growth to the growth of gross output was −0.75 during 2010–2017 which means that 1% growth of gross output on average was equal to 0.75% decrease in the number of employees. At the same time, the ratio of the growth rates of real wages and productivity labor was 0.71, which means that 1% increase in labor productivity was equal to only 0.71% of the growth of real wages.

Such usage of personnel by agrarian enterprises is profitable. Staff is reduced, expenses for its retention are growing slowly, and this rise was relatively insignificant. Then labor productivity is growing significantly. During the investigated period, enterprises adhered to the principle when the growth of labor productivity exceeds rates of wages. This is necessary for expanded reproduction, profit making and growth of profitability. Violation of this principle leads to an overrun of the wage fund, an increase in the production costs, and a reduction of profit.

However, low staff costs are always reflected in the low quality of staff. Nowadays, many agrarian enterprises are faced with the problem of finding highly skilled tractor drivers, combine harvesters, locksmiths, welders, etc. Thus, the quality of work is lost, the terms of its execution are increased, and as a result, the product quality and

**Table 2** Efficiency of staff usage at Ukrainian agrarian enterprises

| Indicators | Years | | | | | | | | Difference between 2017 and 2010 | |
|---|---|---|---|---|---|---|---|---|---|---|
| | 2010 | 2011 | 2012 | 2013 | 2014 | 2015 | 2016 | 2017 | ± | Times |
| Expenses per employee, thousand UAH | 23.3 | 21.5 | 24.3 | 27.2 | 29.7 | 38.4 | 48.0 | 70.4 | 47.1 | 3.0*t* |
| incl. expenses beyond the wage fund | … | … | … | … | … | 0.8 | 1.0 | 1.3 | x | x |
| Share of labor retention (%) | 12.3 | 12.2 | 12.8 | 10.4 | 9.8 | 7.4 | 6.6 | … | | x |
| Share of wages (%) | 9.1 | 9.1 | 9.4 | 7.6 | 7.2 | 5.5 | 5.4 | … | | x |
| The ratio of expenses for personnel retention to income (%) | 8.0 | 7.7 | 8.3 | 6.8 | 6.0 | 3.5 | 5.4 | … | | x |
| Income per employee, thousand UAH | 223.6 | 306.5 | 392.9 | 417.9 | 591.7 | 1070.9 | 980.5 | 1152.7 | 929.1 | 5.2*t* |
| Profit per employee, thousand UAH | 33.3 | 49.6 | 55.1 | 32.0 | 50.3 | 249.7 | 246.6 | 219.3 | 186.0 | 6.6*t* |
| Profit per employee, thousand UAH (in purchasing power in 2017) | 103.4 | 131.6 | 136.8 | 78.8 | 105.0 | 345.1 | 300.4 | 219.3 | 115.9 | 2.1*t* |
| Revenue of labor expenses (%) | 142.9 | 230.7 | 226.5 | 117.6 | 169.5 | 649.5 | 513.6 | 311.7 | 168.7 | x |

quantity are reduced. This problem is connected with the increasing frequency of migration of the most active and productive age group of rural population—youth that leads to increase in the share of disabled and poorly functioning groups. That is why the staff qualification is deteriorating.

The low cost of staff retention in Ukraine affects a relatively higher level of personnel turnover in agriculture than in the economy overall. Thus, the ratio of retired agricultural workers to the average quantity was 56% in 2010–2017 while, in general, in the country—only 30%. Of course, seasonal fluctuations are reflected in this ratio, although low staff stability is always negatively reflected in productivity. Thus, agrarian enterprises do not receive higher output.

## 2.5 *Wages*

The analysis of the wage fund as the main component of staff retention shows that the share of the basic wage fund decreased with the growth of the share of additional wages and other incentive and compensation payments during 2010–2017 (Table 3).

The most significant increase in additional wages was the production results bonuses especially in 2016–2017—up to 7.3%. In absolute terms, it was 3.4 and 5.1 thousand UAH per year per employee (in 2010, it was 0.8 thousand UAH), and in the structure of gross income only 1.2–1.7% (in 2010–1.5%), which indicates the insignificance of the existing level of bonuses as a motivative factor for employees to improve the results of their duties. Other perks and benefits have no actually motivative effect: payment or subsidy for employees' food, material rehabilitation, etc.

The average monthly wage in agriculture remains relatively low despite the fact that in the early 2000s, after the reforms of the collective farms, economic activity in the Ukrainian agrarian sector grew slightly which led to a graduate rise of production volumes and profitability. Despite the periodic increase in the nominal level, the wages of farmers remain almost the lowest in the national economy. As an evidence of that fact is the ratio of the average monthly nominal wage in agriculture to the average monthly nominal wage in the whole country (Fig. 4).

Among the reasons for the low price of labor in agriculture, we should mention the relatively low efficiency of agricultural production and the growth of production volumes in the industry. Another reason for low level of wages in agriculture is a small share of labor costs in gross value added—their share is much lower than in the national economy: in 2010–21.6 and 54.5%, respectively, in 2016—14.7 and 43.2%, respectively.

In general, the dynamics of the average monthly wage in agriculture is positive during 2010–2017: nominal wages increased by 4 times, while real (taking into consideration purchasing power) by 67.9% (Fig. 4).

Growth of wages occurred both in relation to the average level of payment in the national economy and in relation to the minimum wage. If the wages in the industry

**Table 3** The structure of the wage fund at Ukrainian agricultural enterprises (%)

| Indicators | Years | | | | | | | | Difference between 2017 and 2010 (±) |
|---|---|---|---|---|---|---|---|---|---|
| | 2010 | 2011 | 2012 | 2013 | 2014 | 2015 | 2016 | 2017 | |
| Wage fund | 83.7 | 83.7 | 83.8 | 83.6 | 83.3 | 79.2 | 78.9 | 79.2 | −4.5 |
| Fund of additional wages, incl. | 15.0 | 14.7 | 15.0 | 15.4 | 15.4 | 18.8 | 18.6 | 18.0 | 3.0 |
| • Surcharges and surcharges to tariff rates and wages | 3.4 | 3.2 | 3.0 | 3.0 | 2.9 | 3.1 | 3.8 | 4.1 | 0.7 |
| • Production performance bonuses | 4.5 | 4.9 | 5.1 | 5.6 | 5.4 | 7.0 | 7.3 | 7.3 | 2.8 |
| • Accrued for unemployment time | 6.4 | 6.0 | 6.4 | 6.3 | 6.4 | 5.8 | 5.6 | 5.5 | −0.9 |
| • Other | 0.7 | 0.6 | 0.5 | 0.5 | 0.7 | 2.9 | 1.8 | 1.1 | 0.4 |
| Other perks and benefits | 1.3 | 1.6 | 1.3 | 1.0 | 1.3 | 2.0 | 2.5 | 2.8 | 1.5 |
| Total | 100 | 100 | 100 | 100 | 100 | 100 | 100 | 100 | x |

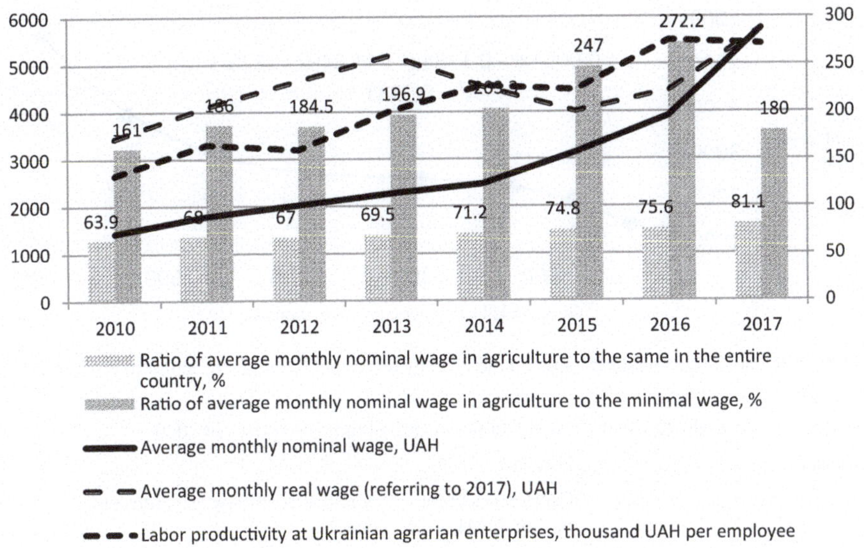


**Fig. 4** Wages in Ukrainian agriculture

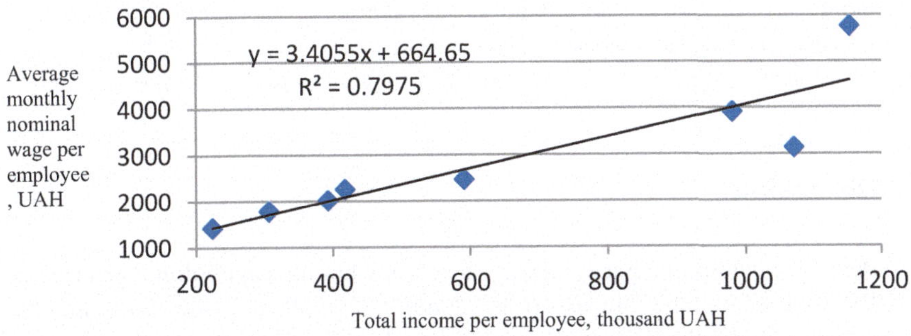


**Fig. 5** Influence of operational incomes on wages at Ukrainian agricultural enterprises

were equal to 63.9% of the average wage and 161.0% of the minimum wage in 2010, then these indicators rose to 81.1 and 180.0%, respectively, by 2017.

Taking into consideration the substantial reduction in real incomes in 2014–2016, the Government of Ukraine decided to raise social standards. From January 1, 2017, the minimum wage in Ukraine was set at UAH 3200 while agrarian enterprises were able to increase nominal wages to UAH 5761.

Obviously, the increase in wages at agricultural enterprises was caused by the growth of labor productivity (Fig. 4). However, as we have researched, more significant impact on wages was caused not only by the growth of production volumes (labor productivity by gross output) during 2010–2017 but by the cost of these products (labor productivity by enterprise incomes and gross value added (Figs. 5 and 6).

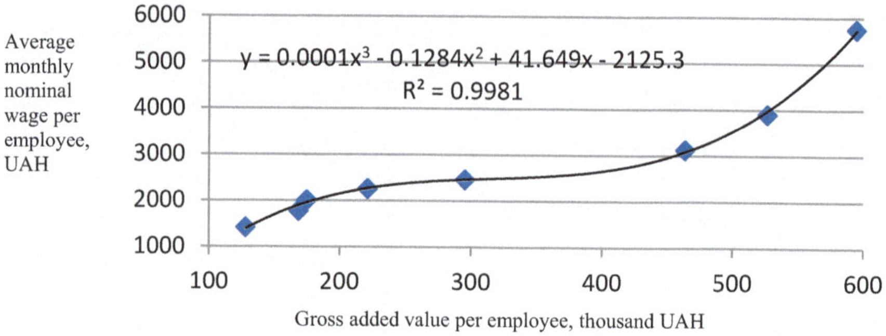


**Fig. 6** Influence of gross added value on the level of wages at Ukrainian agricultural enterprises

**Table 4** Labor productivity forecast at Ukrainian agricultural enterprises in 2020

| Indicators | | | Forecast | | | | | |
|---|---|---|---|---|---|---|---|---|
| | | | Gross production, million UAH | | Number of employees, thousand people | | | |
| Minimum | | | 104114.2 | | 347.1 | | | |
| Point forecast | | | 157130.6 | | 418.7 | | | |
| Maximum | | | 210146.9 | | 490.3 | | | |
| Labor productivity forecast, thousand UAH per employee | | | | | | | | |
| 212.4 | 248.7 | 299.9 | 320.5 | 375.3 | 428.6 | 452.7 | 501.9 | 605.4 |

## 3 Discussion

In order to predict productivity and wage levels at Ukrainian agricultural enterprises in 2020, we executed the forecasts of the number of employees (see above) and the gross value of agricultural products which was done based on trends of gross crop and livestock production:

$$\tilde{y}_t = 51169.8 + 21098.1t - 2861.1t^2 + 143.6t^3, \text{(Crop production)} \quad (14)$$

$$\tilde{y}_t = 23018.3 + 4397.0t - 591.7t^2 + 21.9t^3, \text{(Livestock)} \quad (15)$$

As a result, the following data is received with the probability of 95% (Table 4). So, there are nine variants of labor productivity based on these data that can be achieved with a probability of 95% in 2020.

It is determined that the average level of labor productivity will be 382.8 thousand UAH, median—375.3 thousand UAH, minimum—212.4 thousand UAH, and the maximum—605.4 thousand UAH. Taking into consideration that 2010-labor productivity was an average of UAH 207.1 thousand in 2017, it becomes clear that its level in 2020 is expected to be higher by 1.1–2.9 times.

Prediction of wage levels is more difficult in comparison with labor productivity because its dynamics may be affected by an increase in the general level of prices for goods and services (inflation), legislative changes in case of the minimum wage level, changes in labor demand, etc.

The relationship between prices and wages is relatively well observed in the long term. However, wages and prices may fluctuate relative to each other as a response to temporary effects in the short and medium terms. Competitiveness of enterprises and workers may change; therefore, labor productivity may also change temporarily. Even if these changes do not have long-term tendencies, these events make a complex dynamic regarding the relationship between wages and prices.

During 2010–2017, consumer prices grew by 2.6 times while the average monthly wage in agriculture increased by four times. Such a ratio of inflation and nominal wage growth is the basis for real wage growth of 1.7 times (Fig. 4). The minimum wage increased 3.6 times during this time. Although such growth exceeds the inflation rate, the growth of the minimum wage was not sufficient to cover inflation losses in 2014–2016. Moreover, there was insufficient overall wage growth in 2014–2015 (Table 5), which led to the necessity of raising the minimum wage in 2017.

Thus, inflation has a significant effect on the formation of real wages in agriculture of Ukraine. Therefore, the forecast of nominal wages must include the inflation factor. We have calculated that the average annual growth of consumer prices was 12.8% during 2010–2017. Based on the nominal wage of 5761 UAH/month (2017) and the specified inflation rate, we will calculate the nominal wage in 2020–8265 UAH/month. Consequently, the growth of nominal wages will be 43.5% in 2020, and it will leave real wages of this year at the level of 2017.

It is clear that the proposal for a minimum wage will be responding by workers only in terms of a high labor supply. The trends we have discovered point to the other. Therefore, it is important to estimate the excess of the minimum wage which will form a nominal wage in case of the personnel outflow. In addition, official inflation that we also need to take into consideration affects the growth of real wages. Moreover, the official inflation, which we operate in the calculations, is actually lower than real inflation according to expert estimates.

Real wages in agriculture grew by an average of 7.7% annually during 2010–2017, ranging from $-13$ to $+29.4\%$. According to formula 10, we executed forecasting of nominal wage in 2020, paying attention to inflation and real wage growth:

$$10329.33 = 5761 \cdot 1.128^3 \cdot 1.077^3.$$

Thus, the nominal wage in agriculture will be UAH 10329.33 per month in 2020 based on the annual inflation and average annual growth of real wages observed during 2010–2017.

Based on the parabola equation of the third order (Fig. 6), we determine the level of labor productivity by gross added value ($LP_{GAV}$) which will allow getting nominal wages of 10329.33 UAH/month

**Table 5** Components analysis of changes in nominal wages and their adequacy in case of coverage inflationary expenses for consumer goods

| Indicators | Years | | | | | | |
|---|---|---|---|---|---|---|---|
| | 2010–2011 | 2011–2012 | 2012–2013 | 2013–2014 | 2014–2015 | 2015–2016 | 2016–2017 |
| Change in nominal wage, UAH/month, total, including | 361.00 | 235.00 | 244.00 | 206.00 | 664.00 | 776.00 | 1845.00 |
| • Change of minimum wage | 74.83 | 135.00 | 54.83 | 65.08 | 53.33 | 167.17 | 1761.50 |
| • Change due to the enterprise decision | 286.17 | 100.00 | 189.17 | 140.92 | 610.67 | 608.83 | 83.50 |
| The minimum required wage growth in comparison to the previous year (at the rate of annual inflation), UAH/month | 65.78 | −3.58 | 10.13 | 565.23 | 1072.11 | 389.36 | 536.49 |
| Compensation of the inflation component due to the growth of the minimum wage, UAH/month | 9.05 | 138.58 | 44.70 | −500.15 | −1018.77 | −222.19 | 1225.01 |
| Compensation of the inflation component due to the growth of nominal wages, UAH/month | 295.22 | 238.58 | 233.87 | −359.23 | −408.11 | 386.64 | 1308.51 |

$$10329.33 = 0.0001x^3 - 0.1284x^2 + 41.649x - 2125.3,$$

where $x$ (that is $LP_{GAV}$) = 990.45. Therefore, in order to ensure the indicated increase in nominal wages, labor productivity at gross added value should increase to 990.5 thousand UAH per employee which is 1.7 times higher than in 2017.

At the same time, according to formula 11 the total added value in agriculture should increase in 2020 by an average of 1.4 times comparing to 2017 and be equal to UAH 414.7 million:

$$414.7 = (990.45 \cdot 418.7)/1000.$$

To estimate the accuracy of these calculations, we will determine the expected gross added value in agriculture in 2020 based on the actual gross added value in 2017 (UAH 295.0 billion) and the expected average annual growth rate which is equal, according to our calculations, 12.68%:

$$295.0 \cdot 1.1268^3 = 422.0.$$

Consequently, gross added value which calculated on the basis of average growth rate will be about 422.0 million UAH in 2020 that differs from the forecast based on formula 11 by only 1.7%. This confirms the reliability of the executed calculations, and it can be argued that the level of nominal wages in Ukrainian agriculture will be about 10329.33 UAH/month in 2020 as a continued effect of trends and dependencies that were detected in 2010–2017.

According to our calculations, the share of labor cost in the added value of agriculture will be 12.5% in 2020:

$$\frac{10329.33 \cdot 12 \cdot 418.7}{414.7 \cdot 1000 \cdot 1000} = 12.5.$$

If agricultural companies increase this share in 2020 to the level that was on average in 2010–2017 (19.0%), then the wage fund will be 78.793 billion UAH (414.7 · 0.19) in total, or 15628.07 UAH/month for one employee (((78,793 · 1000)/418.7)/12).

Thus, the wage will possibly range from 10329.33 to 15628.07 UAH/month in 2020. The lower limit characterizes the demand of workers while the upper one is the supply of enterprises. It is clear that the level of wages about $369–558 per month is not very stimulating in comparison with the possibility of earning much more abroad.

## 4 Conclusions

It has been proved that labor productivity at agricultural enterprises increased during the research period, and the main reason for that is the graduate rise of production volumes that exceeded the release of labor. Studies show that according to current trends at agricultural enterprises, the average number of employees is expected to decrease by 30.5% over 2017–2021 comparing to 2010–2016.

Personnel costs per employee increased by almost 3 times during 2010–2016 but still constituted a tiny share among current expenditures. Thus, a small percentage of company revenue was spent on their financing.

It was investigated that 1% of the gross production growth was accompanied by a decrease number of employees by an average 0.75%, while the growth of labor productivity by 1% leads to an increase of real wages by only 0.71%, which confirms the thesis about effective personnel usage by the agrarian enterprises.

Basic wage plays the main role in the structure of the wage fund of agricultural enterprises but the share of bonuses is insignificant that does not stimulate employees to improve the results of their activities.

Growth of wages at agricultural enterprises of Ukraine took place both in relation to the average level of payment in the national economy and in relation to the minimum wage. This increase is mostly connected with the increase in labor productivity which is calculated by the gross added value.

Thus, among the positive changes in the material stimulation of farmers, the next facts can be mentioned as follows: a steady tendency to increase both nominal and real wages; an increase in its share relative to both the average level of payment in the national economy and the minimum social standards; a significant reduction in wage debts, and to the negative ones—relatively low wages and the weakness of the stimulus effect of the fund of additional wages.

It is determined that the projected level of labor productivity by gross product will amount about 382.8 thousand UAH in 2020 ranging from 212.4 to 605.4 thousand UAH per employee or will be higher 1.1–2.9 times in comparison with 2010–2017.

Wages forecasting showed that the nominal wage can be in the range from 10329.33 to 15628.07 UAH/month in 2020 where the lower limit is a characteristic of the demand of employees while the upper one is the supply of enterprises.

## References

1. Bevers, S., McCorkle, D., Hanselka, D.: Farm and Ranch Personnel Management. The Texas A&M System. https://agecoext.tamu.edu/wp-content/uploads/2013/10/rm8-1.pdf (2013)
2. Sajbidorova, M., Lusňakova, Z., Dobisova, M.: Management of human resources in agricultural sector enterprises. Slovak university of agriculture in Nitra. https://spu.fem.uniag.sk/mvd2016/proceedings/en/articles/s3/sajbidorova_dobisova_lusnakova.pdf (2016)
3. Berde, C.: Human resource management in Hungarian agriculture. Jahrbuch der Österreichischen Gesellschaft für Agrarökonomie **15**, 157–164. https://oega.boku.ac.at/fileadmin/user_upload/Tagung/2005/05_Berde.pdf (2006)

4. Bitsch, V.: Personnel Management Research in Agribusiness. Michigan State University. https://ifama.org/resources/files/2009-Symposium/1067_paper.pdf (2009)
5. Syniaieva, L.V., Bocharova, N.O.: Motivation as a function of administrative management and a component of the local system of personnel management. In: Sinyaeva, L.V. (ed) Melitopol Collection of scientific works of Tavria State Agrotechnological University (economic sciences) vol. 1(29), pp. 127–130. http://nbuv.gov.ua/UJRN/znptdau_2015_1_26 (2015)
6. Petryshyn, L.: Improvement of processes of formation and use of personnel of agricultural enterprises as a factor of their economic behavior. Scientific Herald of Uzhgorod National University, Issue 16, part 2, pp. 65–68. http://www.visnyk-econom.uzhnu.uz.ua/archive/16_2_2017ua/16.pdf (2017)
7. Restuccia, D., Tao Yang D., Zhu, X.: Agriculture and aggregate productivity: a quantitative cross-country analysis. J. Monet. Econ., **55**(2), 234–250. http://citeseerx.ist.psu.edu/viewdoc/download?doi=10.1.1.561.4853&rep=rep1&type=pdf (2008)
8. Dharmasiri, L.M.: Measuring agricultural productivity using average productivity index (API). In: International Conference in XXVI th Indian Geographers, Bhopal, January 2006. https://www.researchgate.net/publication/248390836_MEASURING_AGRICULTURAL_PRODUCTIVITY_USING_AVERAGE_PRODUCTIVITY_INDEX_API (2006)
9. Gutierrez, L.: Why is agricultural labour productivity higher in some countries than others? Agricultural Economics Review, Greek Association of Agricultural Economists, **3**(1), 1–15. https://ideas.repec.org/a/ags/aergaa/26431.html (2002)
10. Boghean, C., State, M.: Analysis of the factors affecting the average labour productivity variation in agriculture, forestry and fishing in Romania. USV Ann. Econ. Public Adm. **13**(2)(18), 35–41. https://ideas.repec.org/a/scm/usvaep/v13y2013i2(18)p35-41.html (2013)
11. Zenka, J., Zufan, P., Krticka, L., Slach O.: Labour productivity of agricultural business companies and cooperatives in the Czech Republic: a micro-regional level analysis. Morav. Geogr. Rep. 4/2015, **23**, 14–25. https://www.researchgate.net/publication/288839584_Labour_productivity_of_agricultural_business_companies_and_cooperatives_in_the_Czech_Republic_A_micro-regional_level_analysis (2015)
12. Novotnal, M., Volek, T.: The Significance of farm size in the evaluation of labour productivity in agriculture. Acta Universitatis Agriculturae et Silviculturae Mendelianae Brunensis Number 1, 2016, **64**, 333–340. https://acta.mendelu.cz/media/pdf/actaun_2016064010333.pdf (2016)
13. Bin, P., Vassallo M.: The growth path of agricultural labor productivity in China: a latent growth curve model at the prefectural level. Econ., MDPI, Open Access J. **4**(3), 1–20. https://ideas.repec.org/a/gam/jecomi/v4y2016i3p13-d73090.html (2016)
14. Polyzos, S., Arabatzis G.: Labor productivity of the agricultural sector in Greece: determinant factors and interregional differences analysis. Discuss. Pap. Ser. **11**(12), 209–226. https://www.researchgate.net/publication/228638610_Labor_productivity_of_the_agricultural_sector_in_Greece_determinant_factors_and_interregional_differences_analysis (2006)
15. Herrendorf, B., Schoellman, T.: Why Is Agricultural Labor Productivity so Low in the United States? https://www.economicdynamics.org/meetpapers/2011/paper_1087.pdf (2011)
16. Andreica, M.E., Aparaschivei, L., Cristescu, A., Cataniciu N.: Models of the minimum wage impact upon employment, wages and prices: the romanian case. http://www.wseas.us/e-library/conferences/2010/Iasi/MCBEC/MCBEC-14.pdf (2010)
17. Ponomarenko, I.V.: Average Monthly Wage Forecast in Ukraine. Global and national problems of the economy, Mykolaiv V.O.Sukhomlynskyi National University, Issue 13, pp. 951–964. http://global-national.in.ua/archive/16-2017/190.pdf (2016)
18. Blanchard, O., Katz, L.F.: Wage dynamics: reconciling theory and evidence. Am. Econ. Rev. **89**, 69–74. https://www.nber.org/papers/w6924 (1999)
19. Oliskevych, M.O.: Econometric analysis of the interrelations between labor market indicators in Ukraine Regional Business Economics and Management, vol. 2, pp. 24–29. http://nbuv.gov.ua/UJRN/Rbetu_2013_2_6. (2013)
20. Dupliak, O.M, Yedlichko, A.O.: Analysis approaches to modeling and forecasting wage. scientific herald of Uzhgorod National University, Series: International Economic Relations and World Economy, **6**(1), 117–119. http://nbuv.gov.ua/UJRN/Nvuumevcg_2016_6%281%29_28 (2016)

# Methodical Approaches to Implementation of Financial Bank Stability

**Nataliia Radchenko, Natalia Rubtsova, Iryna Chkan and Inna Yakysheva**

## 1 Introduction

The banking system is one of the key links in the economic system of any country. It should regulate the movement of cash flows, promote the most rational use of financial resources of society and the transfer of capital in those sectors of the economy, where the return on investment will be maximal. The success of the banking system will depend to a large extent on the success of overcoming the economic crisis, reducing inflation and financial stabilization. During the study of various sources of economic literature, most scholars argued that there is no single approach to determining the essence of financial stability of the bank. Very often it is replaced by such economic concepts as solvency and liquidity. It should be noted that these two categories are really important for the bank. Thus, liquidity characterizes the process of transferring financial and tangible assets into cash; solvency characterizes the ability of the bank in a timely manner and in full to pay off its obligations. These are very important components, but financial stability of the bank is a broader concept, which in addition to liquidity and solvency should include such indicators as: balance of assets and liabilities by terms and amounts; the quality of the loan and investment portfolio; reputation of the bank; quality of services; and ability to counteract crisis phenomena effectively.

### *1.1 Literature Review*

Considerable attention is given to the study of this issue by foreign scholars such as Allen and Wood [1], De Nicholas et al. [2], Davis [3], Roy [4] and others.

N. Radchenko (✉) · N. Rubtsova · I. Chkan · I. Yakysheva
Tavria State Agrotechnological University, B. Khmelnitsky ave., 18, Melitopol 72312, Ukraine
e-mail: radchenko80@i.ua

© Springer Nature Switzerland AG 2019

V. Nadykto (ed.), *Modern Development Paths of Agricultural Production*,
https://doi.org/10.1007/978-3-030-14918-5_54

Each scientist who studies the problem of ensuring financial stability of a bank defines this category differently.

The best approach according to Allen and Wood is to define the characteristics of an episode of financial instability first and then define financial stability as a state of affairs in which episodes of instability are unlikely to occur [1].

Davis identifies three generic types of financial instability. The first is centered on bank failures, typically following loan or trading losses, the second involves extreme market price volatility after a shift in expectations and the third being the one that is linked to the second, involves protracted collapses of market liquidity and issuance [3].

Panteleev and Haljava determine financial sustainability as excess of income over costs [5]. In our opinion, this is an incomplete definition, since, the level of profit belongs to the effective indicators of banking activity, which reflect the final result. But in order to assess the financial stability of the bank, you need to know the structure of the sources of profit and the order of their distribution. It is the stability of the bank's income sources that is a hallmark of financial sustainability.

In his scientific works, Kreinina [6] considers stability of the bank as the stability of its activity in the medium and long term, without taking into account the current activity of the bank, which is erroneous. Yu. Maslencnenkov identifies the concept of stability of a bank with the stability of a credit institution in the context of the economic environment in which it operates [7].

## *1.2 Methodology, Variables and Data*

The rating of financial stability of banks is of great importance, since it allows us to assess the current state of the bank and identify factors that negatively affect the results of its activities. Scientists and specialists of the banking sector use various methodological approaches to assess the financial stability of the bank:

- coefficient analysis;
- integral methods;
- mathematical and statistical methods;
- rating methods [8].

Coefficient analysis is based on the calculation of financial indicators that affect the financial stability of the bank (Table 1).

The calculation of coefficients allows us to detail the activities of the bank, as well as to determine the specific features of its activities. However, this approach also has disadvantages, namely the impossibility of generating a generalized integrated assessment of the financial condition of the bank. By means of integral methods, it is possible to assess the financial stability of the bank by calculating a single common indicator. Using integral indicators, a comparative analysis of the financial stability of a large set of banks can be carried out.

**Table 1** The coefficients of financial stability of the bank [9]

| Title | Procedure | Optimal value |
|---|---|---|
| Coefficient of concentration of own equity (independence or autonomy) | *K* (cce) = (Equity/total assets) × 100 | The higher the value, the more stable the state of the bank |
| Coefficient of concentration of attracted funds | *K* (caf) = (Funds raised/total assets) × 100 | Shows how much on 1 UAH own funds invested in assets have attracted funds |
| Coefficient of financial dependence | *K* (fd) = (Total assets/equity) × 100 | Increase in dynamics means an increase in the share of borrowed funds in financing the bank |
| Stability factor | *K* (sf) = (Equity/equity) × 100 | At least 5% |
| Financial risk factor (financial leverage) | *K* (frf) = Funds acquired/equity | Between 1 and 20% |
| Coefficient of protection of profitable assets by own capital | *K* (pac) = Equity/(total assets—non-performing assets) | At least 10% |
| Capital multiplier | *K* (m) = Total assets/authorized capital. | Within 12.0–15.0% |

It is also possible to assess financial stability using mathematical and statistical methods, such as: discriminant analysis, factor analysis, and nonlinear evaluation. Along with integral methods, rating methodological approaches remain equally important. The most famous in the world rating system of banks is the CAMELS system, which is widely used by supervisors of many countries of the world. The CAMELS system is a rating and is based on a combination of accounting and expert judgment. Supervision of banks, based on risk assessments under this rating system, consists in determining the general state of the bank based on common criteria covering all its activities [10].

An assessment of the financial sustainability of Ukrainian banks is carried out by international lenders and financial institutions. So, experts of the World Bank and the IMF, there is an indicator of financial stability of the Z-score. The methodology of this indicator involves assessing the probability of insolvency of the banking system of the country, that is, the probability that the value of assets of banks will be lower than the cost of obligations. The higher (lower) is the value of the Z-score indicator, the lower (higher) is the level of probability of the insolvency of the banking system and, consequently, the higher (lower) level of its financial stability. Depending on the actual value of this indicator, all countries of the world are divided into three groups: the first group: Z-score from 0 to 10 (Greece, Lithuania, Moldova, the Netherlands, Slovenia); second group: Z-score from 10 to 20 (Hungary, Italy, Norway, Poland, Spain, Ukraine); third group: Z-score value more than 20 (Belgium, Czech Republic, Germany, Great Britain, Sweden) [11].

**Table 2** Economic norms established by the NBU [13]

| | Name of the indicator | Normative value |
|---|---|---|
| *Capital ratios* | | |
| I1 | The minimum size of the regulatory capital | Is 500 million UAH |
| I2 | Sufficiency (adequacy) of regulatory capital | Not less than 10% |
| I3 | The ratio of regulatory capital and aggregate assets | Is not less than 9% |
| *Liquidity norms* | | |
| I4 | Instant liquidity | At least 20% |
| I5 | Current liquidity | At least 40% |
| I6 | Short-term liquidity | Not less than 60% |
| *Credit risk standards* | | |
| I7 | The maximum amount of credit risk per one counterparty | Is not more than 25% |
| I8 | Large credit risks | No more than eight times the size of regulatory capital |
| I9 | The maximum amount of loans, guarantees and sureties granted to one insider | No more than 5% |
| I10 | The maximum aggregate amount of loans, guarantees and sureties granted to insiders | No more than 30% |
| *Investment standards* | | |
| I11 | Investing in securities separately for each institution | No more than 15% |
| I12 | Total investment | No more than 60% |

Consequently, the use of one or another methodological approach depends primarily on the subject conducting the assessment. So, for the shareholders the most important are indicators of profitability, for depositors of funds—indicators of liquidity and solvency; for borrowers—indicators of availability of credit resources; for creditors—indicators that characterize the reliability and stability of the bank.

In addition, an active assessment of the financial stability of banks is also carried out by the state, represented by the National Bank of Ukraine. The National Bank of Ukraine (NBU) regulates the financial stability of banks by: defining the norms of mandatory reserve, establishing mandatory economic standards, applying restrictions and requirements to banks, regulating the rates of deductions to provisions for active banking operations [12]. In order to maintain the stability of banking activities, the NBU approved the "Instruction on the Procedure for Regulating the Activities of Banks in Ukraine" [13]. The given instructions contain obligatory economic norms, the values of which must be observed by all commercial banks of Ukraine (Table 2).

The main task of the National Bank in the field of financial stability is to develop a system for timely detection of risks that threaten the stability of the banking and financial system of the state, and the implementation of necessary actions and measures aimed at preventing these risks [14].

The CNB uses stress testing as a tool for assessing the resilience of financial institutions registered in the Czech Republic and of the financial system as a whole. The CNB applies both a top-down macro-aggregate approach and a bottom-up micro-individual approach to stress testing. Macro-stress tests assess the resilience of the banking sector and the pension management companies sector as a whole. The CNB applies two approaches when stress testing the banking sector. In the top-down macro-approach, the CNB performs the tests itself on the basis of the data it has on the banking sector. It uses the results to assess macroprudential aspects of the capital position and liquidity of the banking sector under stress. In the bottom-up micro-approach, the relevant bank conducts the test on the basis of its own data using the methodology and scenarios set by the CNB. The CNB then evaluates the results and uses them in the process of supervisory review of the capital requirements for banks. The banking sector currently undergoes:

1. Solvency macro-stress test.
2. Liquidity macro-stress test.
3. Supervisory stress test (micro-stress test).

Current stress testing methodology:

1. Solvency macro-stress tests.
2. Liquidity macro-stress tests [15].

As part of its prudential supervision, the Reserve Bank undertakes supervisory stress tests. These exercises involve subjecting several financial institutions to a common stress scenario, with oversight from regulators. The Reserve Bank uses supervisory tests to:

1. Investigate and understand the implications of current and emerging risks to financial stability;
2. Assess the resilience of participating banks when subject to severe stress;
3. Support improvements in the use of stress tests by individual banks to identify and manage the risks facing their business [16].

The primary task for domestic banks is to support financial stability and reliability, ensure sufficient solvency and reduce internal risks based on the implementation of modern methodological approaches.

## 2 Descriptive Analysis

When choosing a banking institution, future customers will pay attention to the rating of one or another institution. In Ukraine, the rating of banks until 2016 was determined by the National Bank of Ukraine in the CAMELS system. From January 1, 2017, in accordance with the Resolution of the Board of the National Bank of Ukraine dated November 01, 2016, No. 393-rsh "On Approval of Inspection Procedures," the National Bank changed the procedure for organizing, planning, conducting and finalizing the results of the inspection, as well as the procedure for the determination and approval bank rating based on the CAMELSO rating system. Unlike the previous methodological approach, the new system contains an additional component—operational risk—Operational Risk (O)—the ability of the bank to effectively manage operational and informational risk in order to prevent/minimize financial losses due to the implementation of risks [17].

The disadvantage of the CAMELSO system is that the built rating of banks is confidential information intended for internal use only by the National Bank of Ukraine and is not subject to publication in the media, therefore, access to this information by other participants in the financial market is closed [8].

Other methodological approaches considered in the first section of the work are inconvenient and difficult for the external user because they require special knowledge and can only be used if the bank's accounts are available.

Thus, the task is to develop a methodical approach to assessing banks by the level of financial stability that would be available to customers (the population and business entities) and was used by them when choosing a banking institution.

To solve this problem, we used the method of analytical hierarchy by Saati [18]. Using the features and characteristics of this method, we will try to construct a hierarchy of choice of a banking institution (Fig. 1). In accordance with Fig. 1, alternatives of choice are any banking institutions that provide a standard set of services. To conduct the survey, we selected the four most famous banks in Ukraine: PrivatBank, Alfa-Bank, Oschadbank and Kredi Agricole Bank.

The financial stability of banks is determined by criteria 1 and 2 levels. In our view, as a user of banking services, the criteria for the first level of financial sustainability are: the reliability of a banking institution, financial performance, the price policy of banks, the reputation of banks and the convenience of using banking services.

In turn, indicators of level 1 are formed under the influence of indicators of level 2, a list of which can be seen in Fig. 1. Clusters in this hierarchical structure are: financial performance indicators, reliability indicators, a cluster of pricing policies of a banking institution, a cluster of reputation of a banking institution and a cluster of convenience with the use of banking services.

In T. Saati comparisons, each node of the cluster is compared with the rest of all nodes by the method of pair comparisons using the matrix. On the basis of the results, the priority vector is given—the rating of the nodes, which quantitatively expresses the advantage of one or another node in accordance with the vertices of the cluster.

In case of solving economic, political, etc., paired comparison tasks can be performed using judgments about the relative importance of components. After that, these judgments are expressed numerically using a specially designed scale of relative importance (Table 3). The verification of the quality of the data for consistency is carried out by means of determining the index of consistency and the level of relative matrix compatibility (1–3).

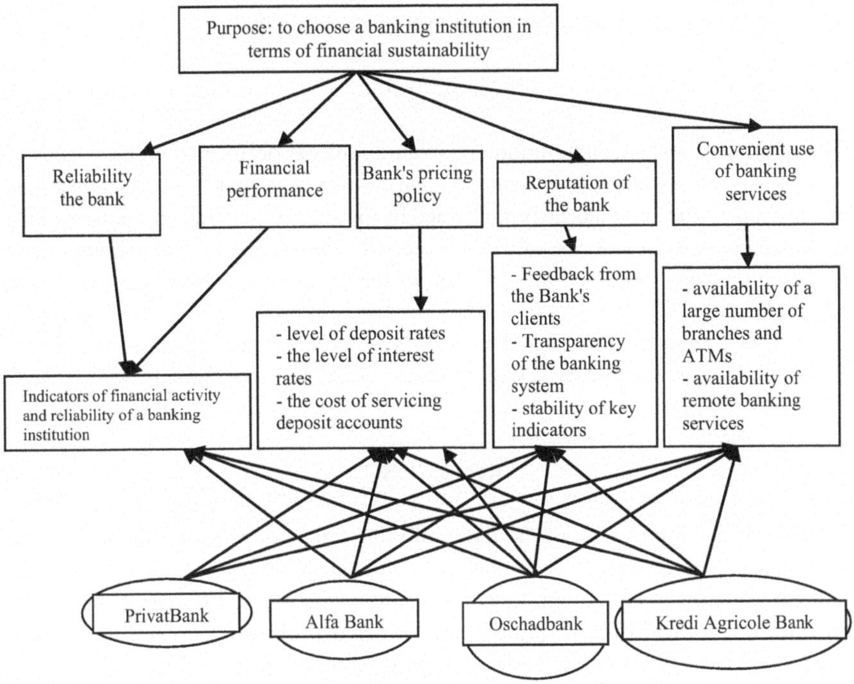


**Fig. 1** Hierarchy of the choice of a potential banking institution for any banking operations

**Table 3** Scale of relative importance [18]

| Intensity of relative importance | Definition |
|---|---|
| 1 | Equally important |
| 3 | Moderate advantage |
| 5 | Strong advantage |
| 7 | Significant advantage |
| 9 | Very strong advantage |
| 2, 4, 6, 8 | Intermediate solutions for neighboring judgments |
| Inverse quantities | If, when comparing one parameter with another, one of the above numbers is obtained, then in the reverse comparison we obtain the inverse value |

*The Coherence Index* (CI) provides information on the extent of the violation of numerical and transitive coherence and is calculated by the following formula:

$$\mathrm{CI} = (\lambda_{\max} - n)/(n-1) \quad (1)$$

$\lambda_{\max}$ the maximum actual value of the matrix;
$n$ is the size of the matrix.

For the symmetric matrix, the equation is always satisfied:

$$\lambda_{max} \geq n \tag{2}$$

The calculation of the maximum eigenvalue λmax is performed on the matrix of pair comparisons as follows: The sum of each column of thoughts is determined, and then, the sum of the first column is multiplied by the value of the first component of the normalized priority vector, the sum of the second column, on the second component, etc. The final step is to determine the sum of the resulting numbers, which is equal to $\lambda_{max}$.

*The level of relative coherence* (RC)—is the ratio of the index of coherence (CI) to its average statistical value (SC) with random selection of coefficients of the matrix of comparisons:

$$RC = CI/SC \tag{3}$$

The average significance of the index of consistency depends on the size of the matrix and is determined in accordance with the scale developed by T. Saati. The matrix is considered sufficiently coherent if the value of the RC is less than 10%; sometimes we will allow 20%. If the RC is more than set limits, then there is a review of expert opinions to improve the level of relative coherence.

## 3 Empirical Analysis

20 respondents were involved in the construction of the priority vectors, among which were customers of banks that already use banking services and potential clients trying to make a choice between banking institutions. Based on the theoretical material [18–20] and respondents' answers, we construct a matrix of pair comparisons and corresponding normalized priority vectors for each cluster separately and write them in Table 4.

$$\lambda_{max} = 5.1458;\ SC = 1.12,\ RC = 0.03$$

As we see, the bank's price policy is the most important for the client (0.4577), the following two positions occupy the reputation of the bank (0.2344) and the convenience of use (0.1677). Less important criteria for potential customers are the financial performance and reliability of the banking institution. The level of relative coherence (RC) of opinions is 0.03 or 3% and meets the recommended requirements; therefore, the coordination of expert opinions is homogeneous and logical.

Further calculations were carried out using the fourth method. The results obtained were listed in Tables 5, 6, 7, 8, 9 and 10 as a synthesis of priorities.

**Table 4** Matrix of paired comparisons for the first level criteria

| First level criteria | Level of reliability of the banking institution | Level of financial performance | Level of the bank's banking policy | Level of reputation of the banking institution | Level ease of use of banking services | Geometric mean elements | Evaluation of components of the vector of priorities |
|---|---|---|---|---|---|---|---|
| The level of reliability of the banking institution | 1 | 1 | 0.2 | 0.33 | 0.33 | 0.4652 | 0.0720 |
| Level of financial performance | 1 | 1 | 0.2 | 0.25 | 0.33 | 0.44 | 0.0682 |
| The level of the bank's pricing policy | 5 | 5 | 1 | 3 | 3 | 2.9542 | 0.4577 |
| The reputation of the banking institution | 3 | 4 | 0.33 | 1 | 2 | 1.5127 | 0.2343 |
| Level ease of use | 3 | 3 | 0.33 | 0.5 | 1 | 1.0823 | 0.1677 |

**Table 5** Synthesis of reliability priorities

| Reliability of the banking institution | Oschadbank | Alfa-Bank | PrivatBank | Kredi Agricole Bank | Geometric mean elements | Evaluation of components of the vector of priorities |
|---|---|---|---|---|---|---|
| Oschadbank | 1 | 3 | 0.5 | 1 | 1.108 | 0.240 |
| Alfa-Bank | 0.33 | 1 | 0.2 | 0.33 | 0.465 | 0.1 |
| PrivatBank | 2 | 5 | 1 | 4 | 2.09 | 0.454 |
| Kredi Agricole Bank | 1 | 3 | 0.25 | 1 | 0.944 | 0.206 |

**Table 6** Synthesis of the priorities of the indicator of financial performance

| Financial performance | Oschadbank | Alfa-Bank | PrivatBank | Kredi Agricole Bank | Geometric mean elements | Evaluation of components of the vector of priorities |
|---|---|---|---|---|---|---|
| Oschadbank | 1 | 4 | 2 | 2 | 2 | 0.44 |
| Alfa-Bank | 0.25 | 1 | 0.5 | 0.5 | 0.5 | 0.11 |
| PrivatBank | 0.5 | 2 | 1 | 1 | 1 | 0.22 |
| Kredi Agricole Bank | 0.5 | 2 | 1 | 1 | 1 | 0.22 |

**Table 7** Synthesis of the priorities of the indicator of price policy

| Bank's pricing policy | Oschadbank | Alfa-Bank | PrivatBank | Kredi Agricole Bank | Geometric mean elements | Evaluation of components of the vector of priorities |
|---|---|---|---|---|---|---|
| Oschadbank | 1 | 0.25 | 2 | 0.33 | 0.634 | 0.13 |
| Alfa-Bank | 4 | 1 | 5 | 2 | 2.51 | 0.5 |
| Privat Bank | 0.5 | 0.2 | 1 | 0.33 | 0.426 | 0.08 |
| Kredi Agricole Bank | 3 | 0.5 | 3 | 1 | 1.456 | 0.29 |

**Table 8** Synthesis of priorities by indicator of reputation

| Reputation | Oschadbank | Alfa-Bank | PrivatBank | Kredi Agricole Bank | Geometric mean elements | Evaluation of components of the vector of priorities |
|---|---|---|---|---|---|---|
| Oschadbank | 1 | 0.5 | 0.33 | 1 | 0.637 | 0.138 |
| Alfa-Bank | 2 | 1 | 0.5 | 2 | 1.19 | 0.26 |
| PrivatBank | 3 | 2 | 1 | 4 | 2.21 | 0.48 |
| Kredi Agricole Bank | 1 | 0.5 | 0.2 | 1 | 0.56 | 0.122 |

**Table 9** Synthesis of priorities on an index of convenience of use

| Ease of use | Oschadbank | Alfa-Bank | PrivatBank | Kredi Agricole Bank | Geometric mean elements | Evaluation of components of the vector of priorities |
|---|---|---|---|---|---|---|
| Oschadbank | 1 | 4 | 0.33 | 6 | 1.68 | 0.29 |
| Alfa-Bank | 0.25 | 1 | 0.2 | 2 | 0.56 | 0.1 |
| Privat Bank | 3 | 5 | 1 | 7 | 3.2 | 0.55 |
| Kredi Agricole Bank | 0.17 | 0.5 | 0.14 | 1 | 0.33 | 0.06 |

**Table 10** Synthesis of global priorities

| Criteria for banking institutions | Level of reliability of the banking institution | Level of financial performance | Level of the bank's banking policy | Level of reputation of the banking institution | Level ease of use of banking services | Global priorities |
|---|---|---|---|---|---|---|
| Candidates | 0.0720 | 0.0682 | 0.4577 | 0.2343 | 0.1677 | X |
| Oschadbank | 0.24 | 0.44 | 0.13 | 0.138 | 0.29 | 0.1877 |
| Alfa-Bank | 0.1 | 0.11 | 0.5 | 0.26 | 0.1 | 0.3212 |
| PrivatBank | 0.454 | 0.22 | 0.08 | 0.48 | 0.55 | 0.289 |
| Kredi Agricole Bank | 0.206 | 0.22 | 0.29 | 0.122 | 0.06 | 0.201 |

Thus, according to the reliability index of the bank in the rating we have built in our first place is PrivatBank and the second—Oschadbank.

According to the financial results, the experts came to the conclusion that Oschadbank will occupy the first position, and Alfa-Bank is the last with low financial results.

The analysis of the pricing policy of the submitted banking institutions made it possible to determine that Alfa-Bank is the most attractive bank for its clients, since it has the most flexible and effective policy for pushing interest rates on banking products. The price policy of Kredi Agricole Bank is also moderate. The most stringent conditions for clients, mainly individuals, appeared to be offered by PrivatBank.

According to experts, the best reputation has PrivatBank and Alfa-Bank. By the indicator of ease of use (Table 9), PrivatBank takes the first position due to the availability of remote customer service through the system of "Privat-24."

Now calculate the global priorities of each banking institution, taking into account the weight of individual indicators, which were calculated in Table 4. The global priority of the Oschadbank is calculated as a result of the calculations:

$$\mathrm{GP} = 0.072 \times 0.24 + 0.068 \times 0.44 + 0.4577 \times 0.13 + 0.2343 \times 0.138 + 0.1677 \times 0.29 = 0.1877$$

Similarly, the results of other banks will be obtained (Table 10).

So, summing up the calculations, it can be argued that Alfa-Bank has the largest global priority due to flexible pricing policies, which bank customers pay the most attention to. It is on this indicator that the vector of financial sustainability accounts for almost 46% (0.4577). On the second position, with a difference of 0.0322 units occupy PrivatBank. It has significant advantages over Alfa-Bank in all other indicators, except for price policy. Thus, for the rating increase, the management of PrivatBank should improve the pricing process to attract more customers. The introduction of such an approach to assessing the bank by level of financial stability, on the one hand, will allow clients to make a careful choice of the institution for servicing, and on the other hand, will encourage banks to more efficient activities aimed at meeting the needs of clients.

## 4 Conclusion

It is proved that overcoming of the crisis phenomena in the banking sector of Ukraine to a large extent depends on the level of trust of the banks on the part of the population and business entities. In this connection, for bank customers, a very relevant choice of bank is being offered, which will offer the most reliable, safe and efficient range of banking services. The methodical approaches we are considering are inconvenient and difficult for the external user, because they require special knowledge and can only be used if the bank reports. The results of the rating system used by the NBU are not freely available to users. Thus, the task was to develop a methodological approach to assessing the banks by the level of financial stability that would be available to clients (the population and business entities) and used them when choosing a banking institution.

To solve this problem, we used the method of analytical hierarchy by T. Saati. In our opinion, the introduction of such an approach to assessing the bank by level of financial stability, on the one hand, will allow clients to make a careful choice of the institution for service, and on the other—will encourage banks to more efficient activities aimed at satisfying, above all, the needs of customers.

## References

1. Allen, W.A., Wood, G.: Defining and achieving financial stability. J. Financ. Stab. **2**(2) (2006)
2. De Nicolo, G.: Size. Charter Value and Risk in Banking: An International Perspective. FRB: International Finance Discussion Paper. No. 689 (2000)

3. Davis, P.E.: Towards a typology for systemic financial instability, Public Policy Discussion Papers 3–20. Economics and Finance Section, School of Social Sciences (Brunel University, 2003)
4. Roy, A.D.: Safety First and the holding of assets. Econometrica **20**(3), 431–449 (1952)
5. Pantelejev, V.P.: Financial stability of the bank: regulating problem. Bank. Ind. **1**, 32–39 (1996). (in Ukrainian)
6. Kreinina, M.N.: Financial stability of enterprise: assessment and decision making. Financ. Manage. **2**, 32–36 (2001). (in Russian)
7. Maslenchenkov, J.S.: Financial management in commercial bank. (M, Perspektyva, 1997) (in Russian)
8. Jaremko, S.A., Bevz, S.V.: The built of rating model of assessing the results of bank's activity. Scientific journa «Computer-integrated technologies: education, science, production», 178–186 (Lutsk, 2015) (in Ukrainian)
9. Kovalenko, V.V.: Anti-crises management of ensuring the financial stability of bank system: monography—Sumy: UABS NBU (2008) (in Ukrainian)
10. Glossary of banking terminology «SYSTEMA CAMELS» .http://www.bank.gov.ua/control/uk/publish/article?art_id=123651 (in Ukrainian)
11. Tkachenko, O.O.: Comparison of financial stability of bank system of European countries. Financial and credit activity: problems of theory and practice, Vyp. **2**, 68–75 (2015) (in Ukrainian)
12. Law of Ukraine «On National Banks of Ukraine» dated by 20.05.1999 No 679-XIV. http://zakon3.rada.gov.ua/laws/show/679–14 (in Ukrainian)
13. Instruction approved by Resolution of National Bank of Ukraine Board «About the order of regulation of bank's activity in Ukraine» dated by 28.08.2001. No 368. http://zakon3.rada.gov.ua/laws/show/z0841–01/page (in Ukrainian)
14. Annual report of National Bank of Ukraine 2015. https://www.bank.gov.ua/ (in Ukrainian)
15. Czech National Bank. https://www.cnb.cz/en/financial_stability/stress_testing/
16. Reserve Bank Of New Zealand. https://www.rbnz.govt.nz/financial-stability/stress-testing
17. Resolution of National Bank of Ukraine Board dated by 01.11.2016 No 393 «On Approval of Inspection Procedures». https://www.bank.gov.ua/ (in Ukrainian)
18. Saati, T.: Making decisions. Method of hierarchy analysis. (M, «Radio ans connecrions», 1993) (in Russian)
19. Bljumyn, S.L., Shujkova, Y.A.: Introduction to the mathematical methods of making decisions (Lypeck, 1999) (in Russian)
20. Ghrechko, A.V.: Onthology of Saaty method of hierarchy analysis. «Artificial intellect». No 3. 746–757 (2005) (in Ukrainian)
21. Beck, T.: *Bank Ownership and Stability: Evidence from Germany*. https://www.fdic.gov/bank/analytical/cfr/2009/june/CFR_SS_2009_beck.pdf

# Accounting and Analytical Methods for Identifying Risks of Agricultural Enterprises' Sustainable Development

**Oleh Sokil, Zhuk Valeriy, Nataliia Holub and Olha Levchenko**

## 1 Introduction

There are many models for diagnosing the risks impact on the financial and economic enterprise state using various indicators calculated on the basis of accounting reports. At the same time, when analyzing the agricultural enterprises' financial condition, such models do not take into account the specific features of sectoral regional specificities and, first of all, the principles of sustainable development, which in normative documents are absent altogether. Given this, based on such an analysis, it is difficult to make an unambiguous conclusion about the agro-industrial enterprises' sustainable development in terms of organizational and technical specifics, goals and strategies, coverage of the market segment, phases of the life cycle, etc. It is necessary to use a complex of different models with the optimal number of coefficients, taking into account the specific nature of the AIC to prevent questionable results of analysis of the impact of risks on sustainable development.

O. Sokil (✉) · N. Holub · O. Levchenko
Tavria State Agrotechnological University, 8 B.Khmelnytsky Ave, Melitopol, Zaporizhia obl.72310, Ukraine
e-mail: oleh.sokil@tsatu.edu.ua

Z. Valeriy
National Science Center «Institute for Agrarian Economics», St. Heroes Oboron, 10, Kiev 03127, Ukraine

© Springer Nature Switzerland AG 2019 
V. Nadykto (ed.), *Modern Development Paths of Agricultural Production*,
https://doi.org/10.1007/978-3-030-14918-5_55

## 2 Methods and Results

### *2.1 Analysis of Common Methods for Assessing the Sustainable Development Risks*

Variety works of investigation the essences, classifications and methods for determining entrepreneurial risk have been conducted by many scientists, in particular: Altman [1], Aman and Rahman [2], Blank [3], Gul et al. [4], Kaminska [5], Khan and Bhatti [6], Khurana and Raman [7], Manfred and Kellezi [8], Smetanko [9], Toffler and Tishaw [10] and other.

The part of the researches is devoted to determining the risk in the audit and the quality of the audit services provision, which can significantly affect on the investors decision, creditors, counterparties, etc. [4, 7]. However, in today's economic realities, there is a need to determine the risk in accounting, since it is the quality of financial reporting data that lead to stakeholders' global decisions [11–14]. At the same time, the impact of risks on decisions made on the basis of accounting and analytical support for the management of agricultural enterprises' sustainable development, almost unexplored.

Analysis of accounting and analytical risks for the indicators formation of assessing the agricultural enterprises' sustainable development should be conducted in several stages: the diagnosis of the probable bankruptcy risk, the assessment of the consequences of risks and the analysis of the cost-effectiveness of anti-risk measures in accounting. On the one hand, the above steps should cover the assessment of risks of an economic, ecological and social nature, and on the other—not to be time-consuming and costly. In addition, the process of analyzing the risks of sustainable development for agricultural enterprises should be based on the following principles: accuracy, measurability and other principles of international counting and financial reporting.

Accounting and analytical model as a basis for the formation of agricultural enterprises' sustainable development reporting depends on the choice of the assessment base and the concept of capital preservation (financial or physical). According to the financial concept of capital, invested funds or the invested purchasing power and the capital are synonymous with the net assets or the equity of the enterprise. According to the physical concept of capital, production capacity and capital are considered as the productivity of an enterprise, based, for example, on the units produced per day [15]. The advantage of the financial concept model, which reflects changes in the capital in financial (monetary) indicators, is the understandability of indicators and the availability of an analysis procedure for managers and stakeholders of agricultural enterprises.

We can use the methods generally accepted in economic analysis to analyze the financial and economic situation, the risk of losses and the likelihood of bankruptcy:

1. Analysis of the break-even sales aims to find out the non-profit turnover and financial strength margin in the form of the difference between the actual and

the threshold sales volume. The method involves calculating the critical level of fixed and variable costs and the marginal price per unit of output. The level of these indicators can be determined graphically or analytically.

2. Analysis of financial ratios provides for the definition of liquidity, financial stability and profitability in order to identify the organization's possibilities for debt recovery, the ratio of own and borrowed capital, the efficiency of using its own working capital, in particular, resources of the enterprise. The quantitative and qualitative filling of the given block of analysis should be carried out taking into account the specific features of the activity of the analyzed enterprise.

The first method covers also an estimation of bankruptcy probability of the enterprise mainly within of discriminant factor models. In international practice, Altman's multifactorial models are widely used to predict bankruptcy, which makes it possible to forecast bankruptcy for a five-year period with an accuracy of up to 70% [16, p. 481].

The first multidimensional model of forecasting bankruptcy was developed by Altman [1] from New York University in the late 1960s. After this pioneering work, a multidimensional approach to forecasting bankruptcy spreads throughout the world among researchers in finance, accounting, banking and credit risk.

It is important to note, that none of the existing models for predicting the bankruptcy probability cannot give an absolutely accurate result, because it is based on the results of a survey of sample data received at enterprises in a certain period of time and certain industries. Therefore, when choosing a model, it is necessary to take into account the specifics of the enterprise's activities, in particular, the agrarian sector with shifting production cycles, seasonality, cyclicality, etc.

## 2.2 The Methodology for Risks Identifying of Agricultural Enterprises' Sustainable Development

The analysis of the risks consequences in accounting for sustainable development contains methods and techniques that allow to identify the impact of each type (or each group) of risks on the performance of the organization and, above all, on capital. The following system of absolute indicators can be used to analyze the consequences of the risks impact on the agricultural enterprise capital (Table 1).

Financial risks that are reflected in accounting can be classified as risks associated with changes in the assets value and risks associated with the implementation of future expenses. The consequences of risks associated with changes in the inventories value, receivables and financial investments are reflected in the accounting records by creating valuation reserves in the event of a decrease in the value (impairment) of the asset, provided that the created reserve increases the amount of other expenses. However, in the future the amount of the created reserve can be restored, which increases the amount of other firm revenues. Despite the fact that the change in the value of an individual object under of such risk influence cannot be favorable, the

**Table 1** Absolute indicators of the change in the capital value under the influence of identified risks

| Type of risks, which are reflected in the sustainability accounting | The economic essence of the indicator | The risk effect characteristics of sustainable development |
|---|---|---|
| The risks of changes in the value of assets inherent in agriculture | The risk characterizes the change in the size of capital in view of changes in the value of assets under the influence of environmental factors | The amount of exchange rate differences, changes in the reserve of doubtful debts, provisions for impairment of financial investments, current market value of financial investments, tangible assets, fixed assets, intangible assets |
| Risks of insufficiency created reserves to cover costs | Characterizes a change in the size of capital due to the insufficiency (excessive) of the reserves of future expenses | The difference between the amount of reserves created by future and actual costs, including the cost of unsecured reserves |
| The risks of property loss | Characterizes a change in the size of capital due to losses, shortage, theft of property | The value of the property losses detected and registered in the accounts |
| Commercial risks | Characterizes the change in the size of capital due to losses in the process of selling products, goods, works, services | Amount of losses from the loss and damage of goods in warehouses, during transportation, losses from substitution of poor quality goods, fines, penalties and penalties for violation of the terms of economic contracts and other arising in the process of selling goods and services |
| Production risks | Characterizes a change in the value of capital as a result of deviations from the normal production process | The amount of losses from lack of production, as well as deviations in the cost of finished products |
| Emergency risks | Characterizes a change in the size of capital under the influence of the effects of emergencies | Difference between loss from emergency and extraordinary income |
| Environmental and social risks | Characterizes a change in the size of capital in connection with causing damage to the environment | Amount of compensation for damage caused to the environment |
| Accounting (information) risks | Characterizes the change in the size of capital under the influence of risks due to the peculiarities of accounting organization | Cost of consequences of risks from changes in accounting policies and risks of distortion of information disclosed in accounting |

**Table 2** Relative indicators of changes in the value of capital under the identified risks influence

| The indicator name | The economic essence of indicator | Method of calculation |
|---|---|---|
| The relative change in the capital amount due to the influence of the *i*-th type of risk in accounting | Characterizes the relative change in the amount of capital under the influence of the *i*-th type of risk consequences in accounting | It is determined by the ratio of the change in the capital amount under the influence of the consequences of the *i*-th type of risk in accounting to the amount of capital |
| The relative change in the amount of capital due to the impact of all types of risks in accounting | Characterizes the relative change in the size of capital due to the impact of the effects of all types of accounting risks | It is determined by the ratio of the change in the amount of capital under the influence of the consequences of all risks in accounting to the amount of capital |
| The share of the change in the amount of capital under the influence of the consequences of the *i*-th type of risk in accounting in the general change in the amount of capital | Characterizes the share of change in the amount of capital under the influence of the consequences of the *i*-th type of risk in accounting in the general change in the amount of capital | It is determined by the ratio of the change in the amount of capital under the influence of the consequences of the *i*-th type of risk in accounting in the overall change in the amount of capital for the reporting period |
| The share of change in the amount of capital under the influence of the consequences of all risks in accounting in the overall change in capital | Characterizes the share of capital change under the influence of the consequences of all risks in accounting in the overall change in the amount of capital | It is determined by the ratio of changes in capital under the influence of the consequences of all types of risks in accounting in the overall change in the amount of capital for the reporting period |
| The share of the change in the amount of capital under the influence of the consequences of the *i*-th kind, the risk in accounting in the change in the amount of capital under the influence of all risks in accounting | Characterizes the share of change in the amount of capital under the influence of the consequences of the *i*-th type of risk in accounting in the change in the amount of capital under the influence of all risks in accounting | It is determined by the ratio of the change in the amount of capital under the influence of the consequences of the *i*-th kind of risk in the accounting of changes in the amount of capital under the influence of the consequences of all risks in accounting |

impact of the risk effects on the amount of capital in different reporting periods can be positive or negative. This is due to the fact that in any periods, the amount of created reserves may exceed the recovered amounts.

The change in the fixed assets value and intangible assets is reflected in the accounting through revaluation. And these results affect to the additional paid-in capital and retained earnings. The result of revaluation can be both an increase and a

**Table 3** Absolute indicators of the analysis from the usage of anti-risk instruments

| Type of anti-crisis measures | The economic essence of indicator | Influence characteristic aimed at reducing the capital risk |
|---|---|---|
| Insurance | Characterizes the change in the capital amount in connection with the insurance implementation | Defined as the difference between the amount of income received (insurance compensation) and the amount of insurance costs incurred |
| Conducting internal and/or external environmental audits | Characterizes the change in the amount of capital in connection with the conduct of internal or external audit | It is defined as the difference between the income from the creation of the internal audit service or, the involvement of external auditors and the costs of creating an internal audit service or the involvement of external auditors |

decrease in the value of fixed assets and intangible assets, and, accordingly, the cost of agricultural enterprise's capital.

The currency risk impact is reflected in the bookkeeping records as currency difference arising as a result of currency exchange fluctuations at the value of monetary assets and liabilities in foreign currency. Positive currency difference leads to an increase in financial income, negative—an increase in financial expenses, except for exchange differences when making contributions to the registered capital in foreign currency, causing changes in additional paid-in capital.

Also, in order to analyze the impact of the risks consequences in accounting and analytical support on the capital amount, it is expedient to calculate the relative indicators (Table 2).

In order to analyze the effectiveness of costs for anti-risk measures in accounting, it is necessary to determine the result from the implementation of certain activities, namely the difference between the income from the activities carried out and the costs for them. Absolute indicators of the analysis from the usage of anti-risk instruments are given in Table 3.

One of the most effective tools for minimizing risks in agriculture is insurance, which allows to transfer the risk of an insurance company for a certain fee, in the event of an insured event, will reimburse the costs of the insured.

In addition, in order to reduce accounting (information) risks, the agricultural enterprise may decide to establish an internal audit service or attract external auditors, which is conditioned by the desire to form and, accordingly, provide better information to users-stakeholders and prevent eco-social imbalances in both middle of the enterprise and beyond. Under such conditions, the effectiveness of such measures is confirmed by a reduction in errors in accounting, which can lead to a

**Table 4** Relative performance analysis of the anti-crisis measures

| The indicator name | The economic essence of indicator | Method of calculation |
|---|---|---|
| Coefficient of costs return for anti-risk measures | Characterizes the degree of reimbursement of expenses for all anti-risk events | Defined as the ratio of the amount of revenue from the implementation of all anti-risk measures to the amount of costs for the implementation of all anti-risk measures |
| Reimbursement ratio for *i*-th type of anti-risk measures | Characterizes the degree of reimbursement of expenses for the implementation of the *i*-th type of anti-risk measures | Defined as the ratio of the amount of income from the implementation of the *i*-th type of anti-risk measures to the amount of costs for the implementation of the *i*-th kind of anti-risk measures |
| Profitability of costs for anti-risk measures | Characterizing the profitability of all anti-risk measures | Defined as the ratio of the effect of the introduction of all anti-risk measures to the amount of costs for the implementation of all anti-risk measures |
| Profitability of expenses for the implementation of the *i*-th kind of anti-risk measures | Characterizing the profitability of the *i*-th kind of anti-risk measures | It is defined as the ratio of the effect from the implementation of the *i*-th kind of anti-risk measures to the amount of costs for the implementation of the *i*-th kind of anti-risk measures |
| The share of the change in the amount of capital from the effective implementation of anti-risk measures | Characterizes the relative change in the amount of capital under the influence of the effect of the introduction of anti-risk measures in the overall change in capital | It is determined by the ratio of the total effect from the introduction of all anti-risk measures to the total change in the amount of capital |

reduction in fines and penalties for violation of legislation, including eco-social. The impact of such consequences actions on the capital amount can be calculated as the difference between the income from the activities carried out (reduction of penalties for violation of the law) and the corresponding costs.

In addition to the above absolute indicators, we can calculate the relative indicators of the analysis of the effectiveness of anti-risk measures which presented in Table 4.

Among such indicators, we can calculate the cost recovery factor for anti-risk measures by types of activities and profitability of expenses for the implementation of anti-risk measures as well as by types of activities. The coefficient of cost recovery for the anti-risk measures is calculated as the ratio of income from the application of anti-risk measures to the cost of carrying out anti-risk measures. The event can be recognized as successful when the cost recovery ratio for it should be at least equal the one, that is, the revenue from the implementation of the event should, at a minimum, cover the costs for it. The profitability of expenses for the implementation of anti-risk measures can be calculated as the ratio of the effect from the introduction of anti-risk measures (the difference between the revenues from the event and the costs of it) to the costs for it. The value of the coefficient shows how many cents of profit from the carrying out of the anti-risk measure are accounted by the dollars costs for its conduct. Such indicators can be defined as for each type of conducted anti-risk measures and for all measures in general.

## 3 Conclusions

A reliable system of management control or external eco-social audit becomes actualized in view of the lack of agricultural enterprises' desire to use insurance tools as a method of preventing risk and reducing their impact on accounting and analytical support for sustainable development.

So, the relative change in the capital amount due to the influence of the *i*-th risk type in accounting depends on the following indicators effect:

- share of the change in the capital amount under the influence of the consequences of the *i*-th kind of risk in accounting in the capital amount change under the influence of risks in accounting;
- share of the change in the capital amount under the influence of the risks effects in accounting in the total capital amount change;
- the growth rate of capital for the period.

Thus, the profitability of costs for implementing anti-risk measures is determined by the following factors:

- shares of the capital amount change under the effect influence of the introduction of anti-risk measures in the overall change in capital;
- growth rate of capital for the period;
- coefficient of capital consolidation;
- the cost of carrying out anti-risk measures per dollar of proceeds from sales.

The article novelty is the analysis combination of standard economic risks and risks of the eco-social direction, taking into account accounting risks. At the same time, the size of the capital of an agricultural enterprise depends on:

- the activity specifics;
- eco-social factors that cause the main risks of sustainable development;
- tax risks;
- risks of land relations.

The practical essence of author's research is the possible use of this method of risk identification by agricultural enterprises in identifying factors affecting the profitability of expenditures under the anti-risk measures to improve sustainable development.

## References


1. Altman, E.I.: Financial ratios, discriminant analysis and the prediction of corporate bankruptcy. J. Finance **23**, 589–609 (1968). [in English]
2. Aman, A., Rahman, E.: Managing relational risks in accounting outsourcing: experiences of small firm. World Appl. Sci. J. **15**(1), 56–62. [online], Available at: http://www.idosi.org/wasj/wasj15(1)11/9.pdf (2011) [in English]
3. Blank, I.A.: Financial Management. Elga, Nika-Tsentr, Kyiv (2005). [in Ukrainian]
4. Gul, F.A., Zhou, G.(S.), Zhu, X.(K): Investor protection, firm informational problems, big n auditors, and cost of debt around the world. AUDITING: J. Pract. Theory **32**(3), 1–30. [online], Available at: http://dx.doi.org/10.2308/ajpt-50462 (2013) [in English]
5. Kaminska, T.H., Kolesnikova, O.M.: Financial risk management accounting. Account. Financ. **1**(63), 20–25. [online], Available at: http://www.afj.org.ua/en/article/139 (2014) [in Ukrainian]
6. Khan, M.M., Bhatti, M.I.: Development in islamic banking: a financial risk-allocation approach. J. Risk Financ. **9**(1), 40–51. [online], Available at: http://dx.doi.org/10.1108/15265940810842401 (2008) [in English]
7. Khurana, I.K., Raman, K.K.: Litigation risk and the financial reporting credibility of big 4 versus non-big 4 audits: evidence from anglo-american countries. Account. Rev. **79**(2), 473–495. [online], Available at: http://dx.doi.org/10.2308/accr.2004.79.2.473 (2004) [in English]
8. Manfred, G., Kellezi, E.: An application of extreme value theory for measuring financial risk. Comput. Econ. **27**, 207–228. [online], Available at: http://dx.doi.org/10.1007/s10614-006-9025-7 (2006) [in English]
9. Smetanko, O.: Improving the process of identification and response by the internal audit service. Ekonomicnij Casopis-XXI (Economic Annals-XXI) 11–12, 135–139 (2014) [in Ukrainian]
10. Toffler, R., Tishaw, H.: Going, going, gone—four factors which predict, Accountancy March 1977, 50–54 (1977) [in English]
11. Nichita, M.: Regression model for risk reporting in financial statements of accounting services entities. SEA—Practical Appl. Sci. **3**(2), 101–107. [online], Available at: http://www.sea.bxb.ro/Article/SEA_8_14.pdf (2015) [in English]
12. Nichita, M., Turlea, C.: Approach regarding a framework for risk reporting in order to enhance the related good practices. Amfiteatru Econ. **40**(17), 1108–1121. [online], Available at: http://www.amfiteatrueconomic.ro/temp/Article_2456.pdf (2015) [in English]
13. Sunder, S.: Risk in Accounting. Abacus **4**(51), 536–548. [online], Available at: http://dx.doi.org/10.1111/abac.12060 (2015) [in English]
14. Schiff, J., Schiff, A., Rozen, H.: Accounting for contingencies: disclosure of future business risks. Manag. Account. Q. **13**(3), 1–8. [online], Available at: https://www.questia.com/library/journal/1G1-298614233/accounting-for-contingencies-disclosure-of-future (2012) [in English]
15. Verhovna Rada Homepage. Conceptual Basis of Financial Reporting, IASB Standard, International document, Adoption of 01.09.2010. [online], Available at: http://zakon3.rada.gov.ua/laws/show/929_009 (2010) [in Ukrainian]
16. Bidyuk, P.I., Verteleckiy, V.V., Zhirova, A.O.: Risk analysis of the enterprises' bankruptcy using clear and fuzzy models. Economic Bulletin of NTUU "KPI" 12, 477–490 (2015) [in Ukrainian]

# Managing of the Living Quality of Population in the Social Sphere

**Lyudmyla Synyayeva, Nataliya Bocharova, Iryna Ahieieva and Anhelina Yarchuk**

## 1 Introduction

Scientific comprehension of the conceptual foundations for ensuring the quality of life of the population of a country allows not only to reveal theoretical knowledge in relation to the multifaceted socioeconomic category, but also to solve a number of scientific and applied problems concerning the improvement of the quality of life in the system of growth productivity.

### *1.1 Analysis of Recent Studies and Publications*

The quality of life as a socioeconomic category was studied in the writings of such prominent foreign economists and sociologists as Whitney C., Galbraith J. Kn., Inglegart R., Turan A., Campbell A., Maslow A., Rogers V., Toffler O., Foster D., Andrus F., Converse F., Milbreight L., McKennel, Michel A., Rogers U., Wright C. and other [1–3]. Nowadays the quality of life is studied by Egorshin A. P. [4], Zhulina E. G. [5], Mazaeva N. [2], Mahmutova A. [6], Mihaylov A. [7], Tsyigankov V. K. [8], Nikiforenko V. G. [9], LIbanova E. M. [10] and other. In their works, attempts have been made to analyze objective socioeconomic, political, cultural, environmental and other conditions of human existence, numerous combinations of objective and subjective characteristics.

L. Synyayeva (✉) · N. Bocharova · I. Ahieieva · A. Yarchuk
Department of Management, Tavria State Agrotechnological University,
B. Khmelnistskiy ave 18, Melitopol 72312, Ukraine
e-mail: ouses@tsatu.edu.ua

© Springer Nature Switzerland AG 2019

V. Nadykto (ed.), *Modern Development Paths of Agricultural Production*,
https://doi.org/10.1007/978-3-030-14918-5_56

## 2 Statement of the Objective and Tasks of the Study

The purpose of the article is to generalize and deepen theoretical and methodological foundations and to substantiate practical recommendations for raising the standard of living of the population in conditions of a transitory economy of Ukraine.

The following tasks were set and solved to achieve the set purpose:

- to study the development of scientific views on the essence of the socioeconomic category of quality of life;
- to specify theoretical and methodological bases for determining criteria and indicators of tools in monitoring the quality of life;
- to analyze the current state and dynamics of indicators of quality of life;
- to substantiate the recommendations aimed at improving the quality of life management of the population.

## 3 Solution

The category "quality of life" is generally accepted in international practice, a highly effective method for assessing the social well-being of the population, and is a comprehensive description of the factors and conditions of the existence of individuality in society.

### *3.1 Methodology, Variables and Data*

In 1961, the United Nations has identified 12 "aspects of living conditions," the state of which can qualitatively characterize the level of development of society. These aspects include health care, education, working conditions, employment, consumption and stocks, transport and communications, housing conditions, recreation and entertainment, social security and personal freedom [11]. Other systems and evaluation methods were also proposed. Under current conditions, the methods used in Ukraine are based on the assessment of the quality of working life in the calculation of the integral index of the human development index and the world's generally accepted comprehensive methods and trends. In particular, the maximum attention is drawn to cognitive control, which is considered as an advanced management technology [12]. Using the cognitive approach with the construction of a cognitive map allows simultaneously to solve two types of tasks: statistical and dynamic. Statistical analysis consists of the allocation and comparison of the ways of influencing some factors to others through the third factors. Dynamic analysis is the generation and analysis of possible scenarios for the development of the object under study in time. The mathematical apparatus supports the theory of sign and fuzzy graphs [13]. The scenario approach (Table 1) analyzing the system makes it possible to determine the influence of the external and internal environment on the efficiency of the formation of the quality of life system, which allows us to identify the parameters that need to be accelerated and require change.

## 3.2 Descriptive Analysis

Considering the practical significance and the assessment of these factors, we can say that almost half of the world's population (3.4 billion people) live below the "poverty line"—less than $5.50 a day [14]. This is announced in the article of the World Bank "Poverty and the general prosperity, common solving the problems of poverty reduction." Three-quarters of Ukrainians live below the poverty line—less than $5.5 a day. At the current rate (UAH 27.6 for $1) 5.5 USD per day—154 UAH, with a monthly average salary of the country 6522 UAH per month (as of the end of 2017—the beginning of 2018), the daily income of ordinary Ukrainians was UAH 217. That is, firstly, this earning is nominal, and secondly, the Ukrainian minimum salary (as of the end of 2017—UAH 3200) has not really crossed the "poverty line." In this case, extreme impoverishment occurs with the existence of less than $1.90. per day (53.2 UAH per day or 1649 UAH per month) [15]. In 2014 below the poverty line lived 15% of the population of Ukraine, today—25%. In fact, Ukraine occupies the second last position of salaries in Europe with an average salary of 278 euro. Below is only Moldova with an average salary of 212 euro [16].

## 3.3 Empirical Analysis

The GDP dynamics for 2010–2017 do not show the existence of economic preconditions for improving the quality of life of the population. The total GDP per capita at comparable prices remains in 2017 below the level achieved in 2011. The GDP growth rates in 2016 and 2017 did not exceed 3%, that is, they were below the rate of decline of the existing population in the corresponding years (3.93% and 4.12%, respectively). This means that GDP growth per capita was ensured not by the expansion of production, but by the negative dynamics of the size of the existing population (Tables 2 and 3) [17]. Consequently, the general (macroeconomic) conditions for improving the quality of life of the population of Ukraine in the period of 2010–2017 did not improve. Accordingly, the significance of distribution factors for determining the dynamics of the quality of life of the population increased [18].

The promiscuity of policies that envisage the predominance of compensatory measures (payment of benefits to the poor), instead of preventive (supporting the growth of demand for highly qualified labor and increasing the availability of goods needed to acquire highly paid competencies), is also evidenced by the dynamics of the spread of poverty in Ukraine (Table 4).

**Table 1** The analysis of the influence of factors on the quality of life system (QLS)

| Factor | Stimulation | Braking | Interpretation factor |
|---|---|---|---|
| X1—productivity | Strongly interacts and has an active influence | Weakly interacts has an active influence | Labor productivity, it has an active influence on QLS and does not depend on changes in this system |
| X2—income of the population | Strong level of interaction and passive influence | Strong interaction and passive influence | Changing incomes is a purpose of management, as well as an indicator and a guarantee of the effective functioning of QLS |
| X3—unemployment | Strongly interacts and has an active influence | Weakly interacts and has a passive influence | It actively affects QLS, which makes it an ideal leverage for QLS control |
| X4—working conditions | Average level of interaction, active influence | Middle level of interaction with moderate activity | Actively affects QLS |
| X5—demography | Weak interaction, passive influence | | At the moment, the factor is weakly affecting the QLS change, but then it can be used as an indicator |
| X6—ecology | Weak interaction, active influence | | Weakly affects the QLS change, but the factor can be used as an indicator |
| X7—social guarantees (state support) | Strongly interacts and has an active influence | Middle level of interaction and the active influence | The factor actively affects the QLS The factor can be applied as the main lever of the QLS contro. |
| X8—education | Strongly interacts and actively influences | Middle level of interaction and active influence | The factor is independent of the QLS change, but it itself can be an ideal leverage of QLS management |

(continued)

**Table 1** (continued)

| Factor | Stimulation | Braking | Interpretation factor |
|---|---|---|---|
| X9—inflation | Weakly interacts, although it has an active influence | | The factor weakly affects the QLS, although it can be an indicator of QLS |
| X10—health care | Strongly interacts, although it has a passive effect | Strong level of interaction with middle level of activity | The factor has a high level of interaction with QLS. Its change is the purpose of managing the QLS |
| X11—leisure | Weakly interacts has a passive influence | | The factor weakly affects the QLS change, but can be applied as an indicator |
| X12—safety of life | Weakly interacts, although it has a passive influence | | The change of this factor is the purpose of the management of the QLS |

**Table 2** Dynamics of changes in the macroeconomic conditions for ensuring the well-being material of households with children

| Indicators | 2010 | 2011 | 2012 | 2013 | 2014 | 2015 | 2016 | 2017 |
|---|---|---|---|---|---|---|---|---|
| The number of the existing population, thousands | 45,962.9 | 45,778.5 | 45,633.6 | 45,553.0 | 45,426.2 | 42,929.3 | 42,760.5 | 42,584.5 |
| GDP per person, nominal, ths. UAH | 23.5 | 28.4 | 30.8 | 32.2 | 34.9 | 46.3 | 55.8 | 70.0 |
| GDP per person, in fixed prices in 2010, ths. UAH/person per year | 23.5 | 24.9 | 25.0 | 25.0 | 23.5 | 22.4 | 23.0 | 23.7 |
| Growth rate of chain (%) | 106.2 | 105.9 | 100.5 | 100.2 | 93.7 | 95.4 | 102.8 | 102.9 |
| Basic (%) | 106.2 | 105.9 | 106.5 | 106.7 | 99.9 | 95.4 | 98.0 | 100.9 |
| Share of final expenditures of households in GDP (%) | 63.0 | 66.1 | 67.6 | 71.5 | 70.6 | 67.0 | 65.8 | 65.7 |
| The share of the state-owned sector (%) | 19.4 | 17.4 | 18.6 | 18.6 | 18.7 | 18.9 | 18.6 | 20.4 |
| Nonprofit organizations serving households (%) | 17.6 | 16.6 | 13.7 | 10.0 | 10.7 | 14.1 | 15.6 | 13.9 |

**Table 3** Dynamics of changes in the indicators of the effectiveness of state compensatory measures to reduce the prevalence of poverty for the period from 2010 to 2017

| Indicators | 2010 | 2011 | 2012 | 2013 | 2014 | 2015 | 2016 | 2017 |
|---|---|---|---|---|---|---|---|---|
| Budgetary Expenditures for Poverty Reduction (Articles on Social Protection and Pension Fund Subsidies),% of GDP | 8.1 | 9.0 | 9.9 | 8.7 | 8.9 | 10.8 | 9.6 | 9.8 |
| The share of the population living below the poverty line | 7.8 | 9.0 | 8.3 | 16.7 | 51.9 | 51.1 | 34.9 | 36.7 |
| Average annual nominal size of MS, UAH/month | 963.1 | 1004.1 | 1134.7 | 1218.0 | 1271.3 | 1378.7 | 1600.0 | 3200.0 |
| Average annual nominal size of average MS, UAH/month | 2648.0 | 3054.0 | 3337.0 | 3619.0 | 4012.0 | 5230.0 | 6475.0 | 8777.0 |
| The ratio of the MS to the average salary (%) | 36.4 | 32.9 | 34.0 | 33.7 | 31.7 | 26.3 | 24.7 | 36.5 |
| Growth rates of MS, chain % of last year | 143.9 | 104.3 | 113.3 | 107.4 | 104.4 | 108.4 | 116.1 | 200.0 |
| Growth average MS, chain % of last year | 118.6 | 115.3 | 109.3 | 108.5 | 110.9 | 130.4 | 123.8 | 135.5 |
| Growth rates of MS, basis % 2010 | 100 | 104.3 | 117.8 | 126.5 | 132.0 | 143.1 | 166.1 | 332.3 |
| Growth average MS, basis % 2010 | 100 | 115.3 | 126.0 | 136.7 | 151.5 | 197.5 | 244.5 | 331.5 |

**Table 4** The dynamics of changes in the indicators of the contribution of education and health sectors to the dynamics of the quality of life of the population in Ukraine for the period from 2010 to 2017

| Indicators | 2010 | 2011 | 2012 | 2013 | 2014 | 2015 | 2016 | 2017 |
|---|---|---|---|---|---|---|---|---|
| Chain indices of dynamics of physical volumes of consumption of goods created in education,% of last year | 100.0 | 99.0 | 110.8 | 100.7 | 91.9 | 98.5 | 97.2 | 120.5 |
| Chains of dynamics of physical volumes of consumption of goods created in health care% of last year | 100.0 | 113.0 | 112.9 | 102.0 | 87.0 | 94.8 | 100.1 | 121.3 |
| Basic indices of dynamics of physical volumes of consumption of goods created in education% 2010 | 100.0 | 99.0 | 109.8 | 110.5 | 101.6 | 100.1 | 97.3 | 117.3 |
| Basic indices of dynamics of physical volumes of consumption of goods created in health care% 2010 | 100.0 | 113.0 | 127.6 | 130.2 | 113.3 | 107.4 | 107.5 | 130.4 |
| Chains of dynamics of the availability of goods created in education,% of last year | 100.0 | 101.05 | 103.45 | 103.89 | 100.03 | 97.17 | 102.14 | 117.49 |
| Chains of dynamics of accessibility of benefits created in health care% of last year | 100.0 | 103.13 | 104.85 | 105.00 | 88.61 | 83.26 | 106.53 | 123.79 |
| Basic indices of dynamics of accessibility of goods created in education% 2010 | 100.0 | 101.05 | 104.54 | 108.60 | 108.63 | 105.56 | 107.82 | 126.68 |
| Basic indices of dynamics of accessibility of benefits created in health care% 2010 | 100.0 | 103.13 | 108.13 | 113.54 | 100.61 | 83.77 | 89.24 | 110.47 |

That is, not a high proportion of budget expenditures directly used to reduce poverty, nor the growth of this share, not only reduces the size of poverty, but also simply suspends its proliferation. For 2011–2017, the share of budget expenditures intended to compensate for the income insufficiency fluctuated between 9 and 10.8% of GDP, but the share of the poor, according to the absolute criterion, increased from 8.6 to 51.1% of the total.

It should be noted that the growth of nominal budget expenditures is low, the inflationary nature is real (taking into account the consumer price index, not the deflator of GDP), the amount of budget expenditures was reduced, which objectively could not affect the real value of all forms of state compensatory measures, but rather a jump-like the growth of the statutory minimum salary (a powerful leverage on the size of official incomes) has significantly reduced the share of poverty—up to 34.9% in 2017 growth of the average salary against the background of a rapid increase in the minimum in 2017 was significantly lower (the real average salary increased in 2017 compared to 2016 by only 33.5% and the minimum—by 2 times). As a result, inter-qualification ratios significantly worsened, once the qualifications of workers were devalued (due to a reduction in the ratio of wages to skilled and unskilled labor—the minimum reached 36.5% of the average against 24.7% in 2016). And again—incentives to invest in human capital have deteriorated. Namely the last resource was and remains the main limitation of growth and development opportunities in Ukraine, and measures to improving the quality of life are limited only by the redistribution of extremely limited economic opportunities between "more or less needy segments of the population."

But this "zero-sum game" has nothing in common with either development or the creation of real possibilities for radical improvement of the positions on the labor market for the general population.

A similar situation is observed with the data on the volume and structure of resource provision of education and medicine (Table 4). From 2010 to 2017, the real (expressed in prices in 2010) volumes of resource support for education in Ukraine decreased by more than 21%. At the same time, the structure of financial resources of education has remained virtually unchanged: The share of budget financing does not deviate by more than 1.5% points from the indicator of 89%, and the share of household expenditures—around the indicator of 10.5%.

Real volumes of healthcare resources up to 2016 continued to decline (to 88.2% of the 2010 level) and in 2017—dynamically increased to 108.3% in 2010. At the same time, if by 2016 inclusive, the increase in the share of household expenditures in the medical sector resources was moderate (to 46% in 2016 vs. 32% in 2010), then in 2017, it increased immediately to almost 52%. The share of budget financing decreased from 67 to 48% in the period from 2010 to 2017.

For education, more peculiar to health, the increase in the availability of goods produced by the industry (the index of availability in 2017 compared with 2010–126.7%, and for health care—110.5%) and less expansion of physical volumes of consumption of goods produced by the industry: the physical volumes of consumption of education services increased in 2017 compared to 2010 by 17.3%, and the growth of physical volumes of households' consumption of health benefits accounted for

30.4% in 2017 p. compared with 2010. Consequently, a more dynamic expansion of the share of non-budget financing peculiar to health has allowed to provide almost twice as large an increase in physical consumption than in education, however, also resulted in lower rates an increase in the availability of goods produced in the industry (110.5% growth rate for the period 2010–2017, compared to 126.7% for education).

## 4 Conclusion

One of the fundamental differences of modern Ukraine from developed countries is expressed in the great dependence of the welfare of the general population on the resource provision of the social policy of the government and the effectiveness of the implementation of targeted social support programs. At the same time, such magnitude and influence of state compensatory measures are perceived as necessary, due to the disadvantages of commercial mechanisms for satisfying the needs of the general population. Therefore, we propose the application of an integrated approach to forming the directions of improving the quality of life of the population on the basics of overcoming the backlog of socioeconomic development and creating a system for motivating the effectiveness of labor activities, improving the system of material incentives, as well as improving the system of change management. The implementation of the proposed measures, activation and optimization of social policy will lead to the strengthening of the processes of market transformation, the formation of a middle class among the population of a powerful, economically independent stratum, the eradication of poverty and raising the level and quality of the population life.

## References

1. Genkin B.M.: Ekonomika i sotsiologiya truda: ucheb. dlya studentov vuzov, obuchayuschihsya po ekon. spets., 5-e izd., dop., M.: Norma,402 s (2007)
2. Mazaeva, N.: Kachestvo trudovoy zhizni—vazhnaya sostavlyayuschaya menedzhmenta personala. Problemyi teorii i praktiki upravleniya., N 3., S. 115–121 (2004)
3. Rayzberg, B.A.: Osnovyi ekonomiki: ucheb. posobie. M.: Infra-M, 407 s (2000)
4. Egorshin A.P., Zaytsev A.K.: Kachestvo zhizni naseleniya regiona. A.P. Egorshin. Narodonaselenie, N 1., S. 14–27. Bibliogr.: s. 27 (2005)
5. Zhulina, E.G.: Sotsialnyie faktoryi formirovaniya kachestva trudovoy zhizni. Grazhdanin i pravo., N 5., S. 64–73., Bibliogr. v snoskah (2007)
6. Mahmutova A.: Nepreryivnoe obrazovanie i kachestvo trudovoy zhizni molodezhi. Chelovek i trud., N 1., S. 53–55 (2007)
7. Mihaylov A. Usloviya truda kak element kachestva trudovoy zhizni. Chelovek i trud., N 2., S. 84–86. - Bibliogr. v snoskah (2006)
8. Tsyigankov, V.K.: voprosu ob otsenke kachestva trudovoy zhizni. Chelovek i trud.,N 2., S. 46–49 (2007)

9. YakIst zhittya naselennya regIonu: analIz, prognozuvannya, sotsIalna polItika/za red. V.G. Nikiforenka. MonografIya/MONMS UkraYini. Odeskiy nats. ekonomIchniy un-t., Odesa, 316 s (2012)
10. Lyudskiy rozvitok v UkraYinI: ModernIzatsIya sotsIalnoYi polItiki: regIonalniy aspekt (kolektivna monografIya)/ za red. E.M. LIbanovoYi; In-t demografIYi ta sotsIalnih doslIdzhen Im. M.V. Ptuhi NAN UkraYini., K., 356 s (2015)
11. Kulinich, A.A.: Kompyuternyie sistemyi modelirovaniya kognitivnyih kart: podhodyi i metodyi. Problemyi upravleniya. .#3. C.2–15 (2010)
12. Kulinich A.A.: Kognitivnaya sistema podderzhki prinyatiya resheniy «Kanva». Programmnyie produktyi i sistemyi., #. 3. S. 25–28 (2002)
13. Kudzh, S.A., SolovYov, I.V., Tsvetkov V.Ya.: Kognitivnyie modeli i metodyi. Kratkiy slovar-spravochnik., M.: Moskovskiy gosudarstvennyiy tehnicheskiy universitet radiotehniki, elektroniki i avtomatiki MGTU MIREA, 95 s (2014)
14. Doklad o chelovecheskom razvitii 2016. Chelovecheskoe razvitie dlya vseh i kazhdogo. Per. s angl.; Programma razvitiya OON., M.: Izdatelstvo «Ves Mir» , 284 s (2017)
15. Sistema sotsIalnogo zahistu ta sotsIalnogo zabezpechennya v UkraYinI. Realniy stan ta perspektivi reformuvannya., K.: Tsentr gromadskoYi ekspertizi, 2009., 104 s., BIblIografIya: s. 104
16. Svitovij bank: Kozhna druga lyudina u sviti zhive za mezheyu bidnosti, https://openknowledge.worldbank.org/bitstream/handle/10986/30418/211330ov.pdf
17. Yarchuk A.V.: Kachestvo zhizni v kontekste teorii i praktiki sotsIalnogo blagosostoyaniya. «Formuvannya kompetentnIsnoYi komponenti fahIvtsIv z oblIku I auditu». MonografIya/ za redaktsIEyu L.V.SinyaEvoYi, O.V.VoronovskoYi., MelItopol: Vidavnichiy budinok MelItopolskoYi mIskoYi drukarnI, 348, S.333–336 (2016)
18. Yarchuk A.V.: ProduktivnIst pratsI yak faktor vplivu na yakIst zhittya naselennya UkraYini. Naukovo-ekonomIchniy ta suspIlno-polItichniy zhurnal «UkraYina: aspekti pratsI», m. KiYiv, 51 s (2017)

# Accounting Essence of Amortization Policy

**Zhuk Valeriy, Trachova Dar'ya, Sakhno Ludmyla and Demchuk Olena**

## 1 Introduction

In the last third of the twentieth century, a new technological revolution, solid computerization, informatization, intellectual property created a fundamentally new situation in society. In order to determine the influence of accounting systems on the processes of forming a depreciation policy, it is necessary to investigate their classification and determine the characteristic features.

### *1.1 Analysis of Recent Studies and Publications*

In the economic encyclopedia, the classification is the division of objects, concepts, names into classes, groups, grades by common feature [1, p. 764]. Pushkar notes that the classification of systems indicates their diversity by structure, origin, purpose and function [2, p. 17]. In the process of economic development of society, as noted by the Russian scientist Sidorov, various signs of classification of accounting systems began to stand out [3, p. 21]. L. Tchaikovska stresses the importance of considering classification system accounting systems. The scientist notes that the accounting system is constantly subject to changes, as all major institutions in different countries, so it is necessary to consider the classification of accounting systems, of course, given the fact that, with the specificity of the historical path of accounting development, its

Z. Valeriy
National Science Center «Institute for Agrarian Economics», St. Heroes Oboron, 10, Kiev 03127, Ukraine

T. Dar'ya (✉) · S. Ludmyla · D. Olena
Tavria State Agrotechnological University, 8 B. Khmelnytsky Ave, Melitopol, Zaporizhia obl. 72310, Ukraine
e-mail: daria.trachova@tsatu.edu.ua

© Springer Nature Switzerland AG 2019

V. Nadykto (ed.), *Modern Development Paths of Agricultural Production*,
https://doi.org/10.1007/978-3-030-14918-5_57

development is subject to a number of general laws [4]. Such a theory, in our opinion, is the institutional (macro-level) and non-institutional (micro-level) economic theory.

According to M. Kozlov, the classification allows to effectively describe and compare the accounting systems of different countries; promotes the development of accounting as a theory and practice; is an integral part of the qualitative training of highly qualified accountants and auditors operating internationally; allows you to solve problems, to predict and prevent their occurrence, based on the experience of other countries that apply a similar model of accounting [5, p. 116]. Leading Ukrainian scientist M. Chumachenko singled out general signs for construction of the classification of systems. For example, in the presence of feedback, the author identified open, in which the elements have a directed action and form a chain sequence; closed, in which the sequence of elements is closed by the feedback; Combined, which include open and closed subsystems [6, p. 105]. The classification symbols proposed by M. Chumachenko, in our opinion, can also be used to construct the classification of accounting systems.

M. Shigun highlights the most problematic questions when constructing the classification of accounting systems, which feature to base classification (type of regulation, influence of national peculiarities, approaches to evaluation); how to take into account the official principles of the regulation of the accounting and reporting system or real practice in the country; how to take into account the economic situation, the type of economy, the peculiarities of the political system; how to take into account the instability of the existing accounting system, which is altered by the influence of legal, political, economic factors, influence of religion and culture within the country and under the influence of world processes; how to reconcile the conflict between national and international business interests that determine the information orientation of accounting [7, p. 296].

Conduct, which is the essence of the institutional approach, as a new scientific methodology, may be an understanding of the essence of accounting theory and practice at a modern, deeper and meaningful level [8, p. 379]. Changes in the accounting system occur under the influence of global globalization processes, interstate economic, political, trade relations. This causes instability of existing accounting systems [Error! Bookmark not defined., p. 295].

The accounting system, as a kind of information system, refers to multidimensional complex-organized systems [Error! Bookmark not defined., p. 108]. Nikandrova notes that the accounting system—a set of interrelated elements, which include types of economic records, which were developed in one or another country in the world due to national, sociopolitical, historical, geographical and other features of its development. Formation of the accounting system at the level of the economic entity is carried out taking into account the goals and objectives of its business, provided by the charter [9, p. 13].

The accounting system should be considered as a separate set of interacting elements, which forms a definite integrity, has certain integral properties, which enables it to perform certain functions [10]. Properties that must be inherent in the accounting system should be considered through the prism of the properties of the system, as categories in general [10, p. 115].

To construct the classification of accounting systems, we are invited to apply a system approach. As the leading national scientist-economist Malyuga observes, the systematic approach to the study of complex systems can reveal an internal mechanism not only of the action of individual components, but also their interaction at different levels. The application of the category of the system, the systematic approach to cognition, in particular, accounting phenomena, has an extremely important theoretical value, since it allows us to investigate the internal integrity of accounting, the organic interconnection and the harmonic interaction of its constituent parts [10, p. 265].

From the standpoint of the evolutionary approach to the development of accounting systems, Bezverhim has identified four main hierarchical units (type-class-family-type) in the systematization of accounting [11]. It is proposed to first determine the essence of such hierarchical units. According to the definitions given in the "Great Dictionary of Contemporary Ukrainian Language" (edited by V. Buzel): the type is a sample, a form, a model, a property that corresponds to a certain group of objects, concepts, phenomena [12, p. 450]; class—a set of objects, phenomena having common features, the same qualities: rank, subdivision [12, p. 544]; family—a set of homogeneous, similar objects, phenomena [12]; view—is a unit that unites a number of subjects on different grounds and is a part of the general higher section of the genus [12, p. 132].

Statement of the objective **and tasks of the study** is to develop the conceptual framework of the information support system for depreciation policy.

## 2 Methods

The research used general scientific and special research methods. Namely: grouping and comparisons for the study of organizational and economic features of the functioning of domestic enterprises, depending on the size and direction of their activities and their impact on the formation of depreciation policy of the enterprise; abstract-logical—during the study of the features of the formation of depreciation policy in the subjects of management; analysis and synthesis—to study the factors of influence of various state regulation measures on the efficiency of asset renewal and the use of the depreciation fund of enterprises.

The theoretical and empirical basis of the study was the development of foreign and domestic scientists, data of statistical observations, regulatory and legal acts regulating the question of formation of accounting and depreciation policy of enterprises.

## 3 The Basic Part of the Study

To determine the historical types of accounting systems, as an essential feature of their structural diversity, a distinguished method of registration.

Types of accounting systems differ in the way of registration of facts of economic life. According to selected features (unified, graphic and polygraphic paradigm), the following historical types of accounting systems are selected as a simple inventory system; accounting information system; integrated accounting system.

The class of the accounting system is characterized by features determined by the levels of building a global economy. From this position, the micro-level (accounting system of the enterprise), meso-level (the unified accounting system of the group) and the macro-level (national accounting system) architecture of the construction of accounting systems are allocated. As O. Petruk correctly observes, in general the accounting system at the national level is characterized by constancy and complexity [13].

The family of accounting sets is given by the specifics of the sectors of the economy and types of economic activity that determine the sectoral accounting standards. It is a record in state (municipal) institutions and accounting in commercial organizations, as well as accounting for types of economic activity: accounting in industry, agriculture, trade, banks, insurance organizations, etc.

The main structural subsystems of the accounting system of the subject of the economy are types of accounting. Each type of accounting has a specific purpose and methodology, there are certain functions that are defined by the purposes of users and generators of accounting and reporting information. Each type of accounting maintains its autonomy in the accounting system and has its own evolutionary trajectory. Each type of account is governed by special rules and requires special knowledge.

Any classification of types of accounting remains subjective, causes controversial judgments and depends on the goals of the researcher. However, the logical process of evolutionary accounting development is characterized by the exchange of accounting and reporting information between types of accounting and the establishment of new interconnections between them.

Thus, for accounting systems characterized by the plurality of states, dynamism, and alternativeness. The indicated signs remain open for refinement and additions. In particular, in each historical period under study, there may be a record of varying degrees of maturity. Thus, by this time the type of simple accounting in households, individual small enterprises, as well as individual entrepreneurs is preserved, while in the medium and large business environment new types of accounting are emerging, namely: (accounting (financial) accounting, internal economy (managerial) accounting, tax calculations, consolidated accounting, accounting according to international standards, social, ecological, etc.).

Each type of accounting is characterized by a set of indicators, but all of them are subject to the organization or regulation of accounting at the national level and represent forms and methods of such regulation [13], p. 36]. In our previous studies, the issue of micro-, meso- and macro-level accounting as a generator of accounting and reporting information for management needs was thoroughly covered [13].

The main task of accounting for depreciation—to provide disclosure in the financial statements of information on the formation and use of funds for the reproduction of fixed assets at all stages of the cycle. Thus, depreciation provides a systematic transfer of the cost of fixed assets to the cost of produced products [14, p. 4].

Among economists, there is no single approach to understanding the essence of the concepts of "depreciation", "wear", "wearing". Between the categories of "depreciation", "wear", "wearing" can be traced to such a relationship: in the use and under the influence of external factors, fixed assets lose their consumer value, which is aggregated by their depreciation. Loss of labor costs is due to transferring it to the finished product (expense period), i.e., through depreciation [15]. The terminological problems of the theoretical basis for accounting for depreciation and amortization of non-current assets are also emphasized in the works of Oshmarina [16]. Therefore, in practice, the reliable definition of either the source of simple reproduction of fixed assets, or their residual value is not ensured, since depreciation is mixed with their depreciation. A problem of great scientific and practical importance is the substantiation of the essence of the depreciation of non-current assets. The article on the economic content of depreciation of fixed assets is devoted to articles by well-known scientists Golov and Tkachenko [17].

At the same time, the problem of mixing it with the depreciation of fixed assets remains beyond the attention of scientists. Only K. Utenkova tried to consider the demolition in the context of regulating the value of non-current assets. But the correlation between these concepts is not investigated by the scientists and therefore in the domestic theory is axiomatic statement: Depreciation of fixed assets—the amount of depreciation of the object of fixed assets from the beginning of its useful use, as interpreted by Standard 7 «Fixed assets». The equation of depreciation to the depreciation of fixed assets is the borrowing of the idea of J. Dumarsche (1874–1946), a French scientist, who, however, at the same time interpreted it as a regulator, but not a reserve [18].

It should be noted that, according to M. Pomazkov, even in the XIV century, were distinguished depreciation and depreciation of fixed assets (this is evidenced by the "Losses and Profits" of the Barcelona department of the company Datini on 13.07.1397–31.01.1399 years), and the latter was determined by the expert assessed and debited from Inventory account for "Profit and Loss" debit.

Nowadays, according to the domestic theory and practice, instead of the methodological substantiation of the essence of the definitions of "depreciation" and "wear" of the basic means, the eclectic approach was used, combining diametrically opposite interpretations of such famous scientists of the late XIX—the beginning of the XX century, as R. Beczman (1870–1936) and E. Sivers (1852–1917).

In particular, the first interpreters of the account amortization as regulating the account "Fixed assets", and the second argued that it is a new fund, so it is a stock account [19].

Based on the study, systematization and generalization of existing views on the definition of depreciation, the difference between the concepts of "depreciation" and "depreciation" of fixed capital is substantiated.

In particular, if wear is a loss in consumer value, and hence the value of labor, then under depreciation is the transfer of the cost of labor to the product being created, followed by their restoration. Both processes, in spite of their differences, are inseparable as two sides of one phenomenon. Therefore, depreciation deductions, reflecting the value of the transferred value, simultaneously show the degree of wear and tear of fixed capital [20].

Critical analysis of theoretical studies and statistical data on the reproduction of capital, the results of enterprises allowed us to conclude that the economy is dominated by linear depreciation and increasing losses of financial opportunities associated with non-linear depreciation.

The study of the historical transformation of accounting components of depreciation policy has revealed the presence of different points of view, discovered in the economic literature. Summarizing a large number of scientists' views on the individual aspects of depreciation, it is possible to note the following features:

(1) Depreciation is the most important moment in the circulation of fixed assets;
(2) Some authors identify the concept of "depreciation" with the concept of "depreciation". But depreciation is the process of transferring the cost of labor resources to the newly created product, and amortization deductions are the cost expression of the lost value of fixed assets in the production process and as a result of the moral wear and tear of labor;
(3) Depreciation simultaneously combines features of cost items that reduce the balance sheet profit and features of income items, while it does not affect the movement of cash.

In the first case, depreciation serves as an element of production costs, not related to the outflow of cash in the form of depreciation deductions, which are included in the cost of production. In the second case, depreciation is a part of the proceeds from the sale of products, which does not bring an additional inflow of funds;

(4) Due to depreciation as an economic process, lost in the process of production, the value of fixed assets is not lost, but stored, accumulate in the depreciation fund;
(5) The theory of depreciation should not oppose one concept of depreciation to another. Each of them is important in terms of ensuring an effective replacement of labor resources.

When establishing the rates of depreciation should be based on economically expedient medium-term operating tools, the need to ensure full compensation for the cost of fixed assets and taking into account their technical and economic aging. The most difficult is the correct definition of the length of the depreciation period (expedient use) of specific means of labor. Usually it is established taking into account many factors, in particular, the overall physical longevity and cost of capital repairs of labor, conditions of their operation, the timing of the onset of technical and economic aging, the possible rates of renewal, etp.

With the development of technology, the improvement of technology and organization of production, the length and nature of the use of certain types of fixed assets change, there is an objective need to reduce the normative terms of their functioning. In this regard, the rates of depreciation should be periodically reviewed and specified.

Depreciation is an extremely complex economic phenomenon that combines the signs of production costs and sources of money, the process of cost movement and the levers of control of reproduction, the compensation of worn out and the accumulation of new means of labor.

In the economic literature, there are three main reasons for easing the role of depreciation in Ukraine: the inadequacy of indexation of fixed assets; long terms of their exploitation; inefficient and inadequate use of accelerated depreciation.

As noted in some literature sources, insufficient use of depreciation deductions and attempts by enterprises to reduce their size is explained by the fact that depreciation affects the increase in the cost of production, if the equipment is loaded less than 50%.

This is explained by the fact that the amount of accrued depreciation depends only on the value of fixed assets, and not on the volumes of manufactured products. If the equipment is not loaded at full capacity, the volumes of the manufactured products are significantly reduced. At the same time, the size of depreciation will not change, but will make a significant share in the cost of production.

At the present stage, the peculiarity of the economic situation in Ukraine is that in order to ensure financial conditions for the implementation of the growth strategy, it is necessary not so much a further increase in the amount of depreciation deductions, such as the maximum liberalization of the depreciation system, which is to enable enterprises to apply flexible methods of accrual depreciation based on the chosen investment strategy.

Depreciation deductions can play a significant role in the development of financial and credit relations. Such funds (depreciation deductions) are kept on a special account of an enterprise in a banking institution. This is most consistent with the principles of financial management, especially as regards the separation of funds for current and strategic use.

Depreciation funds are funds of an exclusively intended purpose, therefore their value is to a certain extent understated. At the same time, it is important that depreciation funds do not go beyond countries. However, depreciation costs may not be used for intended purposes in the event of bankruptcy or a crisis situation. Such use involves the sale of general purpose funds [21].

The modern depreciation system of the main production facilities has a certain number of properties that can be optimized not only by changing the current tax legislation in the field of depreciation norms, applying coefficients and other normative indicators, but also through the implementation of appropriate measures at the enterprise.

Such measures include the ordering of the priority of repairs and restoration of fixed assets for the reporting year. Investigation of the conditions for the depreciation of objects, in accordance with various methods of its calculation in accordance with the requirements of tax legislation and the Regulation (standard) of accounting number 7 "Fixed assets", allows you to determine the economic consequences of using different methods in the accounting for the future [22].

The reproduction of fixed assets and depreciation policy of an enterprise is not only accounting problem. Financing timely of production performance updating does not depend on the accounting interactions of accounts. This is an issue of depreciation process methodology application. The essence of the depreciation policy is determining the necessary amounts of fixed assets for the actual reproduction. This has to form the company's accounting policies.

For the formation of accounting policy, we should consider thoroughly the experience of our country and foreign countries. International standards allow enterprises to apply all methods of national accounting standards, in addition to production and cumulative ones. Enterprises choose the method that best reflects the expected future profit from the operation with assets. This method is constantly used during the reporting period. Method changes take place due to changes in operating the assets or receiving future economic benefits.

Formation of an effective depreciation policy is a common process accounting and technical services company. To justify the need for a radical change in existing approaches to depreciation policy was studied the dynamics of fixed assets and the amount of capital investment in agriculture.

To determine the depreciation policy role in the formation of investment resources of the enterprise, the structure of sources of capital investment companies Ukraine.

Thus, the accounting policies application for the effective depreciation policy formation is very important. The main obstacle to the formation of real depreciation fund of a company is unaccountability and vulnerability of this expenditure.

Optimal depreciation and accounting policies should be based on an integrated approach based on the specific economic conditions of enterprises. Faithfulness and timeliness of depreciation are only checked while calculating the amounts of some income taxes.

The main lever for stimulating the formation of public policy should be the depreciation tax protectionism. It should be implemented as preferential taxation and self-installing your software key components forming the amortization policy economic entities.

## 4 Conclusions

When formulating the methodology of depreciation policy, we must evaluate its components: the terms of use, the need for modernization for innovative development, since this directly affects the efficiency and competitiveness of production and the pace of technological progress.

The essence of this account is the combination of engineering and accounting. To ensure a high-quality information system, it is necessary to determine what can be done by engineers and they are for him. The issue of information and accounting of depreciation policy is a matter of combining engineering, accounting and economics. This information is not only accounting, it is related to other services.

Scientific and technological progress is the main vector that should form the depreciation policy of the enterprise. Determining the direction of development, compiling the list of equipment necessary for this development is the main objective of the engineering and technical service of the enterprise. In determining the prospects for enterprise development, it is necessary to respond promptly to the most up-to-date changes and trends in any industry based on the quality of information support, which allows the exchange and study of best practices and trends in the industry.

Based on these data, engineering and technical services, specialists-economists of specialization must draw up a business plan, the implementation of which, as well as possible sources of funding, need to be planned in conjunction with the specialists of the accounting area, which should make changes in the accounting policies of the company in terms of depreciation.

## References

1. Mochernyy S.V.: Ekonomichna entsyklopediya (Economic Encyclopedia) Part 1. K.: Academy Publishing Center, 864 p (2000)
2. Pushkar M.S.: Rozrobka system obliku (Development of accounting systems). Ternopil′: Kart-blansh, 198 p (2003)
3. Sidorova, M.I.: Razvitiye modeley bukhgalterskogo ucheta v usloviyakh sovremennykh informatsionnykh tekhnologiy (Development of accounting models in the conditions of modern information technologies), 44 p (2013)
4. Chaykovskaya, L.A.: Vliyaniye institutsional′nykh faktorov na sostoyaniye i razvitiye uchetnykh sistem (Influence of institutional factors on the state and development of accounting systems)/Problemy teoriyi ta metodolohiyi bukhhalters′koho obliku, kontrolyu i analizu. Mizhnarodnyy zbirnyk naukovykh prats′. Seriya: Bukhhalters′kyy oblik, kontrol′ i analiz Voll 3 (21), P. 2., pp. 379–384 (2011)
5. Kozlova, M.O.: Klasyfikatsiya natsional′nykh system bukhhalters′koho obliku: analiz naukovykh pidkhodiv ta sproba pereosmyslennya (Classification of national accounting systems: analysis of scientific approaches and attempts to rethink)/Visnyk ZHDTU/Seriya: Ekonomichni nauky. Voll 2 (52), pp. 116–122 (2010)
6. Chumachenko, N.G.: Prinyatiye resheniy v upravlenii proizvodstvom (Decision making in production management). K.: Tehnika., 192 p (1978)
7. Shyhun, M.M.: Rozvytok modelyuvannya systemy bukhhalters′koho obliku: teoriya i metodolohiya (Development of the modeling of the accounting system: theory and methodology). Zhytomyr: ZHDTU, 632 p (2009)
8. Yevdokymov, V.V.: Adaptyvna model′ intehrovanoyi systemy bukhhalters′koho obliku (Adaptive model of the integrated accounting system), Zhytomyr: ZHDTU, 516 p (2010)
9. Nikandrova, L.K.: Formirovaniye uchetnoy sistemy khozyaystvuyushchikh sub"yektov: teoriya, metodologiya i praktika (Formation of the Accounting System of Economic Entities: Theory, Methodology and Practice) (2011)
10. Malyuha, N.M.: Bukhhalters′kyy oblik v Ukrayini: teoriya y metodolohiya, perspektyvy rozvytku (Accounting in Ukraine: theory and methodology, prospects of development). Zhytomyr: ZHDTU, 548 p (2005)

11. Bezverkhyy, K.V.: Vitchyznyanyy oblik yak henerator oblikovo-zvitnoyi informatsiyi dlya potreb upravlinnya diyal′nistyu pidpryyemstv na mikro, mezo- i makrorivnyakh (Domestic accounting as a generator of accounting and reporting information for managing the activities of enterprises at micro, meso- and macro levels). Bukhhalters′kyy oblik, kontrol′ i analiz: problemy teoriyi ta metodolohiyi. K. pp. 40–58 (2013)
12. Busel, V.T.: Velykyy tlumachnyy slovnyk suchasnoyi ukrayins′koyi movy (Great explanatory dictionary of modern Ukrainian language). K.; Irpin′: VTF „Perun", 728 p (2005)
13. Petruk, O.M.: Rehulyuvannya bukhhalters′koho obliku v Ukrayini: teoriya, metodolohiya, praktyka (Accounting Regulation in Ukraine: Theory, Methodology, Practice). Zhytomyr: ZHDTU, 152 p (2006)
14. Hudzyns′kyy, O.H.: Amortyzatsiya v onovlenni osnovnykh zasobiv pratsi (Amortization in updating of basic means of labor). Bukhhalteriya v sil′s′komu hospodarstvi N 7., pp. 4–9 (2008)
15. Duhiyenko, N.O.: Amortyzatsiyna polityka yak instrument derzhavnoho rehulyuvannya protsesu vidtvorennya osnovnoho kapitalu. Zbirnyk tez dopovidey tret′oyi mizhnarodnoyi konf. «Ekonomichnyy i sotsial′nyy rozvytok Ukrayiny v KHKHI stolitti». Part.2., Ternopil′: Ekonomichna dumka., pp. 509–512 (2006)
16. Oshmarin, Y.V.: Amortyzatsiynyy fond: neobkhidnist′ formuvannya ta vidobrazhennya v obliku (Amortization fund: the need for formation and reflection in the account). Visnyk Zhytomyrs′koho derzhavnoho tekhnolohichnoho universytetu.Ekonomichni nauky., Zhytomyr: ZHDTU, **2**(36), 84–91 (2006)
17. Holov, S.: Dyskusiyni aspekty amortyzatsiyi (Discussion aspects of depreciation). Bukhhalters′kyy oblik i audyt., № 5., pp. 10–12
18. Kosova, T.D.: Amortyzatsiya yak investytsiynyy resurs pidpryyemstva (Amortization as an investment resource of the enterprise). Ekonomika pidpryyemstva., № 9., pp. 157–166 (2006)
19. Vasylishyn, S.I.: Napryamy vdoskonalennya obliku amortyzatsiyi osnovnykh zasobiv sil′s′kohospodars′kykh pidpryyemstv (Directions of improvement of accounting of depreciation of fixed assets of agricultural enterprises)/ Biznes Inform., № 12., pp. 383–388 [Electronic resource]. Retrieved from: Режим доступу: http://nbuv.gov.ua/UJRN/binf_2014_12_56., c. 387 (2014)
20. Pyrets′, N.M.: Porivnyal′nyy analiz natsional′nykh ta mizhnarodnykh standartiv obliku osnovnykh zasobiv (Comparative analysis of national and international standards for accounting for fixed assets) [Electronic resource]. Retrieved from: http://nbuv.gov.ua/j-pdf/inek_2013_6_79.pdf (2013)
21. Doskovskiy, V.: Finansovyy mekhanizm nachisleniya amortizatsii i yego teoreticheskaya kontseptsiya (The financial mechanism for calculating depreciation and its theoretical concept), № 5–6, 155 p (2010)
22. Horodyans′ka, L.V.: Vidtvorennya osnovnykh zasobiv na pidpryyemstvakh Ukrayiny: teoriya i prak- tyka obliku ta analizu (Restoration of fixed assets at Ukrainian enterprises: theory and practice of accounting and analysis). K.: KNEU., 224 p (2008)

# Problems and Prospects for Development of Family Households in Ukraine

**Tetiana Yavorska, Yurii Prus, Oksana Lysak and Hanna Zavadskykh**

## 1 Introduction

Agriculture in Ukraine is characterized by the most dynamic rates of development among other sectors of the economy in recent years. Its share in the gross product of the country in 2017 amounted to 10.2%, in exports—33.6%. This situation shows that the state is taking real steps to form an effective agricultural policy. There are many different forms of management in agriculture in Ukraine: agricultural enterprises, farms, households. At different stages of the formation of the agrarian market, the pace of development of these forms is unequal.

At the present stage, the dual organizational structure of the rural economy has developed, which is manifested in the presence of two sectors—corporate and individual. The corporate sector combines agricultural enterprises of various organizational and legal forms, which have the right of a legal entity: state enterprises, private enterprises, economic partnerships, production cooperatives, etc. The individual sector is represented by two types of individual farms: farms and households that carry out agricultural activities. Under the influence of market factors, the corporate and individual sectors occupy their niches in the structure of agricultural production. Joint-stock companies produce the most commercially attractive and export-oriented

T. Yavorska (✉) · Y. Prus · O. Lysak · H. Zavadskykh
Tavria State Agrotechnological University, B. Khmelnitsky ave. 18, Melitopol 72310, Ukraine
e-mail: super_yavort@ukr.net

Y. Prus
e-mail: yurii.prus@tsatu.edu.ua

O. Lysak
e-mail: lysakksana@gmail.com

H. Zavadskykh
e-mail: hannaz2208@gmail.com

© Springer Nature Switzerland AG 2019

V. Nadykto (ed.), *Modern Development Paths of Agricultural Production*,
https://doi.org/10.1007/978-3-030-14918-5_58

types of products: grain, sunflower, rape, etc., and handle more than 90% of the land. In 2017, they produced 47.7% of gross agricultural output.

The individual sector, in particular, households, is focused on self-sufficiency and self-sufficiency in the domestic food market. In the total volume of gross agricultural production, the share of households accounted for 43.6%. They have grown 98.1% of potatoes, 85.5% of vegetables, 83.7% of fruits and berries, 22.6% of grain, 13.4% of sunflower seeds in 2017. Among the types of livestock products, farms have produced 36.9% of meat in slaughter weight, 73.1% of milk, 46.1% of eggs, 87.0% of wool, 98.7% of honey in 2017 [1]. Households also perform important public functions: food security, self-employment, economic base of rural development, social base of reproduction of the peasantry, etc. At the same time, households do not have legal status as producers of agricultural products. The state policy of regulation and support of agricultural production mainly contributes to the strengthening of large-scale production, that is, the corporate sector.

## 2 The Role and Definition of Family Farms in Global Agriculture

World experience shows that the basis of the agrarian sector of developed market economies is family farms. In the world, more than 500 million farms (more than 570 million) are family farms. They account for almost 56% of agricultural production and a significant share of land in cultivation: 83% in North and Central America, 68% in Europe, 85% in Asia, 62% in Africa [2]. Family farms achieve a high level of productivity on their land plots. For example, according to the Food and Agriculture Organization of the United Nations (FAO), in Brazil, family farms account for less than 25% of all agricultural land, and the production of basic crops is almost 40%. In the USA, family farms account for 78% of the country's agricultural land. At the same time, the share of their production of agricultural products is 84% in the amount of 230 billion dollars [3]. In agriculture in the European Union, 85% of all farms are family farms, they handle almost 70% of all agricultural land.

In the world, there is no single definition of the concept of "family farm". For different countries, different definitions are used in the legal field, based on the share of the farmer's own work, on the legal status of the owner, on the amount of land cultivated, on the right to ownership of land and inheritance. According to the FAO definition, the family farm is a way of organizing agricultural, forest, fish, pastoral and aquaculture production, which is carried out by a separate family and is based mainly on the work of family members, both women and men. The family farm includes all types of family farming and local development activities.

Based on a literature review [among others 4–10] derived following elements that define family farming within the current development:

- Both business ownership and managerial control are in the hands of family or near-family members;

- Business ownership and managerial control are transferred within the family over different generations;
- A majority of the labor is provided by the operator and his/her family;
- A substantial part of the capital is furnished by the operator and his/her family;
- The family obtains a major share of its income from farming;
- The principals are related by kinship or marriage; The family lives on the farm.

## 3 Legislative Aspects of the Establishment and Functioning of Family Farms in Ukraine

In Ukraine, the status of a farm is defined by the legislation since 1991. According to the latest draft law No. 1599, "On Amendments to Some Laws of Ukraine Concerning the Promotion of the Establishment and Operation of Family Farms" (2016), it is established that the farm is a form of entrepreneurial activity of citizens who have expressed a desire to produce commodity agricultural products, to process their processing and realization for the purpose of making profit on the land plots granted to them by the ownership and/or use, including lease, for farming, commodity agricultural production, personal farming. Such an economy is subject to state registration as a legal entity or an individual—an entrepreneur.

For the first time in this law, the term "family farming" has been introduced. The status of such an economy can be obtained provided that in its business activities the work of exclusively members of one family is used. Involvement by family farming of other citizens can be carried out exclusively for seasonal and individual work, which are directly related to the activity of the farm and require special knowledge or skills. The head of a family farm may only be a member of the family concerned. It is established that a family farm without a legal entity status is organized by an individual, an entrepreneur, alone or jointly with members of his family, on the basis of a contract for the establishment of a family farmer.

Domestic experts and experts in agriculture believe that for the development and modernization of agricultural production, the introduction of EU standards, all producers of agricultural products should have a single legislative field for the organization of their activities and work on equal tax conditions.

To increase the legal, tax, social status of family farms to a level not lower than other categories of agricultural producers, on August 15, 2018, the Law of Ukraine "On Amendments to the Tax Code of Ukraine and certain Laws of Ukraine on Stimulation of the Formation and Activities of Family Farms". The law provides the possibility for residents of countryside to legalize their activities in the form of entrepreneurial activity for individual entrepreneurs who have organized a farm, including a family. They were given the opportunity to be taxpayers of the single tax of the fourth group, which previously had the right to pay only agricultural producers—legal entities. At the same time, this opportunity can be realized provided the family farm is consistent with the set of such criteria:

- Carrying out activities exclusively within the limits of a farm, registered in accordance with the Law "On Farmers";
- Exclusively engaged in the production, processing and supply of agricultural products;
- Realization of economic activity (except delivery) at the place of tax address;
- Not using the labor of hired persons;
- The members of the farm of such an individual are only members of the family of this person;
- The area of agricultural land owned and/or used, the land of the water fund in the use of members of the farm is not less than 2 ha, but not more than 20 ha.

The law also establishes a preferential procedure for paying a single social contribution for all members of the farm, including its chairmen (additional payment of the single social contribution at the expense of the State budget for 10 years). It is planned to pay the tax on personal income from rent for land (units), to local budgets at the location of such land plots.

Legislative changes are intended to strengthen the social and economic protection of members of family farms, increase readiness and ability of peasants (especially young and middle generations) to self-management on a family basis and will promote increase of incomes of rural inhabitants by the formation of favorable prices for own-produced products. In addition, it will affect the availability of farms to existing agricultural budget support programs and will encourage the introduction of newly designed programs for them.

## 4 Analysis of Households as the Basis for the Establishment of Family Farms

An analysis of the socioeconomic nature, functions, motivations and development of households of the population shows that they can represent in Ukraine the modern stage of evolutionary development of family type of farming in agriculture. Family farms form the basis of the agricultural system of developed countries. Creation of family-owned farms based on households will allow them to acquire forms similar to those existing in other countries over time.

The basis for the establishment of family farms in the countryside of Ukraine is part of households (farms) who own land plots of different sizes and are engaged in the production and sale of crop and livestock products. This is a huge category of agricultural producers (almost 4.8 million units), which foreign experts consider a real "Ukrainian phenomenon" [11].

According to the definitions in the statistical yearbook: "Household (economy of the population)—a set of persons (or one person) living together in one living space (part thereof), have a common household, have common expenses for housing, food, etc.". Households are households that carry out agricultural activities both for self-provision of food products and for the purpose of production of commodity

**Table 1** Characteristics of households in rural areas of Ukraine in 2014–2018

| Indexes | Year | | | | |
|---|---|---|---|---|---|
| | 2014 | 2015 | 2016 | 2017 | 2018 |
| Number of households, ths. | 5173.0 | 4948.7 | 4924.0 | 4900.1 | 4873.6 |
| Of them: have land plots, ths. | 5076.7 | 4834.6 | 4835.7 | 4779.1 | 4799.8 |
| Of them: have land plots (%) | 98.0 | 97.7 | 98.2 | 97.5 | 98.5 |
| Hold livestock, poultry and bees (%) | 76.7 | 76.3 | 76.9 | 75.8 | 76.0 |
| Area of land, total, thousand hectares | 13444,6 | 12163,9 | 14371,7 | 13706,5 | 14020,2 |
| The same, for one household, hundredth of hectare | 265.3 | 251.6 | 297.2 | 286.8 | 292.1 |
| Number of households with an area of land more than 2 ha, thousand units | 1819.3 | 1600.3 | 1832.7 | 1806.5 | 1771.1 |
| Of them: with an area of more than 10 ha | 278.7 | 227.2 | 314.3 | 325.0 | 292.8 |
| The direction of land use by households on average, %: cultivating products only for their own needs | 14.4 | 16.3 | 14.8 | 15.2 | 15.7 |
| For own needs and for sale | 11.7 | 11.5 | 14.3 | 11.7 | 10.0 |
| Leased out | 73.0 | 71.3 | 69.7 | 71.8 | 73.2 |
| Only started to use | 0.9 | 0.9 | 1.2 | 1.3 | 1.1 |

agricultural products. This category includes rural households, urban areas (including collective gardens and gardens), as well as individuals—entrepreneurs who carry on their activities in the field of agriculture without establishing a legal entity.

The total number of households with land plots has decreased somewhat (by 5.5%) in period of 2014–2018 [12]. However, their share in the total number of rural households increased by 0.5%. Increased both the total area of land and on average one farm. According to the statistics of 2018, 36, 9% of Ukraine's households in land use exceeds 2 ha of agricultural land, including 16.5% of them have an area of land more than 10 ha (Table 1).

Increase in land use in households partially affected the volumes of production of certain types of agricultural products. Vector of agricultural production has shifted toward crop production in the period from 2000 to 2017. Production of grain (3.1 times) and sunflower (3.7 times) has increased the most. Vegetable production volumes in 2017 have grown by 48.0% compared to 2000, by 11.0% compared to 2010. Until 2000, livestock farming was an important place in the production of agricultural products. Currently, there is a tendency toward a decrease in the share of households in the production of meat (by 37.8%) and eggs (by 20.1%). At the same time, the share of total milk production in 2017 was 73.1% versus 71.0% in 2000 and 80.3% in 2010 [1]. The share of wool and honey production in the total production of these types of products has increased (Table 2).

Table 2 Production of basic types of agricultural products in households of Ukraine

| Product type | 2000 | | 2010 | | 2017 | |
|---|---|---|---|---|---|---|
| | Thousand tons | Share in total production (%) | Thousand tons | Share in total production (%) | Thousand tons | Share in total production (%) |
| Cereals and legumes | 4494.8 | 18.4 | 9491.6 | 24.2 | 14,011.6 | 22.6 |
| Sugar beets (production) | 1604.7 | 12.2 | 1085.8 | 7.9 | 654.4 | 4.4 |
| Sunflower | 431.7 | 12.5 | 1185.9 | 17.5 | 1638.8 | 13.4 |
| Potato | 19561.4 | 98.6 | 18222.3 | 97.4 | 21778.8 | 98.1 |
| Vegetable crops | 4835.0 | 83.1 | 7157.8 | 88.1 | 7942.4 | 85.5 |
| Fruit and berry crops | 1188.5 | 81.8 | 1459.7 | 83.6 | 1714.2 | 83.7 |
| Meat of all kinds (in slaughtered mass), t | 1224.7 | 73.7 | 924.6 | 44.9 | 835.2 | 36.0 |
| Milk | 8989.2 | 71.0 | 9031.9 | 80.3 | 7514.8 | 73.1 |
| Eggs million pcs. | 5831.3 | 66.2 | 6802.7 | 39.9 | 7140.5 | 46.1 |
| Wool, t | 2089 | 61.4 | 3482 | 83.1 | 1712 | 87.0 |
| Honey, t | 48875 | 93,2 | 69253 | 97,7 | 65384 | 98,7 |

That is farms have a significant place in production of agricultural products in the country and they continue to significantly influence the formation of its offer on the market. In case the area of the land exceeds 2 ha, is not leased or is not dispossession in any other way, and is processed by the own forces of the household, it is essentially an entrepreneurial structure or a family farmer.

## 5 Problems of Family Farms in Ukraine

A significant achievement of Ukrainian agrarian policy is legislative definition of the status of family farms. However, in addition to the legislative definition, a number of problems still need to be resolved for the effective functioning of family farms (Table 3).

**Table 3** Problems of the establishment and functioning of family farms in Ukraine

| Problems | Consequences | Results |
|---|---|---|
| The farm does not act at the same time as the way of life and form of management, as it is in developed countries | Lack of experience in family farming | Slow formation of a layer of family farmers in the countryside |
| Low level on information support | Lack of information and analytical network, which will provide the opportunity to communicate on the issues of creating family farms, purchase and use of productive resources, product sales, participation in support programs, exchange of experience, solving legal, financial and other issues | The lack of awareness of the majority of households about the conditions for the establishment and the benefits of the functioning of family farms |
| Low competitiveness on the market of agricultural products | Lack of established agricultural value-added chains, covering agricultural production, processing, storage, access to profitable markets, etc. | The difficulty of becoming a full member of the agricultural market |
| Low efficiency of control system | Lack of self-regulatory management structures | Limited possibilities for professional associations, lack of representation in state authorities |
| Inefficient state support | Absence of complex conditions for the voluntary transition of households to family farms, by encouraging them to become the official producer of agricultural products and their further functioning | Inhibition of the process of creating family farms |
| Absence of a system of professional training for family management | Lack of relevant knowledge of technological, economic and other levels for farm management | Uncertainty about the expediency of opening your own business and running an effective business |

This pattern has been formed over the centuries for economically developed countries, where family farms are main form of management. The farm here serves both the way of life and the form of management simultaneously. In Ukrainian countryside, the way of life remains as rural system, but family farming is still emerging as a form of management. This is the main problem that affects the formation of a layer of family farmers in the Ukrainian countryside. Accordingly, the disadvantages of family farming in developed countries and in Ukraine are significantly different.

For example, SWOT analysis in Western European agriculture, in particular, in the UK, shows both the advantages and disadvantages of developing modern family farms.

The persistence of family farm depends on the transfer of the farm within the family. The timely designation of a successor is crucial for survival of family farm, and the number of designated successors decreases over time (Calus 2009). On the one hand, there is the movement toward increasing farm size: The economies of scale are used to develop high capital-intensive family farms. On the other hand, there are motivated and skilled young people who are willing to take over the family farm, but there is no viable (family) farm available to take over. New institutions should focus on the combination of viable farms in combination with motivated and skilled successors to develop competitive farms both on the level of farm size and management. Cooperation with external capital providers might be necessary to have a positive development of these kinds of interactions. In order to bring land, management and capital together, a successor bank can be developed.

Most of the problems of family farmers in Ukraine are almost the same for countries in the past as a "socialist camp". For example, family businesses in Georgia are also faced with a number of problems inherent in Ukrainian farms.

Small and diversified land plots are considered to be one of the main challenges and constraints facing Georgian farmers today. Small plots of land located in different places, sometimes not even close to the farmer's house, hinder the expansion and development of farms. Farmers often do not have any legal documents proving ownership of land. The other important byproduct is the lack of knowledge about farming. As was mentioned before, after the collapse of the Soviet Union many families received plots of land but did not know anything about farming. The other important byproduct is the lack of knowledge about farming.

These two major challenges are related and cause many other challenges, such as the lack of access to finance. In many cases, farmers are not bankable, and agriculture is considered to be quite risky. Although the government has initiated an agro-insurance project, which could contribute a lot to the increase in the bankability of farmers, there are still a lot of issues to be resolved.

Limited innovation and low adoption rates of new technologies are another challenge to family farms. In addition, family farms have limited access to markets (e.g., input markets) and services (e.g., veterinary, machinery, etc.). All of which contribute to the weak position of family farms in the value chains. Lastly, it should be mentioned that the quality of services and amenities in rural areas of Georgia are very low. This negatively affects family farms, agricultural enterprises and the rural population in general. Larger enterprises have greater power to cope with those challenges, whereas due to their constraints family farms are quite weak and cannot cope with the lack of rural development without assistance from the government and donor community [13].

One of the main directions of solving problems of farms is the creation of the Ukrainian Foundation for the Development of Family Farm and Villages with the participation of State Authorities, the Association of Farmers and Private Landowners of Ukraine (AFPLU) and donor organizations [14]. Such a foundation will solve most of the problems of family farms listed in Table 3. It will help to create an institutional environment for increasing the capacity of family farming to a greater extent than the adoption of the law on the status of family farming. Accordingly, now it is necessary to concentrate efforts, firstly, on the implementation of support programs for households without requiring the acquisition of the status of a family farm as a condition for obtaining this support; and secondly, to ensure the representation of the interests of households—potential family farmers in the executive [15].

## 6 Conclusions

Today in Ukraine, family farms are almost the main direction of sustainable development of rural areas, an important factor in rural economies. They preserve traditional production, local culture, promote the preservation of agricultural biodiversity and the rational use of natural resources, and so on. Given the full revival of family farms, Ukrainians can not only increase their food security, but also become the largest suppliers of products to foreign markets, as these products have a high export potential.

It is advisable to develop and adopt a strategy to encourage the creation and development of family-owned farms for the period up to 2025, the concept of the relevant State Target Program, as well as the budget program "State Support for the Development of Family farms" to form self-sufficiency and employment of rural population in Ukraine as an economic basis for the development of rural areas. It is advisable to take advantage of the experience of foreign countries in applying the economic instruments of state support to agricultural production in developing these documents, of which the most effective is to improve the mechanism of pricing agricultural products, the provision of state subsidies to farmers, the use of preferential lending and preferential taxation of agricultural producers, the development and implementation of national and sectoral programs development of agriculture, well-balanced foreign trade policy.

In our opinion, without the stimulation of the acquisition of the status of an official producer of agricultural products by high-quality farms, it will not be possible to achieve a significant breakthrough in the creation of family farms. Therefore, an important task of the state agricultural policy should be the formation of sufficient financial incentives for the voluntary transformation of subjects engaged in commodity agricultural production in the status of the economy of the population, as the business unit.

## References


1. Statistical Yearbook “Agriculture of Ukraine” for 2017 State Statistics Service of Ukraine, 245 r. http://www.ukrstat.gov.ua/ (2018)
2. Zubar, I.V.: Development of cooperation of small farms on the basis of world experience. Innov. Econ. **4**, 54–60 (2015)
3. Bezzub, I.: Prospects for the development of family farms in Ukraine. http://nbuviap.gov.ua/index.php?option=com_content&view=article&id=419:agropromislovij-sektor-2&catid=71&Itemid=382
4. Brandth, B., Haugen, M.S.: Gendered work in family farm tourism. J. Comp. Fam. Stud. **38**, 379–393 (2007)
5. Calus, M.: *Factors explaining farm succession and transfer in Flanders. Department of Agricultural Economics*, pp. 229. Ghent University, Ghent (2009)
6. De Haan, H.: Images of family farming in the Netherlands. Sociol. Rural. **33**(1993), 147–166 (1993)
7. Gasson, R., Errington, A.: The farm family business. CAB International, Wallingford (1993)
8. Knutson, R.D., Penn, J.B., Flinchbaugh, B.: Agricultural and Food Policy. Prentice Hall, New Jersey (1998)
9. Loyns, R.M.A., Kraut, M.: The family farm in the next decade: the positive role of part-time farming. Can. J. Agric. Econ. **40**, 591–604 (1992)
10. Small, L.-A.: The influence of “Family” on agrarian structure: revisiting the family farm debate in Bulgaria and southern Russia. J. Comp. Fam. Stud. **36**, 489–503 (2005)
11. Zbarsky, V.K., Alekseeva, Y.Y.: Features of the transformation of rural households in family farms. Econ. Agroindustrial Complex. **7**, 23–29 (2017)
12. Social and Demographic Characteristics of Households of Ukraine [Sotsialno-demohrafichni kharakterystyky domohospodarstv Ukrainy u 2018 rotsi (za danymy vybirkovoho obstezhennia umov zhyttia domohospodarstv)]. Statistical Yearbook. State Statistics Service of Ukraine [Statystychnyi zbirnyk. Derzhavna sluzhba statystyky Ukrainy]. [Online]. Access from: http://www.ukrstat.gov.ua/
13. The Role of Family Farmingin the Sustainable Development of the Agricultural Sector and Poverty Reduction in Georgia [Online]. Access from: http://www.aea.am/files/papers/w1504.pdf. pp. 29–30
14. Borodina, E., Prokopa, I.: Support for family farming: slogans and actions. “Mirror of the week. Ukraine” No. 1, January 15, 2016. http://gazeta.dt.ua/ariculture/pidtrimka-simeynogo-fermerstva-gasla-i-diyi-_.html (2016)
15. Korinets, R.: Family farm as an objective need the Ukrainian farmer, September 2014 [Online]. Access from: http://www.agrotimes.net/journals/article/simejni-fermi-yak-ob%E2%80%99ektivna

# Part IV

# Comprehensive Assessment of the White Roots Aroma

**Iryna Bilenka, Yana Golinskaya, Iryna Kalugina and Liudmyla Kiurcheva**

## 1 Introduction

White roots are a unique raw material in healthy nutrition. Among these root crops, the most common ones are celery and parsnip in the cuisines of the world.

Valuable nutritional and curative properties of celery are provided by a large number of different substances in favorable for the human body combinations. The root crops contain about 40 species of flavor and aromatic substances that improve appetite and digestion [1].

Due to its chemical composition, celery has remarkable healing properties. It positively affects the state of the cardiovascular system, digestive organs, kidneys, and liver, provides calming effect on the nervous system, promotes hematopoiesis and metabolism. With constant consumption of celery, the vitality rises, the body rejuvenates, which is especially important for the elderly. Celery produces an excitatory influence, promotes weight loss, helps to get rid of neuroses, delays deposition of salts in the joints, walls of the vessels, kidneys, liver, prevents allergic rashes on the skin, dermatitis, heals ulcers and wounds. The diuretic effect of celery is due to the high content of potassium. It is known that the associated removal of uric acid from the body is due to the favorable relationship of mineral salts, which makes root crop an effective remedy against gout and rheumatism, as well as in case of propensity to stone formation [2–6].

I. Bilenka (✉) · Y. Golinskaya · I. Kalugina
Odessa National Academy of Food Technologies, Kanatna Str., 112, Odessa 65039, Ukraine
e-mail: irinabelenka@gmail.com

L. Kiurcheva
Tavria State Agrotechnological University, B. Khmelnitsky Av., 18, Melitopol 72312, Zaporizhia Region, Ukraine

© Springer Nature Switzerland AG 2019

V. Nadykto (ed.), *Modern Development Paths of Agricultural Production*,
https://doi.org/10.1007/978-3-030-14918-5_59

## 2 Literature Review

The use of celery enhances the function of the kidneys, the nervous system, and the glands of the inner secretion, promotes clearing of blood and relieves defatigation, increases the displacement of metabolic wastes [5].

Volatile oils of celery roots have disinfectant and antiseptic properties, reduce the activity of putrefactive microorganisms and inflammatory processes in the digestive tract, stimulate the activity of the glands of the inner secretion [7].

Parsnip roots have a pleasant hot-sweet taste and a delicate aroma. Their use in dietary nutrition is useful for gallstones and kidney stone diseases, gout, in case of nervous disorders, pneumonia, and bronchitis. It has been established that when using parsnip, digestion improves, walls of capillary vessels are strengthened, and pain-relieving and expectorative actions are noted. The root crop has tonic properties, promotes appetite, stones, and salts [8].

The conducted information search and experimental studies have shown that white roots are extremely useful raw materials and are characterized by intense specific flavor, which in some cases has a positive effect on the human body, but in some finished dishes, the decrease in the detection of vegetable tones and the overall intensity of the flavor plays a decisive role.

Aroma is one of the most important indicators of food quality. Culinary products are a multicomponent system in which each component contributes to the creation of flavor and aroma properties. The appearance of some volatile compounds is associated with the chemical transformation of substances contained in primary raw materials [9].

It is known that aromatic substances are complex mixtures of various, often related substances, formed by enzymatic means. The main components of the aroma are aldehydes, ketones, ethers, higher alcohols, organic acids, and terpenes. Aromatic compounds change at each technological stage. The biosynthesis of any component of the aroma is a multi-stage process that is carried out under the influence of various enzymes. Fatty alcohols and aldehydes are formed as a result of degradation of phospholipids and fatty acids under the influence of lipoxygenases, hydroperoxidases, isomerases, and dehydrogenases [10, 11].

According to some scientists, the aroma of white roots is due to the presence of alcohols, ethers, volatile acids, hydrocarbons [12].

The aroma of food products should attract the attention of the consumer and, in addition, stimulate appetite and increase the secretion of gastric juice to better food digestion, metabolism improvement. It is believed that volatile substances of aromatic plants have preservative, antiseptic, bactericidal properties [13].

## 3 Research Approach

The intensity of aroma depends on many factors, such as air temperature, degree of fineness, moisture content, and the smell of the finished product also depends on its formulation components.

Given the number of aroma-forming substances, their diversity in chemical nature and the possibility of interconnections, it is difficult to predict the formation of the final bouquet in the finished product. At the same time, the use of some technological methods contributes to the formation of the desired aromatic profile. The purpose of our work was to study the aromatics in both fresh white roots and during their processing.

The free aromatic compounds were determined by solid-phase microextraction (SPME). The fresh raw material was examined and after UFR processing in the chosen mode (650 W power for 1 min). In the analysis, the technique developed in the laboratory of chromatographic investigations of SE "Ukrmetrteststandard" in Kyiv was applied. Determination of aromatic components was performed on the gas chromatograph HP8790B (Agilent Technologies). Gas chromatographic analysis was carried out on the capillary column vf5-ms (Agilent Technologies), phase—5% phenyl-methyl-polysiloxane, length—30 m, internal diameter—0.32 mm, film thickness—0.25 μ. Thread with CarboxenTM/Polydimethylsiloxane coating was applied, film thickness was 75 μm (Supelco, Cat. No. 57318). The sample volume used for extraction is 20 $cm^3$. Extraction time at a temperature of 40 °C—60 min using method of gas-liquid chromatography.

The conditions of the gas chromatographic analysis are given in Table 1.

The number of odor units was determined by the modified method [14]. The basis of the method is the principle of distillation and oxidation of volatile oil with a solution of potassium dichromate in the Vidmark's device at thermostatic conditions, which is based on the ability of the chromic mixture to oxidize the aromatic substances.

Under the insert of the Vidmark's device, the sample weight mark was placed with a precision of 0.01 g of the investigated raw material weighing 0.5 g. We put 2 $cm^3$ of 1% solution of potassium dichromate, dissolved in acid, into a flask of the device. The solution was prepared as follows: 50 g of potassium dichromate was dissolved in 450 $cm^3$ of warm distilled water and 500 $cm^3$ of sulfuric acid, with a

**Table 1** Analysis conditions

| | |
|---|---|
| Injector temperature | 300 °C |
| Injection method | Without stream splitting (Splitless) |
| Carrier gas | Helium |
| Carrier gas velocity | 1.5 $cm^3$/min |
| Thermostat columns temperature | 40 °C (2 min) |
| Heating (5 °C/min) up to 220 °C | 12 min |
| Detector temperature (PID) | 300 |

density of 1.83 g/cm$^3$. Similarly, a control experiment was conducted, but instead of sample weight mark, there was placed distilled water under the insert.

The insert with a stopper was tightly placed in a flask and the device was put in a thermostat with a temperature of 60 °C for 2 h. After thermostating, the flask was cooled for 15 min. After that, the flask was gently opened and rinsed with 25 cm$^3$ of distilled water. The flask was again closed and placed in the refrigerator for 15 min at a temperature of about 0 °C. After that, 1 cm$^3$ of 5% solution of KI was added to the mixture of potassium dichromate. Tithed iodine was titrated in the presence of an indicator (1% starch solution) with 0.1 n sodium thiosulfate solution. In parallel, a control experiment with distilled water was conducted. The number of odor units was calculated according to the following formula:

$$x = (a - b) * K * 100/\mathrm{m} \tag{1}$$

where

- $x$ Number of odor units, RU;
- $a$ Amount of 0.1 n sodium thiosulfate solution, which was sent to titration of the control chromic mixture, cm$^3$;
- $b$ Quantity of 0.1 n solution of sodium thiosulfate, which was sent to titration of chromic mixture, combined with condensate of aroma-forming compounds, cm$^3$;
- К Correction for 0.1 n sodium thiosulfate solution;
- m Sample weight of product, g (cm$^3$).

## 4 Results

As a result of the studies conducted, it was found that the roots of celery and parsnip contain a large number of different volatile compounds. In the roots of celery, there were identified 13 aromatic substances, in the parsnip root—20. It was found that most of the volatile compounds of the white roots are terpene—a class of hydrocarbons, natural organic substances, secondary vegetable metabolites. This is the most widespread category of organic molecules contained in volatile oils. It is known that volatile oils of plants consist largely of mono-, sesqui- and diterpenes and their derivatives, including: α-pinene—the main component of turpentine (pine oil), β-pinene is contained in some types of volatile oils. Both are minor components. Limonene is a component of citrus volatile oils [15].

It is determined that the main aromatic components of the roots of celery are monoterpenes (limonene), sesquiterpenes (β-selenin) and phthalides (butylphthalide).

The background components of the aroma of white roots are represented mainly by monoterpenes, which are products of condensation and oxidation of isoprene. These are monotrenia ($C_{10}H_{16}$), which are the main components of the celery root

**Table 2** Content of aromatic substances of roots of celery, % of the sum

| No. of chromatographic peak | Components | Fresh | After UFR processing |
|---|---|---|---|
| 1 | trans-ocimene | 9.28 | 6.94 |
| 2 | 3-butylphthalide | 3.21 | 3.68 |
| 3 | β-selinene | 20.0 | 23.0 |
| 4 | β-pinene | 6.34 | 6.61 |
| 5 | limonene | 24.6 | 13.9 |
| 6 | myrcene | 2.26 | 0.82 |
| 7 | γ-terpinene (cis-ocimene) | 4.04 | 1.19 |
| 8 | terpinolene | 3.44 | 0.81 |
| 9 | methyl-4-ethyl-hexane | 6.39 | 8.33 |
| 10 | sabinene | 5.08 | 6.67 |
| 11 | 3-methyl butanal | 3.51 | 14.1 |
| 12 | *n*-cymene | 8.95 | 1.49 |
| 13 | α-pinene | 3.07 | 3.35 |

extract. The mass fraction of the aromatic substances of the roots of celery before and after treatment with UFR streams is given in Table 1.

Based on the results of the main aromatic substances, the roots of celery are terpene hydrocarbons—limonene, β-selinene, trans-ocimene, myrcene. Analytical studies of the aroma characteristics of the celery roots components and data in Table 1 show that the aroma is formed from the following components: pine (α-pinene, β-pinene), lemon (myrcene, limonene) and apple (3-methylbutanal—$C_5H_{10}O$). In the samples studied, there is 3-butylphthalide ($C_{12}H_{14}O_2$), which causes the specific aroma of celery. This substance also helps to lower cholesterol and blood pressure, reduces the risk of cardiovascular disease [16, 17].

According to Table 1, it is evident that the content of α-, β-pinene, 3-butylphthalide, sabinene, methyl-4-ethyl-hexane and β-selinene in the roots of celery after UFR processing varies slightly, namely, the increase in the range from 0.27% for β-pinene and up to 3% for β-selinene is observed. The largest changes were observed in limonene, *n*-cymene and 3-methyl butanal, with the content of the first two substances decreased by 10.7 and 7.46% respectively, and the content of the latter increased by 10.59%. After UFR processing, the amount of terpinolene, which causes smoky and woody aromas, decreased by 2.63%.

The qualitative and quantitative content of the aromatic substances of the parsnips roots is given in Table 2.

From Table 2, it can be seen that 22 of the aromatic substances 20 ones are undefined. After analyzing the data obtained, it can be claimed that the main component of fresh and processed parsnips is myristicin ($C_{11}H_{12}O_3$), with its concentration higher after processing with UFR streams and reaches 29.5%. This substance plays a prominent role in the creation of flavor-aromatic sensations, has a subtle, fruity nuance resembling almonds. There are also three asarone isomers in the parsnip root ($C_{12}H_{16}O_3$)—γ-asarone, (Z)-asarone and (E)-asarone. The total content of the three asarone isomers is 10.18%. Asarone has a sedative ability, and under certain conditions it is pain-relieving. It is also known that asarone relaxes unstriped muscles cramps and reduces arterial pressure [7]. There is a significant content of the elimecene in the parsnip roots (16.1%), which has a nutty aroma, and terpionene (14.5%). There is also (Z)-falcarinol in substantial quantities ($C_{17}H_{24}O$)—0.84%, a representative of acetylenic hydrocarbons, which is hardly found in wild plants. This substance is curative, since it prevents the growth of cancer cells. Tests in rats with premalignant tumors have confirmed a significant decline in disease progression already within four months after the start of the study [15–19].

After UFR processing, the major changes in parsnips roots were noted for meta-cymene-8-ol, elemicin and *n*-octyl acetate, their content increased by 15.18, 9.00 and 6.66%, respectively. The content of substances such as terpionene (by 11.29%), *n*-octanol hexanoate (by 6.34%), (Z)-asarone (by 6.96%), and dihydroisocalamendiol (by 3.85%) decreased substantially. For other aromatic components, small changes in the quantitative aspect are characteristic.

The sensory evaluation of celery and parsnips has shown that fresh roots are characterized by a rich, pronounced aroma inherent in this type of food product. After UFR processing, the aroma became more delicate, soft and pleasant, the specific aroma of celery and parsnip roots lost its sharpness and the bright vegetable shade became less pronounced.

The content and composition of the substances forming the aroma vary during the ripening of the plant material, during the fermentative and thermal processes, with the destruction of the tissues of the fruits. When stored and after some technological operations, there is a partial loss of aroma and taste. Traditionally, sensory tasting is used to evaluate the flavor-aromatic bouquet. However, this method is relatively long, requires special skills of the tasters and is not acceptable for studying a large number of objects in mass production, as rather subjective one. Since the substances that cause the aroma belong to different classes of chemical compounds, it is very difficult to express the aroma through a single physicochemical indicator. That is why, as such an integrated index, the number of odor units was compared with the sensory characteristics.

Data on changing the number of odor units of white roots during mechanical and thermal processing are given in Table 3.

It has been established that for any technological processing, the number of white roots' odor units decreases on average from 30 to 62%. The slightest changes occurred in the short-term processing with the UFR stream. The use of such treatment before cleaning raw materials will not only reduce the loss of aroma, but also will be a precautionary measure of undesirable enzymatic processes. Also, the product does not acquire boiled tones that are characteristic for melanoidins (Table 4).

**Table 3** Content of aromatic substances of the parsnips roots, % of the amount

| No. of chromatographic peak | Components | Fresh | After UFR processing |
|---|---|---|---|
| 1 | β-pinene | 0.78 | 0.21 |
| 2 | limonene | 1.12 | 0.50 |
| 3 | β-trans-ocimene | 0.53 | Less than 0.05 |
| 4 | meta-ocimene-8-ol | 1.22 | 16.4 |
| 5 | terpionene | 14.5 | 3.21 |
| 6 | *n*-octyl acetate | 2.01 | 8.67 |
| 7 | bornyl acetate | 5.12 | 2.50 |
| 8 | *n*-oxyl butanoate | 0.76 | 0.52 |
| 9 | myristicin | 27.3 | 29.5 |
| 10 | β-bisabolene | 1.52 | 2.59 |
| 11 | α-calacorene | 3.03 | 5.09 |
| 12 | elemicin | 16.1 | 25.1 |
| 13 | not detected | 0.65 | 0.29 |
| 14 | γ-asarone | 0.77 | 0.79 |
| 15 | not detected | 0.74 | 0.66 |
| 16 | *n*-octanol hexanoate | 8.30 | 1.96 |
| 17 | (Z)-asarone | 7.86 | 0.90 |
| 18 | dihydroisocalamendiol | 4.28 | 0.43 |
| 19 | (E)-asarone | 1.55 | 0.13 |
| 20 | di-n-butyl phthalate | 0.83 | 0.32 |
| 21 | (Z)-falcarinol | 0.84 | 0.17 |
| 22 | 5-docyldehydro 2 (3H)-furanone | 0.14 | 0.05 |

**Table 4** Changing the number of odor units ($x$, $cm^3$ 0.02 n $Na_2S_2O_3$) of white roots during cooking ($n = 5$)

| Type of raw material processing | Celery | Parsnip |
|---|---|---|
| Fresh and unpeeled | 698 | 472 |
| Peeled and chopped | 383 | 266 |
| Juice | 422 | 247 |
| Blanched | 479 | 289 |
| After UFR processing | 491 | 291 |

## 5 Conclusions

Sensory tasting evaluation has shown that the application of certain technological approaches reduces the intensity of unwanted tones of white roots aroma. Thus, the study of aromatic substances of white roots has allowed to carry out quantitative and qualitative estimation of the change of their aroma in the process of technological processing and to determine the directions of consumer quality and nutritional value formation of the finished product.

## References

1. Kokotkiewicz, A., Luczkiewicz, M.: Celery (Apium graveolens var. dulce (Mill.) Pers.) Oils. In Essential Oils in Food Preservation, Flavor and Safety, pp. 325–338. https://doi.org/10.1016/B978-0-12-416641-7.00037-7 (2016)
2. Thompson, M.D., Thompson, H.J.: Botanical diversity in vegetable and fruit intake: potential health benefits. In: Bioactive Foods in Promoting Health pp. 1–17 (2010)
3. Kolarovic, J., Popovic, M., Zlinská, J., Trivic, S., Vojnovic, M.: Antioxidant activities of celery and parsley juices in rats treated with doxorubicin. Molecules **15**(9), 6193–6204 (2010). https://doi.org/10.3390/molecules15096193
4. Ignarro, L.J., Balestrieri, M.L., Napoli, C.: Nutrition, physical activity, and cardiovascular disease: an update. Cardiovasc. Res. **73**(2), 326–340 (2007). https://doi.org/10.1016/j.cardiores.2006.06.030
5. Bialobrzewski, I., Markowski, M.: Mass transfer in the celery slice: effects of temperature, moisture content, and density on water diffusivity. Drying technology, **22**(7), 1777–1789. https://doi.org/10.1081/DRT-200025647 (2004)
6. Popović, M., Kaurinović, B., Trivić, S., Mimica-Dukić, N., Bursać, M.: Effect of celery (Apium graveolens) extracts on some biochemical parameters of oxidative stress in mice treated with carbon tetrachloride. Phytother. Res. **20**(7), 531–537 (2006). https://doi.org/10.1002/ptr.1871
7. Cieślik, E., Gręda, A., Adamus, W.: Contents of polyphenols in fruit and vegetables. Food Chem. **94**(1), 135–142 (2006). https://doi.org/10.1016/j.foodchem.2004.11.015
8. Plehanova, T.: Pasternak—tsennoe ovoschnoe i lekarstvennoe rastenie. Zh. Ovoschevodstvo, (4), 38 (2005)
9. Telezhenko, L.N., Bezusov, A.T.: Biologicheskiaktivnyie veschestva fruktov i ovoschey i ih sohranenie pri pererabotke. Optimus, Odessa, 268 s (2004)
10. Dubova, G.E., Bezusov, A.T.: Nauchnyie osnovyi vosstanovleniya estestvennyih aromatov v pischevyih prorduktah [Tekst] Naukovi pratsi, ONAHT, vipusk 42, tom 2, stor. 33–38
11. Silich M.V.: Opredelenie letuchih soedineniy v pryano-aromaticheskom syire [Teskt], Naukovi pratsi, vipusk 40, tom 2, stor. 39–41
12. Ikeda, R.M., Stanley, W., L., Vannier, S.H., Spitler, E.M.: The monoterpene hydrocarbon composition of some essential oils. J.Food Sciense, **27**(5):455–458 (1962)
13. Johnson, A.E., Nurster, H.E., Williams, A.A. Vegetable volatiles: a survey of components identified: part II, Chemistry and Industry, 23, №10, 1212–1224 (1971)
14. Pat. na korisnu model 95721 Ukrayina, MPK (2015.01) G01 N 30/00, G01 N 31/16 (2006.01). Sposib viznachennya chisla aromatoutvoryuyuchih spoluk/Verhivker Ya.G., Osipova L.A., Iovcheva I.O., Efremov V.V.; vlasnik Odes. nats. akad. harch. tehnologly. -# u201403879; zayavl. 14.04.2014; opubl. 12.01.2015, Byul. # 1
15. Sertel, S1., Eichhorn, T., Plinkert, P.K., Efferth, T.: Chemical composition and antiproliferative activity of essential oil from the leaves of a medicinal erb, Levisticum officinale, against UMSCC1 head and neck squamous carcinoma cells. Anticancer Res. **31**(1):185–91. PMID: 21273597 (2011)

16. Nunes, IS1., Faria, J.M., Figueiredo, A.C., Pedro, L.G., Trindade, H., Barroso, J.G.: Menthol and geraniol biotransformation and glycosylation capacity of Levisticum officinale hairy roots. Planta Med. **75**(4):387–391. https://doi.org/10.1055/s-0028-1112217 (2009). Epub 2009 Jan 20. PMID: 19156598
17. Guzman. J.D1., Evangelopoulos, D., Gupta, A., Prieto, J.M., Gibbons, S., Bhakta, S. Antimycobacterials from lovage root (Ligusticum officinale Koch). Phytother Res. **27**(7):993–998. https://doi.org/10.1002/ptr.4823 (2013). Epub 2012 Aug 16. PMID: 22899555 https://doi.org/10.1002/ptr.4823
18. Schinkovitz, A1., Stavri, M., Gibbons, S., Bucar, F.: Antimycobacterial polyacetylenes from Levisticum officinale. Phytother Res. **22**(5):681–684. https://doi.org/10.1002/ptr.2408 (2008). PMID: 18350523 https://doi.org/10.1002/ptr.2408
19. Efremov V.V. i dr.: Opredelenie chisla aromatoobrazuyuschih soedineniy kak pokazatelya kachestva pryano-aromaticheskogo rastitelnogo syirya (pars) i produktovego pererabotki. Aspirant i soiskatel (6), 38–41 (2014)

# Influence of the Growth Regulator Application Method on Antioxidant Plant System Activity of Winter Wheat (*Triticum Aestivum* L.)

**Zoia Bilousova, Yuliia Klipakova, Victoria Keneva and Serhii Kulieshov**

## 1 Introduction

The successful intensive technology implementation of growing crops is strongly influenced by solving the problem of increasing the plant resistance to adverse environmental factors, both at the stage of the seed germination and during the vegetation period [1, 2]. The influence of abiotic stressors (drought, salinity, etc.) that accompany a plant during the vegetation induces the enhanced formation of active oxygen species (AOS) in chloroplasts and other cellular compartments [3]. The AOS accumulation causes the oxidative stress, which leads to the damage to membrane lipids, nucleic acids, inactivation of enzymes, and fading of pigments [4]. The balance between the formation and the AOS neutralization is crucial for the plant survival under stress.

One of the most effective ways to reduce the negative effects of stress factors on the plant productivity is the use of natural and synthetic growth regulators [5, 6]. The new generation of growth regulators has a triple effect on plants: They increase their own resistance to the adverse factors activity, stimulate physiological processes, and enhance the non-specific plant immunity [7]. Researchers have found that the growth regulator use contributes to an increase of the water-soluble carbohydrate and protein content [8] and increases the antioxidant plant system activity, in particular superoxide dismutase (SOD), catalase, and peroxidase, thereby increasing the plant resistance to the cold and thermal stress [9, 10].

At present, the research findings on the natural growth regulator application for increasing the antioxidant plant system activity of winter wheat have been published, but the influence of synthetic growth regulators on the plant resistance mechanisms to the stressor activity has been insufficiently investigated. In addition, the duration

Z. Bilousova (✉) · Y. Klipakova · V. Keneva · S. Kulieshov
Tavria State Agrotechnological University, B. Khmelnitsky Av., 18, Melitopol 72312, Zaporizhia Region, Ukraine
e-mail: zoiazolotukhina@gmail.com

© Springer Nature Switzerland AG 2019

V. Nadykto (ed.), *Modern Development Paths of Agricultural Production*,
https://doi.org/10.1007/978-3-030-14918-5_60

and effectiveness of this drug group action have been significantly influenced by the application method (the pre-sowing seed treatment or the foliar plant spraying).

The purpose of the work was to research the influence of the AKM synthetic growth regulator application method on the oxidative stress development and the activity of antioxidant enzymes (SOD, catalase) in winter wheat plants.

## 2 Materials and Methods

The experimental part of the work was carried out during 2009–2012 as a stationary experiment of the Department of Crop Production at the Production and Training Center of Tavria State Agrotechnological University, which is located in Lazurne, Melitopol district, Zaporizhia region, Ukraine (46°58′00″N 35°08′00″E, 33 m above the sea level).

The experimental field soil was the southern light-clay black soil. The humus content in the arable layer was increased, the easily hydrolyzed nitrogen content was very low, the movable phosphorus content was increased, the exchangeable potassium content was high, and the interstitial water reaction was close to neutral.

The weather conditions during the research years differed considerably. According to Selyaninov's hydrothermal coefficient (HTC), 2009 and 2012 were characterized as very dry (HTC = 0.5), 2010 was excessively wet (HTC = 1.6), and 2011 was sufficiently wet (HTC = 1.0).

The experiment was conducted for winter wheat of Zolotokolosa variety. The predecessor was bare fallow. The soil cultivation and field preparation for sowing were carried out according to the scheme commonly used for the southern steppe zone of Ukraine. The seeds were sown in the first decade of October in the well-prepared soil using a usual drill sowing method, the burying depth was 5–6 cm, and the seeding rate was 5.0 million of seeds per 1 ha. In the BBCH stage 29, a herbicide was added with the active tribenuron-methyl substance (methyl ester 2-[6-methyl-4-methoxy-1,3,5-triazine-2-yl(methyl) carbamoyl sulfamoil] of benzene carboxylic acid) in the amount of 20 g/ha. In the BBCH stage 37, plants were treated with a fungicide with the active carbendazim substance (N-(benzimidazolyl-2)-O-methyl carbamate) in the amount of 0.5 l/ha. To protect against pests, an insecticide with the active substance dimethoate (O, O-dimethyl-S-(N-methylcarbamidomethyl) dithiophosphate) was used in the amount of 1.5 l/ha.

The scheme of the experiment included four variants:

1. Control (without the AKM growth regulator application);
2. The foliar treatment of plants with the AKM growth regulator (0.33 l/ha) in the BBCH stage 37;
3. The pre-sowing seed treatment with the AKM growth regulator (0.33 l/t);
4. The pre-sowing seed treatment with the AKM growth regulator (0.33 l/t) and the foliar treatment of plants with the AKM growth regulator (0.33 l/ha) in the BBCH stage 37.

AKM is a semi-synthetic film-forming plant growth regulator (PGR) of the antioxidant action, consisting of dimethyl sulfoxide (16–25 g/l), ionol (37.5 g/l), PEG-1500 (540 g/l), and PEG-400 (230 g/l) [11].

The experiment was repeated four times. The total area of the unitary plot was 100 $m^2$, and the accountable area was 50 $m^2$.

For determining the malondialdehyde (MDA) content, the plant material was homogenized in 20% trichloracetic acid and incubated with 0.5% thiobarbituric acid in a boiling water bath for 30 min. In a supernatant obtained after the centrifugation, the MDA content was determined spectrophotometrically at the wavelength of 532 nm and expressed in nM of MDA per 1 g of the dry matter [12].

For determining the SOD activity, the plant tissues were homogenized in 0.1 M phosphate buffer (pH = 7.8) with the addition of the Triton X-100 detergent (the final concentration was 0.2%). The supernatant was analyzed after the homogenate centrifugation (7000 g, 15 min). The overall SOD activity was measured using the method based on the ability of the enzyme to compete for superoxide anions forming as a result of the aerobic interaction of NADH and phenazine metasulfate. The optical density was determined at the wavelength of 530 nm and expressed in conventional activity units per gram of the dry matter [13].

The catalase activity was determined in the following way. The batch of the plant material was homogenized in 0.1 M Tris-HCl buffer (pH = 7.8). The homogenate was centrifuged for 15 min at 7000 g. For the analysis, three glass tubes were taken, 3 ml of supernatant were put into them, 3 ml of 0.3% $H_2O_2$ were added, and then, the content was mixed and incubated for 10 min. Then, the reaction was stopped by adding 1 ml of 4% ammonium molybdate. The optical density of the solution was determined at the wavelength of 410 nm, which was expressed in micromole of $H_2O_2$ per gram of the dry matter • min [14].

The findings were processed statistically by the dispersive analysis method.

## 3 Results and Discussion

The important criterion for assessing the plant resistance to the stress factor action is the level of the oxidative stress development in cells and tissues; one of the indicators is the process of lipid peroxidation (LPO) of cell membranes. The intensity of the LPO has been estimated by the accumulation of its final product—malondialdehyde (MDA). The conducted researches have indicated that the highest MDA content was observed after the spring vegetation renewal during the BBCH 29 plant development stage (Table 1). The high MDA content in the leaves of plants has shown a sharp intensification of free radical processes at this development stage as a result of the plant organism response to the stress caused by low negative temperatures during the overwintering period. Later, by the end of the vegetation, the LPO product content has gradually decreased, which has specified the plant adaptation to changing environmental conditions.

**Table 1** Malondialdehyde content in winter wheat plants depending on the method of the AKM growth regulator application, nM/g of the dry matter (average value in 2009–2012)

| Application method | | The development stage | | | | |
|---|---|---|---|---|---|---|
| Seed pre-sowing treatment | Plant treatment | BBCH 29 | BBCH 37 | BBCH 59 | BBCH 65 | BBCH 73 |
| Control | | 233.59 ± 9.15 | 193.44 ± 9.89 | 142.60 ± 7.73 | 94.72 ± 3.84 | 96.14 ± 5.31 |
| – | AKM | 234.01 ± 9.52 | 194.26 ± 9.85 | 119.32 ± 7.96 | 90.71 ± 5.29 | 87.35 ± 2.54 |
| AKM | – | 199.02 ± 5.95 | 176.36 ± 6.16 | 118.85 ± 7.89 | 86.80 ± 2.11 | 92.05 ± 5.15 |
| AKM | AKM | 192.85 ± 4.60 | 175.84 ± 6.31 | 111.87 ± 6.07 | 83.70 ± 4.35 | 82.14 ± 3.51 |

The AKM growth regulator application in the winter wheat cultivating technology changes the MDA content dynamics, depending on the drug application method. Thus, the pre-sowing treatment of winter wheat seeds with the AKM growth regulator contributed to the MDA content decrease on the average of 11% during the whole vegetation period compared with the control. The foliar plant treatment with the AKM growth regulator at the BBCH 37 development stage has led to the sharp decrease in the MDA level (by 39%) at the next stage of plant development (BBCH 59) compared with the control. MDA content dynamics in the plants of this variant were similar to the pre-sowing AKM seed treatment at subsequent development stages (BBCH 65–73).

The highest efficiency for reducing the intensification of free radical processes in winter wheat plants had the complex AKM growth regulator application both for the pre-sowing seed treatment and for the furrow spraying of plants at the BBCH 37 development stage. This variant of using the growth regulator has led to the MDA content decrease on the average of 15% during the whole vegetation period compared with the control. The findings have provided an opportunity to state the lipid peroxidation process inhibition with the AKM growth regulator.

It is believed that the oxidative stress is caused not so much by the AOS formation, but to a greater extent by a disbalance between their generation and neutralization [15]. The regulation of this process is carried out using an antioxidant protection system, which includes high- and low-molecular components [16]. The important high-molecular components which directly neutralize AOS and are characterized by the high action specificity concerning their certain forms are superoxide dismutase, peroxidase, and catalase [3]. The protection primary line against the oxidative damage is SOD, which breaks up the oxidation of cellular macromolecules at the initiation stage [17].

The conducted researches have shown that the increase in the SOD activity at the BBCH 37 plant development stage (Table 2) was followed by increasing the LPO product content at the previous development stage (Table 1), indicating the increase in this enzyme synthesis under stress factor conditions. The maximum SOD activity was in the period of the transition from vegetative to generative types of plant development (BBCH 59), with the subsequent decrease in its activity due to the organism aging [17, 18].

**Table 2** SOD activity in winter wheat plants depending on the method of the AKM growth regulator application, cu/g dry matter (average for 2009–2012)

| Application method | | The development stage | | | | |
|---|---|---|---|---|---|---|
| Seed pre-sowing treatment | Plant treatment | BBCH 29 | BBCH 37 | BBCH 59 | BBCH 65 | BBCH 73 |
| Control | | 10.63 ± 0.98 | 13.78 ± 0.54 | 18.49 ± 0.53 | 13.59 ± 0.49 | 3.86 ± 0.17 |
| – | AKM | 10.85 ± 0.73 | 13.84 ± 0.43 | 19.99 ± 0.54 | 15.47 ± 0.40 | 4.19 ± 0.10 |
| AKM | – | 13.73 ± 0.37 | 18.15 ± 0.39 | 17.66 ± 0.58 | 14.88 ± 0.23 | 4.22 ± 0.13 |
| AKM | AKM | 14.52 ± 0.50 | 18.50 ± 0.50 | 21.75 ± 0.59 | 21.19 ± 0.39 | 4.92 ± 0.10 |

The AKM growth regulator application for the pre-sowing treatment of winter wheat seeds has facilitated the sharp increase in the SOD activity at the BBCH 29 and 37 development stages—by 29 and 32% correspondingly compared to the control. At the BBCH 59 development stage, there was no significant difference between the check and the given treatment variant, which can be explained by the plant development peculiarities. Thus, in the check without the PGR application, there was faster coming of all plant development phases. It was this factor that determined the sufficiently high SOD activity at the BBCH stage 59 (4-day phase), because this phase duration in the variant with the AKM pre-sowing treatment was 7 days. The shortened plant development period of the control stimulated the higher SOD activity at the BBCH 59 stage with the subsequent rapid depletion of this enzyme amount. Subsequently, during the BBCH 65–73 stages, as a result of the positive effect attenuation of the pre-sowing PGR treatment, the difference between this variant and the check was within 9%.

The foliar plant treatment with the AKM growth regulator at the BBCH 37 development stage contributed to increasing the SOD activity at the following developmental stages (BBCH 59–73) on the average of 10% compared to the control.

The highest effect on the SOD activation in winter wheat plants was noted for the variant with the PGR use both for the pre-sowing seed treatment and vegetative plants. On an average, during the vegetative period in the plants of this variant, the SOD activity was by 34% higher compared to the check.

In general, the SOD activity in the plants of all experimental variants is consistent with the MDA content dynamics.

Since hydrogen peroxide is the result of not only lipid peroxide oxidation, but also the dismutation effect of superoxide anions with the SOD enzyme [17], therefore it is necessary to provide an element of the antioxidant plant protection that will utilize its excessive amount. Such a role in plant organism cells is performed by the catalase enzyme, which restores hydrogen peroxide to water, thereby protecting cells from its toxic effects [19].

**Table 3** Activity of catalase in winter wheat plants depending on the method of the AKM growth regulator application, μmol/g of dry matter • min (the average for 2009–2012)

| Application method | | The development stage | | | | |
|---|---|---|---|---|---|---|
| Seed pre-sowing treatment | Plant treatment | BBCH 29 | BBCH 37 | BBCH 59 | BBCH 65 | BBCH 73 |
| Control | | 66.79 ± 4.78 | 24.04 ± 1.99 | 97.52 ± 4.83 | 92.24 ± 4.61 | 52.36 ± 2.78 |
| – | AKM | 64.48 ± 5.54 | 21.84 ± 0.89 | 125.11 ± 7.07 | 114.08 ± 5.77 | 79.43 ± 4.74 |
| AKM | – | 110.70 ± 6.58 | 50.31 ± 2.87 | 122.40 ± 7.18 | 107.38 ± 6.20 | 75.54 ± 2.90 |
| AKM | AKM | 108.94 ± 6.20 | 51.24 ± 2.63 | 165.61 ± 7.42 | 122.63 ± 5.08 | 111.27 ± 4.44 |

The findings have indicated the rather high catalase activity in the first half of the spring vegetation (BBCH 29) followed by a sharp decreasing in the BBCH 37 stage (Table 3).

The high catalase activity in winter wheat plants in the BBCH 29 stage can be explained by the specific reaction of this enzyme to the chemical stress caused by the application of herbicides at this development stage [20] and requires further researches. The further dynamics of the catalase activity completely coincides with the dynamics of another antioxidant SOD enzyme (Table 2), and this fact confirms the action of these substances as a single enzyme system [3].

The AKM growth regulator application for the pre-sowing treatment of winter wheat seeds contributed to the increase in the catalase activity in the first half of spring vegetation (BBCH 29–37) almost twice compared with the control. Subsequently, there was a gradual decline in the positive effect of the pre-sowing PGR treatment on this enzyme activity (as in the case of SOD), and during the period of generative development (BBCH 59–73), the difference between this variant and the control was within 16–44%.

The foliar treatment of winter wheat plants with the AKM growth regulator at the BBCH 37 development stage increased the catalase activity by 28, 24, and 52% at the following developmental stages (BBCH 59, 65, and 73, respectively) compared to the control.

The integrated PGR application (for the pre-sowing seed treatment and vegetating plants) has had the highest effect on the catalase activity intensification, which has been manifested in the increasing activity of this enzyme on the average of 1.8 times during the vegetation compared with the control.

Also, it should be noted that the growth regulator effect on the catalase activity increase was significantly higher than on the SOD activity stimulation.

## 4 Conclusions

Thus, the AKM growth regulator application in the winter wheat cultivation technology leads to the activity reduction of the lipid peroxidation processes and enhances the functioning of such antioxidant enzymes as superoxide dismutase and catalase.

However, the foliar plant treatment at the BBCH 37 development stage influences positively on the antioxidant system activation only during the generative period of plant development, when all the production elements have been already formed; therefore, it does not have a significant effect on the yield increase [21]. At the same time, the AKM PGR application only for the pre-sowing treatment of winter wheat seeds is characterized by an unstable effect on the antioxidant system stimulation, which also does not guarantee the high plant productivity formation. Therefore, the most efficient way to use the AKM growth regulator is the pre-sowing seed treatment and the subsequent plant spraying at the BBCH 37 development stage, which contributes to the average 15% reduction in the MDA content, the SOD activity increase by 34%, and the catalase growth of 1.8 times during vegetation compared to the control.

## References


1. Osakabe, Y., Yamaguchi-Shinozaki, K., Shinozaki, K., Phan Tran, L-S.: Sensing the environment: key roles of membrane-localized kinases in plant perception and response to abiotic stress. J. Exp. Bot. **64**(2), 445–458 (2013). https://doi.org/10.1093/jxb/ers354
2. Atkinson, N.J., Urwin, P.E.: The interaction of plant biotic and abiotic stresses: from genes to the field. J. Exp. Bot. **63**(10), 3523–3543 (2012). https://doi.org/10.1093/jxb/ers100
3. Singh Gill, S., Tuteja, N.: Reactive oxygen species and antioxidant machinery in abiotic stress tolerance in crop plants. Plant Physiol. Biochem. **48**(12), 909–930 (2010). https://doi.org/10.1016/j.plaphy.2010.08.016
4. Sharma, P., Bhushan Jha, A., Shanker Dubey, R., Pessarakli, M.: Reactive oxygen species, oxidative damage, and antioxidative defense mechanism in plants under stressful conditions. J. Bot. **2012**, 26 (2012). https://doi.org/10.1155/2012/217037
5. Bajguz, A., Hayat, S.: Effects of brassinosteroids on the plant responses to environmental stresses. Plant Physiol. Biochem. **47**(1), 1–8 (2009). https://doi.org/10.1016/j.plaphy.2008.10.002
6. Tassia, E., Pougetb, J., Petruzzellia, G., Barbafieri, M.: The effects of exogenous plant growth regulators in the phytoextraction of heavy metals. Chemosphere **71**(1), 66–73 (2008). https://doi.org/10.1016/j.chemosphere.2007.10.027
7. Kadioglu, A., Saruhan, N., Sağlam, A., et al.: Exogenous salicylic acid alleviates effects of long term drought stress and delays leaf rolling by inducing antioxidant system. Plant Growth Regul. **64**(1), 27–37 (2011). https://doi.org/10.1007/s10725-010-9532-3
8. Santner, A., Calderon-Villalobos, L.I., Estelle, M.: Plant hormones are versatile chemical regulators of plant growth. Nat. Chem. Biol. **5**, 301–307 (2009). https://doi.org/10.1038/nchembio.165
9. Liang, Y., Zhu, J., Li, Z., Chu, G., Ding, Y., Zhang, J., Sun, W.: Role of silicon in enhancing resistance to freezing stress in two contrasting winter wheat cultivars. Environ. Exp. Bot. **64**(3), 286–294 (2008). https://doi.org/10.1016/j.envexpbot.2008.06.005
10. Simova-Stoilova, L., Demirevska, K., Petrova, T., Tsenov N., Feller, U.: Antioxidative protection and proteolytic activity in tolerant and sensitive wheat (*Triticum aestivum* L.) varieties subjected to long-term field drought. Plant Growth Regul., **58**(1), 107–117. https://doi.org/10.1007/s10725-008-9356-6 (2009)
11. Kalytka, V.V., Zhernovyi, O.I., Ivanchenko, O.A., Zolotukhina, Z.V., Yalokha, T.M.: Patent na korysnu model № 58260 «Antystresova kompozytsiia dlia peredposivnoi obrobky nasinnia silskohospodarskykh kultur» [Antistress composition for presowing treatment of farm crop seeds] (2011). (in Ukrainian)

12. Musiienko, M.M., Parshykova, T.V., Slavnyi, P.S.: Spektrofotometrychni metody v praktytsi fiziolohii, biokhimii ta ekolohii roslyn. Fitosotsiotsentr, Kyiv (2001). (in Ukrainian)
13. Chevari, S., Chaba, I., Sekey, I.: Rol superoksiddismutazy v okislitelnykh protsessakh kletki i metod opredeleniya yeye v biologicheskikh materialakh [The role of superoxide dismutase in the oxidative processes of cells and the method of determining it in biological materials]. Lab. Case **11**, 678–681 (1985). (in Russian)
14. Korolyuk, M.A., Ivanova, L.I., Mayorova, I.G.: Metod opredeleniya aktivnosti katalazy [Methods for determination of catalase activity]. Lab. Case **1**, 16–19 (1988). (in Russian)
15. Kolupaev, Y.E.: Aktivnyie formyi kisloroda v rasteniyah pri deystvii stressorov: obrazovanie i vozmozhnyie funktsii [Reactive oxygen species in plants at stressors action: formation and possible functions]. The Bulletin of Kharkiv National Agrarian University. Series Biology, **3**(12), 6–26 (2007). (in Russian)
16. Blokhina, O., Virolainen, E., Fagerstedt, K.V.: Antioxidants, oxidative damage and oxygen deprivation stress. Ann. Bot. **91**, 179–194 (2003)
17. Alscher, R.G., Erturk, N., Heath, L.S.: Role of superoxide dismutases (SODs) in controlling oxidative stress in plants. J. Exp. Bot. **53**(372), 1331–1341 (2002). https://doi.org/10.1093/jexbot/53.372.1331
18. Priss, O.P., Kalitka, V.V.: Effect of heat treatment with antioxidants on oxygen radical scavenging during storage of zucchini squash. Technology Audit and Production Reserves, **6/4**(26), 45–50 (2015). https://doi.org/10.15587/2312-8372.2015.56230
19. Willekens, H., Chamnongpol, S., Davey, M., Schraudner, M., Langebartels, C., Van Montagu, M., Inzé, D., … Van Camp, W.: Catalase is a sink for $H_2O_2$ and is indispensable for stress defence in C3 plants. EMBO J. **16**(16), 4806–4816 (1997). https://doi.org/10.1093/emboj/16.16.4806
20. Mhamdi, A., Queval, G., Chaouch, S., Vanderauwera, S., Van Breusegem, F., Noctor, G.: Catalase function in plants: a focus on Arabidopsis mutants as stress-mimic models. J. Exp. Bot. **61**(15), 4197–4220 (2010). https://doi.org/10.1093/jxb/erq282
21. Kalytka, V.V., Zolotukhina, Z.V.: Formuvannia vrozhainosti ta yakosti zerna ozymoi pshenytsi pid chas zastosuvannia rehuliatora rostu v umovakh sukhoho Stepu Ukrainy [Formation of productivity and quality of grain of the winter wheat at use of the regulator of growth in the conditions of dry Steppe of Ukraine]. Scientific Herald of National University of Life and Environmental Sciences of Ukraine. Series: Agronomy, **183**(2), 67–74 (2013). (in Ukrainian)

# Effects of Different Dietary Selenium Sources Including Probiotics Mixture on Growth Performance, Feed Utilization and Serum Biochemical Profile of Quails

**Volodimir Bityutskyy, Svitlana Tsekhmistrenko, Oksana Tsekhmistrenko, Olexander Melnychenko and Viktor Kharchyshyn**

## 1 Introduction

Poultry farming is an innovative and high-tech meat and egg production sector. In order to provide increased demand for animal protein, improve bird health and product quality, probiotics are an alternative to antibiotics [1]. Birds feedings include probiotics such as Lactobacillus and Bifidobacterium, which have a positive effect on the productivity of broilers, [2–4], regulate lipid metabolism [5] and processes of immunomodulation [6]. Selenium, which is inducted into feed, improves bird productivity and health and has a positive effect on the immune system and fatty acid composition of meat and eggs [7]. There are several different forms of selenium that are used as additives to fodder: inorganic forms (sodium selenite and sodium selenate) and organic forms (selenomethionine). However, new forms of selenium have appeared in recent years, such as 2-hydroxy-4-methyliselobutanoic acid (HMSeBA) and nanoselens that have better bioavailability, efficacy and low toxicity [8]. Selenite and selenate are considered to be the most toxic forms of selenium, and then selenocysteine, whereas methyl compounds of selenium and nanoselenium have the lowest levels of toxicity. The addition of 0.3–0.5 mg/kg of nanoselen proved to be effective and beneficial in improving the protection against oxidative stress, and the maximum dosage of nanoselen should not exceed 1.0 mg/kg [8].

Selenium nanoparticles (SeNPs) are considered as new forms of selenium supplemented with high biological activity and low toxicity. However, the molecular mechanism by which they exhibit biological action for today is not fully understood. SeNPs are known to exhibit biomimetic activity, in particular, oxidase-like, what is also a characteristic of other nanoparticles [8–10].

Regarding the mechanism of action of nanoselen, the work [11] doubts the interpretation that nanoselen has a direct antioxidant effect, since it has been established

V. Bityutskyy (✉) · S. Tsekhmistrenko · O. Tsekhmistrenko · O. Melnychenko · V. Kharchyshyn
Bila Tserkva National Agrarian University, Soborna pl., 8/1, Bila Tserkva 09117, Ukraine
e-mail: voseb@ukr.net

© Springer Nature Switzerland AG 2019

V. Nadykto (ed.), *Modern Development Paths of Agricultural Production*,
https://doi.org/10.1007/978-3-030-14918-5_61

that, as in various forms of selenium, its antioxidant effect is associated with the expression of selenoprotein genes and the synthesis of the corresponding proteins [12]. The studies show the positive effects of nanoselen on the activity of glutathione peroxidase (GSH-Px) and on direct expression of enzyme.

The most prominent biological role of selenium is due to its presence as a functional component in selenium enzymes such as glutathione peroxidase (GPx) and thioredoxin reductase (TrxR). SeNPs, as compared to selenium, selenomethionine, and methyl selenocysteine, have been shown to increase the activity of selenium enzymes in mice and rats more effectively, with significantly lower acute toxicity [13–15].

As the main effects of selenium implemented with selenometionine, selenocysteine and selenoproteins that are not synthesized in humans, animals and birds, this biotransformation is implemented by plants or bacteria [16].

From this point of view, it is interesting to assume that the intestinal microbiota can transform nano-Se into selenite/selenate or recover in $H_2Se$ with further synthesis of selenoproteins [17, 18]. This has been proven that selenium nanoparticles can be oxidized to inorganic selenium oxyanions in the intestine in the presence of bacteria presence for their absorption [17]. Recently, the genes encoding the YedE and YedF proteins are considered as new gene candidates participating in the Se metabolism in prokaryotes, including bacteria [19]. Thus, additional studies are needed to find out the mechanisms for the possible transformation of nano-Se into the gastrointestinal tract of animals and the possible synergy of action by the use of certain strains of probiotics and selenium nanoparticles. The metabolic action of SeNPs depends on the composition of the nanoparticles, including the nanoparticle coating agent [20].

Thus, selenium plays a very important role in feeding birds due to polyfactor action, and its incorporation into compound feed is necessary, especially in areas with soils that are scarce in selenium. In addition, considering the possible toxic effects and the more limited properties of inorganic forms of selenium, there is a need for the development of new compositions, in particular in nanoform.

In order to assess the impact of complex probiotic preparation Probifilact (Probiotic Containing Lactobacillus, Bifidobacterium), which is added to animal feed, along with various forms of selenium on performance, retention and blood biochemical parameters quail conducted the survey data.

## 2 Materials and Methods

In this work there was used a complex of strains of Bifidobacterium animalis VKB, Lactobacillus casei IMV B-7280, Bifidobacterium animalis VKL (Probifilact), what were obtained from the collection of the Institute of Microbiology and Virology named by D.K. Zabolotny National Academy of Sciences of Ukraine. A nanoselenium with spherical shape with a particle size of 7–60 nm (average size of 34 nm) stabilized with citrate was obtained at the Ukrainian State Research Institute "Resurs."

The nanoselenium was tested for purity, morphology and size using an electron microscope and UV spectrophotometry. Electronic microscopy of the samples was performed according to the generally accepted method [21] using an electron microscope JEM-1400 (Institute of Microbiology and Virology named by DK Zabolotny National Academy of Sciences of Ukraine) at 80 kV. The research was conducted at the Bila Tserkva National Agrarian University, Ukraine. According to the scheme, a number of 600 ode-day-old quails were used, of which six groups (with four replication subgroups in each) were formed on the basis of analogues: control and five experimental ones. The research lasted 35 days. The quail diet consisted of full-fodder feed, responsible for the content of energy and other nutrients standards (Standard organization of Ukraine, 2006). Birds of group 1 (control) were fed with standard diet (SD), experimental quails of second group—SD + 0.3 mg of sodium selenite/kg feed, third group—SD + 0.3 mg (SeNPs)/kg feed, fourth group—SD + Probifilact, fifth group—SD + 0.3 mg sodium selenite/kg feed + Probifilact, sixth group—SD + 0.3 mg (SeNPs)/kg feed + Probifilact. The conditions for birds keeping in all groups were the same.

Complex probiotical preparation Probifilact was added to the mixed fodder (150 g/t). Feed intake and body weight were determined at weekly intervals. The weight gain and feed conversion ratios of birds were calculated then. During the experiment records of the preservation of livestock, live weight of quails, feed consumption, calculating average daily and relative increments of live weight, feed costs per 1 kg of growth were kept.

Blood samples were taken from two birds (♂ + ♀) from each replication subgroup at the end of the experiment (35th day), after 4-h-length fasting after decapitation. The samples were held in an oblique position to form a blood clot and then centrifuged at 3000 rpm for 10 min. The serum was poured into sterile test tubes to determine the biochemical parameters. Biochemical blood parameters such as total protein, albumin, uric acid, total bilirubin, cholesterol, triglyceride, creatinine, calcium, phosphorus, AST and ALT were determined in the research laboratory of biochemical methods of research of Bila Tserkva National Agrarian University using standard sets in accordance with the manufacturer's instructions.

The statistical analysis was conducted using the Microsoft Office Excel software. Numerical data were presented as mean arithmetic values and their standard deviations ($M \pm m$). For single comparisons, values of $P$ were determined using Student's $t$ test. Differences between groups were significant at $P < 0.05$.

## 3 Results and Discussion

The results of experimental studies indicate that quails of experimental groups for live weight prevailed in control analogues (Table 1). In the young age, young poultry of the control and experimental groups did not differ significantly by live weight. However, in subsequent age periods, the live weight of quails varied depending on the type and level of feed supplement in the feed. The highest live weight was

**Table 1** Growth performance and feed utilization of supplemented with different Se sources, under probiotics influence

| Group | Parameters/treatments | Final weight (g) after fifth week |
|---|---|---|
| 1 | Standard diet (SD) | 162.2 ± 2.86 |
| 2 | SD + sodium selenite 0.3 mg/kg | 169.5 ± 4.91 |
| 3 | SD + SeNPs 0.3 mg/kg | 172.2 ± 3.9 |
| 4 | SD + Probifilact | 174.5 ± 4.03* |
| 5 | SD + Probifilact + sodium selenite 0.3 mg/kg | 177.6 ± 3.41** |
| 6 | SD + Probifilact + SeNPs 0.3 mg/kg | 188.3 ± 4.23*** |

*$p < 0.05$; **$p < 0.01$; ***$p < 0.001$, the results are probably related to the values in the control part

observed in quail of sixth groups, which consumed feed containing Probifilact + SeNPs 0.3 mg/kg.

Quails of the fourth group that consumed feed from Probifilact practically did not concede to the weight of the fifth group young, probably exceeding the live weight of analogues of control in the period of 35-day age. The quail body weight of the second and third groups was similar, with some advantage of the bird of the third group. At the same time, quails of these groups also surpassed the peers of the control group by weight during the period of 35 days.

In this case, it was shown the reducing of feed intake in experimental groups of quails and poultry mortality during the experiment compared with quails receiving standard feed was shown (Fig. 1).

In studies [22], additives of nano-Se at a diet of 0.3 mg/kg, unlike sodium selenite, improve the growth efficiency, antioxidant and immune properties of broilers grown under high ambient conditions. In researches [23], there were the study of the comparative effects of nanoselenium (Nano-Se) and sodium selenite on growth, bioavailability, antioxidant activity, haematological and biochemical parameters, cellular and humoral immunity of birds. It was found that the body weight of all groups who got Nano-Se (to a dose of 0.3 mg/kg) was significantly ($p < 0.05$) higher than adding sodium selenite and control groups. However, the further increase in Nano-Se content in the feed had a negative impact on the body weight of the birds.

It had been established that selenium deficiency can lead to a decrease in the synthesis of triiodothyronine (T3) and inhibition of bird growth [24]. This is due to the fact that selenium is a component of the key enzyme 5′-deiodination, converting T4 to T3, a hormone that regulates body growth controls the energy and protein consumption. In addition, Se may be part of glutathione peroxidase, which prevents cellular lipid membrane destruction due to oxidative damage, which in general affects the growth rate of the organism [15].

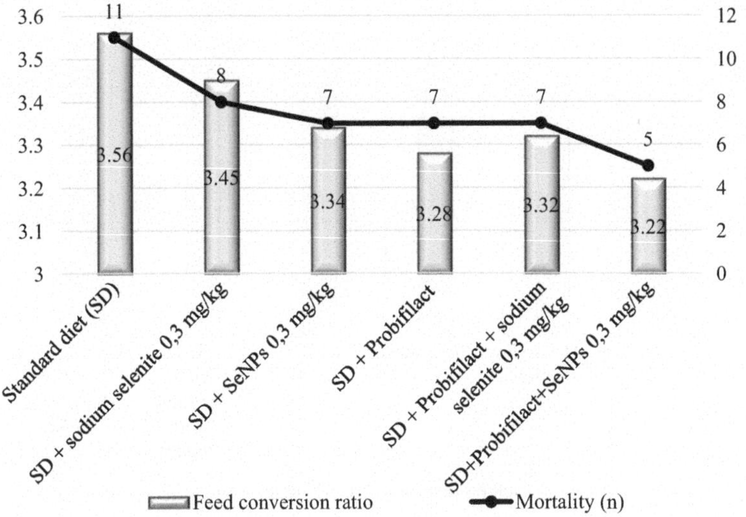


**Fig. 1** Feed conversation ratio and mortality of supplemented with different Se sources, under probiotics influence

Saleh [15] believes that improving the bird growth rates by the use of the probiotic complex and nano-Se may be due to increased metabolizable feed energy. This improvement may be due to the balancing of the microbial population in the gastrointestinal tract, which plays an important role in maintaining the health and energy of bird growth.

We found that quails of the sixth group that consumed feed containing Probifilact + SeNPs 0.3 mg/kg were characterized by the lowest feed costs per 1 kg of body weight gain during the growing season (1–35 days), which was 9.6% less than in control group analogues. The death of quails for the use of various feed additives up to 5 weeks from age varied from 5 to 11% among different groups.

The concentration of total protein, albumin (Fig. 2) and bilirubin (Fig. 3) in serum reflects the main functions of the liver, such as the synthesis of proteins (total protein and albumin), withdrawal of anions and the formation of bile (bilirubin), as well as the degree of damage to hepatocyte membranes (AST, ALT) [25, 26]. Increasing of protein, albumin content, tendency to decrease the activity of AST, ALT (Table 2), absence of probable changes in total bilirubin suggests that the addition of selenium and probiotic compounds to the diet positively influenced the parameters of protein metabolism and had no negative influence on the liver.

Experimental groups (3–6) showed a tendency to decrease the concentration of uric acid and creatinine. The content of uric acid in birds reflects the use of protein, the withdrawal of nitrogen, as well as the antioxidant function of the body [27]. Concentration of uric acid (Fig. 3), as the main product of the catabolism of amino acids and purines in birds [28], correlates inversely with the process of protein degradation and reflects the balance between the consumption, use and degradation of proteins,

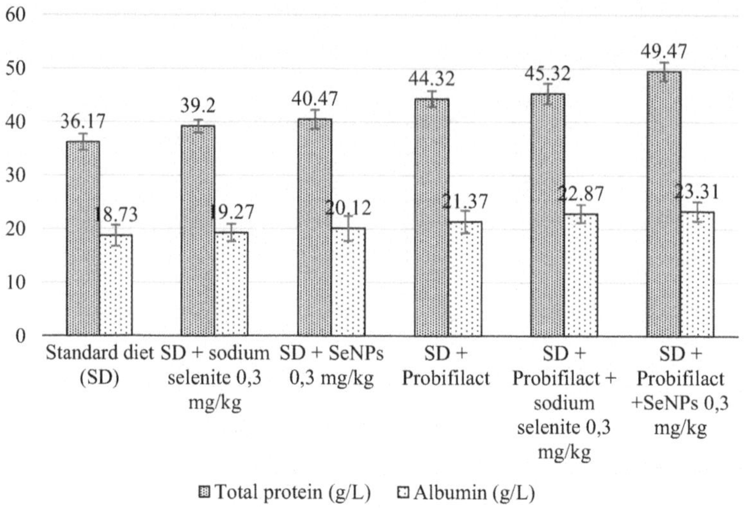


**Fig. 2** The content of total protein and albumin in serum blood of 35-day quail

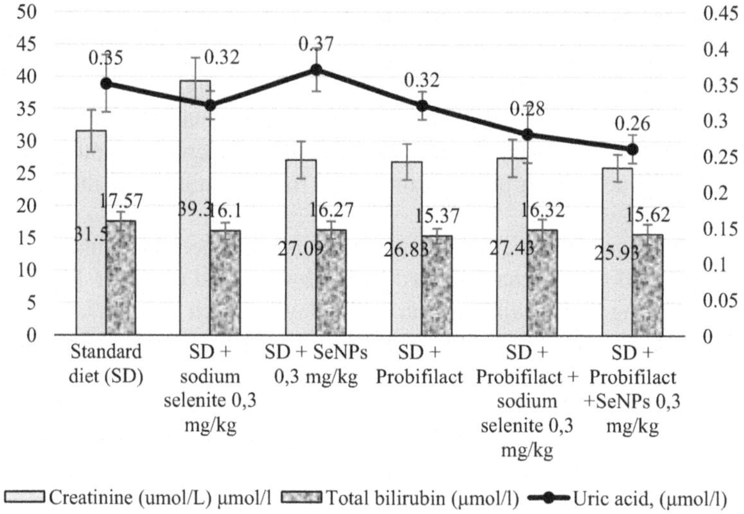


**Fig. 3** Content of creatinine, total bilirubin and uric acid of blood of 35-day quail ($M \pm m$)

**Table 2** Indicators of lipid metabolism of 35-day quail ($M \pm m$)

| Group | Parameters/treatments | Total lipids (g/L) | Cholesterol (mmol/L) | Triglyceride (mmol/L) |
|---|---|---|---|---|
| 1 | Standard diet (SD) | 17.57 ± 1.85 | 6.75 ± 0.72 | 5.63 ± 0.49 |
| 2 | SD + sodium selenite 0.3 mg/kg | 18.11 ± 1.30 | 5.98 ± 1.49 | 4.34 ± 0.61 |
| 3 | SD + SeNPs 0.3 mg/kg | 16.84 ± 2.03 | 4.84 ± 0.74 | 3.9 ± 0.99 |
| 4 | SD + Probifilact | 15.93 ± 1.22 | 4.35 ± 0.43* | 3.61 ± 0.77* |
| 5 | SD + Probifilact + sodium selenite 0.3 mg/kg | 15.27 ± 1.35 | 4.18 ± 0.52* | 3.83 ± 0.41* |
| 6 | SD + Probifilact + SeNPs 0.3 mg/kg | 14.37 ± 1.65 | 4.05 ± 0.95* | 3.46 ± 0.87* |

and the excretion of protein metabolites by the kidneys. Concentration of uric acid in serum is used to evaluate the function of the kidneys, with hyperuricemia (increased serum uric acid levels), often associated with kidney disease [28]. The trend towards lowering the concentration of uric acid and creatinine suggests that the addition of probiotics and various forms of selenium does not have a negative effect on the function of the kidneys.

The analysis of the lipid spectrum of blood established (Table 2) that the content of triacylglycerols and cholesterol decreased ($p < 0.05$) in quails of groups 4, 5 and 6, what is consistent with studies in which the reduction in their content in blood broilers for the introduction of probiotics [5].

It is known that microorganisms in the digestive tract can be involved in the metabolism of cholesterol in the body by acting on the host enzyme cell systems that synthesize endogenous cholesterol. Hypocholesterolemic activity of strains of lactic acid bacteria L. casei IBM-7280 [29] was established in vivo experiments on the model of mice. Researches [15] found that content in cholesterol and triglyceride plasma was lowered, while the number of high-density lipoproteins (HDL) was increased with the addition of a probiotic and selenium. Perhaps this is due to the activation of lipolysis for the supply of selenium [30, 31].

Researches [5] what added probiotics showed a decrease in the content of triglycerols, and the total level of cholesterol in the blood was also lower. There is an assumption that some bacterial probiotic strains can include cholesterol in bacterial cells and also hydrolyse bile salts that inhibit the activity of 3-hydroxy-3-methylglutaryl-coenzyme A (HMG-CoA) reductase, which plays a key role in the synthesis of this sterol, limiting the speed of cholesterogenesis and lowering the total pool of cholesterol in the body [2].

It is believed that probiotics containing specific microorganisms contribute to the transformation and assimilation of nano-Se in the intestine [11].

A significant increase in calcium content was found in experimental groups receiving probiotics and various forms of selenium (Table 3).

**Table 3** The content of calcium, phosphorus and the activity of individual transaminases of blood of 35-day quails (M±m)

| Group | Parameters/treatments | Calcium (mmol/L) | Phosphorus (mmol/L) | AST (U/L) | ALT (U/L) |
|---|---|---|---|---|---|
| 1 | Standard diet (SD) | 2.71 ± 0.22 | 2.59 ± 0.34 | 247.2 ± 19.1 | 48.33 ± 5.74 |
| 2 | SD + sodium selenite 0.3 mg/kg | 2.70 ± 0.51 | 2.62 ± 0.32 | 279.7 ± 21.3 | 58.23 ± 5.04 |
| 3 | SD + SeNPs 0.3 mg/kg | 3.13 ± 0.24 | 2.84 ± 0.41 | 235.5 ± 11.9 | 43.67 ± 3.83 |
| 4 | SD + Probifilact | 3.51 ± 0.37 | 2.78 ± 0.28 | 214.1 ± 29.0 | 40.2 ± 3.03 |
| 5 | SD + Probifilact + sodium selenite 0.3 mg/kg | 3.39 ± 0.26* | 3.05 ± 0.31 | 225.3 ± 18.03 | 45.2 ± 4.15 |
| 6 | SD + Probifilact + SeNPs 0.3 mg/kg | 3.81 ± 0.34* | 3.10 ± 0.24 | 201.0 ± 19.2 | 40.2 ± 3.73 |

In the case of stimulation of the metabolic activity of lactic bacteria and bifidobacteria, low molecular weight fatty acids (acetate, lactate, propionate, butyrate), which lower the pH in the intestine, contribute to the absorption of calcium, magnesium and ferum [32].

It is believed that before this Se form can find itself in commercial poultry/livestock, it is necessary to find out how nanoselenium is converted into active selenoproteins. One of the possible mechanisms for the transformation of nano-Se may be mediated by the action of microbiotics in the intestine, which is capable of converting nano-Se into selenite, $H_2Se$ or Se-phosphate, followed by the synthesis of selenoproteins [11].

## 4 Conclusions

Analysing the data of the experiment, the effectiveness of the application of the studied feed additives in the feeding of quails has been established. The use of probiotics and various forms of selenium in poultry feeding has a positive effect on biochemical parameters, body weight gain, feed conversion rate and livestock retention, and nanoselen alone, and in combination with a probiotic, has a more effective effect compared to sodium selenite. Possibility of involvement of intestinal microbiota in assimilation of nano-Se and metabolism requires further study.

## References


1. Král, M., Angelovičová, M., Mrázová, Ľ.: Application of probiotics in poultry production. Sci. Pap. Anim. Sci. Biotechnol. **45**(1), 55–57 (2012)
2. Kalavathy, R., Abdullah, N., Jalaludin, S., Ho, Y.W.: Effects of Lactobacillus cultures on growth performance, abdominal fat deposition, serum lipids and weight of organs of broiler chickens. Br. Poult. Sci. **44**(1), 139–144 (2003). https://doi.org/10.1080/0007166031000085445
3. Kabir, S.M.L., Rahman, M.M., Rahman, M.B., Rahman, M.M., Ahmed, S.U.: The dynamics of probiotics on growth performance and immune response in broilers. Int. J. Poult. Sci. **3**(5), 361–364 (2004). https://doi.org/10.3923/ijps.2004.361.364
4. Vantsawa, P. A., Umar, T., Bulus, T.: Effects of probiotic lactobacillus acidophilus on performance of broiler chickens. Int. Direct Res. J. Agric. Food Sci. **5**(8), 302–306. (2017). https://doi.org/10.26765/DRJAFS.2017.DRJA16284589
5. Mayahi, M., Razi-Jalali, M., Kiani, R.: Effects of dietary probiotic supplementation on promoting performance and serum cholesterol and triglyceride levels in broiler chicks. Afr. J. Biotechnol. **9**(43), 7383–7387 (2010)
6. Tymoshok, N.O., Lazarenko, L.M., Bubnov, R.V., Shynkarenko, L.N., Babenko, L.P., Mokrozub, V.V., Melnichenko, Y.A., et al.: New aspects the regulation of immune response through balance Th1/Th2 cytokines. EPMA J. **5**(Suppl 1), A134 (2014). https://doi.org/10.1186/1878-5085-5-s1-a134
7. Suchý, P., Straková, E., Herzig, I.: Selenium in poultry nutrition: a review. Czech J. Anim. Sci. **59**(11), 495–503 (2014)
8. Marković, R., Glišić, M., Bošković, M., Baltić, M.Ž.: New scientific challenges—the possibilities of using selenium in poultry nutrition and impact on meat quality. In: IOP Conference Series: Earth and Environmental Science, vol. 85, No. 1, p. 012032. IOP Publishing. https://doi.org/10.1088/1755-1315/85/1/012032
9. Tsekhmistrenko, S.I., Bityutskyy, V.S., Tsekhmistrenko, O.S., Polishchuk, V.M., Polishchuk, S.A., Ponomarenko, N.V.: Enzyme-like activity of nanomaterials. Regul. Mech. Biosyst. **9**(3), 469 (2018). https://doi.org/10.15421/021870
10. Tsekhmistrenko, O.S., Tsekhmistrenko, S.I., Bityutskyy, V.S., Melnichenko, O.M., Oleshko, O.A.: Biomimetic and antioxidant activity of nanocrystalline cerium dioxide. World Med Biol. **14**(63), 196–201 (2018). https://doi.org/10.267254/2079-8334-2018-1-63-196-201
11. Surai, P.F., Kochish, I.I., Velichko, O.A.: Nano-Se assimilation and action in poultry and other monogastric animals: Is gut microbiota an answer? Nanoscale Res. Lett. **12**(1), 612 (2017)
12. Sarkar, B., Bhattacharjee, S., Daware, A., Tribedi, P., Krishnani, K.K., Minhas, P.S.: Selenium nanoparticles for stress-resilient fish and livestock. Nanoscale Res. Lett. **10**(1), 371 (2015)
13. Bao-hua, X., Zi-rong, X., Mei-sheng, X., Cai-hong, H., Yue-song, D., Li, X.: Effect of nano red elemental selenium on GPx activity of broiler chick kidney cellsin vitro. Wuhan Univ. J. Nat. Sci. **8**(4), 1161 (2003). https://doi.org/10.1007/BF02903692
14. Zhang, J., Wang, X., Xu, T.: Elemental selenium at nano size (Nano-Se) as a potential chemopreventive agent with reduced risk of selenium toxicity: comparison with Se-methylselenocysteine in mice. Toxicol. Sci. **101**(1), 22–31 (2007). https://doi.org/10.1093/toxsci/kfm221
15. Saleh, A.A.: Effect of dietary mixture of Aspergillus probiotic and selenium nano-particles on growth, nutrient digestibilities, selected blood parameters and muscle fatty acid profile in broiler chickens. Anim. Sci. Pap. Rep. **32**, 65–79 (2014)
16. Schrauzer, G.N., Surai, P.F.: Selenium in human and animal nutrition: resolved and unresolved issues. A partly historical treatise in commemoration of the fiftieth anniversary of the discovery of the biological essentiality of selenium, dedicated to the memory of Klaus Schwarz (1914–1978) on the occasion of the thirtieth anniversary of his death. Crit. Rev. Biotechnol. **29**(1), 2–9 (2009). https://doi.org/10.1080/07388550902728261
17. Loeschner, K., Hadrup, N., Hansen, M., Pereira, S.A., Gammelgaard, B., Møller, L.H., et al.: Absorption, distribution, metabolism and excretion of selenium following oral administration of elemental selenium nanoparticles or selenite in rats. Metallomics **6**(2), 330–337 (2014). https://doi.org/10.1039/c3mt00309d

18. Bao, P., Chen, S.C., Xiao, K.Q.: Dynamic equilibrium of endogenous selenium nanoparticles in selenite-exposed cancer cells: a deep insight into the interaction between endogenous SeNPs and proteins. Mol. BioSyst. **11**(12), 3355–3361 (2015). https://doi.org/10.1039/C5MB00555H
19. Lin, J., Peng, T., Jiang, L., Ni, J.Z., Liu, Q., Chen, L., Zhang, Y.: Comparative genomics reveals new candidate genes involved in selenium metabolism in prokaryotes. Genome Biol. Evol. **7**(3), 664–676 (2015). https://doi.org/10.1093/gbe/evv022
20. Choudhury, S.R., Goswami, A.: Non-metallic nanoparticles & their biological implications, p. 61. Basic Appl. Aspects, Nanobiotechnology (2017)
21. Golding, C.G., Lamboo, L.L., Beniac, D.R., Booth, T.F.: The scanning electron microscope in microbiology and diagnosis of infectious disease. Sci. Rep. **6**, 26516 (2016). https://doi.org/10.1038/srep26516
22. Mahmoud, H.E.D., Ijiri, D., Ebeid, T.A., Ohtsuka, A.: Effects of dietary nano-selenium supplementation on growth performance, antioxidative status, and immunity in broiler chickens under thermoneutral and high ambient temperature conditions. J. Poult. Sci. **53**(4), 274–283 (2016). https://doi.org/10.2141/jpsa.0150133
23. Mohapatra, P., Swain, R.K., Mishra, S.K., Behera, T., Swain, P., Mishra, S.S., et al.: Effects of dietary nano-selenium on tissue selenium deposition, antioxidant status and immune functions in layer chicks. Int. J. Pharmacol. **10**(3), 160–167 (2014). https://doi.org/10.3923/ijp.2014.160.167
24. Preter, F.S.: Organic selenium: benefits to animal and humans, a biochemist's view. In: Proceedings of Alltech's 16th Symposium, pp. 205–213. Nottingham University Press, Nottingham (2000)
25. Limdi, J.K., Hyde, G.M.: Evaluation of abnormal liver function tests. Postgrad. Med. J. **79**(932), 307–312 (2003). https://doi.org/10.1136/pmj.79.932.307
26. Donaldson, J., Madziva, M.T., Erlwanger, K.H.: The effects of high-fat diets composed of different animal and vegetable fat sources on the health status and tissue lipid profiles of male Japanese quail (Coturnix coturnix japonica). Asian-Aust. J. Anim. Sci. **30**(5), 700 (2017). https://doi.org/10.5713/ajas.16.0486
27. Wright, P.A.: Nitrogen excretion: three end products, many physiological roles. J. Exp. Biol. **198**(2), 273–281 (1995)
28. Kolmstetter, C.M., Ramsay, E.C.: Effects of feeding on plasma uric acid and urea concentrations in blackfooted penguins (Spheniscus demersus). J. Avian Med. Surg. **14**(3), 177–179 (2000). https://doi.org/10.1647/1082-6742(2000)014%5b0177:EOFOPU%5d2.0.CO;2
29. Starovoitova, S.A., Babenko, L.P., Timoshok, N.A., Shynkarenko, L.N., Lazarenko, L.N., Spivak, N. Y.: Cholesterol-lowering activity of lactic acid bacteria probiotic strains in vivo. Microbiologichny zhurnal *Мікробіологічний журнал* **74**(3), 78–85 (2012). http://nbuv.gov.ua/UJRN/MicroBiol_2012_74_3_13
30. Kim, S.H., Yu, D.J., Lee, S.J., Park, S.Y., Ryu, K.S., Lee, D.G.: Effects of feeding aspergillus oryzae ferments on performance, intestinal microfloua, blood serum components and environmental factors in broiler. Korean J. Poult. Sci. **30**, 151–162 (2003)
31. Oppenheimer, J.H., Schwartz, H.L., Lane, J.T., Thompson, M.P.: Functional relationship of thyroid hormone-induced lipogenesis, lipolysis, and thermogenesis in the rat. J. Clin. Invest. **87**(1), 125–132 (1991). https://doi.org/10.1172/JCI114961
32. Cai, S.J., Wu, C.X., Gong, L.M., Song, T., Wu, H., Zhang, L.Y.: Effects of nano-selenium on performance, meat quality, immune function, oxidation resistance, and tissue selenium content in broilers. Poult. Sci. **91**(10), 2532–2539 (2012)

# Influence of Oat Extract on the Antioxidant Status of Geese

Olena Danchenko, Lubov Zdorovtseva, Mykola Danchenko, Oleksandr Yakoviichuk, Tetiana Halko, Elena Sukharenko and Yulia Nicolaeva

## 1 Introduction

Poultry farming is one of the most developed livestock industries in many countries around the world. In Ukraine, the traditional direction of poultry farming is geese breeding. However, recently, for a number of reasons, the number of geese in the country has decreased. So, from 1990 to 2015, the parental population of geese in Ukraine declined by almost 2.5 times [1–5]. However, an analysis of the condition and prospects for the development of the geese population indicates that there are basic prerequisites for the restoration of this industry in Ukraine. This is, first of all, the preserved gene pool of geese and favorable climatic conditions for geese in most regions of Ukraine [6, 7]. The use of innovative technologies and modern technological equipment in the poultry industry and the introduction of scientific achievements in production will contribute to increasing the efficiency of this industry [8, 9].

The use of antioxidants in feeding birds helps to eliminate the harmful effects of negative anthropogenic factors of the existing technologies for its cultivation [10]. Using natural antioxidant impurities has a number of advantages over the traditional synthetic vitamins of the antioxidant group. They are generally available, side effects are minimal or absent, they are not likely to cause toxic organic slags, do not irritate the mucous membrane of the stomach, do not disturb the digestion, and that is why they are well tolerated [11–15].

O. Danchenko (✉) · L. Zdorovtseva · M. Danchenko · E. Sukharenko
Tavriya State Agrotechnological University, B. Khmelnitsky Av., 18, Melitopol 72312, Ukraine
e-mail: nndea@ukr.net

O. Danchenko · O. Yakoviichuk · T. Halko · Y. Nicolaeva
Bogdan Khmelnitskiy Melitopol State Pedagogical University, Getmanska St., 20, Melitopol 72312, Ukraine

© Springer Nature Switzerland AG 2019

V. Nadykto (ed.), *Modern Development Paths of Agricultural Production*,
https://doi.org/10.1007/978-3-030-14918-5_62

**The aim of this work** was to find out the effect of the extract of oat seeding Avéna satíva on the antioxidant activity of the liver tissues of geese of the Italian breed in the pre-slaughter period (from the 35th to the 63rd day).

# 2 Experimental Studies

## *2.1 Research Method and Materials*

The research was carried out on geese of the Italian breed on the basis of the "Victoria" agricultural enterprise of Priazovsky district of Zaporizhzhya region. At the age of 1 day, based on the principle of analogs, two groups of geese (control and experimental) were formed with 26 geese in each group with an average weight (98.5 $\pm$ 4.2) g. Throughout the experiment, the birds of the control group were kept on a standard diet balanced by exchange energy, protein, and vitamins in accordance with the recommendations [16]. The geese of the experimental group from the 35th to the 56th day were given an oat extract. Extraction of flavonoids from oats was carried out with water, the ratio of raw material and extraction is 1:30, extraction time in a boiling water bath is 90 minutes. Slaughter of geese was carried out at 63 days of age, following the standards of the Council of Europe Convention on the protection of animals used in scientific research.

During the experiment, weekly determination of the intensity of lipid peroxide oxidation (LPO) in geese liver tissues was carried out, which was evaluated by the content of peroxidation products reacting with 2-tiobarbituric acid in the homogenates of these tissues (TBCAPout) and after the initiation of $Fe^{2+}$ POL (TBCAPin) [17]. The state of the AOP system was determined using the integral index—the antioxidant activity factor ($C_{AOA}$) [18], which was calculated as the ratio of TBCAPout to TBCAPin, since tissues homogenates contain not only the peroxidation substrate but also AOP components that can inhibit lipid peroxidation. In addition, in the isolated biomaterial, the content of total lipids, vitamins E, A, β-carotene, and antioxidant enzymes: superoxide dismutase (SOD), catalase (CAT), and glutathione peroxidase (HPO) [19–21] was determined. Mathematical processing of research results was carried out by methods of mathematical statistics, including multidimensional correlation and cluster analysis, using the SPSS-13.0 computer software package and MS Excel 2000 software.

## *2.2 Results and Discussion*

The indicated interval of geese ontogenesis is characterized by physiological stress in the body of birds (from the 42nd to the 56th day) caused by the formation of juvenile feathers. This process requires high energy and amino acids, including sulfur-

**Table 1** The content of lipids, lipoperoxidation products, and the antioxidant activity in the geese liver ($M \pm m, n = 6$)

| Age, days, $T$ | Group | TBCAPout ($P_2$), nMol/g | TBCAPin ($P_3$), nMol/g | Lipids ($X$), mg/g | $C_{AOA}$ |
|---|---|---|---|---|---|
| 35 | Control | 62.5 ± 3.8 | 124.7 ± 6.2 | 21.8 ± 1.3 | 0.50 |
| | Experimental | 63.3 ± 3.4 | 126.3 ± 4.9 | 22.7 ± 1.1 | 0.50 |
| 42 | Control | 87.2 ± 4.0 | 182.3 ± 7.5 | 17.3 ± 0.7 | 0.48 |
| | Experimental | 61.7 ± 2.8* | 121.0±5.3** | 18.2 ± 0.4 | 0.51 |
| 49 | Control | 93.5 ± 4.7 | 283.2 ± 12.7 | 14.1 ± 0.8 | 0.33 |
| | Experimental | 84.2 ± 3.8 | 187.1±8.4** | 14.8 ± 0.7 | 0.45 |
| 56 | Control | 71.3 ± 3.5 | 169.8 ± 9.3 | 12.2 ± 0.5 | 0.42 |
| | Experimental | 54.1 ± 2.2** | 102.1±5.7** | 11.9 ± 0.6 | 0.53 |
| 63 | Control | 57.8 ± 2.8 | 118.0 ± 4.7 | 9.8 ± 0.3 | 0.49 |
| | Experimental | 49.8 ± 2.3* | 87.4 ± 3.8* | 10.7 ± 0.5 | 0.57 |

*Note* Here and in Table 2, the difference is probably relative to the control group: *—$p \leq 0.05$; **—$p \leq 0.01$

containing ones. Therefore, even on the background of a balanced diet for energy and protein, the process of juvenile fever formation is accompanied by stress in the AOP system, which is reflected in an increase in the content of lipoperoxidation products in the liver of the 49-day-old geese of the control group compared with the starting value (TBCAPout by 50.0%, TBCAPin—by 2.27 times), while $C_{AOA}$ decreased accordingly by 34.0% (Table 1). At the same time, the nature of the lipid content ($r = -0.895$, $\gamma = 0.002$) is monotonically declining during the experiment.

The physiological tension in the geese was characterized by a decrease in the level of activity of antioxidant enzymes: SOD (by 32.8%) and CAT (by 49.5%), vitamin E (by 26.5%), and β-carotene (by 19.6%) (Table 2). At the same time, the GPO activity from the 35th to the 56th day increased almost twice, and the vitamin A content was maintained at a presumably stable level. The second half of the experiment was accompanied by stabilization of the prooxidant–antioxidant equilibrium, which is confirmed by a decrease in the content of the products of the peroxide oxidation in the liver (TBCAPout at 38.2%, TBCAPin by 2.40 times), and the $C_{AOA}$ during this period increased by 48.5%. However, the activation of antioxidant enzymes of SOD and CAT at the end of the experiment was slowed down, and for 63-day-old geese, these figures on 26.9 and 44.8% inferior to their initial value.

Under the influence of bioflavonoids of oats in the geese of the experimental group, there was a decrease in the average level of TBCAPout by 15.9%, TBCAPin by 28.9%, while the $C_{AOA}$ increased by 15.3%. In the second part of the experiment, a significant activation of CAT and an increase in the content of vitamin E and β-carotene were established in geese liver of the experimental group. However, it is probably more important that activation of the AOP system during physiological stress is observed in 49-day-old geese. It was at this age that the geese of the exper-

**Table 2** The activity of enzymes, the content of vitamins in the liver, and the live weight of geese ($M \pm m$, $n = 6$)

| Age, day, $T$ | Group | Activity of enzymes | | | Vitamin content μg/g | | | Mass of geese ($M$), kg |
|---|---|---|---|---|---|---|---|---|
| | | SOD ($Y_1$), con. un./(min g) | CAT × $10^{-5}$ ($Y_2$), nкat/g | GPO × $10^4$ ($Y_3$), mcmoles/(min g) | A ($V_1$) | E ($V_2$) | β-carotene ($V_3$) | |
| 35 | Control | 11.32 ± 0.63 | 17.09 ± 0.72 | 3.52 ± 0.17 | 4.32 ±0.19 | 13.26 ± 0.53 | 10.86 ± 0.41 | 2.05 ± 0.11 |
| | Experimental | 11.47 ± 0.58 | 17.33 ± 0.81 | 3.52 ± 0.21 | 4.27 ±0.21 | 12.98 ± 0.59 | 10.71 ± 0.63 | 2.12 ± 0.08 |
| 42 | Control | 10.25 ± 0.47 | 11.95 ± 0.73 | 4.81 ± 0.21 | 4.58 ±0.14 | 12.93 ± 0.71 | 10.03 ± 0.42 | 2.46 ± 0.11 |
| | Experimental | 10.97 ± 0.61 | 13.43 ± 0.55 | 3.89* ± 0.19* | 4.02* ±0.17 | 14.29 ± 1.07 | 11.32 ± 0.57 | 2.68 ± 0.09 |
| 49 | Control | 7.61 ± 0.41 | 10.30 ± 0.27 | 5.73 ± 0.27 | 4.17 ±0.25 | 9.74 ± 0.21 | 9.32 ± 0.27 | 2.68 ± 0.14 |
| | Experimental | 8.04 ± 0.32 | 11.05 ± 0.42 | 6.98* ± 0.32 | 3.45* ±0.23 | 10.63* ± 0.17 | 9.97 ± 0.34 | 2.91 ± 0.10 |
| 56 | Control | 9.21 ± 0.43 | 8.63 ± 0.28 | 6.79 ± 0.27 | 3.98 ±0.15 | 10.79 ± 0.40 | 8.73 ± 0.24 | 2.95 ± 0.09 |
| | Experimental | 8.93 ± 0.51 | 10.65* ± 0.47 | 5.47* ± 0.29 | 4.72* ±0.21 | 12.07* ± 0.19 | 9.64* ± 0.17 | 3.36* ± 0.13 |
| 63 | Control | 8.28 ± 0.39 | 9.43 ± 0.37 | 4.37 ± 0.19 | 4.41 ±0.18 | 11.74 ± 0.37 | 9.78 ± 0.32 | 3.24 ± 0.12 |
| | Experimental | 8.65 ± 0.43 | 10.87* ± 0.48 | 5.73* ± 0.31 | 4.63 ±0.25 | 13.95** ± 0.43 | 10.77* ± 0.29 | 3.88* ± 0.13 |

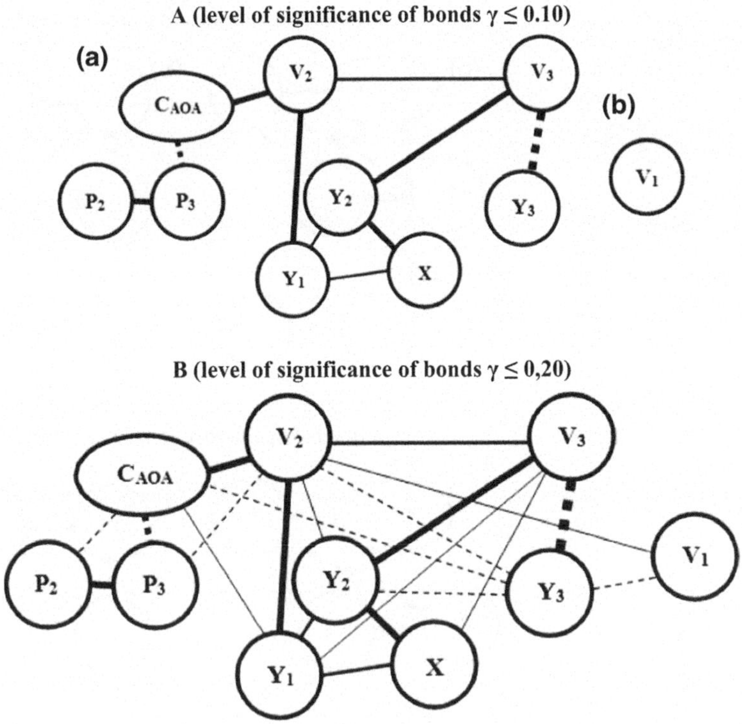


**Fig. 1** Clusters of the investigated parameters of the control group according to the density of their correlation bonds: – or - - - - at the level of significance $\gamma \leq 0.05$; – or - - - - at the significance level $\gamma \leq 0.01$; – or - - - - at the level of significance $\gamma \leq 0.10$; – or - - - - at the level of significance $\gamma \leq 0.20$

imental group exceeded the corresponding index of the control group by 36.4%. In addition, the effect of the extract is stabilization of the level of $C_{\mathrm{AOA}}$, which is confirmed in 1.86 times lower than the control group, the value of the coefficient of variation of the $C_{\mathrm{AOA}}$. A significantly higher mass of 63-day-old geese of the experimental group compared with the control one (19.8%) is an additional confirmation of the activation of the AOP geese system under the influence of oats extract.

A cluster analysis was performed to find out the nature and order of the integrated structure of the investigated parameters of the prooxidant–antioxidant equilibrium. It provides a more visible view of the dependence of the antioxidant activity of liver tissues, quantitatively determined by the $C_{\mathrm{AOA}}$, from the studied parameters. The clustering of these indices of the control group that is the basis of the quantity and density of the correlation links between them at the level of significance $\gamma \leq 0.10$ shows two clusters (Fig. 1A).

Within the base cluster, consisting of the nine indicators, which includes $C_{\mathrm{AOA}}$, a strong influence of vitamin E on the level of $C_{\mathrm{AOA}}$ ($r = 0.910$, $\gamma = 0.032$) attracts attention. All investigated enzymes and β-carotene exhibit a powerful but indirect

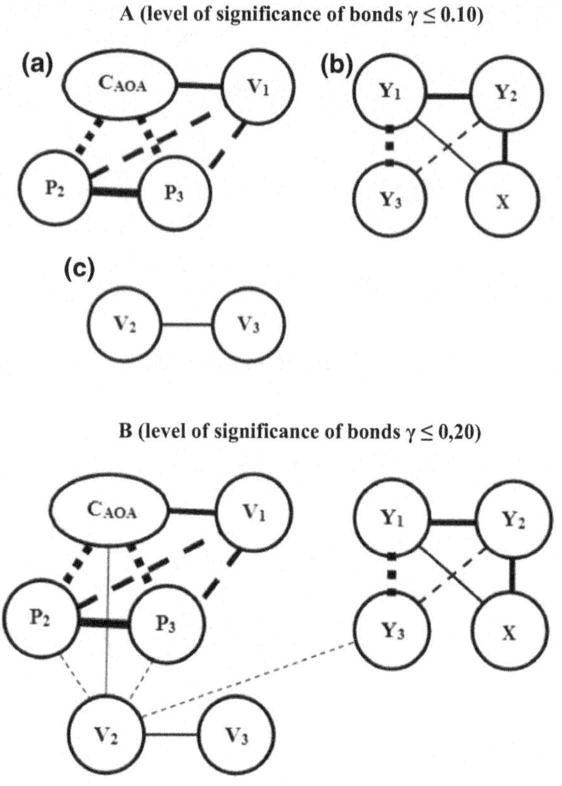


**Fig. 2** Clusters of the studied parameters of the experimental group on the density of their correlation bonds: – or - - - - at the level of significance $\gamma \leq 0.05$; – or - - - - at the significance level $\gamma \leq 0.01$; – or - - - - at the level of significance $\gamma \leq 0.10$; – or - - - - at the level of significance $\gamma \leq 0.20$

influence on $C_{\text{AOA}}$. Meanwhile, vitamin A remains in isolation without any reliable correlation. However, the correlation analysis of the investigated parameters of the control group at the level of significance $\gamma \leq 0.20$ (Fig. 1B) indicates a tendency to link vitamin A content with the GPO activity ($r = -0.709$, $\gamma = 0.180$) and the content vitamin E ($r = -0.688$, $\gamma = 0.199$) and, thus, the indirect ability to influence on $C_{\text{AOA}}$ of liver of geese.

The clusterization of the experimental group ($\gamma \leq 0.10$) showed three clusters (Fig. 2A). Specificity of the functioning of the AOP system of the liver geese of the experimental group is 25.0% lower compared to the control group of the consistency of the studied indicators.

Attention is drawn to an increase in the direct and indirect effects of vitamin A on $C_{\text{AOA}}$. Meanwhile, vitamin E has a distinct position, indicating a weakening of its effect on $C_{\text{AOA}}$. Thus, due to the action of components of the oats extract in the liver tissues of the geese, alternative mechanisms of antioxidant protection, which are characterized by increasing the influence of vitamin A on antioxidant activity ($C_{\text{AOA}}$), are activated. Correlation analysis of the investigated indicators of the experimental group at the level of significance $\gamma \leq 0.20$ (Fig. 2B) proves the presence of a tendency to link the content of vitamin E with the $C_{\text{AOA}}$ ($r = 0.703$,

$\gamma = 0.185$) and with the HPO activity ($r = 0.679$, $\gamma = 0.200$) and thus proves the existence of a single antioxidant system in which, depending on the nature of the influence of exogenous factors, various mechanisms are implemented with the predominant participation of individual components of the AOP system.

## 3 Conclusions

1. Introduction to the diet of geese extract of oat in the pre-slaughter period stabilizes the prooxidant–antioxidant balance in the tissues of their liver, and during physiological stress causes an increase in the antioxidant activity of these tissues, which is accompanied by a significant activation of CAT and an increase in the content of vitamin E and β-carotene in the liver geese.
2. Under the influence of oats extract, alternative mechanisms of antioxidant protection are included, which are characterized by 25.0% lower level of consistency of components of the AOP system.
3. The increase in the mass of the 63-day-old geese of the experimental group compared with the control of almost 20.0% is an additional confirmation of activation of the AOP geese system under the influence of oats extract.

## References

1. Ivko, I.I., Riabinina, O.V., Melnyk, O.V.: Ways to improve the efficiency of native poultry. Efficient Poult. Farming **11**(71), 33–40 (2010) (in Ukrainian)
2. Karpov, V.S.: Breeding geese. Farm **18**, 22 (2011) (in Russian)
3. Khvostyk, V.P.: Geese, geese! Ha… Ha… Ha…/. Agrarian **22**, 20–22 (2014) (in Ukrainian)
4. Melnik, V.A.: Production of waterfowl in the world and in Ukraine [Electronic resource]. For poultry farmers. http://ptitcevod.ru/produkciyapticevodstva/proizvodstvoprodukciivodoplavayushhej-pticy-v-mire-i-v-ukraine.html (2013) (in Russian)
5. Roztalnyy, A., Kuipers, A.: Livestock farming in Central and Easteru Europe and Central Asia. Cattle husbandry in Easteru Europe and China. EAAP Sci. Ser. **135**, 13–36 (2014). https://doi.org/10.3920/978-90-8686-785-1
6. Ishchenko, YuB: Analysis of poultry production in Ukraine and forecasts up to 2020. Modern Poult. Breed. **4**(137), 4–8 (2014) (in Ukrainian)
7. Kyryliuk, O.F.: Development of poultry market. Bullet. Agrarian Sci. **8**(12), 80–82 (2012) (in Ukrainian)
8. Tereshchenko, O.V., Katerinich, O.O., Rozhkovsky, O.V.: Modern trends of poultry development in Ukraine: the state and prospects of scientific support of the industry. Effective Poult. Farming **11**(83), 7–12 (2011) (in Ukrainian)
9. Khvostyk, V.P.: Perspective directions of goose breeding. Mod. Agrarian Technol. **8**, 62–69 (2013) (in Ukrainian)
10. Tsekhmistrenko, S.I., Tsekhmistrenko, O.S.: Biochemistry of meat and meat products. Bila Tserkva (2014) (in Ukrainian)
11. Suray, P., Fisinin, V.I.: Natural antioxidants in embryogenesis of chickens and protection from stress in postnatal development (review). Agric. Biol. **2**, 3–18 (2013) (in Russian)

12. Keriene, I., Mankeviciene, A., Bliznikas, S., Jablonskyte-Rasce, D., Maiksteniene, S., Cesnulevciene, R.: Biologically active phenolic compounds in buckwheat, oats and winter spelt wheat. Zemdirbyste-Agric. **3**(102), 289–296 (2015). https://doi.org/10.13080/z-a.2015.102.037
13. Menshikova, E.B., Lankin, V.Z., Kandalintseva, N.V.: Phenolic antioxidants in biology and medicine. In: Structure, Properties, Mechanisms of Action. LAP LAMBERT Academic Publishing, Saarbrücken (2012) (in Russian)
14. Gangopadhyay, N., Hossain, M.B., Rai, D.K., Brunton, N.P.: A review of extraction and analysis of bioactives in oat and barley and scope for use of novel food processing technologies. Molecules **20**, 10884–10909 (2015). https://doi.org/10.3390/molecules200610884
15. Burtseva, O.V.: Quantitative determination of phenolic compounds Avena sativa L. Ukrainian J. Clin. Lab. Med. **8**(44), 225–228 (2013) (in Ukrainian)
16. Recommendations on the Standardization of Feeding of Poultry. Birky: Poultry Institute UAAS, Kharkiv (2005) (in Ukrainian)
17. Ionov, I.A., Shapovalov, S.O., Rudenko, E.V., Dolgaya, M.N., Ahtyrskij, A.V., Zozulya, Y.A., et al.: Criteria and Methods for Controlling Metabolism in the Organism of Animals and Birds. Institute of animal husbandry NAAS, Kharkiv (2011) (in Russian)
18. Zdorovtseva, L.M., Khromishev, V.O., Danchenko, O.O.: Geese fatty acid composition of brain and heart lipids in hypo-and hyperoxia. Ukrainian J. Ecol. **2**(3) (2012). http://dx.doi.org/10.15421/20122_30 (in Ukrainian)
19. Abreu, I.A., Cabelli, D.E.: Superoxide dismutases—a review of the metal-associated mechanistic variations. Biochim. Biophys. Acta **1804**(2), 263–274 (2010). https://doi.org/10.1016/j.bbapap.2009.11.005
20. Alptekin, O., Tuekel, S., Yildirim, D., Alagoez, D.: Immobilization of catalase onto Eupergit C and its characterization. J. Mol. Catal. **64**(3–4), 177–183 (2010). https://doi.org/10.1016/j.molcatb.2009.09.010
21. Lubos, E., Loscalzo, J, Handy, D.E.: Glutathione peroxidase-1 in health and disease: from molecular mechanisms to therapeutic opportunities. Antioxid Redox Signal. **15**(7), 1957–1997 (2011). https://doi.org/10.1089/ars.2010.3586

# Screening of Agricultural Raw Materials and Long-Term Storage Products to Identify Bacillary Contaminants

**Olena Danylova, Marina Serdyuk, Liudmyla Pylypenko, Victor Pelykh, Inna Lopotan and Antonina Iegorova**

## 1 Introduction and Purpose of the Study

In the production of food from agricultural raw materials, there is a risk of their contamination by microorganisms of the epiphytic microflora, including pathogenic microorganisms which can get on raw materials from soil, water, air, and also from equipment, inventory, and hands of workers during technological processing. Control of feedstock, storage, and types of processing is of particular importance in the manufacturing of food products of plant origin in connection with their diverse biochemical composition, depending on various factors [1, 2]. Microorganisms can remain viable for a long time, and among them may be pathogenic toxin-forming species that are dangerous to human health or microorganisms that alter the organoleptic and, most importantly, nutritional value of the products. The content of certain microorganisms is regulated by industry and international standards, EU regulations in accordance with the Codex Alimentarius and HACCP, which are based on principles and procedures developed to ensure food safety [3–7]. However, this control does not always provide sufficient information for the manufacturers in order to decide on the type of storage or processing of raw materials. *Bacillaceae* family of bacteria includes the microorganisms of the genera *Bacillus*, *Paenibacillus*, *Geobacillus,* and others, among which there are numerous species that can cause toxicoses. This is due to the production of active enzymes (hemolysins, phospholipase, and protease), emetic and diarrheal toxins, at that the toxin-producing ability appeared in previously considered

O. Danylova (✉) · L. Pylypenko · I. Lopotan · A. Iegorova
Odessa National Academy of Food Technologies, Kanatna St., 112, Odesa 65039, Ukraine
e-mail: olenaivdan@gmail.com

M. Serdyuk
Tavria State Agrotechnological University, B. Khmelnitsky Av., 18, Melitopol 72312, Ukraine

V. Pelykh
State Higher Educational Institution, «Kherson State Agricultural University», Stretenskaya St., 23, Kherson 73006, Ukraine

© Springer Nature Switzerland AG 2019

V. Nadykto (ed.), *Modern Development Paths of Agricultural Production*,
https://doi.org/10.1007/978-3-030-14918-5_63

saprophytic species [2, 8–10]. Besides that, more and more attention of producers and consumers of products recently have been attracted to the bacteria of the *Bacillus cereus* group, since they are capable of causing outbreaks of food intoxication. And due to the ubiquity in the environment and the ability to withstand the heating processes, spore-forming bacteria can be found in food, which leads to its spoilage if spores are present in sufficient quantities and in conditions favorable for germination. The microbiological hazard associated with the spoilage of the food product by saprophytic microorganisms and the microbiological danger of the product contaminated with pathogenic microorganisms should be separated. This is due to the fact that most pathogenic microorganisms found in food do not cause damage to them, i.e., a dangerous product can have good organoleptic properties. At the same time, microorganisms that cause food spoilage, not being pathogenic, can produce toxins. That is why it is advisable to study not only the amount and composition of microbiota of raw materials in groups, but also characterize it according to morphotypes and species.

The purpose of this study was to compare the group composition of the epiphytic microbiota and heat-resistant spore-forming microorganisms of plant-grown industrial agricultural raw materials from various regions of Ukraine to predict risks and ensure the safety of the products of its processing.

## 2 Research Methods

For the study, we used medium samples of plant-grown industrial agricultural raw materials from various climatic and geographical regions of Ukraine—Poltava, Odessa, Kherson, Zaporozhye regions. During the research, microbiological, physiological, biochemical, and statistical methods were used [11–15]. The morphological, tinctorial, cultural, and biochemical properties of the isolated cultures of microorganisms were studied by conventional methods on the following features:

- the nature of growth in dense and liquid nutrient media: meat-peptide broth (MPB), meat-peptide agar (MPA), wort agar (WA), MPA with the addition of starch, nitrates, etc.);
- saccharolytic properties (method: seeding in semisolid Giss media);
- proteolytic properties (method: sowing in milk, as well as meat-peptone gelatin (MPG) and pricking into a column of gelatin);
- the possibility of splitting proteins with the excretion of indole (method: control of test paper impregnated with oxalic acid solution);
- catalase production (method: reaction with hydrogen peroxide);
- production of acetoin (method: reaction with $\alpha$-naphthol).

Analogically to studies of epiphytes in raw materials, a study was made of their group composition and of individual, isolated pure cultures: a set of morphological, tinctorial, cultural, and biochemical properties for the products of its processing—vacuum packaged and prepared for canning. To take into account the total number of bacteria, molds, and yeast, the washings were not heated. To isolate heat-resistant aerobic and facultative anaerobic bacteria, the average sample was heated for 20 min at 80 °C, and after cooling to room temperature, plated for meat-peptone agar (MPA), and for obligate anaerobes—in Kitt-Tarozzi media, with further incubation of crops at 30 °C for 24–48 h.

Morphological signs of colonies and bacterial cells were examined after 24–48 h of cultivation on dense and liquid nutrient media (MPB, MPA) and 5 days on WA. Cultural signs of colonies were determined by such indicators: shape, size, surface, profile, color, transparency, edge shape, texture, and consistency. In the broth, the nature of the sediment, turbidity, the presence of the film, and its structure were noted. The ratio to the acidity of the medium was established by the presence of growth of microorganisms at pH values of 3–6 after 24–48 h of cultivation.

The determination of amylolytic activity was carried out by incubating bacterial cultures on MPA medium supplemented with 0.2% starch. The presence of discolored areas in the growth zones after treatment with Lugol's solution indicated a positive test result. The ability to gelatin hydrolyze was determined by the presence of dilution after incubation of the test microorganisms in gelatin tubes at 36 °C for 7 days and cooling to 4 °C. The presence of lecithinase activity was tested on yolk and salt agar. The Voges–Proskauer reaction was carried out by adding 0.6 $cm^3$ of a 5% α-naphthol solution in absolute ethanol to 1 $cm^3$ of the culture of the bacteria in the broth. The mixture was thoroughly mixed, and 0.2 $cm^3$ of a 40% aqueous solution of KOH was added, mixed again, and incubated in an inclined position for better oxygen access. Test tubes were checked after 15 and 60 min. Coloring the surface of the medium in intense red color indicated the presence of acetoin.

Fatty acid analysis was carried out by comparing strains profiles of bacteria with known standards. The automatic system of identification of microorganisms MIDI Sherlock (MIDI, USA) was used on the basis of gas chromatograph Agilent +7890 (Agilent Technologies, USA) with a flame ionization detector and a ULTRA 2 capillary column (25 m × 0.2 mm × 0.33 μm). Hydrogen was used as the carrier gas. Injection of 2 μl of the sample was carried out by distributing the carrier gas stream (split mode) to 40: 1. The temperature of the injector was set at exactly 250 °C, and the column temperature varied from 170 to 270 °C at a rate of 5 °C/min. The concentration of each fatty acid was expressed as a percentage of total fatty acids. Fatty acids, whose concentration was less than 0.2%, were not taken into account.

According to a set of studied features, the species belonging to isolated cultures of microorganisms was determined.

## 3 The Main Result of the Study

In agricultural products, the number of microorganisms and their type depends on the species, as well as on the conditions of growth, collection, storage, and processing of feedstock, so the safety of the product depends on a large number of factors. To ensure the production of safe products with a specified shelf life, it is necessary to determine

**Table 1** Group composition of microorganisms of plant raw materials from various regions ($n = 3, p \leq 0.05$)

| Raw material | Microorganisms (CFU/g) | | |
|---|---|---|---|
| | MAFAnM | Mold fungi | Yeast |
| *Poltava region* | | | |
| Carrot | $(8.0–8.4) \times 10^6$ | 58–70 | $(5.0–5.1) \times 10^4$ |
| Courgettes | $(0.2–1.0) \times 10^5$ | 47–52 | $(2.2–2.5) \times 10^3$ |
| Tomatoes | $(2.4–3.5) \times 10^5$ | 43–49 | $(2.7–3.3) \times 10^3$ |
| Apples | $(3.8–4.5) \times 10^4$ | 27–34 | $(2.6–3.0) \times 10^2$ |
| *Odesa region* | | | |
| Carrot | $(7.6–7.9) \times 10^6$ | 52–63 | $(4.2–4.5) \times 10^4$ |
| Eggplant | $(4.0–4.2) \times 10^5$ | 44–48 | $(9.0–9.3) \times 10^2$ |
| Tomatoes | $(4.5–4.9) \times 10^4$ | 41–46 | $(2.1–2.8) \times 10^3$ |
| Pepper mellow | $(6.5–6.9) \times 10^4$ | 43–45 | $(2.6–2.7) \times 10^2$ |
| Courgettes | $(0.7–1.0) \times 10^5$ | 47–59 | $(2.0–2.4) \times 10^3$ |
| Apples | $(3.3–4.0) \times 10^4$ | 32–39 | $(1.9–2.8) \times 10^2$ |
| Strawberry | $(4.2–5.1) \times 10^5$ | 43–50 | $(3.0–3.9) \times 10^3$ |
| *Kherson region* | | | |
| Carrot | $(2.0–3.0) \times 10^6$ | 39–49 | $(3.1–3.4) \times 10^4$ |
| Eggplant | $(2.8–3.1) \times 10^5$ | 37–48 | $(1.0–1.2) \times 10^3$ |
| Tomatoes | $(4.7–5.1) \times 10^4$ | 43–48 | $(2.5–3.1) \times 10^3$ |
| Pepper mellow | $(7.1–7.4) \times 10^4$ | 39–47 | $(7.9–8.2) \times 10^2$ |
| Courgettes | $(0.6–1.1) \times 10^5$ | 41–49 | $(2.6–3.0) \times 10^3$ |
| Apples | $(3.9–4.2) \times 10^4$ | 31–46 | $(2.2–3.2) \times 10^2$ |
| Strawberry | $(4.6–5.3) \times 10^5$ | 46–54 | $(3.6–4.2) \times 10^3$ |
| *Zaporizhzhia region* | | | |
| Carrot | $(7.5–9.1) \times 10^5$ | 46–51 | $(4.8–5.3) \times 10^4$ |
| Courgettes | $(0.4–1.0) \times 10^5$ | 47–50 | $(2.3–2.9) \times 10^3$ |
| Cherries | $(4.8–5.6) \times 10^4$ | 44–56 | $(2.9–3.8) \times 10^3$ |
| Apples | $(4.1–4.8) \times 10^4$ | 39–45 | $(2.4–2.9) \times 10^2$ |

by microbiological tests the absence of a microbiota that causes food poisoning and specific damage. The study of the group composition of the contaminants of agricultural raw materials was limited to the consideration of bacteria, mycelia, and non-miraculous fungi (Table 1), since exactly heat-resistant bacilli and clostridia and sporiferous microbiota can persist after heat treatment in the preservation and drying processes, and also under anaerobic conditions in a vacuum package.

The generally accepted microbiological criteria for product safety in many countries is the presence or absence of microorganisms and their amount, including

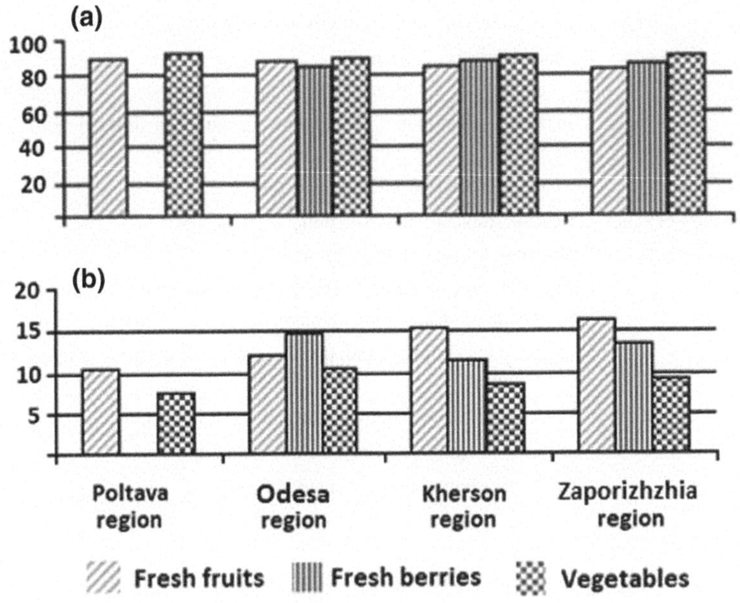


**Fig. 1** The composition of the detected sporiferous microbiota in the samples, %

pathogenic species, as well as the content of toxins or metabolites per unit of mass, volume, and area of product or batch [2, 7–9].

The study showed the presence of high contamination of all types of raw materials, especially from the Poltava and Zaporozhye regions. A significant amount of yeast and fungi was found in samples of raw materials from all the regions studied, especially on berries. The greatest amount of microbiota was on carrots, which is associated with growing conditions and additional pollution by soil microorganisms. Contribution of sporiferous microbiota to total microbial contamination, established by the MAFAnM indicator, had insignificant differences in the studied regions of Ukraine: 48.5, 44.2, 46.3, and 52.3% in Poltava, Odessa, Kherson, Zaporozhye regions, respectively.

Analysis of the composition of the spore-bearing bacterial microbiota (Fig. 1) indicates the prevalence of bacilli—up to 92% (a). Clostridium (b) is present in a much smaller amount (from 7.7 to 16%).

Among the colonies most frequently found on MPA, cultures were identified that comprised nine morphotypes or morphogroups differing in cultural characteristics (Table 2) and a number of morphological properties.

It should be noted that most indicators differ slightly: all the colonies are gray or whitish, and the relief also differs slightly. Representatives of the morphotype I on the MPA formed small grayish shiny colonies, II—slightly blurred, both morphotypes cause liquefaction of gelatin. Two cultures from the third morphogenous group were distinguished by smoother, whitish shiny colonies on MPA growing into the substrate,

**Table 2** Characteristics of the main types of colonies of microorganisms on MPA

| No. | Growth rate | Characteristics of colonies |
|---|---|---|
| I | +++ | Smooth, grayish-white, shiny, growing into a substrate, d up to 1.5–2 mm |
| II | ++ | Small, grayish, blurred, d up to 2 mm |
| III | +++ | Smooth, grayish-white, d 2 mm |
| IV | ++ | Wax-like, grayish, d up to 2 mm |
| V | + | Rounded, grayish-white, with pasty consistency, matte surface, slightly wavy edge, d up to 2 mm |
| VI | ++ | Folded macro relief, round, thick, convex, solid, shiny, mucous, d 2–2.5 mm |
| VII | + | Thin, round, beige, slightly blurry, d up to 1.5–2 mm |
| VIII | ++ | Grayish, shiny, large, slimy, d more than 3 mm |
| IX | ++ | Thin, spreading over the surface, d 2–3 mm |

formation of a thin film on the MPB, and clouding of the medium, peptonization of the milk without clotting, crater-like gelatin dilution during stabbing into the column, lack of amylase and tyrosinase activity, and nitrates reduction ability. Therefore, these cultures, presumably, are referred to the species *Bacillus pumilis*. Microorganisms of morphotype IV cleaved maltose and did not cleave mannitol. They formed spores in 16–18 h. In the early stages of growth on glucose, agar cells contained fat globules. All cultures showed lecithinase activity on the yolk agar, formed acetoin and characteristic ruby colonies on salt agar with 2,3,5-triphenyltetrazolium chloride, as well as indole, which allowed them to be identified as *B. cereus*. The totality of the signs of group V bacilli colony (pasty consistency, matte surface, slightly wavy margin) indicates that this group can be formed by strains of *B. thuringiensis*. On the samples tested, *B. thuringiensis* was found to be a few, except for eggplant and pepper, where they prevailed. An important reference point in the identification of group VI bacilli was the size, folded macrorelief of the colony, which distinguished them from the species described above. The colonies of this group on the MPA are round, thick, convex, whole, shiny, and slimy. With the age of the culture, the substrate turns brown. On the MPB medium, the colonies are scarce in the form of turbidity; on gelatinous media, they form a grayish surface coating. When planting with a stab in the column, a dilution is observed in the form of a crater. These bacilli cause not the folding of milk, but its patronization. In old cultures, fat is found on the MPA during growth. Representatives of the sixth group were identified as *B. megaterium* also presumably since the reactions of tyrosine cleavage and reduction in nitrates varied depending on the age of the culture. All isolates of this morph type have a pronounced mobility in the diurnal culture.

The fight against spore-forming bacteria is one of the main practical tasks in the food industry, including canning industry, in the process of storage and processing of agricultural and food products. Therefore, the total number of microorganisms in the

**Table 3** Total number of microorganisms at different types of processing ($n = 3$, $p \leq 0.05$)

| Raw material | MAFAnM (CFU/g) | |
|---|---|---|
| | After vacuum packaging | Before sterilization |
| Carrot | $(5.6–6.2) \times 10^3$ | $(1.5–1.9) \times 10^4$ |
| Courgettes | $(0.2–1.0) \times 10^3$ | $(0.9–1.2) \times 10^3$ |
| Eggplant | $(1.2–1.9) \times 10^3$ | $(1.4–2.1) \times 10^3$ |
| Pepper mellow | $(1.9–2.2) \times 10^3$ | $(2.7–3.4) \times 10^3$ |
| Tomatoes | $(2.4–3.5) \times 10^2$ | $(1.9–2.7) \times 10^2$ |

samples of products after vacuum packing and storage for 7 days and in the samples of canned food prepared for the sterilization process was determined (Table 3).

The characteristics of acid-forming bacilli, taking into account the basic morphological and biochemical features (the presence of spore that does not modify the vegetative cell, the utilization of sugars and alcohols (arabinose, xylose, mannitol) with acid formation without gas) is presented in Table 4. A total of six morphotypes of acid-forming bacilli have been identified, having the following general properties: medium-sized sticks (0.6–0.8) × (1.5–3.0) – (1.0–1.2) × (3.5–5.0) μm with elliptical spores located centrally, Gram-positive.

The description of the isolated bacilli is given in Tables 4 and 5 on the basis of morphophysiological and biochemical properties.

In Table 5, Gram-positive mobile rods are united, spores of which in diameter exceed the thickness of cells and are located subterminally or terminally. They form catalase, but can grow on MPA under anaerobic conditions, and also hydrolyze starch and casein, reduce nitrates to nitrites, and do not form indole, lecithinase, and tyrosinase. In contrast to the bacilli described in Table 4, when cultivated on media with arabinose, xylose, and mannitol, they form a gas along with the acid. The detection of bacilli of VII–IX morphotypes in raw materials indicates the necessity of their control in products after heat treatment (particularly preserves, canned goods) and when stored in vacuum packaging, as these bacilli are a potential cause of bomb damage. In addition, it is necessary to comply with the storage conditions, especially the temperature and composition of the gas mixture in vacuum packaging. Acid- and gas-forming bacilli on the studied raw materials are represented by a relatively small amount—from 2–3% on tomatoes to 9–14% on carrots from the total number of isolated bacilli.

It should be noted that the morph physiological, cultural, and biochemical properties of the studied cultures were not always convincing. On different media, some R-form colonies passed into S-form colonies, which made it difficult to perform primary identification on cultural features. Because of this, it was not possible to introduce the characteristics of some crops into tables and clearly determine the proportion of isolates studied in the total number of bacilli found on the raw material examined.

**Table 4** Characteristics of the acid-forming bacilli of vegetable raw material

| Indicators | Bacilli properties by morphotypes | | | | | |
|---|---|---|---|---|---|---|
| | I | II | III | IV | V | VI |
| Cell sizes (μm) | (0.7–0.8) × (2.0–3.0) | (0.6–0.8) × (1.5–2.0) | (0.6–0.7) × (2.0–2.5) | (1.0–1.2) × (3.0–4.0) | (1.0–1.2) × (3.0–4.0) | (1.2–1.5) × (2.5–3.0) |
| Growth on MPA in anaerobic conditions | – | + | – | + | + | – |
| Hydrolysis of starch | + | + | – | + | + | + |
| Reduction of nitrates | + | + | – | + | + | + |
| The formation of acetoin | + | + | + | + | + | – |
| Tyrosine decarboxylation | – | – | – | + | +3 Cultures ±1 Culture | + |
| The formation of acid from arabinose, xylose, mannitol | + | Acid formation + Gas formation ± | + | – | – | +1 Culture −1 Culture |
| Prospective species | *B. subtilis* | *B. licheniformis* | *B. pumilis* | *B. cereus* | *B. thuringiensis* | *B. megaterium* |

**Table 5** Characteristics of microorganisms with acid-forming and gas-forming properties

| Indicators | Bacilli properties by morphotypes | | |
|---|---|---|---|
| | VII | VIII | IX |
| Cell sizes (μm) | (0.5–0.6) × (3.0–4.0) | (0.6–0.7) × (2.0–3.5) | (0.7–1.0) × (2.0–3.0) |
| Hydrolysis of casein | – | + | + |
| Gelatin liquefying | + | ± (Weak reaction) | ± |
| The formation of acetoin | – | + | – |
| Tyrosine decarboxylation | – | – | – |
| Prospective species | *P. macerans* | *P. polymyxa* | *B. circulans* |

**Table 6** Specific composition of bacillary contaminants, isolated from raw material and foods of Ukrainian food industry of Poltava region

| Type of microorganisms | Amount of strains | Percentage of identified bacillary contaminants of total microbial contamination |
|---|---|---|
| *B. subtilis-licheniformis* | 54 | 37.76 |
| *B. cereus* | 44 | 30.77 |
| *B. thuringiensis* | 14 | 9.79 |
| *B. pumilis* | 11 | 7.69 |
| *P. polymyxa* | 7 | 4.90 |
| *P. macerans* | 6 | 4.20 |
| *B. megaterium* | 5 | 3.50 |
| *B. circulans* | 2 | 1.40 |

Species belonging to the strains of *Bacillaceae* family isolated from vegetable raw materials were confirmed by carrying out fatty acid analysis, comparing them with known standards. The predominance of branched fatty acids in the fatty acid profile is a characteristic feature of bacteria of the order of *Bacillales* [12, 13]. According to the published data and the results of our studies, the content of branched fatty acids in bacilli was from 54 to 85% of the total fatty acid pool of the cell, including both saturated and unsaturated acids, with predominance of iso-C15: 0 and anti-C15: 0. They also have a high content of anti-C17: 0 and iso-C17: 0 fatty acids. The studied fatty acid composition [15] made possible to refine the species identification of microorganisms in addition to the results obtained by classical methods.

The microbiota of agricultural raw materials is diverse; however, mycelial and non-micellic fungi in thermally processed products are less danger to the consumer than spore-forming bacteria.

Among the bacillus contaminants of the samples studied, the *subtilis–licheniformis* group is the most numerous: about 38% of the total number of bacilli. The amount of *B. pumilis* is 6.6–7.7%, the gas-forming *P. polymyxa* and *P. macerans*—the causative agents of bomb damage—from 4.4 to 4.9% and 3.4 to 4.2%, respectively. The percentage of microorganisms of the group *B. cereus* (*B. cereus* and *B. thuringiensis*) is from 28 to 30% and from 10 to 13%, respectively, in the studied regions of Ukraine (Tables 6, 7, 8 and 9).

The obtained results allow to estimate an essential component of the epiphytic microbiota of plant raw materials of various regions of Ukraine, which forms the so-called residual microbiota of the products of its processing. This is of particular importance for making adjustments to technological processes, both during primary processing to obtain semi-finished products and storage and to ensure compliance of finished products with safety requirements.

**Table 7** Specific composition of bacillary contaminants, isolated from raw material and foods of Ukrainian food industry of Odesa region

| Type of microorganisms | Amount of strains | Percentage of identified bacillary contaminants of total microbial contamination |
|---|---|---|
| *B. subtilis-licheniformis* | 51 | 37.50 |
| *B. cereus* | 42 | 30.88 |
| *B. thuringiensis* | 16 | 11.76 |
| *B. pumilis* | 9 | 6.62 |
| *P. polymyxa* | 6 | 4.41 |
| *P. macerans* | 5 | 3.68 |
| *B. megaterium* | 5 | 3.68 |
| *B. circulans* | 2 | 1.47 |

**Table 8** Specific composition of bacillary contaminants, isolated from raw material and foods of Ukrainian food industry of Kherson region

| Type of microorganisms | Amount of strains | Percentage of identified bacillary contaminants of total microbial contamination |
|---|---|---|
| *B. subtilis-licheniformis* | 56 | 38.36 |
| *B. cereus* | 43 | 29.45 |
| *B. thuringiensis* | 17 | 11.64 |
| *B. pumilis* | 10 | 6.85 |
| *P. polymyxa* | 7 | 4.79 |
| *P. macerans* | 5 | 3.42 |
| *B. megaterium* | 6 | 4.11 |
| *B. circulans* | 2 | 1.37 |

Since contamination by microorganisms of *B. cereus* group is of particular concern due to the possibility of food intoxication, a comparison was made of their presence in samples of vegetables, fruits, and berries (Table 10).

Comparing the obtained results for agricultural raw materials of different regions of Ukraine, it is possible to note practically the same tendencies of detection of *B. cereus*. Its number turned out to be higher in vegetables: 59.1, 56.3, 63.0, 61.8% for the regions of Poltava, Odesa, Kherson, and Zaporizhzhia regions, slightly lower in berries (from 31 to 36%) and the smallest—in fruits (from 17 to 27%).

The average value of contamination by *B. cereus* for samples prepared for processing (before sterilization) is 36%, after vacuum packing and storage for 7 days—about 27%.

**Table 9** Specific composition of bacillary contaminants, isolated from raw material and foods of Ukrainian food industry of Zaporizhzhia region

| Type of microorganisms | Amount of strains | Percentage of identified bacillary contaminants of total microbial contamination |
|---|---|---|
| *B. subtilis-licheniformis* | 57 | 38.26 |
| *B. cereus* | 42 | 28.19 |
| *B. thuringiensis* | 19 | 12.75 |
| *B. pumilis* | 11 | 7.38 |
| *P. polymyxa* | 7 | 4.70 |
| *P. macerans* | 6 | 4.03 |
| *B. megaterium* | 5 | 3.36 |
| *B. circulans* | 2 | 1.34 |

**Table 10** Comparison of the quantities of *B. cereus* bacteria in samples of vegetables, fruits, and berries from different regions of Ukraine

| Raw material | Number of samples | Number of samples containing *B. cereus* | Percentage of samples containing *B. cereus* |
|---|---|---|---|
| *Poltava region* | | | |
| Fresh fruits | 7 | 3 | 42.9 |
| Vegetables | 22 | 13 | 59.1 |
| *Odesa region* | | | |
| Fresh fruits | 12 | 2 | 16.7 |
| Fresh berries | 16 | 5 | 31.3 |
| Vegetables | 48 | 27 | 56.3 |
| *Kherson region* | | | |
| Fresh fruits | 8 | 2 | 25.0 |
| Fresh berries | 14 | 5 | 35.7 |
| Vegetables | 46 | 29 | 63.0 |
| *Zaporizhzhia region* | | | |
| Fresh fruits | 11 | 3 | 27.3 |
| Fresh berries | 9 | 3 | 33.3 |
| Vegetables | 34 | 21 | 61.8 |

# 4 Conclusions

For the first time, the epiphytic microbiota of agricultural raw materials from various regions of Ukraine (Poltava, Odesa, Kherson and Zaporizhzhia regions) has been studied.

It was established that the MAFAnM index is 1–3 orders of magnitude higher than the contamination rate of the studied samples by mycelial and non-micellar fungi.

The dominance of bacilli among heat-resistant spore-bearing microbiota is revealed—up to 92% of the total amount on vegetables and up to 85–87% on berries and fruits.

In the regions of Ukraine, there is an insignificant (by 4.8–8.1%) increase in the number of heat-resistant spore-bearing microorganisms from west to east, which is likely due to the climatic features of the regions and the microbial contamination of their soils.

The carried out researches are actual for forecasting of risks and maintenance of safety of products of processing of industrially grown kinds of agricultural raw materials.

# References


1. Patyka, V.P.: Phytopathogenic bacteria in contemporary agriculture. Mikrobiol. Z. **78**(6), 71–83 (2016)
2. Zhang, Z., Feng, L., Xu, H., et al.: Detection of viable enterotoxin-producing *Bacillus cereus* and analysis of toxigenicity from ready-to-eat foods and infant formula milk powder by multiplex PCR. J. Dairy Sci. **99**(2), 1047–1055 (2016)
3. Codex Alimentarius: http://www.fao.org/docrep/012/i1400e/i1400e.pdf (2010)
4. EU Regulation № 852/2004: https://www.fsvps.ru/fsvps-docs/ru/laws/eu/852-2004.pdf (2004)
5. ДСТУ ISO 7932:2007 Мікробіологія харчових продуктів і кормів для тварин. Горизонтальний метод визначення кількості ймовірно *Bacillus cereus*. Техніка підрахунку за температури 30 °С (2007)
6. Standard CAC/RCP 75-2015: Code of hygienic practice for low-moisture foods (2015)
7. Пилипенко Ю.Д., Мазуренко І.К., Пилипенко І.В., Пилипенко Л.М.: Державні нормативні документи на сировину, напівфабрикати, матеріали та консервовану продукцію. Показники безпечності та якості (Методичні вказівки. Видання офіційне). – К.: Мінагрополітики, 114 с. (2009)
8. Kim, J.-B., Kim, J.-M., Kim, S.-Y., et al.: Emetic toxin producing *B. cereus* Korean isolates contains genes encoding diarrhea related enterotoxins. J. Food Protection **73**(7), 1219–1224 (2010)
9. Forgani, F., Kim, J.B., Oh, D.H.: Profiling of emetic toxin- and enterotoxin-producing *Bacillus cereus*, isolated from food, environmental, and clinical samples by multiplex PCR. Int. J. Food Microbiol. **124**, 224–230 (2008)
10. Biesta-Peters, E.G., Dissel, S., Reij, M.W., et al.: Characterization and exposure assessment of emetic *Bacillus cereus* and cereulide production in food products on the Dutch market. J. Food Protect. **79**(2), 230–238 (2016)
11. Garrity, G.M.: Bergey's Manual of Systematic Bacteriology. The Proteobacteric. Part A., vol. 2, 2nd edn., pp. 317–321. Bergey's Manual Trust Department of Microbiology and Molecular Genetics Michigan State University (2005)

12. Ivanytsia, V.O., Gorshkova, O.G., Korotaeva, N.V., et al.: Fatty acid composition of lipids of strain *Bacillus sp*. O3-5 isolated from oilcontaminated soil of the Zmiiny Island. Microbiol. Biotechnol. **4**(32), 28–36 (2015)
13. Pylypenko, I.V., Paulina, Y.B., Pylypenko, L.N., Yamborko, G.V.: Composition of microbal contaminants of vegetable raw material. Microbiol. Biotechnol. **3**(31), 83–95 (2015)
14. Harley, J.P., Prescott, L.M.: Laboratory Exercises in Microbiology, 4th edn., 466 p. The McGraw Hill Companies, NY (2002)
15. Yamborko, G.V., Ostapchuk, A.M., Sergieieva, Zh, Y., Pylypenko, L.N., Pylypenko I.V.: Chemotaxonomic features and plasmid profiles of aerobic and facultative anaerobic spore-forming bacteria from vegetables. Microbiol. Biotechnol. 1(37), C. 56–72 (2017)

# Development of Formulation and Quality Assessment of Fast-Cooking Grain Composition for Pregnant Women

**Nadya Dzyuba, Liubov Telezhenko, Maryana Kashkano and Oksana Maksymets**

## 1 Introduction

The modern nutrition structure of the population of Ukraine does not correspond to the principles of rational nutrition [1]. Improving the nutritional structure primarily involves increasing the production of food products through the improvement of existing and development of new technologies for functional food products.

Balanced nutrition for pregnant women is one of the most important problems nowadays. Optimal nutrition of a woman during pregnancy and lactation is a prerequisite for giving birth to a healthy baby and its effective growth and development. The full development of the baby in the womb, the construction of new cells, requires a large amount of protein in the diet of a pregnant woman. The recommended ratio of animal and plant proteins is 3:2. Lack of protein can lead to a violation of the synthesis of enzymes, the function of glands, the formation of a negative nitrogen balance, muscle atony, and decrease in the body's resistance to viral diseases. In this regard, it is important to develop a formulation, production technology, and the assessment of the quality of the new dry composition of increased nutritional and biological value.

N. Dzyuba (✉) · L. Telezhenko · M. Kashkano
Odessa National Academy of Food Technologies, Kanatna St. 112, Odessa 65039, Ukraine
e-mail: dzyubanadya282@gmail.com

O. Maksymets
Tavria State Agrotechnological University, B. Khmelnitsky Av., 18,
Melitopol, Zaporizhia Region 72310, Ukraine

© Springer Nature Switzerland AG 2019

V. Nadykto (ed.), *Modern Development Paths of Agricultural Production*,
https://doi.org/10.1007/978-3-030-14918-5_64

## 2 Literature Review

Nutrition of pregnant women should not only support the energy-plastic processes in the body of a mother and her child, but also provide comfortable and energetic activities of women at all stages of pregnancy [2, 3].

Known formulations of porridges with wheat germ cake, finely dispersed shell of chicken eggs [4, 5], balanced in terms of macro- and micronutrient contents, corresponding to the needs of the body of a pregnant woman. The product has a mass fraction of vitamin E, which contributes to the proper functioning of the placenta. The content of folic acid in a cake of wheat germs promotes the regulation of hematopoiesis in the fetus, normalizes the transmission of genetic information, prevents miscarriage, and plays an important role in the formation of nerve fibers of the fetus.

Developed grains line "Vitaminka" for pregnant women [6], the vitamin supplement, which was made on the basis of wheat germ cake and eggshell, previously sterilized and milled to the size of the particles of $40 \times 10^{-6}$ m. Dosage was carried out in quantities of 37 and 4% by weight of the finished product. The main grains ingredients are rice, whole wheat, and buckwheat. Analysis of the chemical composition of the resulting mixtures showed that they satisfy the daily need for iodine, vitamins A and E by 15%.

The scientists of France developed a dry milk mixture "Celia Mama," which includes dry skim milk, whey protein concentrate, lactose, sucrose, inulin, a mixture of vegetable fats and vitamins, and the probiotic culture of bifidobacteria *Bifidobacterium lactis* Bb12 [7]. The mixture "Celia Mama" contains minerals (sodium, potassium, calcium, phosphorus, magnesium, iron, zinc, copper, iodine, manganese, selenium), vitamins (A, D, E, K, $B_1$, $B_2$, $B_6$, $B_{12}$, PP, C, folic acid, pantothenic acid, biotin), and vitamin-like substances (choline, inositol), the level of which corresponds to domestic standards for milk supplements for pregnant and lactating women.

Using soy protein in the diet of pregnant women is an effective way to supplement the diet with essential amino acids, especially lysine [8, 9]. Along with high-quality protein, vitamins, and minerals in the formulation of a specialized product for pregnant women, it is necessary to include natural food fibers, to improve the intestinal motility and prevent constipation [10].

According to the recommendations of the Commission on micronutrients at the Institute of Medicine of the National Academy of the United States (2001), the folate intake is expressed in micrograms of food folate (and not folic acid) and is: for women of reproductive age—400 μg; for pregnant women—600 μg; for nursing mothers—500 μg of food folate per day [11].

Thus, functional food products made from different types of grains, legumes, oilseeds, and other plant material are a new and essential component of the daily diet. In the food industry, there is observed the need to create a product market for functional products based on grain crops [12].

## 3 Composition Development and Quality Assessment of Grain Composition for Pregnant Women

The aim of the work was to develop a recipe and assess the quality of dry grain preparation for pregnant women in the form of grains.

To achieve the goal, the following tasks were set:

- to develop a formulation of a fast-cooking grains composition with a maximum content of protein and folic acid;
- to determine the content of the basic nutrients in the finished mixture;
- to determine the digestibility of the product obtained;
- using the ALST method to determine the storage conditions for fast-cooking compositions.

Optimization of formulations of cereal compositions based on the main parameters of the chemical composition, corresponding to the recommended needs of a pregnant woman, was conducted using mathematical modeling [13]. The implementation of mathematical modeling of grains MIXTypec composition was solved by linear programming using the MS Excel editor [13].

To predict the actual storage time of grains compositions, the ASLT test [14] was used, which allowed to investigate changes in product-related indicators and product safety indicators from the time and temperature of its storage. The main indicators of quality were organoleptic and microbiological indicators. The prepared grains composition was subdivided into 100-g samples stored at the temperature from 10 to 25 °C, with the step of 5 °C, for 6 months with one-month step.

According to DSTU ISO 6658:2005, the organoleptic analysis was carried out by analyzing the descriptive method (the method of profiling) and the method of using scales and categories (assessment using score scale).

Microbiological parameters were determined according to the normative documents: bacteria of the colibacilli (DSTU GOST 30712), mesophilic anaerobic and facultative anaerobic microorganisms (GOST 1044.15), pathogenic microorganisms, including *Salmonella* genus (DSTU IDF 93A), and the number of mold fungi and yeast (GOST 10444.12).

Delphi method was used to develop a fast preparation formulation for pregnant women, which involves obtaining the optimal information of a high degree of reliability in the development of the product. Based on the fact that the composition is a multicomponent system, it allows adjusting the ratio of the main macro- and micronutrients to obtain a balanced product. For the development of the grains composition, both the native composition of the finished product and its product characteristics were taken into account. The main criterion for selecting the raw material for the product was the high nutritional value of the components and the possibility of their combination in order to obtain a product of high biological and nutritional value.

Based on the physiological needs of the body of a pregnant woman, the chemical composition of the selected ingredients, consumers' preferences, and the recipes of the rapid composition of the grains composition were developed by mathemat-

**Table 1** Determination of the grains composition swelling degree

| Time (s) | Swelling stage of the grains composition | | |
|---|---|---|---|
| | 80 °C | 90 °C | 98 °C |
| 20 | 101.2 | 101.5 | 102.3 |
| 40 | 92.4 | 93.4 | 94.1 |
| 60 | 83.7 | 84.2 | 85.3 |

ical modeling. The target function of the mathematical model corresponded to the maximum content of folic acid and protein. The portion was 200 g. Preliminary calculations have shown that it is advisable to introduce the following constraints for component composition for the formulation modeling: mass fraction of wheat—70 to 100 g; mass fraction of rice—60 to 100 g; mass fraction of milk thistle—7 to 10 g; mass fraction of sesame seeds—8 to 10 g; mass fraction of fructose—7 to 10 g; mass fraction of licorice root—5 to 8 g; mass fraction of gluten—6 to 10 g; mass of portion—200 g.

As a result of modeling in MS Excel using the "Solution searching" add-in, a formulation of a balanced grains composition was obtained, which included the above components (g/1 serving): wheat—100, rice—67, milk thistle—7, sesame—8, fructose—7, liquorice root—5, and gluten—6.

Based on the fact that the product was intended for quick cooking, it was necessary to determine the technological parameters in which the dry product would be rapidly reconditioned. The degree of swelling of the grains composition was estimated as an increase in the volume of the product in relation to the initial volume. Initial volume was 100 $cm^3$. Water was used as a reducing agent at a temperature of 80, 90, and 98 °C. Data on the degree of swelling are given in Table 1.

It has been established that the process of swelling and reconditioning of dry mixtures lasts 60 s at a temperature of 98 °C due to the high degree of dispersion of the grain composition.

The tasting of the developed product was conducted with the participation of teachers and auxiliary educational staff of the department of technology of restaurant and health nutrition, as well as students of the 2nd … 5th courses, that is, people of different age groups, different social backgrounds, with different material resources were interviewed. A total of 54 people attended the tasting. An assessment of the organoleptic characteristics of the mixture is given in Table 2. The total estimated value of the finished product, reconditioned by water, was 23.5 points out of 25 possible, indicating high degradation rates of the reconditioned mixture.

According to modern nutrition principles, products must contain a wide range of ingredients that are necessary for the human body, be balanced by nutritional and biological value. A study of the chemical content of grains compositions, the results of which are shown in Table 3.

It has been established that the developed product has a high content of protein substances, which is very important in terms of providing the body with the necessary proteins.

**Table 2** Physicochemical parameters of the grains composition

| Parameter name | Grains composition |
|---|---|
| Appearance | Homogeneous, without any additional impurities, without lumps |
| Consistence | The particles of plant material are evenly distributed throughout the volume |
| Color | From light cream to cream, even throughout the volume |
| Odor | Milky odor with a slightly pronounced fresh smell with a slight flavor of licorice |
| Flavor | Sweet, without foreign flavors |

**Table 3** Chemical composition of a grain composition

| Nutrients | | | g/100 g |
|---|---|---|---|
| Water | | g | 34.405 |
| Proteins | | | 9.14 |
| Fats | | | 2.83 |
| Carbohydrates | General | | 48.99 |
| | Monosaccharides | | 28.34 |
| | Cellulose | | 20.65 |
| Alimentary fiber | | | 4.205 |
| Ash | | | 1.8 |
| Mineral substances | Na | mg | 18.25 |
| | K | | 195.91 |
| | Ca | | 53.1 |
| | Mg | | 140.3 |
| | S | | 41.7 |
| | P | | 209.9 |
| | Cl | | 73.35 |
| | Fe | | 1.805 |
| Vitamins | A | mg | 0.64 |
| | $B_1$ | | 0.239 |
| | $B_2$ | | 0.186 |
| | $B_5$ | | 0.7 |
| | $B_6$ | | 0.325 |
| | $B_9$ | | 31.98 |
| | $B_{12}$ | | 0.275 |
| | E | | 0.27 |
| | PP | | 2.97 |
| | C | | 2.02 |
| Energy value | | kcal | 238.55 |

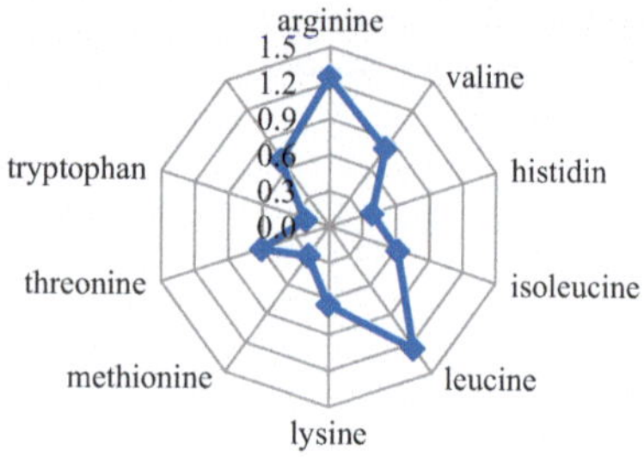


**Fig. 1** Amino acid content of the grain composition

**Table 4** Amino acid score of the grain composition

| Amino acid | Content level recommended by FAO/WHO, mg/1 g protein | Grain composition | |
|---|---|---|---|
| | | mg/1 g protein | AC score |
| Isoleucine | 40 | 53.99 | 134.97 |
| Leucine | 70 | 40.94 | 58.48 |
| Lysine | 55 | 81.56 | 148.30 |
| Methionine | 35 | 42.44 | 121.25 |
| Phenylalanine | 60 | 19.12 | 31.87 |
| Threonine | 40 | 39.36 | 98.39 |
| Tryptophan | 10 | 13.43 | 134.34 |
| Valine | 50 | 45.67 | 91.34 |
| CDAS (%) | | 70.5 | |
| Biological value (%) | | 29.5 | |

Since the grain product contains high protein content, the study of qualitative composition of essential amino acids is given in Fig. 1. The results of the studies prove that nineteen amino acids have been identified in the composition, including all irreplaceable ones. Investigation of the biological value of the developed composition was studied on the basis of the calculation of amino acid score of proteins, which are given in Table 4.

The obtained data showed that the first limiting amino acid is phenylalanine (AC score = 31.87%). The coefficient of difference of the amino acid score (CDAS) shows the average value of the excess amino acid score of essential amino acids in comparison with the lowest level of phenylalanine score. So the CDAS for the developed grains composition is 70.5. Thus, we can conclude that there is an excess of amino acids by 29.5%, but their utilization is defined by the minimum phenylalanine score.

In order to increase the digestibility of the grains composition, it is recommended to use it with nuts or dried fruits, which include a large number of biologically active substances.

**Table 5** Transformation of proteins in the gastrointestinal tract ($n = 5, P \geq 95$)

| Medium | Duration of incubation (h) | | | |
|---|---|---|---|---|
| | 0.5 | 1.0 | 2.0 | 3.0 |
| Gastric juice | 47.15 ± 0.31 | 51.72 ± 0.23 | 67.72 ± 0.37 | 74.15 ± 0.27 |
| Bile | 69.45 ± 0.25 | 73.29 ± 0.13 | 81.72 ± 0.27 | 97.15 ± 0.23 |

By simulating the behavior of the developed grains composition in the human body, the digestion of proteins under the influence of gastric juice (pH 2.1, the content of pepsin—0.25%), at temperature (37 ± 2) °C for three hours, was considered (Table 5). It was shown that proteins of grains composition are digested by 47% after 30 min of gastric juice incubation.

After three hours of incubation in the medium of the gastric juice, the protein of the composition (Table 4) is digested (74.15 ± 0.27%), and after their subsequent incubation in the bile—by (97.15 ± 0.23%). Further incubation leads to almost complete digestion of proteins.

Thus, the study of the digestibility of proteins of the developed grains composition suggests that the use of extruding leads to a decrease in the stability of proteins during digestion and thus increases the degree of digestibility of the product.

The developed grains compositions represent a food product that should be safe according to microbiological parameters. Therefore, microbiological studies of grains in the process of storage and immediately after their production were conducted.

General seeding of experimental specimens and the presence of mushrooms and yeast are determined. Sowing was carried out on meat-peptone agar. The results of microbiological studies and the method of conducting the microbiological analysis are summarized in Table 6. The obtained data allow to state that the developed product meets the sanitary requirements for the microbiota content.

According to the results of the conducted research, it was found that after the processing, the grains compositions meet the standards for the content of microorganisms and in case of proper storage (at low relative humidity and temperature and conditions preventing the product from seeding through the contact with air or the surfaces of the equipment, the quality of the mixture remains high for 4 months of storage at 20 °C in sealed plastic containers).

The organoleptic study of the resulting grains composition was carried out in accordance with GOST 15113.3 "Food concentrates. Methods of determination of organoleptic parameters, readiness of concentrates for use and assessing the dispersion of the suspension." The dynamics of changes in organoleptic parameters in the storage process is shown in Fig. 2.

On the basis of theoretical and experimental studies, a grains composition technology has been developed, which allows to obtain a product that meets the requirements of microbiological, physicochemical, and organoleptic parameters.

The technology includes the primary processing of raw materials, which was carried out as follows. The raw material used for the preparation of the grains com-

**Table 6** Results of microbiological studies of fast-cooking grains composition in storage process

| Name of indicators | Duration of storage, months | | | | |
|---|---|---|---|---|---|
| | 1 | 2 | 3 | 4 | 5 |
| Number of mesophilic anaerobic and facultative anaerobic microorganisms, CFU/cm$^3$, not exceeding | $0.5 \times 10^1$ | $0.7 \times 10^1$ | $1 \times 10^1$ | $1.5 \times 10^1$ | $2.1 \times 10^1$ |
| Coliforms in 1.0 g of the product | Not detected | Not detected | Not detected | Not detected | Not detected |
| Mold fungi in 1.0 g of product, not more | $0.01 \times 10^2$ | $0.08 \times 10^2$ | $0.1 \times 10^2$ | $0.14 \times 10^2$ | $0.2 \times 10^2$ |
| Yeast in 1.0 g of product, not more | $5 \times 10^1$ | $5.1 \times 10^1$ | $5.4 \times 10^1$ | $5.6 \times 10^1$ | $6.1 \times 10^1$ |

**Fig. 2** Dynamics of changes in organoleptic parameters in the process of storage (in scores)

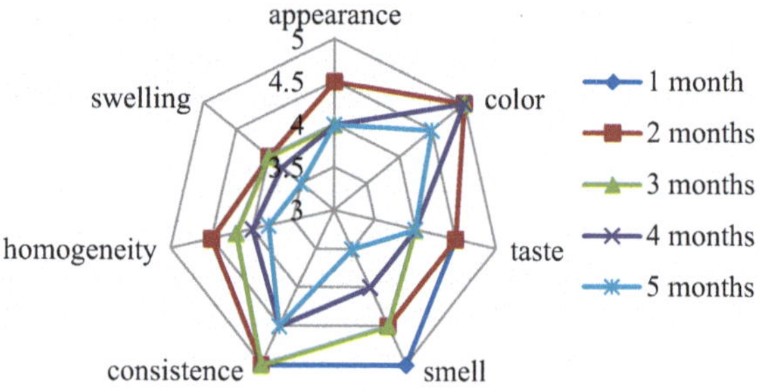


position must comply with TU 9294-001-99621687-07, wheat—GOST R 52554-2006, fructose—TU 9111-011-359-37677-02, thistle—DSTU 7666:2014, sesame seeds—DSTU 7012:2009.

Primary and secondary purification (sorting) of grain consisted of a series of successive operations. Sorting was carried out on air-grid machines, thyristors, pneumatic sorting tables, and other special devices. In addition, thorough purification is needed to improve the safety of the grain, removing foreign impurities. Washing was carried out on washing machines at a temperature of 18–20 °C for 5–7 min, air drying at a temperature of up to 50 °C to a humidity of 14%, which is necessary for the further extruding technological process carried out in the extruder Extruder E-1000 Bronto (Cherkasy city). Then, the milling of the extruded material is carried out up to the diameter of the particles 6–9 mm.

Milk thistle was used as a powder (milled fruits). Seeds of sesame were sorted and milled. The licorice root was sorted, milled, ground, and dried to a moisture content of 15%. Gluten was prepared in the form of powder.

All components were dosed, mixed, and packed in polyethylene bags with the volume of 100 grams.

When grains composition is implemented, it is supplied by weight of the finished dish of 200 g per serving at a temperature (45–50) °C. The finished product must be implemented immediately after the reconditioning.

## 4 Conclusions

Thus, a recipe for a dry composition based on the grain of different cultures has been developed with the introduction of ingredients that enrich dish with the folic acid and contribute to obtaining a product balanced by the amino acid composition. The technologies of preparation of a grains composition with the use of grain extrusion, which promotes the flow of uniform texture of the product, are developed, allows quick cooking, regular nutrition, and increases the level of digestibility of proteins in the human body.

The evaluation of the quality of the grain composition showed its safety, balance, and functionality on the ingredients necessary for pregnant women in the stage of lactation. By the mass fraction of folic acid, one portion of the finished product adequately replaces 30% of daily need and a mass fraction of proteins—25%.

## References

1. Dzyuba, N.A., Zemlyakova, O.V., Pedak, K.G.: ProgresivnI tehnIka ta tehnologIYi harchovih virobnitstv restorannogo gospodarstva I torgIvlI: zb. nauk. prats [redkol.: O.I. Cherevko (vIdpov. red.) ta In.], HarkIv: HDUHT **1**(23), 142–153 (2016)
2. Rodionova, N.S., Manykovskya, M.V.: Development of biotechnology combined foods ystem senriched with plant complexes on the basis of animal origin. J. Eco Agri-Tourism **2**(10), 192–194 (2014)
3. Rodionova, N.S., Alekseeva, T.V., Malikova, T.N.: Materialyi mezhdunar. nauchno-tehn. konf. «Prodovolstvennaya bezopasnost: nauchnoe, kadrovoe i informatsionnoe obespechenie». Voronezh: VGUIT, **1**, 367–368 (2015)
4. Rodionova, N.S., Popov, E.S.: Aktualnaya biotehnologiya **3**(10), 8–11 (2014)
5. Belokurova, E.V., Kolomnikova, Ya.P., Solohin, S.A.: Hleboproduktyi **1**, 56–58 (2015)
6. Alekseeva, T.V., Kalgina, YuO, Shevchenko, V.I., Bokarev, D.A.: Vestnik VGUET **4**, 130–134 (2016)
7. Tumanova, V.A., Serova, O.F., Petruhin i dr., V.A.: Voprosyi ginekologii, akusherstva i perinatologii **3**(3), 101–104 (2004)
8. Vyisotskiy, V.G., Zilova, I.S.: Voprosyi pitaniya **5**(20), 43 (1995)
9. Mendelson, G.I.: Pischevaya promyishlennost **6**, 90 (2004)
10. Spirichev, V.B., Kon, I.Ya., Shatnyuk, L.N., Miheeva, G.A., Spiricheva, T.V.: Voprosyi detskoy dietologii **3**(3), 41–48 (2005)

11. Dietary Reference Intakesf or Thiamin, Riboflavin, Niacin, Folate, Vitamin $B_{12}$, Pantothenicacid, Biotinand Choline, p. 564. Institute of Medicine, National Academy Press, Washington (1998)
12. Ayusheeva, O.G.: Hranenie i pererabotka selhozsyirya **3**, 58–60 (2004)
13. Dzyuba, N., Telezhenko, L., Kashkano, M., Vikul, S., Priss, O., Zhukova, V., Kiurcheva, L., Gaprindashvili, N.: Eureka: LifeSci. **1**, 46–53 (2018)
14. Dzyuba, N.A., Oliiynik, M.I., Bezzodina, A.R.: Science and Society Proceedings of the 1st Internetional Conference, Accent Graphics Communications and Publishing, Hamilton, Canada, pp. 15–20 (2017)

# Innovative Technology of the Scoured Core of the Sunflower Seeds After Oil Expression for the Bread Quality Increasing

**Victoria Evlash, Lydia Tovma, Iryna Tsykhanovska and Nonna Gaprindashvili**

## 1 Introduction

At present, in the diets of the Ukrainian population, there is a deficit of macro- and microelements, protein, polyunsaturated fatty acids, vitamins, food fibers, and biologically active substances. There is an imbalance in their food consumption. This can be explained as available in the processing of inferior raw materials, the deterioration of the environmental situation, violations in the structure of nutrition, that is, the reasons that lead to an increase in the number of people suffering from various diseases, including malnutrition.

Improving the nutritional structure of Ukrainians and ensuring the quality and safety of food products have become the most important priority of our country's domestic policy. The current development of the production sphere of the healthy food products is oriented to the modern nutrition postulates. In particular, with the technology development which allows receiving on the basis of the rational use of natural raw materials, there are the food products that contribute to the nutrition improvement and prevent the diseases associated with the alimentary factor. One of the main tasks of Ukraine food industry is the sustainable production development

V. Evlash (✉)
Kharkiv State University of Food Technology and Trade,
Klochkivska St. 333, Kharkiv 61051, Ukraine
e-mail: evlashvv@gmail.com

L. Tovma
National Academy of the National Guard of Ukraine, Maidan Zakhysnykiv Ukrainy,
3, Kharkiv 61001, Ukraine

I. Tsykhanovska
Ukrainian Engineering-Pedagogics Academy, Universitetska St., 16,
Kharkiv 61003, Ukraine

N. Gaprindashvili
Tavria State Agrotechnological University, B. Khmelnitsky Av., 18,
Melitopol 72312, Ukraine

© Springer Nature Switzerland AG 2019 
V. Nadykto (ed.), *Modern Development Paths of Agricultural Production*,
https://doi.org/10.1007/978-3-030-14918-5_65

of the domestic raw materials and the food in sufficient quantities to provide food products to the country's population using new advanced technologies for the deep and comprehensive processing of food raw materials [1–5].

At present, the secondary resources of oil and fat production are actively used in the food decision, ecological and energy problems being an additional source of substances of natural origin. A significant amount of secondary resources is formed during the sunflower seeds processing—the main oilseed crop in Ukraine—namely sunflower cake and sort. The most valuable properties of sunflower cake are high protein content, low cost, and the absence of toxic and antiviral substances in it. However, sunflower meal and cake are used mainly for the agricultural purposes, in particular as feed additives for poultry and livestock. Analyzing the chemical composition of the kernels of sunflower seeds containing 20% of protein, 50% of lipid, dietary fiber, minerals, and other substances, it can be concluded that they are a good source of nutrients and can be used in food technologies [4, 5].

The valuable nutrients promising source (essential fatty acids, essential amino acids, vitamins, etc.) as well as raw materials improves the functional and technological properties of the food products, and the quality of the finished product is the scoured core of the sunflowers seeds (SCSS). SCSS was made of using the complex innovation technologies as the low temperature of sunflower seeds extrusion. The most valuable properties of SCSS are: high protein content (more than 39% in terms of the dry substance), amino acids, essential polyunsaturated fatty acids, significant amount of antioxidants: vitamin E—15.4% and chlorogenic acid—0.3%; no toxic and anti-nutritional substances and low cost; water and fat-containing ability [6].

The studies' analyses results of the chemical composition showed that the sunflower seeds of modern selection (sorts of Source P-453 Master Buzuluk; hybrids Mercury, Melin, Altair) that is prevalent in the total gross harvest of the sunflower seeds in Ukraine is a prospective raw material for obtaining a range of high nutritional value of the food products, namely oil, lecithin, dietary protein, and natural complex of antioxidants, including chlorogenic acid.

In addition, the minimum content of phenolic compounds in SCSS provides the light color of the scoured core of the sunflower seeds which indicates the possibility of its inclusion in bakery products from the wheat flour without deterioration of crumb color as well as in other food production. All this makes SCSS one of the promising raw materials with a complex action that can be used in food industries [5, 6].

Therefore, it is relevant to use the scoured core of the sunflower seeds in food technology.

## 2 Literature Data Analysis and Problem Statement

Today, the problem of protein nutrients shortage in the diet of Ukraine population reached the maximum value. Economy changes led to the fact that for the last 3 years the purchasing power has decreased significantly. And the consumers economize first of all on food buying the cheap products of dubious quality and insufficient biological and nutritional value.

Indeed, the cost of quality raw materials has increased, and therefore, the use of the new sources of raw materials and enrichment is important today, first of all the mass consumption products—bread and bakery products.

Bread is the basic food consumed daily; therefore, its quality must be subject for all the medical–biological requirements. These figures depend on a number of factors; the main one is the quality of the main and additional raw materials used in the production of bread.

In industry, nearly 50% of the total flour is recycled with reduced properties and the big bakeries use the continuous technologies which have several disadvantages, including the impact on the bread quality.

Not enough is satisfied the need of the population in bakery products for medical–dietary, prophylactic, and functional purpose, especially in the war zones and the ecological trouble. There is also a problem in the production of bread and bakery products for a long storage.

It is therefore necessary to pay close attention to the selection of assortment and creation of new formulations that provide specified consumer properties of finished products.

In Ukraine, the large acreage is sunflower. However, known as oil-bearing crop, it consists of 20% of protein and other valuable substances, and that is why, it can be served as a basis for its use as a polyfunctional component.

The feasibility of secondary products using of oil-bearing seeds processing particularly in SCSS in the bread production is associated with increasing the biological value and improving technological and taste qualities of the products [7, 8].

By now, there are famous works about the improvement in biological value of bakery products by enriching them with the processing products of sunflower protein isolate [8–10]. Its disadvantage is the narrow focus of the action.

The recent work [11–14] has shown that the bread production is enriched with the nutrients of the sunflower seeds and their processing products had certain difficulties.

The introduction of 5% or more of the seed flour or protein isolate from the sunflower seeds or meal using the traditional technology of making bread has led to the deterioration of its quality—eclipse crumb, poorly developed porosity, which significantly reduced the consumer appeal of the finished products [12, 14].

The disadvantages of the other technologies of the sunflower seeds processing and getting the oil are the high-temperature regimes leading to a deep denaturation of seed proteins excludes obtaining possibility from them and the food protein products without additional processing [6, 15].

A complex processing technology of the scoured core of the sunflower seeds (SCSS) is perspective that provides for receiving product—SCSS enriched the physiologically valuable oil and is also very popular in the food technology ingredients like vegetable protein and lecithin.

SCSS also has moisture and the fat-containing ability; the ability to form persistent emulsions has an antimicrobial effect.

Therefore, the scientific rationale for using SCSS for bakery products of high quality and nutritional value has the great practical prospects in the food industry.

## 3 The Purpose and Objectives of the Research

The work purpose is the processing technology development of SCSS for improving the quality of wheat bread and bakery products.

To achieve the goal next tasks were set:

- to investigate the chemical composition and functional–technological properties of the scoured core of the sunflower seeds after the oil expression (SCSS);
- to investigate the SCSS influence on the organoleptical characteristics of the wheat bread prototypes;
- to investigate the SCSS influence on physicochemical characteristics of the experimental samples of the wheat bread;
- to investigate the SCSS influence on the structural and mechanical properties of the experimental samples of the wheat bread;
- to investigate the SCSS influence on the microbiological properties of the experimental samples of the wheat bread.

## 4 Materials, Equipment, and Research Methods of the Scoured Core of the Sunflower Seeds After the Oil Expression

In this work, we investigated the influence on the scoured core of the sunflower seeds (SCSS) after the oil expression on the organoleptical, physicochemical, structural–mechanical, and microbiological characteristics of the wheat grain with the aim of improving its quality, nutritional, and biological value.

The research object is the bread technology from the wheat flour. The research subjects are:

- the scoured core of the sunflower seeds after the oil expression (SCSS) obtained by using the innovative integrated technology for the sunflower seeds processing [the producer of LLC "NAUTECH PLUS," Ukraine];
- pattern 1 the control—wheat bread SSU 7517:2014 quality indicators: humidity crumb—not more than 43.0%; acidity not more than—3.5°; porosity—not less than 57.0%;
- pattern 2—the wheat bread with SCSS in the amount of 2, 5% by weight of flour in powder form;
- pattern 3—wheat bread with SCSS in the amount of 5, 0% by weight of flour in powder form.

To implement the set tasks were used the conventional and standard methods of testing according to SSU-P 4583:2006 and SSU 7517:2014, the modern instrumental methods of biochemical, physicochemical, and microbiological analysis. The contents of crude fat and moisture were determined by using IMR-relaxometry minispec mq-20 (Bruker, Germany) according to AUSS (All Union State Standard) 8.597. Mass fraction of protein was determined by using the system of quantitative identification of N2/DKL 8 protein (VELP Scientifica, Italy) according to AUSS 13496.4. The biological value of the protein complex was studied by experimental determination of amino acid composition with using the system of capillary electrophoresis "KRAPEL-105M," manufacturer Lyumeks (Russia).

Relative biological value (RBV) of protein products was determined by the rapid method by using infusorium Tetrahymena pyriformis in accordance with the recommendation of Ignatiev A. D. et al. Mass fraction of dietary fiber was determined in the apparatus for analysis of fiber Fibretherm FT12 (Gerhardt, Germany) in accordance with AUSS 31675.

The mass fraction of fat in the protein complex was determined on the automatic setup for solid–liquid extraction SOX414a SOXTHERM (Gerhardt, Germany) according to the instructions and AUSS 10857. Fatty acid composition of lipids was determined by the gas chromatograph with a flame ionization detector and integrator "Crystal 5000."

Calcium and magnesium were determined by complexometrically method. The statistical processing of the experiment results was carried out according to the method of student. The organoleptical and physicochemical properties of the wheat bread were determined according to standard methods [16, 17].

The structural and mechanical properties of the crumb of the loaf (elasticity, %; a modulus of elasticity $E$, Pa; shrinkage, %,) were determined by measuring ai elastic properties enabling the penetrameters AP-4/2. With the help of this device was determined the overall deformation of the bread crumb ($\Delta H_{\text{com}}$) that describes its compressibility; plastic deformation ($\Delta H_{\text{plastic}}$) or elasticity; and elastic deformation ($\Delta H_{\text{elasticity}}$) or shrinkage methods [16, 17].

For characteristics of the bread, staling process was used, the determining method of the elasticity modulus E which characterizes the depth of immersion of the indenter cone penetrometer under a load in the bread crumb [16, 17]. Friable was determined in % of the resulting crumbs to take of crumb mass [16, 17]. Dimensional stability of the wheat bread was determined by the formula: $F = H/P$ ($H$—height, cm; $P$—perimeter, cm); specific volume ($V_{\text{pit}}$., $\text{cm}^3/\text{g}$), which was determined by dividing the bread amount on its mass, expressed to the nearest 0.01 $\text{cm}^3/\text{g}$, and the bread volume was measured in $\text{cm}^3$ using the device RZ-BIO, which works on the principle of the volume measuring of loose filler material displaced by the bread (the bread volume was measured three times) [16, 17].

The microbiological parameters list which carried out quality control of the finished bakery products was established on the base of the requirements of SSP 4.4.5.078 and MBR No. 5061-89: the number of mesophilical aerobic and facultative anaerobic microorganisms (NMAFAM, CPU/g); the presence of the bacteria of the intestinal sticks group (BGIS (coliforms) in 0.001 g); the detection of staphylococcus aureus; proteus; and other pathogens (pathogenic microorganisms, including the bacteria of Salmonella, in 25 g) [18–20].

## 5 The Research Results of the Influence of the Unlimited Nucleus of Sunflower Seeds. After Oil Expression on the Quality Indicators of the Wheat Bread Prototypes

The chemical composition of the scoured core of the sunflower seeds (SCSS) was investigated and due to the fact that SCSS used to enhance the nutritional value of the bread and conducted a comparative analysis of the relative chemical composition of the primary raw materials—the wheat flour and the second grade (Table 1).

SCSS studied for GMO content. It was found that SCSS is not genetically modified by DNA has a target sequence of the 35S promoter and NOS-terminator.

Table 2 shows the physicochemical quality indicators of the scoured core of the sunflower seeds (SCSS) after oil expression.

The analysis of the physicochemical quality indicators is presented in Table 2 and shows that SCSS scoured core of the sunflower seeds is characterized by fairly high protein content (42.7%) and fat (25.67%).

Table 3 shows the results of amino acids determination in SCSS and wheat flour.

The analysis showed (Table 3) that the total number of essential amino acids in the SCSS is 2.1–3.1 times higher than in the wheat flour. The first limiting amino acid is lysine.

Due to the fact that in addition to proteins and minerals to physiologically functional ingredients are unsaturated fatty acids and in the work it was done the comparative analysis of fatty acid profile of SCSS and the wheat flour (Table 4).

As shown in Table 4, sample SCSS obtained by innovative resource-saving technologies of complex processing of sunflower seeds (manufacturer LLC "NEWTECH PLUS," Ukraine) meets the requirements of AUSS 30623-98 in terms of quality—fatty acid composition.

**Table 1** Comparative chemical composition of SCSS and the wheat flour of the high grade and the second grade

| Product | Protein (g) | Fat (g) | Sugar (g) | Starch (g) | Fiber (g) | Ash (g) | Ca (mg) | Mg (mg) |
|---|---|---|---|---|---|---|---|---|
| SCSS | 42.7 | 25.7 | 3.7 | 12.5 | 11.9 | 2.9 | 367 | 317 |
| Wheat flour: High grade | 12.0 | 1.3 | 1.9 | 79.7 | 0.12 | 0.58 | 21 | 19 |
| Second grade | 13.6 | 2.1 | 1.0 | 72.8 | 0.7 | 1.3 | 37 | 84 |

From the data in Table 4, it follows that in SCSS a significant content of linoleum acid refers to the essential fatty acids.

In this work, the optimal weight ratio of SCSS was determined experimentally. The weight ratio was established by taking into account the consumer properties of the bread and based on the calculation of the finished products' cost.

Thus, judging by the presented characteristics, SCSS is a very valuable raw material for the bakery industry and the decisive task of which is to create bakery products that have an increased nutritional value.

The indicators of oxidative damage to the scoured core of the sunflower seeds after oil extraction calculated taking into account the oil content are within the normal range; extraneous, metal magnetic impurities and pest infestation are absent. In general, the results of a study of the safety and nutritional values of SCSS allow us to classify the sample of the scoured core of the sunflower seeds after oil expression as a standard food raw material.

Table 5 presents the organoleptic quality control of the experimental samples of the wheat bread with SCSS additive on a 5-point scale. The following indicators were evaluated as appearance: *appearance*: shape, surface condition, color; *the state of the crumb*: got scorched, stirring, porosity; taste; smell.

Table 6 shows the results of the organoleptic analysis of the wheat bread with SCSS additive in the optimum amount of 2.5% to the weight of flour as compared to the control sample.

As follows from the data (Table 6), the prototypes were estimated at 90.2–97.6 points depending on the grade of the wheat flour and the quantity of SCSS. The determining factors for increasing the total score in relation to the control sample were the aroma and taste of bread.

**Table 2** Physicochemical quality parameters of SCSS

| Component name, SCSS | Component quantity (%) |
|---|---|
| Mass fraction of moisture and volatile substances | 8.2 |
| Mass fraction of crude protein in recalculation on s. r. | 42.7 |
| Mass fraction of crude fat in recalculation on s. r. | 25.67 |
| Mass fraction of crude fiber | 11.87 |
| Mass fraction of starch | 12.53 |
| Mass fraction of chlorogenic acid | 0.3 |
| Mass fraction of vitamin E, mg% | 15.4 |
| Allergens:<br>Mass fraction of gluten, mg/kg less than 5 (gluten-free, if the mass fraction of gluten, less than 20 mg/kg | 4 mg/kg |

**Table 3** Content of essential amino acids in SCSS and the wheat flour

| Amino acids | Mass fraction, mg per 100 g of product | | |
|---|---|---|---|
| | SCSS | Wheat flour | |
| | | High grade | Second grade |
| Amino acids amount | 7416 | 3471 | 4105 |
| Including: valin | 1070 | 471 | 525 |
| Isoleucine | 710 | 430 | 560 |
| Leucine | 1343 | 806 | 840 |
| Lysine | 693 | 250 | 330 |
| Methionine + cystine | 785 | 353 | 430 |
| Threonine | 886 | 311 | 365 |
| Tryptophan | 337 | 100 | 130 |
| Phenylalanine + tyrosine | 1592 | 750 | 925 |

**Table 4** Content of fatty acids in SCSS and the wheat flour (%)

| Fatty acid | SCSS | Wheat flour | |
|---|---|---|---|
| | | High grade | Second grade |
| Palmitinum (C 16: 0) | 3.2 | 0.13 | 0.26 |
| Stearic acid (C 18: 0) | 2.1 | 0.01 | 0.02 |
| Oleinov (C 18: 1) | 12.5 | 0.10 | 0.21 |
| Linolev (C 18: 2) | 31.8 | 0.48 | 0.77 |
| Linolenov (C 18: 3) | – | 0.03 | 0.04 |

**Table 5** Organoleptic characteristics assessment of the experimental samples of the wheat bread is enriched with SCSS in comparison with the control sample

| Wheat bread samples | Organoleptic parameters of wheat bread | | | | |
|---|---|---|---|---|---|
| | Taste | Color | Porosity | Appearance and surface of bread | Flavor |
| Sample 1—control | 5 | 4 | 4 | 5 | 4 |
| Sample 2 | 5 | 5 | 5 | 5 | 5 |
| Sample 3 | 5 | 5 | 4 | 5 | 5 |

**Table 6** Results of the organoleptic analysis of the wheat bread with SCSS additive in the optimum amount of 2.5% to the weight of flour as compared to the control sample

| Indicator | Experimental samples of the bread from the wheat flour of the highest grade | |
|---|---|---|
| | Sample 1—control | Sample 2 |
| Shape, surface condition of the plug, $C_W = 2$ | 9.0 | 9.6 |
| Color cork, $C_W = 2$ | 9.4 | 9.6 |
| Color of crumb, $C_W = 2$ | 14.4 | 14.4 |
| Character of porosity $C_W = 2$ | 13.8 | 14.4 |
| Elasticity of crumb $C_W = 23$ | 12.0 | 13.8 |
| Aroma (flavor) $C_W = 23$ | 13.8 | 15.0 |
| Taste $C_W = 23$ | 13.8 | 15.0 |
| Compressibility $C_W = 21$ | 4.0 | 4.8 |
| Total quality index $\Sigma X_i \times C_W$* | 90.2 | 97.6 |

*$C_W$—Weight coefficient, $X_i$—*i*-th quality indicator

**Table 7** Assessment of the influence of the food additive of SCSS on the physicochemical parameters of the experimental samples of the wheat bread

| Experimental bread samples | Physicochemical parameters | | | |
|---|---|---|---|---|
| | Moisture of the crumb, %, not more | Acidity of crumb, degree, not more | Porosity, %, not less | Specific volume ($cm^3/g$) |
| Sample 1 | 42.0 | 3.5 | 57.0 | 4.09 |
| Sample 2 | 42.5 | 3.0 | 65.0 | 5.21 |
| Sample 3 | 43.0 | 2.8 | 64.0 | 5.19 |

To assess the influence of SCSS on the finished products' quality, the physicochemical parameters of the experimental samples of the wheat bread were determined: the mass fraction of crumb moisture, the specific volume, acidity, and porosity of the crumb. Table 7 shows the physicochemical parameters of the experimental samples of the wheat bread, enriched with SCSS as compared to the control sample.

Mass fraction of moisture in the crumb—with this indicator the quality and stability of bread is closely linked during storage as the excess moisture contributes to the flow of the enzymic and chemical reactions for activating the microorganisms' activity including those that lead to the damage of bread, in particular its mold.

The important indicators of the bread quality and bakery products are porosity and acidity. The porosity of bakery products shows the ratio of pore volume to the total volume of the crumb of bakery products and is expressed as a percentage. The dependence of the porosity of the experimental samples of the wheat bread on the amount of SCSS additive is presented in Table 7.

Acidity characterizes the freshness and flavor of bread. Bread and bakery products with the low acidity last longer. It should be noted that the level of acidity of the

**Table 8** Evaluation of the influence of SCSS additive on the crumbling, form stability, and spores contamination of zygomycetes of experimental samples of the wheat bread

| Amount of additive HC, mass % | Quality indices | | | | |
|---|---|---|---|---|---|
| | Slightness of crumb (in 12 days), % | Form stability (in 3 days) | The appearance of mold (at 7° C) through | | |
| | | | 5 days | 10 days | 13 days |
| Sample 1 | 2.5 (6.5) | 0.46 (0.42) | + | + | + |
| Sample 2 | 2.0 (3.5) | 0.60 (0.58) | – | – | + |
| Sample 3 | 1.8 (3.2) | 0.58 (0.56) | – | + | + |

wheat bread and bakery products with the addition of SCSS is lower in comparison with the control (Table 7).

From the data in Table 7, it follows that using SCSS for the wheat bread in an amount of 2.5–5.0% to the flour weight improves the physicochemical characteristics.

The experimental samples of the wheat bread which are enriched with the additive SCSS had the good elasticity and developed the same porosity; the same size pores with thin walls; the crumb did not crumble and did not jam.

Table 8 shows the research data of the influence of the SCSS additive on the crumbling, form stability, and spores contamination of filamentous fungi (Penicillium, Aspergillus, Mucor, etc.) which cause molds of bread and bakery products.

The increase in the specific volume (Table 7) and the form stability (Table 8) of the bread, which enriched with SCSS, is associated with the components' ability of the sunflower seeds (in particular, lipids, glucolipid, and lipoproteins) to complex formation and clustering.

Table 9 shows the microbiological characteristics of the experimental samples of the wheat bread immediately after baking and after storage for 72 h and 10 days (when determining the spores of bacteria, Vas. Subtilis).

The data in Table 9 confirm the antimicrobial action of SCSS and the microbiological safety of developmental prototypes of the wheat bread enriched with SCSS and compliance with the standards established for this type of product [18–20].

Table 10 shows the results of determining the crumb deformation of the samples bread ($\Delta H_{\text{com}}$, mm) during storage for 12 days.

The data analysis in Table 10 shows that the bread staling with SCSS additive passes slower than in the control.

The conducted researches made it possible to develop recipes for new bakery products using SCSS and to determine the preferred way for each of them for dough-making.

Thus, protein sunflower products, in particular SCSS, improve many quality indicators of the wheat bread, significantly increasing its nutritional value.

**Table 9** Influence of SCSS additive on the microbiological parameters of the experimental samples of the wheat bread, which enriched the products of the sunflower seeds processing in comparison with the control during storage

| Indicators name | Standard | Experimental samples of wheat bakery products | | |
|---|---|---|---|---|
| | | Control | Sample 2 | Sample 3 |
| NMAFAnM, CFU/g, immediately/72 h later. | $1.0 \times 10^3$ | $1.0 \times 10^3/1.2 \times 10^3$ | $0.25 \times 10^3/0.30 \times 10^3$ | $0.22 \times 10^3/0.26 \times 10^3$ |
| S. aureus immediately/72 h later | Not allow at 1.0 g | Absent in 1.0 g | Absent in 1.0 g | Absent in 1.0 g |
| BGIS, immediately/72 h later | Not allow 0.001 g | Absent in 0.001 g | Absent in 0.001 g | Absent in 0.001 g |
| Proteus, immediately/72 h later | Not allow at 0.1 g | Absent in 0.1 g | Absent in 0.1 g | Absent in 0.1 g |
| Salmonella, L. monocytogenes, immediately/72 h later | Not allow at 25.0 g | Absent in 25.0 g | Absent in 25.0 g | Absent in 25.0 g |
| Number of bacteria spores Bac. Subtilis, CPU/immediately/72 h later | $0.4 \times 10^3$ | $0.20 \times 10^3/0.26 \times 10^3$ | $0.15 \times 10^3/0.16 \times 10^3$ | $0.13 \times 10^3/0.15 \times 10^3$ |
| Number of bacteria spores Bac. Subtilis, CPU/immediately/after 10 days | $0.4 \times 10^3$ | $0.20 \times 10^3/0.40 \times 10^3$ | $0.15 \times 10^3/0.20 \times 10^3$ | $0.13 \times 10^3/0.17 \times 10^3$ |

**Table 10** Influence of the SCSS additive of the crumb deformation of the experimental bread samples when stored in comparison with the control sample

| Developmental prototype of the bread | The crumb deformation, mm, during storage for | | | |
|---|---|---|---|---|
| | 0 h | 6 h | 72 h | 12 days |
| Sample 1 | 8.0 | 7.0 | 4.5 | 1.5 |
| Sample 2 | 11.5 | 10.8 | 9.7 | 6.5 |
| Sample 3 | 11.5 | 10.8 | 9.7 | 6.5 |

## 6 Discussion of the Investigation Results of the Influence of SCSS on the Wheat Bread Quality

In evaluating of the introducing possibility of the scoured core of the sunflower seeds (SCSS) into the bakery products in order to increase its nutritional and biological value and studied the quality indicators of SCSS, namely: organoleptic indicators, chemical composition and safety indicators. The sample of SCSS has no deviations in taste and smell; it is characterized by the unexpressed with a slight herbaceous tinge flavor, inherent in the sunflower seed without a specific oily taste.

As it can be seen from the data given (Table 1), in SCSS in comparison with the wheat flour a much higher content of fat, mineral substances including calcium and magnesium are observed.

From Table 2, it can be seen that the sunflowers' scoured core contains a significant amount of antioxidants: vitamin E—15.4 mg% and chlorogenic acid—0.3% which effects on the oxalic acid exchange in the human body and prevents gout.

Especially, it would be desirable to pay attention to such property of SCSS as the content of gluten (the mass fraction of gluten, mg/kg less than 5) which allows attributing it to gluten-free products.

When studying the composition of protein of SCSS obtained by the innovative resource-saving integrated technology of the processing sunflower seeds, it was revealed that the developed technology does not lead to a significant change in the native composition of amino acids in the protein part of the sunflower core. The content of essential amino acids in the protein complex of the scoured core of the sunflower seeds after oil expression is more than 36%. In SCSS, the suppressing amino acid is lysine; in addition, a higher content of amino acids leucine and threonine deficient for all varieties of the wheat flour is noted (Table 3).

When evaluating the indices that determine the physiological value of the oil (Table 4) contained in the scoured core of the sunflower seeds, it is established that the lipid part has fat-soluble physiologically valuable nutrients in native form, as well as interchangeable and irreplaceable fatty acids.

It was revealed that, in comparison with the sunflower meal obtained by the pressing method, the efficiency of the technological functional properties of the unlimited nucleus of the sunflower seeds is increased by 15–20%.

The organoleptic analysis of the experimental samples of the wheat bread with the addition of SCSS (Tables 5 and 6) showed that the bread quality from the wheat flour with the additive of the sunflower seeds in the above-mentioned doses differed from the control sample by more pleasant taste, expressed by aroma, well-developed porosity, and large volume. In the experiment process, rational mass fraction of SCSS was determined: for bread from wheat flour of the highest grade—2.5%.

Physicochemical parameters are better for samples with SCSS food additive (Table 7): The humidity of the crumb is increased by 0.5–1.0%; the acidity of the crumb is reduced by 0.5–0.7°; porosity is increased by 2.0–3.0%; the specific volume increases by 1.12–1.17 $cm^3$/h. This is due to the surface-active properties; moisture and fat-sensitive lipids; and lipo-glycoprotein of SCSS. Adding SCSS in an amount of 2.5–5.0% to the weight of the flour helps to reduce the crumbling of the wheat bread by 1.0–1.3% (Table 8) and increase the form stability by 1.0–1.5% (Table 8).

Bread with the addition of SCSS is moldy for 5–8 days later compared to the control (Table 8). This is due to the bactericidal effect of the terpenoid sunflower seeds (in particular, thymol, borneol, camphor), which inhibit growth and destroy pathogenic microbes, fungi, micelles.

Experimental data (Tables 8 and 9) confirm the antimicrobial action of SCSS and the microbiological safety of prototypes of wheat bread enriched with sunflower seeds processing products (SCSS).

When storing the experimental bread samples for 12 days (Table 10), the crumb compressibility is reduced for control—by 5.3 times, for sample 2 and sample 3—by 1.8 times. That is, staling bread with the addition of SCSS is slower.

## 7 Conclusions

1. The chemical composition and functional–technological properties of the scoured core of the sunflower seeds after oil expression were studied. SCSS compared to the wheat flour has a significantly higher content of protein, fat, minerals, including calcium and magnesium.
   Also, SCSS contains a significant amount of antioxidants: vitamin E—15.4 mg% and chlorogenic acid—0.3%, which affects the exchange of oxalic acid in the human body.
   Especially, it would be desirable to pay attention to such property of SCSS as the content of gluten (the mass fraction of gluten, mg/kg less than 5) which allows attributing it to gluten-free products.
   The content of essential amino acids in the protein complex of SCSS is more than 36%. In SCSS, the suppressing amino acid is lysine; in addition, a higher content of amino acids leucine and threonine deficient for all varieties of the wheat flour was noted.
   The technology of obtaining SCSS provides a gentle effect on the lipid part of the scoured core of the sunflower seeds and allows maximally to preserve in the product fat-soluble physiologically valuable nutrients in the native form, as well as interchangeable and irreplaceable fatty acids. It was revealed that, in comparison with the sunflower meal obtained by the pressing method, the efficiency of the technologically functional properties is increased by 15–20%.
2. The data of organoleptic analysis confirm the improvement of the wheat bread quality with SCSS additive by an average of 5.0–7.0% in comparison with the control. SCSS rational dosage for the bread from the wheat flour of superior quality of 2.5% was established.
3. The effect of SCSS additive on the physicochemical parameters of the experimental samples of the wheat bread has been studied.
   SCSS additive contributes to: increase the moisture content of the crumb by 0.5–1.0%; decrease the acidity of the crumb by 0.5–0.7°; increase the porosity

by 2.0–3.0%; increase the specific volume by 1.12–1.17 $cm^3/g$; increase the form stability by 1.0–1.5%; reduce the crumbling by 1.0–1.3%.
4. The influence of SCSS additive on the structural and mechanical characteristics of the experimental samples of the wheat bread in their storage process is studied. When the experimental bread samples are stored for 12 days, the compressibility of the crumb is reduced by 5.3 times for the control, for sample 2 and sample 3—by 1.8 times. That is, staling bread with SCSS additive is slower.
5. The influence of SCSS additive on the microbiological parameters of the experimental samples of the wheat bread during storage has been studied.

The total microbial contamination level of both freshly baked wheat bread and the bread stored for 72 h (and 10 days) at a temperature of 20 °C did not exceed the permissible levels and amounted to: after baking and cooling, the number of microorganisms (NMAFAnM) in 1 g of samples for 2 3 in comparison with the control decreases in 4.0–4.2 times both immediately and through 72 h; the number of bacterial spores Bac. Subtilis after 10 days increases: in control in 2 times, and in samples 2, 3—in 1.3 times. All the prototypes stored for 72 h (and 10 days) at a temperature of 20 °C met the requirements of microbiological standards established for this type of product in Ukraine [18–20].

That is, SCSS has an antimicrobial effect and helps to improve the quality and the shelf life of the wheat bread.

Thus, the results of the conducted studies make it possible to justify the expediency of using the scoured core of the sunflower seeds after oil expression of SCSS by the wheat bread technology.

## References

1. Skurikhin, I.M.: Vso o pishche s tochki zreniya khimika. – M.: Vyssh.shk. 33–40 (1991)
2. Khamagayeva I.S.: Biologicheski aktivnyye dobavki, obogashchennyye essentsial'nymi mikroelementami: Monografiya/I.S. Khamagayeva, A.V. Krivonosova, O.S.Kuznetsova. Ulan-Ude: Izd-vo VSGTU. 178 (2009)
3. Khamagayeva I.S.: Probioticheskiye biologicheski aktivne dobavki: posobiye [Tekst]. Ulan-Ude: Izd-vo VSGUTU. 120 (2009)
4. Lishchenko V.F.: Mirovaya prodovol'stvennaya problema: belkovyye resursy (1960–2005 gg.). M.: DeLi print. 272 (2006)
5. Pashchenko L.P., Zharkova I.M.: Ratsional'noye ispol'zovaniye rastitel'nogo belok-soderzhashchego syr'ya v tekhnologii khleba, p. 239. FGUP IPF, Voronezh (2003)
6. Shcherbakov V.G., Ivanitskiy, S.B.: Proizvodstvo belkovykh produktov iz maslichnykh semyan. – M.: Agropromizdat. 152 (1987)
7. Shchekoldina, T.V., Khristenko, A.G., Ye, A.: Chernikhovets Obespecheniye naseleniya pol-notsennym belkom na puti resheniya voprosov prodovol'stvennoy bezopasnosti. Tekhnologiya i tovarovedeniye innovatsionnykh pishchevykh produktov. **4** (33), 117–123 (2015)
8. Shchekoldina, T.V.K.: voprosu povysheniya biologicheskoy tsennosti khleba i khlebobu-lochnykh izdeliy. Molodoy uchenyy. **5**(85), 111–113 (2015)
9. Shchekoldina, T.V.: Sovershenstvovaniye tekhnologii khlebobulochnykh izdeliy povyshennoy biologicheskoy tsennosti s ispol'zovaniyem belkovogo izolyata podsolnechnogo shrota: diss. … kand. tekhn. nauk: Krasnodar. 169 (2010)

10. Shchekoldina, T.V.: Belkovyy izolyat podsolnechnika – perspektivy ispol'zovaniya dlya povysheniya biologicheskoy tsennosti khlebobulochnykh izdeliy: monografiya. Krasnodar. 164 (2014)
11. Shchekoldina, T.V., Vershinina, O.L., Kudinov, P.I., Ye, A.: Chernikhovets Izucheniye vliyaniya belkovogo izolyata podsolnechnika na svoystva smesi rzhanoy i pshenichnoy muki. Tekhnologiya i tovarovedeniye innovatsionnykh pishchevykh produktov. **1**(30), 20–28 (2015)
12. Vedernikova, Y.I., Knyazhanskaya, B.A., Savrasova, L.L.: Ispol'zovaniye belkov podsolnechnika v khlebopekarnom proizvodstve. M.: TSNIITEIPP. 40 (1980)
13. Patt, V.A., Pyatnitskaya, I.N., Stolyarova, L.F., Yaroshenko, P.A.: Povysheniye pishchevoy tsennosti khlebobulochnykh i sukharnykh izdeliy: Obzor. inform. Ser. Khlebopekarnaya, makaronnaya i drozhzhevaya prom-st'. M.:TSNIITEIPP. 36 (1981)
14. Patt, V.A., Stolyarova, L.F., Paskhina, L.A.: Povysheniye pishchevoy tsennosti khleba putem primeneniya vysokobelkovykh produktov rastitel'nogo proiskhozhdeniya. Tr. VNIIZH. Vyp. **31**, 31 (1974)
15. Lobanov, V.G., Kudinov, P.I., Bochkova i dr, L.K.: Sposob polucheniya pishchevogo belkovogo izolyata iz podsolnechnogo shrota. Patent No. 2340203, A23 J 3/14, 1/14 RF. BIPM. **34**, 4 (2008)
16. Koryachkina, S.Y., Berezina, N.A., Ye. V.: KhmelevaMetody issledovaniya kachestva khlebobulochnykh izdeliy: uch.-metod. pos. Orel: OrelGTU. 166 (2010)
17. Kasymova, C.K.: Laboratornyy praktikum po kursu «Tekhnologiya khleba, konditerskikh i makaronnykh izdeliy», p. 48. Kyrgyzskiy gosudarstvennyy tekhnicheskiy universitet im. I. Razzakova, Bishkek (2012)
18. Mikrobiolohichni normatyvy ta metody kontrolyu produktsiyi hromads'koho kharchuvannya: DSP 4.4.5.078., zatv. MOZ Ukrayiny 7.11.2001r., K.: MOZ Ukrayiny. 17 (2001)
19. Mediko-biologicheskiye trebovaniya i sanitarnyye normy kachestva prodovol'stvennogo syr'ya i pishchevykh produktov. M.: Izd-vo standartov. **5061**, 186 (1990)
20. Produkty pishchevyye. Metod opredeleniya kolichestva mezofil'nykh aerobnykh i fakul'tativno-anaerobnykh mikroorganizmov: GOST 104444.15., K.: Gosstandart Ukrainy. 16 (1996)

# Effect of Living Mulch on Chlorophyll Index, Leaf Moisture Content and Leaf Area of Sweet Cherry (*Prunus avium* L.)

**Tatyana Gerasko, Lyudmila Velcheva, Liudmyla Todorova, Lyubov Pokoptseva and Iryna Ivanova**

## 1 Introduction

Achieving future of safe environment depends on conserving soil, water, and biological resources. Soil is a key component of sustainability. Mulching is one of the methods to protect and enhance the productivity of the soil. Different mulches are widely used in agriculture due to the countless advantages they have [1–4]. There is an increasing interest for use of living mulches in orchards. Based on reports in the literature, we conclude that living mulches increase humification and reduce denitrification and runoff, thus enhancing soil nitrogen availability and water regulation [5–10]. Natural grass cover can act as a living mulch: It is shown that spontaneous vegetation cover enhanced the physical quality of orchard soil [11], improved AM fungal propagules, soil organic carbon, and soil enzyme activities more effectively than did sod culture [12]. Spontaneous grass does not require sowing and irrigation, generates a large biomass, which can remain in place for replenish the soil with organic matter, prevents soil erosion [13], increases biodiversity [14], and provides pollination services [15]. However, it was reported that living mulches decreased soil water availability, impairing apple yields [16], decreased apricot yield, fruit weight, and economic output [17]. That is, trees compete with herbs for water [18, 19] and nutrients [20]. Thus, in an organic orchard, to support the natural biocenosis and to create optimal conditions for the reproduction of the soil fertility, it is necessary to keep soil under living mulch. But the effects of living mulch on the physiological parameters of fruit trees have not been fully explored. To fill this gap, this work aimed at examining the effects of living mulch in the organic orchard on the chlorophyll

T. Gerasko (✉) · L. Velcheva · I. Ivanova
Department of Fruit, Vegetable and Biochemistry, Tavria State Agrotechnological University, B. Khmelnitsky Av., 18, Melitopol, Zaporizhia Region 72312, Ukraine
e-mail: tetiana.herasko@tsatu.edu.ua

L. Todorova · L. Pokoptseva
Department of Plant Production, Tavria State Agrotechnological University, B. Khmelnitsky Av., 18, Melitopol, Zaporizhia Region 72312, Ukraine

© Springer Nature Switzerland AG 2019

V. Nadykto (ed.), *Modern Development Paths of Agricultural Production*,
https://doi.org/10.1007/978-3-030-14918-5_66

index, leaf moisture content, and leaf area of sweet cherry (*Prunus avium* L.) in the conditions of the southern Steppe of Ukraine.

## 2 Materials and Methods

### 2.1 Site Description

The work was conducted from 2017 to 2018 in the southern steppe of Ukraine (Melitopol district, Zaporozhye region) in the experimental orchards of Taurian State Agrotechnological University, which is intended for research purposes. The soil cover of the investigated area is the chestnut soils, which are very low humus. Soils have a weakly alkaline reaction of soil solution (pH varies within 7.1–7.4). On the background of a light granulometric composition, the humus content in the upper humus horizon is 0.6%. The analysis of aqueous extraction revealed that the total content of water-soluble salts does not exceed 0.015–0.024%. Analyzing all physical and agrochemical properties, we can conclude that the soils are suitable for growing sweet cherries. The long-term mean air temperature is +9.6 °C. The summer months (June, July, August) have an average daily air temperature of 20–22 °C. Winters are warm with frequent thaws. The coldest months are January and February. During these months, mean annual air temperature ranges minus 3.7–4.3 °C, but the minimum temperature is reduced to minus 33 °C. Mean annual precipitation for the last 10 years was approximately 350–450 mm.

### 2.2 Orchard Floor Management

Plant material for research was sweet cherry (*Prunus avium* L.) cultivars "Valery Chkalov"/Prunus mahaleb and "Dilemma"/Prunus mahaleb planted in 2011 at 7 m × 5 m. The experiment was designed as a randomized complete block with two treatments, in triplicate. Each experimental plot had an area of 210 $m^2$ (7 m × 30 m). Each plot contained 10 sweet cherry trees. The experiment included two different orchard floor management systems (OFMS): Standard mechanical cultivation—one discing at a 15-cm depth followed by harrowing + manual weeding during the growing season (MC) was compared with living mulch—spontaneous vegetation cover (LM). The natural vegetation of grasses was mowed four times during the growing season, and the clippings were left on the ground for decomposition. Any other management was identical in each treatment. Synthetic fertilizers and chemical plant protection products were not used.

### 2.3 Sampling Measurements and Data Calculation

Leaves for analysis were collected in the first decade of August, with the full development of the leaf surface. The leaf area was determined by the method of cuttings: Ten leaves of each tree were harvested from the middle of the annual shoots on the southern side of the crown and transported to the laboratory for analysis. The leaves were weighed, punched with punch (the area of the cutting was 1 $cm^2$). Cuttings were weighed, and the area of the leaves was calculated by the ratio of the mass of leaves and the mass of the cuttings. The parameters of the water regime of the leaves were determined ravimetrically, as described by G. K. Karpenchuk and A. V. Melnyk: The total moisture content was determined by oven drying (105 °C) until constant weight; the relative turgrescence was calculated as the ratio of the total moisture content to the moisture content after the 24-h saturation in a wet chamber; the moisture deficit—the ratio of moisture absorbed by the leaves (after a 24-h saturation in a wet chamber) to the total moisture content after a 24-h saturation in a humid chamber; the water-retaining ability—the ratio of lost moisture (after a 24-h wilting) to the total moisture content [21]. The content of photosynthetic pigments (chlorophylls *a*, *b* and carotenoids) in the leaves was estimated by determination of visible and near UV light absorption capacity of leaf acetone extracts spectrophotometrically in the biochemical laboratory of the Taurian State Agrotechnological University according to generally accepted methods [22]. Biochemical analyses were conducted in three biological replicates. Means for the treatments were compared using the least significant differences (LSD) and Student's criterion, and significant differences were determined at $P < 0.05$ probability level. All data were analyzed using the Microsoft Excel 2010 [23].

## 3 Results and Discussions

Tables 1 and 2 show data on the content of photosynthetic pigments in sweet cherry leaves. The content of chlorophylls and the sum of chlorophylls *a* and *b* in the leaves of both studied cultivars in the experimental variants did not differ significantly in 2017 or 2018. But it should be noted significantly more carotenoid content in the leaves of the cultivar "Valery Chkalov" in 2017 in LM. The enhanced synthesis of carotenoids is a nonspecific response of plants to stress [24]. Due to the increase in the carotenoid content, the chlorophyll index of this cultivar in LM conditions was significantly lower compared to MC in 2017. At same time (in 2017), for the cultivar "Dilemma," on the contrary, it was noted the increase of the chlorophyll index in LM and a significantly higher content of carotenoids in MC. Further studies of cultivars characteristics are needed to find out the reason for this revealed trend. We hypothesize that this is due to the activity of soil microorganisms, symbiotic mycorrhiza. It is known that LM creates optimal conditions for the development of

**Table 1** The content of photosynthetic pigments in sweet cherry leaves cultivar "Valery Chkalov"

| OFMS | Chlorophyll a content (%) | Chlorophyll b content (%) | Carotenoids content (%) | Sum of chlorophylls $(a + b)$, % | Chlorophyll index $(a + b)/\kappa$ |
|---|---|---|---|---|---|
| *2017 Year* | | | | | |
| MC | $1.77 \pm 0.16$ | $1.73 \pm 0.13$ | $0.17 \pm 0.01$ | $3.49 \pm 0.31$ | $20.4 \pm 0.06$ |
| LM | $2.09 \pm 0.17$ | $1.51 \pm 0.13$ | $0.24 \pm 0.02$* | $3.59 \pm 0.16$ | $14.7 \pm 0.05$* |
| *2018 Year* | | | | | |
| MC | $1.19 \pm 0.12$ | $0.48 \pm 0.05$ | $0.30 \pm 0.02$ | $1.67 \pm 0.17$ | $5.5 \pm 0.04$ |
| LM | $1.12 \pm 0.10$ | $0.52 \pm 0.05$ | $0.27 \pm 0.02$ | $1.65 \pm 0.16$ | $6.2 \pm 0.05$* |

*Note* *—The difference is significant at $P \leq 0.05$

**Table 2** The content of photosynthetic pigments in sweet cherry leaves cultivar "Dilemma"

| OFMS | Chlorophyll a content,% | Chlorophyll b content (%) | Carotenoids content (%) | Sum of chlorophylls $(a + b)$, % | Chlorophyll index $(a + b)/\kappa$ |
|---|---|---|---|---|---|
| *2017 Year* | | | | | |
| MC | $1.86 \pm 0.12$ | $1.39 \pm 0.11$ | $0.25 \pm 0.04$* | $3.25 \pm 0.32$ | $13.3 \pm 0.02$ |
| LM | $1.81 \pm 0.16$ | $1.46 \pm 0.14$ | $0.14 \pm 0.04$ | $3.27 \pm 0.33$ | $24.1 \pm 0.03$* |
| *2018 Year* | | | | | |
| MC | $0.88 \pm 0.12$ | $0.31 \pm 0.05$ | $0.25 \pm 0.02$ | $1.20 \pm 0.12$ | $4.86 \pm 0.23$ |
| LM | $1.22 \pm 0.11$ | $0.80 \pm 0.14$ | $0.25 \pm 0.02$ | $2.02 \pm 0.12$ | $8.21 \pm 0.16$* |

*Note* *—The difference is significant at $P \leq 0.05$

soil symbiotic mycorrhiza [12], but different cultivars may have their own specific features regarding the formation of symbiosis with soil microorganisms.

It should be noted that the weather conditions of 2017 were relatively satisfactory with respect to moisture supply, especially in June, as opposed to 2018, when drought lasted throughout all summer months. Drought was reflected on the physiological state of leaves, which contained significantly less photosynthetic pigment. The content of chlorophylls under such severe conditions in 2018 was lower, and the carotenoids content was higher in both cultivars in both OFMS. It should be noted that the chlorophyll index was higher in LM in both cultivars in 2018, which indicates the positive effect of LM on the physiological state of sweet cherry trees.

Tables 3 and 4 show the parameters of the water regime of sweet cherry leaves. For the cultivar "Valery Chkalov," there was no significant difference between the MC and LM on the total moisture content, relative turgrescence, and moisture deficit. But under hiding conditions, we observed a significant increase in the water-retaining ability of leaves in 2017. This can be associated with an increase in the content of colloids in the tissues of leaves, which is an adaptive plants response to water deficit. Naturally, the deficit of moisture in the leaves was closely negative linked with the relative turgrescence ($r = -0.9$).

**Table 3** Water regime in leaves of sweet cherry cultivar "Valery Chkalov"

| OFMS | Total moisture content (%) | Relative turgrescence (%) | Moisture deficit (%) | Water-retaining ability (%) |
|---|---|---|---|---|
| *2017 Year* | | | | |
| MC | 58.7 ± 0.19 | 23.4 ± 1.38 | 76.6 ± 1.39 | 94.3 ± 0.61 |
| LM | 58.2 ± 0.65 | 21.3 ± 1.36 | 78.7 ± 1.37 | 96.5 ± 0.45* |
| *2018 Year* | | | | |
| MC | 54.3 ± 0.29 | 31.6 ± 0.35 | 68.4 ± 0.55 | 95.8 ± 0.79 |
| LM | 53.6 ± 0.47 | 30.7 ± 0.33 | 69.3 ± 0.42 | 95.2 ± 0.75 |

*Note* *—The difference is significant at $P \leq 0.05$

**Table 4** Water regime in leaves of sweet cherry cultivar "Dilemma"

| OFMS | Total moisture content (%) | Relative turgrescence (%) | Moisture deficit (%) | Water-retaining ability (%) |
|---|---|---|---|---|
| *2017 Year* | | | | |
| MC | 62.1 ± 0.22 | 25.7 ± 2.07 | 74.3 ± 3.08 | 93.3 ± 1.47 |
| LM | 54.8 ± 0.12* | 27.5 ± 1.79 | 72.5 ± 1.79 | 91.7 ± 0.25 |
| *2018 Year* | | | | |
| MC | 55.7 ± 0.35 | 30.9 ± 1.55 | 69.1 ± 1.99 | 94.5 ± 1.41 |
| LM | 51.6 ± 0.43* | 26.6 ± 1.67* | 73.4 ± 1.25* | 97.7 ± 1.44* |

*Note* *—The difference is significant at $P \leq 0.05$

The sweet cherry cultivar "Dilemma" was characterized by a significantly lower total moisture content in the leaves in LM during two years of research; the water-retaining capacity of leaves was increased in 2018. Since the decrease in the total moisture content and relative turgrescence in LM and severe drought in 2018 in the cultivar "Dilemma" was significant (unlike the cultivar "Valery Chkalov"), it could be noted that cultivar "Dilemma" is more vulnerable to drought conditions and competition with herbs. But in the scientific literature, there is a message that an increase in the dysfunction of the hydraulic system in the most vulnerable cultivars may represent a signal for enhancing the delivery of water to fruits, and in this case, water instead of moving into the leaves is delivered to the fruits [25].

To better explain the obtained data, we will analyze the results of determining the sweet cherry leaf area, presented in Table 5.

The leaf area was significantly less in LM for both studied cultivars in 2017. In 2018, there was no significant difference with MC, but the cultivar "Valery Chkalov" showed a tendency to decrease the leaf area in LM, and the cultivar "Dilemma," on the contrary—increased.

**Table 5** Sweet cherry leaf area, $m^2$/tree

| OFMS | Cultivar "Valery Chkalov" | Cultivar "Dilemma" |
|---|---|---|
| *2017 Year* | | |
| MC | 38.2 | 46.8 |
| LM | 29.0 | 32.4 |
| LSD $_{0,5}$ | 3.05 | 3.74 |
| *2018 Year* | | |
| MC | 60.8 | 51.7 |
| LM | 52.8 | 59.3 |
| $LSD_{0.5}$ | 5.17 | 5.04 |

# 4 Conclusions

The content of carotinoids in 2017 was significantly higher in the leaves of the cultivar "Valery Chkalov" in LM and in the leaves of the cultivar "Dilemma" in MC and in 2018 was no significant difference in both cultivars.

The chlorophyll index in LM (compared to MC) was higher in the cultivar "Dilemma" in 2017 and in both cultivars under severe drought in 2018.

The water-retaining ability in 2017 (satisfactory moisture supply) was greater in LM in the leaves of the cultivar "Valery Chkalov," in 2018 (drought)—in the cultivar "Dilemma."

The leaf area was significantly less in LM for both studied cultivars in 2017. In 2018, a significant difference with MC was not noted.

The disadvantage of our study is the lack of data on the state of soil microorganisms, namely the symbiotic mycorrhiza in the rhizosphere of cherry trees, which could explain the revealed trends in the physiological state of the leaves.

# References

1. Chen, Y., Wen, X., Sun, Y., et al.: Mulching practices altered soil bacterial community structure and improved orchard productivity and apple quality after five growing seasons. Sci. Hort. **172**, 248–257 (2014). https://doi.org/10.1016/j.scienta.2014.04.010
2. Granatstein, D., Sanchez, E.: Research knowledge and needs for orchard floor management in organic tree fruit systems. Int. J. Fruit Sci. **9**(3), 257–281 (2009). https://doi.org/10.1080/15538360903245212
3. Bakshi, P., Wali, V.K., Iqbal, M., et al.: Sustainable fruit production by soil moisture conservation with different mulches: a review. Afr. J. Agric. Res. **10**(52), 4718–4729 (2015). https://doi.org/10.5897/ajar
4. Oliveira, M.T., Merwin, I.A.: Soil physical conditions in a New York orchard after eight years under different groundcover management systems. Plant Soil **234**, 233–237 (2001). https://doi.org/10.1023/a:1017992810626

5. Qian, X., Gu, J., Pan, H.-J.: Effects of living mulches on the soil nutrient contents, enzyme activities, and bacterial community diversities of apple orchard soils. Eur. J. Soil Biol **70**, 23–30 (2015). https://doi.org/10.1016/j.ejsobi.2015.06.005
6. Yao, S.R., Merwin, I.A., Bird, G.W., Abawi, G.S., Thies, J.E.: Orchard floor management practices that maintain vegetative or biomass groundcover stimulate soil microbial activity and alter soil microbial community composition. Plant Soil **271**(1/2), 377–389 (2005). https://doi.org/10.1007/s11104-004-3610-0
7. Nikiema, P., Nzokou, P., Rothstein, D.E. et al.: Soil microbial biomass as affected by groundcover management in a Fraser fir (*Abies fraseri* [Pursh] Poir) plantation after 1 year. Biol. Fertil. Soils **48**, 727. https://doi.org/10.1007/s00374-012-0666-5
8. Turrini, A., Caruso, G., Avio, L., et al.: Protective green cover enhances soil respiration and native mycorrhizal potential compared with soil tillage in a high-density olive orchard in a long term study. Appl. Soil. Ecol. **116**, 70–78 (2017). https://doi.org/10.1016/j.apsoil.2017.04.001
9. Huang, J., Wang, J., Zhao, X., et al.: Effects of permanent ground cover on soil moisture in jujube orchards under sloping ground: a simulation study. Agric. Water Manage. **138**, 68–77 (2014). https://doi.org/10.1016/j.agwat.2014.03.002
10. Atucha, A., Merwin, I.A., Brown, M.G., et al.: Soil erosion, runoff and nutrient losses in an avocado (Persea americana Mill) hillside orchard under different groundcover management systems. Plant Soil **368**(1–2), 393–406 (2013). https://doi.org/10.1007/s11104-012-1520-0
11. Fidalski, J., Tormena, C.A., da Silva, A.P.: Least limiting water range and physical quality of soil under groundcover management systems in citrus. Sci. Agric. (Piracicaba, Braz.) **67**(4), 448–453 (2010). https://doi.org/10.1590/s0103-90162010000400012
12. Wang, P., Wang. Y., Wu, Q.S.: Effects of soil tillage and planting grass on arbuscular mycorrhizal fungal propagules and soil properties in citrus orchards in southeast. China Soil Till. Res. **155**, 54–61 (2016). https://doi.org/10.1016/j.still.2015.07.009
13. Garcia-Diaz, A., Bienes, R., Sastre, B.: Nitrogen losses in vineyards under different types of soil groundcover. A field runoff simulator approach in central Spain. Agric. Ecosyst. Environ. **236**, 256–267 (2017). https://doi.org/10.1016/j.agee.2016.12.013
14. Juarez-Escario, A., Conesa, J.A., Sole-Senan, X.O.: Management as a driver of functional patterns and alien species prominence in weed communities of irrigated orchards in Mediterranean areas. Agric. Ecosyst. Environ. **249**, 247–255 (2017). https://doi.org/10.1016/j.agee.2017.07.042
15. Saunders, M.E., Luck, G.W., Mayfield, M.M.: Almond orchards with living ground cover host more wild insect pollinators. J. Insect. Conserv. **17**, 1011–1025 (2013). https://doi.org/10.1007/s10841-013-9584-6
16. Zelazny, W.R., Licznar-Malanczuk, M.: Soil quality and tree status in a twelve-year-old apple orchard under three mulch-based floor management systems. Soil Till. Res. **180**, 250–258 (2018). https://doi.org/10.1016/j.still.2018.03.010
17. Du, S., Bai, G., Yu, J.: Soil properties and apricot growth under intercropping and mulching with erect milk vetch in the loess hilly-gully region. Plant Soil **390**, 431 (2015). https://doi.org/10.1007/s11104-014-2363-7
18. Forey, O., Metay, A., Wery, J.: Differential effect of regulated deficit irrigation on growth and photosynthesis in young peach trees intercropped with grass. Eur. J. Agron. **81**, 106–116 (2016). https://doi.org/10.1016/j.eja.2016.09.006
19. Hammermeister, A.M.: Organic weed management in perennial fruits. Sci. Hort. **208**, 28–42 (2016). https://doi.org/10.1016/j.scienta.2016.02.004
20. Ni, X., Song, W., Zhang, H., Yang, X., Wang, L.: Effects of mulching on soil properties and growth of tea olive (*Osmanthus fragrans*). PLoS ONE **11**(8), e0158228 (2016). https://doi.org/10.1371/journal.pone.0158228
21. Karpenchuk, G.K., Melnyk, A.V.: Uchety, nabliudenyia, analyzy, obrabotka dannykh v opytakh s plodovymy i yahodnymy rastenyiamy: Metodycheskye rekomendatsyy. Uman. s.-kh. yn-t, Uman (in Russian) (1987)
22. Pochynok, H.N.: Metody biokhymycheskoho analyza rastenyi. Nauk, Dumka, Kiev (in Russian) (1976)

23. Lakyn, H.F.: Biometryia. Vysshaia shkola, Moscow (in Russian) (1990)
24. Gross, J.: Carotenoids. In: Pigments in Vegetables. Springer, Boston, MA (1991). https://doi.org/10.1007/978-1-4615-2033-7_3
25. Peschiutta, M.L., Bucci, S.J., Scholz, F.G., et al.: Leaf and stem hydraulic traits in relation to growth, water use and fruit yield in *Prunus avium* L. cultivars. Trees-Struct. Funct. **27**(6), 1559–1569 (2013). https://doi.org/10.1007/s00468-013-0904-y

# Modification of Modeling Method of Toxic Dystrophy of Liver in Rats

**Viktoriya Gryshchenko, Olena Danchenko and Viktoriya Musiychuk**

## 1 Introduction

The liver is the largest digestive gland that plays an important role in providing homeostasis of the internal environment of the body and in the development of adaptive reactions, which is due to its participation in many metabolic processes, anatomical, and functional connections with other organs and systems of the body. Recently, the risk of developing hepatopathy is increasingly associated with irrational pharmacotherapy of the underlying disease [4, 23, 24]. Thus, the administration of tetracycline antibiotics (doxycycline, chlortetracycline, metacycline) causes the occurrence of mitochondrial cytopathies. Toxic hepatitis can occur both on the first day of taking an antibacterial drug and after a few months from the start of treatment. To determine the degree of damage to the liver parenchyma, it is recommended to study the activity of enzymes with various intracellular topographies [7, 11, 16, 19].

Clinical and morphological manifestations of toxic liver lesions are diverse [20]. There are three main types of damage: hepatocellular, cholestatic, and mixed. The target of toxic effects can be hepatocytes (dystrophy, necrosis), bile ducts and tubules (cholestasis), or sinusoidal cells (endothelium). Medication of the liver, in particular parenchyma damage in the form of functional disorders (induction of microsomal enzymes, hyperbilirubinemia), lead to necrosis or apoptosis. Other hepatotoxic effects of pharmacotherapy include the formation of steatosis in the form of acute fat changes, steatohepatitis, cholestasis, granulomatous changes and damage to the vascular system of the liver, etc. [3, 5, 17].

V. Gryshchenko (✉) · V. Musiychuk
National University of Life and Environmental Sciences of Ukraine,
Heroiv Oborony Str., 15, Kiev 03041, Ukraine
e-mail: viktoriya_004@ukr.net

O. Danchenko
Tavria State Agrotechnological University, B. Khmelnitsky Av., 18,
Melitopol, Zaporizhia Region 72312, Ukraine

© Springer Nature Switzerland AG 2019 
V. Nadykto (ed.), *Modern Development Paths of Agricultural Production*,
https://doi.org/10.1007/978-3-030-14918-5_67

A significant compensatory antitoxic reserve of the body neutralizes the negative impact of toxic substances [9, 18]. In the liver, which performs a detoxification function, their inactivation and binding take place for further elimination. The realization of this function during the first phase depends significantly on the activity of cytochrome P450. With the participation of the corresponding enzymatic systems, oxidation, reduction, hydrolysis, hydration, and dehalogenation of toxins occur [10]. In the next phase is registered conjugation of toxic substances and their deactivation due to the processes glucuronidation, acetylation, methylation, binding with amino acids and glutathione. However, the detoxification capacity of the organ is significantly reduced if the influence of harmful factors develops against the background of hepatopathology. The severity and severity of symptoms are determined by the dose of xenobiotics [2].

Morphological heterogeneity of toxic lesions of the liver and preferential localization of necrosis causes metabolic zoning. Behind metabolic activity, hepatocytes are heterogeneous. Their zoning for a difference in metabolic functions determines the selective sensitivity of hepatocytes to various pathological factors. In particular, cells of zone I contain more mitochondria, oxidative processes, gluconeogenesis, synthesis of cholesterol, urea, and bile acids are more intensive. In hepatocytes of the III zone, glycolysis, lithogenesis, cytochrome P450-dependent hydroxylation, and glucuronidation of xenobiotics are most pronounced. The use of tetracycline antibiotics leads to the development of necrosis of hepatocytes predominantly of the III zone [4].

Toxic liver damage due to taking medications presents a certain diagnostic complexity [1, 15]. Sometimes, with prolonged therapy, they proceed without clinical manifestations, which indicate chronic intoxication. This leads to a comprehensive study of the pathogenesis of the development of drug hepatopathology, primarily at the molecular level, the creation of sensitive test systems for early diagnosis, as well as the means of prevention and therapy of such conditions [6]. Therefore, it remains relevant to develop methods for the artificial reproduction of pathological conditions of the liver due to its medicinal damage. We have already developed a method for modeling drug-induced hepatopathy by introducing into the body of rats a non-steroidal anti-inflammatory drug diclofenac sodium, which provokes the development of toxic hepatitis [12].

The purpose of this work was the artificial reproduction in laboratory rats of an acute form of toxic hepatodystrophy without sudden changes in the clinical state, which corresponds to the spontaneous course of this hepatopathology in animals with respect to the complex of general clinical, biochemical, morphological, and pathoanatomical changes, but does not lead to death.

# 2 Experimental Studies

## 2.1 Research Method and Materials

**The Modeling of Toxic Hepatodystrophy**. The experiment involved white laboratory rats (males) who were selected according to the principle of analogues with a body weight of 200–220 g. With the known scheme for modeling fatty hepatosis in laboratory rats by oral administration of 1% solution of tetracycline hydrochloride at a dose of 500 mg/kg, for 5 days, an extremely severe clinical condition is observed, which develops already on the 2–3th day of the experiment, a lethal outcome is possible [13]. In addition, this type of animal with a body weight of 200–220 g is difficult to perceive the recommended volume of tetracycline hydrochloride solution, which borders or even exceeds the physiological capabilities of the stomach, and this in most cases is 5 $cm^3$ or more. This was the basis for work on improving the existing method for modeling the acute form of toxic hepatodystrophy. For this, two groups (control and study) were formed in twenty rats each. The animals of the study group artificially reproduced the acute form of hepatodystrophy by the method [9] of intragastric injection of the 4% solution of tetracycline hydrochloride, modified by us, using a probe at a dose of 250 mg/kg once a day, for 7 days in accordance with the current standardization and quality criteria conducting biological experiments and biomodeling principles [21]. Animals of the study group remained during the experiment without treatment. In the control group, clinically healthy rats were treated, with an equivalent volume of distilled water administered intragastrically by means of a probe.

The rats were separately placed in cages. Before the experiment began, they were in quarantine with a clinical examination for two weeks. The animals were kept on a balanced diet containing all the necessary biologically active and nutritious substances. They had free access to feed and drinking water. Were monitored changes in body weight and animal feed intake. The duration of the experiment was 7 days. Euthanization of animals and selection of biological material were carried out on the 8th day of the experiment.

During the experiment followed the requirements of the "European Convention for the Protection of Vertebrates used for experimental and scientific purposes" (Strasbourg, 1986), the Law of Ukraine "On the Protection of Animals against Cruel Treatment" No. 3447 of 21.02.2006.

**Histomorphological Studies**. After decapitation, a pathological anatomical dissection and selection of liver samples were performed for histomorphological studies. Pieces of liver were selected for microscopic studies and were fixed in a 10% solution of neutral buffered formalin [14], washed in running water, dehydrated in spirits of increasing concentration (70°, 96°), aged in chloroform, and poured into paraffin. The cooled paraffin blocks were attached to wooden cubes. After that, slices were made on a sine microtome, and slices 10 μm thick were obtained. The resulting sections were glued onto slides, stained with hematoxylin Karatsu and eosin, and examined with a light microscope [21].

**Morphological and Biochemical Studies of Blood**. Native blood was collected from rats from the abdominal aorta into test tubes with heparin solution, which was examined for hemoglobin content, absolute number of erythrocytes, leukocytes, and platelets, calculated by leukogram, and thrombocrit was determined on a Micro CC-20 Plus Auto HTI (USA) analyzer. To obtain plasma, it was centrifuged at 1500 rpm for 15–20 min. In blood plasma of rats were investigated biochemical indicators, such as total protein, total and conjugated bilirubin, activity of aspartate-aminotransferase (AST, EC 2.6.1.1), alanine-aminotransferase (ALT EC 2.6.1.2), γ-glutamyl-transpeptidase (γ-GTP, EC 2.3.2.2), alkaline phosphatase (APF, EC 3.1.3.1) on an open biochemical semi-automatic analyzer GBG Stat Fax 1904 Plus (Awareness Technology, Inc., Florida, USA) using DAC-SPECTROMED SRL (Moldova) reagents.

**Statistics**. The results were processed in the package Statistica 6.0. The probability of the difference between the samples was estimated by Student's *t* test, having previously checked the normality of their distribution. The disagreements were considered reliable at $P < 0.05$.

## 2.2 *Results and Discussion*

Regular administration of tetracycline hydrochloride preparation to rats for seven days is accompanied by the emergence of severe clinical symptoms of acute toxic dystrophy of liver that begin to appear in animals on the second or third day of its administration and are characterized by general depression, decreased appetite, polydipsia or thirst reduction, losses the average body weight for the group of 10–15 g, a dull coat, a decrease in elasticity and dry skin, hair loss, a slight increase stomach, liquefaction of stool.

**Pathoanatomical Studies**. Data from the clinical manifestation of toxic hepatodystrophy complement the results of pathoanatomical studies of the body—the liver acquires a light yellow color or has a mottled mosaic pattern, the brownish-red areas alternate with yellow, slightly enlarged, rounded edges, flabby consistency.

**Histomorphological Studies**. Histomorphological examination shows the decomposition of hepatocytes, diffuse placement of fat droplets of different sizes (small, medium, and large droplets) in liver cells, focal histiolymphoid infiltration, expansion, and blood vessel overflow (Fig. 1). In addition to obesity, a pronounced granular dystrophy is revealed. In some places, there are small proliferates of Kupffer cells. Changes in portal tracts are expressed in sclerosis with a slight thickening and inflammatory response—the appearance of small, local histiocytic, and lymphoid elements [22].

**Morphological Indicators of Blood**. As a result of the study of the morphological composition of the blood (Table 1), the development of leukocytosis (27% increase in the number of white blood cells), erythrocytopenia (a decrease in the number of erythrocytes by 27%), and simultaneous manifestation of hypochromaemia (a decrease in hemoglobin by 23%) was established in rats of the experimental group.

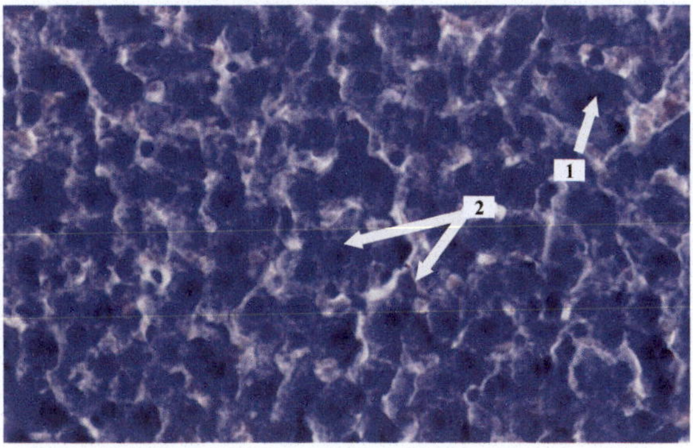


**Fig. 1** Liver of the rat of study group (a dose of tetracycline hydrochloride 250 μg/kg). Hepatocytes in the state of fatty degeneration (1), hyperemia of the vessels of the particle (2). Stained with hematoxylin Karatsu and eosin, ×200

**Table 1** Morphological blood indices of experimental hepatodystrophy rats against tetracycline hydrochloride at a dose of 250 mg/kg ($M \pm m$, $n = 20$, $P < 0.05$)

| Index | Control group | Study group |
|---|---|---|
| Leukocytes, $10^9$/l | 4.8 ± 0.2 | 6.1 ± 0.5* |
| Erythrocytes, $10^{12}$/l | 6.6 ± 0.5 | 4.8 ± 0.3* |
| Hemoglobin (g/l) | 182.7 ± 12.1 | 141.0 ± 3.6* |
| Mean hemoglobin in erythrocyte | 28.3 ± 1.1 | 29.5 ± 1.7 |
| Hematocrit (%) | 35.2 ± 2.3 | 27.2 ± 1.7* |
| Platelets, $10^9$/l | 470.0 ± 18.0 | 515.0 ± 21.0 |
| Thrombocrit (%) | 0.249 ± 0.018 | 0.322 ± 0.016* |
| Lymphocytes (%) | 54.0 ± 1.1 | 52.0 ± 1.1 |
| Stab neutrophils (%) | 8.0 ± 0.2 | 5.5 ± 0.5* |
| Segmented neutrophils (%) | 33.5 ± 0.8 | 36.0 ± 0.4* |
| Monocytes (%) | 2.5 ± 0.1 | 2.5 ± 0.1 |
| Eosinophils (%) | 1.5 ± 0.2 | 3.5 ± 0.5* |
| Basophils (%) | 0.5 ± 0.0 | 0.5 ± 0.1 |

* results were significant for $p \leq 0.05$ when compared to control meanings

At the same time, the average hemoglobin content in the erythrocyte remains unchanged; it may be a compensatory response to the development of anemia, which we also observed when rats were administered toxic doses of diclofenac sodium [8]. Simultaneous reduction of the hematocrit value by 1.3 times is probably a consequence of a significant decrease in the number of erythrocytes ($P < 0.05$) and consequently proves the presence of anemia in patients with rats. The reason for increasing the value of thrombocrit (by 1.3 times), first of all, may be a violation in the hematopoiesis system or the result of the body's reaction to the development of other pathological processes that lead to stimulation of platelet production and change the parameters of the state of the platelet unit. We have diagnosed changes in

**Table 2** Biochemical indicators of blood plasma in rats of experimental tetracycline-induced hepatosis ($M \pm m$, $n = 20$, $P < 0.05$)

| Index | Control group | Study group |
|---|---|---|
| Total protein (g/l) | 68.0 ± 0.1 | 60.5 ± 0.1* |
| Total bilirubin (μmol/l) | 3.2 ± 0.1 | 45.6 ± 2.9* |
| Conjugated bilirubin (μmol/l) | 1.45 ± 0.23 | 27.33 ± 0.81* |
| Alanine aminotransferase (U/l) | 24.12 ± 3.31 | 72.05 ± 2.38* |
| Aspartate aminotransferase (U/l) | 71.25 ± 3.11 | 126.83 ± 4.04* |
| Alkaline phosphatase (U/l) | 250.02 ± 10.02 | 400.10 ± 9.79* |
| γ-Glutamyl transpeptidase (U/l) | 12.32 ± 0.68 | 21.34 ± 1.10* |

* results were significant for $p \leq 0.05$ when compared to control meanings

thrombocrit in the body of animals suffering from toxic hepatodystrophy, probably due to the development of an acute inflammatory reaction in the liver parenchyma, which is confirmed by the presence of leukocytosis in these animals. In the leukogram of sick animals, a significant increase in the relative number of segmented neutrophils in 1.1 times is observed, along with a decrease in the number of stab nuclear (1.5 times), which indicates a shift of the nucleus of neutrophils to the right and, above all, is the result of toxic effects of tetracycline hydrochloride on the bone marrow.

For these animals, an increase in the amount of eosinophils in the blood is 2.3 times, which is explained by their known antitoxic function. They adsorb toxic products of protein nature and destroy them, and in areas of inflammation phagocytic immune complexes, products of tissue decay, although their phagocytic activity is lower than neutrophils.

**Biochemical Indicators of Blood**. In the study of biochemical indicators of blood plasma in patients with toxic hepatodystrophy in rats (Table 2), there is a significant increase in activity of liver-specific enzymes: alanine aminotransferase 3 times, aspartate aminotransferase—81%, alkaline phosphatase—60% and γ-glutamyl transpeptidase 73% with a simultaneous decrease in the total protein content by 11%, an increase in the concentration of total bilirubin by 15 times due to the conjugated fraction, the level of which increased by 19 times compared to the control, indicating destruction changes in hepatocytes, a decrease in the intensity of protein-synthesizing processes, a violation of pigment metabolism, and the development of intrahepatic cholestasis.

Thus, we proposed a method for modeling the acute form of toxic drug-induced hepatodystrophy in laboratory rats with pronounced clinical, pathologic-anatomical, and biochemical changes in both the liver and the level of the whole organism.

This experimental model can be used, first of all, in medicine and veterinary medicine in order to determine the peculiarities of ultrastructural and metabolic changes in the animal organism in the development of toxic liver dystrophy, as well as for the clinical testing of hepatoprotective profile preparations again and introduction of effective treatment regimens in animal husbandry with a similar manifestation spontaneous hepatopathology. The most important result of the development of experimental hepatodystrophy is the obtaining of desired changes in the liver parenchyma with relatively moderate clinical course in rats and their 100% survival rate.

## 3 Conclusions

### *3.1 Tetracycline-Induced Hepatosis*

As a result of intragastric administration of 4% tetracycline hydrochloride solution to laboratory rats using a probe at a dose of 250 mg/kg once a day, for 7 days a complex of symptoms characteristic for acute toxic dystrophy of liver with an average severity of clinical course and absence lethal cases were marked. Regular administration of tetracycline hydrochloride preparation to rats for seven days is accompanied by the emergence of severe clinical symptoms of acute toxic dystrophy of liver that begin to appear in animals on the second or third day of its administration and are characterized by general depression, decreased appetite, polydipsia or thirst reduction, losses the average body weight for the group of 10–15 g, a dull coat, a decrease in elasticity and dry skin, hair loss, a slight increase stomach, liquefaction of stool.

### *3.2 Pathological Changes in the Liver*

In experimental hepatodystrophy—the liver acquires a light yellow color or has a mottled mosaic pattern, brownish-red areas alternate with yellow, slightly enlarged, rounded edges, flabby consistency. Histomorphological examination shows decomposition of hepatocytes, diffuse placement of fat droplets of different sizes in liver cells, focal histophilic infiltration, expansion, and overflow with blood vessels. In addition to obesity, a pronounced granular dystrophy is revealed.

### *3.3 Morphological Indicators of Blood*

For patients with toxic hepatodystrophy in rats, the development of leukocytosis, erythrocytopenia with simultaneous manifestation of hypochromaemia, a decrease in the hematocrit size and growth of thrombocrit is characteristic, which indicates that these animals have an inflammatory process and anemia. In the leukogram of sick animals, there is a shift of the nucleus of neutrophils to the right, which, above all, is the result of toxic effects of tetracycline hydrochloride on the bone marrow. For these animals is characteristic eosinophilia, which is explained by their antitoxic function of eosinophils.

### 3.4 Biochemical Indicators of Blood

For the biochemical profile of the blood plasma of patients with toxic hepatodystrophy in rats, an increase in activity of enzymes relatively specific for the liver is characteristic: aminotransferases, alkaline phosphatase and $\gamma$-glutamyl transpeptidase and hypoproteinemia, hyperbilirubinemia due to the conjugated fraction, which indicates destructive changes in hepatocytes, a decrease in the intensity of protein-synthesizing processes, violation pigment metabolism, and development of intrahepatic cholestasis.

**Finding**. A method for modeling the acute form of toxic medicamentous hepatodystrophy has been developed. It can be used in experimental medicine and veterinary medicine to determine the peculiarities of ultrastructural and metabolic changes in the organism of animals with a similar spontaneous pathology of the liver, as well as for the clinical testing of hepatoprotective profile preparations again.

## References

1. Calitz, C., Hamman, J.H., Fey, S.J., Wrzesinski, K., Gouws, C.: Recent advances in three-dimensional cell culturing to assess liver function and dysfunction: from a drug biotransformation and toxicity perspective. Toxicol. Mech. Methods **28**(5), 369–385 (2018). https://doi.org/10.1080/15376516.2017.1422580
2. Chernova, V.M.: Liver pathology in patients with blood diseases. Mod. Gastroenterol. **3**(89), 105–113 (2016) (in Ukrainian)
3. Choi, Y.J., Lee, C.H., Lee, K.Y., Jung, S.H., Lee, B.H.: Increased hepatic fatty acid uptake and esterification contribute to tetracycline-induced steatosis in mice. Toxicol. Sci. **145**(2), 273–282 (2015). https://doi.org/10.1093/toxsci/kfv049
4. Deng, Z., Yan, S., Hu, H., Duan, Z., Yin, L., Liao, S., Sun, Y., Yin, D., Li, G.: Proteomic profile of carbonylated proteins in rat liver: discovering possible mechanisms for tetracycline-induced steatosis. Proteomics **15**(1), 148–159 (2015). https://doi.org/10.1002/pmic.201400115
5. Fabbrini, E., Magkos, F.: Hepatic steatosis as a marker of metabolic dysfunction. Nutrients **7**(6), 4995–5019 (2015). https://doi.org/10.3390/nu7064995
6. Forsberg, K.J., Patel, S., Wencewicz, T.A., Dantas, G.: The tetracycline destructases: a novel family of tetracycline-inactivating enzymes. Chem. Biol. **22**(7), 888–897 (2015). https://doi.org/10.1016/j.chembiol.2015.05.017
7. Gryshchenko, V.A.: Biochemical properties of the plasma of rats with the experimentally induced hepatitis after oral administration of sodium diclofenac. Regul. Mech. Biosyst. **8**(2), 191–196 (2017). https://doi.org/10.15421/021730 (in Ukrainian)

8. Gryshchenko, V.A.: Hematological profile of rats in experimental diclofenac-induced hepatitis. Ukrainian J. Ecol. **7**(3), 78–83 (2017). https://doi.org/10.15421/2017073 (in Ukrainian)
9. Gryshchenko, V.A., Tomchuk, V.A., Musiychuk, V.V.: Method modeling of fatty liver. Patent UA, 105657 (2016) (in Ukrainian)
10. Madian, A.G.: Redox proteomic investigation of tetracycline-induced steatosis. Proteomics **15**(1), 8–9 (2015). https://doi.org/10.1002/pmic.201400554
11. Markley, J.L., Wencewicz, T.A.: Tetracycline-inactivating enzymes. Front. Microbiol. **9**(1058) (2018). https://doi.org/10.3389/fmicb.2018.01058
12. Melnychuk, D.O., Gryshchenko, V.A.: Method modeling of toxic hepatitis. Patent UA, 105657 (2016) (in Ukrainian)
13. Melnychuk, D.O., Tomchuk, V.A., Yanchuk, P.I., Gryshchenko, V.A., Reshetnyk, E.M., Synelnyk, T.B., Capenko, P.K., Kartyfuzova, ZhV, Govorukha, T.M., Makarchuk, MYu., Veselskyj, S.P.: Research Methods of Liver and Biliar System Functional State. NUBiP Ukraine, Kyiv (2015) (In Ukrainian)
14. Pototsky, M.K.: Fundamentals of Histopathological Engineering: Methodical Instructions. NUBiP Ukraine, Kyiv (2001) (In Ukrainian)
15. Schueller, F., Roy, S., Vucur, M., Trautwein, C., Luedde, T., Roderburg, C.: The role of miRNAs in the pathophysiology of liver diseases and toxicity. Int. J. Mol. Sci. **19**(1), 261 (2018). https://doi.org/10.3390/ijms19010261
16. Schwartz, J., Holmuhamedov, E., Zhang, X., Lovelace, G.L., Smith, C.D., Lemasters, J.J.: Minocycline and doxycycline, but not other tetracycline-derived compounds, protect liver cells from chemical hypoxia and ischemia/reperfusion injury by inhibition of the mitochondrial calcium uniporter. Toxicol. Appl. Pharmacol. **273**(1), 172–179 (2013). https://doi.org/10.1016/j.taap.2013.08.027
17. Szalowska, E., van der Burg, B., Man, H.Y., Hendriksen, P.J.M., Peijnenburg, A.A.C.M.: Model steatogenic compounds (amiodarone, valproic acid, and tetracycline) alter lipid metabolism by different mechanisms in mouse liver slices. PLoS ONE **9**(1), e86795 (2015). https://doi.org/10.1371/journal.pone.0086795
18. Thakkar, S., Chen, M.J., Fang, H., Liu, Z.C., Roberts, R., Tong, W.D.: The liver toxicity knowledge base (LKTB) and drug-induced liver injury (DILI) classification for assessment of human liver injury. Expert Rev. Gastroenterol. Hep. **12**(1), 31–38 (2018). https://doi.org/10.1080/17474124.2018.1383154
19. Tikhonov, S.N., Rotov, K.A., Alekseev, V.V., Snatenkov, E.A., Khrapova, N.P.: Activity of some enzymes in the liver of experimental animals after treatment with the liposomal formulation of tetracycline and streptomycin. Bull. Exp. Biol. Med. **149**(1), 47–49 (2010). https://doi.org/10.1007/s10517-010-0872-9
20. Tikhonov, S.N., Rotov, K.A., Khrapova, N.P., Alekseev, V.V., Snatenkov, E.A., Zamarin, A.A., Simakova, N.A.: Effect of liposomal tetracycline hydrochloride on enzymatic function of the liver. Bull. Exp. Biol. Med. **145**(4), 443–445 (2008). https://doi.org/10.1007/s10517-008-0113-7
21. Vlizlo, V.V., Fedoruk, R.S., Ratych, I.B.: Laboratory Methods of Investigation in Biology, Stock-Breeding and Veterinary. Spolom, Lviv (2012) (in Ukrainian)
22. Volkova, O.V., Eletsky, Yu.K.: Fundamentals of Histology and Histological Techniques. Medicine, Moscow (1987) (In Russian)
23. Yao, X.M., Li, Y., Li, H.W., Cheng, X.Y., Lin, A.B., Qu, J.G.: Bicyclol attenuates tetracycline-induced fatty liver associated with inhibition of hepatic ER stress and apoptosis in mice. Can. J. Physiol. Pharmacol. **94**(1), 1–8 (2016). https://doi.org/10.1139/cjpp-2015-0074
24. Yu, H.-Y., Wang, B.-L., Zhao, J., Yao, X.-M., Li, Y.: Protective effect of bicyclol on tetracycline induced fatty liver in mice. Toxicology **273**(1–3), 60–60 (2010). https://doi.org/10.1016/j.tox.2010.03.010

# Technological Properties of Winter Wheat Grain Depending on the Ecological and Geographical Origin of a Variety and Weather Conditions

**Hrygoriy Hospodarenko, Olena Cherno, Ihor Prokopchuk and Marina Serdyuk**

## 1 Introduction and Purpose of the Article

The quality of winter wheat grain is formed under the influence of weather conditions and biological characteristics of the variety. The thousand-kernel weight, vitreousness, protein content, gluten content, gluten deformation index, and grain unit characterize the technological properties of wheat grain. It is difficult to overestimate the protein importance. It affects the nutritional value, as well as the technological value of bread. The basis of the production of high-quality grain is a variety which should combine high productivity and good quality. The cultivation conditions have a significant effect on the crop formation and, especially on its quality. It is found that even under the conditions of sufficient moisture, a high air temperature can lead to a shortage of crops and excessive precipitation during the period of filling and maturing grain can reduce its quality. Therefore, in order to improve wheat quality, winter wheat requires knowledge of varietal characteristics in specific growing conditions [1, 2, 3]. There are problems with wheat grain quality when we strive to get high yield. The protein content in wheat grain is a crucial indicator characterizing both the nutritional value and the technological value of grain and flour. According to the Department of Agriculture and Food of Western Australia, weather conditions prevail over the overall effect on the protein content of grain (up to 50%) [4]. The situation is complicated by the fact that protein content is a rather complicated trait for breeding improvement. This is due to its low inheritance and high sensitivity to growing conditions [5, 6]. Scientists [7–10] believe that when selecting wheat varieties with

H. Hospodarenko (✉) · O. Cherno · I. Prokopchuk
Uman National University of Horticulture, Institutska Str. 1, Uman, Cherkassy Region 20305, Ukraine
e-mail: pivotbi@ukr.net

M. Serdyuk
Tavria State Agrotechnological University, B. Khmelnitsky Av., 18, Melitopol, Zaporizhia Region 72312, Ukraine

© Springer Nature Switzerland AG 2019

V. Nadykto (ed.), *Modern Development Paths of Agricultural Production*,
https://doi.org/10.1007/978-3-030-14918-5_68

widespread use of intraspecific crossings of high-yielding local varieties with excellent grain quality and valuable varieties from geographically distant regions, one can obtain a large variety of economic and valuable features. Thus, under the conditions of forest-steppe in Ukraine, the significant influence of the variety genotype on the following parameters of the flour quality was by gluten deformation index, raw gluten content, and sedimentation index [11]. Wheat varieties differ significantly in protein content. This important indicator determines the economic expediency of cultivating a certain variety and determines directions of technological use of grain.

Modern agriculture has a number of varieties that are characterized by good adaptive properties which allows you to obtain a high yield of appropriate quality. However, its level and quality of winter wheat grain depend on weather conditions during the period of winter wheat vegetation which can cause a significant decrease. This forces scholars and manufacturers to look for solutions to this problem. The selection of yielded varieties plays an important role in this aspect. The research purpose is to determine the influence of varieties of different ecological and geographical origin and weather conditions on the quality indicators of winter wheat grain. Research tasks

The research tasks include:

- To analyze the quality indicators of grain depending on the variety;
- To determine the effect of weather conditions on the thousand-kernel weight, vitreousness, protein content, gluten content, and gluten deformation index.

The experimental work was performed on the research field of Uman National University of Horticulture. The agricultural technology of winter wheat cultivation is generally accepted for right-bank forest-steppe of Ukraine. Varieties of winter wheat of various ecological and geographical origins were cultivated in the experiment, and the predecessor was peas. The total plot was 50 $m^2$, the registration plot was 36 $m^2$, and the number of replications was three and consistent allocation. In order to assess the quality of winter wheat grain, protein content was determined according to ДСТУ 4117:2007, gluten content and its quality by ДСТУ ISO 21415-1:2009, thousand-kernel weight by ДСТУ4138–2002, and vitreousness by ГОСТ 10987–76. Mathematical processing of experimental data was carried out by the method of dispersion and correlation analysis.

The largest thousand-kernel weight of winter wheat plants is formed under favorable meteorological conditions during the period of filling and maturing grain. It should be within 45–55 g at yields of 7.0–9.0 t/ha [12, 13]. Weather conditions during the research years were contrasting and significantly influenced the quality of winter wheat grain. Crop vegetation in 2012 was in fairly warm weather. The elevated temperature regime caused the reduction of interphase periods and accelerated development of plants. The increase in the air temperature to 30–31 °C suppressed winter wheat plants. Due to the hot weather with a shortage of precipitation in June, there was an increase in air–soil drought, unfavorable agricultural meteorological conditions for the formation of quality winter wheat yields. There was premature maturing grain before usual for almost 2.5 weeks. Spring and summer vegetation of winter wheat in 2013 took place under the conditions of optimal moisture which

**Table 1** Thousand-kernel weight and vitreousness of winter wheat depending on the variety and weather conditions

| Variety | Thousand-kernel weight (g) | | | | Vitreousness (%) | | | |
|---|---|---|---|---|---|---|---|---|
| | 2012 | 2013 | 2014 | Average for three years | 2012 | 2013 | 2014 | Average for three years |
| *Varieties selected in forest-steppe conditions* | | | | | | | | |
| Podolianka (check variant) | 34.7 | 36.1 | 32.3 | 34.4 | 90 | 84 | 52 | 75 |
| Artemisia | 44.8 | 49.1 | 34.8 | 42.9 | 91 | 87 | 77 | 85 |
| Favorytka | 41.2 | 49.0 | 30.1 | 40.1 | 89 | 80 | 69 | 79 |
| *Varieties selected in steppe conditions* | | | | | | | | |
| Misia Odeska | 43.0 | 49.7 | 28.9 | 40.5 | 79 | 70 | 42 | 64 |
| Vdala | 40.1 | 48.9 | 29.3 | 39.4 | 84 | 75 | 50 | 70 |
| $HIP_{05}$ | 2.0 | 2.3 | 1.6 | 0.0 | 4.3 | 4.0 | 2.9 | 2.0 |

ensured the formation of high yield and grain quality. In 2014, the most critical period in the development of winter wheat (from the shooting stage to the ear formation stage) and in the period from grain formation to waxy ripeness was under the conditions of excessive moisture. It negatively affected the yield of winter wheat, as plants were lying down, affected by diseases and decreased grain quality (Table 1).

The thousand-kernel weight was the highest under favorable weather conditions of 2013. Under the conditions of excessive moisture in 2014, it was the smallest. On average, over three years of research, varieties selected under forest-steppe conditions exceeded Podolianka check variant by 5.7–8.5 g. Artemisia variety was characterized by the highest thousand-kernel weight 42.9 g. The varieties selected in steppe zone, although exceeding the check variant by 5.0–6.1 g, but the thousand-kernel weight was smaller compared to Artemisia and Favorytka. It is found that this decrease was significant in varieties selected in steppe zone, due to excessive moisture in the grain filling season (2014). The baking properties of flour depend on grain vitreousness much. The most important conditions determining it are the soil and air humidity and the level of mineral nutrition [14–17]. During the research, grain was half vitreous (in 2014) and vitreous in 2012 and 2013.

On average, over three years of research, grain vitreousness of winter wheat varieties was in the range of 64–85%. Vitreousness of winter wheat grain in varieties selected under the conditions of forest-steppe was higher in comparison with the check variant of 5–13%. Artemisia variety had the highest vitreousness (85%). Grain of varieties selected under steppe conditions had vitreousness of 64–70%, but this figure was (by 5–9%) smaller compared to the check variant.

Protein is one of the components characterizing biochemical parameters of quality and nutritional value of grain [18, 19]. By protein content of 9–10%, the flour has a low quality. The minimum protein in the Ukrainian wheat to provide satisfactory quality should be 12% [1, 2].

**Table 2** Protein content in grain of winter wheat varieties depending on the weather conditions, %

| Variety | 2012 | 2013 | 2014 | Average for three years |
|---|---|---|---|---|
| *Varieties selected in forest-steppe conditions* | | | | |
| Podolianka (check variant) | 14.8 | 13.7 | 12.3 | 13.6 |
| Artemisia | 16.5 | 16.1 | 12.6 | 15.1 |
| Favorytka | 14.0 | 13.8 | 11.1 | 13.0 |
| *Varieties selected in steppe conditions* | | | | |
| Misia Odeska | 11.0 | 11.4 | 11.3 | 11.2 |
| Vdala | 14.6 | 13.5 | 11.9 | 13.3 |
| $HIP_{05}$ | 0.7 | 0.7 | 0.6 | – |

It is established that the protein content of winter wheat grain also depends on the weather conditions and the place where the variety is selected. In the phase of filling grain in dry conditions in 2012 at the temperature of about 31 °C and the absence of precipitation, the protein accumulation and gluten accumulation was better than in other years of research (Table 2). This can be explained by the inhibition of photosynthetic processes under the arid conditions, increased breathing, and increased carbohydrate expenditure which contributes to the protein accumulation. This year, the highest protein content was observed in Artemisia variety and exceeded the check variant by 1.7%. Favorytka variety was inferior to 0.8%.

Similar patterns were observed in 2013. Instead, in 2014, there was 299 mm of precipitation during April–June that is 109 mm higher than the average multi-year indicator. Winter wheat plants were heavily thickened, overgrown, affected by illnesses and lodged. In turn, it affected the grain quality and led to a decrease in the protein content. Grain of Favorytka, Misia Odeska, and Vdala varieties corresponded to the third degree of quality and grain of Podolianka and Artemisia varieties corresponded to the second degree of quality.

On average, over three years of research, the protein content of winter wheat was in the range of 11.2–15.1% depending on the variety. Among varieties selected in forest-steppe, only Artemisia variety had the high protein content (15.1%) which corresponded to the first degree of quality and exceeded check variant by only 1.5 points or by 11%.

Among the varieties selected under steppe conditions, Misia Odeska variety had the low protein content of 11.2% (third degree of quality). Vdala variety had the protein content of 13.3% (second degree of quality) which is 0.3% less than Podolianka variety.

There is a strong correlation relationship ($r = 0.70$) between winter wheat vitreousness and its protein content which is described by the following regression equation:

$$Y = 6.173x - 6.60$$

x is the protein content,% (Fig.1).

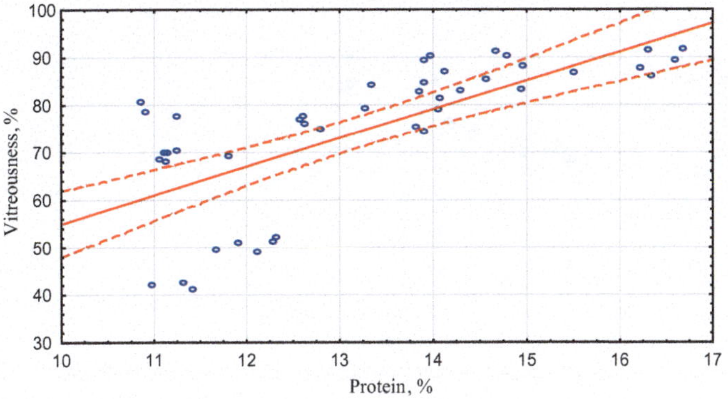


**Fig. 1** Correlation between protein content in and vitreousness, in 2012–2014

**Table 3** Gluten content in winter wheat varieties and gluten deformation index (GDI) depending on the weather conditions

| Variety | Gluten content (%) | | | | GDI, unit | | | |
|---|---|---|---|---|---|---|---|---|
| | 2012 | 2013 | 2014 | Average for three years | 2012 | 2013 | 2014 | Average for three years |
| *Varieties selected in forest-steppe conditions* | | | | | | | | |
| Podolianka (check variant) | 33.1 | 32.7 | 24.3 | 29.2 | 77 | 80 | 98 | 85 |
| Artemisia | 34.2 | 33.4 | 25.4 | 32.0 | 72 | 75 | 80 | 76 |
| Favorytka | 29.2 | 22.7 | 21.2 | 29.0 | 70 | 77 | 88 | 78 |
| *Varieties selected in steppe conditions* | | | | | | | | |
| Misia Odeska | 21.4 | 22.7 | 19.2 | 21.1 | 68 | 79 | 80 | 76 |
| Vdala | 32.1 | 29.8 | 21.3 | 28.2 | 79 | 88 | 94 | 87 |

$y$ is vitreousness, %;
$x$ is the protein content, % (Fig. 1).

On average, over three years of research, the gluten content in winter wheat varieties varied within the range of 21.1–32.0%. Artemisia variety showed the highest gluten content of 32.0% and exceeded check variant by 2.8% among varieties selected in forest-steppe. This indicator of Favorytka variety was at the level of Podolianka variety (Table 3).

The varieties selected in steppe conditions were characterized by a lower content of gluten which was in the range of 21.1–28.2%. It is 1.0–8.1% less than in the check variant. It was particularly low in excessive moisture conditions in 2014 of Misia Odeska variety.

The gluten content in winter wheat grain and its quality can vary considerably. Typically, the gluten content is well correlated with protein content (Fig. 2). How-

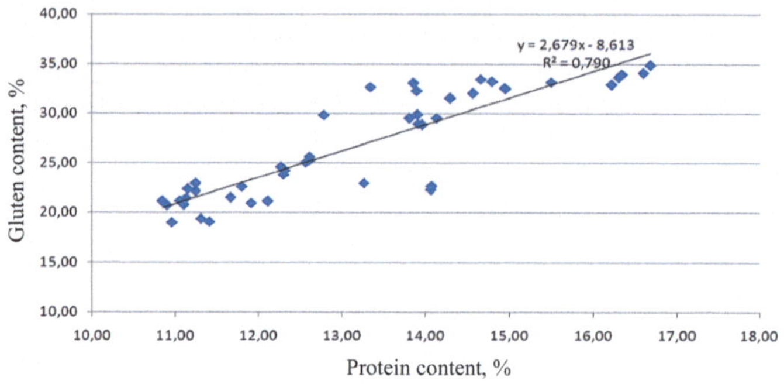


**Fig. 2** Correlation between the gluten content and the protein content in wheat grain (in 2012–2014)

ever, there are cases where the gluten protein may change its physical and chemical properties under the influence of certain factors. This reduces the strength of the hydrated mass of raw gluten, and, accordingly, there is a lower yield of gluten at a slightly higher content of protein in grain. In the first place, this phenomenon can be caused by excessive precipitation [1, 11]. A similar phenomenon was observed in 2014 in Misia Odeska variety when the protein content was lower than the gluten content.

There is a strong correlation ($r = 0.88$) between the gluten content and the protein content in winter wheat grain which is described by the following regression equation:

$$Y = 2.679x - 8.613$$

$y$ is the gluten content, %;
$x$ is the protein content, % (Fig. 2).

The study found that varieties and weather conditions during the period of filling and maturing grain affected its quality in different ways.

In 2012, during hot dry weather and insufficient supply of plants with moisture, during the period of grain maturing, there was the formation of stronger and elastic gluten, but less elastic. Its quality was good and corresponded to the quality group I.

The rainy weather during harvesting in 2014 led to a significant deterioration in the gluten quality which was satisfactorily weak in all varieties.

On average, in 2012–2014, the gluten of winter wheat grain selected in forest-steppe was good in quality and its deformation index was within the range of 76–78 units and responded to quality group I. It was satisfactorily weak in Podolianka and Vdala varieties and responded to quality group II.

## 2 Conclusions

The quality indicators of winter wheat grain vary considerably depending on the variety. Artemisia variety is characterized by the high content of protein selected in forest-steppe. The weather conditions during the grain formation significantly affect its quality. The best quality grain is formed providing plants with moisture, and worse quality grain is formed with excessive moisture during the formation and maturing grain.

## References


1. Rybalka, O.I.: Wheat quality and its improvement. Odesa (2011)
2. Matviyets, V.G.: Problem of soft wheat grain system approach to solving. Bull. Central Sci. Res. Center Kharkiv Reg. **10**, 151–159 (2011)
3. Burdeniuk-Tarasevich, L.V.: Wheat. Depth of genetic potential. Grain **4**(48), 49–51 (2010)
4. Lemon, J.: Nitrogen management for wheat protein and yield in the Esperfnce Port zone. In: Bulletin 4707. Department of Agriculture and Food Australia, pp. 1–27 (2007)
5. Snape, J., Hyne, V., Aitken, K.: Tarqetinq qenes in wheat usinq marker mediated approaches. In: Proceedings of the 8th International Wheat Genetics Symposium, Beijing, China, pp. 749–759 (1993)
6. Jonson, V., Mattern, P., Kuhr, S.: Improvement of wheat protein by traditional breeding and genetic technique. Cereal Chem. **62**, 203–211 (1985)
7. Zhemela, G.P.: Agrotechnical factors of improvement of the quality of winter wheat grains. Sci. Works Poltava State Agrarian Acad. **23**, 115–119 (2005)
8. Deshev, D.: Productivity of selection strategy in durum wheat. In: 5th International Wheat Conference, June 10–14, pp. 19–20. Ankara, Turkey (1996)
9. Dixon, J., Nalley, L., Kosina, P., La, R.: Adaptation and economic impact of improved wheat varieties in the developing world. Rovere Others Agric. Sci. **6**, 489–502 (2006)
10. Lauzon, J., Fallow, D., O'Hallovan, J.: Assessing the temporal stability of spatial parents in crop using combine yield monitor data. Can. Soil Sci. **3**, 439–451 (2005)
11. Demidov, O.A., Vasilenko, N.V., Pravdziva, I.V., Koliuchy, V.T.: Quality indicators of grain of new wheat varieties of soft winter wheat. Myronivska Variety Agro ONE, p. 21 (2017)
12. Cherno, O.D.: Physical and biochemical indicators of grain quality of winter wheat after long fertilizing. Interagency Thematic Sci. Collect. Agric. **1**, 98–102 (2015)
13. Lykhochvor, V.V., Prots, R.R.: Winter Wheat. Ukrainian technologies, Lviv (2002)
14. Hospodarenko, G.M., Cherno, O.D.: The quality of winter wheat grain for long-term use of fertilizers in the field crop rotation. Bull. Uman Natl. Unive. Hort. **1**, 11–15 (2016)
15. Rybalka, O.I., Lytvynenko, M.A.: Wheat quality: Recent genetic aspects of its improvement. Bull. Agrarian Sci. **6**, 25–31 (2007)
16. Podpriatov, G.I., Voitsekhivsky, V.I., Matseiko, D.M.: Fundamentals of Standardization, Quality Management and Certification of Plant Production. Manual. K., Aristei (2004)
17. Shapoval, I.S.: Agrobiological Bases of Formation of Stable Crops of Winter Wheat on Black Chernozems of Left-Bank. Monograph, Chornobai, Forest-Steppe of Ukraine (2012)
18. Linina, A., Ruza, A.: Variety by environment interactions for winter wheat quality traits. Agronomijas vēsti **8**, 122–126 (2005)
19. Krejčirova, L., Capouchova, I., Petr, J., Bicanova, E., Kvapil, R.: Protein composition and quality of winter wheat from organic and conventional farming. Zemdirbyste Agric. **93**(4), 285–296 (2006)

# Multicriteria Optimization of Quality Indicators of Sweet Cherry Fruits of Ukrainian Selection During Freezing and Storage

**Iryna Ivanova, Iryna Kryvonos, Liudmila Shleina, Galina Taranenko and Tatyana Gerasko**

## 1 Introduction

The South of Ukraine is characterized by the great natural opportunities for industrial cultivation of fruit crops. One of the most recognizable features of the Zaporizhia region is the sweet cherry crop. The sweet cherry fruits are widely popular not only as an early fruit crop that opens the fruit season, but also as a source of easily digestible monosaccharides, a powerful immunostimulating complex of many phytonutrients [1–3]. The latter one, together with pectin, creates a double barrier for cholesterol [3, 4]. A wide range of nutrients causes the need for evenly usage in a diet of feed of sweet cherry fruits throughout the year [5].

From 70–80 thousand tons of sweet cherry, which is supplied annually to the markets in Ukraine, 25% are the fruits varieties developed at the Horticultural Research Station named after M. F. Sydorenko [3, 6]. The selection lists of this station have delivered more than 90 promising varieties of sweet cherries to the StateTest [6, 7].

Reducing the losses of the cultivated biological sweet cherry harvest, increasing the efficiency of processing enterprises, and solving the problem of the annual supply of the population by fruits and products of their processing are possible with the use of quick-freezing process. The freezing process of fruits and berries is one of the best ways of their long-term storage with minimal changes in their biochemical composition and quality. One of the factors of the development hindering of production of quick frozen sweet cherry is the inadequate degree of studying of the modern domestic variety of this crop [8, 9]. In most fruit crops, there is a pronounced influence of the variety on the quality change during freezing. Qualitative properties of raw materials, determined by the genetic factor, are corrected by the influence of ecological and agrotechnical conditions of cultivation [10]. Numerous studies have

I. Ivanova (✉) · I. Kryvonos · L. Shleina · G. Taranenko · T. Gerasko
Tavria State Agrotechnological University, B. Khmelnitsky Av., 18, Melitopol, Zaporizhia Region 72312, Ukraine
e-mail: irynaivanova2017@gmail.com

© Springer Nature Switzerland AG 2019

V. Nadykto (ed.), *Modern Development Paths of Agricultural Production*,
https://doi.org/10.1007/978-3-030-14918-5_69

revealed a clear pronounced effect of varietal features of plant objects on the quality of quick frozen products [11].

In recent years, the attention of scientists attracts an evaluation of the suitability for freezing of sweet cherries, especially for the high harmony of fruit flavors. It was found that the best for freezing are the sweet cherries varieties with an intense dark color of a cellular juice, a dense flesh, a small stone, and a thin peel [11, 12]. These include selection varieties of Horticultural Research Station named after M. F. Sidorenko—Natkhnennia, Zustrich, Solidarnist, Amulet, Prykmetna [13].

Analyzing the foregoing and taking into account that the large area of plantations of the investigated culture is located in Ukraine, it is relevant to continue the research work on the determination of suitability of the sweet cherry fruits of new zoned varieties of different ripening periods for freezing and long-term storage.

## 2 Analysis of Literary Data and Problem Statement

Production experience and scientific research have determined that effective in terms of maintaining the quality of frozen fruits is considered to be rapid freezing at a temperature of minus 30–35 °C, using air circulation and storing frozen products at a temperature not higher than minus 18 °C [5, 6].

Over the years, scientists have been conducting studies to evaluate the suitability of different types of plant raw material for freezing [1, 8]. In most of the fruit crops within the kind, there is an effect of the variety on the change of quality during freezing [11, 12]. It is determined that cherry is one of the optimal stone kind for freezing [3, 7]. A number of general requirements for sweet cherry varieties for low-temperature storage have been formed: The fruits should have a uniform, intense-red color, large size, dense pulp, a small stone, a thin skin, and be sweet to taste [10]. As a result of studies of sweet cherry hybrids, the dukes found that varieties with an intense dark coloration of cellular juice are preferable to freezing [4, 13].

Studies have shown that sweet cherry is little used for industrial freezing [3, 8]. The reason for the above fact is that the kind belongs to those fruits whose suitability for quick freezing is determined by the characteristics of the variety [12]. The insufficient degree of the studied modern sweet cherry variety causes the problem of a comprehensive evaluation of the suitability of new varieties to a quick freezing and storage [2, 6].

The problem of choosing an optimal sweet cherry variety for freezing and long-term storage remains unresolved. The varieties of late ripening present the particular interest; the one of the key problems in their evaluation is the lack of developed ranges of values of chemical composition [5, 6].

For the analysis of frozen products, it is important to determine not only physical and biochemical parameters, but also the ability to carry out a comprehensive evaluation of the studied varieties by many incompatible criteria (physical and biochemical parameters of sample varieties) [3, 5], which leads to a scientifically based evaluation of the suitability of sweet cherry varieties for freezing and storage at low temperatures.

## 3 Purpose and Tasks of the Research

The research purpose was to evaluate the effect of quick freezing separated fruits and long-term storage on the quality of parameters of the sweet cherry fruits of new zoned varieties of late-ripening period of time.

In accordance with the purpose, the task was set and solved:

- To establish a complex of parameters of physical-biochemical properties of fruits which will be better for freezing and long-lasting 6-month storage of late sweet cherry variety.

## 4 Materials and Methods of Research

The research was conducted during 2013–2017. The sweet cherry fruits, taken for research, was grown in the south of the Zaporizhzhya region of Ukraine in the Horticultural Research Station named after M. F. Sydorenko. The sweet cherry varieties selected for our research were grown in the climate of a region characterized by insufficient rainfall, uneven distribution over periods of the year, high temperatures in the summer, and also low humidity of air. In the dry steppe zone, there are dry winds almost every two years.

The scheme of the research—the influence of varietal characteristics and freezing on the preservation of physical and biochemical indicators of sweet cherry fruits of a late-ripening period—is presented in Fig. 1.

For the research, the sweet cherry samples of new zoned varieties of late-ripening period were stored for six months.

The average fruit sample is 2.5 kg. The freezing process was carried out in polyethylene packages of 0.5 kg capacity at $t = -30 \pm 1$ °C. The freezing was considered to be complete when the temperature in the center of the fruit reached $-18 \pm 1$ °C.

The research of the chemical and technological characteristics of each sweet cherry variety was carried out on the frozen samples and was stored for six months.

### *4.1 Research Methods*

The evaluation of fruits quality indices was carried out in triple repetition. The amount of juice loss (AJL) was determined by the difference in the mass of frozen fruits

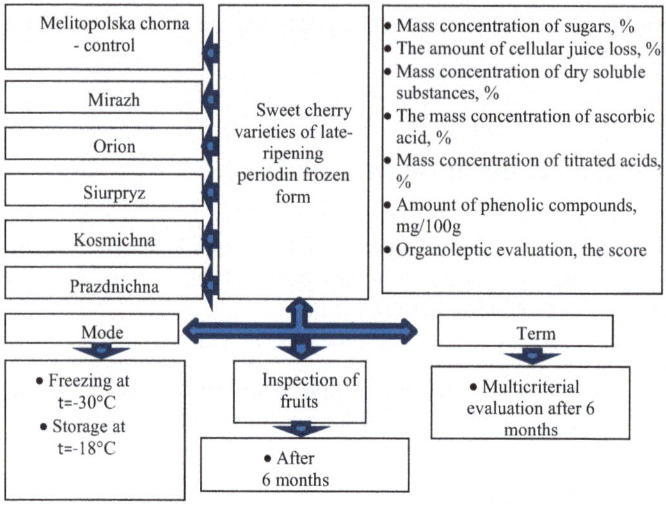


**Fig. 1** Scheme of research on impact of varietal characteristics and freezing on the preservation of physical-biochemical and organoleptic characteristics of sweet cherry fruits of late-ripening period

and defrosted sample varieties with a percentage calculation. Defrostation of frozen samples was carried out by air at $t$ = 18–21 ± 1 °C [3].

The mass concentration of dry soluble substances (DSS) was determined by refractometric method. With a help of press, we squeezed juice from the average test variety sample, and the further juice researches on the content of DSS were performed on a dispersive refractometer MRF-454B2M. The resulting value of the refractive index of the light beam in the transition of a light beam from one medium to the juice of sweet cherry fruits was conducted on a scale in percent content of dry substance. In the calculation of the refractive index, the values of corrections for temperature were taken into account [3].

Determination of the mass fraction of sugars (S), %—by the permanganate method by Bertrand [3]. The method is based on the ability of sugar carbonyl groups to recover copper oxide (II) in copper oxide (I) in alkaline medium. During dissolution, the iron is formed by ammonium oxide iron oxide copper (I), oxidized to copper oxide (II), restores iron (III) into iron (II), the amount of which is determined by titration with a solution of potassium manganese oxide.

The mass concentration of titrated acids (TA) was determined by titerometric method—titration 0.1 H with NaOH solution [3]. By Formula 1:

$$X = \frac{M \cdot K \cdot O_{\mathrm{w}} \cdot 100}{M_{\mathrm{w}} \cdot O_{\mathrm{s}}} \tag{1}$$

where $X$—total acidity, % (100 g); $M$—quantity of 0.1 $n$ of a solution of alkali spent on titration, cm$^3$; $K$—conversion factor for malic acid 0.0067; $O_{\mathrm{w}}$—the volume to which the weight has been proved, ml; $M_w$—weight of the test substance, g; $O_s$—volume of solution taken for titration, ml.

The determination of the mass fraction of ascorbic acid (AA) was carried out by titerometric method, Tillman's paint [3]. The basis of the method for determining the mass fraction of vitamin C is the restoration of the Tillmans reagent (2,6-dichlorophenol-indophenol). The amount of spent on titration of a reagent is calculated by the content of ascorbic acid in the extractors of the Formula 2:

$$X = \frac{M_1 \cdot K \cdot V_1 \cdot 0.088 \cdot 100}{M_2 \cdot V_2} \quad (2)$$

where $X$—mass fraction of ascorbic acid, mg/100 g; $M_1$—number of reagents that went into titration, ml; $K$—correction to a titre of 0.001 $n$ of a solution of a paint; $V_1$—total volume of extraction, ml; $M_2$—weight loss, g; $V_2$—volume of extractor, taken for titration, ml; 0.088—coefficient of conversion of the amount of reagent to ascorbic acid.

The content of polyphenols was determined with a Folin–Denis reagent. The method involves conducting a reaction of complexation of polyphenols with a Folin—Denis reagent and the formation of colorants with the following definition of optical density. The mass fraction of the phenol field was calculated by the Formula 3:

$$X = \frac{c \cdot V \cdot a \cdot 100}{m \cdot V_1} \quad (3)$$

where $c$—mass concentration of polyphenolic compound (routine), which is determined according to the calibration schedule, mg/cm$^3$; $V$—volume of extract, cm$^3$; $V_1$—volume of extract for analysis, cm$^3$; $m$—mass of product weight gain, g (fruits of sweet cherry); 100—coefficient of determination of the polyphenols content per 100 g of product; $a$—is the coefficient of the degree of dilution of the sample extract.

To establish a complex of physical-biochemical parameters of the fruits better for freezing and storage of the sweet cherries of late-ripening period, the method of multicriteria optimization (geometric convolution of the criteria) was applied.

At the heart of the method is the application of the decision-making mechanism according to many criteria, which eliminates the influence of the units of the indicators measurement of the physical and biochemical properties of the fruits, as well as the values of the intervals of the permissible values of each indicator for the choice of the variety (target function).

The physical-biochemical indicators (criteria $A_j$) of the sweet cherry fruits of the late-ripening period during freezing and storage are the amount of juice loss ($A_1$); the dry soluble substance ($A_2$); sugars ($A_3$); titrated acids ($A_4$); ascorbic acid ($A_5$); the amount of phenolic compounds ($A_6$).

To exclude the influence of the measurement units of physical-biochemical and organoleptic indicators of fruits of different varieties, we carry out an operation of rationing that will allow to translate the values of physical-biochemical and organoleptic indicators into dimensionless quantities ($f_j \to \widehat{f}_j$).

Before starting the operation of rationing, we need to set:

(1) Maximum ($f_j^+$) and minimum ($f_j^+$) of the $j$ criterion of the researched varieties ($x_i$);
(2) The optimal value of $j$ criterion by the following rule:

– If the evaluation criterion ($f_j$) tends to a minimum value

$$(f_j^{\text{opt}} \to \min), \text{ then} = f_j^{\text{opt}} = f_j^-;$$

– If the evaluation criterion ($f_j$) tends to a maximum value

$$(f_j^{\text{opt}} \max), \text{ then} = f_j^{\text{opt}} = f_j^+.$$

The aspiration for the optimal value of $j$ criterion ($f_j^{\text{opt}}$ min; $f_j^{\text{opt}}$ max) is taken into account when choosing Formulas (4) or (5) for the operation of rationing

$$\widehat{f}_j(x_i) = \begin{cases} \frac{(f_j(x_i) - f_j^-)}{(f_j^+ - f_j^-)}, \text{ if } f_j^{\text{opt}} \to \max \\ \frac{(f_j^+ - f_j(x_i))}{(f_j^+ - f_j^-)}, \text{ if } f_j^{\text{opt}} \to \min \end{cases} \tag{4, 5}$$

– The value of the $j$ criterion in the normalized look for the $i$ variety;
– The value of $j$ criterion for the $I$ variety in the corresponding units of measurement;
– The area of admissible values of the $j$ criterion of the comparable varieties.

After the operation of rationing, the values of the target function ($\varphi$) for each variety ($x_i$) are calculated according to the Formula (6):

$$\varphi(x_i) = \sum^{n} \left| \widehat{f}_j(x_i) - \widehat{f}_j(x^u) \right| \to \min, \quad \text{where } 0 \le \widehat{f}_j(x_i) \le 1;$$

$$\widehat{f}_j(x^u) = 1 \tag{6}$$

$\varphi(x_i)$ the target function of the $i$ varieties;
$N$ the number of criteria.
$\widehat{f}_j(x_i)$ the value of the $j$ criterion in the normalized look for the $i$ variety;
$\hat{f}_j(x^u)$ the value of the $j$ criterion in the normalized look for the ideal variety;
$x^u$ the ideal variety (with optimal criteria values)

The proof that $\widehat{f}_j(x^u) = 1$.

If $f_j^{\text{opt}} \to \max$, then according to the Formula (1), the value of the $j$ criterion in the normalized look for the ideal variety for the verification we can calculate by the Formula (4):

$$\widehat{f}_j(x^u) = \frac{f_j(x^u) - f_j^-}{f_j^+ - f_j^-}, \text{ as } f_j(x^u) = f_j^{\text{opt}} = f_j^+, \text{ then}$$

$$\widehat{f}_j(x^u) = \frac{f_j^+ - f_j^-}{f_j^+ - f_j^-} = \frac{1}{1} = 1 \tag{7}$$

If $f_j^{\text{opt}} \to \min$, then according to the Formula (2), the value of the $j$ criterion in the normalized form for the ideal variety for verification we can calculate by the Formula (5);

$$\widehat{f}_j(x^u) = \frac{f_j^+ - f_j(x^u)}{f_j^+ - f_j^-}, \text{ as } f_j(x^u) = f_j^{\text{opt}} = f_j^-, \text{ then}$$

$$\widehat{f}_j(x^u) = \frac{f_j^+ - f_j^-}{f_j^+ - f_j^-} = \frac{1}{1} = 1 \tag{8}$$

The choice of the best variety is determined from the conditions of maximally approximating its target function to the target function of the ideal variety, which is equal to zero.

We will prove that $\varphi(x^u) = 0$.

According to the Formula (3), $\varphi(x^u) = \sum^n \left| \widehat{f}_j(x^u) - \widehat{f}_j(x^u) \right| = \sum^n |1 - 1| = 0.$

Consequently, if the smaller quantity of the target function of the variety in the range of the criteria values of the researched varieties, the more suitable it is to freezing and long-term low-temperature storage. On this principle, the construction of the rank row and the choice in the range of values of the indicators of comparable varieties of the complex of parameters of the physical-biochemical properties of the fruits better for freezing and the long-term low-temperature storage of the sweet cherry variety within the limits of the researched ripening period is established.

## 5 Research Results

The variation of the mean values of experimental data for years of research on the size of juice loss immediately after defrosting of fruits after 6 months of storage occurs in the range of 12.9–16.7% (Table 1). The minimum value of the juice loss was recorded in the variety sample Mirazh (12.9%) and Prazdnichna (13.1%), which is significantly less than the value of this indicator in the control variety Melitopolska chorna 14.1% ($HIP_{05} = 0.7\%$). The difference between the content of the researched index in the fruits of Orion variety is not reliable, is equal to control, and is 14.8%.

The study of the amount of dry soluble substances in the sweet cherry fruits of analyzed varieties showed that the index varies in the range of 16.4–18.9%. The maximum safety of dry soluble substances was noted in Kosmichna variety (18.9%). Prazdnichna and Orion varieties have a stable minimum content of the researched

**Table 1** Value results of the target functions $\varphi(x_1)\ldots\varphi(x_6)$ when selecting the optimal sweet cherry variety for quick freezing and storage after six months

| Alternatives | | Criteria, $A_j$ | | | | | | | | | | | | The values of target functions, $\varphi(x_i)$ | Rank |
|---|---|---|---|---|---|---|---|---|---|---|---|---|---|---|---|
| Variety | | The amount of juice loss (%), $A_1$ | | Dry soluble substances (%), $A_2$ | | Sugars (%), $A_3$ | | Titrated acids (%), $A_4$ | | Ascorbic acid (mg/100 g), $A_5$ | | The sum of phenolic compounds (mg/100 g), $A_6$ | | | |
| | | $f_1$ | $\hat{f}_1$ | $f_2$ | $\hat{f}_2$ | $f_3$ | $\hat{f}_3$ | $f_4$ | $\hat{f}_4$ | $f_5$ | $\hat{f}_5$ | $f_6$ | $\hat{f}_6$ | | |
| $x_1$ | Melitopolska chorna - control | 14.1 | 0.68 | 17.8 | 0.56 | 12.0 | 0.04 | 0.52 | 0.35 | 5.9 | 0.27 | 510.0 | 1.00 | 2.90 | 3 |
| $x_2$ | Mirazh | 12.9 | 1.00 | 18.5 | 0.84 | 14.5 | 1.00 | 0.74 | 1.00 | 7.2 | 0.77 | 410.7 | 0.62 | 0.77 | 1 |
| $x_3$ | Orion | 14.8 | 0.50 | 17.2 | 0.32 | 11.9 | 0.00 | 0.40 | 0.00 | 6.2 | 0.38 | 430.1 | 0.70 | 4.60 | 6 |
| $x_4$ | Siurpryz | 14.9 | 0.47 | 18.1 | 0.68 | 13.1 | 0.46 | 0.44 | 0.12 | 6.1 | 0.35 | 240.9 | 0.00 | 3.92 | 5 |
| $x_5$ | Koamichna | 16.7 | 0.00 | 18.9 | 1.00 | 12.1 | 0.08 | 0.45 | 0.15 | 5.2 | 0.00 | 337.0 | 0.36 | 3.41 | 4 |
| $x_6$ | Prazdnichna | 13.1 | 0.95 | 16.4 | 0.00 | 12.1 | 0.08 | 0.55 | 0.44 | 7.8 | 1.00 | 469.9 | 0.84 | 2.69 | 2 |
| | $HIP_{05}$ | 0.7 | | 0.4 | | 0.8 | | 0.11 | | 1.1 | | 37.1 | | | |
| $f_j^-$ | | 12.9 | | 16.4 | | 11.9 | | 0.40 | | 5.2 | | 240.9 | | | |
| $f_j^+$ | | 16.7 | | 18.9 | | 14.5 | | 0.74 | | 7.8 | | 510.0 | | | |
| $f_j(x^u)$ | | | 1 | | 1 | | 1 | | 1 | | 1 | | 1 | | |
| $f_j^{opt}$ | | 12.9 (min) | | 18.9 (max) | | 14.5 (max) | | 0.7 (max) | | 7.8 (max) | | 510.0 (max) | | | |

index of 16.4 and 17.2%, respectively. The changes in the analyzed index in the sweet cherry fruits of the researched varieties immediately after freezing are statistically significant.

The sugars content in the researched sweet cherry varieties varies in the range of 11.9–14.5%. Fruits of the varieties Prazdnichna (12.1%), Kosmichna (12.1%), Orion (11.9%), and Melitopolska chorna (12.0%) have no statistically significant difference in sugar content. The fruits of Mirazh variety are characterized by the highest sugar content of 14.5%.

The titrated acid content in frozen sweet cherry fruits after the storage of 6 months for the Kosmichna and Prazdnichna varieties is at the Melitopolska chorna standard level, and the difference between the indexes is not statistically significant. Mirazh variety is marked with a maximum content of the researched index of 0.74%, which is statistically significant in terms of six varieties of late-term ripening.

The mass concentration of ascorbic acid varies in frozen variety samples from 5.2 to 7.8 mg/100 g. The content of the indicator in the fruits of Mirazh and Prazdnichna varieties is 7.2 mg/100 g and 7.8 mg/100 g, which is significantly higher relative to the Melitopolska chorna control variety—5.9.

With a long 6-month storage in late varieties of sweet cherries, the variation in the amount of phenolic compounds occurs in a wide range and makes 240.9–510.0 mg/100 g. The varieties fruits Mirazh, Orion, Siurpryz and Prazdnichna inferior to the content of the researched index for the control variety after 6 months of storage, which is statistically reliable.

Based on the analysis of the experimental material, it should be noted:

- Maximum preservation of cellular juice, sugars, titrated acids after freezing, and low-temperature storage for six months was noted in the zoned variety Mirazh—12.9, 14.5, 0.74%—respectively;
- The frozen variety samples of late sweet cherries Kosmichna when stored are characterized by a maximum dry soluble substances content (18.9%) and exceed this index in relation to all varieties, including the Melitopolska chorna control variety (17.8%).
- Relatively less oxidation of ascorbic acid after 6 months of storage is noted in the zoned variety Prazdnichna—7.8 mg/100 g.
- After the concentration of the amount of phenolic compounds after six months of storage, it should be noted the Melitopolska chorna variety of late-ripening period of time (510.0 mg/100 g).

Taking into account the wave-like decreasing nature of the change in physical-biochemical indicators of sweet cherry fruits during storage up to 9 months for obtaining a quick frozen late-ripening sweet cherries, with a maximum content of dry soluble substances, sugars, titrated acids, the sum of phenolic compounds, ascorbic acid, and the minimum values of the juice loss index long-term storage of fruits should be limited to six months. On the basis of determining the optimal shelf life of sweet cherry fruits of the researched varieties samples, the choice of the optimal sweet cherry varieties for quick freezing and long-term storage by multicriteria optimization was carried after 6 months.

## 6 Discussion of Research Results

The analysis of the experimental data requires a comprehensive comparative evaluation of the researched varieties in all quality indices, so selecting the optimal sweet cherry variety for quick freezing and long-term storage by multicriteria optimization was carried out after 6 months.

In the values analysis of the target functions, a range of varieties is ranked according to the degree of suitability for freezing and 6-month storage (Table 1). According to the data, the prevailing number of researched modern zoned sweet cherry varieties in the southern Steppe of Ukraine by a set of qualitative indicators of quick frozen fruits exceeds the control variety—Melitopolska chorna. Within the researched group of late varieties, the new zoned variety Mirazh (1 rank) was found to be the best for freezing and 6-month storage $-\varphi(x_2) = 0.77$. Melitopolska chorna as a control variety has the rank 3 by the value of the target function—$\varphi(x_1) = 2.9$, and the zoned variety Prazdnichna by the complex of physical-biochemical indicators has received the value $\varphi(x_6) = 2.69$ and holds the rank 2. The value of the target functions of variety samples Kosmichna ($\varphi(x_5) = 3.41$) and Siurpryz ($\varphi(x_4) = 3.92$) has given us an opportunity to evaluate frozen products in a complex way and get ranks 4 and 5 accordingly. The maximum value of the target function $\varphi(x_3) = 4.60$, and the frozen Orion variety fruits have received rank 6.

## 7 Conclusions

The comparative evaluation of the researched varieties according to many incommensurate criteria (physical and biochemical parameters of the fruits) by the method of multicriteria optimization (geometric convolution criterion) allowed to exclude the influence of the measurement units of qualitative indicators, as well as the size of the intervals of the permissible values of each indicator for the target function—$\varphi(x_i)$. The values results of target functions $\varphi(x_1)\ldots\varphi(x_6)$ when selecting the optimal sweet cherry variety for quick freezing and storage for six months, it was found that the best fruits for freezing in the specified optimal term are the fruits of Mirazh variety. Based on the values of physical-biochemical indicators of Mirazh variety, a set of parameters has been developed that allows scientifically to predict the greatest suitability for freezing and storage of sweet cherry late varieties: the amount of juice loss immediately after freezing—11.5%; the initial concentration of dry soluble substances—19.3%; sugars—15.2%; titrated acids—0.79%; ascorbic acid—8.1 mg/100 g; the sum of bioflavonoids—530.7 mg/100 g.

## References

1. Ruliev, V.A.: Gardening of Ukraine south (in Ukraine). Wild Field, Zaporizhzhya (2003)
2. Carranca, C., Brunetto, G., Tagliavini, M.: Nitrogen nutrition of fruit trees to reconcile productivity and environmental concerns. Plants (Basel) **7**, 4 (2018). https://doi.org/10.3390/plants7010004
3. Statista: http://www.statista.com/statistics/264001/worldwide-production-of-fruit-by-variety (2016)
4. Fageria, N.K., Baligar, V.C., Li, Y.C.: The role of nutrient efficient plants in improving crop yields in the twenty first century. J. Plant Nutr. **31**, 1121–1157 (2008). https://doi.org/10.1080/01904167.2016.1249798
5. Granatstein, D., Sanchez, E.: Research Knowledge and Needs for Orchard Floor Management in Organic Tree Fruit Systems. Inter. J. Fruit Sci. **9**(3), 257–281 (2009). https://doi.org/10.1080/15538360903245212
6. Bakshi, P., Wali, V.K., Iqbal, M., et al.: Sustainable fruit production by soil moisture conservation with different mulches: a review. Afr. J. Agric. Res. **10**(52), 4718–4729 (2015). https://doi.org/10.5897/AJAR
7. Juarez-Escario, A., Conesa, J.A., Sole-Senan, X.O.: Management as a driver of functional patterns and alien species prominence in weed communities of irrigated orchards in Mediterranean areas. Agric. Ecosyst. Environ. **249**, 247–255 (2017). https://doi.org/10.1016/j.agee.2017.07.042
8. Pochynok, H.N.: Metody biokhymycheskoho analyza rastenyi (in Russian). Nauk, Dumka, Kiev (1976)
9. Karpenchuk, G.K., Melnyk, A.V.: Uchety, nabliudenyia, analyzy, obrabotka dannykh v opytakh s plodovymy i yahodnymy rastenyiamy: Metodycheskye rekomendatsyy (in Russian). Uman. s.-kh. yn-t, Uman (1987)
10. Taiz, L., Zeiger, E.: Plant Physiology, 5th ed. (pp. 67–86). Sinauer Associates, Sunderland, MA, USA (2010)
11. Lishchenko, V.F.: Mirovaya prodovol'stvennaya problema: belkovyye resursy (1960–2005 gg.). M.: DeLi print. 272 (2006)
12. Mittler, R.: Oxidative stress, antioxidants and stress tolerance. Trends Plant Sci. **7**(9), 405–410 (2002)
13. Serdyuk, M., Stepanenko, D., Baiberova, S., Gaprindashvili, N., Kulik, A.: Eur. J. Enterp. Technol. **4/11**(82), 62–68 (2016). https://doi.org/10.15587/1729-4061.2016.76235

# Effect of Preparations Methyure (6-Methyl-2-Mercapto-4-Hydroxypyrimidine) on Corn (*Zea Mays* L.) Biological Productivity Under Saline Soil Conditions

**Maksym Kolesnikov, Yuliia Paschenko, Halyna Ninova, Maryna Kapinos and Anastasiia Kolesnikova**

## 1 Introduction

Climate change, which has recently been observed, increases the influence of abiotic environmental factors on plants [1]. Soil salinization has become a typical factor for the southern regions of Europe. It leads to a significant loss of crop yields. Droughts and salinization of soils are spreading rapidly, especially on irrigated fields, and discontinue use of big squares of arable land annually. It is projected that by the middle of the twenty-first century only half of them will preserve, but their fertility will be halved [2]. According to the international standards, the soil is considered saline, if the electrical conductivity of their water extraction is more than 4.0 $dSm^{-1}$, corresponding to 40 mM NaCl [3]. The suitability of such soils for agricultural use depends on their water–salt regime and ionic composition, in which the most dangerous ion is $Na^{+}$ due to its toxicity to plant organisms. Only certain plant species adapted to the existence in the saline environment by means of various methods for ion accumulation preventing. In Ukraine, according to the Land Cadastre Administration, there are about 4 million ha of salt soils, including 2.7 million ha of arable land, and these squares are increasing every year.

Attempts to reduce the salinization of arable soils by washing or the use of phytoremediators were ineffective, and gypsum of soil is almost not carried out. Salt stress leads to violations of physiological and biochemical processes in plant organism. It is noticed changes in cell division, stomatal conductance, carbon assimilation, absorption of mineral nutrients enzymatic activities of plants. The concentration of activated Oxigene forms and peroxidation products in cells increased, and the func-

M. Kolesnikov (✉) · Y. Paschenko · H. Ninova · M. Kapinos
Tavria State Agrotechnological University, B. Khmelnytsky Av., 18,
Melitopol, Zaporizhia Region 72312, Ukraine
e-mail: maksym.kolesnikov@tsatu.edu.ua

A. Kolesnikova
Taras Shevchenko National University of Kyiv, Hlushkova Av., 2, Kiev 03127, Ukraine

© Springer Nature Switzerland AG 2019

V. Nadykto (ed.), *Modern Development Paths of Agricultural Production*,
https://doi.org/10.1007/978-3-030-14918-5_70

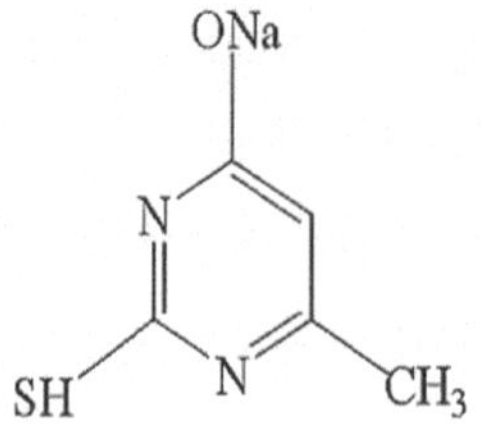

Methyure (6-methyl-2-mercapto-4-hydroxypyrimidine)

**Fig. 1** Structural formula of preparations Methyure (Mr = 165 g/mol)

tion of the antioxidant system of the plant organism activated under salt stress. High concentrations of salts, like many other stressors, inhibit plant growth [4].

Therefore, the direction of research was aimed at increasing of the plant salt tolerance. However, the creation of salt-resistant varieties by classical selection methods proved to be impossible. These methods did not guarantee the preservation of a single cell salt-resistant in the whole organism during ontogenesis [5].

It suggested a method to increase corn bioproductivity on saline soils using synthetic phytohormones of 6-benzylaminopurine, Ivin preparation [6], products of thermophilic methane fermentation [7]. However, their positive effect was observed only during the vegetation period of young plants in boxes with the presence of 0.1 M NaCl in soil, the results were obtained on different varieties, and hybrids varied significantly. It showed positive effect of gibberellic acid (0.3 mM) and kinetin (0.05 mM) [8], brassinosteroids [9], salicylic acid [10], jasmine acid and ascorbate [11] on the growth regulation of plants under various types of soil salinity and other abiotic stresses [12].

The preparations Methyure (6-methyl-2-mercapto-4-hydroxypyrimidine) has a low toxicity of $LD_{50} = 4000$ mg/kg (Fig. 1) [13].

It was found that Methyure stimulated the growth of corn seedlings and roots, while these exposed in salt medium. Methyure treatment ($10^{-7}$ M) of seeds caused a decrease in the content of the ABA hormone by 51%, an increase in the content of IOA—by 2.21 times and CK—by 6% compared to untreated plants on the salt medium [14].

The action of Methyure was directed at the accumulation of roots mass, at the formation of generative organs, at the increasing grain yield under corn cultivation on salt soils. The influence of various forms of Methyure on the biometric parameters of corn seedlings in the early stages of ontogeny was demonstrated [15].

Corn is one of the most valuable crops. In world agriculture, maize takes the third place after wheat and rice in terms of sown area and gross harvest of grain. According to the USDA, world corn production in 2016/2017 was 980 million tons, with an average yield of 5.0–5.2 t/ha. Ukraine is among the five largest corn producers in the world with a volume of up to 30 million tons. Much of the corn crops in our country are located in the areas with insufficient humidification, high summer air temperatures, soils of varying degrees of salinity, which significantly reduces grain yield [16]. Consequently, the study of the corn salt-protective effect of the Methyure under the conditions of south of Ukraine saline soils is relevant and has a great practical importance.

The aim of the work was to determine the influence of the preparations Methyure on the corn growth, development and yield formation under the conditions of the semiarid region saline soils of Ukraine.

## 2 Material and Methods

### *2.1 Materials*

The work was laid on the experimental field of Tavria State Agrotechnological University (Melitopol, Zaporizhia region, Ukraine). The corn seeds of the hybrid Monsanto DKC 5143 were used.

It used preparations Methyure (6-methyl-2-mercapto-4-hydroxypyrimidine) synthesized at the Institute of Bioorganic Chemistry of the National Academy of Sciences of Ukraine and kindly given by Palladina T. A. (D.Sc., Professor of the M. G. Kholodny Institute of Botany).

### *2.2 Soil Description*

Field plot is located on the Haplic Kastanozem with a pH value of 7.3, humus (by Tyurin)—2.6%, nitrogen (according to cornfield)—111.3 mg/kg, mobile phosphorus (by Chirikov)—153.7 mg/kg, exchangeable potassium (by Chirikov)—255 mg/kg, salinity type—chloride, salinity level—slightly saline [17].

### *2.3 Design of Experiment*

The maize seeds of the experimental variants processed semi-volatile by the solutions of Methyure at concentrations ($10^{-3}$, $10^{-4}$, $10^{-6}$, $10^{-8}$ M) and with the addition of Liposam adjuvant (5 ml/L). Sowing is carried out in a well-prepared soil. The seed rate was 80,000 seeds/ha. The accounting area of one plot was 3 $m^2$. All experiments were performed in four repeats, and variants were placed systematically [18].

The cracking stage was marked for 12 days after sowing. The first foliar treatment of crops was carried out at the phase of 6–7 leaves (16–17 BBCH code) and the second treatment—at the beginning of tassel emergence (51 BBCH code). The plant samples (ten plants) were taken from each plot using the randomization method at the stages of 5–6 leaves unfolded (15–16 BBCH code), stem elongation (30–31 BBCH code), tassel emergence (5–57 BBCH code) [19].

### 2.4 Determination of Seed Germination

At the end of cracking stage, the field germination of corn seeds was determined. It counted the number of seedlings in the area of 1 $m^2$ and calculated the field germination in relation to the number of sown seeds in a mentioned area [18].

### 2.5 Determination of Leaf Area Index (LAI)

After leaf collection, leaf area was calculated by means of gravimetric techniques. The gravimetric method correlates weight of leaves from ten plant samples to weight of leaf die-cuts ($D = 10$ mm). Knowing the area leaf die-cuts, the area of the leaf surface was calculated and resulted in 1 $m^2$ of sowing area [20].

### 2.6 Determination of Chlorophyll Content

Total chlorophylls were determined in leaves fluorometrically (N-tester, "Yara", Japan). The measurement point should be in the middle of the first developed leaf. Thirty random measurements in the plot performed using the usual "*W*" scheme give an average value. The results were presented in conventional units [21].

### 2.7 Determination of Netto Photosynthesis Productivity (NPP)

The netto photosynthesis productivity is the ratio of the growth of plant dry matter mass (g) for a certain period of time (days) to the unit of leaf area ($m^2$). NPP was determined periodically by sampling plants. Calculation of NPP was made using the formula (1):

$$\mathrm{NPP} = \frac{B_2 - B_1}{0.5(S_1 + S_2)n} \tag{1}$$

where

NPP netto photosynthesis productivity, g/$m^2$ * day;

$B_1$ and $B_2$ — plant dry matter mass in the beginning and at the end of the accounting period, g;
$(B_2 - B_1)$ — the growth of the plant dry matter mass in $n$ days, g;
$S_1$ and $S_2$ — leaves' area at the beginning and at the end of the accounting period, $m^2$;
$0.5(S_1 + S_2)$ — average working leaves' area during the experiment;
$n$ — period between two observations, days [22].

## 2.8 Determination of Corn Biological Yield

The determination of biological yield elements (average number of ears per plant, average weight of 1 kernel, weight of 1000 seeds, seed moisture, grain nature, individual grain productivity, biological yield) was provided according to the generally accepted agrobiology techniques [18]. Harvesting was carried out manually.

All obtained results were analyzed statistically, and their reliability was checked by the Student test and the least significant difference using Statistica 6.0 software. Statistical tests with $p = 0.05$ level were considered significant, and the null hypothesis was rejected.

## 2.9 Calculating of Energetic Factor

The energy factor is calculated as the ratio of the energy received from the crop to the total amount of expended anthropogenic energy using the formula (2):

$$K_e = \frac{GE_Y}{E_a} = \frac{Q_c \times Y \times C_d}{E_a} \qquad (2)$$

where

$K_e$ — energy factor, MJ;
$GE_Y$ — output of gross energy from 1 ha, MJ
$Q_c$ — total energy in 1 kg of dry matter, MJ/kg;
$Y$ — yield of agricultural crop t/ha;
$C_d$ — content of dry matter, %;
$E_a$ — anthropogenic energy, MJ.

When calculating anthropogenic energy, we took into account the energy of fuel, electricity, energy consumption for the production and application of fertilizers, pesticides, seeds and other substances, labor costs, depreciation costs, equipment repair, etc. [23].

# 3 Results and Discussion

Pre-sowing treatment of agricultural crop seeds with complexes of fungicides, trace elements, inoculants, antistresant can improve the production efficiency. Thus, it observed a positive effect on the germination of the seeds after pre-sowing treatment of maize seeds with Methyure in all concentrations. Seed treatment with Methyure at the concentration ($10^{-3}$ M) was maximally affected on the germination of maize and reached 81.7%, while in the control variant this indicator was 75.8%.

It is previously shown that the main reason for the salt-protective effect of Methyure is to increase the removal of $Na^+$ from the cytoplasm of cells by activating the work of $Na^+/H^+$-anteners. The positive effect of pre-sowing seed treatment with Methyure on the growth of seedlings is due to the inhibition of peroxidation processes in their tissues, due to the activation of the endogenous protective system [24].

Preparation Methyure ($10^{-3}$ M) the most efficiently increased LAI of maize at the phase of 5–6-leaves (BBCH code 15–16) (Table 1).

In the further ontogeny, it observed an increasing LAI at 36–40% of corn crops under the influence of Methyure at the range of $10^{-3}$–$10^{-6}$ M. At the experimental maize plants in the tassel stage was noted the exaggeration of LAI at 12% compared with this indicator in control crops. The lower concentration of Methyure ($10^{-8}$ M) did not significantly affect the leaf area index of maize crop.

The data on the chlorophyll content determined by fluorimetric method using the N-tester in corn plants of the hybrid DKC 5143 under the influence of Methyure are presented in Table 2. During the vegetation period, it was found that the preparation Methyure was promoted to the biosynthesis of chlorophyll. At the same time, the results obtained are ambiguous and statistically insignificant. In general, the content of chlorophyll increased by 2.5–10.3% under the effect of various concentrations of preparation compared with control (Table 2).

There is a link between the production process and the photosynthetic parameters [25]. However, it is often difficult to find a quantitative correlation between the intensity of photosynthesis and the productivity of plants in crops; i.e., first and

**Table 1** Leaf area index ($m^2/m^2$) of corn hybrid DKC 5143 under the Methyure effect

| Experiment factor | Code of BBCH phenological scale | | |
|---|---|---|---|
| | 15–16 | 30–31 | 51–57 |
| Control | 0.187 ± 0.025 | 0.972 ± 0.084 | 2.764 ± 0.154 |
| Methyure ($10^{-3}$ M) | 0.247 ± 0.028* | 1.321 ± 0.105* | 3.091 ± 0.187 |
| Methyure ($10^{-4}$ M) | 0.207 ± 0.015 | 1.345 ± 0.125* | 3.060 ± 0.193 |
| Methyure ($10^{-6}$ M) | 0.192 ± 0.020 | 1.356 ± 0.098* | 3.107 ± 0.205* |
| Methyure ($10^{-8}$ M) | 0.163 ± 0.018 | 1.164 ± 0.119 | 2.772 ± 0.214 |

*—the difference is significant at the $p \leq 0.05$ level

**Table 2** Chlorophyll content (c.un.) in corn leaves hybrid DKC 5143 under the Methyure effect

| Experiment factor | Code of BBCH phenological scale | | |
|---|---|---|---|
| | 15–16 | 30–31 | 51–57 |
| Control | 459.3 ± 15.6 | 546.8 ± 27.6 | 561.3 ± 20.1 |
| Methyure ($10^{-3}$ M) | 483.0 ± 2.6 | 559.8 ± 23.0 | 565.4 ± 19.8 |
| Methyure ($10^{-4}$ M) | 468.7 ± 11.5 | 548.3 ± 9.7 | 564.0 ± 12.0 |
| Methyure ($10^{-6}$ M) | 460.7 ± 2.1 | 558.8 ± 10.4 | 572.6 ± 15.3 |
| Methyure ($10^{-8}$ M) | 459.3 ± 18.0 | 567.5 ± 22.3 | 575.2 ± 18.1 |

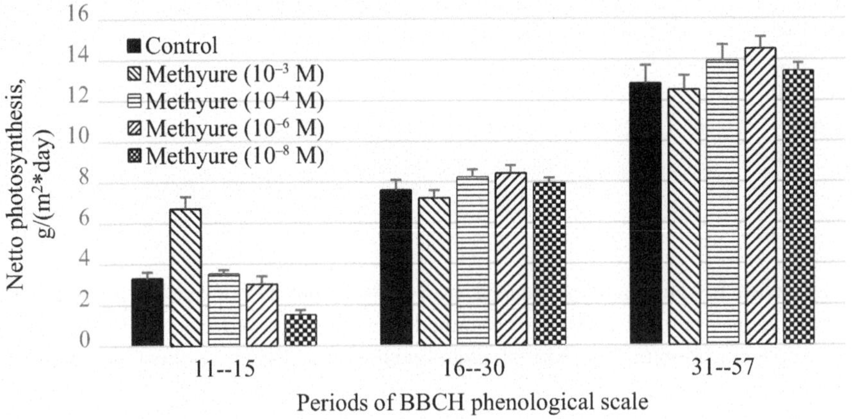


**Fig. 2** Changes of netto photosynthesis productivity of corn hybrid DKC 5143 under the Methyure effect

foremost, they depend on the conditions of the environment. It was found that the investigated preparation increased the photosynthesis activity during corn vegetation (Fig. 2).

In the early stages of the vegetation, the netto photosynthesis productivity of the maize crops exceeded the control value by almost 2 times only under the effect of Methyure at concentration of $10^{-3}$ M. However, it noticed the excretion of corn crop netto photosynthesis at 4.0–13.3% under the foliar treatments by Methyure (from $10^{-4}$ M to $10^{-8}$ M) in subsequent periods of corn vegetation. Methyure ($10^{-6}$ M) showed the most expressive effect on corn plants at tassel stage.

Table 3 presents the data about effect of Methyure treatment on bioproductivity of corn crops.

It is known that the adaptive strategy of plants is some anatomic–morphological changes under the influence of abiotic factors [26]. In particular, plant height refers to such indicators. It was found that Methyure caused a decrease in plant height by 8.8%, which suggests a better adaptation of plants to the effect of osmotic stress. In addition, there was a constant amount of leaves on plants due to shortening interstices.

**Table 3** Effect of Methyure on biological yields of corn hybrid DKC 5143

| Experiment factor | Index | | | | | | |
|---|---|---|---|---|---|---|---|
| | Plant altitude (cm) | Weight of 1000 seeds (g) | Average number of ears per plant, unit | Weight of 1 kernel (g) | Grain nature (g/L) | Individual grain productivity (g) | Biological yield (c/ha) |
| Control | 215.0 | 233.2 | 1.00 | 17.2 | 887.0 | 83.6 | 36.2 |
| Methyure ($10^{-3}$ M) | 203.8 | 237.2 | 1.06 | 20.3 | 904.0 | 99.9 | 45.8 |
| Methyure ($10^{-4}$ M) | 201.1 | 241.1 | 1.05 | 15.5 | 910.0 | 94.4 | 42.0 |
| Methyure ($10^{-6}$ M) | 206.0 | 238.8 | 1.04 | 17.4 | 887.0 | 86.7 | 40.5 |
| Methyure ($10^{-8}$ M) | 196.0 | 230.2 | 1.04 | 15.3 | 835.0 | 81.1 | 36.8 |
| $LSD_{0.05}$ | 12.8 | 6.7 | 0.04 | 4.1 | 15.0 | 14.7 | 5.7 |

Methyure insignificantly increased (by 4–6%) the number of ears per plant. It was noted that the weight of 1000 seeds obtained from maize crops treated with Methyure in the concentration of $10^{-4}$ M (calculated on the basic 14% humidity) increased by 3.4%.

As a result of these changes, there was an increasing individual grain under the influence of Methyure. Thus, the average weight of grain obtained from one plant under the action of Methyure in concentrations of $10^{-3}$ and $10^{-4}$ M increased by 13–19% compared with control plants of corn. Precisely, these concentrations of the preparation had an effect on the formation of the grain, as indicated by increasing grain nature by 2.6%.

The climatic conditions were dry for forming and ripening corn kernel during the growing season. Therefore, the biological yield of corn grown without irrigation was not high enough. The biological yield of control corn crops was 36.2 c/ha. Methyure treatment at concentrations of $10^{-3}$ and $10^{-4}$ M affected the yield, which increased up to 45.8–42.0 c/ha, so yield of experimental variants of corn exceeded control one by 16–26%.

The data presented in Table 4 show that the energy factor of all variants is within the permissible limits ($E_k > 1$). The highest energy factor corresponded to the variants with Methyure treatment at concentrations $10^{-3}$ and $10^{-4}$ M. In variants where Methyure was used at the above-mentioned concentrations, the energy factor exceeded the control index by 25.9 and 15.4%, respectively. This fact suggests that Methyure application will make corn growing more cost-effective and its technology will become more resource saving. The largest component in the structure of the costs of agricultural crop growing is fertilizers, pesticides and seeds, which in all cases exceeds 50%. The use of Methyure allows to lower the rate of fertilizers and pesticides while corn growing.

**Table 4** Calculation of the energy factor of corn-growing technology under the Methyure application

| Experiment factor | Total energy of dry matter ($Q_c$), MJ/kg | Coefficient of dry matter content ($C_d$), % | Output of gross energy ($GE_y$), MJ/ha | Total anthropogenic energy ($E_a$), MJ/ha | Energy factor ($E_K$) |
|---|---|---|---|---|---|
| Control | 17.6 | 0.86 | 54,792.3 | 52,721.8 | 1.04 |
| Methyure ($10^{-3}$ M) | 17.6 | 0.86 | 69,322.9 | 52,856.5 | 1.31 |
| Methyure ($10^{-4}$ M) | 17.6 | 0.86 | 63,571.2 | 52,786.4 | 1.20 |
| Methyure ($10^{-6}$ M) | 17.6 | 0.86 | 61,300.8 | 52,786.3 | 1.14 |
| Methyure ($10^{-8}$ M) | 17.6 | 0.86 | 55,700.5 | 52,608.0 | 1.06 |

## 4 Conclusion

The use of the preparation Methyure in corn-growing technology positively influenced the field germination of seeds, the formation of crop photoassimilation apparatus, the adaptation of plants to semiarid agro-climatic conditions and salt soil. It was proved that Methyure ($10^{-3}$ and $10^{-4}$M concentrations) improved the formation of some elements of corn crop structure.

As a result, the inclusion of the preparation Methyure in the above-mentioned concentrations to the technology of corn hybrid DKC 5143 growing allowed to increase the biological productivity by 26.5–16.0% respectively and was compared with control.

## References

1. Hansen, J.W.: Realizing the potential benefits of climate prediction to agriculture: issues, approaches, challenges. Agric. Syst. **74**, 309–330 (2002)
2. Wang, W., Vinocur, B., Altaian, A.: Plant responses to drought, salinity and extreme temperature toward genetic engineering for stress tolerance. Planta **218**, 1–14 (2003)
3. Hasegawa, P.M., Bressan, R.A., Zhu, J.-K., Bohnert, H.J.: Plant cellular and molecular responses to high salinity. Plant Physiol. **51**, 463–499 (2000)
4. Munns, R., Tester, M.: Mechanisms of salinity tolerance. Annu. Rev. Plant Biol. **59**, 651–681 (2008)
5. Flowers, T.J.: Improvig crop salt tolerance. J. Exp. Bot. **55**, 307–319 (2004)
6. Gurmani, A.R., Bano, A., Khan, S.U., Din, J., Zhang, J.L.: Alleviation of salt stress by seed treatment with abscisic acid (ABA), 6-benzylaminopurine (BA) and chlormequat chloride (CCC) optimizes ion and organic matter accumulation and increases yield of rice (*Oryza sativa* L.). Aust. J. Crop. Sci. **5**, 1278 (2011)

7. Kalinina, N.A.: Stimulyatsiya preparatom PTMB prorastaniya semyan kukuruzyi v usloviyah hloridnogo zasoleniya. Growth and development regulator of plant. In: V International Conference, Moscow (1999) (in Russian)
8. Tuterev, S.L.: Fiziologo-biohimicheskie osnovyi upravleniya stresso-ustoychivostyu rasteniy v adaptivnom rastenievodstve. Vestnik zaschityi rasteniy **1**, 11–35 (2000) (in Russian)
9. Sharma, I., Ching, E., Saini, S., Bhardwaj, R., Pati, P.K.: Exogenous application of brassinosteroid offers tolerance to salinity by altering stress responses in rice variety Pusa Basmati-1. Plant Physiol. Biochem. **69**, 17–26 (2013)
10. Miura, K., Tada, Y.: Regulation of water, salinity, and cold stress responses by salicylic acid. Front. Plant Sci. **5**, 4 (2014)
11. Sheteawi, S.A.: Improving growth and yield of salt-stressed soybean by exogenous application of jasmonic acid and ascobin. Int. J. Agric. Biol. **9**, 473–478 (2007)
12. Priss, O., Yevlash, V., Zhukova, V., Kiurchev, S., Verkholantseva, V.: Effect of abiotic factors on the respiration intensity of fruit vegetables during storage. Eastern Eur. J. Enterp. Technol. **6**(11), 27–34 (2017). https://doi.org/10.15587/1729-4061.2017.117617
13. Prikazchikova, L.P.: Nitrogene content fertilizer. Patent No. 2116727, Russia Federation (1997) (in Russian)
14. Palladina, T.A., Ribchenko, J.I., Konturska, O.O.: Zalezhnist adaptogennogo efektu preparatu Methyure na roslini za umov solovogo stresu vid yogo molekulyarnoYi strukturi. Biotechnology **5**, 115–119 (2012) (in Ukrainian)
15. Kovalenko, N.O., Palladina, T.A.: Gene expression of H+ -pumps in plasma and vacuolar membranes of corn root cells under the effect of sodium ions and bioactive preparations. Ukrainian Biochem. J. **88**(2), 89–97 (2016). http://dx.doi.org/10.15407/ubj88.02.089
16. Shpaar, D., Ginapp, K., Dreger, D., Zaharenko, A., Kalenska, S.: Corn. Growing, Harvesting, Preserving and Using. Alfa-Stevia LTD, Kyiv (2009) (in Russian)
17. National Standard of Ukraine. Soil Quality. Indicators of Soil Fertility: DSTU 4362:2004. Kyiv, Ukraine (2006) (in Ukrainian)
18. Yeshchenko, V.O., Kopytko, P.H., Kostohryz, P.V., Opryshko, V.P.: Fundamentals of Research in Agronomy. TD Edelweiss, Vinnitsa (2014) (in Ukrainian)
19. Meier, U.: Growth stages of mono-and dicotyledonous plants. In: BBCH Monograph, 2nd edn. Federal Biological Research Centre for Agriculture and Forestry (2001)
20. Jonckheere, I., Fleck, S., Nackaerts, K., Muys, B., Coppin, P., Weiss, M., Baret, F.: Review of methods for in situ leaf area index determination: Part I. Theories, sensors and hemispherical photography. Agric. For. Meteorol. **121**, 19–35 (2004)
21. Lamb, J.J., Eaton-Rye, J.J., Hohmann-Marriott, M.F.: An LED-based fluorometer for chlorophyll quantification in the laboratory and in the field. Photosynth. Res. **114**, 59–68 (2012)
22. Hrytsaienko, Z.M., Hrytsaienko, A.O., Karpenko, V.P.: Methods of Biological and Agrochemical Studies of Plants and Soils. Nichlava, Kyiv (2003) (in Ukrainian)
23. Medvedovskiy, O.K., Ivanenko, P.I.: Enerhetychnyi analiz intensyvnykh tekhnolohii v silskohospodarskomu vyrobnytstvi. Urozhai, Kyiv (1988) (in Ukrainian)
24. Kovalenko, N.O., Bilyk, ZhI, Palladina, T.A.: Effect of adaptogenic preparations on $Na^+/H^+$-antiporter function in plasma membrane of corn root cells under salinity conditions. Ukrainian Biochem. J. **86**, 134–141 (2014)
25. Zelitch, I.: The close relationship between net photosynthesis and crop yield. Bio-Science **32**, 796–802 (1982)
26. Araus, J.L., Serret, M.D., Edmeades, G.: Phenotyping maize for adaptation to drought. Front. Physiol. **3**, 305 (2012)

# Sweet Ices with High Nutritional Value

**Juliya Kozonova, Victoria Stepanova, Alla Salavelis and Alina Kulyk**

## 1 Introduction and Purpose of the Article

One of the sweet products types is the ice cream without use of milk base—fruit or sweet ice. These products on a natural basis can be considered as modern sweet eco-products with low sugar and caloric content and with a high content of biologically active substances. Nowadays, people are paying a lot of attention to the nutrition, emphasizing that it should be diverse. Consequently, the development of new products on a vegetable natural basis without the addition of artificial components is the topic of current interest. The latest trends in the sweet products market require the use of natural raw materials, the removal of preservatives and flavor enhancers from the recipes and maximal reduction of sugar and, accordingly, the caloric content of products [1, 2]. These top trends are reflected in the scientific works of many authors. Thus, there is a known technology for obtaining a low-calorie dessert in which instead of the milk basis, the soy or wheat protein isolate was used and the sugar was replaced by fructose and non-starch polysaccharides [3]. The authors Rodriguez and Campderrós [4] investigated the replacement of sugar by a combination of 50% stevia and 50% sucrose and also selected inulin as a fat substitute. There are also studies about the use of honey with a combination of other natural product compounds as sugar replacement [5]. It is known the technology of making ice cream on the base of normalized milk and pumpkin puree to enrich the product with mineral substances and food fibers, as well as it is proposed the production of tea ice on the basis of Ivan tea extract, which is recommended to use for the nervous system normalization [6]. The food market research testifies the considerable demand and consumers interest

J. Kozonova (✉) · V. Stepanova · A. Salavelis
Odessa National Academy of Food Technologies, Kanatna Str 112, Odessa 65039, Ukraine
e-mail: kozonova5@gmail.com

A. Kulyk
Tavria State Agrotechnological University, B. Khmelnitsky Av., 18, Melitopol, Zaporizhia Region 72312, Ukraine

© Springer Nature Switzerland AG 2019

V. Nadykto (ed.), *Modern Development Paths of Agricultural Production*,
https://doi.org/10.1007/978-3-030-14918-5_71

**Table 1** Recipe composition of the sweet ice «happiness»

| Raw material | Recipe amount (g) |
|---|---|
| Black currant | 19 |
| Apple | 19 |
| Carrot | 25 |
| Honey | 4 |
| Topping from walnuts | 13 |
| Product output | 80 |

to the new sweet products with improved raw material composition, but the analysis of literary sources show a small assortment of such products; therefore, the development of such products is appropriate and relevant. The purpose of the article is to develop the technology of high nutritional value sweet ice with topping on the base of walnut as a source of fatty component with balanced fatty acid composition. The sweet ice topping is the dispersion of the walnut in water, so the main tasks are to study the stability of the system and to determine the emulsion sizes of particles that affect the organoleptic characteristics of the product, in particular the sensation in the mouth and the presence of grainy.

## 2 The Recipe of Sweet Ice and the Main Features of Its Production

A complicated environmental situation, high level of stress, and intense work are just a few factors that have an impact on almost every person today. Under such conditions, the increase of the useful substances in food products, along with the use of natural raw materials—the leading task of scientists nowadays. Continuing the trend, as an alternative to milk-based ice cream, we offered sweet ice, which contains fruit and vegetable puree sweetened with honey and topped with emulsion from walnuts.

An analysis of the vegetable raw materials chemical composition was conducted, and the sweet ice recipe was chosen; the data are given in Table 1.

For the sweet ice processing, the fruit and vegetable raw materials were turned into puree. The use of natural brightly colored containing bioflavonoids raw materials, such as, for example, black currant, can provide fruit puree with attractive color without the use of artificial colorants [7]. It is possible to introduce the developed product technology on the existing equipment of canning plants. The primary process of fruit and vegetable semi-finished products preparation includes the following technological operations: sorting, washing, inspection, crushing, welding, and rubbing. Then, they mix fruit and vegetable puree with honey according to the recipe insert a wooden stick and freeze in the GyroFreeze apparatus. Then, they cover the

**Table 2** Recipe of walnut topping

| Raw material | Recipe amount (g/100 g) |
|---|---|
| Drinking water | 85.7 |
| Kernels of walnuts | 14.3 |

product with toppings of walnut emulsion and conduct the additional processing in GyroFreeze.

The fatty base is a key ingredient that affects the quality of the finished product, in particular such a dessert as ice cream [8]. As a source of fatty component, we suggested to use the walnut emulsion. New experiments confirm the positive effect of this raw material on the human body [9–11], and the extracts from it contain polyphenols and balanced fatty acids, which have a positive effect on lipid metabolism [12]. Usually, emulsions on the base of plant material, including walnut, are not stable, and during their storage, the stratification is observed. Chinese scientists [13] investigated the positive influence of tea polyphenols on its stability and proved the need of raw materials fine grinding to increase the dispersion stability. The proposed walnut topping has a homogeneous consistency without the sensation of product particles and pale cream color, and thanks to the fine grinding, it remains stable during the manufacturing process of sweet ice. The fineness of the nut topping provides the formation of small crystals of ice, which positively reflects on the quality and organoleptic properties of the finished frozen sweet ice. The recipe of walnut topping is presented in Table 2.

The topping for sweet ice is a dispersion of the walnut kernel in water, which is made at a ratio of raw materials and drinking water 1:7. For the production of such a disperse system, the purified kernels of walnuts are dosed and soaked in water up to 10 h at a GM 1:1 and at a temperature of (20 ± 3) °C. Then, the walnut kernels are washed for 180 s, poured by boiling water and subjected to the wet-thermal treatment at a temperature (98–100) °C for 120 s. Water is drained, and the walnut kernels are poured with water at GM 1:7 with temperature (20 ± 3) °C. The mixture is grinded up by grinding machine with a high capacity. After infusion of the crushed mixture for 1800 s, the mixture is homogenized using a homogenizer with a power of at least 300 Wt for 180 s. The resulting mixture is filtered through nylon or metal sieves with a holes minimum diameter.

## 3 The Stability Assessment of the Walnut Topping Disperse System

To assess the stability of the system and to determine the walnut dispersion particles sizes that affect on the product organoleptic characteristics, in particular the sensation in the mouth and the presence of grainy, the study of the light absorption capacity of the walnut dispersion was carried out using a photoelectrocalorimeter KFK 3-01

**Table 3** Indices of the walnut dispersion light-absorbing ability

| Wavelength λ (nm) | 320 | 340 | 400 | 440 | 490 | 540 | 590 | 670 | 750 | 870 | 980 |
|---|---|---|---|---|---|---|---|---|---|---|---|
| Optical density D | 0.063 | 0.148 | 0.500 | 1.047 | 1.402 | 1.630 | 1.689 | 1.753 | 1.626 | 1.406 | 1.166 |

**Table 4** Disperse system characteristics

| Indicator | Indicator value |
|---|---|
| Refraction index | 1.3999 |
| Relative particle radius (m) | $14.38 \times 10^{-6}$ |
| Turbidity of complexes systems ($sm^{-1}$) | 0.192973 |
| Number of particles per volume unit ($sm^{-3}$) | $13211 \times 10^{-4}$ |
| Weight concentration (%) | 0.0626 |

at a wavelength of 320–980 nm [14]. The investigation of the particles sizes in the obtained walnut dispersions was carried out by the method of the turbidity spectrum on the base of disperse system light scattering [15–17]. Using the experimental data obtained by measuring the dependence of the dispersions optical density from the wavelength, they constructed logarithmic curves and calculated the wave exponent as a tangent of the slope of the curve to the abscissa with a negative sign. In turn, the wave exponent allows us to calculate the particles sizes and their number in disperse systems and to investigate the formation of supramolecular structures. Using the disperse system light-scattering characteristic functions tables [16], the dispersion characteristic indices were calculated. The obtained data are presented in Table 3.

The obtained data show that the maximum value of the optical density of the manufactured walnut dispersion falls on the wavelength in the range of 540–670 nm; it can be explained by the presence of polyphenols in the dispersion [5].

On the base of the experimental disperse system optical density at a wavelength of 540–670 nm, using the turbidity spectrum method, the radius of the particles was calculated. To do this, it is necessary to construct a graphical dependence of the disperse system optical density change from the length of the wave dispersion in the system lg D-lg λ. Experimental points constructed in the coordinate system lg D-lg λ are on the line in the range ΔD—0.2 to 0.3 μ. In such cases, the wave exponent n is defined as a tangent of the slope of the straight line lg D to lg λ with the opposite sign [15].

Using the disperse systems light scattering characteristic functions tables [16], by the wave exponent n were found tabular values that characterize the dispersion systems and were calculated the relative particles refraction index in the dispersion system, the particle radius in the dispersion system, the turbidity of the disperse system, the number of particles per volume unit, weight concentration of dispersed particles (Table 4).

Taking into account the low indicator of the particles weight concentration in the walnut dispersion sample, it can be argued that the dispersion stability is high and

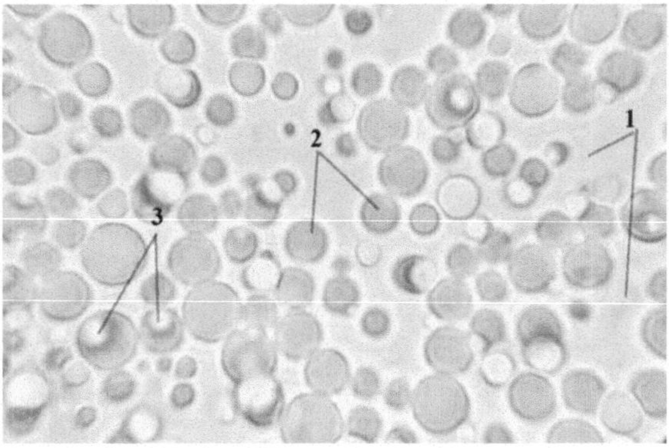


**Fig. 1** Microscopic image of walnut dispersion (2000 times increased in size)

the dispersion fraction is evenly spreading in a dispersion environment. The relative particle radius of the system is about $14.38 \times 10^{-6}$ m, which is typical for fine-grained systems. Such size ensures the system homogeneity, and for taste sensations, it is not characterized by the presence of inclusions and all that positively affects the organoleptic assessment of the finished product. The obtained results were confirmed by the sedimentation analysis. The results of the study indicate that the most active disperse system sedimentation occurs in the first 540 s of the experiment. On the base of the sedimentation analysis data, it was determined the relative mass of fractions which fall precipitated in a certain period of time and investigated the integral and differential curves of particle distribution in radius. Based on the calculations, it was found that the particle size of the disperse system is $(26–30) \times 10^{-6}$ m and is about 70% of the whole system.

To investigate the stability of the walnut dispersion, to prevent its stratification and to establish the type of disperse system, we were carried out microscopic examination of the walnut disperse system samples. The results are presented on Fig. 1.

Microstructural studies of the walnut dispersion under an electron microscope testify the heterogeneity of the system. The largest fractional compound is the water component (1) in which there are fat inclusions, which are presented as fatty balls (2). According to research, their size ranges from 0.5 to 17 mkm, most of which are within the range of 1–10 mkm, which indicates the homogeneity of the dispersion. In the middle of some fatty balls, there are some inclusions that are water covered with fat shells (3); therefore, it can be predicted that the obtained dispersion is a multiple emulsion in which the external system is a direct emulsion and the internal one is a reverse emulsion [18, 19]. Due to this phenomenon, the dispersion system is stable to the stratification and there is no separation of the hydrophobic fatty particles from the continuous environment.

The content of biologically active nutrients in the finished topping composition varies depending on the period of raw material use and the conditions of its storage. Walnut kernels contain more biologically active nutrients in the autumn–winter period, compared with the spring–summer period, and, therefore, toppings from walnuts with a shorter storage time after harvesting contain more nutrients.

**Table 5** Macronutrient composition of walnut topping ($n = 2, p \leq 0.05$)

| Indicator | The mass fraction | |
|---|---|---|
| | (g/100 g) | On dry weight (g/100 g) |
| Nutrients (g) | | |
| Proteins | 2.275 | 19.78 |
| Lipids | 5.000 | 43.48 |
| Carbohydrates, of | 3.570 | 31.04 |
| which food fibers | 1.570 | 13.65 |
| Ash | 0.655 | 5.65 |
| Water | 88.500 | – |

The largest proportion of nutrients up to 43.5% falls on fats, about 19.8% make proteins and 31%—carbohydrates. As a result of the chromatographic analysis, it was found that there are 50% linoleic, 23.8% oleinic, and 12.7% linolenic fatty acids from all fatty acids of the walnut topping, which indicates the evenly extraction of the lipid component from the walnut into the dispersion environment. To determine the usefulness of using walnut topping in sweet ice, its chemical composition has been investigated (Table 5).

The correlation of fatty acids ω-3 and ω-6 remains at 1:4 levels, which meets the requirements of FAO/WHO for better assimilation. Vitamins are dominated by B group (B1, B3, B6), PP, C and E, and from microelements potassium, phosphorus and magnesium are presented in the topping.

## 4 Conclusions

To extend the range of sweet products on a natural basis without the use of artificial substances, we offered and developed the technology of sweet ice «Happiness». The basis of the product are fruit and vegetable puree with the addition of honey as a sweetener, the product is covered with walnut topping. The developed topping has similar rheological properties to cow's milk, but has a nutty aroma and is different in color: Cow's milk is white-colored, while the walnut topping is painted as pale cream. As a result of the experiment, it is possible to confirm the high stability of the walnut topping dispersion and the uniform distribution of the dispersion fraction in the dispersion environment. The relative radius of the system particles is about $14.38 \times 10^{-6}$ m, which is typical for fine-grained systems. Such sizes ensures homogeneity of the system and for taste sensations is not characterized by the presence of inclusions, which positively affects the organoleptic characteristics of the finished product. In addition, the disperse system is resistant to the stratification of the hydrophobic fatty particle from the whole environment. Therefore, the developed sweet ice, due to carefully selected raw materials, may be an alternative to conventional milk based milk ice cream for people who have lactose intolerance, as well as for vegetarians.

## References


1. Alija, J., Talen, C.: New concept of desserts with no added sugar. Int. J. Gastronomy Food Sci. **1**, 116–122 (2012). https://doi.org/10.1016/j.ijgfs.2013.06.002
2. Bucher, T., Van der Horst, K., Siegrist, M.: Fruit for dessert. How people compose healthier meals. Appetite **60**, 74–80 (2013). https://doi.org/10.1016/j.appet.2012.10.003
3. Nepovinnykh, N.V., Kliukina, O.N., Ptichkina, N.M., Bostan, A.: Food hydrocolloids. Available https://www.sciencedirect.com/science/article/pii/S0268005X17316612 (2018).Accessed 22 March 2018
4. Furlán, L.T.R., Campderrós, M.E.: The combined effects of Stevia and sucralose as sugar substitute and inulin as fat mimetic on the physicochemical properties of sugar-free reduced-fat dairy dessert. Int. J. Gastronomy Food Sci. **10**, 16–23 (2017). https://doi.org/10.1016/j.ijgfs.2017.09.002
5. Moriano, M.E., Alamprese, C.: Honey, trehalose and erythritol as sucrose-alternative sweeteners for artisanal ice cream. A pilot study. LWT, **75**, 329–334 (2017). https://doi.org/10.1016/j.lwt.2016.08.057
6. Sevostyanova, A.M., Zemlyakova, E.S.: O vozmozhnosti ispolzovaniya rastitelnogo syirya Kaliningradskoy oblasti v proizvodstve morozhenogo i chaynogo lda. Vestnik molodezhnoy nauki **1**, 14–17 (2015)
7. Dubova, G.E.: Rozrobka tehnologiyi slivovogo ta chornosmorodinovogo sokIv z vikoristannyam metodu tsentrifuguvannya, PhD thesis, Odesa (2000)
8. Smith, K.W.: Specialty oils and fats in food and nutrition. In: Food Science, Technology and Nutrition, Woodhead Publishing Series, pp. 271– 284 (2015). https://doi.org/10.1016/c2014-0-01770-4
9. Thangthaeng, N., Poulose, S.M., Fisher, D.R., Shukitt-Hale, B.: Walnut extract modulates activation of microglia through alteration in intracellular calcium concentration. Nutr. Res. **49**, 88–95 (2018). https://doi.org/10.1016/j.nutres.2017.10.016
10. Byerley, L.O., Samuelson, D., BlanchardIV, E., Luo, M., Lorenzen, B.N., Banks, S., Ponder, M.A., Welsh, D.A., Taylor, C.M.: Changes in the gut microbial communities following addition of walnuts to the diet. J. Nutr. Bioch. **48**, 94–102 (2017). https://doi.org/10.1016/j.jnutbio.2017.07.001
11. Choi, Y., Abdelmegeed, M.A., Song, B.J.: Preventive effects of dietary walnuts on high-fat-induced hepatic fat accumulation, oxidative stress and apoptosis in mice. J. Nutr. Bioch. **38**, 70–80 (2016). https://doi.org/10.1016/j.jnutbio.2016.08.013
12. Liang, X., Chen, D., Cao, L., Zhao, S.: Effects of pressed degreased walnut meal extracts on lipid metabolism in postnatally monosodium glutamate-induced mice and 3T3-L1 preadipocytes. J. Funct. Foods **31**, 89–96 (2017). https://doi.org/10.1016/j.jff.2017.01.026
13. Liu, S., Liu, F., Xue, Y., Gao, Y: Evaluation on oxidative stability of walnut beverage emulsions. Food Chem. **203**, 409–416 (2016). https://doi.org/10.1016/j.foodchem.2016.02.037
14. Thomas, O., Junqua, G., Thomas, M.-F.: UV-Visible Spectrophotometry of Water and Wastewater, pp. 349–378. Elsevier B.V (2017)
15. Klyonin, V.I., Schegolev, S.Y., Lebedeva, L.G.: Opt. Spektrosk. **6**, 1161 (1973)
16. Klyonin, V.I., Schegolev, S.Y., Lavrushin, V.I.: Harakteristicheskie funktsii svetorasseyaniya dispersnyih sistem. Izd. Sarat. Un-ta, Saratov (1977)
17. Dyakonova, A., Stepanova, V., Shtepa, E.: Preparation of the core of walnut for use in the composition of soft drinks. Food Sci. Technol. **11**(3), 71–79 (2017). https://doi.org/10.15673/fst.v11i3.609

18. Nicolai, T., Murray, B.: Particle stabilized water in water emulsions. Food Hydrocolloids **68**, 157–163 (2017)
19. Shestakov S.D.: Kavitatsionnyiy reaktor, kak sredstvo prigotovleniya i stabilizatsii emulsiy dlya hlebopekarnoy promyishlennosti. Hranenie i pererabotka selskohozyaystvennogo syirya, **3**, 5–6 (2003)

# Nitrogen in Soil Profile and Fruits in the Intensive Apple Cultivation Technology

**Tetiana Maliuk, Natalia Pcholkina, Liliia Kozlova and Oksana Yeremenko**

## 1 Introduction

Perennial fruit crops occupy no more than 1% of the world's agricultural land. Despite this, they are of great economic importance in the world trade and the economy of many regions [1]. The south of Ukraine is referred to such regions due to the soil and climatic conditions, which are beneficial for deciduous fruit crops of temperate climate.

The apple tree (Malus x domestica Borkh.) ranks second among fruit trees in fruit harvesting (89.33 million tons was produced in 2016) all over the world after banana [2]. According to the information of S. Pereira-Lorenzo et al., who have compiled a large array of information on apple cultivation in the world, Ukraine ranks one of the leading places in Europe in terms of the land area occupied by this crop [3]. Modern apple cultivation technologies in the south of Ukraine provide for the planting of 2500 trees/ha and more, the use of clonal stocks, small crown forms, mandatory irrigation, etc. [4]. The optimum supply of fruit plants with nitrogen and moisture in such technologies is of particular importance [1]. It is connected with the peculiarity of the root system of trees location on clonal stocks, occupying a small amount of soil [5–7].

Recently, problems have become more and more important to increase the efficiency of fertilizers and reduce the environmental load on the soil [8, 9]. This applies to nitrogen fertilizers (N-fertilizers) in the first place. They determine a significant increase in crop yields, but they have a high energy intensity and high ability to pollute the environment. Focusing solely on increasing the yield of gardens often leads

T. Maliuk (✉) · O. Yeremenko
Tavria State Agrotechnological University, B. Khmelnitsky Av., 18, Melitopol, Zaporizhia Region 72312, Ukraine
e-mail: agrochim.ios@ukr.net

T. Maliuk · N. Pcholkina · L. Kozlova
Melitopol Research Fruit Growing Station Named After M.F, Sydorenko of Institute of Horticulture, Vakulenchuka St 99, Melitopol, Zaporizhia Region 72311, Ukraine

© Springer Nature Switzerland AG 2019 
V. Nadykto (ed.), *Modern Development Paths of Agricultural Production*,
https://doi.org/10.1007/978-3-030-14918-5_72

to the use of excessive doses of N-fertilizers and pollution of ground and surface waters with nitrates, to the accumulation of toxic substances in products, and health problems [9, 10].

Developing reliable recommendations for fertilizing with nitrogen (N) of apple trees is very difficult, because perennial crops have a specific nature of substances absorption and accumulation. It is associated with the special structure of wood and its ability to accumulate and then use the nutrient reserves, physiological stages of growth and development of trees, the structure of trees root system [1, 11]. De Angelis et al. [11] confirmed that the time and method of applying N-fertilizers can affect not only the plants in the year of application, but also the yield of the next year, and can also affect various forms of nitrogen accumulation and mobilization.

What to consider as excessive doses of nitrogen is also a controversial issue. Neilsen et al. [12] report that an apple tree on an M.9 vegetative stock with a density of 3300 trees per hectare absorbed no more than 8.8–44 kg N/ha. Ernani et al. [13] calculated that the maximum N value required for an apple tree is always less than 50 kg N/ha, even in years with high yields.

Doses of nitrogen, which are often used nowadays in Ukrainian apple orchards, range from 60 to 150 kg N/ha. Although, if nitrogen is not absorbed by the roots of plants, it is lost with surface runoff or leached down the soil profile along with water, evaporation, and denitrification [14, 15].

The reaction of fruit trees to the application of N-fertilizers is expressed in various manifestations of plant functions. These are changes in the magnitude of the vegetative growth, layings of fruit buds, the fall of fruits, the average weight of a fruit, the total amount of fruit on a tree, the change in the quality of fruits [11, 16–19]. That is, the effect of nitrogen (positive or negative) consists ultimately of the complex effect on yield.

Apple tree's need for N depends on its content in the soil. Despite the large amount of information on the effect of N-fertilizers on plant nutrition, there is no consensus on its effect on a fruit tree. A number of experiments have shown low efficiency of N-fertilizer application in gardens, especially in the first years of fruiting [19, 20]. Others note the positive effect of exceptionally moderate doses of N in the soil [21, 22].

Inorganic compounds N ($NH_4^+$, $NO_2^-$, and $NO_3^-$) are the main form of this element absorbed by plants [23]. The supply of these forms is very limited and constitutes no more than 1–5% of the total nitrogen in the soil, which forces farmers to use large doses of N-fertilizers to increase yields. The formation and accumulation of inorganic nitrogen forms in the soil depend on many factors, including the characteristics of crops, the type of fertilizers, soil properties, temperature, and water content [19, 24, 25]. The size of N losses at high doses of fertilizers also depends on these factors.

Mineral nitrogen ($N_{min}$) is used most often for the diagnosis of plant nutrition. Such measurements of $N_{min}$ may include the total amount of nitrates N–$NO_3$ plus ammonium nitrogen (N–$NH_4$). Only N–$NO_3^-$ can be used, since this form is usually dominant in the soil [26]. N–$NO_3$ is a form of nitrogen that is most suitable for direct uptake by plant roots. It also has the greatest ability to move along the soil profile [27].

Based on publications in the literature, the main factors of accumulation and movement of $N–NO_3$ along the profile are periodic rains of high intensity, irrigation, excessive doses of N-fertilizers, loess soils with a rough texture, replacement of crops for gardens that require strong fertilizing, and features of the hydrothermal soil regime [28, 29]. The keeping of gardens under summer fallow influences the accumulation of $N–NO_3$ in the soil and, sometimes, heavy precipitation even in enough dry areas can also contribute to the downward movement of $N–NO_3$ into the soil profile [30]. In the same studies, the authors report that under the gardens there is the highest potential for leaching $NO_3$ compared to forests and pastures.

The accumulation of nitrates in fruits, as in the soil, depends on three main factors, including the type and dose of N-fertilizers application, the type of soil, and environmental conditions [31, 32]. Most of the $NO_3^-$ (on average, about 85%) enters the organism with vegetables, meat products, and fruits, including fresh ones and their processed products [32, 33]. Although in general, fruits are considered the products with a low ability to accumulate $NO_3^-$ [32], nitrogen fertilizers can determine the amount of $NO_3$ accumulation in fruits [31]. Ziarati, quoting various sources, reports that in the countries of Europe the acceptable nitrate content for apple tree is very different and ranges from 24 to 108 mg/kg [34]. The maximum permissible concentration of $NO_3$ in apple fruits in Ukraine is 60 mg $NO_3$/kg of the raw substance [35].

## 2 Formulation of the Problem

In general, the dependence between the doses of nitrogen fertilizers, the content of various forms of nitrogen in the soil and the nitrogen nutrition of fruit trees remains insufficiently studied. This is consistent with the opinion of a number of scientists [7, 12, 19, 21]. The study of the correlation between the different forms of nitrogen content in the soil, the hydrothermal regime of the soil, nitrogen doses, the effect on apple tree yields and the accumulation of nitrates in fruits are central to understanding the role of N-fertilizers in horticulture [8, 9, 12]. However, there are a number of difficulties and inaccuracies with respect to the mechanisms for controlling the responses of trees to nitrogen, which presumably vary depending on the forms of nitrogen, weather conditions, crop load, natural characteristics of the soil, and external sources of nitrate. More complete information about the factors that control the accumulation of nitrogen in the soil, the influence of different forms of nitrogen on the yield of apple trees, and some indices of fruit quality will help identify limitations in developing more efficient nitrogen fertilization systems and reduce the ecological load on the soil and plants.

# 3 Purpose of the Study

The purpose of the study is to study the dynamics and accumulation of nitrogen mineral forms in the soil and the factors affecting it, to establish the size of the nitrates downward migration in the soil profile in a fruiting apple orchard fertilized with nitrogen, to study the connection of increasing doses of N-fertilizers with the yield and accumulation of nitrate compounds in fruits in the long field experiments. This, in its turn, can increase the efficiency of N usage in the interests of the environment and the productivity of gardens.

# 4 Materials and Methods

## *4.1 Characteristics of the Research Region*

During 2006–2015, the studies were carried out in Melitopol research fruit growing station named after M.F. Sydorenko of Institute of Horticulture (46°50′36.2″N 35°21′22.6″E), Melitopol, Ukraine.

The research region belongs to the steppe arid zone of the south of Ukraine. The climate is moderately continental. The average annual temperature was 10.1–12.1 °C over the years of research, and the annual precipitation was 363–669 mm. The duration of the frost period is 85–100 days; the duration of the active growing season (the number of days with air temperature more than +10 °C) is 170–185 days. The sum of active air temperatures >10 °C is 3400–3700 °C. Meteorological data were used from the institution "Melitopol Meteorological Station," located directly on the territory of the Horticulture Station (Table 1).

## *4.2 Characteristics of Plantations*

An apple garden (Malus Domestica Borkh.) was planted in the study in 2003 using the varieties "Aidared" and "Florina," stock M.9. Tree planting scheme is 4 × 1 m (2500 trees per 1 ha). There is drip irrigation; soil moisture was maintained at 75–80% of the full moisture saturation of the soil with a free outflow of water. All agrotechnical activities were carried out as it is customary everywhere in this region for an apple tree, as recommended [4].

The experiment included such doses of nitrogen fertilizer: $N_0$ (control without fertilizer); $N_{30}$ (30 kg N/ha); $N_{60}$ (60 kg N/ha); $N_{90}$ (90 kg N/ha); and $N_{120}$ (120 kg N/ha). Fertilizers were applied evenly to the crown projection area. The trees were fertilized annually. Nitrogen in the form of ammonium nitrate was applied 2 weeks before the expected flowering of the apple tree as traditionally used in the gardens of the region. In addition, 50 kg of $P_2O_5$ in the form of superphosphate and 30 kg of $K_2O$ in the

**Table 1** Meteorological indices in the research region, 2006–2015

| Year | Index | | | |
|---|---|---|---|---|
| | The average annual air temperature (°C) | Annual precipitation (mm) | Precipitation April–September (mm) | Precipitation October—March (previous year–current year) (mm) |
| 2006 | 10.2 | 550 | 334 | 290 |
| 2007 | 12.1 | 370 | 142 | 187 |
| 2008 | 11.0 | 383 | 265 | 194 |
| 2009 | 11.1 | 486 | 161 | 193 |
| 2010 | 11.9 | 669 | 299 | 344 |
| 2011 | 10.3 | 471 | 333 | 273 |
| 2012 | 11.7 | 337 | 169 | 176 |
| 2013 | 11.7 | 363 | 154 | 182 |
| 2014 | 11.3 | 543 | 364 | 164 |
| 2015 | 11.9 | 458 | 204 | 259 |

form of potassium sulfate were applied annually under all trees. Each variant of the experiment consisted of 6 trees and was distributed randomly within 4 repetitions (blocks), that is, the variant included 24 trees in total. Between the repetitions, there was a minimum 2 protection tree.

## 4.3 *Characterization, Sampling, and Soil Analysis*

The soil of the study sites is Haplic Chernozems, ID 14-2 according to [36], loam clay (the soil has a heavy loamy texture). Parent material of the soil is Carboniferous loess clay (parent rocks of these soils consist of carbonate loess-like clays). Since the root system of the apple tree on the stock M9 has a surface location and its main mass is up to a depth of no more than 60 cm [5–7], the soil characteristic is given for this layer (0–60 cm) on average: The organic matter content is 2.33%, pH (water) 7.8, N total—0.17%, $CaCO_3$—4.5–6%.

Soil samples were taken 4 times a year: before fertilization, during the period of active vegetative growth, during the period when fruit buds were laid, and at the end of the growing season. The samples were taken during the growing season to a depth of 60 cm. In addition, in the spring and autumn of each year, the samples were taken to a depth of 150 cm to observe the movement of nitrates along the soil profile. After 10 years of research in 2015, samples were taken for nitrate content to a depth of 500 cm. Soil samples were selected for each variant in three replications at a distance of 0.5 m from the trunk of a tree in a row and in the row spacing. The sampling was carried out with a soil drill with a cylinder diameter of 3 cm. The drill has the ability to increase the length due to additional inserts for sampling from deep layers of the profile.

In the soil samples, the content of N–$NO_3$ and N–$NH_4$ was determined using a spectrophotometer as described in [37, 38]. The sum of nitrate and nitrite corresponded to the nitrate content; that is why, nitrites in the soil were not separately determined, which corresponds to [37, 38]. Mineral nitrogen ($N_{min}$) in the soil was defined as the sum of nitrates and ammonia. To convert mg N/kg to kg N/ha, we multiplied the indicator in mg/kg by the layer depth (cm) to the apparent density (g/$cm^3$) and divided by 10.

## 4.4 Selection and Analysis of Fruits, Yield Accounting

The average sample of the fruits was taken from each repetition of the experiment variant, 3 fruits from the southern and northern sides of the tree. Preparation of the fruit for homogenization consisted in the fact that they were cleaned with water, wiped dry, and cut crisscross into 4 equal parts. For analysis, we took ¼ from each fruit without a pedicle and seed nests, which were cut out. The determination of N–$NO_3$ in the fruits was carried out three times as they ripened using an ion-selective electrode, as described in [39].

The yield of apples from each tree (kg) was determined annually by taking into account the number of permanent fruits on the tree and multiplying by the average weight of the fruit. The average weight of the fruit was determined as the average of 100 fruits; the weighing of the fruit was carried out with an accuracy of $\pm 0.01$ g. The yield of each tree was then multiplied by the number of trees per hectare to find out the yield per hectare (t/ha). The accountings were conducted separately for each repetition of the variant. For each repetition of the variant, then the average indices were calculated, which were further used in mathematical processing.

## 4.5 Statistical Analysis

The data were subjected to ANOVA to check for significant differences based on the least significant difference (LSD) between doses of fertilizers for average values of the mineral nitrogen content in the soil, nitrates in fruits, and apple tree yield. Probability levels of less than 0.05 were indicated as significant and very significant. Descriptive statistics was used to analyze a sample of nitrogen content in the soil. Correlation and regression analysis was used to describe the connection between different forms of nitrogen in the soil, doses of fertilizers, indices of the hydrothermal regime of the soil, and the yield of apple trees. They were performed using software packages, including MS Excel, Statistica 7.0.

**Table 2** Content of mineral nitrogen in the soil (mg/kg) (average over the years of research)

| Dose N | The beginning of the growing season (before fertilizing) | | | Active vegetative growth | | |
|---|---|---|---|---|---|---|
| | $N_{min}$ | $N–NO_3$ | $N–NH_4$ | $N_{min}$ | $N–NO_3$ | $N–NH_4$ |
| $N_0$ | 8.1c | 4.2c | 3.9a | 16.6d | 10.4d | 6.2b |
| $N_{30}$ | 9.1bc | 6.2b | 2.9a | 33.3c | 21.3c | 12.0ab |
| $N_{60}$ | 10.8b | 6.8b | 4.0a | 42v5bc | 30.6c | 11.9ab |
| $N_{90}$ | 10.9bc | 7.2b | 3.2a | 59.7b | 41.2b | 18.5ab |
| $N_{120}$ | 13.9a | 9.1a | 4.8a | 74.3a | 56.8a | 17.5a |
| LSD | 1.8 | 1.7 | 1.3 | 12.8 | 9.2 | 4.4 |
| Dose N | Differentiation of fruit buds | | | The end of the growing season | | |
| | $N_{min}$ | $N–NO_3$ | $N–NH_4$ | $N_{min}$ | $N–NO_3$ | $N–NH_4$ |
| $N_0$ | 11.9d | 6.7c | 5.2a | 7.2d | 4.0c | 3.7a |
| $N_{30}$ | 18.3c | 10.3c | 8.0ab | 11.8c | 7.6b | 4.2b |
| $N_{60}$ | 26.6b | 17.5b | 9.1ab | 14.8b | 9.2ab | 5.6bc |
| $N_{90}$ | 27.2b | 17.3b | 9.9b | 16.9b | 9.8a | 7.1 cd |
| $N_{120}$ | 39.7a | 28.4a | 11.3a | 19.5a | 10.8a | 8.7d |
| LSD | 5.5 | 4.4 | 1.9 | 2.3 | 1.8 | 1.5 |

## 5 Results and Discussion

Observations on the dynamics of the nitrogen mineral forms in the soil for the period of 10 years made it possible to determine the values of nitrate nitrogen ($N–NO_3$), ammonia nitrogen ($N–NH_4$), and mineral nitrogen ($N_{min}$) in the soil during the apple growing season (Table 2). The dose of nitrogen fertilizer (N-fertilizer) significantly influenced these indices. Increasing the dose of nitrogen linearly enlarged the content of all studied forms of nitrogen in the soil.

The maximum content of $N_{min}$ was during the period of active vegetative growth of apple trees (June)—16.6–74.3 mg/kg. This index decreases greatly in the second half of the growing season to 11.9–39.7 mg/kg. This is due to the absorption of nitrogen by plants, including the forming of fruits, as well as possible loss of nitrogen. It can be lost by leaching and denitrification from the soil due to excessive doses of fertilizers [9, 14, 15, 30]. There are fewer differences between the doses of N-fertilizers at the beginning and end of the growing season. Not high level of nitrogen mineral forms content can be associated with low microbiological activity at the beginning of the growing season, when there is no optimal temperature for this, and washing out of nitrogen during the cold period of the year.

In addition, a statistical analysis was conducted for a sample of 160 observations for each variant during 10 years of research. Data included measurements at all

**Table 3** Nitrogen mineral forms content in the soil layer 0–60 cm during 10 years of research

| Dose N | Form N | x | s | V | Confidence interval |
|---|---|---|---|---|---|
| $N_0$ | $N_{min}$ | 11.2 | 5.9 | 52.7 | 9.3–13.1 |
| | $N–NO_3$ | 6.1 | 4.6 | 75.4 | 4.8–7.4 |
| | $N–NH_4$ | 5.4 | 3.0 | 55.6 | 3.8–5.6 |
| $N_{30}$ | $N_{min}$ | 18.0 | 12.3 | 68.3 | 14.9–21.1 |
| | $N–NO_3$ | 10.7 | 9.3 | 86.9 | 8.0–13.4 |
| | $N–NH_4$ | 6.9 | 4.4 | 63.8 | 5.8–8.0 |
| $N_{60}$ | $N_{min}$ | 23.9 | 16.8 | 70.3 | 19.9–27.9 |
| | $N–NO_3$ | 15.3 | 13.2 | 86.3 | 12.2–18.4 |
| | $N–NH_4$ | 8.3 | 5.9 | 71.1 | 7.0–9.6 |
| $N_{90}$ | $N_{min}$ | 27.9 | 22.5 | 80.6 | 21.4–34.4 |
| | $N–NO_3$ | 18.6 | 17.4 | 93.5 | 13.6–23.6 |
| | $N–NH_4$ | 9.4 | 6.0 | 63.8 | 8.1–10.7 |
| $N_{120}$ | $N_{min}$ | 37.3 | 31.3 | 83.9 | 30.2–44.4 |
| | $N–NO_3$ | 26.2 | 25.1 | 95.8 | 20.0–32.4 |
| | $N–NH_4$ | 11.2 | 7.2 | 64.3 | 9.4–13.0 |

X—the average, s—standard deviation, V—coefficient of variation

sampling times. The boundaries and sizes of nitrogen forms variation in the soil at different doses of nitrogen were determined (Table 3).

The coefficient of variation is a manifestation of variability as a mathematical characteristic. Taking this into account, relative variability was the greatest for $N–NO_3$. That is, this form of nitrogen mostly changes over time. The $N–NH_4$ fluctuations were at the $N_{min}$ level. The great variability of the nitrogen mineral forms in time and space and a large dependence on many factors are noted in a number of works [14, 15, 24, 30].

The studies have shown that the main part of $N_{min}$ is represented by nitrate compounds. These values are closely connected linearly at $R^2 = 0.95$, and the deviation from the regression is ±4.6 mg/kg (Fig. 1). It is possible to determine only nitrate nitrogen in the soil in order to establish the availability of soil mineral forms of nitrogen in general. We do not claim that this is generally suitable for Haplic Chernozems, but it is possible to use the condition of summer fallow and irrigation in gardens to simplify the analytical work for the 0–60 cm soil layer. The priority of nitrogen nitrate form in mineral nitrogen is often noted, including in [26]. In addition, there is evidence that the nitrate form in the field conditions has an advantage for the diagnosis of nutrition and is closely related to the feedback on the application of N-fertilizers [27].

Among the influence factors on the variability and accumulation of nitrogen in the soil, there are land use, the term of land use, plant specificity, the supply of mobile

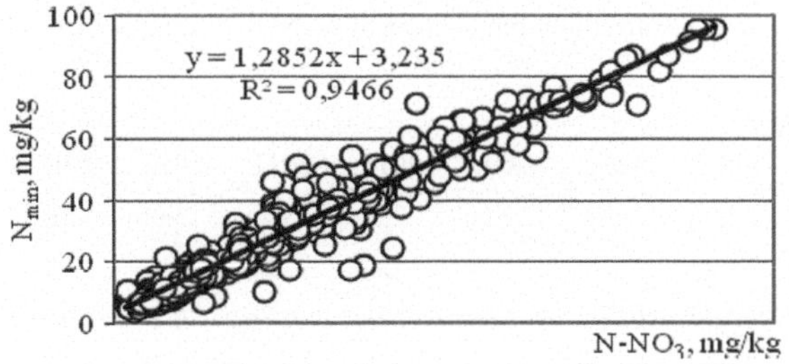


**Fig. 1** Content of $N_{min}$ (y) in the soil with different content of N–$NO_3$ (x)

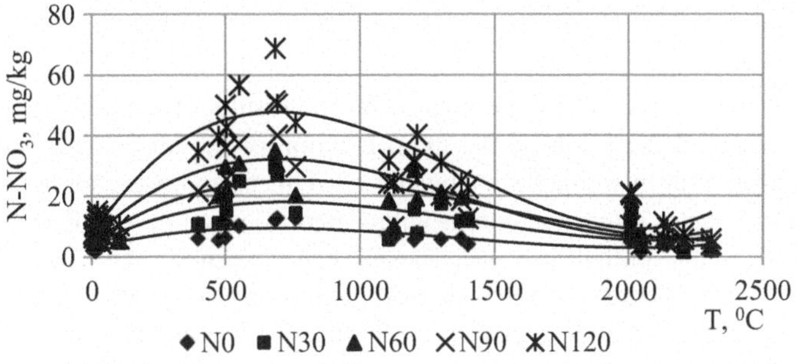


**Fig. 2** Trend lines for the dependence of N–$NO_3$ on the sum of soil temperatures >10 °C

organic substances, fertilizer doses, and hydrothermal soil regime [14, 15, 19, 22, 24, 30]. The initial content of nitrogen mobile forms and agrotechnical measures were the same in the study, and the humidity was also constant through the irrigation. Therefore, we determined the dependence degree of the priority form N–$NO_3$ on temperature and fertilizer doses. The temperature regime determines the variability of N–$NO_3$ by 34–42%, the dose of fertilizers by 46–52%, their mutual interaction was less significant. The differences are significant at 0.1% significance level. The dependence of N–$NO_3$ from the sum of the soil temperatures accumulation above 10 °C due to the large influence of the temperature regime was determined. These indices are curvilinearly connected (Curvilinear dependence) at $R^2 = 0.68–0.87$. The nature of the trend lines is similar at the control and the application of nitrogen.

The largest accumulation of N–$NO_3$ was in the period of active apple tree growth (June) with the accumulation of soil temperature amounts T = 550–760 °C, which coincides with the period of active vegetative growth of apple tree (Fig. 2). Lowering of the nitrates concentration occurs in the layer of 0–60 cm to the phase of fruit buds differentiation and further to the end of the growing season.

So, the research has shown that N-fertilizers lead to the accumulation of a large amount of very mobile N–$NO_3$ in the upper layer of soil. In this regard, it is important to determine its movement along the profile, that is, the potential threat of groundwater pollution.

**Table 4** Influence of N-fertilizers on the content of N–$NO_3$ in a layer of 60–150 cm in spring (mg/kg)

| Dose N | Year | | | | | | | | | |
|---|---|---|---|---|---|---|---|---|---|---|
| | 2006 | 2007 | 2008 | 2009 | 2010 | 2011 | 2012 | 2013 | 2014 | 2015 |
| $N_0$ | 3.5a | 2.6c | 3.2c | 2.0c | 2.6c | 3.4c | 3.9b | 1.1b | 1.6c | 4.3c |
| $N_{30}$ | 3.4a | 3.9c | 3.6c | 1.9c | 3.8c | 2.6c | 7.4b | 1.9b | 4.2c | 6.6c |
| $N_{60}$ | 3.2a | 10.2c | 10.7b | 4.0bc | 5.9c | 6.8bc | 4.2b | 2.4b | 15.3b | 6.8c |
| $N_{90}$ | 3.8a | 26.9b | 14.7ab | 10.3b | 11.7b | 10.8b | 28.4a | 5.5a | 14.9b | 16.5b |
| $N_{120}$ | 3.3a | 39.7a | 19.1a | 24.0a | 16.9a | 19.7a | 34.7a | 6.4a | 24.8a | 24.0a |
| LSD | 0.7 | 6.4 | 5.3 | 5.3 | 3.9 | 4.38 | 7.0 | 2.5 | 7.1 | 6.1 |

The movement of N–$NO_3$ is confirmed by the results of the soil layer analysis 60–150 cm beyond the limits of the main location of the apple trees roots. They were held every year in spring before bud break. This period was chosen to trace the movement of nitrates after the autumn-winter period, when moisture accumulation occurs mainly in the south of Ukraine. The amount of washed out nitrogen depended on the dose of N-fertilizers and the meteorological conditions of the year. The largest supply of nitrates was in this layer with the addition of $N_{90}$ and $N_{120}$ and in the years with high rainfall during the growing season and/or the autumn-winter period of tree rest—up to 39.7 or 429 kg/ha (Table 4). The application of $N_{30}$ and $N_{60}$ leads rarely to a significant accumulation of residual nitrogen in the soil below 60 cm.

It is necessary to recognize that the content of N–$NO_3$ in the soil in the spring was back to the yield obtained that year. The higher was the harvest, the less excess nitrogen remained. This happened after the harvest years of 2008, 2009, 2010, and 2012. As a result, a large amount of N–$NO_3$ indicates that the doses of N-fertilizers $N_{90}$ and $N_{120}$ often used in the gardens of the south of Ukraine were much more than trees require even in the years with relatively high yields. Increasing the dose of N-fertilizers did not always lead to a significant increase in yield. The data in Tables 5 and 6 show that there is almost always a significant difference between control without fertilizers and with N-fertilizers even in the years with low yields. But doses of more than N60 led to an increase in significant yield only in 2008 and 2010 according to "Aidared" and in 2006 and 2012 according to "Florina" or in 20% of cases.

In 2015, samples were taken to a depth of 5 m. The purpose of this experiment was to determine the effect of the total nitrogen dose for 10 years on the movement of nitrates below a 150 cm layer, which was studied annually. The use of N-fertilizers led to the accumulation of a large amount of nitrates at a depth of 250–300 cm. This was the second layer in terms of N–$NO_3$ accumulation after the top one (0–150 cm). The amount of nitrogen at this depth was with the annual introduction of $N_{30}$–4.1 mg/kg, $N_{60}$–6.1 mg/kg, $N_{90}$–10.8 mg/kg, and $N_{120}$–15.2 mg/kg (Fig. 3). This is 1.7–6.3 times more than in the control (2.4 mg/kg). We assume that the reason for the concentration of N–$NO_3$ at this depth may be related to the groundwater level. The wells, made at

**Table 5** Yield of apple tree "Aidared" depending on the doses of N-fertilizers

| Dose N | Year | | | | | | | | | |
|---|---|---|---|---|---|---|---|---|---|---|
| | 2006 | 2007 | 2008 | 2009 | 2010 | 2011 | 2012 | 2013 | 2014 | 2015 |
| $N_0$ | 11.2c | 20.3b | 30.7a | 41.9a | 41.3c | 19.5b | 53.8c | 18.0a | 19.9c | 11.4a |
| $N_{30}$ | 14.0bc | 25.8a | 33.8bc | 43.2a | 49.3b | 21.3ab | 58.7bc | 20.0a | 23.5bc | 13.0a |
| $N_{60}$ | 18.7a | 27.0a | 33.3bc | 43a | 51.4ab | 24.4a | 64.2a | 22.0a | 30.4a | 11.5a |
| $N_{90}$ | 15.6ab | 26.2a | 39.6a | 42.9a | 49.3b | 21.5ab | 60.5ab | 20.3a | 27.3ab | 15.4a |
| $N_{120}$ | 15.6ab | 28.4a | 36.6ab | 43.9a | 53.6a | 23.8a | 58.1bc | 21.0a | 30.7a | 13.2a |
| LSD | 3.8 | 4.3 | 3.4 | 3.0 | 2.7 | 2.5 | 4.3 | 3.5 | 4.0 | 2.8 |

**Table 6** Yield of apple tree "Florina" depending on the doses of N-fertilizers

| Dose N | Year | | | | | | | | | |
|---|---|---|---|---|---|---|---|---|---|---|
| | 2006 | 2007 | 2008 | 2009 | 2010 | 2011 | 2012 | 2013 | 2014 | 2015 |
| $N_0$ | 15.5a | 17.0b | 24.3b | 35.3b | 37.0c | 30.1c | 41.2c | 23.1c | 26.5b | 9.8b |
| $N_{30}$ | 12.3b | 20.6ab | 21.9b | 38.4ab | 41.8b | 33.3b | 44.4b | 26.5b | 32.3a | 10.2b |
| $N_{60}$ | 11.5b | 18.9b | 28.9a | 37.8ab | 43.0a | 30.3ab | 47.5ab | 31.8a | 30.1a | 10.5b |
| $N_{90}$ | 16.1a | 19.1ab | 25.8ab | 39.7a | 42.5a | 33.3a | 46.0ab | 31.0a | 29.7a | 11.7b |
| $N_{120}$ | 15.1a | 21.7b | 23.7b | 37.7ab | 41.8a | 30.6ab | 49.0a | 33.0a | 30.7a | 13.5a |
| LSD | 2.1 | 2.9 | 3.9 | 2.4 | 1.9 | 2.7 | 3.2 | 3.6 | 2.5 | 1.4 |

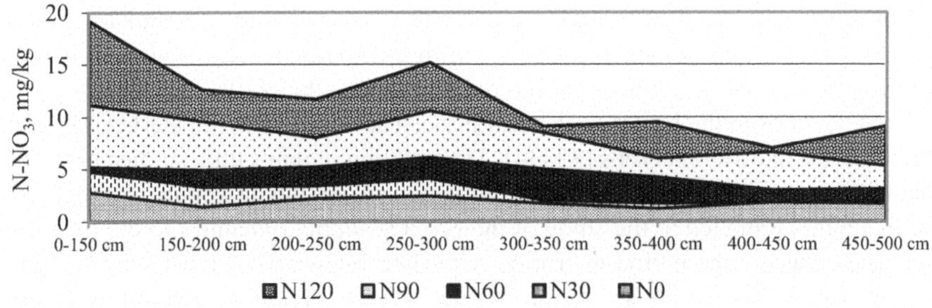


**Fig. 3** Migration of N–$NO_3$ in the soil profile under the influence of N-fertilizers, spring 2015

the site to a depth of 5 meters, showed that it ranged from 2.5 to 4.5 m depending on the time of year.

The accumulation of N–$NO_3$ at the maximum surveyed depth of 450–500 cm depended on the total dose of nitrogen. Its increase from 300 kg ($N_{30}$ annually) to 1200 kg ($N_{120}$ annually) led to an increase in N–$NO_3$ from 23.9 to 84.6 kg/ha. The control was 17.6 kg/ha. The intake of N–$NO_3$ ranged from 0.63 to 6.7 kg/ha relatively to annual one to this layer. This index was calculated by dividing N–$NO_3$ (kg/ha) in a layer of 450–500 cm minus N–$NO_3$ (kg/ha) on the control by the number of fertilizing years. Under such conditions, groundwater may be contaminated with N–$NO_3$. This means that it is necessary to solve the problem of excessive doses of

**Table 7** Content of N–$NO_3$ in the fruits of apple trees, average 2006–2015

| Dose N | Period of fruit development | | | | | |
|---|---|---|---|---|---|---|
| | Growth | | Beginning of ripening | | Ripeness | |
| | A | F | A | F | A | F |
| $N_0$ | 41.3c | 52.4b | 29.4d | 33.1c | 22.6b | 39.8b |
| $N_{30}$ | 44.7c | 53.1b | 30.4d | 35.1c | 23.0b | 39.5b |
| $N_{60}$ | 46.1c | 56.6b | 35.7c | 32.6c | 24.3b | 40.6b |
| $N_{90}$ | 72.9b | 67.4a | 60.8b | 43.1b | 35.9a | 41.4b |
| $N_{120}$ | 88.9a | 73.5a | 68.2a | 65.3a | 40.2a | 44.9a |
| LSD | 9.6 | 7.5 | 4.4 | 10.4 | 6.3 | 3.5 |

A—"Aidared," F—"Florina"

nitrogen, dividing the total amount of nitrate fertilizers into several doses, to reduce the summer fallow and the term of keeping the soil under the summer fallow.

That is, a high potential ability is observed in the intensive apple orchard to wash out N–$NO_3$ even on the soil with high clay content. This coincides with the data about high risk of nitrate washing out in the gardens that require intensive nitrogen fertilizing and abundant irrigation plus occasional heavy rains and moderate temperature. There is also information about the increased drainage function of tree roots [40].

But in the end, most of the nitrogen N–$NO_3$ in our studies accumulates in the upper 0–150 cm soil layer and is connected with soil texture features and periodically droughty periods of the region. This coincides with the data [30].

It is known from the literature that nitrates can accumulate in the marketable part of the harvest of different crops [32–34]. The question remains little studied as to the effect of nitrogen fertilizers on the accumulation of nitrates in apples. Ambiguous results on this issue led to the study of the N–$NO_3$ supply dynamics to the fruits as they ripen. During the period of fruit growth (late July–early August), the N–$NO_3$ content was 41.3–83.9 mg/kg unripe weight of fruits (Table 7). According to the gradations for Ukraine with doses of $N_{90}$ and $N_{120}$, there were more of them than it was allowed. Nitrogen doses affected this index significantly.

The amount of N–$NO_3$ decreased greatly in the early September at ripening by 11.2–42.4%, which is connected with the intensive use of nitrogen-containing compounds in the synthesis. The content of N–$NO_3$ in fruits and soil during this period is closely connected at $r = 0.72 \pm 0.03$. By harvesting fruit (end of September–October, depending on the year and variety), N–$NO_3$ was below the maximum permissible concentration at all doses of N-fertilizers. In general, the N–$NO_3$ content decreases as the fruit ripens to the levels safe for human. The dose of nitrogen directly affects this index in all terms of fruit selection. But the difference is mainly between the control without fertilizers and other variants of experience.

That is, the potential threat of nitrates MPC exceeding is in apple fruits in intensive plantations only if they are gathered before the time and doses above $N_{60}$ are applied.

## 6 Conclusions

1. The use of N-fertilizers in increasing doses led to the accumulation of mineral forms of nitrogen in the soil profile in an intensive apple orchard.
2. The first place was for the nitrate form of nitrogen, which we recommend for determining the supply of soil with all available nitrogen if it is necessary.
3. In general, the most N–$NO_3$ was found in the upper soil layer 0–150 cm, but there is migration to a depth of 500 cm despite the loamy texture of the soil and occasionally dry periods of the year. The amount of mineral forms of nitrogen at $N_{90}$ and $N_{120}$ was very high outside the main root-bearing layer of apple trees on the dwarf stock.
4. Our results demonstrated that the use of N-fertilizers leads to a significant increase in yield. Large doses cannot always increase the yield of apple trees, but they can affect negatively the properties of the soil and the quality of fruits. That is why the attention should be focused on optimizing the nitrogen fertilizers for the most efficient use of nitrogen in horticulture.

## References

1. Carranca, C., Brunetto, G., Tagliavini, M.: Nitrogen nutrition of fruit trees to reconcile productivity and environmental concerns. Plants (Basel) **7**, 4 (2018). https://doi.org/10.3390/plants7010004
2. Statista 2016. http://www.statista.com/statistics/264001/worldwide-production-of-fruit-by-variety
3. Pereira-Lorenzo, S., Ramos-Cabrer, A., Fischer, M., M. Gradziel, T.: Breeding Apple (Malus x Domestica Borkh). In: Breeding Plantation Tree Crops: Temperate Species, pp. 33–81. (2009). https://doi.org/10.1007/978-0-387-71203-1_2
4. Ruliev, V.A.: Gardening of Ukraine south (in Ukraine). Wild Field, Zaporizhzhya (2003)
5. Robinson, T.: Advances in apple culture worldwide. Rev. Bras. de Fruticultura, Jaboticabal, vol. especial, pp. 37–47. (2011). https://doi.org/10.1590/s0100-29452011000500006
6. Akbari, M., Maejima, T., Otagaki, S., Shiratake, K., Matsumoto, S.: Efficient rooting system for apple M9 rootstock using rice seed coat and smocked rice seed coat. Int. J. Agron. **1**, 1–4 (2015). https://doi.org/10.1155/2015/107906
7. Neilsen, G.H., Forge, T.A., Angers, D.A., Neilsen, D., Hogue, E.J.: Suitable orchard floor management strategies in organic apple orchards that augment soil organic matter and maintain tree performance. Plant Soil **378**(1–2), 325–335 (2014). https://doi.org/10.1007/s11104-014-2034-8
8. Fageria, N.K., Baligar, V.C., Li, Y.C.: The role of nutrient efficient plants in improving crop yields in the twenty first century. J. Plant Nutr. **31**, 1121–1157 (2008). https://doi.org/10.1080/01904167.2016.1249798
9. Martínez-Alcántara, B., Quiñones, A., Forner-Giner, M.A., Domingo, J.I., Primo-Millo, E., Legaz, F.: Impact of fertilizer-water management on nitrogen use efficiency and potential nitrate leaching in citrus trees. Soil Sci Plant Nutr. **8**(5), 659–669 (2012). https://doi.org/10.1080/00380768.2012.733678
10. Zahoor, W.A., Khanzada, H., Bashir, U., Aziz, K., Zahir, S., Faheem, A.K.: Role of Nitrogen Fertilizer in Crop Productivity and Environmental Pollution. J. Agric. For. **4**(3), 201–206 (2014). https://doi.org/10.5923/j.ijaf.20140403.09

11. De Angelis, V., Sánchez, E., Tognetti, J.: Timing of nitrogen fertilization influences color and anthocyanin content of apple (*Malus domestica* Borkh. cv 'Royal Gala') fruits. Int. J. Fruit Sci. **11**(4), 364–375 (2011). https://doi.org/10.1080/15538362.2011.630298
12. Neilsen, G.H., Neilsen, D., Herbert, L.C.: Nitrogen fertigation concentration and timing of application affect nitrogen nutrition, yield, firmness, and color of apples grown at high density. HortScience **44**(5), 1425–1431 (2009). https://journals.ashs.org/hortsci/abstract/journals/hortsci/44/5/article-p1425.xml
13. Ernani, P.R., Rogeri, D.A., Proenca, M.M., Dias, J.: Addition of nitrogen had no effect on yield and quality of apples in a high density orchard carrying a dwarf rootstock. Rev. Bras. Frutic. **30**, 1113–1118 (2008). https://doi.org/10.1590/S0100-29452008000400044%5bonline%5d
14. Liu, C-W, Sung, Y., Chen, B-C, Lai, H-Y: Effects of nitrogen fertilizers on the growth and nitrate content of lettuce (Lactuca sativa L.). Int. J. Environ. Res. Public Health **11**(4), 4427–4440 (2014). https://doi.org/10.3390/ijerph110404427
15. Pole, V., Missa, I., Rubauskis, E., Kalva, E., Kalva, S.: Effect of nitrogen fertiliser on growth and production of apples in the conditions of Latvia. In: Proceedings of the Latvian Academy of Sciences. Section b, vol. 71, No. 3(708), pp. 115–120. (2017). https://doi.org/10.1515/prolas-2017-0020
16. Muhammad, S., Sanden, B., Saa, S., Lampinen, B.D., Smart, D., Shackel, K.A, Dejong, T., Brown, P.: Optimization of nitrogen and potassium nutrition to improve yield and yield parameters of irrigated almond (Prunus dulcis (Mill.) D. A. webb). Sci. Hortic. **228**, 204–212 (2017). https://doi.org/10.1016/j.scienta.2017.10.024
17. Milić, B., Čabilovski, R., Keserović, Z., Magazin, N., Tarlanović, J.: Interactions between nitrogen fertilization and apple fruit thinning with NAA. Acta Sci. Pol. Hortorum Cultus. **16**(3), 55–65 (2017). https://doi.org/10.24326/asphc.2017.3.6
18. Dechen, N.G., Roque, A.: Long-term annual fertilization with nitrogen and potassium affect yield and mineral composition of 'Fuji' apple. Sci. Agricola **66**(3), 377–385 (2009). https://doi.org/10.1590/S0103-90162009000300013
19. Kowalczyk, W., Wrona, D., Przybyłko, S.: Content of minerals in soil, apple tree leaves and fruits depending on nitrogen fertilization. J. Elem. **22**(1), 67–77 (2017). https://doi.org/10.5601/jelem.2016.21.1.1011
20. Neto, C., Carranca, C., Clemente, J., Varennes, A.: Nitrogen distribution, remobilization and re-cycling in young orchard of non-bearing 'Rocha' pear trees. Sci. Hortic. **118**, 299–307 (2008). https://doi.org/10.1016/j.scienta.2008.06.023
21. Stefanelli, D., Goodwin, I., Jones, R.: Minimal nitrogen and water use in horticulturae: Effects on quality and content of selected nutrients. Food Res. Int. **43**(7), 1833–1843 (2010). https://doi.org/10.1016/j.foodres2010.04.022
22. Malyuk, T., Pcholkina, N., Pachev, I.: Diagnostics of parameters of interrelations of mineral nutrition and formation of yield of fruit crops for intensive technologies of their cultivation. Banat's J. Biotechnol. **9**, 41–44 (2014). https://doi.org/10.7904/2068-4738-V(9)-41
23. Taiz, L., Zeiger, E.: Plant physiology, 5th ed, pp. 67–86. Sinauer Associates Sunderland, MA, USA, (2010)
24. Watts, D.B., Torbert, H.A., Prior, S.A., Huluka, G.: Long-term tillage and poultry litter impacts soil carbon and nitrogen mineralization and fertility. Soil Sci. Soc. Amer. J. **74**, 1239–1247 (2010). https://doi.org/10.2136/sssaj2008.0415
25. Dessureault-Rompré, J., Zebarth, B.J., Burton, D.L., Sharifi, M., Cooper, J., Grant, C.A., Drury, C.F.: Relationships among mineralizable soil nitrogen, soil properties, and climatic indices. Soil Sci. Soc. Amer. J. **74**, 1218–1227 (2010). https://doi.org/10.2136/sssaj2009.0213
26. Zhang, S., Gao, P., Tong, Y., Norse, D., Lu, Y., Powlson, D.: Overcoming nitrogen fertilizer over-use through technical and advisory approaches: A case study from Shaanxi Province, northwest China. Agr. Ecosyst. Environ. **209**, 89–99 (2015). https://doi.org/10.1016/j.agee.2015.03.002
27. Pole, V., Missa, I., Rubauskis, E., et al.: Effect of nitrogen fertiliser on growth and production of apples in the conditions of Latvia. In: Proceedings of the Latvian Academy of Sciences. Section B. Natural, Exact, and Applied Sciences, **71**(3), 115–120 (2017). https://doi.org/10.1515/prolas-2017-0020

28. Salem, M., Al-Ethawi, L.: A study of the presence of residual of nitrogenous fertilizer nitrate (NO3-) in some soils of Brack—Ashkada agriculture project. J. Life Sci. Technol. 84–88 (2013). https://doi.org/10.12720/jolst.1.1.84-88salem
29. Di, H.J., Cameron, K.C.: Nitrate leaching in temperate agroecosystems: sources, factors and mitigating strategies. Nutr. Cycl. Agroecosyst. **64**(3), 237–256 (2002). https://doi.org/10.1023/A:1021471531188
30. Fan, J., Hao, M., Malhi, S.S.: Accumulation of nitrate-N in the soil profile and its implications for the environment under dryland agriculture in northern China: A review. Can. J. Soil Sci. **90**, 429–440 (2010)
31. Susin, J., Kmecl, V., Gregorcic, A.: A survey of nitrate and nitrite content of fruit and vegetables grown in Slovenia during 1996–2002. Food Addit. Contam. **23**, 385–390 (2006). https://doi.org/10.1080/02652030600573715
32. Anjana, S.U., Iqbal, M.: Nitrate accumulation in plants, factors affecting the process, and human health implications. A review. Agron Sustain Dev **27**, 45–57 (2007). https://doi.org/10.1051/agro:2006021
33. Ysart, G., Miller, P., Barrett, G., Farrington, D., Lawrance, P., Harrison, N.: Dietary exposures to nitrate in the UK. Food Addit. Contam. **16**, 521–532 (1999). https://doi.org/10.1080/026520399283669
34. Ziarati, P., Mohammad-Makki, F.M.: Removal of nitrate and nitrite from tomato derived products by lemon juice. BBRA **12**(2), 767–772 (2015). https://doi.org/10.13005/bbra/2258
35. State Sanitary Regulations and Norms 4.4.2.030–99 (Decree № 30, 1999)
36. IUSS Working Group WRB (2015). World Reference Base for Soil Resources 2014, Update 2015. International Soil Classification System for Naming Soils and Creating Legends for Soil Maps. World Soil Resources Reports No. 106, Rome: FAO
37. Houba,V.J.G., Novozamsky, I., Lexmond, Th. M., Van der Lee, J.J. Applicability of 0.01 M $CaCl_2$ as a single extraction solution for the assessment of the nutrient status of soils and other diagnostic purposes. Commun. Soil Sci. Plant Anal. **21**(19–20), 2281–2290 (2008). https://doi.org/10.1080/00103629009368380
38. Houba, V.J.G., Novozamsky, I., Huybregts, A.W.M.: Cite as Comparison of soil extractions by 0.01 M $CaCl_2$, by EUF and by some conventional extraction procedures. Plant Soil **96**, 433–437 (1986). https://doi.org/10.1007/BF02375149
39. Consalteri, A., Rigato, A., Clamor, L., Giandon, P.: Determination of nitrate in vegetables using an ionselective electrode. J. Food Compos. Anal. **5**, 252–256 (1992). https://doi.org/10.1016/0889-1575(92)90044-K
40. Ghestem, M., Sidle, R.C., Stokes, A.: The influence of plant root systems on subsurface flow: implications for slope stability. Bioscience **61**, 869–879 (2011). https://doi.org/10.1525/bio.2011.61.11.6

# Technological Indices of Spring Wheat Grain Depending on the Nitrogen Supply

**Larysa Novak, Vitalii Liubych, Serhii Poltoretskyi and Mykola Andrushchenko**

## 1 Introduction

Production of food grain has always been the main problem of agriculture. Nowadays, it is of particular importance because there is an acute shortage of grain, including spring wheat, with restriction of energy reserves, and the demand for it in Ukraine and foreign market increases. Besides, grain of spring wheat is the main raw material for the production of flour, high-quality cereals, bread, and pastry [1].

## 2 Analysis of Reference Sources Data and Problem Statement

Ensuring needs in grain of spring wheat is possible by expanding the planted area of crops and improving the technology of its cultivation and, above all, by fertilization [2–4].

Solving these problems can only be based on the rational use of land resources, introducing a scientifically based system of agriculture in each farm, enhancing soil fertility, and applying intensive technologies of growing cereal crops [5–8].

It is known that spring wheat uses fertilizer elements and their aftereffect well [9–11]. Modern varieties of spring wheat are characterized by high grain yield and

L. Novak (✉) · V. Liubych · S. Poltoretskyi
Uman National University of Horticulture, Institutska St. 1, Uman, Cherkassy Region 20305, Ukraine
e-mail: lnovak590@gmail.com

M. Andrushchenko
Tavria State Agrotechnological University, B. Khmelnitsky Av., 18, Melitopol, Zaporizhia Region 72312, Ukraine

© Springer Nature Switzerland AG 2019

V. Nadykto (ed.), *Modern Development Paths of Agricultural Production*, https://doi.org/10.1007/978-3-030-14918-5_73

protein content. Therefore, the study of the influence of fertilizer norms on the productivity of spring wheat and grain quality formation is important.

The problem is to determine the yield and quality of spring wheat grain, depending on the nitrogen fertilizer.

## 3 Research Objectives and Tasks

The research objective was to analyze the technological properties of spring wheat grain, depending on the conditions of nitrogen fertilizer.

Research tasks were meant to establish productivity of wheat grain by determining the yield, vitreous, protein content.

## 4 Materials and Methods of Research

The study was conducted under conditions of research fields in Uman NUH, in Cherkasy region, Ukraine, Fig. 1. The soil was the loamy podzolized chernozem. Studies were conducted in 2012–2014. The experiment was carried out according to the schedule presented in Tables 1, 2 and 3. Phosphate and potassium fertilizers were applied during the primary tillage, and nitrogen fertilizers were applied in spring during the pre-sowing cultivation.

As shown in Fig. 1, the area of studies is Ukraine, Cherkasy region, research fields of Uman National University of Horticulture.

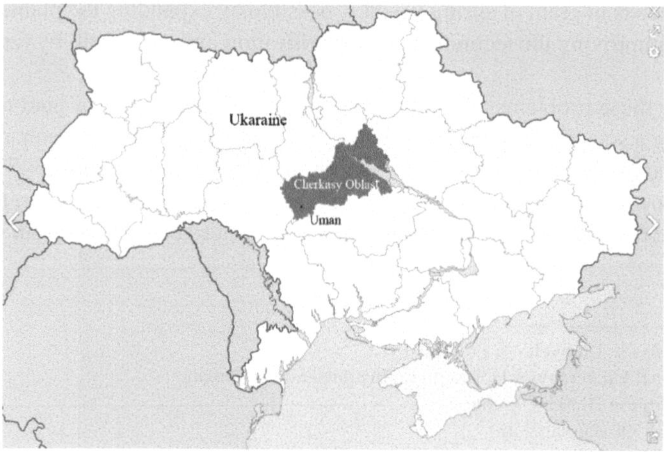


**Fig. 1** Area of studies is Ukraine, Cherkasy region, research fields of Uman National University of Horticulture

**Table 1** Yield of spring wheat under different norms of nitrogen fertilizers and ground (t/ha)

| Variant of the experiment | Year of the study | | |
|---|---|---|---|
| | 2012 | 2013 | 2014 |
| Without fertilizers (check variant) | 2.79 ± 0.14* | 5.01 ± 0.17* | 5.37 ± 0.50* |
| Ground $P_{60}K_{60}$—Ground I | 3.06 ± 0.18* | 5.25 ± 0.22* | 5.59 ± 0.18* |
| Ground I + $N_{30}$ | 3.43 ± 0.18* | 5.81 ± 0.46* | 6.05 ± 0.23* |
| Ground I + $N_{60}$ | 4.20 ± 0.09* | 6.35 ± 0.47* | 6.64 ± 0.40* |
| Ground I + $N_{90}$ | 4.43 ± 0.33* | 6.60 ± 0.38* | 6.99 ± 0.28* |
| Ground I + $N_{120}$ | 4.51 ± 0.37* | 6.67 ± 0.27* | 7.08 ± 0.31* |
| Ground I + $N_{150}$ | 4.67 ± 0.32* | 6.63 ± 0.28* | 7.14 ± 0.40* |
| Ground I + $N_{180}$ | 4.55 ± 0.34* | 6.71 ± 0.53* | 7.13 ± 0.29* |
| Ground I + $N_{210}$ | 4.51 ± 0.24* | 6.73 ± 0.20* | 7.12 ± 0.39* |
| Ground $P_{120}K_{120}$—Ground II | 3.30 ± 0.30* | 5.30 ± 0.37* | 5.88 ± 0.26* |
| Ground II + $N_{30}$ | 3.55 ± 0.33* | 5.95 ± 0.35* | 6.12 ± 0.39* |
| Ground II + $N_{60}$ | 4.20 ± 0.36* | 6.52 ± 0.59* | 6.83 ± 0.32* |
| Ground II + $N_{90}$ | 4.51 ± 0.38* | 6.82 ± 0.34* | 7.12 ± 0.51* |
| Ground II + $N_{120}$ | 4.72 ± 0.46* | 7.02 ± 0.41* | 7.32 ± 0.33* |
| Ground II + $N_{150}$ | 4.80 ± 0.31* | 7.12 ± 0.42* | 7.39 ± 0.32* |
| Ground II + $N_{180}$ | 4.81 ± 0.34* | 7.12 ± 0.38* | 7.38 ± 0.14* |
| Ground II + $N_{210}$ | 4.80 ± 0.33* | 7.12 ± 0.36* | 7.39 ± 0.31* |

*Values are expressed by mean ± SD of four independent observations, with significance $p$ value $< 0.005$

Farming equipment for spring wheat cultivation is common for Right-Bank Forest-Steppe of Ukraine. In the experiment, after spring barley Kolektyvna 3 spring wheat variety was grown. Agricultural technology included the application of phosphate and potash fertilizers in autumn according to the schedule of the experiment, and then primary tillage in 1–2 stages and under-winter plowing. In early spring, the soil surface was leveled by harrowing, and then, applying nitrogen fertilizers, pre-sowing cultivation was carried out according to the depth of seeding spring wheat. The total area was 72 $m^2$ and the registration plot was 40 $m^2$. The experiment replication was triple and the plot placement was consistent.

Grain yield was determined in sections by direct harvesting with a combine.

Protein content was determined by GOST 4117:2007 and grain vitreousness was determined by GOST 10987–76 to assess the quality of spring wheat grain.

The experiment replication was four-time and randomized. Data samples of each repetition were checked for adequacy and reproducibility. Since all samples were unevenly distributed, their statistical analysis was carried out using the methods of descriptive and nonparametric statistics.

Weather conditions during the period of studies were unstable compared to average long-term indicators.

**Table 2** Protein content in spring wheat grain under different norms of nitrogen fertilizers and ground (%)

| Variant of the experiment | Year of the study | | |
|---|---|---|---|
| | 2012 | 2013 | 2014 |
| Without fertilizers (check variant) | 15.2 ± 0.16 | 16.0 ± 0.49 | 14.5 ± 0.00 |
| Ground $P_{60}K_{60}$—Ground I | 15.2 ± 0.34 | 16.4 ± 0.29 | 14.6 ± 0.00 |
| Ground I + $N_{30}$ | 15.5 ± 0.40 | 17.4 ± 0.40 | 14.9 ± 0.44 |
| Ground I + $N_{60}$ | 15.7 ± 0.28 | 17.1 ± 0.33 | 15.1 ± 0.41 |
| Ground I + $N_{90}$ | 15.8 ± 0.35 | 17.5 ± 0.33 | 15.3 ± 0.18 |
| Ground I + $N_{120}$ | 16.0 ± 0.38 | 17.4 ± 0.34 | 15.4 ± 0.21 |
| Ground I + $N_{150}$ | 16.1 ± 0.27 | 17.1 ± 0.37 | 15.7± 0.48 |
| Ground I + $N_{180}$ | 16.1 ± 0.29 | 17.5 ± 0.20 | 15.7 ± 0.39 |
| Ground I + $N_{210}$ | 16.1 ± 0.24 | 17.5 ± 0.18 | 15.8 ± 0.68 |
| Ground $P_{120}K_{120}$—Ground II | 15.4 ± 0.37 | 16.0 ± 0.21 | 14.7 ± 0.37 |
| Ground II + $N_{30}$ | 15.4 ± 0.60 | 16.5 ± 0.42 | 14.8± 0.26 |
| Ground II + $N_{60}$ | 15.7 ± 0.53 | 16.8 ± 0.35 | 15.1 ± 0.45 |
| Ground II + $N_{90}$ | 16.0 ± 0.27 | 16.9 ± 0.42 | 15.4 ± 0.36 |
| Ground II + $N_{120}$ | 16.1 ± 0.40 | 17.0 ± 0.50 | 15.5 ± 0.36 |
| Ground II + $N_{150}$ | 16.2 ± 0.37 | 17.5 ± 0.41 | 15.6 ± 0.25 |
| Ground II + $N_{180}$ | 16.4 ± 0.41 | 17.4 ± 0.25 | 15.8 ± 0.11 |
| Ground II + $N_{210}$ | 16.5 ± 0.41 | 17.6 ± 0.30 | 15.9 ± 0.31 |

*Values are expressed by mean ± SD of four independent observations, with significance $p$ value <0.005

Weather conditions in 2012 were more favorable for the growth and the development of spring wheat. However, during its growing period, there was precipitation up to 184.1 mm that was 1.5 times less than average long-term indicators.

Weather conditions in 2013 were characterized by uneven distribution of rainfall during the growing season of spring wheat and inactive increase in heat at the beginning of its vegetation. In general, weather conditions were favorable for obtaining high yields of spring wheat. However, in the period of April–July, there was precipitation up to 173.6 mm that is 1.6 times less than the average long-term indicator.

Weather conditions in 2012 were characterized by sufficient rainfall. Thus, during the period of April–July, there was precipitation up to 294.3 mm that is 1.1 times more than the average long-term indicator [3]. In 2014, there was 140.8 mm of rainfall during the stem elongation. The temperature also affected the growth and the development of plants of spring wheat. Thus, in a period of intense stem growth (stem elongation—heading) in 2013, it was unfavorable compared to the optimal temperature (9–16 °C) and it was 18–21 °C. During this period, for other years of studies, the air temperature was optimal.

**Table 3** Vitreousness of spring wheat grain under different norms of nitrogen fertilizers and ground (%)

| Variant of the experiment | Year of the study | | |
|---|---|---|---|
| | 2012 | 2013 | 2014 |
| Without fertilizers (check variant) | 89 ± 0.25 | 79 ± 0.41 | 84 ± 0.41 |
| Ground $P_{60}K_{60}$—Ground I | 89 ± 0.50 | 79 ± 0.41 | 85 ± 0.41 |
| Ground I + $N_{30}$ | 90 ± 0.41 | 80 ± 0.41 | 90 ± 0.41 |
| Ground I + $N_{60}$ | 92 ± 0.41 | 82 ± 0.41 | 95 ± 0.41 |
| Ground I + $N_{90}$ | 93 ± 0.41 | 83 ± 0.41 | 95 ± 0.25 |
| Ground I + $N_{120}$ | 93 ± 0.25 | 83 ± 0.41 | 97 ± 0.41 |
| Ground I + $N_{150}$ | 95 ± 0.41 | 85 ± 0.41 | 97 ± 0.41 |
| Ground I + $N_{180}$ | 96 ± 0.25 | 86 ± 0.41 | 98 ± 0.41 |
| Ground I + $N_{210}$ | 97 ± 0.71 | 87 ± 0.41 | 98 ± 0.58 |
| Ground $P_{120}K_{120}$—Ground II | 91 ± 0.71 | 81 ± 0.58 | 90 ± 0.41 |
| Ground II + $N_{30}$ | 93 ± 0.25 | 82 ± 0.71 | 92 ± 0.41 |
| Ground II + $N_{60}$ | 96 ± 0.41 | 86 ± 0.41 | 95 ± 0.41 |
| Ground II + $N_{90}$ | 95 ± 0.41 | 85 ± 0.71 | 96 ± 0.41 |
| Ground II + $N_{120}$ | 97 ± 0.71 | 87 ± 0.41 | 97 ± 0.41 |
| Ground II + $N_{150}$ | 98 ± 0.71 | 88 ± 0.41 | 97 ± 0.41 |
| Ground II + $N_{180}$ | 98 ± 0.41 | 88 ± 0.41 | 98 ± 0.41 |
| Ground II + $N_{210}$ | 97 ± 0.58 | 87 ± 0.41 | 98 ± 0.41 |

*Values are expressed by mean ± SD of four independent observations, with significance $p$ value <0.005

## 5 Results and Discussion

The research has found that weather conditions of the growing season and nitrogen fertilizer norms had a significant impact on the productivity of spring wheat as shown in Table 1.

The yield of spring wheat increased considerably in variants with the application of nitrogen fertilizers compared to $P_{60}K_{60}$ variant within three years of the research. On average, over three years of the research, spring wheat yield in unfertilized areas was 4.38 t/ha and increased to 6.13 t/ha in the variant with the highest rate of nitrogen fertilizers ($N_{210}$), Fig. 2.

However, it varied considerably over the years of research. Thus, in 2012, in the check variant, the yield was 2.72 t/ha and increased to 4.51 t/ha or 64% when applying $N_{210}$ shown in Table 1. In 2013, this figure also significantly increased in accordance from 5.01 to 6.73 t/ha or 34%; in 2014, it changed from 5.37 to 7.12 t/ha, or 33%. It should be noted that the increase in nitrogen fertilizer norm to $N_{120\text{-}210}$

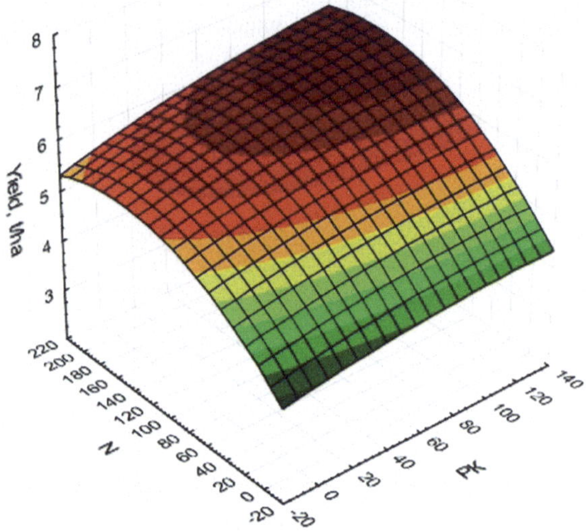

**Fig. 2** Average wheat grain yield

hardly differed from the variant with the application of 90 kg/ha of nitrogen fertilizers speaking about grain yield.

A similar pattern was observed when growing wheat on the ground $P_{120}K_{120}$. On average, over three years of the research, the yield increased from 4.41 to 6.46 t/ha. Variants with the application of $N_{120\text{-}210}$ provided a significant increase in grain yield. Data analysis of the productivity shows the inappropriate application of $P_{120}K_{120}$ for spring wheat.

Spring wheat grain is characterized by a high content of protein which significantly changed depending on weather conditions of the growing season and nitrogen fertilizer norms as shown in Table 2.

On average, over three years of the research, this figure in the variant without fertilizers amounted to 15.2% and increased to 15.9–16.5% by applying $N_{30\text{-}210}$ in the ground $P_{60}K_{60}$, Fig. 3. In 2012 and 2013, the protein content was higher compared to 2014 because this year was characterized by higher rainfall compared to other years of the research.

A similar pattern was observed when applying nitrogen fertilizers in the ground $P_{120}K_{120}$, and on average, over three years of the research, the protein content increased from 15.2 to 16.7% Fig. 3.

Also, applying nitrogen fertilizers significantly increased vitreousness of spring wheat grain as shown in Table 3.

On average, over three years of the research, vitreousness of spring wheat in the variant without fertilizers accounted for 84% and increased to 87–94% in variants with the application of 30–210 kg/ha of nitrogen fertilizers on the ground $P_{60}K_{60}$.

**Fig. 3** Average protein content in wheat grain

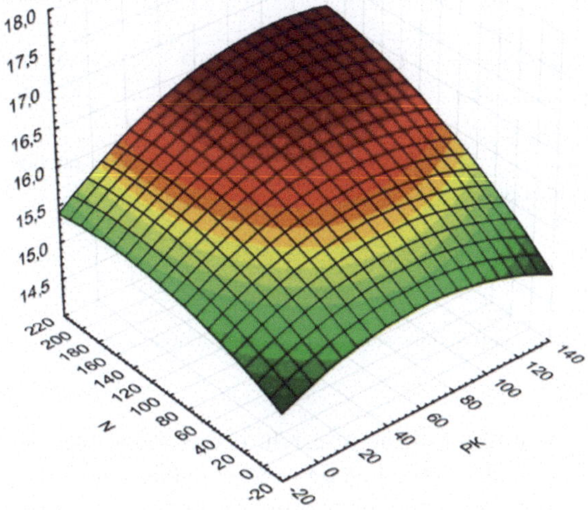


**Fig. 4** Average grain vitreousness index

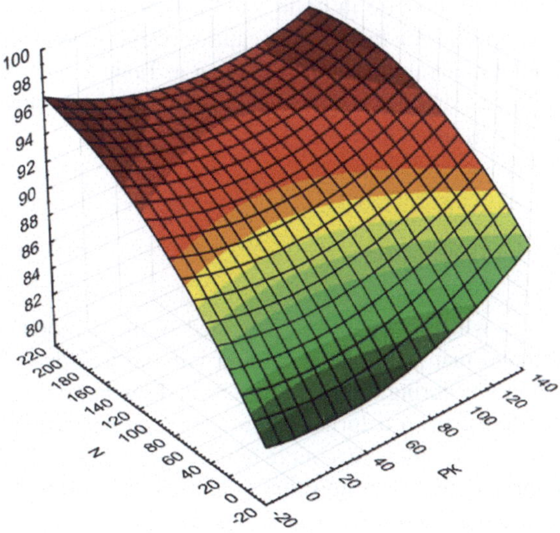


However, this figure also varied over years of the research. Thus, in 2012, it was 89–98%; in 2013, it was 79–87%; and in 2014, it was 84–98%.

Vitreousness of spring wheat after applying nitrogen fertilizers was similar when growing it on the ground $P_{120}K_{120}$. The figure did not almost change and was 85–94%, Fig. 4.

**Table 4** Spearman rank order correlations (2013)

| | Yield (t/ha) | Protein content (%) | Grain vitreousness (%) | PK | N |
|---|---|---|---|---|---|
| Yield (t/ha) | 1.000000 | 0.924970* | 0.891238* | 0.440606 | 0.900570* |
| Protein content (%) | 0.924970* | 1.000000 | 0.969550* | 0.242631 | 0.988259* |
| Grain vitreousness (%) | 0.891238* | 0.969550* | 1.000000 | 0.256519 | 0.973757* |
| PK | 0.440606 | 0.242631 | 0.256519 | 1.000000 | 0.164764 |
| N | 0.900570* | 0.988259* | 0.973757* | 0.164764 | 1.000000 |

*Marked correlations are significant at $p$ <05000 $e$ <0.005

As a result of the research, it was found that the yield of spring wheat varied significantly depending on weather conditions of the growing season. Spring wheat responds best to applying nitrogen fertilizers compared to phosphate and potassium ones. However, it responds best to applying $N_{60\text{-}90}$ on the ground $P_{60}K_{60}$.

A further increase in the nitrogen fertilizer norm and ground does not provide a significant increase in grain yield and protein content. Strength of the linear relationship between the studied features (yield, protein content, grain vitreousness, and dose of mineral fertilizers) is shown in Table 4.

Table 4 shows that indices of yield and protein content, yield and nitrogen fertilizer doses correlate closely.

## 6 Conclusions

Spring wheat grain is characterized by high protein content and vitreousness. It is found that protein content and vitreousness depend on the features of weather conditions during the growing season. Low humidity, high temperature, and lack of moisture in the soil during the growing season contribute to protein content increasing in the grain compared to more humid-growing season. These figures can be improved optimizing nitrogen nutrition conditions, for example, in ground I when applying $N_{180}$ and ground II when applying $N_{180}$.

## References

1. Saiko, V.F.: Scientific systems of grain farming. Higher school, **336**, (1994)
2. Panchenko, I.A.: Influence of nutrition backgrounds on the soft wheat grain quality. Bulletin of Kharkiv State Agricultural University, **1**, 207–210 (2001)

3. Petrenko, V., Liubich, V., Bondar, V.: Baking quality of wheat grain as influenced by agriculture systems, weather and storing conditions. Rom. Agric. Res. 69–76 (2017)
4. Ruisi, P., Frangipane, B., Amato, G., Frenda, A.S., Plaia, A., Giambalvo, D., Saia, S.: Nitrogen uptake and nitrogen fertilizer recovery in old and modern wheat genotypes grown in the presence or absence of interspecific competition. Front Plant Sci. **6**, 185–193 (2015)
5. Pan, D., Kong, F., Zhang, N., Ying, R.: Knowledge training and the change of fertilizer use intensity: Evidence from wheat farmers in China. J Environ. Manage. **197**, 130–139 (2017)
6. Goucher, L., Bruce, R., Cameron, D.D., Lenny Koh, S.C., Horton, P.: The environmental impact of fertilizer embodied in a wheat-to-bread supply chain. Nat Plants. **3**, 170–182 (2017)
7. Zhen, S., Deng, X., Zhang, M., Zhu, G., Lv, D., Wang, Y., Zhu, D., Yan, Y.: Comparative phosphoproteomic analysis under high-nitrogen fertilizer reveals central phosphoproteins promoting wheat grain starch and protein synthesis. Front Plant Sci. **8**, 67 (2017)
8. Dustin, A., Douglas, L., William, L., Agron, J.: Economically optimal nitrogen fertilization for yield and protein in hard red spring wheat. **96**, 116–123 (2004)
9. Gary, E.: Crop rotation and nitrogen effects on normalized grain yields in a long-terms study. Agron. J. **92**, 938–941 (2000)
10. Zhang, W., Liu, D., Liu, Y., Chen, X., Zou, C.: Overuse of phosphorus fertilizer reduces the grain and flour protein contents and zinc bioavailability of winter wheat (Triticum aestivum L.). J. Agric. Food Chem. **65**, 1473–1482 (2017)
11. Yang, L., Zhang, L., Yu, C., Li, D., Gong, P., Xue, Y., Song, Y., Cui, Y., Doane, T.A., Wu, Z.: Nitrogen fertilizer and straw applications affect uptake of 13C,15N-Glycine by soil microorganisms in wheat growth stages. **12**, 169–177 (2017)

# Use of Alternative Types of Fuel for Grain Drying

**Nina Osokina, Hennadii Tkachenko, Yana Yevchuk and Olena Hryhorenko**

## 1 Introduction

The largest energy expenses of after-gathering processing of corn, sunflower, and soybean grain are spent for drying [1].

Every year prices for the traditional types of fuel (natural gas, diesel fuel, mazout, coal) are rising, and therefore, it is reasonable to use alternative types of fuel in the process of grain drying. The most widespread types of them are organic waste from cultivation and processing of agricultural crops (straw, grain cleaning waste, sunflower husk), as well as wood (logs, wood shaving, fuel chips, granules) [2, 3].

## 2 Analysis of Reference Sources Data and Problem Statement

It is better to use the straw left on the field as organic fertilizer, and the intensity of its combustion in bales is difficult to control. The most primitive technology of burning one-meter-long logs (Brazilian driers Kepler Weber [4]) requires considerable labor costs. Significant disadvantage of using this type of fuel is its high humidity which is difficult to remove. Thus, one-meter-long logs by air-and-sun drying for several months are mostly above 30% in humidity [5, 6].

N. Osokina · H. Tkachenko (✉) · Y. Yevchuk
Uman National University of Horticulture, Instytutska Str., 1, Uman, Cherkasy Region 20305, Ukraine
e-mail: tkachenkogenady@gmail.com

O. Hryhorenko
Tavria State Agrotechnological University, B. Khmelnitsky Av., 18, Melitopol, Zaporizhia Region 72312, Ukraine

© Springer Nature Switzerland AG 2019

V. Nadykto (ed.), *Modern Development Paths of Agricultural Production*,
https://doi.org/10.1007/978-3-030-14918-5_74

Wood for accelerating its drying is crushed on fuel chips up to 30 mm. The latter can be received by branches cutting of fruit and forest plantations remained after preventive pruning. Screw, belt, scraper conveyors, and grain cores are suitable for their transportation. Fuel chips are dried in drum driers [7].

Granulation of biofuel increases its friableness and caloric content and reduces transportation costs. Heat generators for combustion of crushed fuel are equipped with mechanized loading systems [8].

Products of biofuel combustion with high content of ash substances are given to the grain mass in the driers of Brazilian technology; in addition, grain absorbs the smell of smoke. Driers with a heat exchanger are deprived of this lack. Air temperature does not exceed 120 °C at the outlet that positively affects the quality of grain, but it can reduce the productivity of grain driers designed for higher temperatures of the drying agent. It is difficult to keep a stable temperature of the drying agent operating under such grain driers, especially when burning of nonuniformity in caloric content and fuel humidity (grain cleaning waste, fuel chips) [9].

Technical and economic grounds, technological scheme and working projects of a heat generator for combustion of fuel chips, pellets, and a small amount of waste of wood processing industry (with knots and nails that cannot be crushed to chips) were developed. It is necessary to increase the power of a heat generator and fuel consumption by almost twice in case of using heat exchanger. Effective operation of a heat exchanger is possible only by highly skilled, disciplined personnel together with engineering solutions—automation of the combustion process, protection against electricity irregularity, etc.

Construction project provides purification of flue gases in cyclones and verification of safety indicators of soybean seeds for remains of incomplete combustion substances: dioxins, heavy metals, polycyclic aromatic hydrocarbons (benzopyrene, etc.).

## 3 Research Objectives and Tasks

Purpose and objectives of the study consist in finding optimal types of alternative fuel and machines for their preparation for combustion in a heat generator of a grain drier—DSP-32*2 M.

## 4 Research Materials and Methods

Our study was conducted on the basis of "Vidrodzhennia" LLC of Hromada village of Liubar district, Zhytomyr region.

The study determined: productivity of log splitters by dividing the volume of wood delivered by one car (indicated in the consignment note) for the time necessary for its processing; chipping machines by marking the time required to fill the operational

**Table 1** Basic technical and technological indexes of the machines for wood splitting

| Indicator | KG(КГ)-1000 | KM(КМ)-2000 |
|---|---|---|
| Maximum tree length (mm) | 1000 | 2000 |
| Maximum tree diameter (mm) | 700 | 1000 |
| Actual maximum diameter (mm) | 300 | 500 |
| Maximum productivity ($m^3$/h) | 2–9 | 9–20 |
| Actual productivity ($m^3$/h) | 1 | 3 |
| Number of blades (pieces) | 2 | 1 |
| Engine power (kW) | 11 | 22 |
| Weight (kg) | 310 | 1000 |

bunker with volume of 6 $m^3$; mass fraction of a large fraction of fuel chips by weighing of manually selected chips in the process of reloading from the operational (volume of 64 $m^3$) into the accumulation bin of a heat generator. Such repetition was done three times.

## 5 Results and Their Discussion

It is necessary to split that wood which has a larger diameter than maximum permissible size of a chopping machine. You should have a hydraulic log splitter KG(КГ)-1000 together with a chopping machine PL(ПЛ)-160. We designed a mechanical log splitter KM(KM)-2000 for a chopping machine MRNP(МРНП)-30 (Table 1).

It was established that the actual maximum diameter of wood for both machines was lower than in the passport because there was wood of non-rectilinear form for combustion, with a large number of branches of fibrous structure. However, excess of load resulted in carriage bending of overheating of hydraulic fluid (KG(КГ)-1000) or chain breaking (KM(KM)-2000), and lack of mechanical supply significantly reduced the productivity of a log splitter (KM(KM)-2000) to 3 $m^3$/h.

Optimal modes of chopping machines operation expected the use of wet wood, drying and storing of fuel chips. Using such scheme requires additional operations for transporting fuel chips to storehouses for their keeping. We chose the following scheme for this purpose which includes: air-and-sun drying of wood, its transportation, chopping, and burning. Blades of chopping machines were sharpened on the TChN(ТЧН) 21–5 bench; angle for blades sharpening was chosen according to the wood moisture content.

Wood chopping was performed using a stationary chopping machine MRNP (МРНП)-30 and a mobile machine—PL(ПЛ)-160. The drive of a chopping machine PL(ПЛ)-160 could be carried out both from the electric motor and from the power shaft of capacity takeoff of a tractor of MTZ(МТЗ)-82 class. The latter option allowed chopping the crowns of trees directly in the forest (Table 2).

**Table 2** Basic technical indexes of chopping machines PL(ПЛ)-160 and MRNP(МРНП)-30

| Indicator | PL(ПЛ)-160 | MRNP(МРНП)-30 |
|---|---|---|
| Maximum tree diameter (mm) | 160 | 250 |
| Maximum productivity ($m^3$/h) | 9 | 30 |
| Number of blades (pieces) | 3 | 16 |
| Engine power (kW) | 30 | 90 |
| Wood supply | Hydraulic | Gravitation |
| Weight (kg) | 860 | 5750 |
| Actual productivity ($m^3$/h) | 3 | 21 |

Supply of fuel chips to the heat generator was carried out by screw conveyors, and therefore, one of the important indicators in the work of chopping machines was mass fraction of chips that could not be given by a screw (length—more than 100 mm; thickness—more than 20 mm).

Chopping machine PL(ПЛ)-160 is completed with an electric motor with a power of 30 kW by the factory manufacturer that is much less than the power of a tractor engine of MTZ(МТЗ)-82 class. Studies showed that the use of electric drive lowered actual productivity of a chopping machine to 3 $m^3$/h. Problems with exceeding the temperature of hydraulic fluid also caused forced stops. The lack of mechanized wood supply to the chopping machine MRNP(МРНП)-30 was the main reason for reducing the actual productivity to 21 $m^3$/h.

Chopping machine MRNP(МРНП)-30 is located below the shop floor which makes gravitation wood supply easier. Chips were given into a bunker of fuel receiving from vehicles with centrifugal force, for example, from a mobile chopping machine PL(ПЛ)-160. Contents of the bunker were unloaded onto the belt conveyor. Scraper conveyor was used to lift fuel chips into a storage bunker. Fuel chips were given to the heat generator TPG(ТПГ)-5 (project of Martsun O. M., Skoblenko V. P., Yaroshenko V. V., Tkachenko G. V.) with the help of four screw conveyors.

It is found that qualitative combustion of fuel on a sloping grate grid was carried out by the linear size of chips within the limits: width and height—10 to 25 mm; length—25 to 100 mm. Fuel pieces of smaller size fell through the holes in a grate grid, and they damaged it with high temperatures burning under a grid. Pieces with a high surface area and low weight were raised by air flow, did not have time to burn in the fuel spray zone. and increased the amount of sparks and products of incomplete combustion in flue gases.

It was revealed that a large fraction of fuel was difficult to transport with screws, and the duration of its drying on the sloping part of a grate grid was longer. A large fraction had a smaller total surface area; therefore, the intensity of the oxidation process was slowing down. The mass fraction of a large fraction of chips is shown in Fig. 1.

In 2017, about 40 thousand tons of soy to 8% of humidity [10, 11] without the use of natural gas was dried at the enterprise. A gas burner with minimum fuel

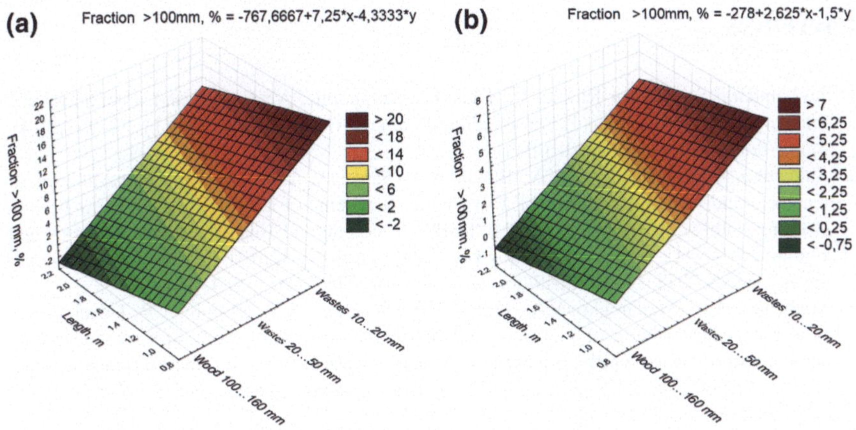


**Fig. 1** Mass fraction of chips with a length of more than 100 mm under processing of wood and wastes from woodworking with the length of 1 and 2 m by chopping machines **a** ПЛ(ПЛ)-160 and **b** МРНП(МРНП)-30

consumption was turned on at night only for drying individual batches with a seed moisture content of more than 25%. Technical documentation for building the second stage of an elevator with grain drier with a capacity of 100 t/h on alternative types of fuel was developed on the accumulation of ascorbic acid in black currant fruits.

## 6 Conclusions

1. Actual productivity of a log splitter KG(КГ)-1000 and chopping machine PL(ПЛ)-160 was lower than passport index under the conditions of uninterrupted operation at a large enterprise of 1 and 3 $m^3$/h, respectively.
2. Lack of mechanized supply of raw materials reduced actual productivity of a log splitter KG(КГ)-2000 to 3 $m^3$/h and a chopping machine MRNP(МРНП)-30 to 21$m^3$/h.
3. Greatest output of a large fraction of fuel chips (20%) was observed under use of a chopping machine MRNP(МРНП)-30.
4. Output of a large fraction of fuel chips was observed by processing of raw materials in the length of 1 m by 60–100% higher than by 2 m in length.

## References

1. Jayasundara, S., Wagner-Riddle, C., Dias, G., Kumudinie, A.: Energy and greenhouse gas intensity of corn (Zea mays L.) production in Ontario. Can. J. Soil Sci. **94**(1), 77–95 (2014). https://doi.org/10.4141/cjss2013-044
2. Shubenko, A.L., Maliarenko, V.A., Babak, N.Y., Senetsky, A.V., Sarapin, V.P.: Estimation of technical-and-economic indexes of a cascade heat scheme of a turbine plant on low-boiling working bodies. In: Bulletin of NTU "KhPI". Energetic and Heat-Technical Processes and Equipment—NTU, Kharkiv, "KhPI", **11** (1233), pp. 6–15 (2017) Bibliography: 20 references.—ISSN 2078-774X. https://doi.org/10.20998/2078-774x.2017.11.01
3. Shubenko, A.L., Maliarenko, V.A., Babak, N.Y., Senetsky, A.V., Sarapin, V.P.: Development of a cascade heat scheme of a turbine plant on low-boiling working bodies for rural energetic. In: Bulletin of NTU "KhPI". Energetic and Heat-Technical Processes and Equipment. NTU, Kharkiv "KhPI", **10**(1232), pp. 13–24 (2017)—Bibliography: 26 references. ISSN 2078-774X. https://doi.org/10.20998/2078-774x.2017.10.02
4. Bocca, L.S., Galves, M.L.: Multicriteria decision aid to implement an on-farm storage system for soybeans. Eng. Agríc. **36**(6) Jaboticabal (Oct/Dec 2016). https://doi.org/10.1590/1809-4430-eng.agric.v36n6p1250-1260/2016
5. Lokshtanov, B.M., Bacherikov, I.V.: Fuel chip and its properties. Urgent directions of scientific researches of XXI century: theory and practice, Voronezh. **3**(8–2), (19–2), pp. 274–278 (2015). https://doi.org/10.12737/15535
6. Golovchuk, A.F.: Improvement of the heat supply system of the campus of National University of horticulture. Bull. Dnipropetrovsk Nat. Univ. Railway Transp. **3**(63), 76–96 (2016). https://doi.org/10.15802/stp2016/74720
7. Nguen, F.: Machines systems for logging in energy plantations. StudArctic Forum. **1**(1) (2016). https://doi.org/10.15393/j102.art.2016.104
8. Yermochenkov M.G., Yevstigneev A.G.: Change in calorific value of wood fuel under torefication. Forest. Bull. Lestnoy vestnik **21**(1), 64–68 (2017). https://doi.org/10.18698/2542-1468-2017-1-64-68
9. Sokhansanj, S., Webb, E.: Evaluating industrial drying of cellulosic feedstock for bioenergy: a systems approach. Biofuel Bioprod. Biorefin. **10**, 47–55 (2016). https://doi.org/10.1002/bbb.1619
10. Li, Y., et al.: Effect of extruding parameters on soybean oil yield of enzyme-assisted aqueous extraction processing. Adv. Mater. Res. **148–149**, 7–18 (2011). https://doi.org/10.4028/www.scientific.net/AMR.148-149.7
11. Martsun, O.M., Osokina, N.M., Yaniuk, T.V., Yaroshenko, V.V., Tkachenko, G.V.: Method of soybean seeds preparation by flattening for oil removal by extrusion followed by pressing: license 110507 Ukraine, MPK C11B 1/04 A23P 30/20/—No.u 2016 04102; docum. 14.04.2016; valid since 10.10.2016, Bulletin No.19

# Effect of Seed Sowing Period on Antioxidant Protection of Basil (*Ocimum basilicum* L.) Under Greenhouse Conditions

**Olesia Priss, Iryna Korotka, Galina Simakhina, Victoria Koliadenco and Tatiana Kolisnychenko**

## 1 Problem Statement

The industrial production of green crops, including basils, in Ukraine is limited by a number of factors. One of the main limiting factors for the effective cultivation of basil in the inter-season period is lack of well-grounded technologies of the cultivation on protected soils. Elements of technology that facilitate the receipt of earlier harvests with the most complete return on the funds invested into the cultivation are of particular attention in the greenhouse industry. Therefore, determination of optimum timing for sowing basil in these protected soil conditions is an important task. However, at the current level of development of vegetable production, ensuring high yield of agricultural crops, including green ones, is practically impossible without the assessment of physiological state of the plant during the vegetation period.

## 2 Analysis of Recent Studies and Publications

It is known that the plant reacts quickly to the action of stress factors with generating reactive oxygen species (ROS) [1, 2]. Usually, the amount of ROS in the cell is maintained at a physiologically normal, background level due to the presence in the biochemical composition of plants of a large number of biologically active substances (BAS) having antioxidant properties and, in general, constituting a multi-

O. Priss (✉) · I. Korotka · V. Koliadenco
Tavria State Agrotechnological University, B. Khmelnitsky St., 18, Melitopol 72310, Ukraine
e-mail: olesyapriss@gmail.com

G. Simakhina
National University of Food Technology, Volodymyrska St., 68, Kiev 01601, Ukraine

T. Kolisnychenko
Oles Honchar Dnipro National University, Gagarin Av., 72, Dnipro 49010, Ukraine

© Springer Nature Switzerland AG 2019

V. Nadykto (ed.), *Modern Development Paths of Agricultural Production*,
https://doi.org/10.1007/978-3-030-14918-5_75

level antioxidant defense system. The balance between the generation of ROS and the antioxidant activity of BAS is an important condition for ensuring normal functioning of the plant organism [3]. However, any external influence is accompanied with an increase in free radical processes and an equilibrium shift in the direction of accumulation of ROS, which in turn induces rearrangements in the protective antioxidant system, in particular, changes in the activity of antioxidant enzymes and a pool of low molecular weight antioxidants [4]. This process is one of the initial stages that lead to the condition of stress. In response to various stresses in plant cells, there is an increase in the content of malonic dialdehyde (MDA), which is associated with the activation of free radical reactions under the indicated conditions [5, 6]. Thus, the content of MDA can serve as an indicator of the activity of oxidation processes caused by oxygen radicals [7].

Biochemical protection systems play the pivotal role in the adaptation of plants to the effects of adverse environmental factors. The main utilizers of reactive oxygen species, generated in the respiratory chain and other metabolic processes, are antioxidant enzymes, namely, superoxide dismutase [8]. Superoxide dismutase (SOD) is one of the key enzymes for the protection of cells and tissues from oxidative degradation. It is the only enzyme to catalyze the dismutation of the superoxide anion radical ($O^{-\bullet 2}$) to $O_2$ and $H_2O_2$, thus regulating the intracellular concentration of free oxygen radicals. The activity of SOD on condition of adverse environmental factors, when the formation of active forms of oxygen in plant organisms increases, can vary in different ways, which depend on the duration and intensity of the stress factor, as well as on the stability of the organism, the stage of plants development, etc. [9]. So, some increase in the activity of SOD can be explained by the response to stress during the cultivation of agricultural crops.

## 3 Materials and Methods of Research

The studies were held in 2014—2016 in the conditions of plastic film greenhouses with technical heating. The research used varieties of basils of Ukrainian selection, namely: Badioryi (control) and Rutan that have a green color of leaves, Philosopher and Purpurova Zorya with a purple color and Siayvo with predominant green and the patches of anthocyanin. The optimum timing of sweet basils sowing included the following study options: 1—sowing seeds in the 3rd ten-day period of February, 2—sowing seeds in the 2nd ten-day period of March, 3—sowing seeds in the 2nd ten-day period of April.

Seeds were planted in boxes in the rows 5 cm wide. The temperature regime during germination of seeds was maintained at the level of 22–25 °C. When the first pair of true leaves was formed, the plants were diced in portions of $6 \times 6$ cm. The seedlings were planted after the formation of 3 pairs of true leaves. The area of control was 2 $m^2$, with five times repetition.

The development of oxidative damage was estimated with the content of malonic dialdehyde (MDA), which is considered to be a marker of oxidative stress. The determination of the MDA content in plant cells is based on its reaction with 2-thiobarbituric acid, resulting in a colored product with a maximum absorption at a wavelength of 532.

The sample weight of the investigated material (0.5 g) was triturated in a porcelain mortar with 3 ml of distilled water. 3 ml of trichloroacetic acid was added thereto and the mixture was homogenized for the second time. Two samples of 2 ml were taken from the obtained homogenate into the test tubes. 2 ml of a 20% solution of trichloroacetic acid was added to one of the samples. Subsequently, this test was used as control in spectrophotometry. 2 ml of 0.5% solution of thiobarbituric acid was added to the second test. After incubating the samples for 30 min in a boiling water bath, they were cooled and centrifuged for 10 min at 3000 rpm. The supernatant was collected with a syringe into a test tube and measured on a spectrophotometer at a wavelength of 532.

*Determination of superoxide dismutase activity.* The activity of superoxide dismutase in the plant cells is determined according to their ability to suppress the reaction of autoxidation of adrenaline in the alkaline medium. The rate of reaction is evaluated spectrophotometrically by the magnitude of the accumulated product optical density of the auto-oxidation of adrenaline absorbed at a wavelength of 347 nm.

The sample weight of the test material (0.5 g) was triturated in a porcelain mortar on ice, with addition of 5 ml of a phosphate buffer (pH 7.8). The homogenate was transferred to centrifuge tubes, where 0.3 ml of chloroform and 0.6 ml of alcohol were added and centrifuged for 20 min at 8000 rpm. The supernatant was collected with a syringe in a test tube and measured on a spectrophotometer at a wavelength of 347 nm.

Control Measurement: 1 ml of 0.1% adrenaline hydrochloride was added to 20 ml buffer pH 10.65. Measurements were held in 3 min against the buffer pH 10.65 with no adrenaline addition.

Experimental Measurement: 1 ml of the experimental specimen and 1 ml of 0.1% adrenaline hydrochloride were added to 20 ml buffer pH 10.65. Measurements were held in 3 min against pH buffer 10.65 with addition of 1 ml of the test sample.

SOD activity is calculated in accordance with the formula:

$$\frac{1 - D_{\text{experimental}}}{D_{\text{control}}} \cdot 100\%, \tag{1}$$

where

$D_{\text{experimental}}$ shall mean the results obtained with experimental measurements;
$D_{\text{control}}$ shall mean the results obtained with control measurements.

# 4 Results

Studies have shown that the cultivation of basil in different periods was accompanied with changes of the content of malonic dialdehyde in leaves, indicating different adaptation of plants to this factor and the intensity of the process of the active forms of oxygen generating in plants. The level of malonic dialdehyde in basils ranges from 13.04 nmol/g for the Philosopher to 15.90 nmol/g for the Syaivo variety (Table 1).

The significant increase in the level of MDA is observed in all breeds of February sowing period, on average, it is up to 18.74 nmol/g. The plants which were sown in March had a level of MDA lower by 32.7%, those of April sowing period had 36.6%, indicating more favorable conditions for growth and development of a plant. The

**Table 1** Content of malonic dialdehyde in the basil greenery at the beginning of budding phase, M $\pm$ m, $n = 5$ (the average for 2014–2016)

| Breed | Terms of planting | Malonic dialdehyde (nm/g) |
|---|---|---|
| Badioryi (control) | 3rd ten-day period of February | 20.90 $\pm$ 1.77 |
| | 2nd ten-day period of March | 12.66 $\pm$ 0.44 |
| | 2nd ten-day period of April | 12.18 $\pm$ 0.58 |
| Average (A) | | 15.25 |
| Rutan | 3rd ten-day period of February | 19.87 $\pm$ 0.78 |
| | 2nd ten-day period of March | 12.08 $\pm$ 0.78 |
| | 2nd ten-day period of April | 10.82 $\pm$ 0.21 |
| Average (A) | | 14.26 |
| Philosopher | 3rd ten-day period of February | 15.85 $\pm$ 0.44 |
| | 2nd ten-day period of March | 12.22 $\pm$ 0.81 |
| | 2nd ten-day period of April | 11.05 $\pm$ 0.54 |
| Average (A) | | 13.04 |
| Purpurova Zorya | 3rd ten-day period of February | 16.58 $\pm$ 0.32 |
| | 2nd ten-day period of March | 12.40 $\pm$ 1.07 |
| | 2nd ten-day period of April | 11.89 $\pm$ 0.66 |
| Average (A) | | 13.63 |
| Syaivo | 3rd ten-day period of February | 20.51 $\pm$ 0.86 |
| | 2nd ten-day period of March | 13.73 $\pm$ 0.73 |
| | 2nd ten-day period of April | 13.45 $\pm$ 0.73 |
| Average (A) | | 15.90 |
| Average (B) | 3rd ten-day period of February | 18.74 |
| | 2nd ten-day period of March | 12.62 |
| | 2nd ten-day period of April | 11.88 |
| $HIP_{0.05}$ (A) | | 0.5 |
| $HIP_{0.05}$ (B) | | 0.5 |

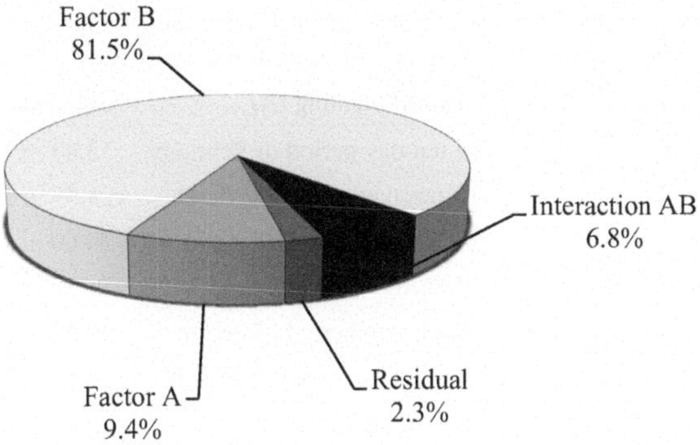


**Fig. 1** Share of influence of the factors on the level of malonic dialdehyde in the greenery of basils: ▩—breed (A); □—terms of seeds planting (B); ■—interaction AB; ▩—residual

planting time in particular was the determining factor in the growth of the level of malonic dialdehyde for sweet basils, which is confirmed by a two-factor dispersion analysis (Fig. 1).

However, the effect of stress factors is accompanied with the activation of protective mechanisms of the plant, aimed at the elimination of active forms of oxygen. The level of superoxide dismutase of all breeds of basils fluctuated within the range of 30.35–32.59 provisional units; the true difference between the breeds was not detected (Table 2).

However, the effect of stress factors is accompanied with the activation of protective mechanisms of the plant, aimed at the elimination of reactive oxygen species. The level of superoxide dismutase for breeds of basils fluctuated within the range of 30.35–32.59 provisional units; the true difference between the breeds was not detected (Table 2).

The influence of the breed factor is insignificant (Fig. 2).

Consequently, the aforementioned evidence suggests that the best sowing period of the seeds of basils with green and mixed color in conditions of plastic film greenhouses with technical heating is the second ten-day period of March and the second ten-day period of April. For plants with purple leaves, the optimum sowing period is the second ten-day period of March. This is evidenced with the lowest level of MDA in the leaves of all breeds—11.88 to 12.62 nM/g of raw matter, and the level of SDA which is 70.3–76.1% less compared to February sowing period.

**Table 2** Superoxide dismutase activity in leaves of basils depending on the dates of sowing at the beginning of the budding phase, M ± m, $n = 5$ (average for 2014-2016)

| Breed (A) | Terms of planting (B) | SDA, provisional units |
|---|---|---|
| Badioryi (control) | 3rd ten-day period of February | 53.09 ± 1.23 |
| | 2nd ten-day period of March | 21.57 ± 2.93 |
| | 2nd ten-day period of April | 20.60 ± 2.49 |
| Average (A) | | 31.75 |
| Rutan | 3rd ten-day period of February | 60.48 ± 1.63 |
| | 2nd ten-day period of March | 15.26 ± 3.31 |
| | 2nd ten-day period of April | 18.03 ± 1.28 |
| Average (A) | | 31.26 |
| Philosopher | 3rd ten-day period of February | 68.86 ± 0.82 |
| | 2nd ten-day period of March | 8.44 ± 0.81 |
| | 2nd ten-day period of April | 15.30 ± 0.38 |
| Average (A) | | 30.87 |
| Purpurova Zorya | 3rd ten-day period of February | 68.91 ± 0.96 |
| | 2nd ten-day period of March | 10.89 ± 1.78 |
| | 2nd ten-day period of April | 17.96 ± 1.78 |
| Average (A) | | 32.59 |
| Syaivo | 3rd ten-day period of February | 54.97 ± 2.82 |
| | 2nd ten-day period of March | 17.06 ± 2.08 |
| | 2nd ten-day period of April | 19.03 ± 1.14 |
| Average (A) | | 30.35 |
| Average according to the terms of seeds sowing | 3rd ten-day period of February | 61.26 |
| | 2nd ten-day period of March | 14.64 |
| | 2nd ten-day period of April | 18.18 |
| $HIP_{0.05}$ (A) | | 2.3 |
| $HIP_{0.05}$ (B) | | 1.4 |

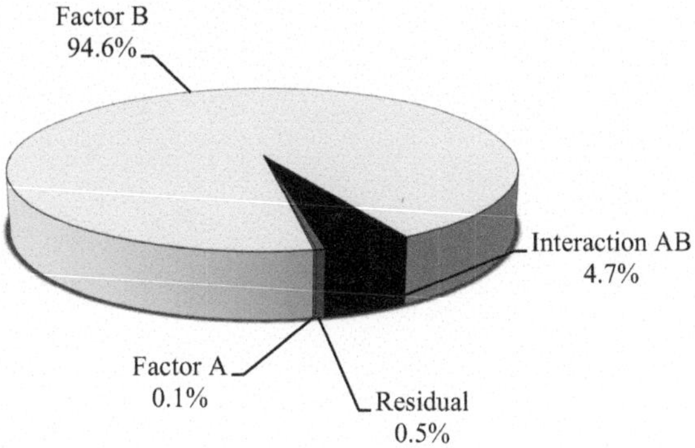


**Fig. 2** Share of influence of the factors on the level of superoxide dismutase activity in the greenery of basils: ▒—breed (A); □—terms of seeds planting (B); ■—interaction AB; ▓—residual

## References


1. Xia, X.J., Zhou, Y.H., Shi, K., Zhou, J., Foyer, C.H., Yu, J.Q.: Interplay between reactive oxygen species and hormones in the control of plant development and stress tolerance. J. Exp. Bot. **66**(10), 2839–2856 (2015). https://doi.org/10.1093/jxb/erv089
2. Choudhury, F.K., Rivero, R.M., Blumwald, E., Mittler, R.: Reactive oxygen species, abiotic stress and stress combination. Plant J. **90**(5), 856–867 (2017). https://doi.org/10.1111/tpj.13299
3. Das, K., Roychoudhury, A.: Reactive oxygen species (ROS) and response of antioxidants as ROS-scavengers during environmental stress in plants. Front. Environ. Sci. **2**, 53 (2014). https://doi.org/10.3389/fenvs.2014.00053
4. Mittler, R.: Oxidative stress, antioxidants and stress tolerance. Trends Plant Sci. **7**(9), 405–410 (2002)
5. Meng, Q.M., Zou, J., Zou, J.H., Jiang, W.S., Liu, D.H.: Effect of $Cu^{2+}$ concentration on growth, antioxidant enzyme activity and malondialdehyde content in garlic (*Allium sativum* L.). Acta Biol. Cracov. Bot. **49**(1), 95–101(2007)
6. Savicka, M., Škute, N.: Effects of high temperature on malondialdehyde content, superoxide production and growth changes in wheat seedlings (*Triticum aestivum* L.). Ekologija **56**(1), 26–33. (2010). https://doi.org/10.2478/v10055-010-0004-x
7. Del Rio, D., Stewart, A.J., Pellegrini, N.: A review of recent studies on malondialdehyde as toxic molecule and biological marker of oxidative stress. Nutr. Metab. Cardiovasc. Dis. **15**(4), 316–328 (2005)
8. Karuppanapandian, T., Moon, J.C., Kim, C., Manoharan, K., Kim, W.: Reactive oxygen species in plants: their generation, signal transduction, and scavenging mechanisms. Aust. J. Crop Sci. **5**(6), 709 (2011)
9. Alscher, R.G., Erturk, N., Heath, L.S.: Role of superoxide dismutases (SODs) in controlling oxidative stress in plants. J. Exp. Bot. **53**(372), 1331–1341 (2002)

# Determining the Risks of the Production Environment of an Agricultural Enterprise

Yurii Rohach, Oleh Yatsukh and Mykhailo Zoria

## 1 Introduction and Purpose of the Article

A properly designed and consistently applied risk management system protects and enhances corporate values. Therefore, the risk aspects should not play the role of only individual functions and processes of the company. Risk assessment should be fixed in the corporate strategy of the enterprise, which considers the planning and implementation of it in all production processes. Only this guarantees optimal use of chances in the market and greatly increases the competitiveness of the enterprise.

All agricultural enterprises in the process of implementing their production and financial activities face a particular type of risk that is characteristic of the agrarian sector as a whole, as well as for a certain type of activity (operational, financial, etc.), or for a particular culture [1].

It is important that agricultural commodity producers know how dangerous this or that risk is in the threat. Managers of enterprises have to deal with different types of risks, including with production risks. Occupational risks arise mainly due to uncertainty about revenues and services [2]. In crop farming production risks arise during sowing and plant care. The effect of production risks, mainly manifested in reducing the volume of agricultural production. The reasons for the occurrence of production risks are, first of all, the negative effects of such unpredictable factors as weather conditions [3], which in recent years have a steady tendency to deterioration, the spread of diseases and pests of plants and animals, non-compliance with agronomic terms of implementation of basic field work, unbalanced feeding of animals, lack of feed, etc.

Industrial risks in the activities of agricultural enterprises are essential factors in their effectiveness. This is explained not only by the high dependence of agricultural production, first of all on plant production, on climatic conditions that are

Y. Rohach · O. Yatsukh (✉) · M. Zoria
Tavria State Agrotechnological University, B. Khmelnitsky Avenue, 18, Melitopol, Zaporizhia Region 72310, Ukraine
e-mail: jazuch73@ukr.net

© Springer Nature Switzerland AG 2019

V. Nadykto (ed.), *Modern Development Paths of Agricultural Production*,
https://doi.org/10.1007/978-3-030-14918-5_76

uncontrolled and unpredictable, but also by considerable spatial dispersion, seasonal production, long production cycles, different soil quality [4, 5].

The natural and climatic conditions have a significant influence on the output of gross agricultural products, in particular: soils, relief, climate, water resources. Extreme weather conditions, such as hailstones, storms, floods, frost, and drought, are the greatest danger, as they can lead to serious loss of income for a very short time [6]. Often, it depends on them: yield of agricultural crops and their assortment; technology and organization of works in the crop production industry; the formation of machine-tractor and automobile park and the effectiveness of their use; the magnitude and nature of the costs of production.

The general objective of our study is to better understand the causes of the risks that agricultural enterprises experience during their activities. In this study, we focus on finding the most effective methods for identifying farm production risks and reducing them using a modern risk management methodology.

## 2 Discussion

Problems of minimizing occupational risks are dedicated to research [7–9]. The paper [7] is devoted to the study of the legal framework of Ukraine on occupational safety and the establishment of interrelationships between the quality of normative documents and the emergence of professional risks. Proposed ways of solving certain problems proposed by the authors concern improvement of the normative and legal basis of Ukraine on labor protection and creation of a unified automated system of analysis and updating of normative acts on occupational safety and directed, respectively, to minimization of a specific group of occupational risks.

In the study [8], based on the comprehensive analysis of statistics on occupational injuries, the legislative base of Ukraine on occupational safety, training programs for students of higher education institutions of Ukraine and the EU, problems have been identified that directly and indirectly contribute to raising the level of occupational risk in enterprises in Ukraine and affect the human security in the conditions of production. The authors proposed the priority ways to eliminate (minimize) identified shortcomings that are organizational and legal and do not foresee the introduction of practical measures to minimize occupational risks.

The study [9] is devoted to a comprehensive analysis of occupational injuries in the world and the search for ways to minimize occupational risks. The authors emphasize that the rapid development of science and technology, global computerization of production contribute to the emergence of new and transformation of known professional dangers. The main direction for finding effective ways to minimize occupational risks, the authors consider the way to improve the production culture at the national level. Such an approach is quite global and definitely effective, but it is not able to provide an acceptable level of professional security, since it does not involve the use of an integrated approach to solving this urgent problem.

An analysis of recent scientific studies shows that there are no practical tools available for use in occupational health and safety management systems of enterprises and organizations, which would allow to effectively minimize the risks associated with the negative impact on the worker of hazardous and harmful production factors of psychophysiological nature.

## 3 Research of Yield and Estimation of Risks in the Cultivation of Sunflower

According to the results of the study of the dynamics of sunflower yield for 2007–2017 and applying the method of analytical alignment, in particular the function of direct ($\tilde{y}_t = a + bt$) calculate the cumulative effect on it of two groups of factors: natural climatic and agronomic.

To simplify the calculation of parameters at the beginning of the time frame ($t = 0$) we take the middle of a series of dynamics. For an odd number of levels, we obtain the following values $t$ (Table 1).

In our case $\sum t = 0$ and therefore the system of equations takes the form [10]:

$$\begin{cases} \dfrac{\sum y = an,}{\sum yt = b\sum t^2.} \end{cases} \quad (1)$$

From here:

$$a = \frac{\sum y}{n}, \quad (2)$$

$$b = \frac{\sum yt}{\sum t^2}. \quad (3)$$

The results of calculations are presented in Table 2.

According to Table 2, we calculate the unknown parameters of the equation a and b (formulas 2 and 3) [10]:

$$a = \frac{\sum y}{n} = \frac{197.9}{11} = 18.0;$$

$$b = \frac{\sum yt}{\sum t^2} = \frac{99.7}{110} = 0.9.$$

**Table 1** Serial number of the year

| Year | 2007 | 2008 | 2009 | 2010 | 2011 | 2012 | 2013 | 2014 | 2015 | 2016 | 2017 |
|---|---|---|---|---|---|---|---|---|---|---|---|
| $t$ | −5 | −4 | −3 | −2 | −1 | 0 | 1 | 2 | 3 | 4 | 5 |

**Table 2** Output data for the analytical equalization of the dynamic range of sunflower yields for direct ($\tilde{y}_t = a + bt$)

| Serial number of the year ($n$) | Yield, centners per ha ($y$) | $y^2$ | $T$ | $t^2$ | $y * t$ | $\tilde{y}_t$ | $(\tilde{y}_t - \bar{y})^2$ |
|---|---|---|---|---|---|---|---|
| 1 | 12.2 | 148.84 | −5 | 25 | −61.0 | 13.5 | 20.25 |
| 2 | 15.3 | 234.09 | −4 | 16 | −61.2 | 14.4 | 12.96 |
| 3 | 15.2 | 231.04 | −3 | 9 | −45.6 | 15.3 | 7.29 |
| 4 | 15.0 | 225.00 | −2 | 4 | −30.0 | 16.2 | 3.24 |
| 5 | 18.4 | 338.56 | −1 | 1 | −18.4 | 17.1 | 0.81 |
| 6 | 16.5 | 272.25 | 0 | 0 | 0 | 18.0 | 0 |
| 7 | 21.7 | 470.89 | 1 | 1 | 21.7 | 18.9 | 0.81 |
| 8 | 19.4 | 376.36 | 2 | 4 | 38.8 | 19.8 | 3.24 |
| 9 | 21.6 | 466.56 | 3 | 9 | 64.8 | 20.7 | 7.29 |
| 10 | 22.4 | 501.76 | 4 | 16 | 89.6 | 21.6 | 12.96 |
| 11 | 20.2 | 408.04 | 5 | 25 | 101.0 | 22.5 | 20.25 |
| Total | 197.9 | 3673.39 | 0 | 110 | 99.7 | 198.0 | 89.10 |

*Source* Calculated by the author according to the data of the State Statistics Service of Ukraine

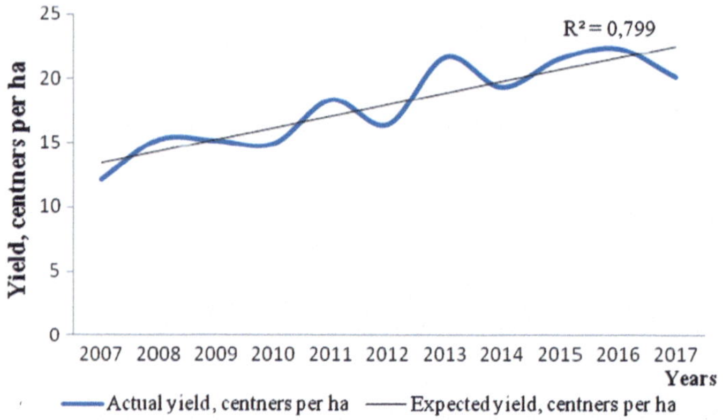


**Fig. 1** Dynamics of sunflower yield in Ukrainian enterprises (all categories of farms), c/ha. *Source* Built by the author according to the data of the State Statistics Service of Ukraine

The straight line equation will look like this:

$$\tilde{y}_t = 18.0 + 0.9t, \quad (4)$$

where $\tilde{y}_t$—Theoretical value of crop change over the analyzed period, centners per ha.

Thus, for the analyzed period, the yield of sunflower was increased annually by 0.9 c/ha (Fig. 1).

The value of the approximation reliability is one of the most important indicators, which can be judged on the validity of using the trend line equation: The value of 0.799 indicates that this function can be used to predict the result. The closer to 1 square of the correlation coefficient, the more reliable the selected model.

The total deviation of the yield of sunflower was formed under the influence of two groups of factors—natural climatic and agrotechnical. The dispersion of theoretical values of yield characterizes its variation under the influence of agrotechnical factors [10]:

$$\sigma^2 = \frac{\sum (\tilde{y}_t - \bar{y})^2}{n} = \frac{89.1}{11} = 8.1. \quad (5)$$

where ($\bar{y}$)—average yield for the analyzed period, c/ha.

Calculations show that the change in yield by 73.6% ((8.1:11) * 100) was formed due to agrotechnical factors, and the remaining variations belong to the natural and climatic factor (26.4%), that is, the agrotechnical factor was decisive in increase in yields for the analyzed period (2007–2017).

The effectiveness of the functioning of agricultural enterprises is influenced by many risk factors, but the question arises which of these factors have a greater impact on the size (level) of risks, and which is less. Due to the fact that the choice of the main risk factors cannot be made only on the basis of technological cards, accounting and financial statements of the enterprise, it is expedient to use the expert estimation method for the assessment of the significance of the matter [11]. For this we conducted a questionnaire survey of agricultural enterprises of various organizational and legal forms of the Zaporizhzhya region. The survey attracted leading specialists, specialists from a particular industry, who are better informed about the specifics of the enterprise's business. As a rule, an expert group of three to four people involved the head of the company, chief accountant, economist, and other competent specialists.

Each member of the expert group provided a list of risks that threatened the sustainability of the functioning of agricultural enterprises. The specific composition of the risks proposed for evaluation by experts included: production, commercial, financial, and investment risks, each of which contained a list of factors that negatively affect the efficiency of production and economic activity of the enterprise. The assessment of the probability of occurrence of risk events is proposed to be guided by the following system of assessments:

0—the risk is considered insignificant;
25—risk is not likely to be realized;
50—it is impossible to say anything about the onset of the event;
75—the risk is likely to manifest itself;
100—the risk will arise unambiguously.

The summarized results of the expert evaluation of risks are reflected in Fig. 2.

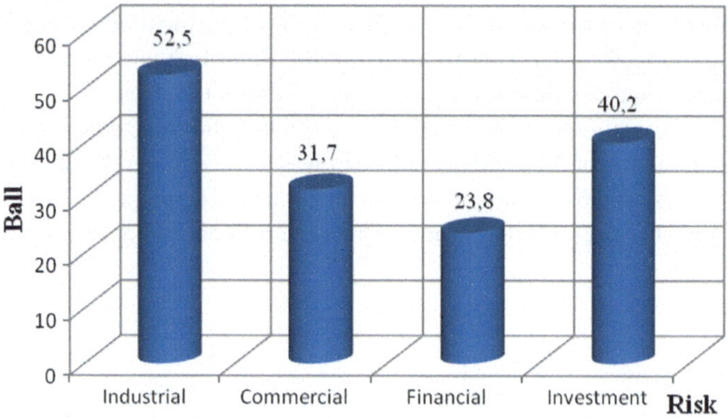


**Fig. 2** Weight of risk groups based on the results of expert evaluation. *Source* Built by the author

**Table 3** Balance assessment of the consequences of occurrence of risk events

| Degree (degree) of gravity | Balance score | Characteristic |
|---|---|---|
| 1. High (catastrophic) risk | more than 7 | The resulting losses will exceed the value of the assets of the agricultural enterprise |
| 2. Moderate (critical) | 4–6 | The resulting losses of the enterprise correspond to the size of its capital |
| 3. Weak (admissible) | 1–3 | The losses incurred by the economy are comparable to the funds in the current account and production inventories |

*Source* [11]

An analysis of experts' assessments of the importance of individual risk factors showed (Fig. 2), in the first place is occupational risk (52.5 points). The investment, commercial and financial risks are placed in the order of reduction.

Consequently, the results of the analysis of expert assessments showed that the most significant are production and investment risks, which confirms the relevance of the study.

The purpose of the expert assessment of risk factors is to determine the possible consequences (in monetary terms) and the probability of the events that led to this. On the basis of the score of the consequences of events (Table 3), it is possible to determine, for each of the factors, which losses (losses) can be incurred by the enterprise in the event of production risks.

Thus, with a detailed examination of the initial data obtained by expert estimates for each of the four groups of assessed risk factors, it has been established that among the production risks the most significant, according to experts, factors that have a negative impact on the results of the enterprise's business is the growth of monetary expenses (7.5 points), physical and moral wear of equipment (6.5 points), decrease of crop yields (5.7 points).

## 4 Conclusions

The research establishes a significant dependence of the results of the production and economic activity on the risks, which is a prerequisite for justifying the need for their minimization and improvement of risk management.

The results of research on the search for effective mechanisms for managing industrial risks in agricultural enterprises will help to reduce losses that arise under the influence of adverse (risky) events. In addition, we want to use the obtained results to work on improving the modern risk management methodology, which would meet the market requirements and ensure sustainable development on the basis of timely identification of risks and the choice of the best ways to prevent, minimize, or eliminate negative consequences.

## References

1. Wojnalowich, A.V., Capko, W.G.: Ocena ryzyka zawodowego w branzach rolniczych Uкrainy, Ocena ryzyka zawodowego w rolnictwie. Praca zbiorowa pod redakcja Agnieszki Buczaj, Leszka Soleckiego, Lublin, Instytut medycyny wsi, pp. 239 –246. (2010)
2. Berg, E.: Integriertes Risikomanagement: Notwendigkeit und Konzepte für die Landwirtschaft. In: Deiter, J. (Hrsg) Agrarökonomie im Wandel, Bonn, pp. 53–67. (2005)
3. Mußhoff O., Odening, M., Xu, W.: Zur Bewertung von Wetterderivaten als innovative Risikomanagementinstrumente in der Landwirtschaft. In: Agrarwirtschaft **54**. Jg, pp. 197–209 (2005)
4. Frentrup M., Hezder, M., Theuvsen, L.: Risikomanagement in der Landwirtschaft, Frankfurt a. Main (2010)
5. Gleißner, W., Romeike, F.: Risikomanagement: Umsetzung—Werkzeuge—Risikobewertung. Freiburg im Breisgau (2005)
6. Lehrner. Risiken in der Landwirtschaft. – S. 13–15. (2002)
7. Bochkovskyi, A.P., Sapozhnikova, N.Y.: Promising direction for improving regulatory legal framework on labour protection for enterprises production food and beverages. Environ. Saf. Sustain. Resour. Manage. **2**(12), 85–93 (2015). https://doi.org/10.13140/rg.2.1.4156.3927
8. Bochkovskyi, A.P., Sapozhnikova, N.Y., Gogunskii, V.D.: Legal and organizational issues of improving the labor protection an industrial safety level at the Ukrainian enterprises. Sci. Bull. Nat. Min. Univ. **5**(161),100–108 (Scopus) (2017). https://doi.org/10.13140/rg.2.2.33613.23528
9. Kim, Y., Park, J., Park, M.: Creating a culture of prevention in occupational safety and health practice. Saf. Health Work. **7**(2), 89–96 (2016). https://doi.org/10.1016/j.shaw.2016.02.002
10. Savytskaia, H.V.: Analyz khoziaistvennoi deiatelnosty predpryiatyia: Navch. posibhyk. 7-e vyd, vypr. Nove znannia, Minsk, (Ekonomichna ocvita). 704 p.(2002)
11. Saprykina, O.M.: Analiz vyrobnychykh ryzykiv silskohospodarskykh pidpryiemstv. PhD thesis, Melitopol, TSATU. (2014). Avaliable http://nbuv.gov.ua/UJRN/znptdau_2014_2_42. Accessed 10 Dec 2018

# Development of Emulsion Sauce Technology for Preventive Nutrition

Liubov Telezhenko, Gennadiy Diduch, Svitlana Kolesnichenko and Valentina Zhukova

## 1 Introduction and Purpose of the Article

The emulsion-type sauces are the leaders among a variety of sauces used in restaurants as they perfect the taste and aroma of food and have a nice buttery consistency. While on the one hand the fat base has high commercial specifications and contributes to the history of a number of physiological processes in the human body and fat-soluble vitamin absorption, on the other hand, the processes of fat bioconversion depend on the availability of various emulsifiers and lipoid substances. It is possible to eliminate contradictions between all these factors and provide the cells of the body with fat components that are conductive to building of the liquid crystal cell membranes by careful selection of component composition, especially in the presence of lecithin. It is known that excessive mass fraction of fat is not so much providing the body with essential ingredients as it is deposited as spare material under the skin and even on the internal organs of the human body. At the same time, the mass fraction of fat in the sauce can be reduced while maintaining its high organoleptic characteristics. Fat replacers are referred to the promising ingredients that can reduce the sauce caloric content with no loss of its sensory characteristics. In recent years, special attention is paid to the development of the recipes for low-caloric emulsion sauces with the inclusion of protein ingredients. Particularly promising in this respect are whey protein concentrates as fat simulators, composition and properties of which are modified under the influence of the microparticulation process [1].

L. Telezhenko · G. Diduch (✉) · S. Kolesnichenko
Odessa National Academy of Food Technologies, Kanatna Str., 112, Odessa 65039, Ukraine
e-mail: genad69@gmail.com

V. Zhukova
Tavria State Agrotechnological University, B. Khmelnitsky Av., 18, Melitopol, Zaporizhia Region 72312, Ukraine

© Springer Nature Switzerland AG 2019

V. Nadykto (ed.), *Modern Development Paths of Agricultural Production*,
https://doi.org/10.1007/978-3-030-14918-5_77

**Table 1** Chemical composition of sucrose whey

| Serum composition | Number |
|---|---|
| Dry substance, % , including | 4.5–7.2 |
| Butterfat | 0.3–0.5 |
| Protein nitrogen compounds | 0.5–0.8 |
| Lactose | 3.9–4.9 |
| Ash | 0.52 |
| pH (active acidity) | 6.0 |

According to Professor Petrovsky's definition, the biological value of lactoserum can be characterized by the following formula: "a minimum of caloric at the maximum biological value." This allows us to consider the lactoserum and food products based on it as biologically complete ones having dietary and even medicinal properties.

Whey proteins are a source of amino acids with a balanced amino acid composition (Table 1). Their introduction into the composition of foodstuffs stimulates muscle protein synthesis, which promotes the growth of muscle tissue.

Some amino acids (especially leucine), which are part of whey proteins, when ingested, affect glucose homeostasis, have insulinotropic action (they stimulate the transport of glucose to muscle tissue) and stimulate the resynthesis of glycogen in muscles and liver.

The colloidal structure of whey proteins determines their easy accessibility and good digestibility by proteolytic enzymes; assimilation of whey protein amounts to 97%. The valuable composition of whey has contributed to the development of its processing technology. One of the modern and promising methods of processing is the thermomechanical method of producing microparticulate.

It is known from the literature that the technology of microparticulate production and its introduction in the manufacture of food products has been worked through and implemented [2].

The whey protein microparticulate can be obtained from cheese whey by the ultrafiltration method, and in the conditions of restaurant management—by the method of boiling (heat denaturation) after separation of fat and casein dust [3].

When the colloidal system of thermally coagulating proteins is heated, molecules that are part of it denaturate (unfold) and begin to gather into one whole. Instead of forming a jellylike spatial grid of coagulated protein, microparticulate whey proteins form microparticles and never create a gel. This process of molecule formation can be compared with the process of "winding pasta on a fork" [4].

In the process of obtaining the microparticulate, the soluble protein molecules are denatured and aggregated under strictly controlled conditions. Due to this, the microparticles represent a very stable form of whey protein that is not able to agglomerate or jellify when heated. The product maintains its functional properties at high pasteurization temperatures and aseptic production (patent) [2]. The process of molecular aggregation starts with dimers and being heated further may continue until all the molecules combined. However, the spontaneous tendency to aggregate may be terminated within the nano-metric limits by means of applying a shearing load of sufficient intensity in the heating process [4].

## 2 Microparticulate Microparticle Technology

The Restaurant and Healthy Catering Technology Department, the Odessa National Academy of Food Technologies, has obtained the microparticulate by the method of whey protein thermal coagulation and application of a strong mechanical shear by means of the dispergator under the following technological modes:

- Boiling the whey protein at a temperature of 87–90 °C, pH: 4.6 for 40 min;
- Cooling and settling of proteins for 1 h at a temperature of 20–25 °C;
- Separation of permeate from whey protein by drainage;
- Separation of protein concentrate by filtration;
- Washing the protein with water three times to remove lactose and bring pH to 6.3–6.5;
- Separation of protein concentrate by filtration;
- Obtaining the whey protein concentrate with a solid mass fraction of 24%; and
- Microparticulating the whey protein concentrate at a temperature of 95–97 °C under the action of the dispergator with a rotation speed of about 30,000–40,000 $min^{-1}$ [2].

As a result of microparticulation, protein particles smaller than 2 μm are formed, which give the product an oily and viscous consistency. The particles whose size exceeds 2 μm give the product flour taste and sandy consistency.

The content of indispensable amino acids in the microparticulate as compared with the scale of FAO/WHO is given in Table 2.

**Table 2** Content of irreplaceable amino acids in microparticulates

| The name of the amino acid | Number in microparticulates (g/100 g) | Scale FAO/WHO |
|---|---|---|
| Throne | 5.47 | 4.0 |
| Valine | 3.81 | 5.0 |
| Leucine | 9.74 | 7.0 |
| Isoleucine | 4.90 | 4.0 |
| Lizin | 5.26 | 5.5 |
| Tryptophan | 2.53 | 1.0 |
| Methionine + cystine | 8.50 | 3.5 |
| Fenilalanine + tyrosine | 7.62 | 6.0 |

## 3 The Technology of Making Sauce

To make high-quality emulsion sauces, it is necessary to consider some features of component introduction. In order to obtain the emulsion, emulsifier, stabilizer or thickener should first be dissolved in water, and then oil should be added.

Soybean lecithin was selected as the emulsifier. The information search suggests that when creating foodstuffs of prophylactic design, the use of vegetable lecithins can be considered as a physiological functional nutrient with physiological effect of high efficiency. Lecithin is a multifunctional prescription component, which allows to substantially improve the organoleptic and physicochemical properties of foodstuffs [5, 6], and their use enables solving a number of problems of technological nature and ensuring predetermined consumer properties of the food products. Besides, lecithin is placed among natural antioxidants, which exhibits antioxidant activity, showing synergy in relation to other natural antioxidants, such as tocopherols [7, 8]. Lecithin inactivates heavy metal ions, which enter the body from the atmosphere and food.

Pre-emulsification of fat in the presence of lecithin was carried out by obtaining lecithin–water–oil food composition. The composition consisted of granulated soy lecithin (20%), oil of camelina sativa (20%), olive oil (20%) and water (40%). A suspension of lecithin was injected in drinking water (1/2 of the total quantity) and heated to (80–85) °C, and then the solution was kept for 40–60 min and mixed in a blender with sequential addition of camelina sativa oil, water and olive oil with a temperature of (35–40) °C until obtaining of viscous consistency [9, 10].

The use of camelina sativa oil enables enriching the sauce with omega-3 and omega-6 fatty acids; besides, this oil has high stability and resistance to oxidative transformations due to the natural combination of tocopherols, sterols and compounds of phenolic nature. Shelf life of camelina sativa oil is significant; it is approximately two years [8].

The aim of this work was the development of the emulsion sauce composition of increased biological value on the basis of microparticulate and lecithin composition.

We have chosen the cold method of the emulsion sauce preparation.

To ensure the necessary consistency and finely dispersed particles in the mixture, the microparticulate was introduced in the food composition by small doses. After mixing the basic ingredients, mustard and salt were added, the mixture was stirred and then (the last thing) flavor ingredients sugar and vinegar according to the recipe were introduced.

The discrete method of preparing emulsion mixtures has significant advantages: relatively low cost of equipment, flexibility and stability of small enterprises and the ability of the restaurant industry to produce the low-caloric emulsion sauce for their needs.

The flow chart of production of emulsion sauce based on microparticulate is shown in Fig. 1.

Compared with the most common recipe of mayonnaise sauce "Provencal," the recipe of the developed low-caloric sauce has a number of characteristic features, but the main distinction of the developed sauce is its qualitative difference from

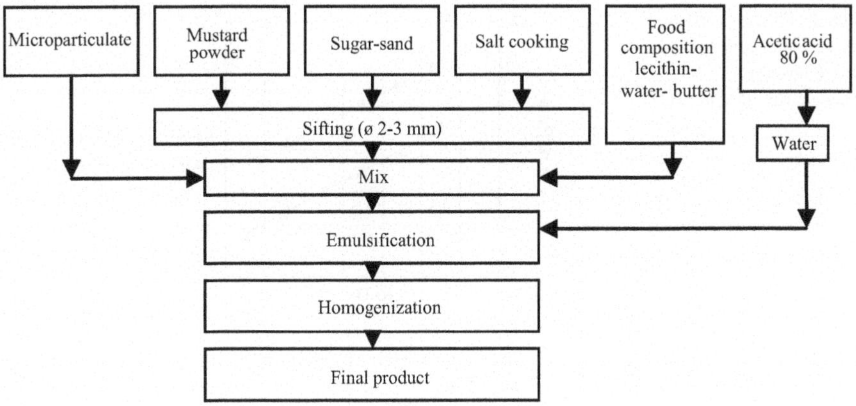


**Fig. 1** Technological scheme of production of emulsion sauce on the basis of microparticulate

other products of the same type. This is primarily qualitative composition of a PUFA product, which certainly transfers it to the level of healthy foods.

Recipe for making emulsion sauce

| | |
|---|---|
| Microparticulate | 47.6 |
| The food composition is lecithin–water–butter | 46.6 |
| Mustard powder | 1.0 |
| Sugar sand | 2.0 |
| Salt cooking | 1.0 |
| Acetic acid 80% | 1.8 |
| Together | 100.0 |

The standard formula of mayonnaise Provencal (67% fat)

| | |
|---|---|
| Vegetable oil refined, deodorized | 65.4 |
| Egg powder | 5.0 |
| Dried, skimmed milk powder | 1.6 |
| Mustard powder | 0.75 |
| Soda food | 0.05 |
| Sugar sand | 1.5 |
| Salt kitchen | 1.0–1.3 |
| Acetic acid 80% -a | 0.55–0.75 |
| Water | 24.15–23.65 |
| Total | 100.0 |

Positive consumer perception of traditional sauces is primarily caused by the consistency of the product. In order to determine how a change in the recipe will affect

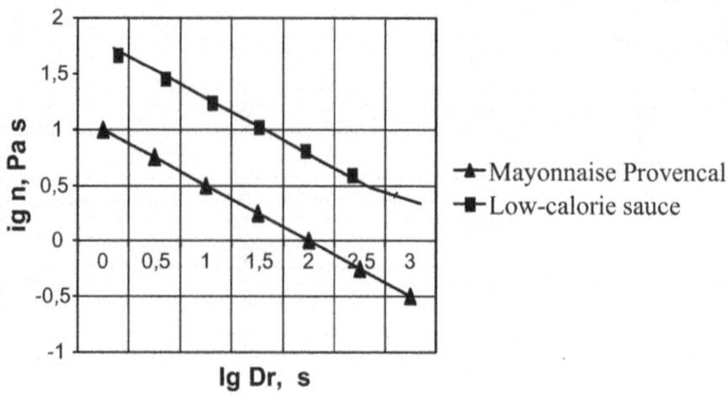


**Fig. 2** Fluidity curves of mayonnaise and low-caloric sauce based on microparticulate at 20 °C

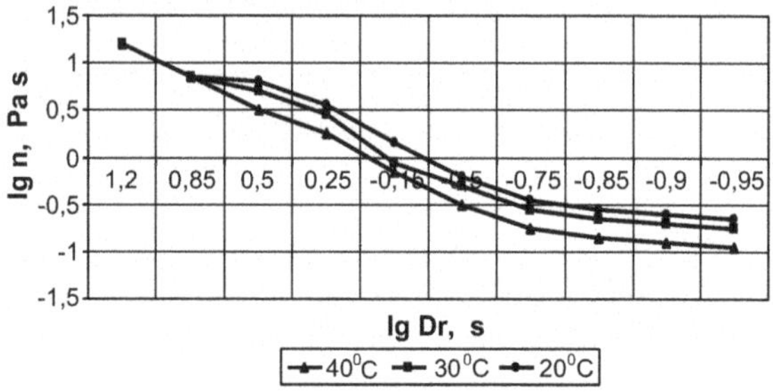


**Fig. 3** Curves of volatility of Provencal mayonnaise at different temperatures

the sauce rheological properties, the sauce fluidity and viscosity were investigated (Figs. 2 and 3).

It was established that the nature of fluidity of the sauce pilot sample is similar to that of mayonnaise sauce "Provencal."

Temperature is one of the most important factors that affects the properties of mayonnaise emulsions; therefore, the effect of the sauce supply temperature on the rheological properties of sauces was studied. The nature of the viscosity change, the data of its absolute values and other physical characteristics of non-Newtonian systems at different temperatures allow the technological process of sauce production to be rationalized as well as the modes of technological equipment operation to be optimized.

As the sauce, which is being devised, is intended for the decoration of both cold (not below 15 °C) and hot (not above 60 °C) dishes, then according to the current TU, it was appropriate to choose a range of temperatures within an interval of 15–60 °C for the study of viscosity characteristics.

It is shown that the curves of viscosity $\eta$ depending on the gradient of the shearing rate *Dr* have the same form, which is characteristic of viscoelastic materials. At temperatures (15, 20 and 30 °C), the dynamic viscosity decreases when the shearing rate increases, especially at low speeds of shear deformation.

The change of rheological properties of the studied systems with a temperature increase can be explained by the restructuring of the suspension texture. When the temperature increases, the interaction between the dispersion system particles weakens, the viscosity of the product decreases and destruction of the "physical grid,"

**Table 3** Organoleptic evaluation of emulsion sauce

| Indicator | Characteristics of the product |
|---|---|
| Appearance, consistency | Homogeneous cream mass |
| Taste and smell | Taste milk and cream, slightly spicy |
| Color | Cream, homogeneous throughout the mass |

which was formed in the suspension aqueous phase by whey protein molecules, takes place.

The outcomes of organoleptic evaluation of the developed sauce are shown in Table 3.

It was established through microbiological studies that the level of QMAFAnM contamination with bacteria does not exceed 3·103 CFU per gram in six hours after production, which is significantly less than the norm.

The recommended period of the developed sauce usage is six hours of production, which is traditional for similar dishes in restaurants.

## 4 Conclusions

Developed was the fabrication method of an emulsion sauce based on whey protein microparticulate and food lecithin composition. It is shown that it is expedient to use the microparticulate of whey proteins for partial replacement of fat in emulsion sauces. The main difference between the developed sauces is the characteristic of the product lipid qualitative composition and the presence of PUFA in it, which certainly puts the sauce on the level of healthy foods. The development in question has the following advantages:

- Increasing efficiency and environmental friendliness of the secondary raw materials recycling and
- Expanding the range of emulsion sauces of a healthy diet.

## References

1. Diduch, G.: Food Sci. Technol. **2**(31), 52–56 (2015)
2. Didukh, G., Goosak, Y., Shklovskaya, Y.L.: Patent nakorisnumodel 86713 Ukraine, IPC A23S 13/00 (2013.01) , №2013 08048, appl . 25/06/2013, publ. 10.01.2014, Bull. Number 1
3. Didukh, G.: Food Sci. Technol. **2**, 3–8 (2017)
4. Hramtsov, A.: Phenomen molochnoy syvorotki. SPb. Professiya, 804. (2011)
5. Kolesnichenko, S.,L. Telezhenko, L.K.: Chimiya, bio- i nanotechnologii, ekologiya i ekonomika v pischevoy i kosmetischeskoy promischlenosti, Charkov, 35–38, 2016
6. Kolesnichenko, S.: Archivarius, **6**(10), 5–8. (2016)

7. Hosomi, R., Fukunaga, K., Arai, H.,etal.: Eur. J. Lipid Sci/Technol. **112**. 537–544 (2010)
8. John V.L. Henry, Peter J. Fryer, William J. Frith, Ian T. Norton,// Food Hydrocolloids 24, 66–71 (2010)
9. J. Chem. Soc. Pak. **37**(04), 824–829 (2015)
10. Henry, J.V.L., Fryer, P.J., Frith, W.J., Norton, I.T.: Food Hydrocolloids, **24,** 66–71. (2010)

# The Influence of AKM Growth Regulator on Photosynthetic Activity of Oilseed Flax Plants in the Conditions of Insufficient Humidification of the Southern Steppe of Ukraine

**Oksana Yeremenko, Svitlana Kalenska, Lyubov Pokoptseva and Liudmyla Todorova**

## 1 Introduction

Zone of South Steppe of Ukraine is distinguished by unstable weather conditions. A weather risk factor substantially influences the productivity of agricultural cultures and is one of objective and the least predictable [1, 2]. Therefore, the productivity of oilseed flax seeds in the conditions of South Steppe of Ukraine is unstable. Productivity of plants largely depends on functioning of their photosynthetic apparatus [3, 4]. In stress conditions, there are the structural-functional alterations of photosynthetic membranes, connected with the peculiarities of structure and metabolism of lipid components that are directed to support the homoeostasis of a plant organism. The changes of lipid constituents of chloroplast membranes are an important link in forming of adaptive reactions as a result of that resistance of the living system to external influences rises and the productivity increases [5, 6].

Increasing plant resistance to stresses under differentiated use of plant growth regulators (PGRs) at the various growth stages is an effective way to increase yield [7, 8].

Raising crop productivity under the PGR influence is associated with the increase in plant adaptation to the growing conditions and intensification of cell activity in the plant organism due to stimulation of biochemical processes thereby leading to the optimization of nutrition, respiration and photosynthesis processes [9]. PGRs facilitate the higher realization of genetic potential of plants [10].

O. Yeremenko (✉) · L. Pokoptseva · L. Todorova
Tavria State Agrotechnological University, B. Khmelnitsky Av., 18, Melitopol, Zaporizhia Region 72312, Ukraine
e-mail: ok.eremenko@gmail.com

S. Kalenska
National University of Life and Environmental Sciences of Ukraine, Heroiv Oborony St., 15, Kiev 03041, Ukraine

© Springer Nature Switzerland AG 2019 
V. Nadykto (ed.), *Modern Development Paths of Agricultural Production*,
https://doi.org/10.1007/978-3-030-14918-5_78

Mohsen Janmohammadi claims that silicon nanoconcentrations ($nSiO_2$) increase leaf surface area by 48% and stimulate growth processes of safflower under adverse weather conditions and against various organic and mineral nutrition [11].

According to a group of scientists including Sibgha Noreen, Muhammad Ashraf, Mumtaz Hussain, Amer Jamil, Kazem Ghassemi-Golezani and Ayda Hosseinzadeh-Mahootchi, the use of salicylic acid to grow sunflower and safflower reduces the negative impact of stress factors by increasing activity of antioxidant enzymes. In addition, the processes of growth and photosynthetic activity of plants are also activated [12, 13].

The well-known growth regulator, which constituent is DMSO (dimethylsulphoxide), is fumar (10% solution of dimethyl ether of aminofumar acid in dimethylsulphoxide) [14], which during preplanting encrustation by 0.0001% solution of sunflower seeds promotes their field germination (by 8.2–9.2%) and the productivity (by 2.6–3.0 q/ha). In the absence of system data in relation to this chemical influence on passing of physiologo-biochemical processes in a plant organism, it is difficult to judge, what its effect is caused by.

In the works of Kalytka V. V. and Pokoptseva L. A., it is proved that antioxidant chemical of distinol (reactant ionol and DMSO) had positive influence not only on the productivity of sunflower but also on keeping of sowing qualities of seeds in a storage period [15, 16].

Growth-stimulative efficiency of PGRs AKM is shown, in which distinol is united with filming agents [17–19]. It is proved that AKM has the expressed antioxidant effect: Seeds of agricultural cultures, treated by AKM, have lower level of malonic dialdehyde (MDA), than untreated. The increase of antioxidant enzymes activity, content of phospholipids, vitamin E, carotenoids and sum of chlorophylls a and b is marked also in the plants of soya during the vegetation under the influence of AKM [20].

The achievement of positive effect from the application of growth-stimulative substances is possible only during optimal concentration of working solution of a chemical, as the majority of biologically active substances works as stimulators in low doses and in high doses—as inhibitors. Except that, the action of plant growth regulators is stipulated by the display of weather conditions of the year in a certain agro-climatic zone of growing and biological features of culture [21, 22].

But all these chemicals need the deeper study of their influence on plants photosynthetic activity of oil-bearing cultures in the conditions of South Steppe of Ukraine. Therefore, the aim of the work was to define the influence of growth regulator AKM on plants photosynthetic activity of oilseed flax.

## 2 Materials and Methods

### 2.1 *Materials*

Alkalis, acids, solvents, Mohr's salt and other reagents are purchased at "Himlabor-reaktyv" company (Kyiv, Ukraine).

### 2.2 *Methods*

Soil preparation: soil sample, which was taken from the experimental plot, was air dried. Then all the roots are carefully removed, the ground was rubbed in agate mortar and sieved it through a sieve with 0.25 mm holes.

The humus content in the soil was determined using I.V. Turin method [23], which is based on the oxidation of organic matter of the soil by 0.4 n solution of bichromate of potassium ($K_2Cr_2O_7$) till the formation of carbon dioxide. Reaction occurs by the equation:

$$3C + K_2Cr_2O_7 + H_2SO_4 = 2K_2SO_4 + 2Cr_2(SO_4)_3 + 8H_2O + 3CO_2$$

Unused for oxidation residue of chromic mixture was titrated by Mohr's salt (double salt) of ammonium sulfate and iron oxide sulfate $[(NH_4)_2SO_4\ FeSO_4]\cdot 6H_2O$. By the number of consumed Mohr's salt, residue of chromium compounds was determined, and by the difference between the original amount (idling determination result), and the remainder amount of chromium compounds, which went to the oxidation of humus was determined.

Definition of hydrolyzed nitrogen by Cornfield [24]. The principle of the method is that the soil is hydrolyzed by treating it with alkali. As a result, nitrogen of exchange ammonium, amide, aminosugars and other compounds is released from the soil in the form of $NH_3$, which is caught by boric acid.

Determination of mobile forms of phosphorus and potassium in the soil using Chirikov method [25]. The method is based on "extraction" of phosphorus and potassium from the soil by 0.5 normal acetic acid at a ratio soil: solution (1:25) followed by phosphorus determination as molybdenum blue on photoelectric colorimeter, potassium—on flame photometer.

Content of dry substance was determined by the thermostatic-weighting method [26].

The net productivity of photosynthesis was determined by basic phase-transfer periods of oilseed flax development (g/m$^2$ per twenty four hours) by test plants sampling, in which the general mass, mass of separate organs and leaves area were determined and calculated by the formula [27, 28].

Photosynthetic potential (PP) (mln.m$^2$ * days/ha) was determined by the basic phase-transfer periods of plants development by the formula of Quidet, West and Brigs [29, 30].

The concentration of pigments was determined, mg/g of raw mass in acetone extractions (100%) in spectrophotometric way at the wavelength of 662 and 644 nm (chlorophylls a and b) [31] and 440.5 nm (sum of carotenoids) [32]. The calculation of pigments concentration was conducted by a next formula (1–5):

$$C_a = 9.784 * \mathrm{D}_{662} - 0.990 * \mathrm{D}_{644} \tag{1}$$

$$C_b = 21.426 * \mathrm{D}_{644} - 4.650 * \mathrm{D}_{662} \tag{2}$$

$$C_{a+b} = 5.134 * \mathrm{D}_{662} + 20.436 * \mathrm{D}_{644} \tag{3}$$

$$C_{\mathrm{car.}} = 4.695 * \mathrm{D}_{440.5} - 0.268 * (\mathrm{C}_{a+b}) \tag{4}$$

where $C_a$, $C_b$, and $C_{\mathrm{car}}$—chlorophylls concentration *a*, *b* and carotenoids, mg/l.

D optical density at the wavelength in formula

Then the content of pigments *A* in plant material, mg/g of raw mass is calculated:

$$A = \mathrm{VC}/1000\mathrm{P}, \tag{5}$$

C pigments concentration, mg/l;
V extraction volume, ml (25 ml);
P band-and-hook hinge of plant material, g (0.1–0.2 g) and pigment percentage (% of pigment) for raw mass by formula (6):

$$\%\,\mathrm{of\,pigment} = A/10, \tag{6}$$

*A* pigment content in 1 g

The productivity of chlorophyll functioning was calculated as ratio of mass increase of plant dry substance toward the mean value of chlorophyll content in leaves [33–35].

## 2.3 Equipment

In order to receive quantitative indicators during experiments on the definition of soil quality and seeds of sunflower, the following equipment was used: analytical scales ANG220 (AXIS, Poland), centrifuge Multifuge (Thermo/Heraeus, Germany), two-channel flame photometer CL22 (Russia), scanning spectrophotometer UV-2800-UNICO (Russia).

**Table 1** Hydrothermal conditions of the growing season of oilseed flax during research years

| Indexes | 2013 | 2014 | 2015 | 2016 |
|---|---|---|---|---|
| Rainfall during the growing season, mm | 122 | 233 | 155 | 192 |
| The sum of active (above +10 °C) temperatures (°C) | 2996 | 2869 | 2756 | 2872 |
| CHU (Crop Heat Units—CHU) [37] | 3519 | 3375 | 3225 | 3369 |
| Hydrothermal coefficient | 0.41 | 0.81 | 0.56 | 0.67 |
| Minimum relative air humidity during flowering (%) | 61.8 | 36.9 | 45.8 | 35.5 |

## 2.4 *Soil and Climatic Conditions of Research Conduction*

The study was conducted in 2013–2016 in Veselovskom district of Zaporizhzhya region and in the laboratory of monitoring soil and crop products quality research institute "Agrotechnologies and Ecology" of Tavria state agrotechnological university.

Soil of the research sites is southern chernozem with average weighted humus content of 3.4%, easily hydrolyzed nitrogen (by Cornfield)—91 mg/kg, mobile phosphorus (by Chirikov)—119 mg/kg and exchangeable potassium (by Chirikov)—127 mg/kg of the soil. The results were compared with established by DSTU standards of indexes of soil fertility for agricultural lands [36].

Soil moisture conditions in the research years differed, both by the rainfall amount and by its uniformity (Table 1). Almost the same amount of precipitation during the growing season was observed in 2013—122 mm, while in 2014 times more precipitation fell—233 mm. At the same time, 2013 were distinguished by irregular rainfall, high temperatures and large soil drought during the period from germination to seed maturation. HTC indexes varied within the limits of 0.4–0.8 over the years. Hydrothermal conditions in 2014 compared to 2013, 2015 and 2016 were more optimal both by amount and by uniformity of rainfall. Together with it, year of 2016 was characterized by the least humidity of air in a period of sunflower flowering (35.5%) comparatively with the investigated years.

## 2.5 *Scheme of Field Experiment*

In order to address stated objectives and goal was laid three-factor field experiment was laid (Table 2). Where is factor *A*—sort of oilseed flax. The factor *B*—seed treatment with AKM plant growth regulator [38]. The factor *C*—hydrothermal conditions of the year of the research. Seed treatment was carried out 1–2 days before sowing by incrustation method at the rate of 10 l of working solution per 1 ton of seeds.

**Table 2** Scheme of field experiment

| Variant | Preparation consumption rate (l/*t*) | Concentration of active substance in the working solution (g/l) |
|---|---|---|
| 1 (C) | Seed treater—Krujzer 10.0 and Maksim XL 5.0 | – |
| 2 | C + AKM, 0.330 | Ionol + dimethyl sulfoxide, 0.0015 |
| 3 | C + AKM, 0.330 | Ionol + dimethyl sulfoxide, 0.015 |

### 2.6 Studied Preparation Characteristics

AKM is a film-forming semi-synthetic plant growth regulator of antioxidant action allowed for seed treatment and spraying of grain, oilseed, legume, vegetable crops and hops. The aqueous solution consists of dimethyl sulfoxide, ionol, PEG-1500 and PEG-400. The consumption rate depends on the crop and level of agricultural background.

### 2.7 The Technology of Oilseed Flax Cultivation in the Experiment

Researches were conducted on the sorts of oilseed flax of home selection: Evryka and Orfei [39]. Oilseed flax seeds were sown early in the third ten-day period of April, and the rate of sowing was 4.5–5.0 million seeds/ha with 15 cm row spacing. The predecessor was winter wheat. Care of crops, accounting and monitoring of growth and development of plants, yield structure formation of oilseed flax were carried out according to [27]. Mathematical analysis of the results was carried out by Student's test [40] and licensed AgroStat computer program.

## 3 Results and Discussion

The seeds of oilseed flax have substantial differences from the structure of other oil-bearing cultures seeds. A seed coat fits closely to the kernel. The cells of seed coat contain mucous substances that strongly swell in water. Therefore, we conducted a laboratory experience in determination of different concentration of AKM influence (with reactant) on energy of germination and laboratory likeness.

Reliable difference in variants between energy of germination and laboratory likeness was not determined by us. In all experiment, variants PGRs AKM promoted these indexes comparatively with control (Fig. 1).

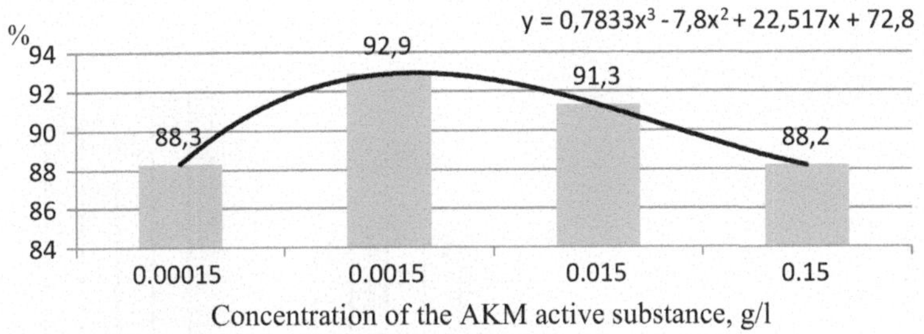


**Fig. 1** Dependence of laboratory germination of oilseed flax seeds on concentrations of the AKM active substance:—approximating polynomial curve of the 4th order

Laboratory likeness of oilseed flax in a control variant was 87.8%. As we see from Fig. 1, high (0.15 g/l) and low (0.00015 g/l) concentrations of AKM do not have substantial influence on the processes of seed germination. Plant growth regulator AKM in a concentration of 0.0015 g/l had the most influence on laboratory likeness of oilseed flax (92.9%). Between variants with the use of PGRs AKM in a concentration 0.0015 and 0.015 g/l, there was not registered the reliable difference, that is why in the field conditions, the research was conducted exactly with these concentrations. Photosynthetical activity of sowing is a founding for prognostication of the agricultural cultures productivity. Abiotic factors and investigated elements of technology of growing influenced the forming of assimilatory surface of oilseed flax. A negative reaction to the reduction of assimilatory surface on a plant is conditioned by shading of understory leaves; as a result, they turn yellow and die off.

The greatest productivity of the reproductive sowing is realized on the leaf area within the limits of 40–50 thousand $m^2$/ha. The further increase of leaf area results in the substantial decline of rates of reproductive organs formation and productivity of seeds.

Processes of formation and accumulation of organic substance are the integral index of all physiology and biochemical processes that takes place in a plant organism. The accumulation of dry above-ground biomass in plants during the researches was determined by the basic phases of growth and development of oilseed flax plants. The researches showed that as there were growth and development of oilseed flax so was increase of dry mass output. The accumulation of dry substance by the oilseed flax plants during vegetation took place unevenly and depended both on the hydrothermal conditions of the year and on the concentration of reactant of PGRs AKM (Fig. 2).

In the stage of plants BBCH code 19–20 development, the content of dry substance of oilseed flax sowing ranged from 54.61 to 99.50 g/$m^2$. The accumulation of this index directly depended on the agrometeorological conditions of growth and development of plants in this period. Maximal content of dry substance was marked in oilseed flax sowing in 2014 and 2016. We did not define any sort peculiarities in dry substance accumulation by plants in this phase of development. PGRs AKM in both investigated concentrations promoted the increase of this index, and on the

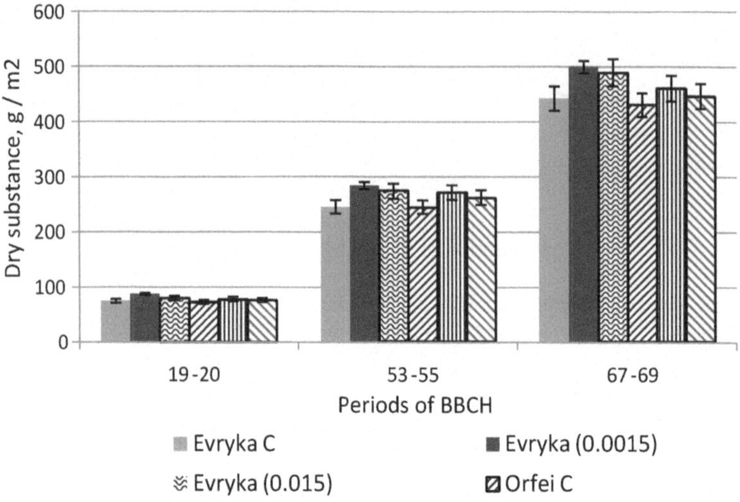


**Fig. 2** A dynamics of content changes of dry substance is in the plants of oilseed flax with action of growth regulator AKM (g/m$^2$)

average, it was from 2.1 to 7.6%, in comparison with control. But in 2013 (droughty), this index for the sort Evryka was by 27.0% more than control one and for the sort Orfei—by 12.1%.

Due to active growth and development of plants in a period BBCH code 19–20—BBCH code 53–55, the content of dry substance increased almost in four times. Fluctuation of this index was within the limits from 227.6 to 300.5 g/m$^2$.

Use of PGRs AKM for preplanting treatment of oilseed flax seeds promoted the accumulation of dry substance in the phase of budding in the average during years by 13.9% for the sort Evryka and by 9.0% for the sort Orfei. Maximal influence on this index was marked in sowing in 2014 and 2016.

Plants of oilseed flax of the sort Orfei in the phase of development of the complete flowering had less content of dry substance during the investigated years. In our view, it is related to the sort peculiarities of plants.

With the preplanting treatment of oilseed flax seeds PGRs AKM, the content of dry substance in the plants of the sort Evryka increased within the limits from 9.7 to 13.2% and in the plants of the sort Orfei—from 3.3 to 6.9%.

Between content of dry substance and plants leaf area of oilseed flax in a phase of development "herringbone" and budding was defined correlation connection of weak force $r = 0.448$ and $r = 0.472$, accordingly. While in the phase of the active flowering of plants, this correlation dependence was of middle force and it was $r = 0.681$.

Part of hydrothermal conditions of year influence on the accumulation of dry substance in the plants of oilseed flax was 23.6% and PGRs AKM—7.9%.

Forming of harvest depends not only on the size of leaf area, but also on the time of its functioning. Photosynthetic potential unites these indexes. This index of oilseed flax changed as follows (Table 3). With the increase of leaf area, there is an increase of photosynthetic potential index.

PGRs AKM in the concentration of reactant 0.0015 g/l increased the PP of oilseed flax plants, while in the concentration of reactant 0.015 g/l this influence was not defined.

Photosynthetic potential of plants in a budding period—flowering ranged in limits from 422.2 to 484.4 thousand $m^2$ * days/ha. The most increase of photosynthetic potential is marked during the treatment of seed sort Evryka PGRs AKM reactant 0.0015 g/l. Also close correlation connection between the productivity and photosynthetic potential ($r = 0.931$) is defined.

The most effective work of assimilatory apparatus was with realization of preplanting treatment of oilseed flax seeds PGRs AKM. In this variant due to the considerable improvement of organization of agrocoenosis morphostructure, the net productivity of photosynthesis is the highest. NPP depends, first of all, on the amount of the carbon dioxide g/$m^2$ $CO_2$ acquired in the process of photosynthesis for twenty-four hours [41]. Thus, in an interphase period a "herringbone"—budding and budding—flowering the use of PGRs of antioxidant type provides the increase of the photosynthesis net productivity index.

The highest indexes of the net productivity of photosynthesis were defined at the plants of oilseed flax in a period "herringbone"—budding. Fluctuation of this index was within the limits from 5.10 to 8.22 g/$m^2$. Close correlation connection of high force between the productivity and net productivity of photosynthesis $r = 0.912$ was defined. We did not define a distinct sort difference in the work of assimilatory apparatus of oilseed flax plants.

Undertaken studies showed that the content of chlorophylls a, b, and their sums in the leaves of control plants did not depend on sort peculiarities. So, chlorophyll content fluctuation in the investigated sorts was within the limits to 5% (Table 4). In 2015, maximal content of chlorophyll was discovered in the plant leaves of both investigated sorts. Fluctuation of this index was within the limits from 1.896 to 1.973 mg/g of raw substance.

Application of PGRs AKM for preplanting treatment of oilseed flax seeds had positive influence on the accumulation of chlorophyll content. So, in the concentration of reactant 0.0015 g/l, the specimen increased this index on the average during the years of researches at the plants of the sort Evryka by 2.7% and at the sort Orfei—by 10.4%. Thus, it is possible to assert that for the plants of the sort Evryka, the adaptation to the unfavorable agrometeorological conditions of vegetation is more characteristic.

At the same time in a period of the active flowering (in 2014), there was an increase of carotenoids content by 6.8% in the variant with the use of AKM (reactant 0.0015 g/l), comparatively with control one. Taking into account the fact that carotenoids have protective properties due to participating in redox reactions [42], it is possible to assert that the use of growth regulator AKM promotes the best adaptation of plants to the unfavorable conditions of growing, which are very often

**Table 3** Photosynthetic and net productivity of oilseed flax plants at preplanting treatment of seeds PGRs AKM

| Sort (factor A) | PGRs (factor B) | Period | | | | | | | |
|---|---|---|---|---|---|---|---|---|---|
| | | BBCH code (19–20)–(53–55) | | | | BBCH code (53–55)–(67–69) | | | |
| | | Year (factor C) | | | | | | | |
| | | 2013 | 2014 | 2015 | 2016 | 2013 | 2014 | 2015 | 2016 |
| PP, thousand $m^2$ * days/ha | | | | | | | | | |
| Evryka | C | 231.5 | 288.5 | 287.2 | 281.5 | 449.6 | 478.5 | 474.4 | 435.5 |
| | AKM[a] | 247.7 | 283.5 | 299.9 | 280.7 | 482.2 | 484.4 | 483.1 | 449.5 |
| | AKM[b] | 231.1 | 281.2 | 284.1 | 275.5 | 451.8 | 474.3 | 465.7 | 438.0 |
| Orfei | C | 227.2 | 272.5 | 278.4 | 262.6 | 438.4 | 465.8 | 476.4 | 427.6 |
| | AKM[a] | 243.9 | 271.4 | 288.5 | 258.1 | 475.6 | 456.2 | 477.9 | 428.0 |
| | AKM[b] | 224.1 | 263.6 | 268.7 | 253.9 | 440.5 | 448.2 | 460.2 | 422.2 |
| $LSD_{05}$ A | | 2.03 | 1.98 | 2.21 | 2.04 | 1.74 | 2.17 | 3.64 | 3.19 |
| B | | 2.31 | 2.05 | 1.76 | 1.68 | 1.62 | 1.99 | 2.51 | 3.27 |
| NPP, $g/m^2a$ 24 h | | | | | | | | | |
| Evryka | C | 8.01 | 5.10 | 6.41 | 5.87 | 3.75 | 5.04 | 3.85 | 4.52 |
| | AKM[a] | 7.90 | 7.29 | 6.75 | 7.16 | 4.14 | 4.93 | 4.38 | 4.76 |
| | AKM[b] | 8.22 | 6.92 | 7.20 | 6.81 | 4.38 | 5.12 | 4.14 | 5.18 |
| Orfei | C | 7.79 | 5.56 | 6.97 | 6.32 | 4.17 | 4.68 | 3.13 | 4.55 |
| | AKM[a] | 7.82 | 6.39 | 7.21 | 7.87 | 3.97 | 4.75 | 3.46 | 4.36 |
| | AKM[b] | 8.08 | 6.48 | 7.45 | 7.49 | 4.27 | 4.50 | 3.59 | 4.34 |
| $LSD_{05}$ A | | 0.11 | 0.23 | 0.26 | 0.31 | 0.09 | 0.14 | 0.18 | 0.18 |
| B | | 0.14 | 0.15 | 0.18 | 0.24 | 0.16 | 0.22 | 0.34 | 0.24 |

*Note* [a]Concentration AKM reactant 0.0015 g/l
[b]Concentration AKM reactant 0.015 g/l

**Table 4** State of pigmental complex in the plants of oilseed flax depending on the action of PGRs AKM in flowering phase (BBCH code 67–69)

| Year (factor C) | Sort (factor A) | PGRs (factor B) | Chlorophyll, mg/g of raw substance | | | Carotenoids mg/gr of raw substance | $\frac{\text{Chlorophyll } a}{\text{Chlorophyll } b}$ | $\frac{\text{Chlorophylls}}{\text{Carotenoids}}$ |
|---|---|---|---|---|---|---|---|---|
| | | | a | b | a + b | | | |
| 2013 | Evryka | C | 0.889 | 0.378 | 1.267 | 0.579 | 2.35 | 2.19 |
| | | AKM[a] | 0.825 | 0.429 | 1.254 | 0.505 | 1.92 | 2.48 |
| | | AKM[b] | 0.861 | 0.384 | 1.245 | 0.538 | 2.24 | 2.31 |
| | Orfei | C | 0.907 | 0.421 | 1.328 | 0.499 | 2.15 | 2.66 |
| | | AKM[a] | 1.044 | 0.464 | 1.508 | 0.493 | 2.25 | 3.06 |
| | | AKM[b] | 0.965 | 0.398 | 1.363 | 0.497 | 2.42 | 2.74 |
| 2014 | Evryka | C | 1.238 | 0.455 | 1.693 | 0.411 | 2.72 | 4.12 |
| | | AKM[a] | 1.356 | 0.468 | 1.824 | 0.439 | 2.89 | 4.15 |
| | | AKM[b] | 1.324 | 0.448 | 1.772 | 0.426 | 2.96 | 4.16 |
| | Orfei | C | 1.108 | 0.397 | 1.505 | 0.397 | 2.79 | 3.79 |
| | | AKM[a] | 1.386 | 0.469 | 1.855 | 0.415 | 2.96 | 4.47 |
| | | AKM[b] | 1.297 | 0.457 | 1.754 | 0.382 | 2.84 | 4.59 |
| 2015 | Evryka | C | 1.466 | 0.436 | 1.902 | 0.555 | 3.36 | 3.43 |
| | | AKM[a] | 1.478 | 0.426 | 1.904 | 0.499 | 3.47 | 3.82 |
| | | AKM[b] | 1.472 | 0.424 | 1.896 | 0.560 | 3.47 | 3.39 |
| | Orfei | C | 1.396 | 0.538 | 1.934 | 0.531 | 2.59 | 3.64 |
| | | AKM[a] | 1.452 | 0.521 | 1.973 | 0.489 | 2.79 | 4.03 |
| | | AKM[b] | 1.401 | 0.512 | 1.913 | 0.481 | 2.74 | 3.98 |
| 2016 | Evryka | C | 0.946 | 0.483 | 1.429 | 0.518 | 1.96 | 2.76 |
| | | AKM[a] | 0.936 | 0.498 | 1.434 | 0.478 | 1.88 | 3.00 |
| | | AKM[b] | 0.945 | 0.462 | 1.407 | 0.501 | 2.05 | 2.81 |
| | Orfei | C | 0.889 | 0.429 | 1.318 | 0.504 | 2.07 | 2.62 |
| | | AKM[a] | 0.876 | 0.478 | 1.354 | 0.399 | 1.83 | 3.39 |
| | | AKM[b] | 0.861 | 0.482 | 1.343 | 0.474 | 1.79 | 2.83 |
| $LSD_{05}$ A | | | 0.048 | 0.027 | 0.051 | 0.024 | | |
| B | | | 0.042 | 0.041 | 0.072 | 0.031 | | |
| C | | | 0.085 | 0.062 | 0.097 | 0.035 | | |

*Note* [a]Concentration AKM reactant 0.0015 g/l
[b]Concentration AKM reactant 0.015 g/l

characterized by an air and ground drought. The evidence of it is an increase of pigmental index at variants with the use of AKM.

In the period of the active flowering, we can observe the influence of growth regulator on the pigmental complex of oilseed flax plants. It attests to fading of positive effect from the preplanting treatment of seeds. So, there was the increase of sum of chlorophylls in the plants of the experimental variant that on the background of low content of carotenoids resulted in the increase of pigmental index almost by 10% (on the average), comparatively with the control variant. Worsening of lighting conditions for the plants of the experimental variant is related to the increase of leaf area, comparatively with control.

## 4 Conclusions

Under different concentrations of active substances, the incrustation of oilseed flax seeds with AKM leads either to stimulation or inhibition of germination.

The increase of pigments concentration in the leaves of experimental plants of all sorts is coordinated with strengthening of leaf area growth and higher intensity of photosynthesis.

Correlation dependence between pigments content and index of NPP, force of which depended both on the sort and on the influence of growth regulator AKM, is defined. So, for the control variant of the sort Evryka a strong correlation connection was marked between chlorophyll content and NPP ($r = +0.76$), for the sort Orfei—between the chlorophyll index and NPP ($r = +0.84$).

Thus, the increase of the photosynthesis productivity with the actions of AKM for the oilseed flax plants took place due to the increase of assimilative leaves area and amount of green pigments.

## References

1. Quiroga, A.R., Díaz-Zorita, M., Buschiazzo, D.E.: Safflower productivity as related to soil water storage and management practices in semiarid regions. Commun. Soil Sci. Plant Anal. **32**, 2851–2862 (2001). https://doi.org/10.1081/CSS-120000967
2. Bezditko, O.E.: Influence of weather risk factors on the crop yielding capacity. Bull. Zhytomyr Nat. Agroecological Univ. **2**(1), 374–382 (2011) (in Ukrainian)
3. Andrianova, Y.E., Tarchevskiy, E.A.: Hlorofill i produktivnost rasteniy, 135 (2000) (in Russian)
4. Shadchyna, T.M., Hulyayev, B.I., Kiriziy D.A.: Rehulyatsiya fotosyntezu i produktyvnist'roslyn: fiziolohichni ta ekolohichni aspekty. Ukr. fitosotsiotsentr, 384 (2006). (in Ukrainian)
5. Kordyum, E.L., Syitnik, K.M., Baranenko, V.V.: Kletochnyie mehanizmyi adaptatsii rasteniy k neblagopriyatnyim vozdeystviyam ekologicheskih faktorov v estestvennyih usloviyah. Nauk. dumka, 273 (2003) (in Russian)
6. Mokronosov A.T., Gavrilenko, V.F., Zhigalova, T.V.: Photosynthesis. Physiological and ecological and biochemical aspects, 2nd ed, Academy (2006) (in Russian)

7. Svetlova, N.B., Kalenska, S.M., Pantaliyenko, A.V., Okanenko, A.A., Musienko, M.M.: Plant growth regulators and formation of adaptive reactions of plants to drought. Sci. J. Nat. Univ. Life Environ. Sci. Ukraine **58**, 11–17 (2002) (in Ukrainian)
8. Priss, O., Kalitka, V.: Effect of heat treatment with antioxidants on oxygen radical scavenging during storage of zucchini squash. Eastern European J. Enterp. Technol. **6/10** (77), 47–53 (2015). https://doi.org/10.15587/1729-4061.2015.561
9. Klymenko, I.I.: Effect of plant growth regulators and fertilizers on seed yield of lines and sunflower hybrids. Sel. Seed Sci. **107**, 183–188 (2015) (in Ukrainian)
10. Kalytka, V.V., Zolotuhina, Z.V.: Productivity of winter wheat in case of pre-sowing seed treatment by anti-stress composition. Sci. J. Nat. Univ. Life Environ. Sci. Ukraine **162**(1), 93–99 (2011) (in Ukrainian)
11. Janmohammadi, M., Amanzadeh, T., Sabaghnia, N., Ion, V.: Effect of nano-silicon foliar application on safflower growth under organic and inorganic fertilizer regimes. Bot. Lithuanica **22**(1), 53–64 (2016). https://doi.org/10.1515/botlit-2016-0005
12. Noreen, S., Ashraf, M., Hussain, M., Jamil, A.: Exogenous application of salicylic acid enhances antioxidative capacity in salt stressed sunflower (*Helianthus annuus* L.) plants. Pak J Bot **41**, 473–479 (2009). https://www.cabdirect.org/cabdirect/abstract/20093118627
13. Ghassemi-Golezani, K., Hosseinzadeh-Mahootchi, A.: Improving physiological performance of safflower under salt stress by application of salicylic acid and jasmonic acid. WALIA J. **31**(S1), 104–109 (2015). https://www.researchgate.net/publication/275659831
14. . Perelik pestitsidiv i agrohimikativ dozvolennih do vikoristannya v Ukraini (2014) (in Ukrainian)
15. Pokoptzeva, L.A., Kalytka, V.V.: Vplyv peredposivnoho obrobitku nasinnia soniashnyku dystynolom na yakist nasinnia pry zberihanni. Visnyk bilotserkivskoho derzh. ahr. universytetu **30**, 139–147 (2004) (in Ukrainian)
16. Pokoptzeva, L.A., Ivanchenko, O.A.: The use of the multi-objective sunflower seeds optimization method for optimal variant substraiation of presoving treatment of antioxidant preparation distinol. Bull. Agrarian Sci. Black Sea Reg. **4**, 163–169 (2011) (in Ukrainian). http://elar.tsatu.edu.ua/bitstream/123456789/3282/1/16.pdf
17. Polviy, A.M.: Silskogospodarska meteorologiya (2012) (in Ukrainian)
18. Yeremenko, O., Kalenska, S., Kalytka, V.: Safflower productivity depending on seed treatment by AKM plant growth regulator and level of mineral nutrition. Agric. For. **64**(1), 65–72 (2018). https://doi.org/10.17707/agricultforest.64.1.08
19. Ieremenko, O.A., Kalitka, V.V.: Influence of oilseed flax seed treatment by AKM plant growth regulator on sowing qualities during storage. Nauchnyiy vzglyad molodyih: poiski, innovatsii v APK. Mezhdunarodnaya nauchno-prakticheskaya konferentsiya molodyih uchenyih, Kazahstan, Alma-Ata, 19–24 (2017) (in Russian). http://www.kaznau.kz/page/conference/
20. Kalytka, V.V., Zolotuhina, Z.V.: Forming the yield of winter wheat in low moisture Steppe zone of Ukraine. Naukovi i praktychni aspekty aghropromyslovogho vyrobnyctva ta rozvytku siljsjkykh reghioniv: materialy Mizhnarodnoi naukovo-praktychnoi forumu [Scientific and practical aspects of agro-industrial production and rural development. Proc. Int. Sci. Practical Forum] 50–54. Ljviv, Ukraine, Sep 22–24 (2010) (in Ukrainian)
21. Shevchenko, A.O., Taracenko, V.O.: Reguljatory rostu. Pryncypovo novyj vysokoefektyvnyj element sil's'kogospodars'kyh tehnologij. Zahyst roslyn, **1**, 17–19 (1998) (in Ukrainian)
22. Eremko, L.S., Sidorenko, A.V., Olepir, R.V., Agafonova, S.O.: Produktyvnist okremykh silskohospodarskykh kultur za zastosuvannia rehulatoriv rostu roslyn. Bull. Poltava State agrarian academy **1**, 43–45 (2009). (in Ukrainian). https://www.pdaa.edu.ua/sites/default/files/visnyk/2009/01/09_r1_roslinnictvo_1_2009.pdf
23. Soil quality. Methods of determination of organic matter: DSTU 4731:2007. Ukraine, 11 (2007) (in Ukrainian)
24. Gerkial, O.M., Gospodarenko, G.M., Kolarkov, Y.V.: Agrochemistry: Study Guide, Uman, 300 (2008) (in Ukrainian)
25. Soils. Determination of mobile compounds of phosphorus and potassium using modified Chirikov method: DSTU 4115–2002. - [Valid from 2002-07-18]. K.: Derzhspozhivstandart of Ukraine, 13 (2002). (in Ukrainian)

26. Rozhkov, A.O., Puzik, V.K., Kalensjka, S.M., Puzik, L.M., Popov, S.I., Muzafarov, N.M., Bukhalo, V.Ja., Kryshtop, Je.A.: Doslidna sprava v aghronomiji, [Pilot case in agronomy]. **1**(2). Majdan, Kharkiv (2016) (in Ukrainian)
27. .Gritsaenko, Z.M., Ponomarenko, S.P., Karpenko, V.P., Leontyuk, I.B.: Biologichno aktivni rechovini v roslinnitstvi, p. 352 (2008) (in Ukrainian)
28. Kazakov Y.O.: Metodolohichni osnovy postanovky eksperymentu z fiziolohiyi roslyn. Fitosotsiotsentr, p. 272 (2000) (in Ukrainian)
29. Nichiporovich, A.A.: Fotosinteticheskaya deyatelnost rasteniy v posevah, p. 186 (1961) (in Russian)
30. Kiriziy, D.A.: Fotosintez i rost rasteniy v aspekte donorno-aktseptornyih otnosheniy, Logos, p. 191 (2004) (in Russian)
31. Musienko, M.M., Pershikova, T.V., Slavnyiy, L.S.: Spektrofotometricheskie metodyi v praktike fiziologii, biohimii i ekologii rasteniy, p. 199 (2001) (in Russian)
32. Maslova, T.G., Popova, I.A.: Krytycheskaia otsenka spektrofotometrycheskoho metoda kolychestvennoho opredelenyia karotynoydov. Fiziologiya rasteniy **39**(6), 615–619 (1986) (in Russian)
33. Tretyakov, N.N., Koshkin, E.I., Mokrushkin, N.M.: Fiziologiya i biohimiya selskohozyaystvennyih rasteniy, Kolos, p. 640 (2000) (in Russian)
34. Nychyporovych, A.A.: Fotosyntez y produktsyonniy protsess: nauchnoe izdanye. Yn-t fyzyolohyy rastenyi ym. K.A.Tymyriazeva, Nauka, p. 277 (1988) (in Russian)
35. Cuperman, F.M., Podolnyiy, V.Z.: Eksperimentalnyiy morfogenez. MGU **1**, 296–306 (1963) (in Russian)
36. Soil quality. Indices of soil fertility: DSTU 4362:2004. Ukraine, p. 19, (2006) (in Ukrainian)
37. Brown, D.M., Bootsma, A.: Crop heat units for corn and other warm season crops in Ontario. Factsheet Ministry of Agriculture, Food and Rural Affairs, 32–41 http://www.omafra.gov.on.ca/english/crops/facts/93-119.htm. (1993)
38. Kalytka, V.V., Zolotuhina, Z.V., Ivanchenko, O.A., Yaloha, T.M., Zhernovy, O.I.: Anti-stress composition for pre-sowing seed treatment of agricultural crops, Patent. 58260 Ukraine, MPK[51] A01C 1/06, A01 N 31/00. №201010482; publ. 11.04.2011, Bull.**7** (2011) (in Ukrainian)
39. State register of plant varieties suitable for dissemination in Ukraine, p. 377 (2015) (in Ukrainian). http://vet.gov.ua/node/919
40. Dospekhov, B.A.: Methods of field experiment (with the fundamentals of statistical processing of the results of research), Agropromizdat, p. 351 (1985) (in Russian)
41. Janishevskiy, L.I., Matsiychuk, V.M.: Center for Sci. Provision of Agrobusiness in the Kharkov region **15**, 167–172 (2013)
42. Lebedeva, T.C., Syitnik, K.M.: Pigmentyi rastitelnogo mira, p. 325 (1986) (in Russian)

# Investigation of the Grinding Mode of the Enriched Wheat Products in the Rolling Mill 1-Grinding System of the Milling Mill of Wheat Grinding

**Olena Yeremeeva, Yevgen Kharchenko, Hennadii Tkachenko, Iryna Shapoval and Olena Hryhorenko**

## 1 Introduction

In accordance with the accepted structure of wheat grinder, the first grinding system is sub-products of the first quality. This system produces the largest amount of high-grade flour, so the study of the grinding regimes of this system is important for obtaining the highest yield of high-grade flour. The operating mode of the grinding systems is determined by the product of a higher-grade flour.

Investigations of the grinding regimes in the drowning process have established that there are regimes of grinding of grain and ladder products of the drown process in rollers, which achieve the greatest output of circular wheat, which are then sent to enrichment. The largest yield of wheat products that can be obtained in the drowning process enables to increase the yield of flour of high grades, which are obtained on the first three grinding systems. Enriched grains and dunes are fed into grinding systems, which are ground to a particle size of less than 130 μ, forming the highest quality flour. Therefore, the question arises about the efficiency of grinding modes of enriched wheat products in roller mills grinding systems in order to obtain the largest output of flour.

In Ukraine, OLIS LLC has started mass production of ESM marker enthlators-dismemblers, which are structurally different from the typical entolyators of P3-BER;

O. Yeremeeva (✉) · H. Tkachenko · I. Shapoval
Uman National University of Horticulture, Institutska Str, 1, Uman, Cherkassy Region 20305, Ukraine
e-mail: eremeeva.elena1961@gmail.com

Y. Kharchenko
National University of Food Technology, Volodymyrska Str., 68, Kiev 01601, Ukraine

O. Hryhorenko
Tavria State Agrotechnological University, B. Khmelnitsky Av., 18, Melitopol, Zaporizhia Region 72312, Ukraine

© Springer Nature Switzerland AG 2019 
V. Nadykto (ed.), *Modern Development Paths of Agricultural Production*,
https://doi.org/10.1007/978-3-030-14918-5_79

therefore, the use of new ESD-1.5 dispensers' entolyators is promising as they have a greater technological efficiency of crushing of first quality products in flour.

## 2 Analysis of Literary Data and Problem Statement

The cruddust products obtained in the drowning process undergo a process of enrichment and are fed into the grinding process for maximum grinding them into flour [1]. The structure of the milling stage is simpler than the drowning stage.

With the introduction of BUHLER grinding technologies after rollers, additional entrays P3-BER and detergents A1-DBG are used for additional grinding and destruction of particle conglomera. The P3-BER enolators are mainly used on the first three grinding systems that process first-grade grinding products with low ash content.

Several researchers at Campbell G. M., Fistes A., O. Vereschinskii and other researchers took part in the research on grain milling in rollers. Campbell G. M., Fang C., Muhamad I. I., Webb C., Bunn P. J., Hook S. C. W., Sadhukhan J., Mateos-Salvador F. [2–7, 11]. On these grinding systems, the highest amount of high-grade flour is obtained. Anthotics RZ-BER represent a kind of machines of the so-called disintegrator type and detachers—machines of shock-erosion action [3–7]. Studies of domestic and foreign researchers [3–6, 8–10] confirmed the expediency of additional grinding of intermediate grinding products in milling mills of durum wheat milling.

The grinding process is less studied, since the results of its work have less effect on the quality of the output of the flour than the process. In the production conditions, the grinding process is also paid less attention than drained.

Company OLIS LLC has developed enteroliters-dismembers ECM-1.5 [7], which differ from the enthtores RZ-BER with the intensity of action on the crushed product. A similar entolyoresis-dismembler is manufactured by the Dutch company OTTE-VANGER. Anthotics-dismembranes are promising for use in the flour mill industry due to more intense action on intermediate grain milling products than typical RZ-BER entolyores. Proceeding from this, it is expedient to investigate the influence of grinding regimes of the "roller machine—entolyator-dismembler" system.

In the literary and informational sources, there are no data on mathematical models of description of granulometric characteristics from the grinding regimes in the rolling mills of the most important first three grinding systems of the first quality [2–6, 8–10] on which the highest amount of flour of high varieties is obtained. These issues are important for designers in developing the balance of grinding new flour mills, as well as for technologists of existing mills. Proceeding from this, one of the tasks of our research is the study of the modes of crushing cropodundstovyh products in rollers and enteroliths-dismemblers, as well as the effectiveness of their joint action.

## 3 The Purpose and Task of Research

The purpose of the work is to increase the competitiveness of products of domestic flour mill enterprises by improving the technological process of milling grain of wheat into flour and reducing energy consumption in the process of its production.

During the research, wheat grain was processed with the following quality indices: grain weight—791 g/l, grain moisture content on the I drought system—16.2%, vitreousness—38%, garbage impurity 0.4%, grain impurity—2.3%.

## 4 Materials and Methods of Research

The research of quality indicators of grain and grinding products was carried out in the production laboratory in the production conditions of a flour mill with a productivity of 330 T per day.

To study the system of "roller mill—entolyator-dismembler", the first three first-class grinding systems were selected, on which enteroliths-dismembIers are installed for the maximum grinding of enriched products into flour.

Changing the grinding regimes and selecting the samples on the first grinding system was carried out as follows: During the work of the grinding unit on the specified system, a product was withdrawn throughout the roll length by means of a tray in the quantity up to 400–500 g. As the grinding products of the first grinding system are fed by pneumatic transport from the roller in the entolyores-dismembIers, and then in the cyclone discharger, then the sampling at different grinding regimes was carried out through the hatches after the cyclone discharger. After selecting the product with a helmet, equipped with rollers, the gap between the rollers was changed and the product was re-selected throughout the roller length and after the entleoter-dismemberer.

The grinding products were sown on a set of sieves No. 27PA-120, No. 33/36PA, No. 41/43PA and No 49/52PA. The mode of operation of the rolling mills of the first three milling systems was determined by the output of the higher-grade flour, since for high-quality wheat mills, the output of the higher-grade flour has the highest value and the main product of the milling at the mill stage is the output of the flour. The evaluation of the working conditions of the crushing equipment of the milling stage shall be carried out on the product of a higher-grade flour.

The output of the high-grade flour was calculated taking into account the under-development by the formula:

$$V_{f.h.g.} = \frac{m_{f.h.g.}}{m_z} * 100 - H \tag{1}$$

where $V_{f.h.g.}$—output of high-quality flour,%; $m_z$—the total mass of the sample after the roller or entleiter-dismembler, g; $m_{f.h.g.}$—mass of flour of the highest grade obtained after sieving, s; $H$—shortcomings, determined by sifting the selected sample of the product entering the first grinding system, %.

The obtained results of the output of separate fractions of grinding products were graphically depicted in the coordinates of "the total product of flour—the yield of fraction".

On the first grinding system installed microshock rollers. The rotor speed of the entolyator-dismembler rotors was 50 s–1 (3000 rpm) and during the research did not change.

## 5 Results and Discussion

### 5.1 *Modes of Grinding of Cereal-Rich Products Enriched in Roller Mills of 1 Grinding System*

In order to substantiate the reduction of the milling stage in flour, we investigated the grinding regimes of enriched rounds (large ones) in the roller mill and the enteroliter-dismemberer of the first grinding system in accordance with the generally accepted structure, which includes a consistent combination of the roller machine and the entolyator-dismemberer ESM-1.5. On the basis of our research, it was found that with an increase in the total product of the flour in the rolling mill of the 1st grinding system from 7.3 to 28.7%, the output of the mixture of stair products (medium and small grains) from 80.9 up to 34.9% by curvilinear dependence. At the same time, the output of hard dunst increases, which increased from 5.4 to 10.4% and the output of soft dunst, which increased from 5.1 to 21.3%. In the range of general product flour, the yield of flour of grade 1 increased by a linear dependence of 1.3–5.1%. The results of the research are shown in picture 1. Analysis of data in picture 1 shows that with the increase of the general product of high-quality flour in the rolling mill of the 1st grinding system, not only the grinding of enriched large grains but also the excessive crushing of medium and small grains occurs, as evidenced by the decreasing nature of the corresponding curve 1 in Fig. 1.

1. stair product (a mixture of small and medium cereals);
2. hard dunst;
3. soft dunst;
4. 1st-grade flour.

The increase in the yield of flour did not exceed 28.7%, since the actual capabilities of the existing roller coaster did not allow this to be carried out and could lead to the outlet of the electric motor of the machine. In Fig. 2, the averaged results of the research of the output of stair products and flour of the 1st grade after shredding in the roller machine and the ESM-1.5 enthooler-dismemberer of the 1st grinding system are presented.

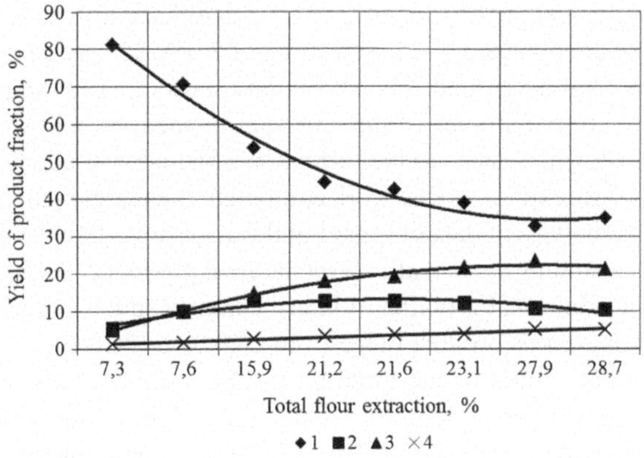


**Fig. 1** Output of stair products after grinding in a roller machine 1 of a grinding system taking into account lack of sight

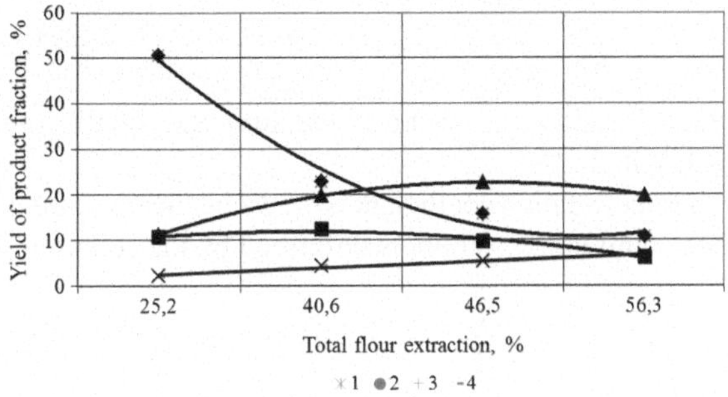


**Fig. 2** Averaged output of stair products on 1 grinding system (roller machine + entolyator-dismembler) taking into account lack of access

1. stair product (a mixture of small and medium cereals);
2. hard dunst;
3. soft dunst;
4. 1st-grade flour.

Analysis of Fig. 2 indicates that the highest grade flour is mainly produced by grinding on the 1st grinding system of medium and small grains, as evidenced by the declining nature of curve 1, as well as the increase of the general product of high-quality flour leads to a decrease in the yield of soft and rigid dunes In the general product of high-quality flour on the system of 25.2%, the average yield of the mixture of medium and small grains was 50.5%, the yield of hard dunes—10.6%, soft dunes—1.3%, flour 1st grade—2.3%.

With an increase in the intensity of crushing of enriched turntables on the 1st grinding system from 25.2 to 56.3%, the yield of the mixture of medium and small grains decreased from 50.5 to 10.7%, rigid duns from 10.6 to 6.3%, soft dunes from 11.3 to 20.0%, 1st-grade flour from 2.3 to 6.7%.

From the analysis of Figs. 1 and 2 shows that the nature of the output of products after the roller machine and after the additional passage of the product through the entolyator-dismemberer has not changed significantly, the values of the output of circular dandruff products and flour are different. In addition, the comparative analysis shows that in the ESM-1.5 distilled entolyator, the medium and small grains that are obtained by the screening of 27PA-120 are shredded the most, and the dunes are slightly ground. The above data shows that in the ESM-1.5 distilled entolyator, the larger particles of the endosperm (medium and small grains) are shredded the most, which leads to the formation of a higher-grade flour from them. Smaller parts are less exposed to the work equipment than larger particles and their conglomerates.

Analysis of data of Figs. 1 and 2 also shows that the use of a roller machine on the 1st grinding system allows the general product of high-quality flour to reach 28.0%, and additional grinding of intermediate products in the entolyator-dismemberer can increase this product twice from 28.0 up to 56.0%.

The output of the grinding products on the 1st grinding system, depending on the general product of the flour of the highest grade Vb.versus, is described by the following equations: yield of a mixture of medium and small cereals, Vsr.kr + dr.kr:

$$V_{m.gr.+sm.gr.} = 0.0343\, V_{f.h.g.}^2 - 4.08\, V_{f.h.g.} + 131.61 \quad (2)$$

where $V_{m.gr.+sm.gr}$—yield of medium and small grains,%; $V_{f.h.g.}$—yield of high-quality flour, %.

The correlation coefficient is 0.99, and the standard deviation is 0.27%. The output of the hard dunst, Wrath, is described by the equation:

$$V_{h.d.} = -0.0136\, V_{f.h.g.}^2 + 0.963\, V_{f.h.g.} - 4.86 \quad (3)$$

where $V_{h.d.}$—the output of hard dunst, %; $V_{f.h.g.}$—yield of high-quality flour, %.

The correlation coefficient is 0.97, and the standard deviation is 0.56%. The output of the soft dunst, Vmd, is described by the equation

$$V_{s.d.} = -0.0207\, V_{f.h.g.}^2 + 1.98\, V_{f.h.g.} - 25.58 \quad (4)$$

where $V_{s.d.}$—output of soft dunst, %; $V_{f.h.g.}$—yield of high-quality flour, %.

The correlation coefficient is 0.98, and the standard deviation is 0.74%.

The output of the first-grade flour, Vb.1c, is described by the equation:

$$V_{f.1g.} = 0.143\, V_{f.h.g.} - 1.35 \quad (5)$$

where $V_{b.1g.}$—output of hard dunst,%; $V_{f.h.g.}$—yield of high-quality flour, %.

The correlation coefficient is 0.99, and the standard deviation is 0.08%.

In the process of milling, the entolyators are auxiliary crushing equipment, which is installed after the roller. The grinding mode of the rolls in the roller machine affects the grinding mode of the grooves in the ESM-1.5 dismemberer entolyator. Researches have shown that the increase of the product of the high-quality flour in the roller machine from 7.5 to 28.3% leads to an increase in the yield of flour and in the entolyator from 17.7 to 28.0%. Dependence is linear.

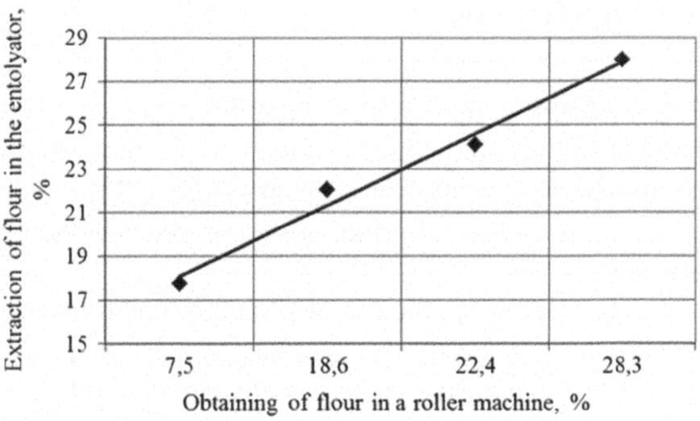


**Fig. 3** Dependence of the output of high-quality flour in the enterolter-dismemberer ECM-1.5 from the output of the flour in the roller machine on the 1st grinding system, taking into account lack of time

It can be concluded that the particles of the grinding products that were not crushed to the size of the flour particles in the roller saw, but due to the mechanical effect, reduced the strength of the bonds further grinded through entolyores-dismemblers. Thus, not crushed grains and dunes are further grinded through the impact machines. In addition, the conglomeration of particles formed after passing circular dentate products through the rollers are also destroyed by passing through the entolyores-dismembranes (Fig. 3).

Rolling machine tools of the first three milling systems worked under the following regimes: the ratio of rotation speeds of rollers, K equaled 1.25, speed of rotation of the speed roll, V, −5 m/s, rollers with a micro-hollow surface.

The results of the research show that in order to intensify the grinding process of enriched circular dentistry products, it is necessary to reduce grinding regimes in the rolling mill of the 1st grinding system to 30% of the total product of the grinding, which positively affects the efficiency of the entolyores-dismemberers. During the laboratory study of crushed intermediate products of the 1st grinding system on the fraction, there was a presence of large particles of the nucleus of yellowish color in the east of the screen number 1.0 in the amount of 0.1–0.2% of the total mass of the product in the form of flakes. In the fraction obtained by the passage of the screen number 1.0 and the screen of the screen No. 12PCH-240, there was also a presence of small particles of the embryo, but it was not possible to separate the embryos and shells separately, and the total yield of this fraction of the product varied from 1.6 to 2.9%.

In the framework of the conducted researches to establish the optimal modes of grinding, the 1st grinding system failed. Although it is known from reference sources that an excessive reduction in the size of the gap between the rollers of the grinding systems leads to the splitting of the grinding products. In the production conditions, such modes were not achieved. For production, it is possible to recommend the general product of flour in a roller machine on 1st grinding system 28.0%.

## 6 Conclusions

The modes of grinding are determined. and the mathematical dependences of the product of flour and wheat products of the first grinding system with the additional equipment by the entolyator-dismemberer ESM-1.5 are proposed. The largest product of flour on the 1st grinding system reaches the regimes of grinding on a roller machine—28%.

At the same time, the use of ESD-1,5 dental emulator allows to increase the total product of flour on the 1st grinding system up to 56%. It has been experimentally established that the dependence between the product of the highest grade flour in the roller mill of the grinding system and the output of the high-grade flour in the ESM-1.5 dental emulator entolyarator has a linear dependence. The results of experimental studies are described by mathematical dependencies, which allow them to be used to calculate the quantitative balance of flour.

## References

1. Kharchenko, Y., Sharan, A., Yeremeeva, O.: L. Novak Ukr. Food J. **6**(4), 603–617 (2017)
2. Mateos-Salvador, F., Sadhukhan, J., Campbell, G.M.: Powder Techn. **237**, 107–116 (2013)
3. Mateos-Salvador, F., Sadhukhan, J., Campbell, G.M.: Powder Techn. **208**, 144–157 (2011)
4. Fistes, A., Tanovic, G., Mastilovic, J.: J. Food Engin. **85**, 296–302 (2008)
5. Fistes, A., Rakic, D., Takaci, A., Brdar, M.: Chem. Eng. Sci. **102**, 346–353 (2013)
6. Campbell, G.M., Sharp, C., Wall, K., Mateos-Salvador, F., Gubatz, S., Hutty, A., Shewry, P.: J. Cereal Sci. **55**, 415–425 (2012)
7. Vereshchynsky, A.P.: NUFT, Kiev (2013)
8. Fistes, A., Rakic, D., Takaci, A., Brdar, M.: Powder Techn. **268**, 412–419 (2014)
9. Storage and processing of grain, **4**, 45–48 (2013)
10. Fistes, A., Rakic, D., Takaci, A., Brdar, M.: Powder Tech. **268**, 412–419 (2014)
11. Kharchenko, Y., Perehuda, N., Shnipko, I.: Grain Storage Proc. **6–7**, 62–64 (2015)

# Correction to: Mathematical Model Changing the Value of the Process of Leakage Current in 0.38 kV Networks

**Viacheslav Gerasymenko, Volodymyr Kozyrskyi, Natalia Maiborodina, and Oleksandr Kovalov**

## Correction to:
## Chapter 35 in: V. Nadykto (ed.), *Modern Development Paths of Agricultural Production*, https://doi.org/10.1007/978-3-030-14918-5_35

In the original version of the chapter, reference 8 was incorrect in the initially published version. It has been corrected as below;

Ovcharov, S., & Strebkov, A. (2015). Active energy losses research in an asynchronous electric motor in operating terms. Eastern-European Journal of Enterprise Technologies, 2(8(74), 22–28. https://doi.org/10.15587/1729-4061/2015/39026. The book and the chapter have been updated with the changes.

The updated version of this chapter can be found at
https://doi.org/10.1007/978-3-030-14918-5_35

© Springer Nature Switzerland AG 2024

V. Nadykto (ed.), *Modern Development Paths of Agricultural Production*,
https://doi.org/10.1007/978-3-030-14918-5_80

The manufacturer's authorised representative in the EU is Springer Nature Customer Service Centre GmbH, Europaplatz 3, 69115 Heidelberg, Germany. If you have any concerns regarding our products, please contact ProductSafety@springernature.com

Printed and bound by CPI Group (UK) Ltd, Croydon, CR0 4YY
30/09/2026
02225440-0019